Geography of the Physical Environment

The *Geography of the Physical Environment* book series provides a platform for scientific contributions in the field of Physical Geography and its subdisciplines. It publishes a broad portfolio of scientific books covering case studies, theoretical and applied approaches as well as novel developments and techniques in the field. The scope is not limited to a certain spatial scale and can cover local and regional to continental and global facets. Books with strong regional focus should be well illustrated including significant maps and meaningful figures to be potentially used as field guides and standard references for the respective area.

The series appeals to scientists and students in the field of geography as well as regional scientists, landscape planners, policy makers, and everyone interested in wide-ranging aspects of modern Physical Geography. Peer-reviewed research monographs, edited volumes, advance and undergraduate level textbooks, and conference proceedings covering the major topics in Physical Geography are included in the series. Submissions to the Book Series are also invited on the theme 'The Physical Geography of…', with a relevant subtitle of the author's/editor's choice. Please contact the Publisher for further information and to receive a Book Proposal Form.

More information about this series at https://link.springer.com/bookseries/15117

Pravat Kumar Shit • Biswajit Bera •
Aznarul Islam • Sandipan Ghosh •
Gouri Sankar Bhunia
Editors

Drainage Basin Dynamics

An Introduction to Morphology, Landscape and Modelling

Springer

Editors
Pravat Kumar Shit
PG Department of Geography
Raja N. L. Khan Women's College
(Autonomous)
Midnapore, West Bengal, India

Biswajit Bera
Department of Geography
Sidho-Kanho-Birsha University
West Bengal, India

Aznarul Islam
Department of Geography
Aliah University
Kolkata, India

Sandipan Ghosh
Department of Geography
Chandrapur College
Barddhaman, West Bengal, India

Gouri Sankar Bhunia
Randstad India Pvt. Ltd
New Delhi, India

ISSN 2366-8865 ISSN 2366-8873 (electronic)
Geography of the Physical Environment
ISBN 978-3-030-79633-4 ISBN 978-3-030-79634-1 (eBook)
https://doi.org/10.1007/978-3-030-79634-1

Cover image by Sonja Weber, München

This Springer imprint is published by the registered company Springer Nature Switzerland AG
The registered company address is: Gewerbestrasse 11, 6330 Cham, Switzerland

Dedicated to

Young Scholars in the field of Geomorphology and Environment

Foreword

I am extremely pleased to introduce this inspiring and impressive volume on *Drainage Basin Dynamics: An Introduction to Morphology, Landscape and Modelling* edited by a group of young and energetic geomorphologists. This volume is the collection of excellent articles contributed by reputed as well as promising geomorphologists. The articles are academically rich, technically sound and contain outputs of very sincere and robust research. The articles are arranged by the editors in an orderly sequence leading to a comprehensive and complete understanding of the topic concerned.

Examining the conservation of mass and energy in landscape evolution in a way to understand the self-regulatory nature of the landscape in the drainage basin units is the key to geomorphic knowledge. Drainage basins of different spatial orders are arranged in the nested hierarchy where both spatial and temporal scales are coupled. Geomorphologists are interested in the working of rivers under different boundary conditions and the effectiveness of formative events in landscape shaping depending on event size and distribution of magnitude-frequency.

From this perspective, the present effort is the brilliant and exceptional instance of academic pursuit of the editors in promoting academic quest, advancing the geomorphic understanding, fostering the interest in exploring the process-form relationship among the learners and encouraging fellow colleagues in a similar honest and genuine endeavour.

Prof. Ramkrishna Maiti
Department of Geography
Vidyasagar University
Midnapore, India

Preface

Drainage basin is often called the cradle of civilizations as it offers many vital resources such as water, land, and the forest that shape the development of ecological, economical, cultural, and geopolitical issues. In the era of twenty-first century, these issues have become more pertinent in the context of 'Anthropocene'. The river flows at its own whim. Nature's whimsy changes and the games of breaking are played. People depend on the river since the very inception of the civilizations. The evolution of civilization centred on the river is fast-changing. People are constantly trying to tame the river. As a result, the river is increasingly getting metamorphosed in one way and influences the socio-economic fabric of human civilization in another way. Consequently, increased river floods, ecological degradation, economic loss and damage of the infrastructure, etc. are clearly noted in recent times. Therefore, taming the river is not important, rather we need to understand the dynamism of drainage systems, morphological characteristics, and landscape features towards sustainability using modern geospatial technology.

This book is composed of 24 chapters associated with spatial modelling in drainage basin evolution, system approach, morphology, drainage basin hydrology and sedimentology, connectivity and interaction, human interference, river engineering, natural and anthropogenic hazards, and various management techniques. The contemporary researches in this field have been amalgamated to comprehend the various methods for the evolution of geomorphic landforms and combating the hydro-geomorphic hazards, and disasters with proper mitigation strategies.

We are very much thankful to all the authors who have meticulously completed their documents on a short announcement and played a vital role in building this edifying and beneficial publication. We do believe that this will be a very convenient book for geographers, geologists, ecologists, river scientists, and others working in the field of water resources management including research scholars, environmentalists, and policymakers. We also acknowledge our deep gratitude to the publisher Springer Nature especially to Doris and her team for contracting with us for such a timely publication.

Midnapore, India | Pravat Kumar Shit
West Bengal, India | Biswajit Bera
Kolkata, India | Aznarul Islam
Barddhaman, India | Sandipan Ghosh
New Delhi, India | Gouri Sankar Bhunia

Acknowledgments

The preparation of this book has been guided by several geomorphologic pioneers. We are obliged to these experts for providing their time to evaluate the chapters published in this book. We thank the anonymous reviewers for their constructive comments that led to substantial improvement in the quality of this book. Because this book was a long time in the making, we want to thank our family and friends for their continued support. This work would not have been possible without constant inspiration from our students, knowledge from our teachers, enthusiasm from our colleagues and collaborators, and support from our family. Finally, we also thank our publisher Springer and its publishing editor for their continuous support in the publication of this book.

Disclaimer

The authors of individual chapters are solely responsible for the ideas, views, data, figures, and geographical boundaries presented in the respective chapters of this book, and these have not been endorsed, in any form, by the publisher, the editor, and the authors of forewords, preambles, or other chapters.

Contents

Editors and Contributors

About the Editors

Dr. Pravat Kumar Shit is an Assistant Professor at the PG Department of Geography, Raja N. L. Khan Women's College (Autonomous), West Bengal, India. He received his M.Sc and Ph.D. degrees in Geography from Vidyasagar University and PG Diploma in Remote Sensing and GIS from Sambalpur University. His research interests include applied geomorphology, soil erosion, groundwater, forest resources, wetland ecosystem, environmental contaminants and pollution, and natural resources mapping and modelling. He has published 16 books (Springer-12, Elsevier-02, CRC Press-01, and others) and more than 70 papers in peer-reviewed journals. He is currently the editor of the GIScience and Geo-environmental Modelling (GGM) book series, Springer Nature.

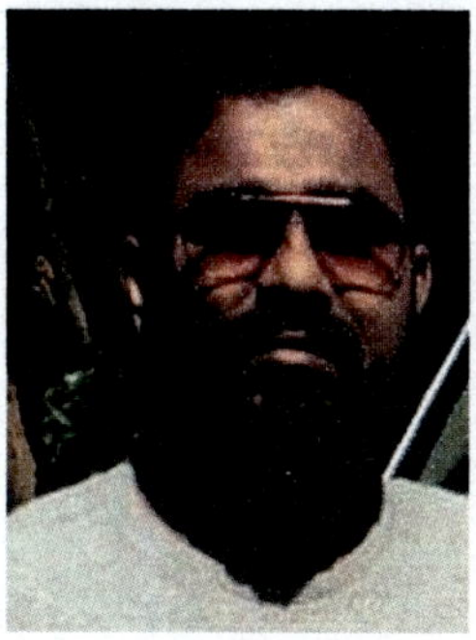

Dr. Biswajit Bera is an Assistant Professor in the Department of Geography, Sidho Kanho Birsha University, West Bengal (India). He did M.Sc and Ph.D. from the University of Calcutta, West Bengal (India). He has done a P.G. Certificate course in groundwater hydrology from IISWBM, Kolkata. He has a rich experience of teaching in different core areas of Geography in various Colleges and Universities over the last 10 years. He is the recipient of the prestigious Young Geomorphologist Award and International Young Geomorphologist scholarship. Dr. Bera has published five books for school, college, and university students and several research articles in both National and International peer-reviewed journals. His research interests include fluvial Geomorphology, Hydrology, and Environmental Geography. Dr. Bera regularly appears in debate/talk in both National print and electronic media. Recently, he has completed the ICSSR Major Research Project awarded by the Ministry of HRD, Govt. of India.

Dr. Aznarul Islam is an Assistant Professor in the Department of Geography, Aliah University, Kolkata, India. He did Master of Science in Geography from Kalyani University, India, and M. Phil and Ph.D. in Geography from the University of Burdwan, India. He has already published more than twenty-five research papers in different national and international journals, edited volumes, and conference proceedings. He is an editorial board member of five international journals and also acting as the reviewer of eight international journals. He is an editor of '*Neo-Thinking on Ganges Brahmaputra Basin Geomorphology*', Springer International Publishing and '*Quaternary Geomorphology in India—Case Studies from the Lower Ganga Basin*', Springer Nature. He has already completed an ICSSR Major Research Project funded by the Ministry of HRD, New Delhi, Government of India. His principal area of research includes geomorphology of the Bengal basin especially riverbank erosion, channel migration, flood, anthropo-geomorphology, and channel decaying.

Dr. Sandipan Ghosh holds the post-graduate degrees of M.Sc, M.Phil, and Ph.D. from the University of Burdwan, West Bengal, India. He has published four books and more than 40 international and national research articles in various geography and geosciences journals. His principal research field includes Fluvial Geomorphology, Regolith Geology, and Quaternary Geomorphology. Currently, he is working as an Assistant Professor in the Department of Geography, Chandrapur College (Barddhaman, West Bengal).

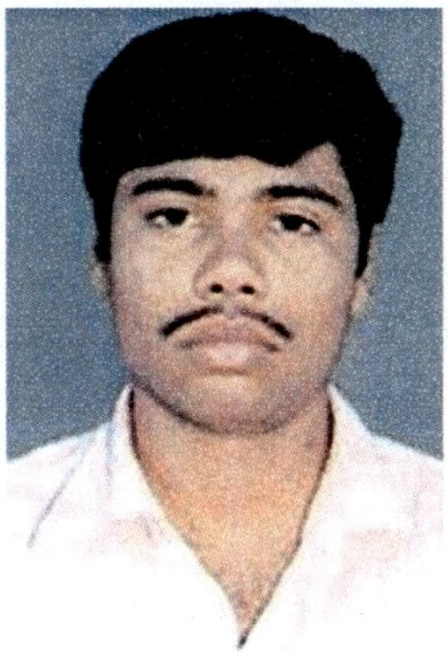

Dr. Gouri Sankar Bhunia received Ph.D. from the University of Calcutta, India, in 2015. His Ph. D. dissertation work focussed on environmental control measures of infectious disease using Geospatial technology. His research interests include environmental modelling, risk assessment, natural resources mapping and modelling, data mining, and information retrieval using Geospatial technology. Dr. Bhunia is associate editor and on the editorial boards of three international journals in Health GIS and Geosciences. Dr. Bhunia has published more than 60 articles in various journals in Scopus indexed. He is currently the editor of the GIScience and Geo-environmental Modelling (GGM) book series, Springer Nature.

Contributors

S. Adelabu Deparment of Geography, University of the Free State, Bloemfontein, South Africa

Jatisankar Bandyopadhyay Department of Remote Sensing and GIS, Vidyasagar University, Midnapore, India

Monali Banerjee Department of Geography, University of Calcutta, Kolkata, West Bengal, India

Biswajit Bera Department of Geography, Sidho-Kanho-Birsha University, Purulia, West Bengal, India

Ujjwal Bhandari Department of Geography, University of Calcutta, Kolkata, India

Rajkumar Bhattacharya Department of Geography and Environment Management, Vidyasagar University, Midnapore, West Bengal, India

Gouri Sankar Bhunia Department of Geography, Seacom Skills University, Santiniketan, Bolpur, West Bengal, India

Manash Jyoti Bhuyan Department of Geography, Nowgong Girls' College, Nagaon, India

Ashok Kumar Bora Department of Geography, Gauhati University, Guwahati, India

Debashree Borah Department of Geography, Arya Vidyapeeth College, Guwahati, India

Nilanjana Das Chatterjee Department of Geography and Environment Management, Vidyasagar University, Midnapore, West Bengal, India

Kousik Das Department of Geography and Environment Management, Vidyasagar University, Midnapore, West Bengal, India

Souvik Das Department of Geography, The University of Burdwan, Burdwan, West Bengal, India

Manasi De Department of Geography, Lady Brabourne College, Kolkata, India

Nityananda Deka Department of Geography, Gauhati University, Guwahati, India

Someswar Devulapalli Independent Researcher, Amberpet, Hyderabad, Telangana, India

Gour Dolui Department of Geography, Panskura Banamali College (Autonomous), Panskura, Purba Medinipur, West Bengal, India

O. A. Fashae Department of Geography, University of Ibadan, Ibadan, Nigeria

Debabrata Ghorai Independent Researcher, Tajpur, Paschim Medinipur, West Bengal, India

Debasis Ghosh Department of Geography, University of Calcutta, Kolkata, West Bengal, India

Sandipan Ghosh Department of Geography, Chandrapur College, Chandrapur, Purba Barddhaman, West Bengal, India

Supriya Ghosh Department of Geography, Sidho-Kanho-Birsha University, Purulia, West Bengal, India

MD Hasanuzzaman Department of Geography, Raja N.L.Khan Women's College (Autonomous), Midnapore, West Bengal, India

Aznarul Islam Department of Geography, Aliah University, Kolkata, India

Subrata Jana Department of Geography, Belda College, Belda, Paschim Medinipur, India

Shuvasish Karmokar Department of Geography, Lady Brabourne College, Kolkata, India

Sadik Mahammad Department of Geography, Aliah University, Kolkata, India

Anukul Chandra Mandal Department of Geography, Seacom Skills University, Bolpur, West Bengal, India

Mrinal Mandal Department of Geography, Sidho-Kanho-Birsha University, Purulia, West Bengal, India

Md. Mofizul Hoque Department of Geography, Aliah University, Kolkata, India

Biraj Kanti Mondal Department of Geography, Netaji Subhas Open University, Kolkata, India

Ismail Mondal School of Oceanographic Studies, Jadavpur University, Kolkata, India

Indrasish Mukherjee Department of Geography, Netaji Subhas Open University, Kolkata, India

Jayesh Mukherjee Department of Geography, Presidency University, Kolkata, West Bengal, India;
Centre for the Study of Regional Development, School of Social Sciences, Jawaharlal Nehru University, New Delhi, India

Rameswar Mukherjee Department of Geography, Samsi College, Samsi, West Bengal, India

Uttam Mukhopadhyay Department of Geography, Vidyasagar College, Kolkata, India

Senjuti Nandy Department of Geography, Visva-Bharati, Santiniketan, India

Binod Kumar Nath Assam Survey & Settlement Training Centre, Guwahati, India

A. O. Olusola Department of Geography, University of Ibadan, Ibadan, Nigeria;
Deparment of Geography, University of the Free State, Bloemfontein, South Africa

O. D. Onafeso Department of Geography, Olabisi Onabanjo University, Ago-Iwoye, Ogun-State, Nigeria

Subhadip Pal Department of Geography, University of Calcutta, Kolkata, West Bengal, India

Subodh Chandra Pal Department of Geography, The University of Burdwan, Burdwan, West Bengal, India

Padmini Pani Centre for the Study of Regional Development, Jawaharlal Nehru University, New Delhi, India

Priyank Pravin Patel Department of Geography, Presidency University, Kolkata, West Bengal, India

Ashis Kumar Paul Department of Geography and Environment Management, Vidyasagar University, Midnapore, India

Suvendu Roy Department of Geography, Kalipada Ghosh Tarai Mahavidyalaya, Bagdogra, Darjeeling, West Bengal, India

Sanchita Saha Department of Geography, Netaji Subhas Open University, Kolkata, India

Sumantra Sarathi Biswas Centre for the Study of Regional Development, Jawaharlal Nehru University, New Delhi, India;
Department of Geography, Sukumar Sengupta Mahavidyalaya, Keshpur, West Bengal, India

S. Satheeshkumar Department of Geology, Periyar University, Salem, India

Pravat Kumar Shit Department of Geography, Raja N.L. Khan Women's College (Autonomous), Midnapore, West Bengal, India

S. Venkateswaran Department of Geology, Periyar University, Salem, India

1 Introduction to Drainage Basin Dynamics: Morphology, Landscape and Modelling

Pravat Kumar Shit, Biswajit Bera, Aznarul Islam, Sandipan Ghosh, and Gouri Sankar Bhunia

Abstract

Drainage Basin Dynamics—An Introduction to Morphology, Landscape and Modelling book offers a flexible introduction to the study of drainage basin evolution, system approach, morphology, drainage basin hydrology and sedimentology, connectivity and interaction, human interference, river engineering, natural and anthropogenic hazards and various management techniques. This book enormously covers the responsible factors of sediment yield and rate of delivery through tributaries to the main streams within the zone of production. Subsequently, it highlights the various hydro-geomorphic processes and signatures within river channels as well as on flood plains or specifically in the zone of transportation and zone of accumulation. The evolution of individual components of drainage basins has been systematically discussed and how they are adjusted in the geomorphic time scale over millions of years is also discussed. In the last two or three decades, huge anthropogenic stress such as the construction of linear embankments, revetment, dam, barrage, sluice gate and bridge and execution of different developmental projects on flood plain has been intensified within the river basin. As a result, various geomorphological and hydrological hazards have been significantly amplified particularly in tropical and sub-tropical regions. The main aim of this book is to highlight the magnitude of natural dynamicity and human intervention within the drainage basin and to examine the micro to macro level landforms evaluation through RS, GIS modelling and machine learning algorithms. A total of 24 chapters have been incorporated and almost all geomorphic, hydrological and sedimentological aspects along with methods have been precisely integrated in this book. Similarly, the natural and anthropogenic hazards and their mitigation strategies have also been included in different chapters of this book. This valuable edited volume will definitely help the students, researchers, scholars and policymakers for

P. K. Shit
Department of Geography, Raja N.L. Khan Women's College (Autonomous), Gope Palace, Medinipur, West Bengal 721102, India

B. Bera (✉)
Department of Geography, Sidho-Kanho-Birsha University, Ranchi Road, Purulia, West Bengal 723104, India

A. Islam
Department of Geography, Aliah University, 17 Gora Chand Road, Kolkata 700014, India

S. Ghosh
Department of Geography, Chandrapur College, Purba Bardhaman, Chandrapur, West Bengal, India

G. S. Bhunia
Department of Geography, Seacom Skill University, Santiniketan, West Bengal, India

P. K. Shit et al. (eds.), *Drainage Basin Dynamics*, Geography of the Physical Environment,
https://doi.org/10.1007/978-3-030-79634-1_1

acquiring knowledge and integrated sustainable river basin or watershed management.

Keywords

Drainage basin dynamics · Morphology · Modelling · Landforms evolution · Sustainable basin management

1.1 Introduction

The drainage basin is a unique hydrological or geomorphological unit and it is hierarchically connected with other drainage basins at a lower elevation as the sub-drainage basin. A drainage basin or river basin simply means a topographical area where all forms of water drain through multiple outlets and ultimately these smaller outlets connect with a big outlet. Rivers are the great shapers and integrators in the terrestrial landscape. Naturally, the drainage basin has multiple physical, chemical and biological components, and thus, these components are interconnected with each other. Rivers are not only channelized flow of water rather than a momentum flow of water, sediments, solutes and organisms. A drainage basin is basically a powerful dynamic open system. This integrated complex or complicated geomorphic system operates under the specific environment through the input of energy and output of materials. The critical balance or equilibrium is still maintained through the input and output of energy and materials. Subsequently, the integration of the geomorphic system is performed through processes and landforms dynamics. This system is interlinked with many sub-systems within the drainage basin. This dynamic open system is being interrupted and interacted by surrounding drainage basin or external components. Whatever the scale is considered, the drainage basin is a complex open system due to multiple components, sub-systems, interactions, structures and inputs of materials from the surrounding environment. The interaction between the variables includes thresholds, lag time and feedbacks. The geomorphic feedbacks within the river system signify the various degree or magnitude and different dimensions of interaction at different stretches of the river channels. Both the positive and negative feedbacks regulate directly the river channel components and indirectly the terrestrial flood plain features. Fluvial thresholds connote the abrupt changes in the river courses through the internal or external inputs on different scales. The rivers are not only a channelized flow but various phrases have been used to portray the characteristics and significance of the river and these idioms are popularly applied such as 'the river corridor' or 'the fluvial system' or 'the river system' or 'the river ecosystem'. The ancient history explicitly documented that many old civilizations or settlements established along the river side or on the flood plains and directly or indirectly human society received numerous benefits from different dimensions like water supply, aquatic plants and animals, transport, energy and power, fertile soils and riparian vegetations. The material inputs like water, sediments, solutes and organisms are naturally mobilized from the terrestrial flood plain and transferred into the river channel. These inputs are partially stored on the channel beds and ultimately these materials are transferred through the natural river corridor into the ocean. However, river engineering interrupts the connectivity within the fluvial system. Similarly, human actions (such as soft and hard engineering works in the river system and on the flood plains) alter the natural river flow, sediment supply and channel morphology. The interactions between inputs and valley context bring spatial heterogeneity, non-linear behaviour, connectivity and integrity in the drainage basin. The applied contemporary research and simulation models comprehensively recognize the importance of such fluvial connectivity. The fluvial connectivity explicitly defines the integration of different components within the system and their multifaceted relationship among the variables (Brunsden and Thornes 1979). Hydrologic connectivity refers to the movement of fluid from the sources through the hill slopes into the river channels and eventually along river networks (Pringle 2001; Bracken and Croke 2007). River connectivity simply defines the flow of association of water, sediments, solutes and organisms through tributary and distributaries channel networks (Ward 1997). Sediment connectivity overtly describes the movement and deposition of sediments on channel beds through multiple river channel links (Harvey 1997; Fryirs et al. 2007a, b; Kuo and Brierley 2013; Bracken et al. 2015). Biological connectivity clearly explains the mobility and dispersal of plant and animal communities throughout the river course for their biological needs and survival. Landscape connectivity can refer to the integration of water, sediments and other materials between terrestrial landscape and river system (Brierley et al. 2006). Structural connectivity also defines interlinks among the landforms units within the drainage basin. Functional connectivity basically illustrates the ability of assimilation between the process-specific structural units in the system (Wainwright et al. 2011). In general, the connectivity describes the competency or capacity of material transfer within the drainage basin between the system components like material mobilization from the flood plain into the river channels or networks (Wohl et al. 2019). These networks are the integration of geomorphic compartments (hill slopes, flood plains, etc.), links (confluences) and nodes (multiple channel junctions) in spatial and temporal scale (Kupfer et al. 2014; Heckmann et al. 2015; Passalacqua 2017). The multiple dimensions of connectivity are more important for landform evolution as well as the survival of both terrestrial and aquatic species. It is very important to realize how human alterations affect the fluvial system and broadly the evolution of river basin morphology (Kondolf 2006). Diverse anthropogenic activities significantly reduce river connectivity within the river basin or watershed (Jaeger and Olden 2012). Dam, sluice gate and culvert across the fluvial system efficiently disrupt hydrological and sediment connectivity along the longitudinal courses of the river (Burchsted et al. 2010; Magilligan et al. 2016; Mould and Fryirs 2017). River bank stabilization and construction of levees effectively interrupt the lateral connectivity between river channels and flood plain (Florsheim and Dettinger 2015; Lininger and Latrubesse 2016). Sometimes, flow diversion may amplify the connectivity within the river networks for the endurance of exotic animals (Zhan et al. 2015). Various processes of channelization like dredging and straightening of river courses boost up the longitudinal river connectivity; as a result, the sediments, solutes and organisms can transfer easily through the channel networks. However, the various aspects such as system, feedbacks and thresholds are significantly regulating the drainage basin dynamics. Dynamics simply defines the constant change and progress within the system and the energy or power comes from internal and external forces. Dynamics are different types like water or fluid dynamic, sediment dynamic, off-take and confluence dynamic whereas broadly flood plain dynamic refers to the spatiotemporal evolution of individual landforms on diverse geomorphic platforms. Lithology and structure, tectonic forces, pedological characteristics, geomorphological setup, environmental aspects and socio-cultural entity greatly control the magnitude of dynamics in different segments within the drainage basin. Subsequently, human actions have been intensified in the natural landscape as well as within fluvial systems or river corridors. As a result, various consequences like sediment yield, sediment delivery ratio and accumulation of sediments have been increased in different stretches in the river channels. Through the evolution of space and time, the river channel geometry has been completely transformed and tremendously modified and it is termed as river metamorphosis. A contemporary research highlighted that the morphology of the meandering channel has been totally converted into the braided channel in different pockets of the Himalayan foothills (Bera et al. 2019a). The significant transformation of channel beds of the Chota Nagpur Plateau region of India has been influenced by tectonic events and channel bed structural imprints (Bera et al. 2019b). With the advancement of science and technology, various types of models such as RS and GIS, mathematical, statistical and machine learning algorithms have played a significant role in the study of drainage basin evolution as well as watershed management. The evolution of morphometric parameters needs an examination of linear, aerial and relief factors of the channel network to assess the hydrological system characteristics of the basin area (Horton 1932; Magesh et al. 2011; Majumdar 1982; Nag and Chakraborty 2003; Nautiyal 1994). Furthermore, the formation of flood sediment yields, erosion rates, surface and sub-surface runoff, and flow regime of the drainage system are being estimated applying the morphometric parameters (Gardiner 1990; Ozdemir and Bird 2009). Various software packages are accessible for analysis of drainage basin parameters in Digital Elevation Models (DEMs) and raster grids of satellite images. All these platforms and algorithms are applied for the extraction of morphometric parameters of river networks and the design of various thematic hybrid maps (Harvey and Eash 1996; Lin et al. 2008; Shahzad and Gloaguen 2011a, b; Thomas et al. 2012). The various contemporary model-based studies such as flood prediction, river bank erosion, channel cut-off, sediment yield, channel avulsion, flood plain evolution, watershed management and hydrological modelling have properly assessed the spatiotemporal evolution of drainage basin and subsequently, these models have brought a significant role for the management of contemporary fluvio-hydrological hazard and disaster along with policymaking.

1.2 Key Aims of the Book

Drainage Basin Dynamics—An Introduction to Morphology, Landscape and Modelling book provides a versatile introduction to the study of drainage basin evolution, morphology, drainage basin hydrology and sedimentology, human interference, natural and anthropogenic hazards and various management techniques. This book enormously offers the responsible factors of sediment yield and their absolute and specific growth and rate of delivery through tributaries to the main streams. Rivers are important geomorphic agents which reflect an amazing variety of forms and behaviour, showing the wide range of natural environments in which they are originated. The drainage system evolution and spatial network development within the dynamic nature are being discussed and how they are adjusted in the geomorphic time scale over millions of years is also discussed. This book shows how drainage systems function and react to change and why this thoughtful is required for flourishing integrated basin management. In tropical and subtropical countries, population pressures as well as different developmental projects are being executed in the drainage basin without proper planning. Today scientists consider the drainage basin as an administrative unit during the implementation of regional projects. In this context, this book will carry a benchmark for scholars and young scientists.

In addition, it will provide supporting material and fundamental knowledge for both undergraduate and postgraduate students. More specifically, this book will definitely help to think of a new research outline, various methods and research metadata for scholars and earth scientists. Since Drainage Basin Morphology, Landscape and Modelling is a speedily intensifying, multidisciplinary scope, the book will also be of curiosity to those who want a wide impression of the subject, such as hydrologists, geologists, ecologists and engineers.

Most of the Physical geography or Geomorphology textbooks are currently available in the market only for undergraduate and postgraduate students or beyond. Similarly, most of the students and scholars are struggling to know the basic concepts and mathematical applications. Different chapters are being organized to build on itself throughout the book for the interest of students, scholars and researchers. Chapter-wise concepts, widely accepted methods and techniques as well as case study with examples should be highlighted. New terms and methods should be illustrated using bold text and an extensive glossary at the end of the book. Selected articles will be incorporated to bring a new shape and signature and to fulfill the overall impression of the book as well as to comprehend the demand of the students and scholars through the changing paradigm.

1.3 Organization of This Volume

The title of this book '*Drainage Basin Dynamics—An Introduction to Morphology, Landscape and Modelling*' reflects the intrinsic connections among drainage basin dynamics, morphology, landscape and modelling. The contemporary researches in this field have been amalgamated to comprehend the various methods for the evolution of geomorphic landforms and combating the hydro-geomorphic hazards and disaster.

Chapter 2 covers the morphological dynamics, erosion potential and morphogenesis of lateritic badlands in Bengal Basin, India. The current study examines the required conditions for gully formation and morphological diversity of gullies to comprehend the geomorphic interpretation (viz., anomalies of geomorphic indices, inherent and extrinsic changes, triggers and geochronology of gully initiation) and quantitative inference of hydro-geomorphic processes (viz., rainfall-runoff simulation, sediment yield and erosion indices) in the lateritic badlands. These are developed as elevated interfluves (i.e., *Rarh* Bengal) of Neogene—Early Quaternary laterites, dissected by the west–east-flowing peninsular rivers and streams in the western shelf zone of the Bengal Basin.

Chapter 3 covers the plan shape geomorphology of the alluvial valley in the middle-lower and deltaic courses of the Subarnarekha river basin, India. Significant techniques like geospatial techniques, repeated field observations, total station survey and sedimentological analysis of bank facies have been considered to investigate the spatial diversity of plan shape geomorphology in different sections of the studied river course. The study divulges that the course of the river sections brings diverse geometry of meander properties with discontinuous straight courses and wider valleys. The presence of palaeo-shorelines, ancient delta-fan lobes and cut and fill terraces along the courses of the river valley shows the role of dynamic marine and fluvio-marine environment within the river basin.

Chapter 4 highlights the quantitative assessment of channel planform dynamics and meander bend evolution of River Ramganga (Ganga Basin), India. This study focuses on the causes and dimensions of planform dynamics through quantitative assessment and the study tried to bring the human actions for the modifications of channel planform as well as the evolution of channel corridors.

Chapter 5 focuses on the change of flow regime, impacts on the peak flood discharge, sedimentation process, seasonal flow pattern and morphology of River Barakar. It is the main tributary river of Damodar River in the Chhotanagpur plateau fringe region of India and the flow characteristics are altered by human interventions. In general, the trend of water discharge has been decreased from 1980 to 2013 at both the gauge stations due to the result of human intervention through the modification of water discharge from the dams.

Chapter 6 presents landscape characterization using geomorphometric parameters for a small sub-humid river basin of the Chotanagpur Plateau, Eastern India. The statistical method like principal component analysis has been used to demarcate distinct physiographic/landscape entities. Authors have used the classical and modern morphometric methods along with a higher resolution Digital Elevation Model (DEM) to examine the terrain characteristics of the Kharkai River Basin. Multi-temporal land cover and land use layers extracted from Landsat datasets were overlain on the extracted terrain units to estimate changes in the same across different landscape types. The scientific study shows that the terrain units are strongly correlated with the lithology and here, the local slope and drainage system are largely controlled by underlying lithology.

Chapter 7 emphasizes on River Raidak-I migration dynamics within the Himalayan Foreland Basin applying quaternary sedimentological bank facies and geospatial techniques. Natural processes such as rapid sedimentation due to low channel gradient, high discharge during monsoon and non-cohesive bank materials primarily regulate the channel migration and erosion–accretion dynamics of the river system within the quaternary geological site of the sub-Himalayan alluvial flood plain of West Bengal. In addition, human interference in the fluvial setting also influences the channel dynamics in this area. The current study portrays how the fluvial system or river corridor is being disconnected from the flood plain and longitudinal sediment, and structural connectivity has been partially affected by the anthropogenic stress within fast-changing landscape.

Chapter 8 covers the spatiotemporal variation of morphological characteristics of the River Bhagirathi within Murshidabad district, West Bengal (India). The different flood plain geomorphic features like active river channel / main river course, tributary river channel, the seasonal stretch of waterways, meander scar, paleochannel, shallow depression, deep depression, oxbow lake and wet sand have been identified applying multi-temporal Landsat satellite data (1990–2017) and Normalized Difference Water Index.

Chapter 9 portrays the sedimentation and shifting of the lower Mundeswari and Rupnarayan River, West Bengal, India. Applying the modern fluvio-hydrological, pedological and sedimentological techniques, the authors have highlighted the natural as well as anthropogenic factors behind unidirectional shifting tendency along the active DamodarFault line. This field-based study also brought how unscientific human activities are playing a significant role to invite flood vulnerability within the lower stretch of this river.

Chapter 10 covers the role of controlling factors for the development of drainage networks around Rajmahal Hills, Jharkhand and West Bengal state of India. The field and the geospatial-based study provides ample evidence about the development of drainage networks and here, geology and geomorphology along with high rainfall play a vital role in the evolution of landforms as well as drainage networks. This chapter also shows the important association between lithology and landforms evolution.

Chapter 11 focuses on the role of morphometric attributes for the evolution of micro landforms and drainage networks in the Kopai river basin of West Bengal, India. Despite the application of classical techniques, geospatial and statistical techniques have been used to identify the micro landforms as well as changing paradigms of hydro-geomorphological processes.

Chapter 12 highlights the impact assessment of the check dam in the pappiredipatti watershed (South India) using LULC and NDVI signatures. Geospatial techniques have been used to assimilate spatial information pertaining to functional characteristics of land cover within buffering distances in the proposed site of watersheds. The results of the Landsat data specify a decline in the dense vegetation area that has led to widespread drought conditions in some parts of the watershed.

Chapter 13 covers the morphometric analysis of sub-watershed and prioritization of coastal regions of Tamil Nadu in India. In this current study, authors have tried to comprehend the hydro-geomorphological characteristics using Digital Elevation Model (DEM) which is obtained from the Shuttle Radar Topographic Mission (SRTM) with 30 m spatial resolution. The geospatial techniques extensively assist to identify the drainage basin characteristics for sustainable watershed management.

Chapter 14 highlights the spatiotemporal variation of channel migration and vulnerability assessment of River Bhagirathi of West Bengal, India. Satellite images of Landsat 5 Thematic Mapper (TM) of 1987, 1997, 2008, and Landsat 8 Operational Land Imager (OLI) image of 2018 have been used to detect the changing bank lines. All the data are processed on ArcGIS 10.3.1 platform to prepare the maps of the river course. This current study deeply investigated the natural as well as anthropogenic causes of river bank migration and fluvial dynamics of river Bhagirathi applying fluvio-hydrological and geospatial techniques.

Chapter 15 describes the landscape transformation of an alluvial channel of the Himalayan Foothill during 1987–2020 using geospatial technology. The current study brings significant evidence of spatiotemporal geomorphic evolution within fast-changing landscape of the

Himalayan foreland basin. This chapter provides various types of avulsion mechanisms of river Kaljani with their tributaries and also focussed on geomorphic instability along with sustainable relocation of settlements on the vulnerable flood plain.

Chapter 16 highlights the river flow, sediment flux, anthropogenic impacts and river dynamics of South-East Asian dams and Indian rivers. Mean monthly discharge data of 270 months (Jan, 1998 to June, 2020) from Dartmouth Flood Observatory (DFO) with ten gauge stations of major Indian rivers have been used to estimate discharge volume at different return periods (2.3 years (Y), 5Y, 10Y, 20Y, 50Y, 75Y, 100Y, 125Y and 200Y) using Log Pearson III flood frequency method. The hydro-geomorphic study highly focuses on the role of the dam to alter the characteristic of sediment and hydrology of a drainage basin.

Chapter 17 covers the contemporary techniques and integrated approach of river health assessment especially based on physico-chemical parameters for the River Subarnarekha, India. Rapid alterations of land use and land cover like urbanization, industrialization, intense agriculture, etc. have been intensified for the deterioration of the river health. The authors have tried to establish how the river ecosystem has been damaged in different stretches of the above-mentioned river. The field-based scientific study has strongly recommended some relevant mitigation measures to restore river health.

Chapter 18 introduces an automatic approach to classify the river network based on Strahler's order using Python language which is an object-oriented, general-purpose and high-level programming language. The approach is an end-to-end execution wherein the Python program needs only digital river network data as input and output the same digital river network data with added a field of Strahler's order. This chapter fundamentally highlights Strahler's stream order computing on digital stream network dataset for wide applications like flood risk assessment, water resource management, flood inundation mapping, watershed management, etc.

Chapter 19 focuses on the morphometry-based sub-watershed prioritization for flood potentiality analysis of the Gumani river basin (India) using the TOPSIS model. Various hydro-geomorphological parameters have been considered to identify the flood potentiality analysis, and the relative weights of the sub-watersheds have been computed using the Technique for Order of Preference by Similarity to Ideal Solution (TOPSIS) which ranks the alternative based on the closest distance to the ideal solution and the farthest distance to the negative-ideal solution. The field-based applied study also highlights the mitigation strategies and micro-planning for the sustainable livelihood of the local people.

Chapter 20 covers the NRCS-Curve Number Method and geospatial techniques for runoff estimation of the Kolong river basin in Assam. Subsequently, stream flow, Hydrologic Soil Groups (HSGs), slope and land use land cover maps have been prepared using satellite images on RS and GIS environments. This scientific study strongly suggested about the availability of hydrological resources and land use and land cover characteristics of the river basin for sustainable water resources management and agricultural development.

Chapter 21 focuses on the role of drainage basin for geomorphological processes-forms dynamics especially of third-order basin in Southern Nigeria. Topographic properties have been extracted using System for Automated Geoscientific Analysis (SAGA) version 4.0.0, GIS software. Rockware software was used in creating rose diagrams for lineament patterns and channel networks. Long profile is also prepared from Digital Elevation Models, DEMs (30 m) within sub-basins in Upper Ogun River Basin. This geospatial study showed the inherent relation between lithology and geomorphic attributes for the evolution of individual landforms.

Chapter 22 covers multi-criteria-based morphometric prioritization for soil erosion susceptibility and denudation rate assessment of Purulia District, India. Erosion susceptibility and denudation rate have been measured based on multi-criteria morphometric prioritization and hierarchical anomaly index using remote sensing and GIS techniques. All the morphometric parameters divulge the interrelations between hydrological processes and their related geomorphic landforms. This chapter brings the varying degree of susceptibility along with appropriate hydro-geological factors which have been responsible for spatiotemporal variations under the Precambrian dominant granitic rock.

Chapter 23 highlights the geomorphic appraisal of active tectonics and fluvial anomalies in the Peninsular Rivers of the Bengal Basin (West Bengal, India). Many scientific studies documented that the peninsular river system (viz., Brahmani, Dwarka, Mayurakshi, Ajay, Damodar, Dwarkeswar, Silai and Kasai river basins) are directly influenced by the underlying geological structures and *en echelon* faults. The present study attempts to comprehend the significant role of tectonic elements, geomorphometric anomalies and soft-sediment deformation structures on the alluvial river valleys and Quaternary flood plains using seismic information, proxy data and geomorphic indices of active tectonics, thematic mapping and stratigraphic analysis of depositional facies.

Chapter 24 emphasizes the geomorphic evolution of the Bhagirathi flood plain in West Bengal, India, using geospatial technology. The result of this study achieved from investigating satellite imagery and Survey of India (SoI) toposheets and its propositions is a valuable data source to know the channel planform dynamics and will assist as a decision-making tool for implementing drainage development works. This geospatialbased scientific study recommended soft and hard engineering techniques for the mitigation of hydro-geomorphic hazards and disasters.

References

Bera B, Bhattacharjee S, Ghosh A, Ghosh S, Chamling M (2019a) Dynamic of channel potholes on Precambrian geological sites of Chhota Nagpur plateau, Indian peninsula: applying fluvio-hydrological and geospatial techniques. SN Appl Sci 1(5):494. https://doi.org/10.1007/s42452-019-0516-2

Bera B, Bhattacharjee S, Roy C (2019b) Estimating stream piracy in the lower Ganga plain of a Quaternary geological site in West Bengal, India applying sedimentological bank facis, log and geospatial techniques. Curr Sci 117(4):662–671. https://doi.org/10.18520/cs/v117/i4/662-671.

Bracken LJ, Croke J (2007) The concept of hydrological connectivity and its contribution tounderstanding runoff-dominated geomorphic systems. Hydrol Process 21:1749–1763

Bracken LJ, Turnbull L, Wainwright J, Bogaart P (2015) Sediment connectivity: a framework for understanding sediment transfer at multiple scales. Earth Surf Proc Land 40:177–188

Brierley GJ, Fryirs KA, Jain V (2006) Landscape connectivity: the geographic basis ofgeomorphic applications. Area 38:165–174

Brunsden D, Thornes JB (1979) Landscape sensitivity and change. Trans Inst Br Geogr 4(4):463–484

Burchsted D, Daniels M, Thorson R, Vokoun J (2010) The river discontinuum: applying beaver modifications to baseline conditions for restoration of forested headwaters. Bioscience 60:908–922

Florsheim JL, Dettinger MD (2015) Promoting atmospheric-river and snowmelt-fueledbiogeomorphic processes by restoring river-floodplain connectivity in California's Central Valley. In: Hudson PF, Middelkoop H (eds) Geomorphic approaches to integrated floodplain management of lowland fluvial systems in North America and Europe. Springer, New York, pp 119–141

Fryirs KA, Brierley GJ, Preston NJ, Kasai M (2007a) Buffers, barriers and blankets: the(dis)connectivity of catchment-scale sediment cascades. CATENA 70:49–67

Fryirs KA, Brierley GJ, Preston NJ, Spencer J (2007b) Catchment scale (dis)connectivity in sediment flux in the upper Hunter catchment, New South Wales, Australia. Geomorphology 84:297–316

Gardiner V (1990) Drainage basin morphometry. In: Goudie AS (ed) Geomorphological techniques. Unwin Hyman, London, pp 71–81

Harvey AM (1997) Coupling between hillslope gully systems and stream channels in the Howgill Fells, northwest England: temporal implications. Geomorphologie: Relief Processus Environ 3:3–19

Harvey CA, Eash DA (1996) Description, instructions, and verification for Basin soft, a computer program to quantify drainage basin characteristics, U.S. Geologic Survey Water Resources Investigations Report 25:95–4287

Heckmann T, Schwanghart W, Phillips JD (2015) Graph theory—recent developments of its application in geomorphology. Geomorphology 243:130–146

Horton RE (1932) Drainage-basin characteristics. Trans Am Geophys Union 13:350–361

Jaeger KL, Olden JD (2012) Electrical resistance sensor arrays as a means to quantify longitudinal connectivity of rivers. River Res Appl 28:1843–1852

Kondolf GM (2006) River restoration and meanders. Ecol Soc 11(2):42

Kuo CW, Brierley GJ (2013) The influence of landscape configuration upon patterns of sediment storage in a highly connected river system. Geomorphology 180–181:255–266

Kupfer JA, Meitzen KM, Gao P (2014) Flooding and surface connectivity of Taxodium-Nyssastands in a southern floodplain forest ecosystem. River Res Appl 31:1299–1310

Lininger KB, Latrubesse EM (2016) Flooding hydrology and peak discharge attenuation along the middle Araguaia River in central Brazil. CATENA 143: 90–101

Lin WT, Chou WC, Lin CY, Huang PH, Tsai JS (2008) Win basin: Using improved algorithms and the GIS technique for automated watershed modeling analysis from digital elevation models. Int J Geogr Inf Sci 22 (1):47–69

Magesh NS, Chandrasekar N, Soundranayagam JP (2011) Morphometric evaluation of Papanasam and Manimuthar watersheds, parts of Western Ghats, Tirunelveli district, Tamil Nadu, India: AGIS approach. Environ Earth Sci 642:373–381

Magilligan, F.J., Graber, B.E., Nislow, K.H. et al. (2016). River restoration by dam removal: enhancing connectivity at watershed scales. Elem: Sci Anthropocene, 4. https://doi.org/10.12952/journal.elementa.000108

Majumdar JP (1982) Morphometric analyses of the 4th order drainage watersheds of the Khowai river basin, Tripura, India—some preliminary results and observations. J Indian Soc Remote Sens 10(3):49–53

Mould S, Fryirs K (2017) The Holocene evolution and geomorphology of a chain of ponds, southeast Australia: establishing a physical template for river management. CATENA 149:349–362

Nag SK, Chakraborty S (2003) Influence of rock types and structures in the development of drainage network in hard rock area. J Indian Soc Remote Sens 31(1):25–35

Nautiyal MD (1994) Morphometric analysis of a drainage basin, district Dehradun, Uttar Pradesh. J Indian Soc Remote Sens 22(4):251–261

Ozdemir H, Bird D (2009) Evaluation of morphometric parameters of drainage networks derived from topographic maps and DEM in point floods. Environ Geol 56:1405–1415

Passalacqua P (2017) The Delta connectome: a network-based framework for studying connectivity in river deltas. Geomorphology 277:50–62

Pringle CM (2001) Hydrologic connectivity and the management of biological reserves: a global perspective. Ecol Appl 11:981–998

Shahzad F, Gloaguen R (2011a) TecDEM: A MATLAB based toolbox for tectonic geomorphology, part 1: Drainage network preprocessing and stream profile analysis. Comput Geosci 37:250–260

Shahzad F, Gloaguen R (2011b) TecDEM: A MATLAB based toolbox for tectonic geomorphology, part 2: Surface dynamics and basinanalysis. Comput Geosci 37:261–271

Thomas J, Joseph S, Thrivikramji KP, Abe G, Kannan N (2012) Morphometrical analysis of two tropical mountain river basins of contrasting environmental settings, the southern Western Ghats India. Environ Earth Sci 66(8):2353–2366

Wainwright J, Turnbull L, Ibrahim TG et al (2011) Linking environmental regimes, space and time: interpretations of structural and functional connectivity. Geomorphology 126:387–404

Ward JV (1997) An expansive perspective of riverine landscapes: pattern and process across scales. River Ecosyst 6:52–60

Wohl E, Brierley G, Cadol D et al (2019) Connectivity as an emergent property of geomorphic systems. Earth Surf Proc Land 44:4–26

Zhan A, Zhang L, Xia Z, et al (2015) Water diversions facilitate spread of non-native species. Biol Invasions 11:3073–3080

2 Morphological Dynamics, Erosion Potential and Morphogenesis of Badlands in Laterites of the Bengal Basin, India

Sandipan Ghosh and Suvendu Roy

Abstract

The landforms of badlands, mainly gullies, record accelerated water erosion into variable regolith or soft sediments or colluviums by integrated processes of overland flow, mass-wasting, channel erosion and pipe flow, mostly observed in the semi-arid and tropical wet–dry climatic regions. Gully can be considered as an ideal geomorphic unit, and badlands come into sight to offer the geomorphic laboratory of model landscape (with a shortened temporal scale), the active fluvial processes and resultant ravine landforms which may be influenced by human activities in the Anthropocene. This chapter examines the controlling or triggering factors (intrinsic and extrinsic changes) for gully formation and morphological diversity of gullies (anomalies of geomorphic indices) to realize the geomorphic evolution (geochronology of gully initiation) and quantitative inference of hydro-geomorphic processes (rainfall–runoff simulation, sediment yield and erosion indices) in the lateritic badlands of West Bengal. These terrains are developed as elevated interfluves (i.e. *Rarh* Bengal) of Neogene—Early Quaternary Laterites, dissected by the west–east-flowing peninsular rivers and streams in the western shelf zone of the Bengal Basin. The study has analysed that tropical weathering of laterite profiles, mass-wasting and water erosion processes on the badland slopes exhibit complex spatial variability of drainage network (i.e. topologically distinct channel networks and fractal dimension of drainage networks) and morphological attributes (e.g. hypsometric integral, stream-length gradient index, basin shape indices, relief ratio, gully density, length of overland flow and constant of channel maintenance), and the forms and processes are quantitatively analysed to deal with the rate of erosion and factors of gully initiation. The study investigates the critical topographic condition, expressed by local slope gradient (S) and drainage area (A), controlling the development and position of gully heads in the landscape. Here, the model of landscape connectivity is introduced as a crucial determinant of badland morphology and evolution, considering a theoretical energy-utilization perspective of an open geomorphic system.

Keywords

Gully · Badlands · Topology · Connectivity · Threshold · Sediment yield

S. Ghosh (✉)
Department of Geography, Chandrapur College, Purba Bardhaman, Chandrapur, West Bengal, India

S. Roy
Department of Geography, Kalipada Ghosh Tarai Mahavidyalaya, Darjeeling, West Bengal, India

P. K. Shit et al. (eds.), *Drainage Basin Dynamics*, Geography of the Physical Environment,
https://doi.org/10.1007/978-3-030-79634-1_2

2.1 Introduction

During the past 55 years, many studies have documented the magnitude of soil erosion problems in different parts of the world (Table 2.1), expressed as billions of tons of eroded soil or billions of dollars of erosion and sedimentation damage in each year (Ahmad 1968; Narayana and Babu 1983; Lal 1990; Singh et al. 1992; Kothyari 1996; Wasson 2003; Vente and Poesen 2005; Pimentel 2006; Reddy and Galab 2006; Thakkar and Bhattacharyya 2006; Casali et al. 2009; Kumar and Pani 2013; Pimentel and Burgess 2013; Sharda et al. 2013; Sharda and Dogra 2013; Froehlich 2018; Sharma 2018; Poesen 2018; Pennock 2019). It is found that the ravines or badlands along the banks of the Yamuna, Chambal, Mahi, Tapti and Krishna Rivers revealed soil losses exceeding 18 t ha^{-1} $year^{-1}$, and the mean erosion rates on the alluvial Indo-Gangetic Plains of Punjab, Haryana, Uttar Pradesh, Bihar and West Bengal were 5–10 t ha^{-1} $year^{-1}$ (Singh et al. 1992). Interestingly, gully erosion represents a key sediment yield process in the landscape, generating between 10 and 95% of total eroded mass at basin scale, whereas the networks of the gully often occupy less than 5% of the total basin area (Poesen 2011; Sinha and Joshi 2012). The reservoirs and dams of India are losing about 1.3 billion m^3 of storage capacity each year due to high sediment yield of watersheds and siltation. That should be alarming enough for everyone as at today's rates, the creation of 1.3 billion m^3 storage capacity would cost Rs. 1448 crores (Thakkar and Bhattacharyya 2006). In this country, major rainfed crops suffer an annual production loss of 13.4 Mt due to water erosion which amounts to a loss of Rs. 205.32 billion in monetary terms (Bawa 2017; Sharda et al. 2019). Now, the major point of research interest is the laterite terrain of West Bengal, eastern India (known as *Rarh* Bengal, i.e. the land of red soil), which is severely dissected by the dense network of rills and gullies, developing miniature forms of watersheds and badlands which refers to regions that soft and poorly consolidated material outcrops, limited vegetation, reduced or no human activity and a wide range of geomorphic processes, such as weathering, gully erosion, mass movements and piping (Bryan and Yair 1982; Martinez-Murillo and Nadal-Romero 2018). Here, realistic assessment of badland morphology, gully erosion risk or soil loss rate of lateritic badlands thus constitutes the first step for understanding the ground reality of erosion and raising awareness among governmental and other stakeholders in a given region to adopt appropriate strategies for erosion protection.

Linear erosion process by concentrated overland flow is a natural phenomenon that has accelerated in Anthropocene due to climate changes, transformation of land use and land management practices with downbeat impact on agriculture. Gully erosion is a significant and intensive process of soil erosion, which accounts for 10–94% of watershed sediment export by flowing water (Poesen et al. 2003; Dube et al. 2020). The morphology of water erosion are either rills or gullies, with rills corresponding to the smaller structures with incision depths below 0.25 m and gullies corresponding to features deeper than 0.25 (Poesen 2018; Dube et al. 2020). A gully is normally defined as a severely incised channel on a valley side or hillside, generally cut by flow water, and often containing an ephemeral flow, and also gullies are generally isolated fluvial features that cut into a smoother pre-existing surface, and typically rectangular or *U*-shaped or *V*-shaped in cross-section, having one or many head cuts and kinckpoints in channel profile (gully cross-section area may be greater than 928 cm^2) (Bull and Kirkby 1997; Poesen et al. 2003; Kirkby and Bracken 2009). Gullies or ravines or badlands are getting more importance in recent decades because these changes in landscapes have severe economic consequences and may disrupt land productivity and livelihoods. According to Torri and Poesen (2014), the lines or fluvial networks of preferential connection is observed in between uplands and the main channel or floodplain, and this system has enough ability to modify water and sediment connectivity during intense rainstorms.

Table 2.1 Important research works on the gully geomorphology and badland evolution

Title of article/Book chapter/Book	Authors	Remarks
Distribution and Causes of Gully Erosion in India	Ahmad (1968)	One of earliest articles on the genesis and evolution of gullies in Indian terrains
Rill and Gully Erosion in the Subhumid Tropical Riverine Environment of Teonthar Tahsil, Madhya Pradesh, India	Singh and Agnihotri (1987)	One of earliest attempts to address gully erosion in the terrain of India; and developing gully erosion research as part of Indian fluvial geomorphology
Drainage evolution in a badland terrain at Gangani in Medinipur District, West Bengal	Bandyopadhyay (1988)	Geomorphic explanation of badland evolution in the Gangani badlands
Gully processes and modelling	Bull and Kirkby (1997)	Exploring the processes of gully erosion and modelling the flow erosion processes
Gully Erosion and Management: Methods and Application	Singh and Dubey (2002)	An important book having key chapters devoted to outlook of research design, morphological techniques of erosion assessment and field measurements of soil erosion
Geomorphological Investigation of Badlands: A Case Study at Garhbeta, West Medinipur District, West Bengal, India	Sen et al. (2004)	Analysing spatio-temporal extent of gullies, morphological dynamics of gullies and evolutionary stages of badland development in the Gangani badland, Garhbeta (Paschim Medinipur)
Gully Erosion: Impacts, factors and Control	Valentin et al. (2005)	Key geomorphic aspects of gully erosion across the world, components of gully, topographic threshold and controlling factors of erosion
Soil erosion science: reflections on the limitations of current approaches	Boardman (2006)	Indentifying the research achievements, advances and gaps in the study of soil erosion
Connectivity as a crucial determinant of badland morphology and evolution	Faulkner (2008)	Introducing the concept of coupling and landscape connectivity to explain the evolution of badlands in the semi-arid landscape
Principles of Soil Conservation and Management	Blanco and Lal (2008)	Evaluating the various aspects of water erosion, effect of erosion, quantitative modelling and strategies of soil conservation sustainable for farmers
Progress of Researches in Ravines and Gullies Geomorphology in India	Sharma (2009)	Depicting the various ideas and concepts related to ravines of Chambal Valley and analysing the theoretical models of ravine development in India
Degraded Lateritic Soils Cape and Land Uses in Birbhum District, West Bengal, India	Jha and Kapat (2011)	Emphasizing the distinctiveness of laterite soils, morphology of gullies and land degradation processes in Birbhum District of West Bengal
Handbook of Erosion Modelling	Morgan and Nearing (2011)	Important empirical and physical-based erosion model used in various spatial and temporal scales
A review of topographic threshold conditions for gully head development in different environments	Torri and Poesen (2014)	Examining the slope–area threshold model to understand the critical condition of gully initiation in different parts of the world

Due to the importance of gullies or ravines or badlands as a sediment source of watershed, gully erosion needs to be better understood, managed and its effects mitigated (Torri and Poesen 2014). Rills and gullies are primarily initiated through two key hydrological processes: (1) surface flow transport soil particles, and (2) subsurface flow resulting in piping. The concentration of surface flow is a function of rain intensity, infiltration rate, land use type and soil properties that govern water infiltration rate and runoff rate mostly in the catchments of gullies (Kirkby and Bracken 2009). In contrast, piping can be a result of abrupt retardation of infiltration rate by a less permeable sub-soil layer, which facilitates high subsurface lateral flow, and the tunnels eventually collapse leading to the initiation of the linear features (Dube et al. 2020). The gully systems, once formed, tend to grow larger and are very difficult and costly to eradicate. It is therefore important to understand the processes of gully formation and the sensitivity of areas to gullying, in order to recommend and adopt the preventive measures that will minimize the risks of new gullies and slow or reverse the growth of existing systems (Bull and Kirkby 1997; Kirkby and Bracken 2009; Poesen 2018).

Gully is an active geomorphic unit in surface evolution and is widely distributed in the world, mostly in the badlands. Given the special position of gullies on the earth's surface, the spatial distribution, influential factor and the evolutionary process of gully landforms have been a focus in geomorphology and erosion research (Erskine 2005; Li et al. 2019). The morphology of gullies or other concentrated flow erosion channels (e.g. length, width, depth, surface area of gully top and volume) found all over the world vary highly in respect of sediment characteristics and climatic characteristics (Bocco 1991; Ionita 2003; Poesen et al. 2003; Valentin et al. 2005; Torri et al. 2006; Vanmaercke et al. 2012; Wells et al. 2016; Poesen 2018). Previous literature reviews on linear erosion have considered the main mechanisms of channel incision and development and the main factors of control and modelling approaches (Poesen et al. 1998; Valentin et al. 2005; Casali et al. 2006; Torri et al. 2006; Vanmaercke et al. 2012; Poesen 2018; Li et al. 2019). Many geographers and geomorphologists (Niyogi et al. 1970; Biswas 1987; Bandyopadhyay 1988; Sen et al. 2004; Sen 2008; Ghosh and Maji 2011; Shit and Maiti 2012; Ghosh 2013; Shit et al. 2013; Roy and Sahu 2016; Ghosh and Guchhait 2017 and 2020; Shit and Adhikary 2020; Islam et al. 2020; Saha et al. 2020; Sarkar et al. 2020; Mahala 2020) have also explored and analysed the morphodynamics and erosion processes in the badlands of West Bengal which are mostly developed over the Early–Middle Quaternary secondary laterites (i.e. *Rarh* Bengal) of the Bengal Basin western shelf zone. These studies yielded important scientific and, to our knowledge, systematic compilation of the morphological characteristics of gully erosion channels found under different environments worldwide and West Bengal has been made, but there are a need for new-fangled or innovative research outputs in soil erosion study to explore the morphogenesis and erosion potential of lateritic badlands using drainage basin morphometry, hydrological models, geo-statistical tools and advanced geomatics. It is desirable to stress more quantitative information on gully morphology (indentifying the forms, anomalies and regional patterns) for a better understanding of controlling factors and landform evolution, because a gully can be considered as a geomorphic unit of study, and the badlands can be recognized as a model landscape to investigate the process–form relationship. The main objective of this chapter is to provide the quantitative hydro-geomorphic expressions of gully topology, morphology, geochronology and channel erosion potential (i.e. topographic thresholds), and finally, evolutionary model of *Rarh* badlands which are fundamentals to achieve an improved understanding of the formation mechanism of gullies and resultant landform anomalies.

2.2 Geomorphic Identities of Study Area

In West Bengal, the lateritic uplands or upland red soil groups (Singh et al. 1998) occur along a NE–SW trending belt parallel to the western margin of Bengal Basin. This unique geomorphic region (i.e. western shelf zone of Bengal Basin) is designated as *Rarh* Bengal by Bagchi and Mukherjee (1983). The zones of laterites with Neogene gravel deposits are the main spatial unit of study, bounded by the latitude of 21°30′ to 24° 40′ N and longitude of 86°45′ to 87°50′ E (Fig. 2.1). The duricrusts, ferruginous gravels and kaolinite deposits (from Rajmahal Basalt Traps to Subarnarekha Basin) appeare as interfluves (uplifted terrain in between 90 and 40 m from msl) and border this province to make the transitional diagnostic tropical weathering surface and distinct sedimentary Early Quaternary lithofacies in between the Archaean–Gondwana litho-unit at the west and the Late Quaternary alluvial litho-unit of Bengal Basin at the east (Niyogi et al. 1970; Niyogi 1975; Biswas 1987; Das Gupta and Mukherjee 2006). The west–central part of West Bengal is neo-tectonically influenced by the basin margin faults systems (i.e. Chota Nagpur Foothill Fault, Garh Moyna–Khandoghosh Fault and Pingla Fault), and the zone of laterites is placed in between the fault systems as an uplifted block.

It is observed that The lateritic *Rarh* region and plateau fringe (covering districts of Purulia, Bankura, Paschim Bardhaman, Birbhum and Paschim Medinipur) show lower T value (soli-tolerance level) ranging from 2.5 to 5.0 Mg ha^{-1} year^{-1} (Mondal and Sharda 2011; Lenka et al. 2014), signifying high susceptibility to rill and gully erosion. A parallel west–east-flowing (due to general west to east trending slope of

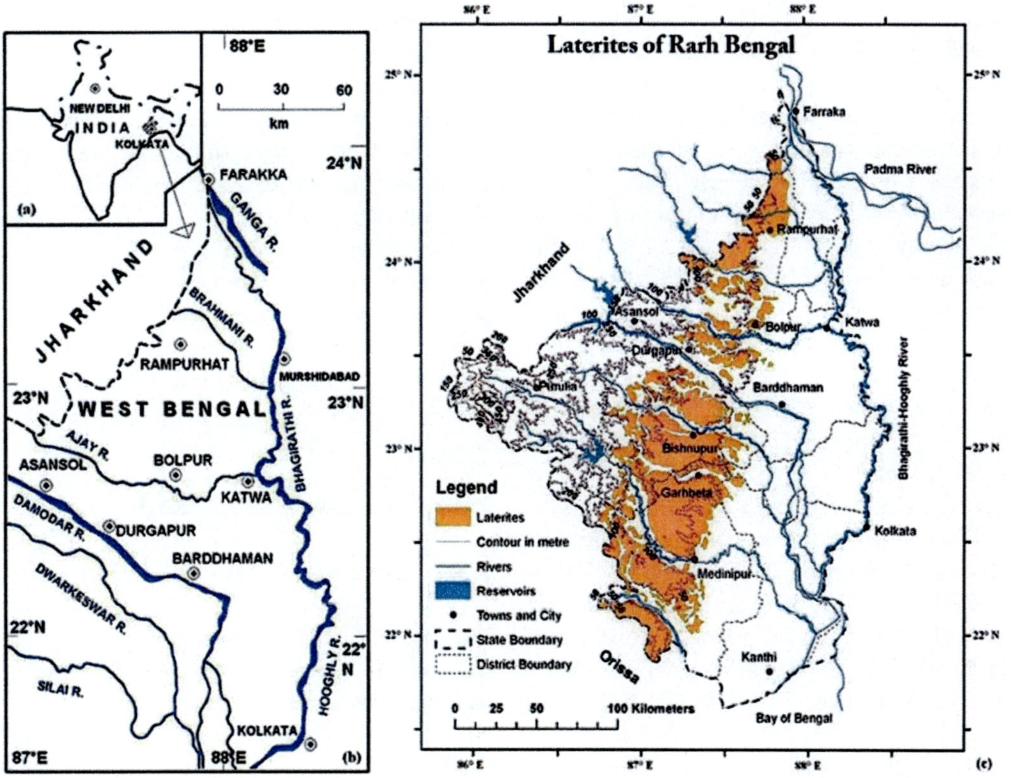

Fig. 2.1 Distribution of *Rarh* Laterites and the dissection patches (interfluves in between the peninsular west–east flowing rivers) of badlands in the western shelf zone of Bengal Basin

underlying structure) peninsular drainage system (viz., Brahmani, Dwarka, Mayurakshi, Ajay, Damodar, Dwarakeswar, Silai, Kangsabati and Subarnarekha rivers) dissects the lateritic *Rarh* region into patches of badlands and tropical deciduous forests of West Bengal (an area of 7,700 km^2). The tropical plant species include *Shorea robusta*, *Madhuca indica*, *Terminalia chebula*, *Eucalyptus globulus*, *Tectona grandis* and *Acacia auriculaeformis* (Ghosh and Guchhait 2015). This region bears the characteristics of a tropical hot and sub-humid type of monsoon climate (mean annual rainfall of 1200–1600 mm), controlled mainly by proximity to the Bay of Bengal in the south and the alignment of the Himalayas in the north (Singh et al. 1998).

The spatial scale of geomorphic study is the watershed or basin of a gully in the badlands. The drainage basin is an ideal geomorphic unit, and it may be defined as the area which contributes water to a particular channel or set of channels (Leopold et al. 1992). It is the source area of precipitation eventually provided to the stream channels by various paths. As such, it forms a convenient unit of the earth's surface within which basic climatic quantities can be measured and characteristics landforms described, and a system within which a balance can be struck in terms of inflow and outflow of moisture and energy (Leopold et al. 1992). So, in this study gully basins of six badlands (laterite terrain) are selected to analyse the topology, morphology and erosion processes. The geographical extensions of these badlands are depicted as (Figs. 2.2, 2.3 and 2.4) follows: (1) Bhatina Badlands of Birbhum (24°12′26″ to 24°08′57″ N, 87°39′52″ to 87°44′35″ E), (2) Surul Badlands of Birbhum (23°40′50″ to 23°40′04″ N, 87°38′29″ to 87°39′35″ E), (3) Radhamohanpur Badlands of Paschim Bardhaman (23°35′42″ to 23°34′45″ N, 87°29′41″ to 87°30′57″ E), (4) Birsingha Badlands of Bankura (23°12′26″ to 23°09′50″ N, 87° 25′36″ to 87°27′07″ E), (5) Gangani Badlands of Paschim Medinipur (22°51′58″ to 22°51′12″ N, 87°20′15″ to 87°21′06″ E) and (6) Bagnada Badlands of Paschim Medinipur (22°53′25″ to 22°51′46″ N, 87°08′49″ to 87°10′16″E).

In these badlands 11 basins of permanent gullies and seventy gully head cuts are selected for the morphometric study. The land use–land cover (LULC) classification of these badlands (Fig. 2.5) reveals five major classes which can be detected as mainly natural forest cover, shrub land, barren surface, built-up area, arable land and water bodies. The important finding of LULC classification is that (1) natural forest cover of badlands varies from 20.83 to 30.60% of total area, (2) shrub land and barren surface varies from 32.13 to 42.12% of total area, (3) road networks and built-up areas varies from 7.27 to 14.13% of total area and (4) the areal coverage of arable land varies from 16.10 to 32.66% of total area, respectively (Fig. 2.3). These LULC classes reveal the role of human activity on the landscape through agriculture, roads and settlements. Generally, the thin solum is loamy-skeletal and hypothermic (sandy loam to sandy clay loam texture of red soils) in nature developing on the barren lateritic wastelands with sparse bushy vegetation and grass. The dark reddish to brown coloured sandy clay loam of 0–16 cm (A_o horizon, maximum grass root zone) is developed over the fragmented secondary laterites. These ferruginous soils have weak fine crumb and granular structure (slightly hard, friable and slightly sticky), 2–5 mm size of manganese nodules, >2 mm size of ferruginous nodules with goethite cortex, 30–40% travels and pebbles, excessive drained surface and ${}_pH$ of 5.4–5.7. The annual soil erosion rates of these regions range from 10.415 to 17.605 t ha^{-1} $year^{-1}$.

2.3 Materials and Methods

2.3.1 Secondary Data Collection

The key sources of the main secondary data are regional soil report, geology report and other physical environmental reports published by NBSS and LUP (National Bureau of Soil Service and Land Use Planning), Census of India, district gazetteer, Irrigation and Waterways Dept. of

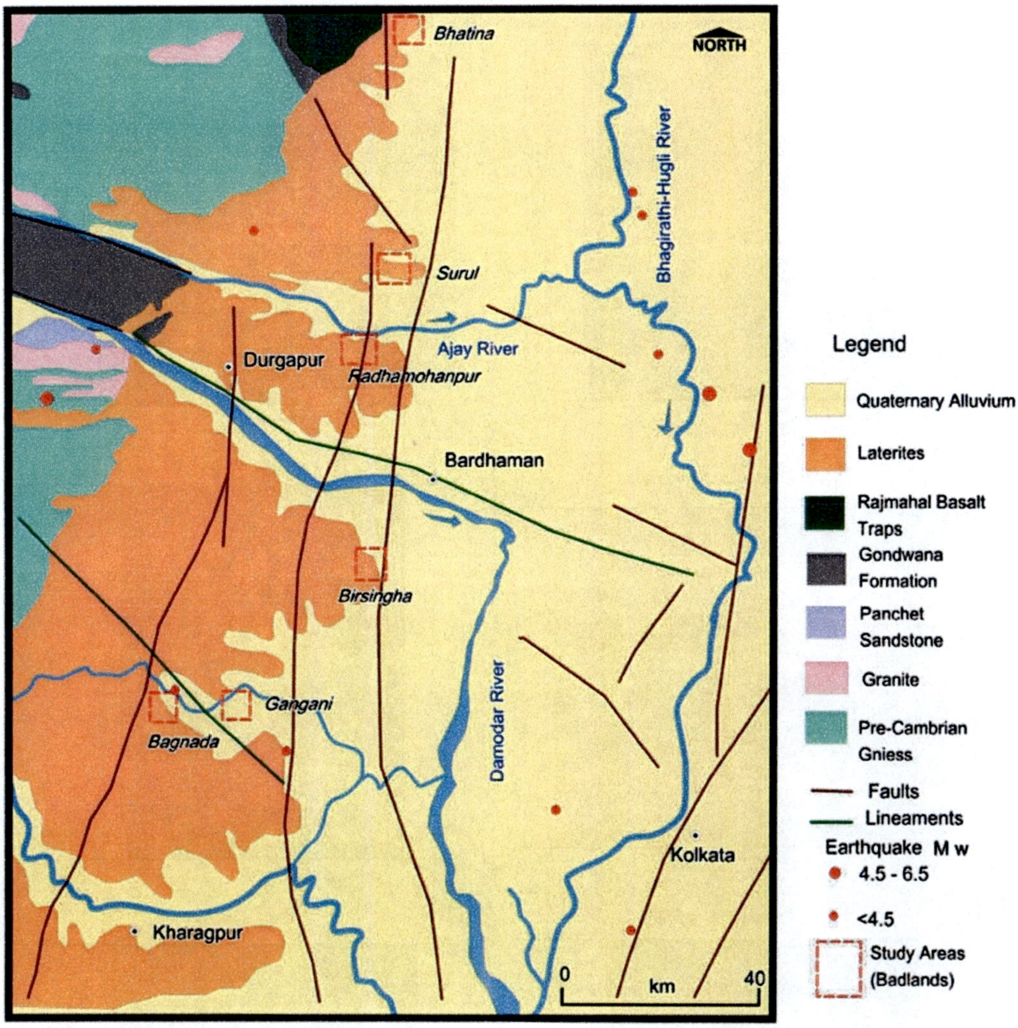

Fig. 2.2 Locations of selected badlands (study areas) in association with geological units and major faults of the Bengal Basin

Govt. of West Bengal (IWD), Geological Survey of India (GSI), related e-books and e-journals. In this case, we have gathered the daily rainfall data from six IWD automatic rain-gauge stations (https://wbi.wd.gov.in) at (1) Rampurhat (24° 10′13″ N, 87°46′50″ E), (2) Shyambati (234141 N, 874,054 E), (3) Satkahania (233544 N, 872957 E), (4) Sonamukhi (231757 N, 872458 E), (5) Amlagora (225044 N, 872003 E) and (6) Simlapal (225458 N, 870413 E) which are situated in and around the study areas, having an areal distance of 18–25 km. From the database, we have selected the rainfall events (10 days) of greater than 50 mm from June to August 2020, and these events vary from 52 to 144.5 mm of rainfall in 2020 for the SCS-CN (Soil Conservation Service Curve Number) modelling. The relevant soil data was collected from the official website of BHOOMI geoportal (www.bhoomigeoportal-nbsslup.in), prepared by Indian Council of Agricultural Research (ICAR) and National Bureau of Soil Survey and Land Use Planning (NBSS&LUP).

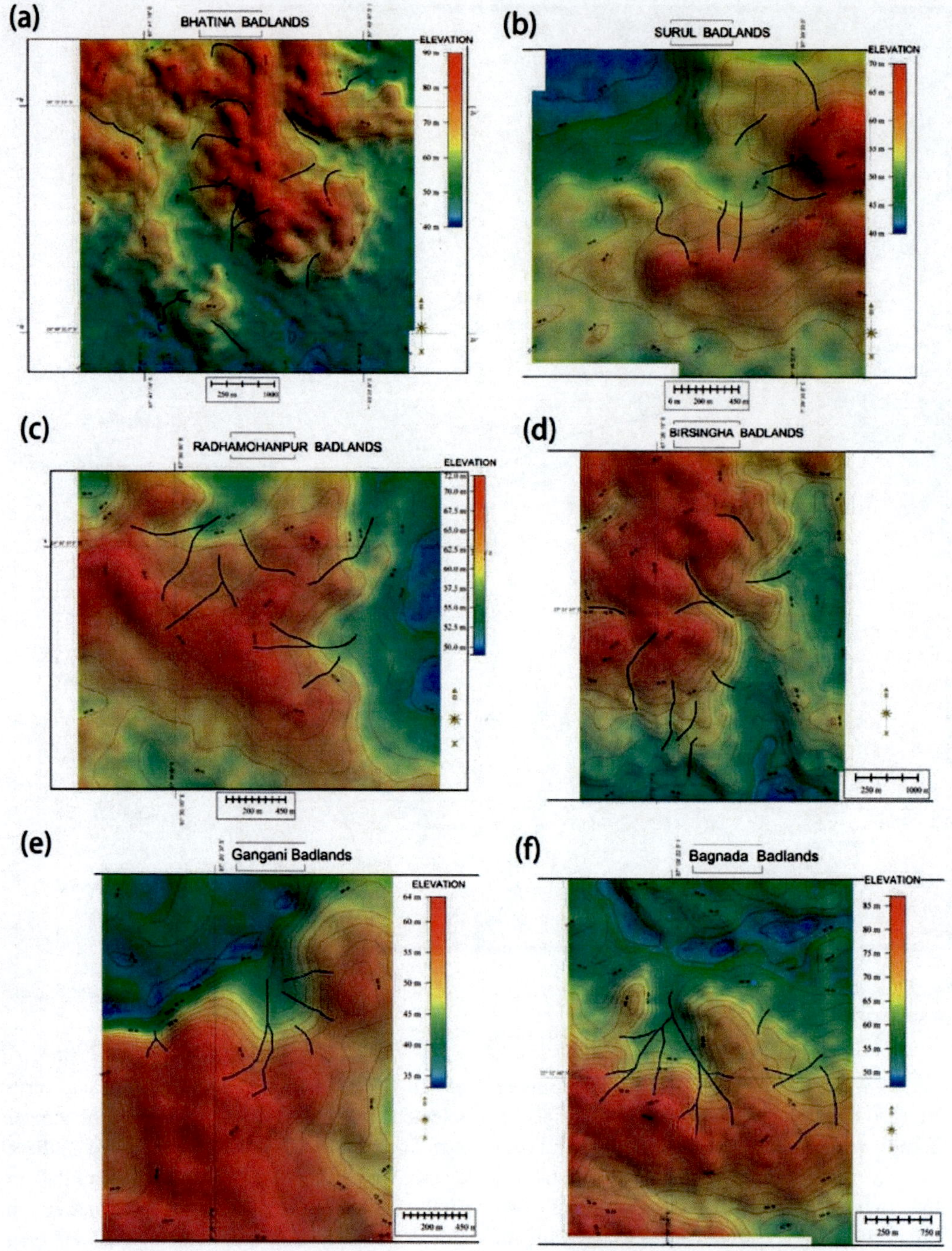

Fig. 2.3 Major drainage lines (generated using 3D flow accumulation modelling of Global Mapper 21.0) of trunk gullies super-imposed on DEM (SRTM 30 m) of (**a**) Bhatina badlands, (**b**) Surul badlands, (**c**) Radhamohanpur badlands, (**d**) Birsingha badlands, (**e**) Gangani badlands and (**f**) Bagnada badlands

Fig. 2.4 Panoramic view: Badlands of *Rarh* laterites having potential for geomorphosites at (**a**) Bhatina, Birbhum, (**b**) Garhbeta, Paschim Medinipur, (**c**) Radhamohanpur, Paschim Bardhaman and (**d**) Surul, Birbhum

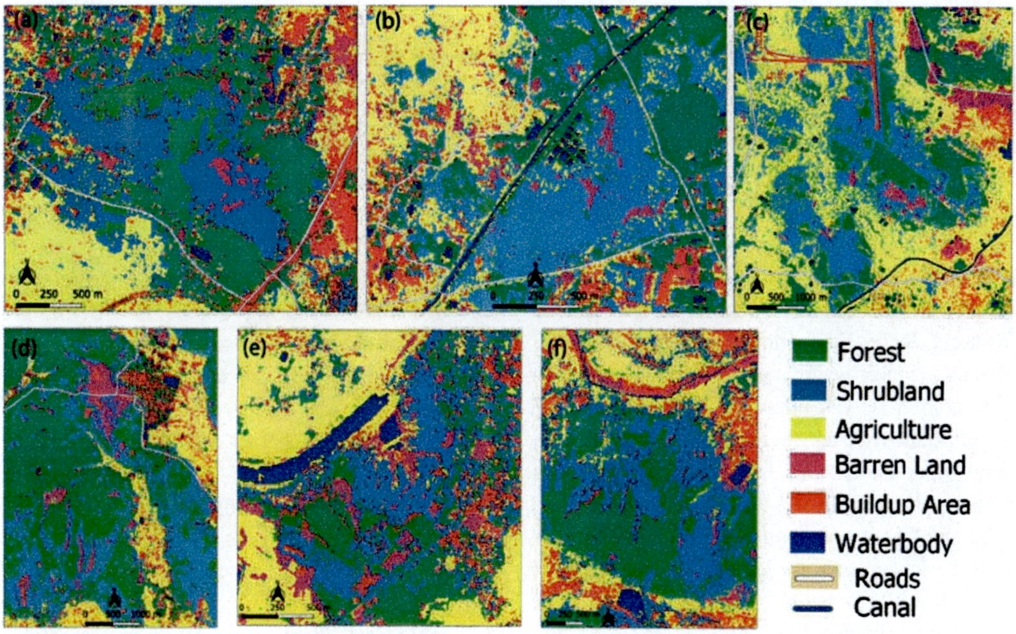

Fig. 2.5 The spatial distribution of selected land use–land cover classes in the (**a**) Radhamohanpur badlands, (**b**) Surul badlands, (**c**) Bhatina badlands, (**d**) Birsingha badlands, (**e**) Gangani badlands and (**f**) Bagnada badlands

For the LULC classification of badlands the Sentinel 2A (10-m resolution), images of tile numbers T45QWF and T45QWG (date of acquisition: 31 December 2019) were collected from the website GloVis of USGS (https://glovis.usgs.gov), having 12 numbers of bands (range of central wavelength: 0.443–2.190 μm). To get relief and drainage information, the SRTM (Shuttle Radar Topography Mission, 30-m resolution or 1 arc-second) digital elevation model (DEM) data (vertical datum of EGEM96 and horizontal datum of WGS84), prepared by the National Aeronautics and Space Administration (NASA) and the National Geospatial-Intelligence Agency (NGA), is used here using ERDAS Imagine 9.1, ArcGis 9.1 and Global Mapper 21.0 software. The SRTM data was collected from the official website Earth Explorer of United States Geological Survey (USGS, https://earthexplorer.usgs.gov).

2.3.2 Digital Topographic Analysis

To create a watershed, the Global Mapper 21.0 software allows the user to perform a 'watershed analysis tool' on the loaded terrain data to find the stream paths as well as delineate the watershed areas that drain into a given stream section. The watershed calculation used the eight-direction pour point algorithm (D-8) to calculate the flow direction at each location, along with a bottom-up approach for determining flow direction through flat areas and a custom algorithm for automatically filling depressions in the terrain data of SRTM 30 m. The 'stream threshold section' of Global Mapper 21.0 controls how much water must flow to a particular cell before it is considered a part of a stream. Larger values of cell count will result in only more water flow areas being delineated, while smaller values will cause more minor water flows to be marked as streams or permanent gullies. Each stream segment (i.e. portion between inflow and outflow) is detected in this analysis, and the indentified drainage lines of Google Earth Pro Web-GIS (*.kml* files) are super-imposed on the DEM-derived stream segments to a detailed network of streams or gullies at a micro-basin scale. After that, the option of 'calculate Strahler stream order' is used to generate a Strahler attribute containing a value classifying each stream of the watershed. In addition, the flow direction point creation option of the software creates a point feature (3D view) with the direction (flow angle) and magnitude (flow accumulation the count of cells flowing to a given cell) attribute at each point, then enabling the layer point style for the quiver plot. The options of 'create contour' and 'path profile tool' are widely used to display a vertical profile along the specified path using loaded DEM data. The 'path profile' or 'line of sight' dialog displays the 3D path profile and provides several essential information of relief, slope, elevation differences, dissection, cut-fill volume and long profile of elevation with changing gradient related to the profile. The 3D path profile is derived by selecting a buffer of 2000 m path, containing the mean elevation of counted cells.

2.3.3 Topology of Drainage Network

Spatial distribution of links (branches) in a drainage basin having a systematic relationship with slope, discharge, rock or soil types and type of exposure is to be studied with due importance for the evolution of drainage in the ravines or badlands (Maiti 2016). This branching system develops due to the bifurcation at the source and the development of numerous rills and gullies on the valley side slopes. The Horton–Strahler system of stream ordering does not provide the hierarchical magnitude of streams from source to outlet. So, in this part the schemes (Shreve 1966 and 1967; Scheidegger 1967; Ranalli and Scheidegger 1968; James and Krumbein 1969; Smart and Wallis 1971; Mock 1971 and 1976; Dacey and Krumbein 1976; Zaliapin et al. 2010; Zanardo et al. 2013) of stream magnitude and link branching system are applied in the badlands.

Topological identity means that the planimetric projections of two-channel networks with the same number of sources can be rotated and deformed within the projection plane so as to become congruent (Mock 1976). The infinite, topologically random model as stated by Shreve (1966) is "…in the absence of geological controls a natural population of channel networks will be topologically random", and later (Shreve 1967) "….in the overall network which will be termed as infinite topologically random channel network, all topologically distinct sub-networks with the same number of sources occur with frequency". Only two properties are necessary to define topological identity—the number of sources and the arrangement by which they are linked together to form a network (Mock 1971). Most of the common geomorphic descriptors and parameters are independent of topology. The number of topologically distinct arrangements increases rapidly with increasing network size. Since Shreve (1967) introduced the topologically random model, a sizeable body of work has accumulated in which natural channel networks have been examined within the predictive framework of the model.

Three important aspects of this study deal only with gully network development—(1) development of a numerical hierarchical network, (2) the law of stream numbers and (3) the law of stream lengths (Strahler 1952). The ultimate tributaries in a gully network are designated as order 1. Wherever two streams of the same order, Ω, join, the resulting stream is of order Ω + 1. Wherever two streams of unequal order join, the succeeding streams are of the higher order (there is no change of order in the trunk stream). A complete stream consists of the entire reach of channels from the formation of order Ω to the point where it terminates in a higher order. If we let N_{Ω} and $N_{\Omega+1}$ be the number of streams or order Ω and Ω + 1, respectively, then R_b, the bifurcation ratio (Strahler 1952), is defined as

$$R_{\mathrm{b}} = N_{\Omega}/N_{\Omega+1}$$

The empirical rule, given by Horton (1945), is that R_b tends towards through a range of R_b which is used to formulate the law of stream numbers, namely

$$N_{\Omega} = R_{\mathrm{b}}^{k-\Omega}$$

where k is the order of the master stream (or drainage basin).

Horton (1945) stated the law of stream lengths:

$$L\prime_{\Omega} = L_1 R_L^{\Omega-1}$$

where L_1 is the mean length of the first order, L'_{Ω} is the mean length of Ω order streams, R_L is stream-length ratio ($R_L = L'_{\Omega}/L'_{-1}$) and $L'_{\Omega-1}$ is the mean length of lower order streams.

In the gully network, exterior links (sources) are those that extend from the gully heads to the first junction, and interior links are the channel segments extending between two junctions or any junction and outlet (Maiti 2016). A network with n sources has $(n-1)$ nodes and $(2n-1)$ links, of which n are exterior links and $(n-1)$ are interior. James and Krumbein (1969) and Dacey and Krumbein (1976) distinguished bifurcation from branching and two types of links, viz., 'cis' and 'trans' links. Cis links are formed by two successive tributaries joining the main stream on the same side of the channel. Trans links are formed by two successive tributaries joining the main stream from the opposite sides. By means of the concept of magnitude, it is possible to define several types of interior links and two types of exterior links (Mock 1971):

- *Exterior Links*—There are two types of exterior links. The S (source) link is a magnitude 1 link that joins another magnitude 1 link at its downstream fork. The TS (tributary source) link is a magnitude 1 link that joins a link of a magnitude greater than 1 at its downstream fork.
- *Interior Links*—The B (bifurcating) link is a link of magnitude μ that is formed at its upstream fork by the confluence of two links, each of a magnitude of 1/2 μ and that flows at its downstream fork into a link of a magnitude of less than 2μ. The TB (tributary bifurcating) link is a link of magnitude μ that is formed at its upstream fork by the confluence of two links, each of a magnitude of $1/2\mu$ and that

flows at its downstream fork into a link of magnitude greater than or equal to 2μ. The T (tributary link is a link of magnitude μ that is formed at its upstream fork by the confluence of two links of unequal magnitude and that flows at its downstream fork into a link of a magnitude greater than or equal to 2μ. The CT (cis–trans) link is a tributary link if stream magnitude becomes μ which is formed at its upstream fork by the confluence of two links of unequal magnitude, and CT link flows at its downstream fork into a link of a magnitude of less than 2μ.

The last downstream link for any system under study, i.e. the highest magnitude link, is arbitrarily defined as either T or TB link depending on the magnitude relationships at its upstream fork (Mock 1971). Let M signify the magnitude of a network, and the magnitude of individual links. A network of magnitude M contains M exterior links ($\mu = 1$) and $M - 1$ interior links ($\mu > 1$) (Mock 1976). The total number of links is $2M - 1$ for a network of magnitude of M. Let $N(M)$ be the number of topologically distinct arrangements possible for a network of magnitude M. $N(M)$ is given by

$$N(M) = 1/2M{-}1\,(2M{-}1\,M)$$

Mock (1971 and 1976) defined the probability (fraction between 0 and 1 or in percentage) of drawing exterior ($P_{\mu ext}$), interior ($P_{\mu int}$), source ($P_{\mu s}$) and tributary source links ($P_{\mu ts}$) from a topologically random population of magnitude M:

$$P_{\mu ext} = M/(2M{-}1)$$

$$P_{\mu int} = (M{-}1)/(2M{-}1)$$

$$P_{\mu s} = M(M-1)/(2M-1)(2M-3)$$

$$P_{\mu ts} = (M/2M{-}1)\,(M{-}2/2M{-}3)$$

2.3.4 Morphometric Attributes

Morphometry is defined as the quantitative measurement of landscape shape. At the simplest level, landforms can be characterized in terms of their size, elevation and slope. Quantitative measurements allow geomorphologists objectively to compare different landforms and to calculate less straightforward parameters (geomorphic indices) that may be useful for identifying a particular characteristic of an area (its level of erosion intensity, stage of landform evolution and also tectonic activity). Some geomorphic indices have been developed as basic reconnaissance tools to identify areas experiencing active erosion and landform dissection in response to terrain relief and slope variations. In this analysis of gully erosion, some of the geomorphic indices are found to be quite useful to detect the form–process relationship and anomalies in relief and slope, such as Length of Overland Flow (Horton 1945), Hypsometric Integral (Strahler 1952) (Fig. 2.6a), Basin Circularity Ratio (Miller 1953), Basin Elongation Ratio (Schumm 1956), Constant of Channel Maintenance (Schumm 1956), Relief Ratio (Schumm 1956), Sinuosity Index (Schumm 1956), Dissection Index (Dov Nir 1957), Stream-Length Gradient Index (Hack 1973) and Fractal Dimension of Drainage Network (Barbera and Rosso 1989) (Table 2.2).

2.3.5 SCS-CN Method and Runoff–Sediment Yield Coupling

The Soil Conservation Service Curve Number (SCS-CN) rainfall–runoff simulation model (Mishra and Singh 2003; Mishra et al. 2006; Bhunya et al. 2014; Gajbhiye et al. 2014; Srivastava and Imtiyaz 2016; Meshram et al. 2017) is based on the water balance equation and two fundamental hypotheses. The first hypothesis equates the ratio of the actual amount of direct surface runoff (Q) to the total rainfall (P) (or maximum potential surface runoff) to the ratio of the amount of actual infiltration (F) to the amount of the potential maximum retention (S). The second hypothesis relates the initial abstraction (I_a) to the potential maximum retention. The direct runoff (Q) can be derived as follows (Mishra and Singh 2003):

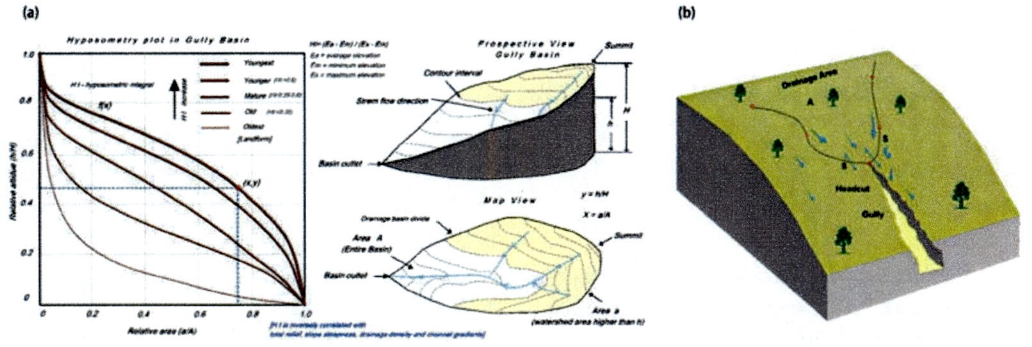

Fig. 2.6 **a** A conceptual model to understand the hypsometric integral of a basin, and **b** a conceptual model of the block diagram to show the threshold components of a gully

Table 2.2 Geomorphic indices used to determine morphometric signatures of gully basins

Indices	Computed by	Formula	Application to geomorphology
Length of overland flow (L_g)	Horton (1945)	$L_g = 1/2D$; D = drainage density	Choudhari et al. (2018)
Hypsometric integral (HI)	Strahler (1952)	HI = $(E_{mean} - E_{min})/(E_{max} - E_{min})$; E_{mean} = mean elevation of the selected basin; E_{max} = maximum elevation of the selected basin; E_{min} = minimum elevation of the selected basin	Hassen et al. (2014)
Constant of channel maintenance (C_m)	Schumm (1956)	$C_m = A_b/\Sigma L$; A_b= basin area; ΣL = total length of stream	Choudhari et al. (2018)
Basin elongation ratio (E_R)	Schumm (1956)	$E_R = 2\sqrt{(A_b/\pi)}/L_b$; A_b = basin area; L_b = maximum basin length; π = 3.14	Asthana et al. (2015)
Basin circularity ratio (R_c)	Miller (1953)	$R_c = A_b/P$; A_b = basin area; P = area of circle of same perimeter	Asthana et al. (2015)
Fractal dimension of drainage network (FD)	Barbera and Rosso (1989)	FD = $\ln R_b/\ln R_L$; R_b = bifurcation ratio; R_L = stream-length ratio	Mahmood and Gloaguen (2014)
Stream-length gradient index (SL)	Hack (1973)	SL $=(h/l)L_s$; h = changes in elevation of the reach around the selected point; l = length of the channel in between the selected elevation contour; lL_s = longest length of the channel from the drainage divide to the selected point of the reach	Joshi et al. (2013)
Dissection index (D_I)	Nir (1957)	$D_I = (E_{max} - E_{min})/E_{max}$; E_{max} = maximum elevation of the selected basin; E_{min} = minimum elevation of the selected basin	Bhunia et al. (2012)
Relief ratio (R_r)	Schumm (1956)	$R_r = H/L_b$; H = elevation difference between maximum and minimum height; L_b = maximum basin length	Choudhari et al. (2018)
Sinuosity index (SI)	Schumm (1956)	SI = AL_0/EL_0; AL_0 = actual or observed length of the stream; EL_0= expected length of the stream	Joshi et al. (2013)

$$Q = (P - I_a)^2/(P - I_a + S)$$

The equation is valid from $P \geq I_a$; $Q = 0$, otherwise. According to Mishra and Singh (2003), the equation can be written as

$$Q = (P - 0.2\,S)^2/(P + 0.8\,S)$$

In practice, S is derived from a mapping equation expressed in terms of the curve number (CN) (Mishra et al. 2006):

$$S = 1000/CN - 10(\text{where } S \text{ in inch})$$

The equation is transformed to SI units:

$$S = 25400/CN - 254(\text{where } S \text{ } in \text{ cm})$$

The CN is a dimensionless value (30–100) to reflect the runoff yield for a specific event-based rainfall in respect of land use–land cover, soil type (selecting hydrologic soil group B), hydrologic condition and antecedent moisture condition (AMC), computed in Indian context (Meshram et al. 2017). The CN value is adjusted to the local slope of the gully basin, considering the effect of slope on runoff (Meshram et al. 2017). The weighted CN values are derived based on the areal coverage of land use and land cover at each badlands.

One measure of geomorphic activity is sediment yield, which is defined as the amount of sediment per unit area removed from a watershed by runoff during a specified period of time. Sediment yield can be defined as the amount of sediment reaching or passing a point of interest in a given period of time and sediment yield estimates are normally given as tonnes per year. Coupling the SCS-CN method and the Universal Soil Loss Equation (USLE), a new model is proposed for the estimation of the rainstorm-generated sediment yield from a watershed (ungauged sites of badlands). Both the SCS-CN method and the USLE method share a common characteristic in that they account for watershed characteristics, albeit differently. It is therefore conjectured that by coupling these two methods, one can compute the sediment yield from the knowledge of rainfall, soil type, land use and antecedent soil moisture conditions. It is a scaling down approach of hydrology where the annual basis of soil erosion rate estimation comes down to daily basis sediment yield at each rainfall event. The methodology of SCS-CN and USLE coupling (i.e. rainfall–runoff erosion modelling) is adapted from the works of Mishra and Singh (2003), Mishra et al. (2006), Bhunya et al. (2014), Gajbhiye et al. (2014) and Srivastava and Imtiyaz (2016). The estimation of runoff is derived from the following equation (Mishra and Singh 2003):

$$Q = (P - I_a)^2/(P - I_a + S_o)$$

where Q is runoff (mm), P is total rainfall (mm), I_a is the initial abstraction (mm) and S_o is the potential maximum retention (mm).

The USLE estimates the potential soil erosion (sheet and rill erosion) from upland areas, and it is expressed as follows (Wischmeier and Smith 1978; Ghosh and Guchhait 2020):

$$E_T = R_e \cdot K_d \cdot L_S \cdot C_f \cdot P_f$$

where E_T is the annual potential soil loss per unit area (t ha^{-1} year^{-1}), R_e is the rainfall and runoff factor (MJ mm ha^{-1} year^{-1}), K_d is the soil erodibility factor, L_S is the slope-length and slope-steepness factor, C_f is the cover and management factor and P_f is the support practice factor.

Taking $I_a = 0.2\ S_o$ which is standard practice, the equation of sediment yield can be recast as (Mishra et al. 2006)

$$S_y = (P - 0.2S_o)E_T/P + 0.8S_o$$

which suggests that sediment yield (S_y, t ha^{-1}) reduces with the increasing initial abstraction and vice versa.

2.3.6 Quantitative Measures of Gully Erosion Potential

With the methodological framework of geomorphology and hydrology the discipline of soil erosion science is tried to find out the following research questions: (1) Where is erosion occurring? (2) Why is it happening and who is to blame? (3) How serious is it? (4) How does it affect? (5) Where do gullies develop in the landscape? (6) Is there any role of active tectonics in badland development? (5) What should be the response? (6) Can we assess it? (Poesen et al. 1998; Boardman 2006; Joshi and Nagare 2013; Borrelli et al. 2020). In this study the quantitative measures, used by Begin and Schumm (1979), Vandaele et al. (1996), Poesen et al. (1998 and 2002), Joshi and Nagare (2013), Torri and Poesen (2014), Rossi et al. (2015), Yibeltal et al. (2019), Ghosh and Guchhait (2020), Geeter et al. (2020), Kariminejad et al. (2020) and Sidorchuk (2020), are applied to estimate the erosion potential of gully and topographic threshold of gully initiation.

Topography is an important factor controlling gully erosion rates. The study of self-enhancing feedback mechanisms generally is encouraged by a threshold approach, whereas the study of self-regulating feedback mechanisms generally is encouraged by an equilibrium approach (Bull 1980). A threshold is a turning point or boundary condition that separates two distinct phases of an inter-connected process, a dynamic system that is powered by the same energy source (Ghosh and Guchhait 2020). The possibility of predicting the spatial occurrence of gullies from topographic attributes using a power function has been investigated by various researchers around the world (Torri and Poesen 2014; Rossi et al. 2015). Usually, the topographic threshold conditions for gully heads are reported as double logarithmic plots of upslope area (*A*) and slope gradient (*S*), where *A* (ha) is the area of the catchment draining towards the gully head and *S* (tangent, m m^{-1}) is the local slope of the soil surface at the gully heads (Fig. 2.4b) (Patton and Schumm 1975; Poesen et al. 2002; Torri and Poesen 2014). The relation between critical valley slope and drainage basin area ($S = a\ A^{-b}$, where a = coefficient and b = exponent of relative area) is used as a predictive model to locate those areas of instability within alluvial valleys where gullies will form:

$$S = aA^{-b}$$

A threshold line is drawn through the lower limit of scatter of the points, and this line represents, for a given area, a critical value for valley slope above which entrenchment of the laterite should occur. A theoretical division of the landscape into process regimes in terms of log S (*X*-axis) and log A (*Y*-axis) signifies different geomorphic thresholds to gully erosion, and the resultant critical threshold line is demarcated as the Montgomery–Dietrich (MD) envelope, through A–S threshold (Vandekerckhove et al. 1998; Moeyersons 2003; Montgomery and Dietrich 1994; Samni et al. 2009). Patton and Schumm (1975) and Begin and Schumm (1984) began modelling gully erosion as a threshold process and suggested that the equation defining threshold could be derived from the fact that concentrated overland flow should produce flow shear stresses in excess of a critical value to erode a gully channel (Torri and Poesen 2014). The final expression of the equation is (Torri and Poesen 2014; Rossi et al. 2015; Bartley et al. 2020)

$$SA^b \geq K \text{ or } S \geq KA^b$$

where *K* is a sort of black box coefficient or constant which depends on local climate (rainfall), soil and land use (reflecting water infiltration rate, soil shear strength and hydraulic roughness). To initiate gully head sinus of slope gradient must be greater than the *K* value (Sin $S \geq K$). Torri and Poesen (2014), Rossi et al. (2015) and Kariminejad et al. (2020) give an expression of *K* as follows:

$$K = 0.73ce^{1.3RFC}(0.00124S_{0.05} - 0.037)A^{-0.38}$$

where $S_{0.05}$ (in mm) is the potential rainfall abstraction and infiltration during a rainfall as used in the SCS-CN method (AMC-II condition), RFC is the percentage of rock fragments, c is assumed equal to 1 and A is the upslope contributing area (ha).

Runoff is the main parameter in assessing the potential triggering of gully heads, but a comprehensive investigation of such erosion should consider local conditions such as morphology, land use, soil characteristics and vegetation. To account for all these aspects, the LANDPLANNER (Landscape, Plants, Landslide and Erosion) model applied a potential erosion index (*PEI*) defined by the following equation (Kariminejad et al. 2020):

$$\mathrm{PEI} = \alpha\left[(Q_r \cdot \mathrm{Sin}S)^{\beta}/S_{0.05}\right]$$

$$S_{0.05} = 25.4\left[1.33\,(1000/CN{-}10)^{1.15}\right]$$

where Q_r is daily runoff (mm) calculated by the SCS-CN method, CN is the curve number, and the coefficients α and β are assumed to be 1, but further calibration is needed to identify the value of the parameters or at least their ranges.

2.4 Results and Interpretation

2.4.1 Stream Ordering System in Gullies

The Strahler (1952 and 1957) number of stream ordering (bifurcation ratio, R_b) is a numerical measure of its branching complexity. In the application of the stream ordering method at basin scale of the gully, each gully of a basin is treated as a node in a tree structure in Global Mapper 21.0 version, with the next downstream as its parent. The important part of this method is that when a 2nd-order or N_{n+1} order of gully combines with another 2nd- or N_{n+1} order of gully, then it becomes a 3rd or numerically next higher order of gully. But a gully or a stream of lower order joining a higher order gully, the ordering system does not change into higher order level. In the practical field, the joining of lower order gullies into higher order gullies make a profuse impact on gully size, sediment yield and discharge. Therefore, that problem was rectified by Sherve (1966 and 1967) with the introduction of the stream magnitude (*M*) concept. With the joining of different orders of gullies, the main order of master gully will increase downstream.

The bifurcation ratio (R_b) signifies the number of lower order streams or gullies required to maintain the highest order of gully in respect of the present fluvial system. A high value of R_b means a high level of discharge out from the fluvial system to the river or floodplain. The study of stream ordering principle suggests that there are dominance of mostly 2 order and 3 order gully basins, though having one 5 order and seven 4 order basins in the six badlands. In these basins R_b value is estimated as (Fig. 2.7): (1) 2.76–3.21 (Bhatina badlands), (2) 3.85–4.16 (Surul badlands), (3) 2.59–4.0 (Radhamohanpur badlands), (4) 3.08 (Birsingha badlands), (5) 4.38 (Gangani badlands) and (6) 3.22–4.12 (Bagnada badlands) (Table 2.3). The highest 5th-order of the gully basin (R_b of 3.86) is observed in the Radhamohanpur badlands. So, the relatively high (>3.0) R_b of these badland gullies reflects a dense concentration of lower order gullies in the upper catchment, generating high runoff discharge and sediment yield into the main system.

2.4.2 Fractal Dimension of Drainage Network (FDDN)

Mandelbrot (1982) introduced the concept of fractals to describe the complexity of nature's shapes, which differ from those of ordinary geometry. For self-similar fractal objects, the fractal objects, the fractal dimension characterizes the scaling properties of the object, and it provides a similarity exponent, since it tells how the associated measure (e.g. mass or length) changes after a change of scale (Barbera and Rosso 1989). The geometric pattern of the stream network of a drainage basin can be viewed as 'fractal' with a fractional dimension. The drainage patterns present irregularities with self-similar (fractal) characteristics in space. The space-filling nature of drainage network is a strong marker of the area vulnerable to intensive erosion and tectonic deformation (Mahmmod

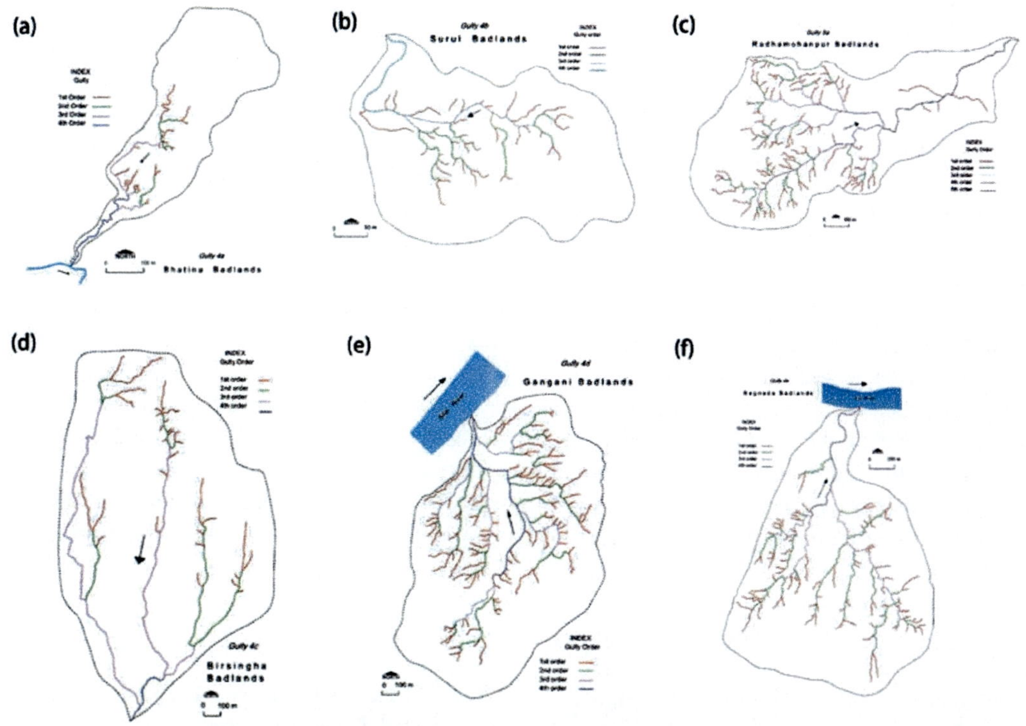

Fig. 2.7 Drainage network of gullies according to Strahler ordering system in the **(a)** Bhatina badlands, **(b)** Surul badlands, **(c)** Radhamohanpur badlands, **(d)** Birsingha badlands, **(e)** Gangani badlands and **(e)** Bagnada badlands

Table 2.3 Basic property of drainage network in the gullies of selected badlands

Badlands	Stream order	R_b	R_L	FD	Gully channel length (m)	Basin area (m^2)	Maximum Basin length (m)	Elevation range (m)	Average basin slope (%)
Bhatina	4th	2.76	1.37	3.172	1440.23	207860.20	833.65	66–48	2.65
Surul	3rd	3.83–4.16	1.29–1.35	4.477–5.550	1212.22–1371.66	67567.11–82645.92	309.34–363.52	66–55	2.30–2.40
Radhamohanpur	4th and 5th	2.49–4.00	1.24–1.29	3.037–6.183	1302.79–4116.32	75418.62–229327.30	466.10–656.75	67–54	2.20–2.56
Birsingha	4th	3.08	1.38	3.486	5036.44	1757813.00	2084.62	86–62	2.35
Gangani	4th	4.39	1.20	7.968	8873.0	75522.50	1137.00	64–34	2.55
Bagnada	3rd and 4th	3.22–4.12	1.24–1.25	4.976–7.109	2227.16–10452.97	162246.83–2183615.07	858.50–1951.36	77–52	1.75–2.55

Note R_b = bifurcation ratio, R_L = stream-length ratio and FD = fractal dimension of drainage network

and Gloaguen 2014). The highly steep regions cause the drainage network to reorganize and linearize which is evidence of uplifted conditions. For an ordered drainage system, Barbera and Rosso (1989) first proposed to derive the fractal dimension from Horton's laws of stream number (bifurcation ratio, R_b) and stream lengths (stream-length ratio, R_L). This measure of fractal dimension (FD) can help to understand fluvial network geometry and composition. Barbera and

Rosso (1989) derived the following equation of *FD*:

$$\mathrm{FD} = \log R_{\mathrm{b}} / \log R_{\mathrm{L}} (\text{when } R_{\mathrm{b}} > R_{\mathrm{L}})$$

$$\mathrm{FD} = 1\,(\text{when } R_{\mathrm{b}} < \text{or} = R_{\mathrm{L}})$$

The FD quantifies the degree of irregularity or segmentation of an object or spatial pattern. The drainage system or network loses its dendricity to become linearized as topographic threshold triggered by external factors that modify the flow network geometry (Mahmmod and Gloaguen 2014). Composite branching architecture and dendritic shape of drainage network make the impression of a fractal-like structure. Superimposition of diverse physical and geological processes governing the development of rivers and streams, including also random components, produces fractal river networks. The linear analyses chiefly focus on the secondary parameters (contributing drainage area, channel lengths, channel slope and elevation) and completely overlook the fractal behaviour of drainage network. The measure of FD is a powerful tool as patterns with dissimilar space-filling characteristics can easily distinguish those areas, which yield some signatures using the conventional linear analysis (Mahmood and Gloaguen 2014). There is a close inter-relationship between R_{b} and R_{L} in the sample basins of gullies, having correlation coefficient (*r*) value of (–) 0.485. With increasing R_{b} of gully basin, the R_{L} is decreasing, following a negative trend ($R_{\mathrm{L}} = 1.5583 - 0.0713R_{\mathrm{b}}$, $R^2 = 0.4936$). It may be argued that with an increasing number of 1st-order gullies, the resultant mean R_{L} is decreasing. Next, FD provides an idea about the complexity and self-similarity of drainage network at micro-scale, i.e. gully basins of badlands. The FD of badlands varies from 3.176 to 8.543 which signify intensive dendritic nature and high complex geometry of network in space. The FD value close to 1 reflects linearity (strong effect of tectonic deformation on rocks), and FD greater than 1 means increasing dendricity of fluvial networks (more dominance of surface processes on alluvium). In the study area, the structural weakness of laterites (i.e. high erodible nature and susceptible to fluvial erosion) is minimal to generate a dendritic network of gully systems (in which trunk gullies run sub-parallel). The FD suggests that a dense network of lower order gullies feeds a master gully channel that matches and is strongly accordant to the overriding gradient of the land. In the lateritic badlands with increasing R_{L} of basins, the FD is progressively decreased (8–3), following a negative trend line ($\mathrm{FD} = 3.621 - 4.19R_{\mathrm{L}}$, $R^2 = 0.8703$). On another side, with increasing R_{b} of basins the *FD* is steadily increased (3–8), showing a positive trend line ($\mathrm{FD} = 2.274R_{\mathrm{b}} - 3.062$, $R^2 = 0.7475$) (Fig. 2.8). It can be concluded that with increasing numbers of lower order gullies (source links), the fractal dimension of drainage network is turned into a very complex network system and less self-similarity in space.

2.4.3 Model of Stream Magnitude and Channel Link

A classification of channel links in stream networks (Shreve 1966; Mock 1971) has provided a fundamental probabilistic basis for statistically evaluating the topologic properties of stream works. The power of this approach has been shown by the fact that the Horton–Strahler law of stream numbers is derivable as a maximum likelihood event (Mock 1971). By introducing the concept that links are basic elements of stream networks and by assigning a magnitude to each link, one makes available the necessary elements for systematic analysis of real networks in terms of an available theoretical model.

Two items (differ from the Horton–Strahler ordering system) worth noting here are (1) 1st-order streams and magnitude 1 links are equivalent, and (2) the focus shifts from streams to links as basic entities in the fluvial system (Mock 1976). Shreve (1967) introduced the concept of stream magnitude in the ordering system of fluvial networks. Shreve (1966) defined terms in the following manner:

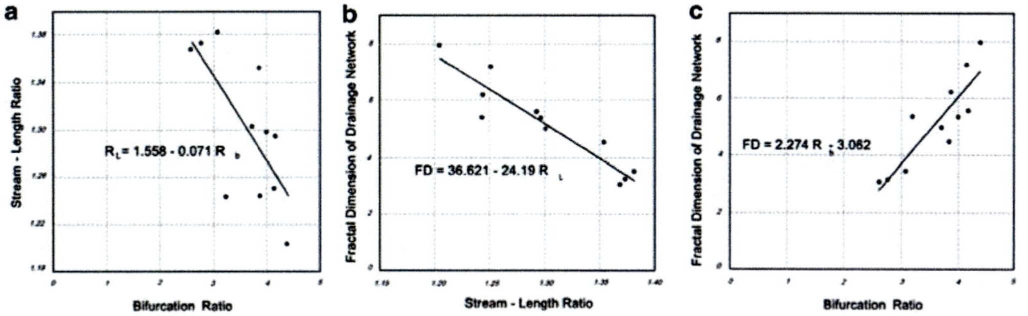

Fig. 2.8 The scatter diagrams and linear regressions showing the inter-relationship between (**a**) bifurcation ratio and stream-length ratio, (**b**) stream-length ratio and fractal dimension of drainage network and (**c**) bifurcation ratio and fractal dimension of drainage network

- The points farthest upstream in a channel network are termed *sources*.
- The point of confluence of two channels is a *fork*.
- The term *link* will refer to a section of channel reaching without intervening forks from either a fork or the outlet at its upstream end to either a fork or the outlet at its downstream end.
- Links are subdivided into *exterior* links, which head at a source, and *interior* links, which head at a fork.
- The magnitude of a link is equal to the total number of sources ultimately tributary to it, where all sources are assigned a magnitude (μ) of 1. Magnitude is an additive property; two links joining at a fork produce a third link whose magnitude is the sum of the magnitudes of the first two.
- The magnitude of a stream network (*M*) is defined as being equal to the magnitude of the outlet link, which is defined by the investigator.

The present paper introduces the magnitude of the gully network, and a classification of channel links in the selected badlands by type presents the necessary quantitative expressions and illustrates some of the potential uses and applicability of this model or system. *M* signifies the magnitude of drainage network and μ the magnitude of individual links. Shreve (1967) ordering system of stream magnitude is applied in the gully basin to number the final order of gully outlet into a floodplain or river. It is completely missing in the Strahler number, because same order basins have many morphologic aspects of dissimilarity which needs to be incorporated in the study. The magnitude of a gully network is defined as being equal to the magnitude of the outlet link, which is defined by the investigators. The main focus of the topology shifts from stream to links as basic entities in the fluvial system of the gully. In this study, link topology and magnitude of the channel are assigned in each basin of the gully to understand the number of topologically distinct channel networks (TDCN) and the probability of link numbers. Each basin has a unique probability of interior and exterior links in respect of final stream magnitude.

The gully basins of the badlands are identified as 3rd-, 4th- and 5th-order drainage basins according to the Strahler method, but those that have no similarity with same order basins of other areas in terms gully morphology and drainage development. This problem is rectified by the magnitude concept of Shreve (1967) and the link concept (Mock 1971), because each gully basin or any stream is quite distinct and different from another due to complex development of drainage network. Now, the following section provides a brief summary of topologically distinct drainage development in the badlands (Fig. 2.9 and Table 2.4):

- *Bhatina Badlands*—The magnitude (*M*) of Bhatina badlands is assigned as *M*-22, having

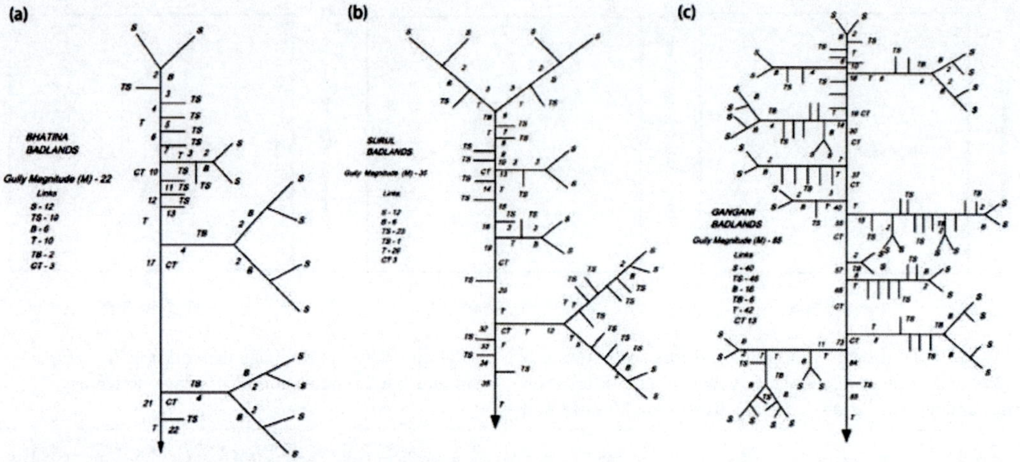

Fig. 2.9 The topological presentation of ideal channel links and stream magnitude in the gully basins of **(a)** Bhatina Badlands, **(b)** Surul Badlands and **(c)** Gangani Badlands

Fig. 2.19 (continued)

Table 2.4 Topologically distinct channel network (TDCN) of badlands

Gully Basin of Badlands	Magnitude (M)	TDCN (Ambilateral Class)	Numbers of link type sets					
			N_S	N_{TS}	N_B	N_T	N_{TB}	N_{CT}
Bhatina	22	44	12	10	6	10	2	3
Surul	35	70	12	23	6	26	1	3
Radhamohanpur	88	88	56	48	26	37	11	19
Birsingha	30	60	12	16	5	17	–	5
Gangani	85	85	40	45	16	42	6	13
Bagnada	81	81	36	42	13	42	3	12

12 S links (source links), 10 TS links (tributary source links), 6 B links (bifurcating links), 10 T links (tributary links), 2 TB links (tributary bifurcating links) and 3 CT links (cis–trans links). The expected probability of exterior ($P_{\mu ext}$) and interior ($P_{\mu int}$) is estimated as 52.1% and 48.9%, respectively, in the fluvial system. Next, the expected probability of

S link and TS link is calculated as 26.21 and 24.95% in the Bhatina Badlands.

- *Surul Badlands*—The 4th-order gully basin of Surul badlands, The magnitude (*M*) of Bhatina badlands is assigned as *M*-35, having 12 S links, 23 TS links, 6 B links, 26 T links, 1 TB links and 3 CT links. The expected probability of exterior ($P_{\mu ext}$) and interior ($P_{\mu int}$) is estimated as 50.54% and 49.46%, respectively, in the fluvial system. Next, the expected probability of S link ($P_{\mu s}$) and TS link ($P_{\mu ts}$) is calculated as 25.55% and 24.94%, respectively, in the Bhatina Badlands.
- *Radhamohanpur Badlands*—The selected gully basin of this badlands is identified as 5th order, but it is recognized as Magnitude (*M*) − 88 by the Shreve method. The network of the gully is assigned as 56 S links, 48 TS links, 26 B links, 37 T links, 11 TB links and 19 CT links. The expected probability of exterior ($P_{\mu ext}$) and interior ($P_{\mu int}$) is estimated as 50.28% and 49.72%, respectively, in the fluvial system. Next, the expected probability of S link ($P_{\mu s}$) and TS link ($P_{\mu ts}$) is calculated as 25.29% and 26.05%, respectively.
- *Birsingha Badlands*—The 4th-order basin of the gully is recognized as *M*-30, having 12 S links, 16 TS links, 5 B links, 17 T links and 5 CT links. The expected probability of exterior ($P_{\mu ext}$) and interior ($P_{\mu int}$) is estimated as 50.9% and 49.1%, respectively, in the Birsingha badlands. Next, the expected probability of S link ($P_{\mu s}$) and TS link ($P_{\mu ts}$) is calculated as 25.93% and 24.96%, respectively.
- *Gangani Badlands*—The 4th-order basin of the gully is defined as *M*-85, and it has 40 S links, 45 TS links, 16 B links, 42 T links, 6 TB links and 13 CT links. The expected probability of exterior ($P_{\mu ext}$) and interior ($P_{\mu int}$) is estimated as 50.2% and 49.8%, respectively, in the Gangani badlands. Next, the expected probability of S link ($P_{\mu s}$) and TS link ($P_{\mu ts}$) is calculated as 24.94% and 25.24%, respectively.
- *Bagnada Badlands*—The 4th-order basin of the gully is assigned as *M*-81, having 36 S links, 42 TS links, 13 B links, 42 T links, 3 TB links and 12 CT links. The expected probability of exterior ($P_{\mu ext}$) and interior ($P_{\mu int}$) is estimated as 50.31% and 49.69%, respectively, in the Bagnada badlands. Next, the expected probability of S link ($P_{\mu s}$) and TS link ($P_{\mu ts}$) is calculated as 25.37% and 24.99%, respectively.

In summary, it can be concluded that the topologically distinct channel network (TDCN) probability of generating S links and TS links is very much similar to all six badlands. Except for Gangani and Bagnada badlands, the other four badlands have a high probability of generating TS links in the system. The gully–river coupling of Gangani and Bagnada badlands—has a strong base level control on the development of drainage network. These basins yield a high magnitude of gully system (*M*-81–85). Though the gully of Radhamohanpur badlands (*M*-88) do not show gully–river coupling at present, it is situated very near to the Ajay River active floodplain (distance about 1.43 km), persisting local base level control on the badlands since Late Holocene. Another important finding is that the same 4th-order basins show different magnitude scales of drainage development in the badlands, and the difference is persisted in the dissection level and morphological diversity of gullies.

2.4.4 Linear Aspects of Gullies

The Horton law of stream lengths (Horton 1945; Strahler 1957) states that a geometric positive relationship exists between the average length of streams of a given order and the corresponding order, having a parameter of stream-length ratio (R_L) (Horton 1945). The R_L values of six badlands are estimated as (1) 1.27–1.39 (Bhatina badlands), (2) 1.29–1.35 (Surul badlands), (3) 1.24–1.36 (Radhamohanpur badlands), (4) 1.38 (Birsingha badlands), (5) 1.20 (Gangani badlands) and (6) 1.24–1.30 (Bagnada badlands). In general, the R_L values reflect that the cumulative length of lower order gullies is R_L factor (i.e. 1.20–1.39) lower than the next higher order gullies. The cumulative lengths of gullies are estimated in six badlands as (1) 1440.23 m

(Bhatina badlands), (2) 1212.22–1371.16 m (Surul badlands), (3) 1302.77–4116.32 m (Radhamohanpur badlands), (4) 5096.44 m (Birsingha badlands) and (6) 1483.0–3072.31 m (Badnada badlands). The areal coverage of gullies varies from 75,997.5 to 1757,813.0 m^2, having a maximum basin length range of 363.52–2084.62 m. The geometric positive relationship between basin area (A_B) and basin length (L_B) is established in the six badlands, i.e. $L_B = 3.616 A_B^{0.437}$, R^2 of 0.914. This empirical relation suggests the geometric progress of growth of gully length in respect of increasing growth of basin drainage area, having high slope value ($b = 0.437$) of trend line.

The gully density (G_d) of unit area (km km^{-2} or m m^{-2}) is a measure to assess the degree of fluvial network generation or crowding and fluvial incision level in the badlands (Table 2.5). In general, all over the world there is a high drainage density in the badlands (Bryan and Air 1982). Similarly, this study reveals a similar fact. The average G_d of six badlands is nearly about 11.264 km km^{-2} which is relatively high in the alluvial floodplain of Bengal Basin. The values of G_d are estimated in six badlands as (1) 6.928 km km^{-2} (Bhatina badlands), (2) 6.93–14.667 km km^{-2} (Surul badlands), (3) 17.274–22.470 km km^{-2} (Radhamohanpur badlands), (4) 2.86 km km^{-2} (Birsingha badlands), (5) 11.748 km km^{-2} (Gangani badlands) and (6) 4.78–13.726 km km^{-2} (Bagnada badlands). Except Birsingha badlands, other five badlands show high G_d, having high drainage texture, high rate of valley incision and high fluvial discharge also.

Next to G_d, the average length overland flow (L_g; Horton 1945) is an important measure to assess gully erosion at hillslope. Overland flow is defined as water that flows over the land surface as either diffuse sheet flow (laminar or mixed laminar flow) or concentrated flow (turbulent flow) in rills and gullies. The value of L_g reflects the flow path which will become concentrated at a distance from the water divide. Through transport and scouring, the overland flow causes rill and inter-rill erosion. The values of L_g are estimated in six badlands as (1) 72.16 m (Bhatina badlands), (2) 34.08 m (Surul badlands), (3) 22.47–28.94 m (Radhamohanpur badlands), (4) 174.51 m (Birsingha badlands), (5) 42.55 m (Gangani badlands) and (6) 36.42–104.43 m (Bagnada badlands). The low value of L_g (<100 m) means the shortest length of the flow path to gully head, having high vulnerability of head cut retreat. The high value of L_g (>100 m) reflects the longest length of the flow path to gully head, having low vulnerability of head cut

Table 2.5 Linear and relief properties of gully basins in the badlands

Badlands	Gully Basins	G_d (km km^{-2})	L_g (m)	C_m (m^2)	R_h	R_c	E_R	HI	S_I
Bhatina	Gully 1a	6.93	72.16	144.32	0.021	0.63	0.31	0.43	1.39
Surul	Gully 2a	14.66	34.08	68.17	0.019	0.78	0.44	0.4	1.64
	Gully 2b	6.94	72.06	79.12	0.023	0.67	0.37	0.43	1.47
Radhamohanpur	Gully 3a	17.27	28.94	57.89	0.019	0.58	0.33	0.57	1.22
	Gully 3b	17.95	27.85	55.71	0.013	0.57	0.29	0.47	1.38
	Gully 3c	22.47	22.25	44.5	0.012	0.54	0.31	0.47	1.12
Birsingha	Gully 4a	2.86	174.51	349.01	0.011	0.83	0.35	0.43	1.16
Gangani	Gully 5a	11.75	42.55	85.11	0.026	0.67	0.43	0.53	1.37
Bagnada	Gully 6a	13.73	36.42	72.84	0.013	0.57	0.26	0.58	1.42
	Gully 6b	10.39	48.13	96.25	0.017	0.54	0.29	0.52	1.36
	Gully 6c	4.78	104.43	208.87	0.018	0.72	0.42	0.53	1.32

Note G_d = gully density; L_g = length of overland flow; C_m = constant of channel maintenance; R_h = relief ratio; R_c = circularity ratio; E_R = elongation ratio, HI = hypsometric integral; S_I = sinuosity index retreat. Due to high slope steepness, many gullies require the shortest path of overland flow in the upstream drainage area.

The constant of channel maintenance (C_m; Schumm 1954) is a very useful measure in gully erosion because the value of C_m reflects a quantitative expression of the minimum or limiting drainage area (m^2 or km^2) required for the development of unit m or km length of gully channel (Table 2.4). So, the basin with high C_m means low erosion vulnerability, because to develop a unit length gully there is a requirement of high drainage area (i.e. high runoff) upstream. In six badlands C_m is estimated as (1) 144.32 m^2 (Bhatina badlands), (2) 68.17 m^2 (Surul badlands), (3) 55.71–57.89 m^2 (Radhamohanpur badlands), (4) 349.01 m^2 (Birsingha badlands), (5) 85.11 m^2 (Gangani badlands) and (6) 72.84–208.88 m^2 (Bagnada badlands). Except Birsingha badlands, the other five badlands show low C_m, signifying relatively low drainage area requirement to develop one unit metre of the gully. The sinuosity index (S_I) is applied here to analyse the degree of sinuous nature in the gullies. The S_I value of 1.0 means perfectly straight and >1.5 means meandering course of the channel. In these badlands the channel S_I varies from 1.12 to 1.64, having a mean value of 1.35. It reveals that the courses of the gully channel is situated in between straight and meandering nature, reflecting the high energy of the channel with a dominance of erosion over deposition in most of the landscape.

2.4.5 Slope and Relief Aspects of Gullies and Statistical Inference

The DEMderived average slope of gully basin varies from 2.16 to 2.8%, and the maximum valley side slope varies from 4.7 to 6.4% in the badlands. Gully head slope (between head cut and water divide) is a key aspect of erosion threshold and it has a geometric negative relationship with drainage area. In this study, the gully head slope ranges from 0.0138 m m^{-1} to 0.0931 m m^{-1}, having a mean of 0.035 m m^{-1}. On average with an increase of one metre downstream length from elevated water divide, the drop of elevation is nearly about 0.035 m. The channel slope (from outlet to the gully head cut), measured along the thalweg (or deepest part of gully channel), is measured in master gullies of six badlands as (1) 0.12–4.04% (Bhatina badlands), (2) 0.91–4.98% (Surul badlands), (3) 0.59–3.88% (Radhamohanpur badlands), (4) 1.56–4.44% (Birsingha badlands), (5) 1.82–4.63% (Gangani badlands) and (6) 0.38–4.17% (Bagnada badlands). In general, considering all channel slopes (S_a) and lengths (L), it is found that the slope is increasing from outlet to the gully head cut following a positive trend line, i.e. $S_a = 0.367\ \ln(L) + 0.117$. Now, considering each gully slope profile, it is understood that initially the slope from the outlet towards the head cut becomes minimum and after a point it is escalating towards the head cut. Here, the 2nd-order polynomial curve (main parabola; $S_a = aL^2 + bL + c$) fits with the scatters with a maximum coefficient of determination (R^2 = 0.439–0.896) (Fig. 2.10). Observing the six scatter diagrams and trend lines, it is concluded that slope near the outlet is minimum due to flat fan sediments of gully deposits, but due to an eventual rejuvenation of gully incision at fan deposits, the slope gets higher at the lowest part of the profile.

The stream length gradient index (S_L; Hack 1973) is calculated along each main gully (from head cut to outlet) and used to evaluate the erosional resistance of the available laterite deposits and relative intensity of fluvial erosion and active tectonics also. SL correlates to the total stream power, available at a particular reach of the gully which is an important hydrologic variable, because it is related to the ability of a gully to incise its valley and transport sediments. The minimum and maximum values of S_L are estimated in the gully elevation profiles of six badlands as (Fig. 2.11) (1) 4.39–32.59 m (Bhatina badlands), (2) 2.37–20.669 m (Surul badlands), (3) 0.80–34.8 m (Radhamohanpur badlands), (4) 0.81–32.59 m (Birsingha badlands), (5) 11.58–35.32 m (Gangani badlands) and (6) 4.83–47.32 m (Bagnada badlands). The

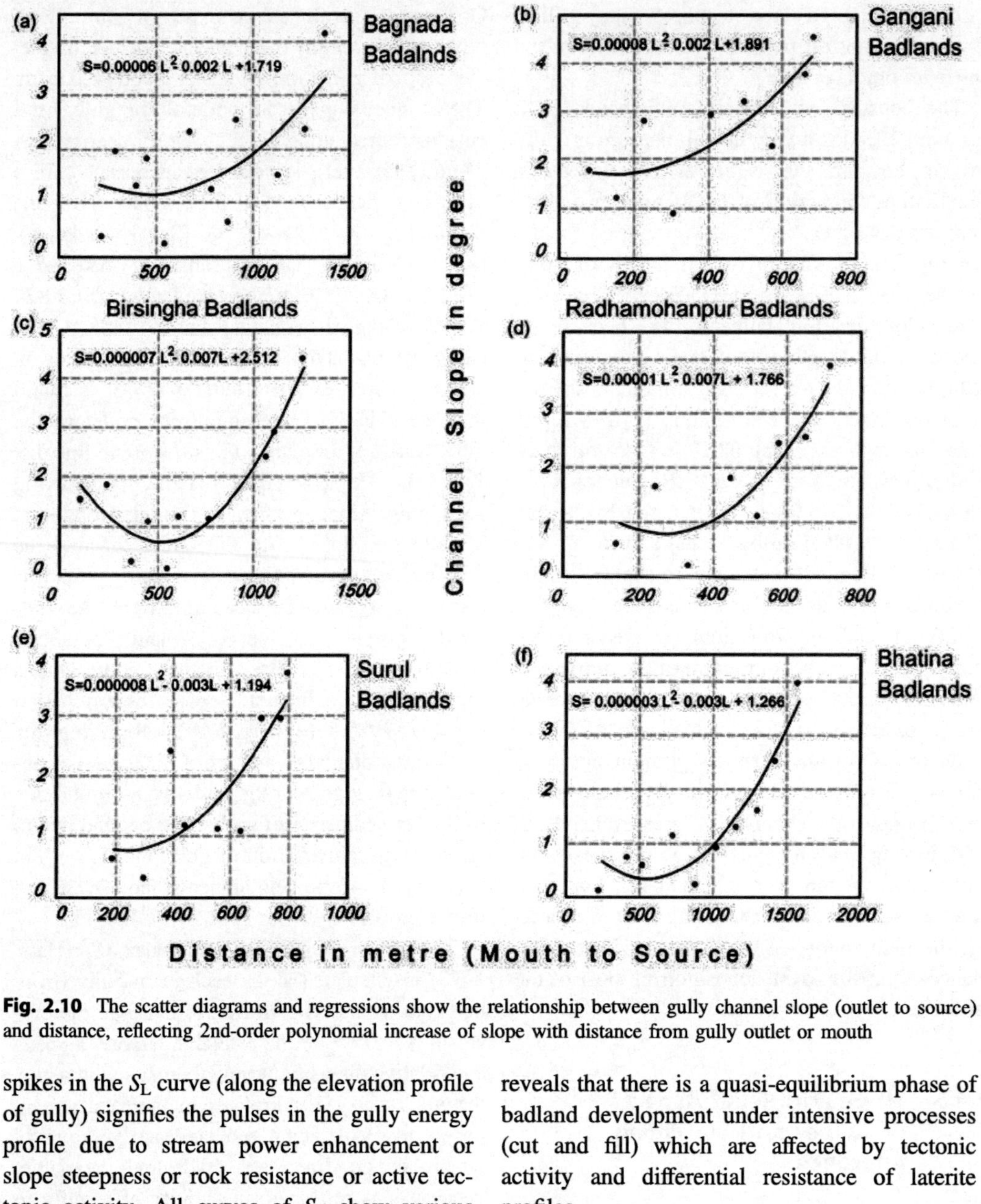

Fig. 2.10 The scatter diagrams and regressions show the relationship between gully channel slope (outlet to source) and distance, reflecting 2nd-order polynomial increase of slope with distance from gully outlet or mouth

spikes in the S_L curve (along the elevation profile of gully) signifies the pulses in the gully energy profile due to stream power enhancement or slope steepness or rock resistance or active tectonic activity. All curves of S_L show various spikes from a head cut to the outlet. The developments of several knickpoints or incised channels generate those spikes due to the enhancement of stream energy. This enhancement is occurred due to sudden tilting of the surface and increasing kinetic energy of flow or tectonic upliftment and increasing slope steepness in the region of *Rarh* laterites. This analysis reveals that there is a quasi-equilibrium phase of badland development under intensive processes (cut and fill) which are affected by tectonic activity and differential resistance of laterite profiles.

Relief (difference between the maximum and minimum elevation, R_r) is another parameter to assess the range of dissection in the landscape, and high relief is associated with deep valley incision at a mature–youthful stage of landscape development. The relief of the basin is estimated in six badlands as (1) 18 m (Bhatina badlands), (2) 7 m (Surul badlands), (3) 8–10 m

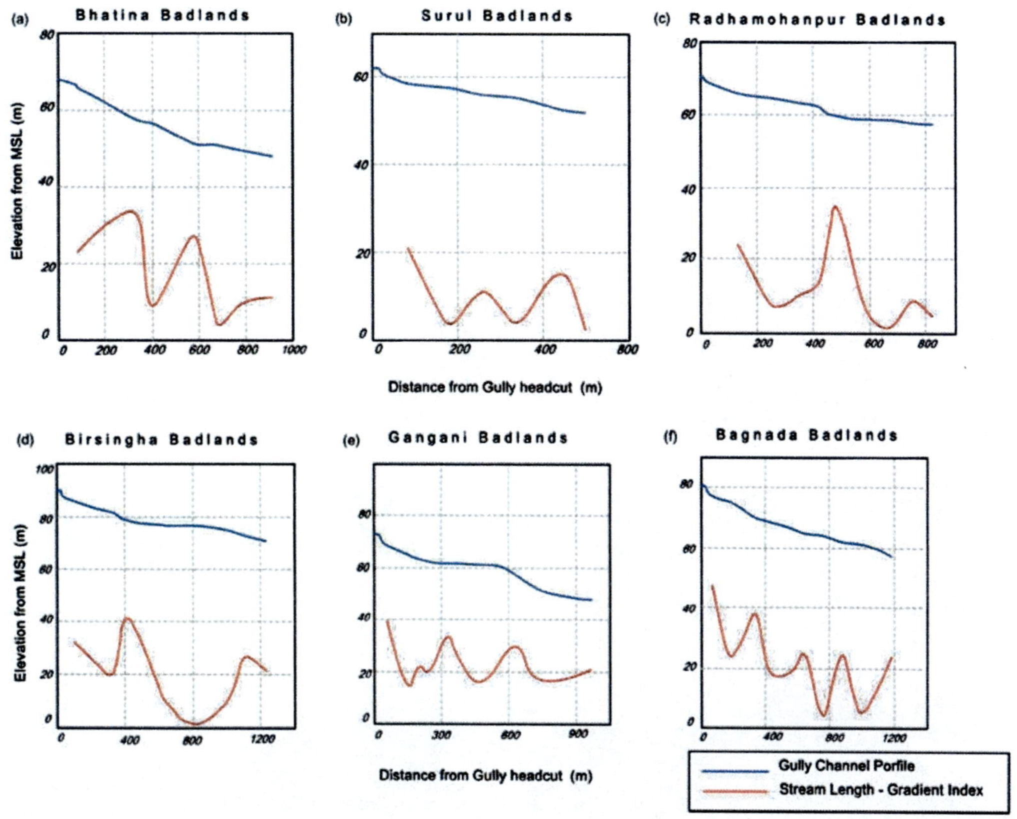

Fig. 2.11 The combined association of gully channel elevation profile and stream-length gradient index (*SL*) showing the variable energy pluses (cut-and-fill sequences) of *SL* index from gully head to outlet, signifying the role of active tectonics, slope-controlled stream power and active channel erosion against bed resistance

(Radhamohanpur badlands), (4) 24 m (Birsingha badlands), (5) 30 m (Gangani badlands) and (6) 12–27 m (Bagnada badlands). The high relief of range 18–30 m reveals the potential energy of the system to erode the landscape and to generate high drainage density. Relief ratio (R_H) is another parameter to get information on the grade condition of stream and slope (i.e. fall of elevation in unit length). The value of R_H varies from 0.011 to 0.026 m m^{-1} in six badlands. The badlands of Gangani and Bhatina yield a high value of R_H (i.e. 0.021–0.026 m m ^{1}), and the average is nearly about 0.017 m m^{-1} which reflects that the gullies are not in a grade condition (i.e. very close to zero). In a grade condition, the relief will be minimum so that there will be a delicate balance between erosion and deposition, but this condition does not prevail in the badlands. Another parameter, dissection index (D_I; Dov Nir 1983), is applied to examine the roughness of the surface. D_I – 0 means absence of dissection and old stage of landscape development, whereas D_I – 1 means vertical cliff or deep valley incision and younger stage of landscape development (Fig. 2.12). A high value of D_I (>0.25) signifies an increasing roughness of the terrain with the development of numerous gullies. In six badlands the range of D_I is estimated as (1) 0.178–0.401 (Bhatina badlands), (2) 0.101–0.213 (Surul badlands), (3) 0.057–0.184 (Radhamohanpur badlands), (4) 0.341–0.474 (Birsingha badlands), (5) 0.06–0.267 (Gangani badlands) and (6) 0.137–0.273 (Bagnada badlands). It is found that except badlands of Surul and Radhamohanpur, all other badlands show relatively high D_I, i.e. greater than 0.25.

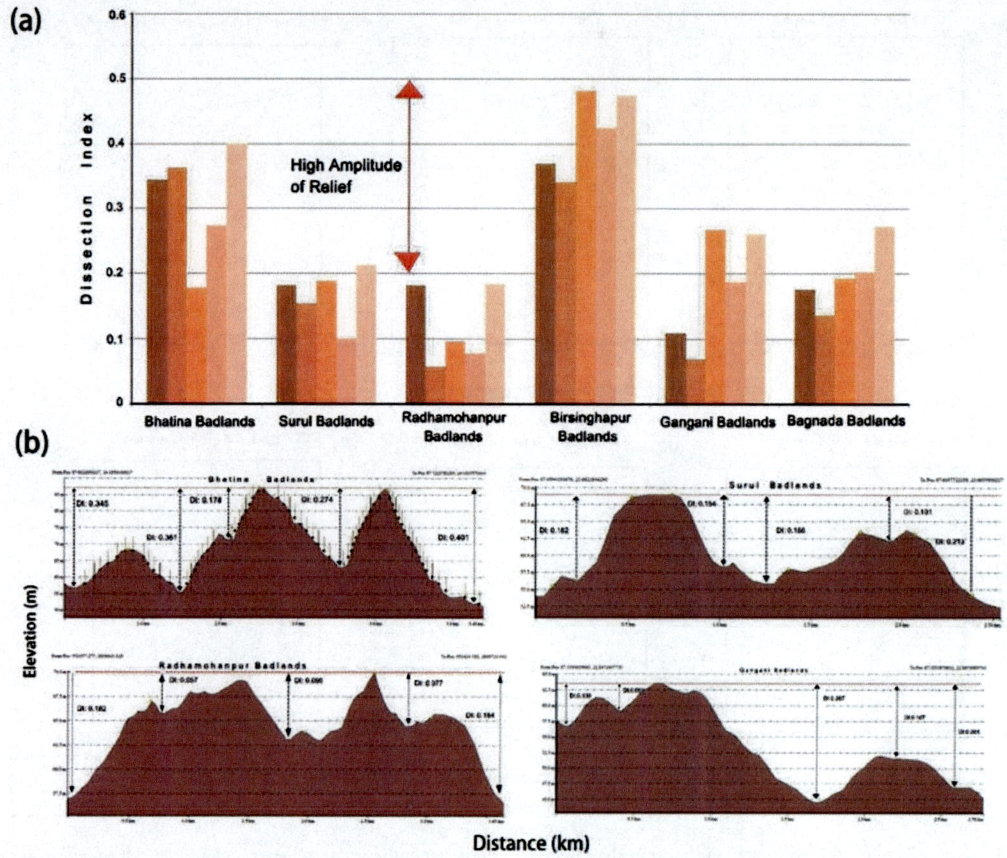

Fig. 2.12 (**a**) Variations of dissection index in the badlands of *Rarh* laterites (*DI* 0.2–0.5 means high amplitude of relief and depth of erosion), and (**b**) cross-sectional elevation profiles of badlands showing regional dissection and relief variation due to gully erosion

Hypsometric integral (H_I; Strahler 1952) is a significant parameter to judge the stage landscape development and the active tectonic nature of the landscape in respect of relief—area variation. The average value of H_I is 0.493 which reflects a youthful–mature landscape (minimum entropy of system) in a quasi-equilibrium condition (>0.35 – < 0.6). In six badlands H_I is estimated as (1) 0.436–0.631 (Bhatina badlands), (2) 0.404–0.437 (Surul badlands), (3) 0.478–0.571 (Radhamohanpur badlands), (4) 0.436 (Birsingha badlands), (5) 0.538 (Gangani badlands) and (6) 0.527–0.583 (Bagnada badlands). It can be concluded that the badlands have an influence on active tectonics and local base level control of erosion, as the H_I value varies between 0.4 and 0.6. The landform is still situated in a condition of high erosion rate than deposition, and the terrain is very much susceptible to tectonic uplift and watershed erosion. Now, the two indices of basin shape (circularity ratio, R_C and elongation ratio, E_R) are applied to indentify the geometric structure of the gully basin. The value of R_C varies from 0.548 to 0.836, and the value of E_R varies from 0.292 to 0.446 in six badlands. R_C value below 1 and E_R value below 0.5 signify a semi-oval to elongated shape of the basin. In this badlands most of the basins or sub-basins of gullies have an elongated shape. It is found that basin with $E_R < 0.5$ or $R_C < 0.7$ signify a tectonically active and high level of gully incision.

In the above discussion, the individual parameter of linear and areal aspects of basin morphometry is scrutinized to get new fangled geomorphic information on the badlands of *Rarh* laterites. Now, the statistical inter-relationship (Fig. 2.13) is needed to get the correlation or connection between the parameters using linear to non-linear regression. H_I of all gully basins is greater than 0.35, i.e. mature to youthful stage of landscape development and tectonically active terrain. It is found that with increasing S_I of basins, H_I is reducing ($H_I = 0.647–0.114S_I$). It means that relatively youthful–mature stage of badlands to tectonically active region restricts the sinuous nature of gully channel, promoting active channel erosion straight course. It is established that with increasing HI, the ER of basins is decreasing, signifying the youthful–mature stage of development ($E_R = 0.492–0.288H_I$). High H_I promotes more elongation of the basin. On the other side with increasing SI, the ER of basins is increasing ($E_R = 0.169 + 0.135S_I$). It means the highly sinuous nature or oscillation of gully channels enhances the basin shape from elongation towards more oval. It is established that with increasing HI, the RC of basins is reducing ($R_C = 1.061 - 0.835H_I$), reflecting more elongation, less sinuosity and more valley incision. On another side with increasing SI, the RC of the basin is also increasing ($R_C = 0.493 + 0.113S_I$). It means the gully channel with more sinuous nature is producing an oval-shaped basin rather than elongation. At last, it is established that with increasing HI, the RH of basins is increasing ($R_H = 0.012 + 0.009H_I$). It means that the youthful–mature stage of the landscape develops high relief and dissection in the badlands and it increases HI and resultant RH also, i.e. more potential energy to erosion (minimum entropy of system).

2.4.6 Topographic Threshold of Gully Initiation

The slope (m m^{-1}) and drainage area (ha) ($S = a\,A^{-b}$) determine the topographic threshold condition of gully initiation in the lateritic badlands of West Bengal. Based on the database of 70 permanent gullies, it is finally established that there is a negative or inverse inter-relationship between the slope of catchment and catchment drainage area above the gully head cut. It means that the gullies which have a larger drainage area (high potential yield of runoff volume) have required a minimum angle of slope steepness to trigger a gully head cut through a high concentration of overland flow (providing kinetic energy to erosion over the resistance of laterite). The model of *S*–*A* relationship can provide a clue about the landscape evolution and erosion vulnerability. The present situation of six badlands is quantitatively described by the empirical equation, $S_{cr} = 0.0207A^{-0.2784}$, where the critical slope of gully drainage area is defined as S_{cr} (i.e. minimum amount slope required to initiate gully) (Fig. 2.14a). Now, S_{cr} is calculated in the gullies of six badlands to assess the vulnerability of laterite terrain to gully incision (Table 2.6):

(1) Bhatina Badlands—S_{cr} 0.2023–0.0566 m m^{-1} (mean—0.0367 m m^{-1});
(2) Surul Badlands—S_{cr} 0.2104–0.0346 m m^{-1} (mean—0.0283 m m^{-1});
(3) Radhamohanpur Badlands—S_{cr} 0.0335–0.0499 m m^{-1} (mean—0.0407 m m^{-1});
(4) Birsingha Badlands—S_{cr} 0.0174–0.0221 m m^{-1} (mean—0.0222 m m^{-1});
(5) Gangani Badlands—S_{cr} 0.0196–0.0325 m m^{-1} (mean—0.0272 m m^{-1});
(6) Bagnada Badlands—S_{cr} 0.0194–0.0343 m m^{-1} (mean—0.0252 m m^{-1}).

Now, it can be confirmed from the topographic threshold model that the secondary laterites of *Rarh*, having a mean critical slope range of 0.0222–0.0407 m m^{-1}, are very susceptible to rill and gully erosion, particularly in the upstream deforested catchment of the barren surface. The estimated MD envelope (Montgomery and Dietrich 1994) distinguishes mass movement dominated gullies from hydraulic erosion dominated gullies in these badlands (Fig. 2.14b). In this study (taking sample counts of 70 gully head cuts), 38.57% of gullies are influenced by overland flow erosion threshold and 35.71% of gullies are affected by landslide erosion threshold. The thresholds for gully head position in a landscape traditionally take into consideration local slope angle (λ) and gully

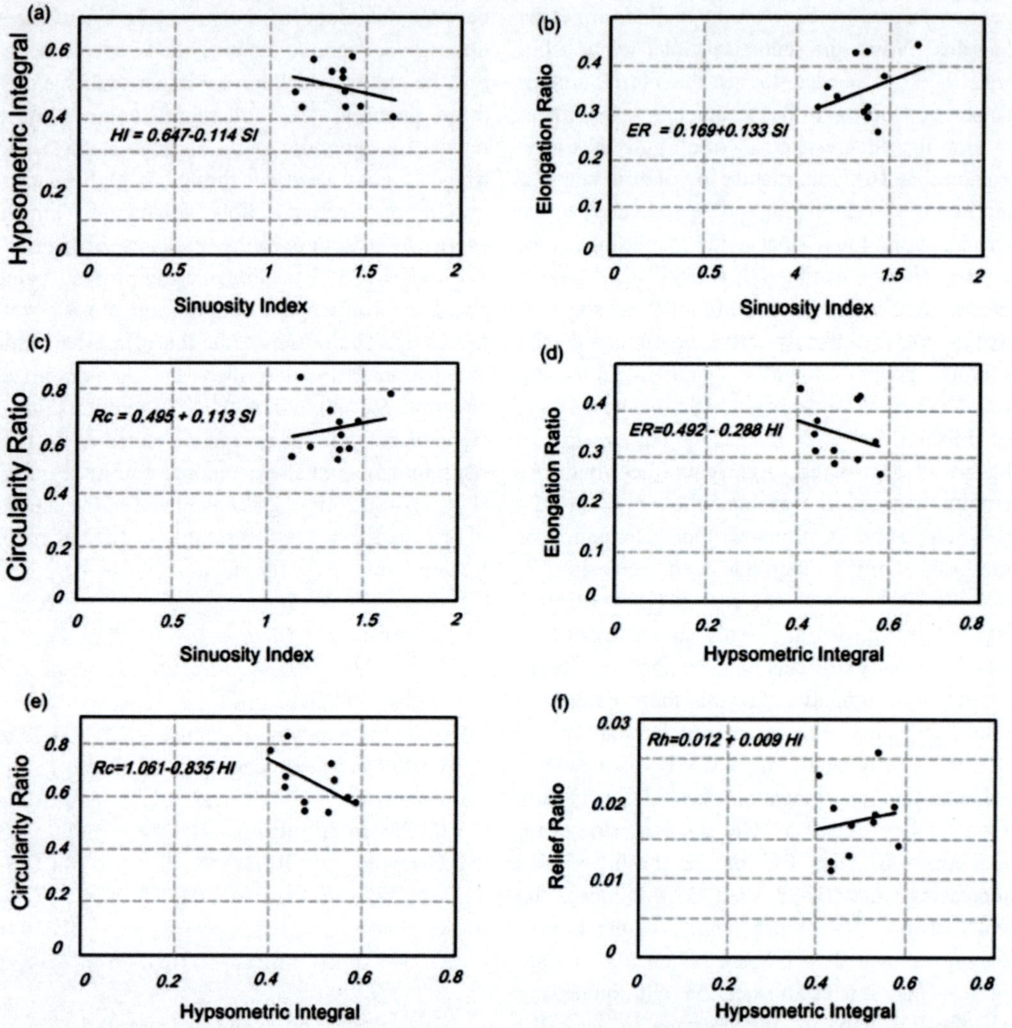

Fig. 2.13 The scatter diagrams and regressions showing (**a**) negative relationship between hypsometric integral and sinuosity index, (**b**) positive elongation ratio and sinuosity index, (**c**) positive relationship circularity ratio and sinuosity index, (**d**) negative relationship between elongation ratio and hypsometric integral, (**e**) negative relationship between circularity ratio and hypsometric integral and (**f**) positive relationship between relief ratio and hypsometric integral in the gully basins of badlands

head drainage area (*A*) by Patton and Schumm (1975) who expressed it through an equation:

$$\sin(\lambda)A^b \geq K$$

where *K* is a sort of black box coefficient (determining threshold), filled with effects of a series of factors such as soil resistance and hydraulic roughness and treated as a local constant of terrain; the exponent *b* deriving from the equations linking unit discharge to the runoff shear stresses. The *b* value, greater than 0.2, usually signifies the dominancy of overland flow erosion in the gully initiation. It is explored that b depends on the typology of flows: it increases from laminar sheet flow ($b = 0.5$) to turbulent motion ($b = 0.857$) in the gullies of the world. Torri and Poesen (2014) have established that *b*, 0.38–0.40, is determined for rough turbulent flow during peak runoff.

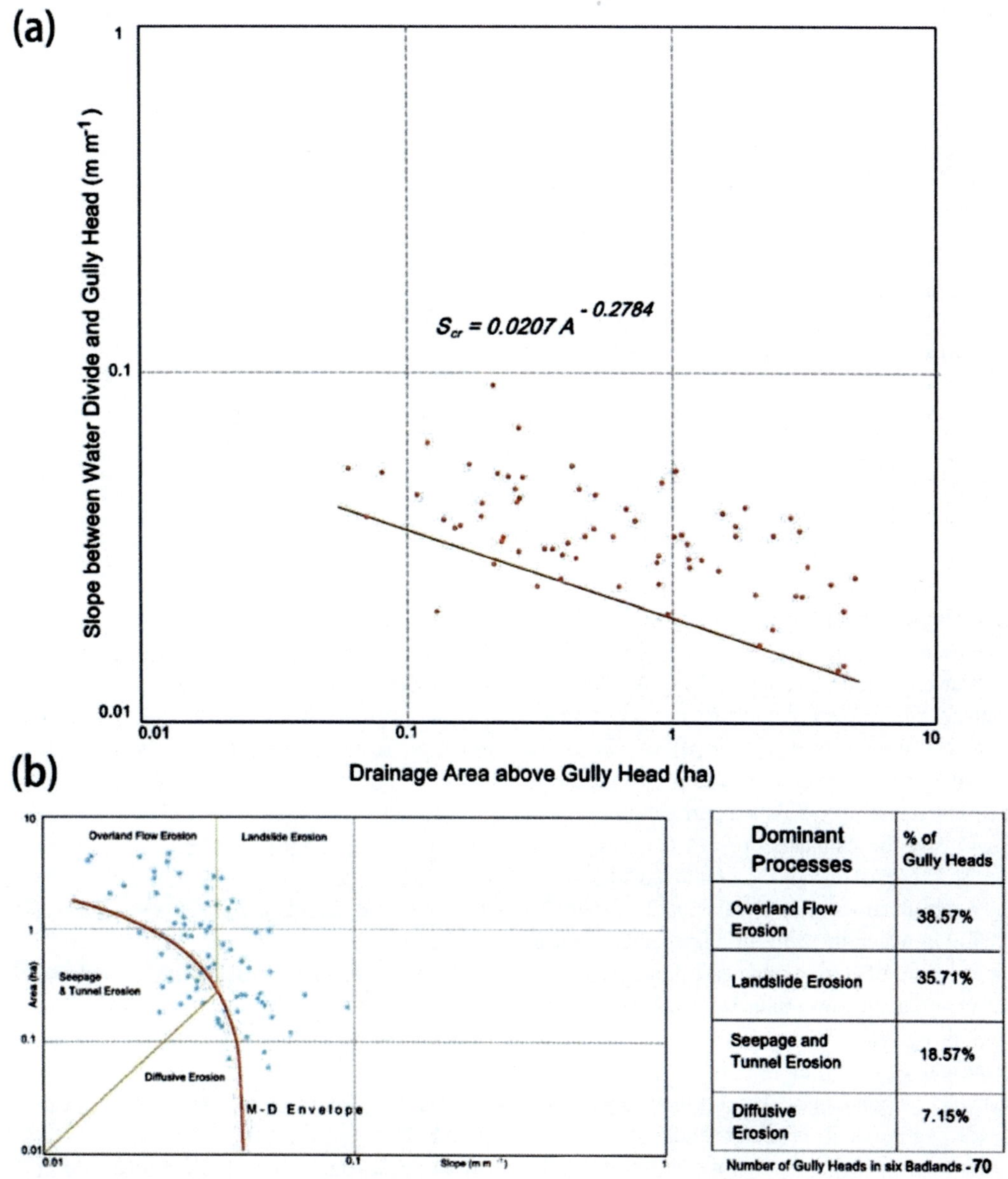

Dominant Processes	% of Gully Heads
Overland Flow Erosion	38.57%
Landslide Erosion	35.71%
Seepage and Tunnel Erosion	18.57%
Diffusive Erosion	7.15%

Fig. 2.14 (**a**) Establishing critical slope–area threshold relation for the permanent gullies of *Rarh* laterites ($S_{cr} = 0.0207A^{-0.2784}$) on the basis of intrinsic threshold, and (**b**) the diagram showing log A (ha) and log S (m m^{-1}) scatter plot in Montgomery–Dietrich Envelope (i.e. brown colour curve) to depict dominant erosion processes in gullies

The study has found the *b* value of 0.2784 which is very much lower than the above values. It signifies the dominant laminar overland flow and high surface roughness due to relatively good vegetation growth in the drainage area. The exponent value of the *S–A* model suggests that the vegetation of gully basins exerts a strong control on water movement and flow concentration, obliging the overland flow to be laminar also as high Reynolds number ($b \sim 0.25$). To justify the validation of the *b* value, the standard error of estimate (S_E) is used with a 0.05

Table 2.6 Summary database of gully head drainage area, critical slope (S_{cr}) and threshold coefficient (K) in the badlands

Badlands	Gully head drainage area (ha)	Mean slope of drainage area (λ, m m^{-1})	sin λ	Mean critical slope for gully initiation (S_{cr}, m m^{-1})	Mean K value
Bhatina	0.06–3.01	0.0284–0.0931	0.0567–0.1851	0.0367	0.395
Surul	0.41–2.43	0.0204–0.0337	0.0407–0.0673	0.0283	0.262
Radhamohanpur	0.11–0.46	0.0444–0.0340	0.0188–0.0888	0.0407	0.482
Birsingha	2.04–4.48	0.0142–0.0245	0.0283–0.0489	0.0222	0.189
Gangani	0.51–3.12	0.0226–0.0441	0.0451–0.880	0.0272	0.231
Bagnada	0.42–3.23	0.0181–0.0535	0.0361–0.1067	0.0252	0.182

confidence interval of judgement in a normal distribution. The value of product-moment correlation coefficient (r) is estimated as (–) 0.5137 to inter-relate the variables of S and A. The t-test statistics of r value is calculated as 5.1128 which is much more than $t_{0.05}$ (2.000) of two-tailed statistics at 68 degrees of freedom. So, it reflects a strong statistically satisfied correlation between S and A in the badlands of *Rarh* laterites. Then, the S_E range of b value (confidence interval at 0.05 significance level) is calculated as 0.2652–0.2909 in which the zero limit does not exist. So, the b value of topographic threshold is statistically significant to emphasize the role of overland flow. Now, it can be concluded that the empirical equation of S_{cr} and A has a strong statistical judgement of geomorphic analysis, and it can be applied in other badland areas of lateritic terrain to recognize the vulnerable hillslope (susceptible to gully erosion).

K coefficient of the threshold is very dynamic in range, and it primarily depends on soil resistance, flow hydraulic roughness, runoff peak discharge and land use character. A gully head cut will initiate when the flow resistance overcome K threshold, depending on the catchment characteristics (Torri and Poesen 2014). The K value function (Torri and Poesen 2014: Torri et al. 2015) is a product of potential rainfall abstraction of the SCS-CN method and exponential function of rock or very coarse fragment cover. In the badlands the K value should be equal or greater than the sinus function of a gradient. The initial abstraction ($S_{0.05}$) of six badlands varies from 112.25 to 154.40 mm which is based on the AMC-II CN (72–79) of badlands. Finally, the calculated K coefficients of six badlands are presented as follows:

(1) Bhatina Badlands—0.158–0.713 (mean—0.395);
(2) Surul Badlands—0.152–0.352 (mean—0.262);
(3) Radhamohanpur Badlands—0.362–0.643 (mean—0.482);
(4) Birsingha Badlands—0.142–0.352 (mean—0.189);
(5) Gangani Badlands—0.141–0.292 (mean—0.231);
(6) Bagnada Badlands—0.124–0.281 (mean—0.182).

The further analysis suggests that in the study areas, the K value of permanent gullies is two to ten times higher than sin λ. Especially in the badlands of Gangani, Bagnada and Bhatina (having direct gully–river coupling and landscape connectivity), the K coefficient value is only one and half times of sin λ (i.e. flow energy or shear stress overcome easily the surface resistance), signifying the high rate of gully

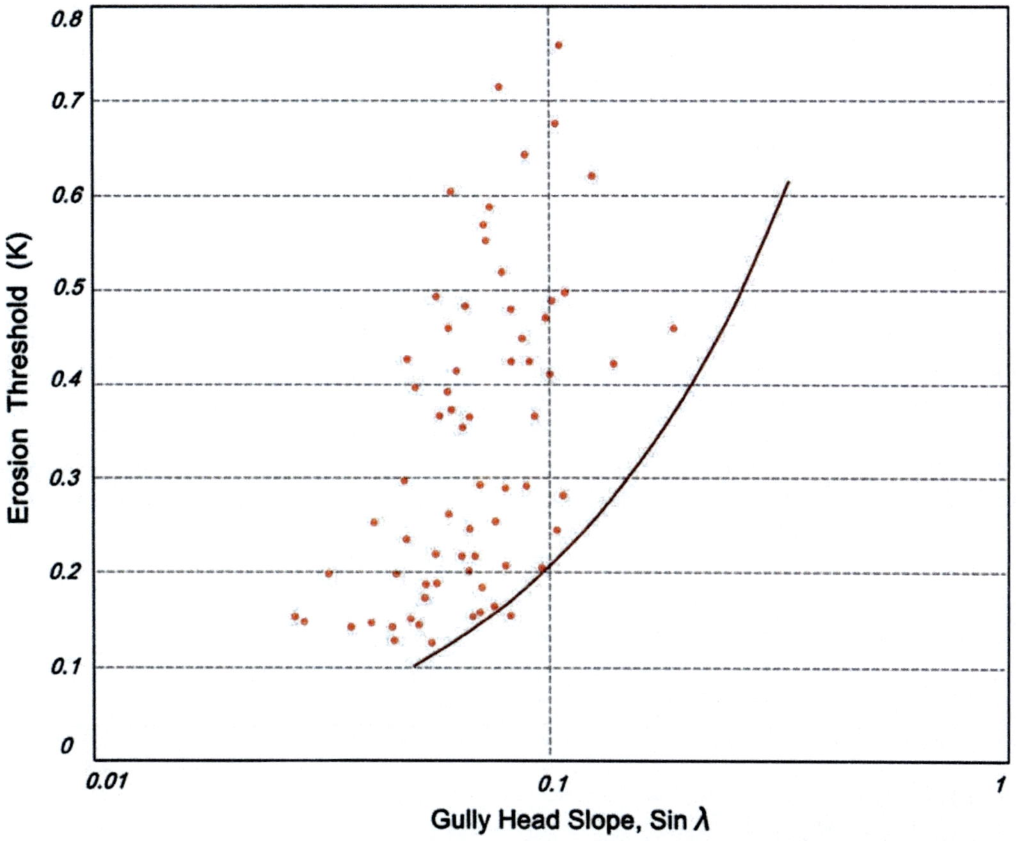

Fig. 2.15 Erosion threshold coefficient (K) of gullies is geometrically progressed with an incessant increase in sine function of gully drainage area gradient ($\sin \lambda = 2.5381K^{0.7973}$)

incision and initiation in the region of S and TS links. The final empirical equation, based on K and $\sin \lambda$, is established as follows:

$$\sin \lambda = 2.5381K^{0.7973}$$

The equation is also considered as the marginal critical condition of gully initiation in the lateritic badlands, showing the positive influence of K threshold coefficient (Fig. 2.15).

2.4.7 Rainfall–Runoff Simulation and Sediment Yield

The daily rainfall (greater than 50 mm per day) of six rain-gauge stations (viz., Rampurhat, Shyambati, Satkahania, Sonamukhi, Amlagora and Simlapal), around the periphery of selected badlands, were recorded in during June–August 2020, as follows: (1) Bhatina Badlands—57.2–100.4 mm, (2) Surul Badlands—55.2–99.8 mm, (3) Radhamohanpur Badlands—55.0–103.6 mm, (4) Birsingha Badlands—55.0–144.5 mm, (5) Gangani badlands—52.0–116.5 mm and (6) Bagnada Badlands—55.3–111.3 mm, respectively. Based on the areal coverage of land use–land cover classes and hydrological soil group B (sandy loam texture of red soils), the weighted CN (modified using slope function) and mean $S_{0.05}$ (AMC-II condition) are estimated as using the SCS-CN method (Table 2.7): (1) Bhatina Badlands: CN – 75.29 and $S_{0.05}$ – 132.55 mm; (2) Surul Badlands: CN – 77.88 and $S_{0.05}$ – 112.25 mm; (3) Radhamohanpur Badlands: CN – 76.85 and $S_{0.05}$ – 120.11 mm;

(4) Birsingha Badlands: CN – 72.74 and $S_{0.05}$ – 154.40 mm; (5) Gangani badlands: CN – 79.85 and $S_{0.05}$ – 119.41 mm; and (6) Bagnada Badlands: CN – 76.98 and $S_{0.05}$ – 120.65 mm, respectively. The estimated daily runoff amount of the selected rainfall events is presented as follows: (1) Bhatina Badlands: 15.31–53.65 mm; (2) Surul Badlands: 27.3–66.49 mm; (3) Radhamohanpur Badlands: 11.41–40.98 mm; (4) Birsingha Badlands: 11.89–79.48 mm; (5) Gangani badlands: 12.0–57.93 mm; and (6) Bagnada Badlands: 12.79–52.44 mm, respectively. The scatters of daily rainfall (P) and runoff (R) values show a positive correlation with an increasing trend in the rainfall–runoff simulation. The established relation is expressed as $R = 0.658P - 20.74$ ($R^2 = 0.89$) in these badlands (Fig. 2.16). The most interesting finding of this SCS-CN analysis depicts that two distinct trends of scatters are recognized in the regression diagram, because due to differential CN of gully basins, the runoff yield is very high in one basin than other in the same rainfall amount. In the badlands of Gangani, Bagnada and Bhatina, the runoff yield is quite high (two to three times higher) than the other due to the high percentage of the barren lateritic surface at the catchment scale.

The organization of NBSS and LUP has already estimated the mean annual soil erosion rate (using USLE) in different physiographic regions of West Bengal (Table 2.7). In the study areas, the rate of erosion varies from 10.415 to 17.605 t ha^{-1} $year^{-1}$. The main task is to find out the particular sediment yield of each badlands due to daily runoff events. The coupling of USLE and SCS-CN methods has measured the potential sediment yield (S_y) of the selected rainfall events (June–August 2020) as follows (Table 2.8): (1) Bhatina Badlands—5.36–8.31 t ha^{-1};

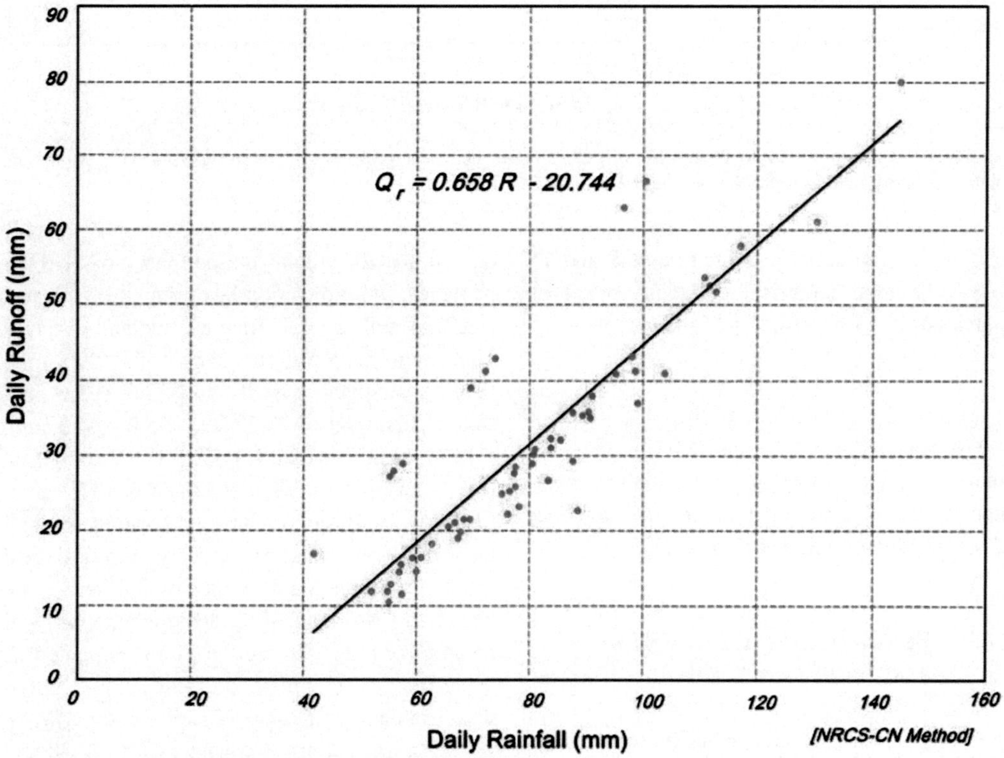

Fig. 2.16 SCS-CN method establishing a linear positive relationship between daily rainfall (R) and daily runoff (Q_r), developing an empirical relation ($Q_r = 0.658R - 20.744$) in the un-gauge watersheds of badlands

Table 2.7 Input parameters of the SCS-CN method and estimation of daily runoff, sediment yield and potential erosion index in the badlands

Badlands (area under consideration of SCS-CN)	Principal land use–land cover classes	Areal coverage (%)	Weight CN	Weight CN slope adjustment	$S_{0.05}$ in AMC II (mm)	R_d (mm)	Q_r (mm)	S_y (t ha^{-1})	PEI
Bhatina (total area: 6.13 km^2)	Forest Cover	28.14	69.92	77.29	132.55	57.2–110.4	15.31–53.65	5.36–8.31	0.078–0.573
	Shrubland	30.3							
	Agricultural Land	24.77							
	Barren Land	2.26							
	Buildup Area	14.5							
Surul (total area: 5.81 km^2)	Forest Cover	30.51	66.65	73.38	112.25	52.0–99.8	27.01–66.49	4.0–7.31	0.137–0.538
	Shrubland	41.99							
	Agricultural Land	17.29							
	Barren Land	4.93							
	Buildup Area	5.26							
Radhamohanpur (total area: 23.86 km^2)	Forest Cover	19.39	69.24	76.85	120.11	57.0–130.1	11.41–61.2	1.27–7.26	0.089–0.482
	Shrubland	22.38							
	Agricultural Land	50.04							
	Barren Land	4.46							
	Buildup Area	3.8							
Birsingha (total area: 21.77 km^2)	Forest Cover	41.71	62.54	67.74	154.4	55.0–144.5	11.89–79.48	1.91–4.70	0.066–0.444
	Shrubland	26.03							
	Agricultural Land	15.47							
	Barren Land	9.78							
	Buildup Area	6.99							
Gangani (total area: 3.73 km^2)	Forest Cover	21.97	69.32	76.94	119.41	52.0–116.5	12.0–57.9	5.1–9.1	0.107–0.418
	Shrubland	27.03							
	Agricultural Land	34.44							
	Barren Land	9.04							
	Buildup Area	7.49							
Bagnada 11.72	Forest Cover	28.24	67.67	71.78	119.41	55.3–111.3	12.79–52.42	4.6–8.4	0.096–0.394
	Shrubland	31.3							
	Agricultural Land	24.87							
	Barren Land	2.67							
	Buildup Area	12.9							

Note CN = curve number; $S_{0.05}$ = potential rainfall abstraction and infiltration during a rainfall as used in the SCS-CN method (AMC-II condition); R_d = recorded daily rainfall greater than 50 mm per day (June–August 2020); Q_r = estimated daily runoff using SCS-CN method; S_y = sediment yield from the basin; PEI = potential erosion index

Table 2.8 Input parameter of Universal Soil Loss Equation (USLE) and estimation of mean annual erosion rates in the badlands

USLE Parameters and Erosion rate	Bhatina Badlands	Surul Badlands	Radhamohanpur Badlands	Birsingha Badlands	Gangani Badlands	Bagnada Badlands
R_e	624.23	596	587	587	623	585
K_d	0.463	0.52	0.52	0.40	0.42	0.41
P_f	0.5	0.3	0.5	0.5	0.5	0.5
L_S	0.366	0.425	0.457	0.401	0.612	0.367
C_f	0.28	0.3	0.2	0.2	0.2	0.4
E_T (t ha^{-1} year^{-1})	14.809	11.854	14.163	10.415	16.013	17.605

Note E_T = potential annual soil erosion rate; R_e = rainfall Erosivity index; K_d = soil Erodibility index; L_S = slope-length factor; C_f = crop cover factor; P_f = soil protection cover factor

(2) Surul Badlands—4.03–7.11 t ha^{-1};
(3) Radhamohanpur Badlands—4.44–7.20 t ha^{-1};
(4) Birsingha Badlands—2.15–4.71 t ha^{-1};
(5) Gangani Badlands—5.70–9.10 t ha^{-1};
(6) Bagnada Badlands—4.63–8.45 t ha^{-1}.

From the result, it can be revealed that within the rainfall range of 52–144 mm, the gully network of badlands can generate sediment yield of 4–9 t ha^{-1} (400,000–900,000 kg km^{-2}) from the basins. The results of S_y are well depicted in the scatter diagrams, showing an increasing trend (but differentially due to the difference in *CNs*) in between runoff (Q_r) and sediment yield. This relation is well established as $S_y = a\ln Q_r - b$, with a high coefficient of determination (>0.9) (Fig. 2.17). The geometric progression of daily runoff can generate a geometric increase of sediment yield in these badlands. Except for Birsingha badlands all these badlands shows a high slope (b) of trend lines, i.e. ranging from 2.5 to 3.6. It means these regions are very susceptible to gully erosion in respect of increasing runoff or rainfall in present geo-climatic settings. Especially, the badlands of Bhatina, Gangani and Bagnada shows a maximum value of S_y (>8.0 t ha^{-1}) within the recorded rainfall range.

The potential erosion index (*PEI*) (Torri and Poesen 2014; Kariminejad et al. 2020) is estimated based on the daily runoff, initial abstraction and basin slope and the value 1 means the highest level of erosion in the basin of the gully. The results show that *PEI* range varies from 0.078 (least susceptible to erosion) to 0.554 (maximum susceptible to erosion) in the badlands. Similarly, the badlands of Bhatina, Gangani and Bagnada shows the high value of *PEI*. In this analysis, two valid empirical equations (exponential function) are established among *PEI*, Q_r and S_y (Fig. 2.18):

$$\text{PEI} = 0.0721e^{0.0309\text{Qr}}$$

$$\text{PEI} = 0.0847e^{0.1517\text{Sy}}$$

Interestingly, in this analysis again two different trends of scatters are recognized: (1) PEI is high due to high S_y of basins in respect of high $S_{0.05}$, and (2) PEI is high due to high Q_r in respect of high *CN*. It is confirmed that daily exponential increase of runoff generates high sediment yield in the basins, and at last *PEI* is drastically increased in the basins of gullies. The stream magnitude of gullies is well associated with *PEI*. The stream magnitude of 81–88 (e.g. Radhamohanpur, Gangani and Bagnada badlands) shows a relatively high PEI of 0.47–0.55.

2.5 Discussion

Gully erosion occurs under different forms (and names), as described by Poesen et al. (1998), Casali et al. (2006), Poesen (2018) and Bartley et al. (2020), and the most common types of gully erosion are observed and recognized in the study areas. 'Ephemeral gullies' form where overland flow concentrates (mostly in cultivated

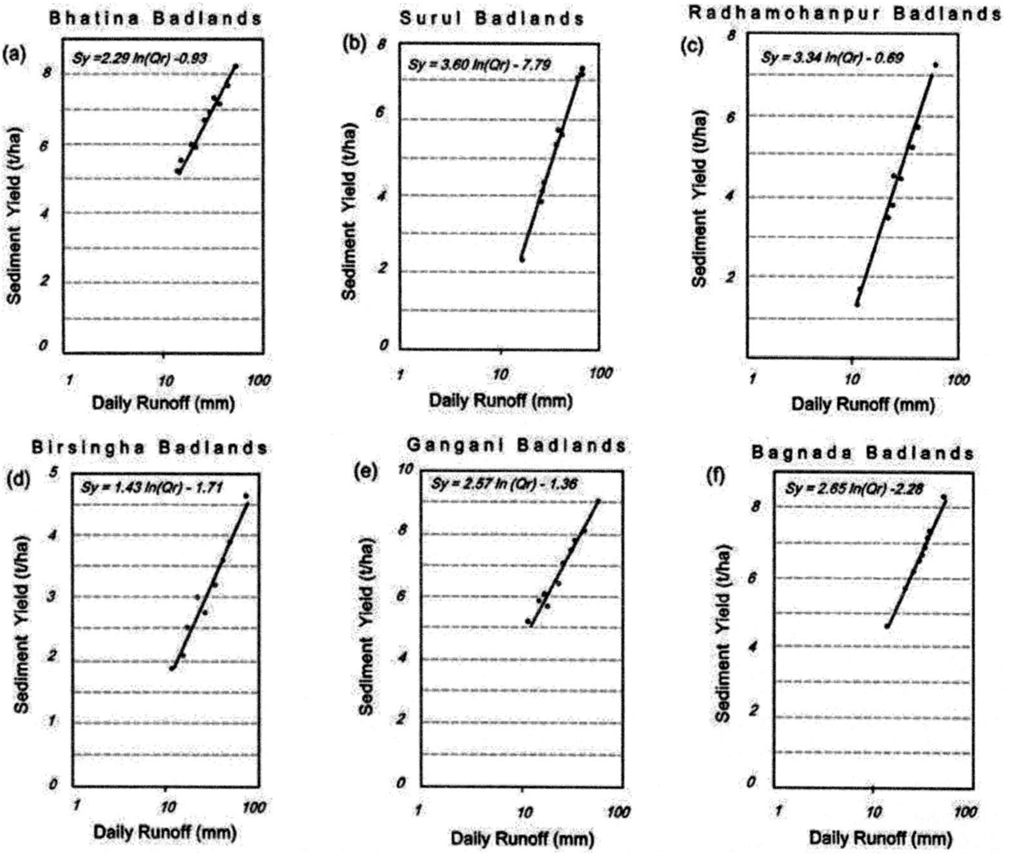

Fig. 2.17 The scatter diagrams and regressions ($S_y = b\ln Q_r - a$) showing a positive increment relation between daily runoff and sediment yield in the basins of gullies

landscapes), i.e. either in natural drainage lines (thalweg of zero-order basins or hollows) or along (or in) linear landscape elements such as drill lines, dead furrows on parcel borders or at the limit of headlands and tractor routes or unpaved access roads. When overland and shallow subsurface flow meet the gully head, energy dissipation from the resulting waterfall causes the headwall to mitigate upslope, leading to the term head cut. 'Hillslope gullies' are often defined from an agricultural perspective as croded channels too deep to ameliorate easily with ordinary farm tillage equipment. Typically, their depths range from 0.5 to as much as 25–30 m. Permanent hillslope gullies are most commonly found in concentrated flow zones of non-cultivated land on colluvial red soils, including natural drainage pathways. These gullies are formed by hydraulic erosion due to excess flow shear stress at the channel head and on the channel bottom caused principally by vegetation degradation or removal. 'Alluvial or bank gullies' may rapidly develop at or below the soil surface by fluvial scour, piping and mass movement processes. Once initiated, these gullies retreat by head-cut migration into the more gentle sloping soil surface of the adjacent alluvium and further into low-angled pediments, river or agricultural terraces. Urban sprawl into sloping land often leads to a dramatic increase of runoff coefficients, decreases in concentration times and a concomitant increase in peak runoff discharges. Under certain conditions (e.g. intensive rainfall regimes, relatively steep slopes, erodible soils

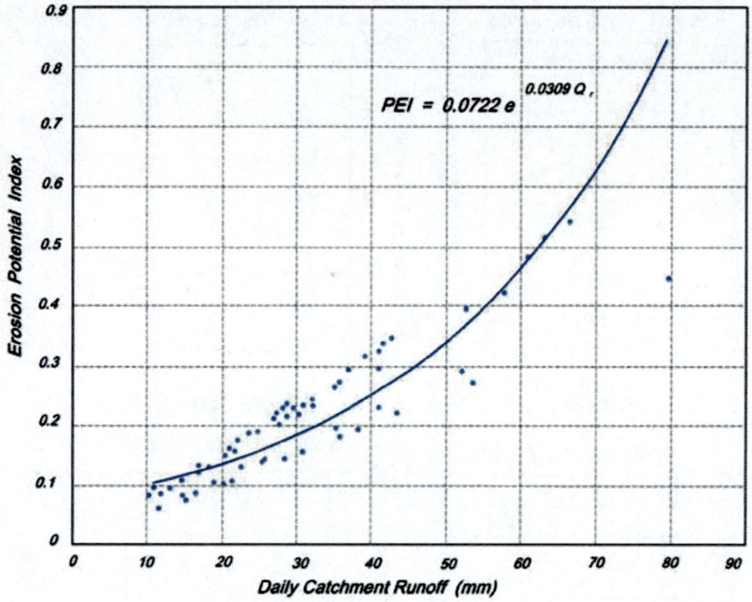

Fig. 2.18 The diagram showing a daily exponential increase of runoff generates high sediment yield in the basins, and at last *PEI* is drastically increased in the basins of gullies

and lack of adequate drainage infrastructure), this may lead to the formation of large 'urban gullies' (observed in Gangani badlands). 'Badlands' are deeply dissected erosional landscapes, formed in soft lithology. Badlands, also referred to as 'ravine lands', are a distinct landform type rather than the gullies, but they contain gullies and coalesced gully features, dominating by overland flow, mass movement and piping. Now, in the following sub-sections, the various factors or triggers of gully initiation, timing of gully erosion (e.g. geochronology) and evolution of gullied landscape are discussed to un-earth the morphogenesis of *Rarh* badlands.

2.5.1 Triggers of Gully Development

Gullies are a highly visible form of soil erosion, with steep-sided, incised, drainage lines greater than 30 cm deep (differentiate from rill). In lay terms, the word 'gully' is often used to describe any drainage line flowing towards a stream; in soil conservation, it is a section of a drainage line that is unstable, with visible evidence of soil removal. In Anthropocene, gully erosion is considered as both a natural and a human-induced active process in the landscape. Natural gully erosion plays a major role in landscape evolution. Gullies, and the streams that they feed into, help carve out valleys and supply alluvial sediment to fill floodplains. Gullies can occur anywhere in a natural drainage line as runoff flows from the most remote part of a catchment to its outlet. However, a natural drainage line is not a prerequisite for a gully to occur. Gully erosion is determined by many watershed and anthropogenic factors, and some factors intensify the rate of gully erosion and expansion (Fig. 2.19). There are extrinsic and intrinsic thresholds responsible for gully erosion. Extrinsic thresholds are those where external variable changes progressively and eventually trigger an abrupt failure within the system, e.g. deforestation and road construction. Intrinsic thresholds are within the system and change independently of the external variables, e.g. piping. In the study area, a few anthropogenic factors (e.g. deforestation, overgrazing, mining, road and associated construction) and watershed factors (e.g. rainfall, overland flow, subsurface flow, soil profile and lithology) are very visible and dominant to induce rill and gully erosion.

Any management practice and built-up areas that leads to runoff concentration have the potential to cause a gully at any location in the

Fig. 2.19 Different components and forms of gully erosion in the laterite terrain of Birbhum and Paschim Bardhaman: (**a**) intensive vertical incision of trunk alluvial gullies, (**b**) development rills and growth of bank gullies, (**c**) mass movement in gully bank, (**d**) evidence of sheet erosion (region of overland flow path) at gully catchment, (**e**) migration and growth of head cuts towards divides, (**f**) parallel growth of bank gullies, (**g**) development of gully fan at downstream with declivity of divides and (**h**) downstream ephemeral flow in gully channel showing excessive deposition at bed

landscape. This includes cultivation furrows, roads, tracks, stock pads, fence and aerodrome. Gullies can sometimes affect most of the landscape, creating what is referred to as 'badland' erosion which is extensive in vulnerable landscapes of *Rarh* laterites. Wherever possible, it is best to prevent gullies from starting rather than attempting to control them once a gully has formed. Therefore, it is needed to understand the triggers of gully development (Fig. 2.20). Any change in a land use or practice that reduced rainfall infiltration results in shorter times of concentration of runoff, increased volumes and velocities of runoff, and therefore higher risks of erosion. Gullies can be created by local, upstream or downstream influences, and often there is more than one factor involved. In some circumstances, the local or downstream influences may be just as important as those upstream.

2.5.2 Local Triggers

Local triggers of gully formation usually occur on land that is not within a natural watercourse, but may divert and/or concentrate overland runoff flows to land that is vulnerable to erosion. It includes (1) roads and tracks, (2) built-up area (settlement and large construction), (3) stock pads, (4) fences, (5) firebreaks, (6) furrows that can develop into rills that become gullies, (7) failed contour banks and waterways and (8) sinkholes on alluvial plains or large cracks on the soil surface and Diversion banks that direct runoff to an incised stream.

2.5.3 Upstream Triggers

Runoff first commences as overland flow when the rate of rainfall exceeds the rate of infiltration. This situation may occur as a result of raindrops impacting bare soil and reducing its infiltration rate. This occurs more quickly on bare soil compared to covered soil, but can occur when rainfall rate exceeds infiltration rate, regardless of soil cover conditions. An effective ground cover allows rain to soak into the soil, until it becomes so saturated that the rainfall becomes overland runoff. On sloping land, overland runoff can concentrate after a certain distance, perhaps only a few metres. Upstream triggers include increased runoff from changed land use such as tree clearing, overgrazing, cultivation, burning and urban developments. Overgrazing results in degraded surface vegetation—a reduction in biomass, ground cover and vegetal basal area—leading to reduced rainfall infiltration, increased runoff and initiation of the erosion processes.

Heavy rains concentrated in a short time, regardless of the total annual amount, can cause severe gullying. The thunderstorms of the tropics are characterized by high-intensity peaks greater than 50 mm hr^{-1}. Intense rains of tropical

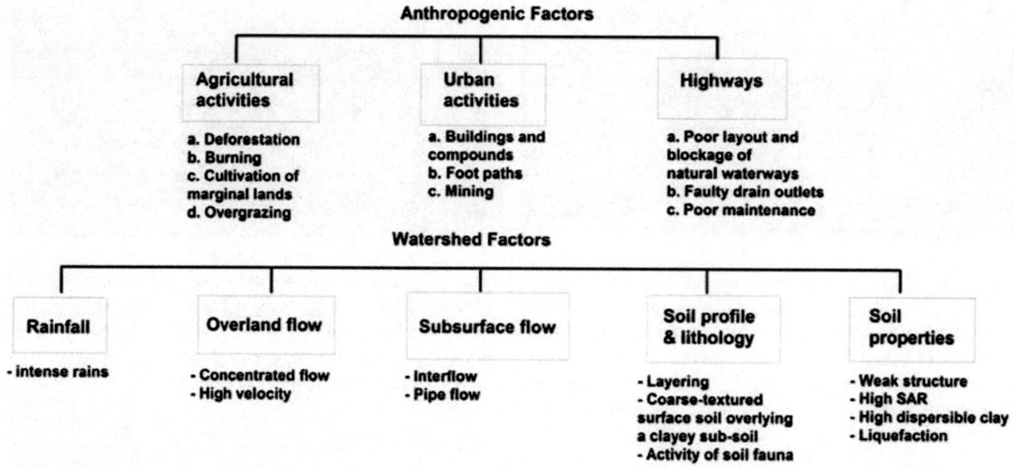

Fig. 2.20 Possible triggers or key factors of gully erosion

cyclones promote less infiltration on the saturated ground and trigger excessive runoff which is responsible for gully head migration and bank failure. Intense rains, coupled with laterite soils prone to slaking and crusting, generate high runoff volume and the concentrated flow affects the lateritic land having the bareness or vegetation break.

Increased runoff from hillslopes will concentrate in a natural drainage line and create a risk for gully erosion to occur. A change in the equilibrium of the stream may occur when a channel needs to modify its shape to allow for an increase in flood levels. The increase in runoff is said to destabilize the equilibrium of the drainage line. Gully erosion is developed by complex hydro-geomorphic processes, whereby the removal of soil is characterized by large incised channels in the badlands of *Rarh* laterites. The gully formation processes include (Bergsma et al. 1996) (a) scour by concentrated runoff, (b) dispersion in sub-soils around seepage lines, followed by tunnel erosion, (c) mass wasting and collapse of overlying soil into tunnel or gully flow, (d) headward erosion gullies and (e) gully widening (Fig. 2.21).

2.5.4 Downstream Triggers

This form of gully initiation is common but rarely mentioned in most texts. An example of a downstream influence is when the outlet of a stream is lowered, resulting in the overall length of its bed coming steeper and lowering of local base level. In the past geological eras, an increase in stream gradients may have occurred as a result of a fall in mean sea levels or a rise in the catchment landscape. When the stream bed erodes, the stream can become more incised. Drainage lines flowing into the stream then become eroding gullies over-falling into the stream. This set off a system of advancing gullies throughout a catchment. Such gullies may eat their way back to the very top of their catchments even though their contributing areas can be very small. Where a gully has virtually no catchment, raindrop impact in the gully itself can become the

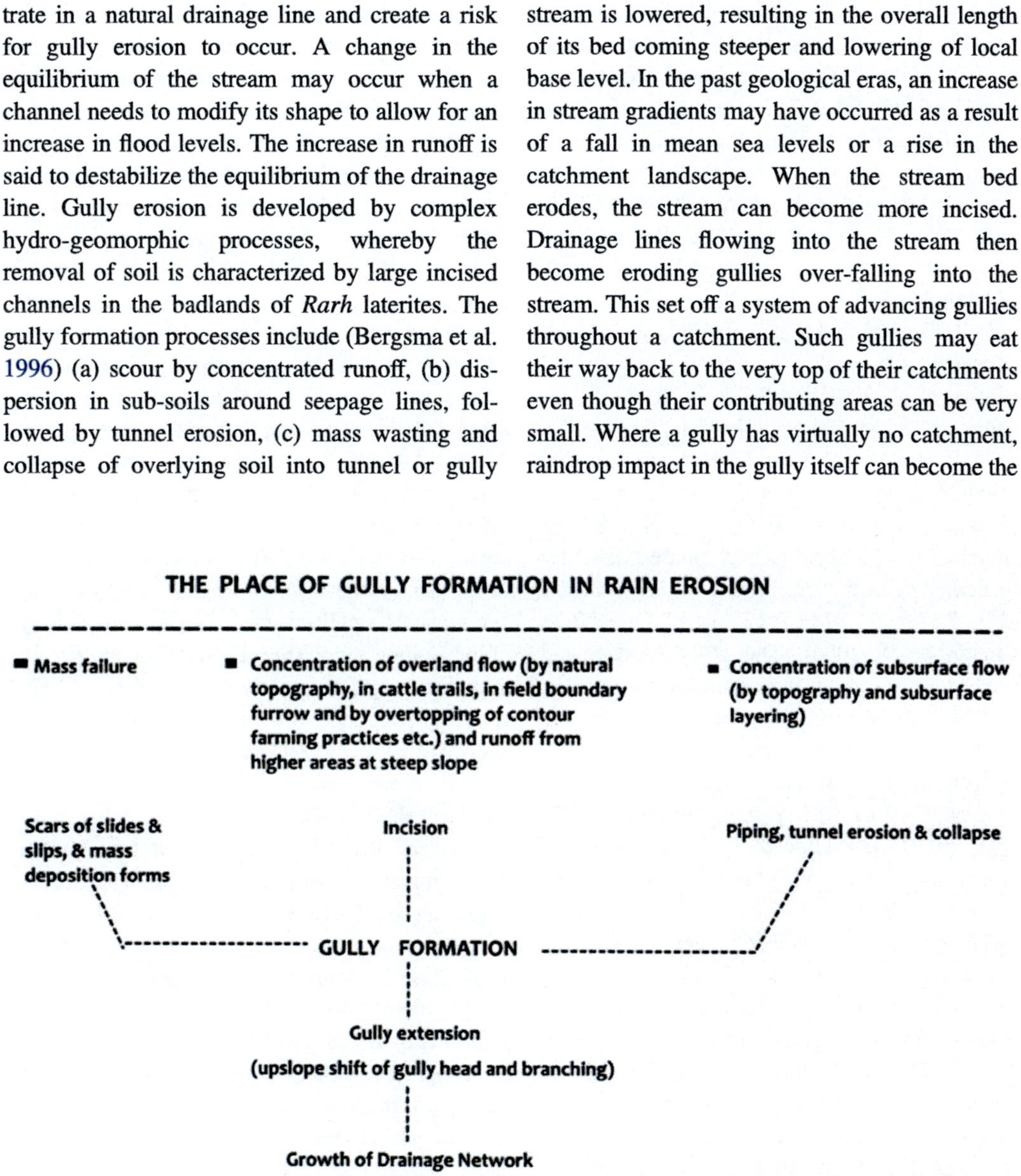

Fig. 2.21 Progressive stages of gully formation and network growth during the prolonged rainstorm event, with a dominance of mass movements, incision and tunnel erosion

main source of runoff, soil loss and subsequent downstream sediment.

2.5.5 Geochronology of Gully Initiation

The badlands along the Yamuna–Chambal Valley, Mahi Valley and Belan Valley represent the vigorous case of gully erosion in India, and these badlands are now recognized as important geomorphosites for fluvial research and geotourism. These badlands are believed to have developed due to past active tectonics and strengthening of Indian Summer Monsoon (ISM) in Late Quaternary (Mishra and Vishwakarma 1999; Gibling et al. 2005; Tandon et al. 2006; Ranga et al. 2015). After Optically Stimulated Luminescence (OSL) dating the gully fill sediments of Chambal Valley, it is explored that the youngest gully fill sediments are dated about ca. 35 ± 4 ka old (Gibling et al. 2005; Ranga et al. 2015). These sediments are actually gully fills, carried by gullies and ravines, must formed prior to that period (Tandon et al. 2006). OSL and radiocarbon dating of cliff-top strata and valley fills of northern India reflect various phases of gully incision, viz., ca. 3.39 ± 0.1 ka, 4.74 ± 0.08 ka, 9.3 ± 0.2 ka, 11.87 ± 0.12 ka and 25 ± 2 ka (Tandon et al. 2006). The gully erosion and ravine development were intensified in the river basins of north-central India due to strengthening of the south-west monsoon (SWM) in Early Holocene (Fig. 2.22): (1) Ganga–Yamuna interfluve at 13–9 ka, (2) Yamuna Valley at 53–10 ka, (3) Belan Valley at 31–21 ka, (4) Mahi Basin at 14–11 ka and (5) Luni Basin at 11–9 ka (Sinha et al. 2020).

OSL dating of badland sediments gives proxy data or clues about the age of fluvial deposition on gully floor in the past, reflecting an idea about the time frame of gully initiation. The dense network of gullies exposes a vertical profile of almost 12–14 m thick package of alternative sandstone–siltstone sequence with an overlying ferruginized sequence of fining upward fluvial sediments at the Gangani Badlands. The section reflects multi-level lateritic weathering, duricrust formation and deposition from Early–Late Pleistocene, and subsequent gully erosion of Holocene exposes this profile. The OSL dating of two gully fan deposits (up to 4.2 m and 2.5 depth) yields an age of ca. 10.6 ± 0.8 ka and 5.3 ± 0.6 ka, respectively, in the Gangani Badlands (Chakraborti 2011). The result justifies that the erosion started around ~10,600 years ago (Early Holocene) and ~5,300 years ago (Mid-Holocene), respectively, due to intensification of SW monsoon rainfall or any tectonic uplift of the Bengal Basin. These badlands are located in between two major basement faults (Chota Nagpur Foothill Fault and Medinipur–Farraka Fault), as an uplifted block, showing coarse drainage texture and centrifugal drainage development. In the badlands of the Bishnupur region, three distinct depositional episodes are indentified in three layers of gully fan deposits (Chakraborti 2011): (1) younger top alluvium (0.9 m depth) of age ca. 1.5 ± 0.2 ka, (2) subsurface deposited older alluvium (2.2 m depth) of age ca. 2.2 ± 0.1 ka and (3) ferruginous coarse deposits with Fe-nodules (3.4 m depth) of age ca. 3.5 ± 0.7 ka. This result identifies that the gully erosion started ~3,500 years ago with occasional intensification ~2,200 and ~1,500 years ago (Late Holocene). The geochronology of the Bolpur region (Surul badlands) signifies an initiation of gully erosion at ~8,400 years ago, and it got another momentum at ~4,100 years ago (Mid-Holocene) (Chakraborti 2011).

The recent earthquakes of Bengal Basin western shelf zone around the faults have emphasized the role of active tectonics on the development of lateritic badlands. The recent important earthquakes of this region were recorded as (1) M_W 5.2 Sagar Island (15 April 1964), (2) M_W 5.0 in the south of Sundarbans (23 June 1975), (3) M_W 4.7 in Ganges Canyon, south of Sundarbans (28 November 2005), (4) M_W 4.5 in Hooghly (28 August 2018), (5) M_W 4.5 in Musabani, Jharkhand (28 July 2019), (6) M_W 4.7 in Bankura (26 May 2019) and (7) M_W 4.1 in Nadia (26 August 2020). The badland terrains of *Rarh* laterites are also genetically linked with

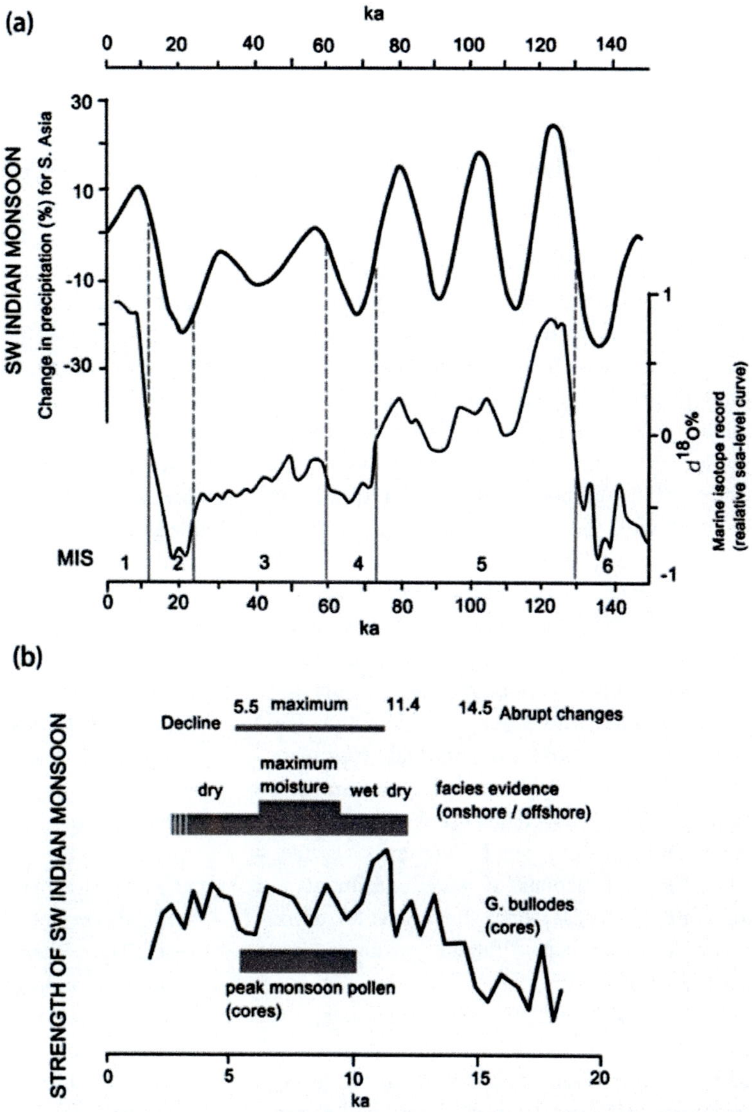

Fig. 2.22 Proxy and dating data showing variations of south-west Indian monsoon: (**a**) modelling resulting of past 150 ka and (**b**) major trends of precipitation for the past 20 ka, showing a relatively high strengthening of monsoon in between 10 and 5 ka (modified from Gibling et al. 2005 and 2008; Roy et al. 2012; Sinha et al. 2020)

inversion of relief and active tectonics (Ghosh and Guchhait 2020). Inversion of relief refers to an episode in landscape evolution when a former valley bottom becomes a ridge, bounded by newly formed valleys on each side (Ollier and Pain 1996; Ollier and Sheth 2008). Since Early Quaternary the unit between Medinipur-Farakka Fault (MFF) and Chota Nagpur Foothill Fault (CFF) started to uplift due to re-activation of basin basement faults during the occasional Himalayan upheaval and active tectonics of Bengal Basin (Ghosh and Guchhait 2020). Then during 6–7 ka, the eastern unit of Tectonic Shelf (between MFF and Damodar Fault) is subjected to subsidence, and the western lateritized unit (i.e. *Rarh Bengal*) is subjected to relief inversion due to neotectonic uplifts and consecutive fluvial erosion in Holocene times. Increased precipitation during ~15 to 5 ka period of peak monsoon recovery probably increased discharge and promoted incision and widespread badland formation (Fig. 2.22). As fluvial erosion proceeds, the valley floor becomes a ridge and interfluves (i.e. laterites of *Rarh Bengal*) bounded by newly formed Late Quaternary valleys on each side.

2.5.6 Connectivity Model of Badland Evolution

The way in which landscape compartments fit together in a catchment influences the operation of biophysical fluxes, and hence the ways in which disturbance responses are mediated over time (Brunsden and Thrones 1979; Brierley et al. 2006). These relationships reflect the connectivity of the landscape. A nested hierarchical framework that emphasizes differing forms of dis (connectivity) in catchments is proposed by Brierley et al. (2006). Analysis of the character and behaviour of landscape compartments, how they fit together (their assemblage and pattern) and the connectivity between them provides a platform to interpret the operation of the geomorphic process in the system of badlands (Brierley et al. 2006). Longitudinal (upstream–downstream and tributary–trunk stream relationship), lateral (slope–channel and channel–floodplain relationship) and vertical (surface–subsurface interactions of water, sediment and nutrients) linkages reflect the operation of different processes at different positions in a catchment of a gully.

The way in which threshold behaviour, landscape sensitivity and connectivity operate together at the meso- and macro-scale in badlands is initially considered from a theoretical energy-utilization perspective (Harvey 2001; Thomas 2001; Faulkner 2008). The argument is developed that intrinsic changes to process domain dominance can result in progressive shifts in process connectivity without the need to invoke external climatic or tectonic change. From this premise, a meso-scale closed system model for the evolution of connected states in regionally isolated badlands is developed, applicable to systems evolving towards a fixed base level within a less erosive host landscape (Faulkner 2008). The model shows how badlands disconnected from regional drainage can develop from a wide range of initial states into convergent quasi-equilibrium, followed by a period of complex response as energy is dissipated across thresholds, giving rise to a range of equifinal maximum entropy landscapes (Faulkner 2008).

This study has introduced the 'connectivity–sensitivity model' (Faulker 2008) as a crucial determinant of badland morphology and evolution in the semi-arid landscapes (Fig. 2.23). It is assumed that it may be useful for geomorphologists to consider the scales and settings within which these 'top-down connectivity' models of hillslope process operation cease to be useful to the interpretation of *Rarh* badland geomorphology. It is assumed that all sediment transport within the landscape occurs at the detachment-limited rate, and it is essentially a 'closed system' model. The connectivity model comprises four evolutionary stages (Faulkner 2008):

- *Initial State*—In the 'initial state', for the point (x,y,z), the disconnected nature of adventitiously developing discontinuous geomorphic elements in the laterite terrain has been developed. At this stage, the badlands are started to extend into the water divides of terrain, and the energy available at (x,y,z) will stabilize or increase drainage area as the badlands extend laterally, with no loss of relief. The badland system at this stage will be young, expanding and aggressive. The actual power of the small developing system is close to the critical power threshold for change, producing cut-and-fill features and increasing connectivity.
- *Quasi-Equilibrium State*—Once the host terrain is used up, the badlands divides will meet, extension will cease and the increasingly integrated network will develop to a quasi-equilibrium stage (Fig. 2.23b) in which all the events are effective and erosion operates in the same direction. This is the most organized stage that the landscape experiences; the 'information' in the landscape (Horton's laws of drainage composition is at a maximum and entropy minimized). It is a relatively insensitive landscape; the robust organization of the convergent landforms

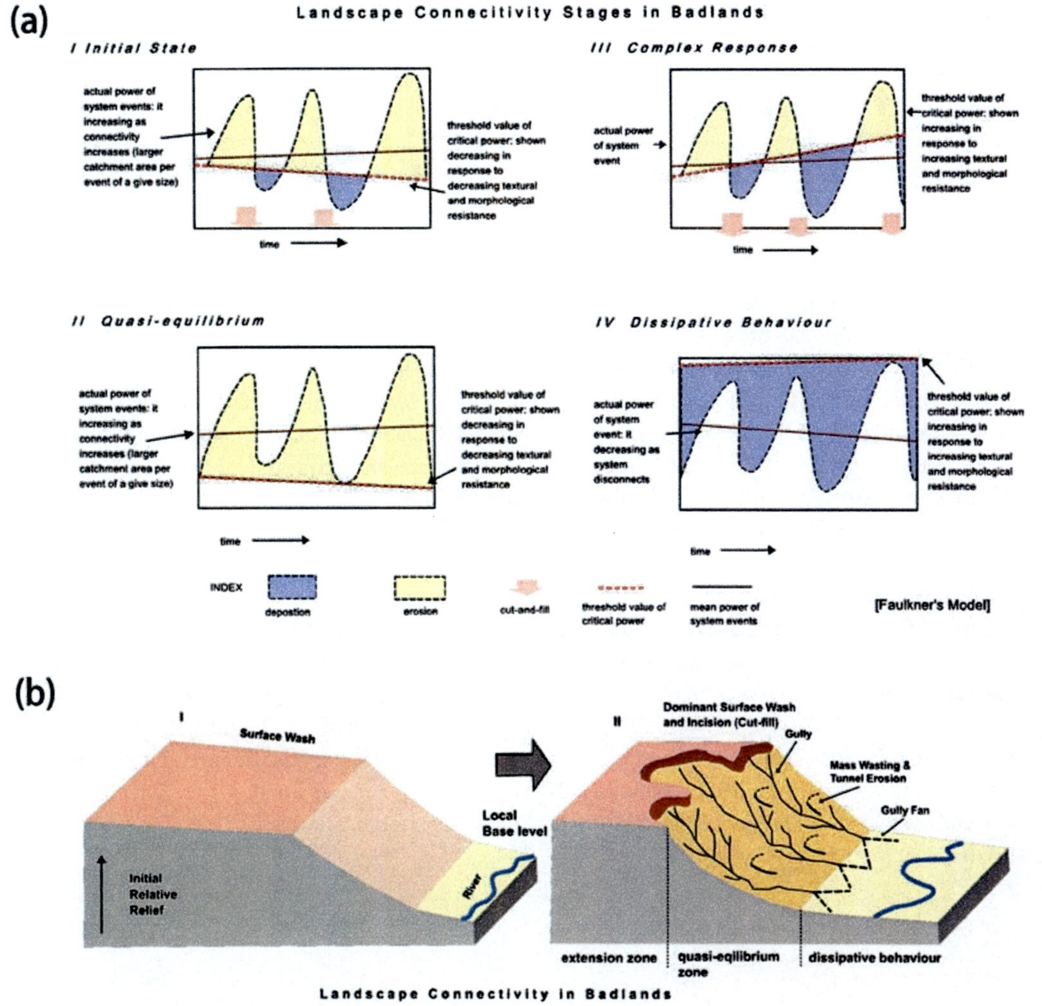

Fig. 2.23 (**a**) Connectivity model of badland evolution: relationship between actual power of system events (mean is solid line) and the threshold value of critical power (suggested as a straight dotted line) at four stages in badlands, and (**b**) a conceptual model of landscape connectivity stages in badland evolution, and development of landscape connectivity after uplift and surface wash in the *Rarh* badlands

ensures that intrinsic and extrinsic threshold ‘avulsions’ are less likely. Within-net deposition will eventually result from a progressive loss of relief, thus, the nature of equilibrium shifts (hence quasi-equilibrium).

- *Complex Response State*—Quasi-equilibrium may be a short-lived stage. The loss of relief inevitably decreases slope angles, and this increases morphological resistance so that the critical threshold for sediment removal starts to rise just as the actual power of the events generated in the reduced relief landscape starts to drop. A point will be reached where once again critical threshold power and the actual range of power of events become straddled in a reversal stage of network organization. The stage reflects complex response (viz., incision, aggradation and planform change), because this increasing lack of flow competence during more frequent events is in a climatic setting still dominated by extreme but infrequent events. A range of sensitive sites must be produced and retrenched over and over again in the badlands, giving rise to many cut-and-fill cycles (a series of ramped or divergent trajectories or episodic erosion).

- *Dissipative Behaviour State*—The progressive loss of relief means that event runoff becomes less integrated: the disconnection of system parts due to loss of flow on transmission and deposition means that the system loses 'information' by disconnecting. The overall context is therefore one of increasing entropy, which increasingly involves both morphological stabilization (slope too gentle for any erosion) and geochemical stabilization (weathering dominates, associated with the removal of dispersive agents from the top layers due to leaching). From the closed system model, 'auto-dissection' is argued to be characteristics of early badland extension, and 'auto-stabilization' as an aspect of late-stage energy dissipation. Both states are common in badland settings.

Many researchers and geomorphologists (Biswas 1987; Niyogi et al. 1970; Bandyopadhyay 1988; Sen et al. 2004; Sen 2008; Shit et al. 2013) have explored the Gangani Badlands of Garhbeta, Paschim Medinipur (about 0.75 km^2) which is now recognized as one of the important geomorphosites in West Bengal. With time, faster erosion in the concave portion of the river bend has lead to the migration of the meander towards the lateritic upland to form an elevated escarpment of Pleistocene laterites (more than 15 m height) along the Silai River's concave side (Bandyopadhyay 1988). Numerous interconnected gullies have developed with time, and further incised into the scarp, escalting fast retreat of an escarpment section and forming an elliptical form of slope. Due to regional variability of laterite thickness, exposure of hard and soft laterites, groundwater, effective rainfall, land use and land cover, soil erodibility, bareness of soil, livestock grazing, local gradient and drainage area, there is some variability of rill and gully formation. Consecutive field investigations reveal seven stages of rill and gully development on the laterites of the study area—(1) Initiation of sub-parallel rills, (2) Cross-grading and Micro piracy, (3) Initiation of gullies, (4) Start of accelerated erosion, (5) Retardation of vertical erosion, (6) Lateral erosion and widening and (7) Gully stabilization (Fig. 2.24).

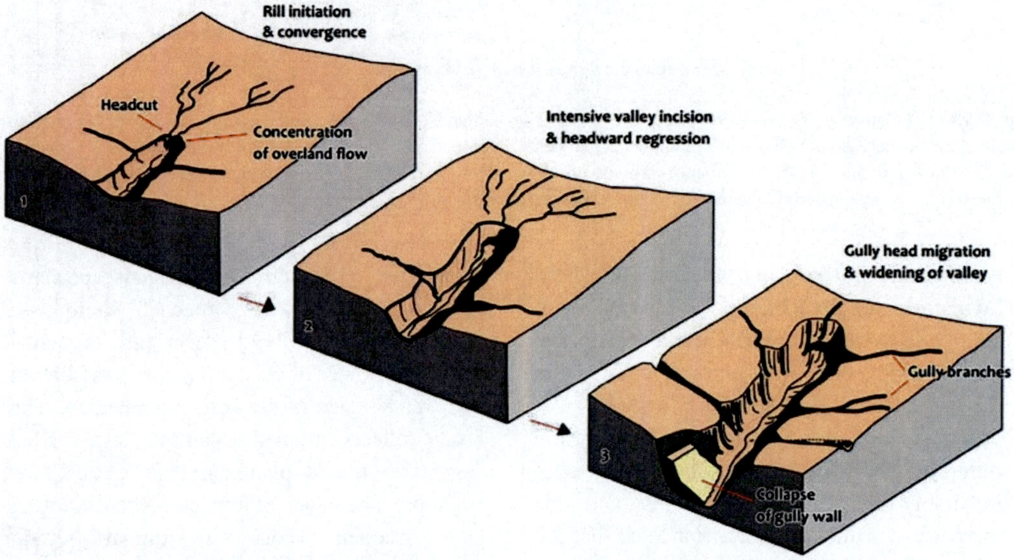

Fig. 2.24 A possible conceptual model showing three prime stages: (1) rill initiation and convergence, (2) valley incision and head cut retreat and (3) head migration and valley widening of rill and gully erosion in the laterite terrain of Bengal Basin

It is now explored that the gully–river coupling of Gangani, Bagnada and Bhatina badlands shows a good example of landscape connectivity model of badland evolution on the *Rarh* laterites (Figs. 2.25 and 2.26). As gully erosion proceeds in the Gangani Badlands (similarly in other badlands), the top-level gullies drain the surface of the laterite upland; middle-level channels formed due to the localized retreat of the cliff line cut into the lower horizons; and the bottom-level gullies dissect the planation left exposed by the retreating cliff line. Along with this denudational operations like slab failure in gully sidewalls, rockfall from river-side duricrusted escarpments, granular disintegration in the top-level duricrust, basal sapping, plunge pool erosion and undercutting are active in gully heads and side walls. In the top-level highly indurated Fe-duricrust surface up to the gully head, sheet erosion is a rather significant surface wash during high-intensity rainfall events of thunderstorms or peak monsoon. The varying cohesiveness, permeability and Erodibility in the different horizons of a lateritic profile produce occasional breakaways. The key erosional factors, responsible for evolution and progress of the landform assemblage, are recognized as: (a) sub-surface hard layer resisting erosion, (b) extent of exposition to laterite weathering, (c) differential waxing slope, (d) erodible debris slope, (e) concave basal slope, (f) convex gully head slope, with convergent rill systems, and (g) gully-side slopes feeding laterally into a gully channel. The River Silai acts as a local control on the base level of erosion since Early Holocene, and the fluvial regime of Holocene SWM monsoon period triggered the various episodes of incision and deposition in the badlands, maintaining a quasi-equilibrium stage of connectivity. It may be interpreted that the active tectonics of Bengal Basin can initiate extrinsic threshold to the badland connectivity system (providing low entropy), and still at present the gully incision phase is dominated over deposition, generating high sediment yield (5–9 t ha^{-1}; within rainfall range of 50–150 mm per day) into the floodplains or rivers.

2.6 Concluding Remarks

The above analysis suggests that the stream ordering system of Strahler is not well explained by the variable drainage network growth of gullies, but the Shreve stream magnitude and channel link topological model treat each gully basin as a distinct network of channel arrangement with increasing stream power. The gullies of study areas are usually identified as 3rd-, 4th- and exceptionally 5th-order of fluvial system (according to Strahler ordering system), whereas the maximum and minimum stream magnitude of gullies are identified as *M*-88 (highest number of channel links—175) and *M*-22 (lowest number of channel links—43), respectively. The *S* Link and *TS* Link probability of TDCN is estimated at about 24.94–26.21% and 24.95–26.05%, respectively, in the badlands. The fractal dimension of drainage network is very high (>1) in the lateritic badlands, varying from 3.037 to 7.968, showing high complexity of drainage network and dense dendritic pattern of network growth. The density of gully network ranges from 2.86 to 13.76 km km^{-2}, having an average length of the overland flow of 22.47–174.51 m. It is learned that to maintain one metre of gully channel, there is a requirement of low drainage area (55.71–349.01 m^2), reflecting high susceptibility of gully erosion. The low value of overland flow path (<100 m) means the shortest length of surface flow towards gully heads, showing high vulnerability of concentrated flow to initiate channels. The stream-length gradient index varies widely throughout the gully channel profiles of six badlands, having a range from 0.81 m to 47.32 m. The index reflects the pulses in the gully energy profile due to stream power enhancement, slope steepness, resistance of underlying sediments and active tectonics of the Bengal Basin. Alongside hypsometric integral values of gully basins, ranging from 0.4 to 0.6, reflect an influence of base-level control to active erosion in the lateritic badlands of *Rarh* Bengal, maintaining low entropy of fluvial system and development of mature–youthful landform stage.

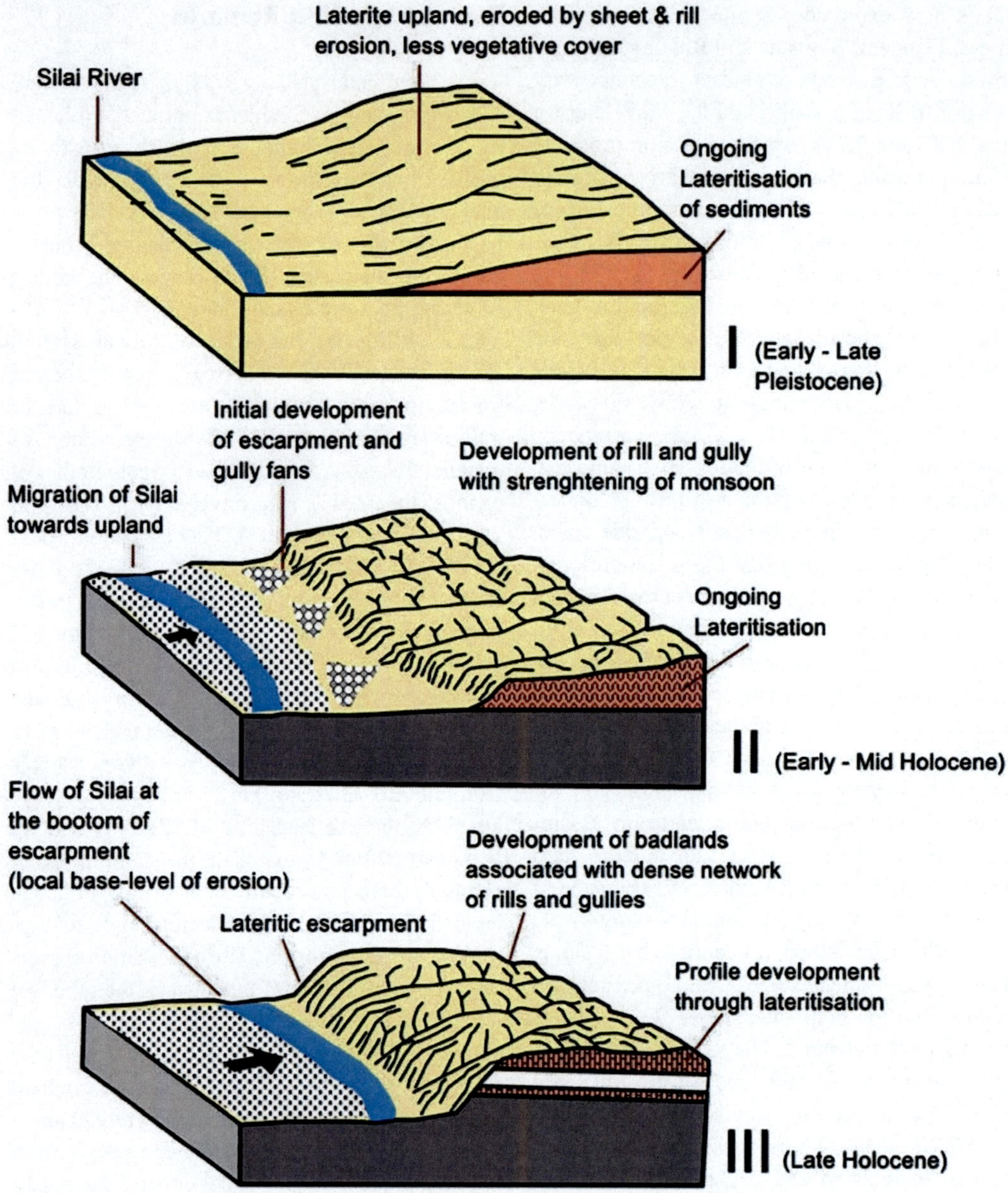

Fig. 2.25 An example of connectivity model in the badlands of *Rarh* laterites, showing the drainage network growth of gullies (in association with consecutive changes in lateritization process) since Late Pleistocene, and increasing connectivity of badlands (got momentum in Early Holocene) with the main river through sediment inputs to develop a quasi-equilibrium to complex state of landform evolution (Late Holocene to Anthropocene)

The present threshold condition of gully initiation is defined as the critical slope (in respect of gully head drainage area) of laterite terrain, varying from 0.0222 to 0.0407 m m^{-1}, and the present sine function of the slope is always greater than the threshold coefficient K (0.182–0.482) in these gullies. In the gullies, the sediment yield is estimated at about 2.15–9.10 t ha^{-1}, in response to daily runoff (11.41–66.49 mm per day). The potential erosion index (0.078–0.554)

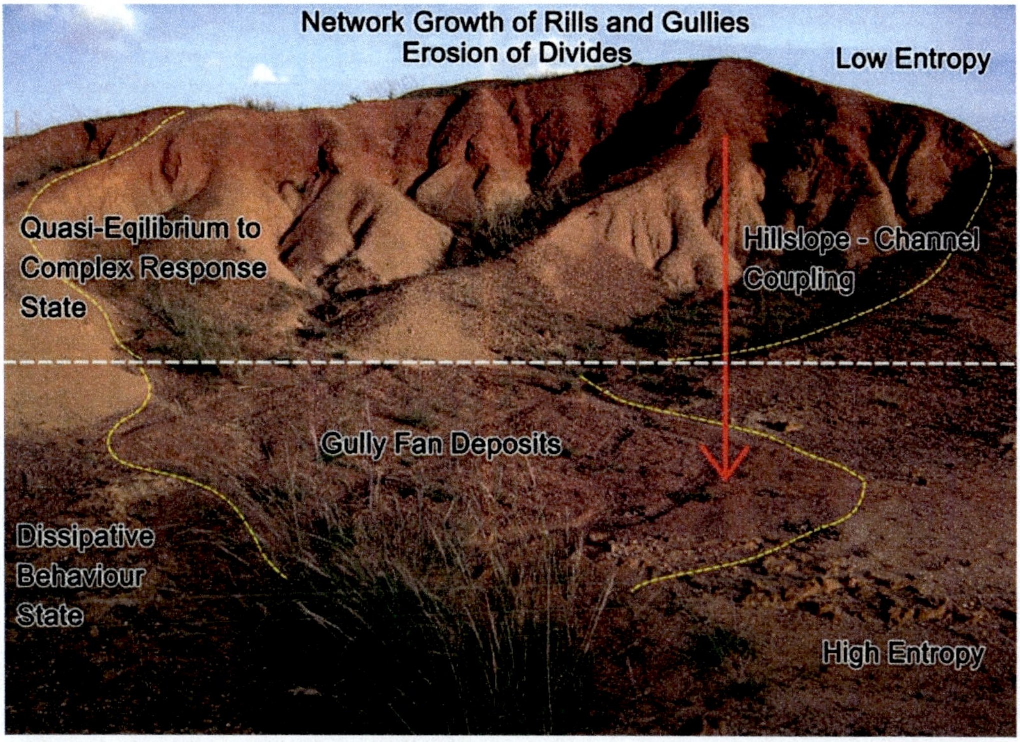

Fig. 2.26 A field example of connectivity model in badland evolution of laterite terrain (Gangani badlands), showing landforms associated with quasi-equilibrium to complex response state at upstream and dissipative state at downstream

is found to be exponentially increased with escalating daily runoff and sediment yield.

It is understood that changes in land use practices, strengthening of south-west monsoon, high rainfall of extreme climatic event and dynamic active tectonics of Bengal Basin are the key factors of badland development, and current anthropogenic activities (viz., roads, settlements, tree clearing, overgrazing and cultivation) act as catalyst to active erosion phase. Geochronology affirms that much proxy data is recorded in gully fan deposits, and these badlands show various phases of active incision since Early Holocene, i.e. 10.6 ka, 8.4 ka, 5.3 ka and 4.1 ka, respectively. Interestingly, these badlands are located over the basin margin fault system of Bengal Basin western shelf zone, and recent earthquakes (M_W 4.1 − 5.2) may have a direct role in changing micro-relief-slope condition as regional tilting towards the east. The important finding of this analysis finally relates with longitudinal and lateral connectivity or linkage of badland evolution which reflects threshold behaviour, landscape sensitivity and process connectivity, operating together at the micro-scale of badlands. The evolutionary stages of badland development are well explained by the connectivity model, and the hillslope–gully–river coupling system in the Gangani, Bagnada and Bhatina badlands (Initial state → Quasi-equilibrium state → Complex response state → Dissipative state). At present time, the laterite terrain may be reached to the quasi-equilibrium to complex state of landform evolution, and it poses an integrated network of gullies which may develop to meet badland divides, maintaining actual increasing power of fluvial system. Consequently, connectivity overcomes the threshold value of critical stream power, promoting more erosion than deposition. In spite of vast researches and present analysis work, a better understanding of these aspects is further needed at a range of temporal and spatial scales in various types of environments. It is undoubtedly the case that there is a strong connection between data availability and an interest in the where, how much, by what means and when gully erosion occurs. Particularly, one of the big questions that bedevilled erosion studies is the issue of the influence of imperceptible change or catastrophic events or anthropogenic factors. Based on the above empirical study, it is realized that there is a need for more research attention for (1) improved understanding of both natural (e.g. climate change and active tectonics) and anthropogenic soil erosion processes and their interactions, (2) scaling up soil erosion processes and rates in space and time with intensive field study and instrumentation, (3) predicting the on-site and off-site impacts of soil erosion and (4) innovative techniques to reduce soil erosion rates.

Acknowledgements We are very much indebted to Dr Pravat Kumar Shit (Assistant Professor, Department of Geography, Raja N. L. Khan Women's College) and Dr Aznarul Islam (Assistant Professor, Department of Geography, Aliah University) their valuable articles, field photographs and suggestions in the field-based data collection and analysis. We are thankful to Dr Sanat Kumar Guchhait (Professor, Department of Geography, The University of Burdwan) for the guidance and valuable suggestions.

References

Ahmad E (1968) Distribution and causes of gully erosion in India. 21st International Geographic Congress (section IV), New Delhi

Asthana AKL, Gupta AK, Luirei K, Bartarya SK, Rai SK, Tiwari SK (2015) A quantitative analysis of the Ramganga drainage basin and structural control on drainage pattern in the fault zones, Uttarakhand. J Geol Soc India 88:9–22

Bagchi K, Mukherjee KN (1983) Diagnostic survey of Rarh Bengal (Part II). University of Calcutta, Calcutta

Bandyopadhyay S (1988) Drainage evolution in a badland terrain at Gangani in Medinipur district West Bengal. Geogr Rev India 50(3):10–20

Barbera PL, Rosso R (1989) On the fractal dimension of stream network. Water Resour Res 25(4):735–741

Bartley R, Poesen J, Wilkinson S, Vanmaercke M (2020) A review of the magnitude and response times for sediment yield reductions following the rehabilitation of gullied landscapes. Earth Surf Process Landf. https://doi.org/10.1002/esp.4963

Bawa AK (2017) Mitigating land degradation due to water erosion. National Academy of Agricultural Science, New Delhi, Policy paper 88, pp 1–19

Begin ZB, Schumm SA (1979) Instability of alluvial valley floors: a method for its assessment. American society ofagricultural engineers. Trans ASAE 22 (2):0347–0350

Begin ZB, Schumm SA (1984) Gradational thresholds and landform singularity: significance for Quaternary morphology. Geol Soc Am Bull 56(3):267–274

Bergsma E, Charman P, Gibbons F, Hurni H, Moldenhauer WC, Panichapong S (1996) Terminology for soil erosion and conservation. Wageningen, International Soil Reference and Information Centre

Bhunia GS, Samanta S, Pal B (2012) Quantitative analysis of relief characteristics using space technology. Int J Phys Soc Sci 2(8):350–365

Bhunya PK, Jain SK, Singh PK, Mishra SK (2014) A simple conceptual model of sediment yield. Water Resour Manag 24:1697–1716

Blanco H, Lal R (2008) Principles of soil conservation and management. Dordrecht, Springer

Biswas A (1987) Laterities and Lateritoids of Bengal. In: Datye VS, Diddee J, Jog SR, Patial C (eds) Exploration in the Tropics. K.R. Dikshit Felicitation Committee, Pune, pp 157–167

Boardman J (2006) Soil erosion science: reflections on the limitations of current approaches. CATENA 68:73–86

Bocco G (1991) Gully erosion: processes and models. Prog Phys Geogr 15(4):392–406

Borrelli P, Robinson DA, Panagos P, Lugato E, Yand JE, Alewell C, Wuepper D, Montanarella L, Ballabio C (2020) Land use and climate change impacts on global soil erosion by water (2015–2070). Proc Natl Acad Sci USA 117(36):21994–22001

Brierley GJ, Fryirs KA, Jain V (2006) Landscape connectivity: the geographic basis of geomorphic applications. Area 38:165–174

Brunsden D, Thrones JB (1979) landscape sensitivity and change. Trans Inst Br Geogr 4:463–484

Bryan R, Yair A (1982) Perspectives on studies of badland geomorphology. In: Bryan R, Yair A (eds) Badland Geomorphology and Piping. Geo Books, Norwich, pp 1–12

Bull LJ, Kirkby MJ (1997) Gully processes and modelling. Prog Phys Geogr 21(3):354–374

Bull WB (1980) Geomorphic thresholds as defined by ratios. In: Coates DR, Vitek JD (eds.) Threshold in geomorphology. George Allen & Unwin, New York, pp 259–263

Casali J, Campo JLMA, Santisteban LMD, Alvarez-Mozos J (2006) Accuracy of methods for field assessment of rill and ephemeral gully erosion. CATENA 67:128–138

Casali J, Gimenez R, Bennett S (2009) Gully erosion processes and modelling. Earth Surf Process Landf 34 (14):1839–1840

Chakraborti S (2011) Final report on Quaternary laterites in the western districts of West Bengal—their geomorphology, stratigraphy, genesis and implications for climate change. Geol Surv India East Reg, Kolkata, pp 1–88

Choudhari PP, Nigam GK, Singh SK, Thakur S (2018) Morphometric based prioritization for groundwater potential of Mula river basin, Maharashtra, India. Geol, Ecol Landsc 2(4):256–267

Dacey MF, Krumbein WC (1976) Topological properties of disjoint channel networks within enclosed regions. J Int Assoc Math Geol 8:429–461

Das Gupta AB, Mukherjee B (2006) Geology of N.W. Bengal Basin. Geological Society of India, Bangalore

Dube HB, Mutema M, Muchaonyerwa P, Poesen J, Chaplot V (2020) A global analysis of the morphology of linear erosion features. Catena 190https://doi.org/10.1016/j.catena.2020.10452

Erskine WD (2005) Gully erosion. In: Lehr J, Keeley J, Lehr J, Kingery III (eds) Surface and agricultural water (water encyclopaedia). Wiley-Interscience, New Jersey, pp 183–188

Faulkner H (2008) Connectivity as a crucial determinant of badland morphology and evolution. Geomorphology 100:91–103

Froechlich DC (2018) Estimating reservoir sedimentation at large dams in India. In: E3S web of conference, vol 40, pp 1–8

Gajbhiye S, Mishra SK, Pandey A (2014) Relationship between SCS-CN and sediment yield. Appl Water Sci 4:363–370

Geeter SD, Poesen J, Vanmaercke M (2020) Does the topographic threshold concept explain the initiation points of sunken lanes in the European loess belt? CATENA 192:1044586. https://doi.org/10.1016/j.catena.2020.104586

Ghosh S (2013) Evolution and morphology of gully erosion of lateritic lands of Paschim Medinipur West Bengal. Unpublished Ph.D. Thesis, Tripura University, https://shodganga.inflibnet.ac.in/handle/10603/180836

Ghosh S, Guchhait SK (2020) Geomorphic threshold and SCS-CN-based runoff and sediment yield modelling in the gullies of Dwarka-Brahmani interfluve, West Bengal, India. In: Shit PK, Pourghasemi HR, Bhunia GS (eds) Gully Erosion Studies from India and Surrounding Region, Springer Nature, Switzerland, pp 45–68

Ghosh S, Maji T (2011) Pedo-geomorphic analysis of soil loss in the lateritic terrain of Rampurhat I block of Birbhum district, West Bengal and Shikaripara block of Dumka district Jharkhand. Int J Environ Sci 1 (7):1734–1750

Ghosh S, Guchhait SK (2017) Estimation of geomorphic threshold in permanent gullies of lateritic terrain in Birbhum, West Bengal India. Curr Sci 113(3):478–485

Ghosh S, Guchhait SK (2015) Characterization and evolution of primary and secondary laterites in northwestern Bengal Basin, West Bengal India. J Palaeogeogr 4(2):203–230

Gibling MR, Sinha R, Roy NG, Tandon SK, Jain M (2008) Quaternary fluvial and Aeolian deposits on the Belan River, India: palaeoclimatic setting of Palaeolithic to Neolithic archeological sites over the past 85,000 years. Quat Sci Rev 27(34):391–410

Gibling MR, Tandon SK, Sinha R, Jain M (2005) Discontinuity-bounded alluvial sequences of the southern Gangetic plains, India: aggradation and degradation in response to monsoonal strength. J Sediment Res 75(3):369–385

Hack JT (1973) Stream profile analysis and stream gradient index. J Res US Geol Surv 1(4):421–429

Harvey AM (2001) Coupling within hillslopes and channels in upland fluvial systems: implications for landscape sensitivity, illustrated from the Howgill Fells, northwest England. CATENA 42:225–250

Hassen MB, Deffontaines B, Turki MM (2014) Recent tectonic activity of the Gafsa fault through morphometric analysis: southern atlas of Tunisia. Quat Int 338:99–112

Horton RE (1945) Erosional development of the streams and their drainage basins: hydrophysical approach to quantitative morphology. Geol Soc Am Bull 56 (3):275–370

Ionita I (2003) Hydraulic efficiency of the discontinuous gullies. CATENA 50:369–379

Islam A, Sarkar B, Das BC, Barman SD (2020) Assessing gully asymmetry based on cross-sectional morphology: a case of Gangani Badland of West Bengal, India. In: Shit PK, Pourghasemi HR, Bhunia GS (eds) Gully erosion studies from india and surrounding region, Springer Nature, Switzerland, pp 69–92

James WR, Krumbein WC (1969) Frequency distribution of stream link lengths. J Geol 77(5):544–565

Jha VC, Kapat S (2011) Degraded lateritic soils cape and land uses in Birbhum district, West Bengal, India. Soc Natureza 23(3):545–556

Joshi VU, Daniels MJ, Kale VS (2013) Morphology and origin of valley-side gullies formed along the watersheds of Deccan Province, India and the Rangeland of Colorado, USA. Transactions 35(1):103–122

Joshi VU, Nagare VB (2013) Badland formation along the Pravara River, western Deccan, India. Can neotectonics be the cause? Zeitschrift fur Geomorphologie 57(3):349–370

Kariminejad N, Rossi M, Hosseinalizadeh M, Pourghasemi HR, Santosh M (2020) Gully head modelling in Iranian Loess Plateau under different scenarios. Catena 194: 104769, https://doi.org/10.1016/j.catena.2020.104769

Kirkby MJ, Bracken LJ (2009) Gully processes and gully dynamics. Earth Surf Process Landf 34(14):1841–1851

Kothyari UC (1996) Erosion and sedimentation problems in India. IAHS Publication No. 236, pp 531–540

Kumar H, Pani P (2013) Effects of soil erosion on agricultural productivity in semi-arid regions: the case of lower Chambal valley. J Rural Dev 32(2):165–184

Lal R (1990) Soil erosion and land degradation: the global risk. In: Lal R, Stewart BA (eds) Soil degradation volume II soil advances in soil science. Springer, New York, pp 129–172

Lenka NK, Mondal D, Sudhishri S (2014) Permissible soil loss limits for different physiographic regions of West Bengal. Curr Sci 107(4):665–670

Leopold LB, Wolman MG, Miller JP (1992) Fluvial Processes in Geomorphology. Dover Publication Inc, New York

Li J, Xiong L, Tang G (2019) Combined gully profiles for expressing surface morphology and evolution of gully landforms. Front Earth Sci 13:551–562

Mahala A (2020) Land degradation processes of Silabati River Basin, West Bengal, India: a physical perspective. In: Shit PK, Pourghasemi HR, Bhunia GS (eds) Gully erosion studies from India and surrounding region. Springer Nature, Switzerland, pp 265–278

Mahmood SA, Gloaguen R (2014) Appraisal of active tectonics in Hindu Kush: insights from DEM derived geomorphic indices and drainage analysis. Geosci Front 3(4):407–428

Maiti R (2016) Modern approaches in fluvial geomorphology. Primus Books, New Delhi

Mandelbrot BB (1982) The fractal geometry of nature. W. H. Freeman and Co., San Francisco

Martinez-Murillo JF, Nadal-Romero E (2018) Perspectives on badland studies in the context of global change. In: Nadal-Romero E, Martinez-Murillo JF, Kuhn NJ (eds) Badlands dynamics in the context of global change. Elsevier, Amsterdam, pp 1–25

Meshram SG, Sharma SK, Tignath S (2017) Application of remote sensing and geographical information system for generation of runoff curve number. Appl Water Sci 7:1773–1779

Miller VC (1953) A quantitative geomorphic study of drainage basin characteristics in Clinch Mountain area, Virginia and Tennessee. Issue 3 of Technical Report, Columbia University

Mishra MN, Vishwakarma LL (1999) Morphotectonics of the Chambal and the Yamuna valleys in the eestern Marginal Gangetic alluvial plains. Geological Survey of India, https://www.scribd.com/document/61292761/Morphotectonics-Chambal-Yamuna

Mishra SK, Singh VP (2003) Soil conservation service curve number (SCS-CN) methodology. Kluwer Academic Publishers, Dordrecht

Mishra SK, Tyagi JV, Singh VP, Singh R (2006) SCS-CN based modelling of sediment yield. J Hydrol 324:301–322

Mock SJ (1971) A classification of channel links in stream networks. Water Resour Res 7(6):1558–1566

Mock SJ (1976) Topological properties of some trellis pattern channel networks. CRREL Report 76-46, Crops of Engineers (US Army), New Hampshire

Moeyersons J (2003) The topographic thresholds of hillslope incisions in south-western Rwanda. CATENA 50:381–400

Mondal D, Sharda VN (2011) Assessment of permissible soil loss in India employing a quantitative bio-physical model. Curr Sci 100(3):383–390

Montgomery DR, Dietrich WE (1994) A physically based model for the topographic control on shallow landsliding. Water Resour Res 30(4):1153–1171

Montgomery DR, Dietrich WE (1994) Landscape dissection and drainage area-slope thresholds. In: Kirkby MJ (ed) Process Models and Theoretical Geomorphology. Wiley, Chichester, pp 221–246

Morgan RPC, Nearing MA (2011) Handbook of Erosion Modelling (ed). Wiley-Blackwell Publishing Ltd, New York

Narayana DVV, Babu R (1983) Estimation of soil erosion in India. J Irrig Drain Eng 109(4):419–434

Nachtergaele J, Poesen J, Gerard G (2002) Ephemeral gullies. A spatial and temporal analysis of their characteristics, importance and prediction. Belgeo 2:1–23

Nir D (1957) The ratio of relative and absolute altitudes of Mt. Carmel: a contribution to the problem of relief analysis and relief classification. Geogr Rev 47 (4):564–569

Nir D (1983) Man, a geomorphological agent: an introduction of anthropic geomorphology. OKeter, Jerusalem and Reidel, Dordrecht

Niyogi D (1975) Quaternary geology of the coastal plain in West Bengal and Orissa. Indian J Earth Sci 2:51–61

Niyogi D, Mallick S, Sarkar SK (1970) A preliminary study of laterites of West Bengal, India. In: Chatterjee SP, Das Gupta SP (eds) Selected papers physical geography, vol 1. 21st Inter national Geographical Congress, Calcutta, National Committee for Geography, pp 443–449

Ollier CD, Sheth HC (2008) The high deccan duricrusts of India and their significance for the laterite issue. J Earth Syst Sci 117(5): 537–551

Ollier CD, Pain CF (1996) Regolith, Soils and Landforms. John Wiley & Sons, New York.

Patton PC, Schumm SA (1975) Gully erosion, northwestern Colorado: a threshold phenomenon. Geology 3:88–90

Pennock D (2019) Soil erosion: the greatest challenge for sustainable soil management. FAO, Rome

Pimentel D, Burgess M (2013) Soil erosion threatens food production. Agriculture 3:443–463

Pimentel D (2006) Soil erosion: a food and environmental threat. Environ Dev Sustain 8:119–137

Poesen J (2011) Challenges in gully erosion research. Landf Anal 17:5–9

Poesen J, Nachtergaele J, Verstraeten G, Valentin C (2003) Gully erosion and environmental change:

importance and research needs. CATENA 50(2–4):91–133

Poesen J, Vandaele K, Wesemael B (1998) gully erosion: importance and model implications. In: Boardman J, Favis-Mortlock D (eds) Modelling Soil Erosion by Water. Springer, Berlin, Heidelberg, pp 285–311

Poesen J, Vandekerckhove L, Nachtergaele J, Oostwoud Wijdenes DJ, Verstraeten G, Wesemael B (2002) Gullyerosion in dryland environments. In: Bull LJ, Kirkby MJ (eds) Dryland Rivers: Hydrology and Geomorphology ofSemi-arid Channels. John Wiley & Sons, Chichester, pp 229–262

Poesen J (2018) Soil erosion in the Anthropocene: research needs. Earth Surf Process Landf 43(1):64–84

Ranalli G, Scheidegger AE (1968) A test of the topological structure of river nets. Hydrol Sci J 13(2):142–153

Ranga V, Mohapatra SN, Pani P (2015) Geomorphological evolution of badlands based on the dynamics of palaeo-channels and their implications. J Earth Syst Sci 124(5):909–920

Reddy VR, Galab S (2006) Looking beyond the debt trap. Econ Polit Wkly 41(19):1838–1841

Rossi M, Torri D, Santi E (2015) Bias in topographic thresholds for gully heads. Nat Hazards 79:S1–S69

Roy NG, Sinha R, Gibling MR (2012) Aggradation, incision and interfluve flooding in the Ganga Valley over the past 100,000 years: testing the influence of monsoonal precipitation. Palaeogeogr, Palaeoclim, Palaeoecol 356–357:38–53

Roy S, Sahu AS (2016) Effect of land cover on channel form adjustment of headwater streams in a lateritic belt of West Bengal (India). Int Soil Water Conserv Res 4:267–277

Saha A, Ghosh M, Pal SC (2020) Understanding the morphology and development of a rill-gully: an empirical study of Khoai Badland, West Bengal, India. In: Shit PK, Pourghasemi HR, Bhunia GS (eds) Gully erosion studies from India and surrounding region. Springer Nature, Switzerland, pp 147–161

Samni AN, Ahmadi H, Jafari M, Boggs G, Ghoddousi J, Malekian A (2009) Geomorphic threshold conditions for gully erosion in south-western Iran (Boushehe—Samal watershed). J Asian Earth Sci 35(2):180–189

Sarkar T, Mishra M, Chatterjee S (2020) On detailed field-based observations of laterite and lateritization: a study in the Paschim Medinipur lateritic upland of India. J Sediment Environ 5:219–245

Scheidegger AE (1967) On the topology of river nets. Water Resour Res 3(1):103–106

Schumm SA (1954) The relation of drainage basin relief to sediment loss. Int Assoc Hydrol 1:216–219

Schumm SA (1956) The evolution of drainage system and slopes in badlands at Perth Amboy, New Jersey. Geol Soc Am Bull 67(5):597–646

Sen J (2008) Geomorphology of Garhbeta badlands West Medinipur district West Bengal. Unpublished Ph.D. Thesis, University of Calcutta, https://shodhganga.inflibnet.ac.in/handle/10602/155878

Sen J, Sen S, Bandyopadhyay S (2004) Geomorphological investigation of badlands: a case study at Garhbeta, West Medinipur district, West Bengal, India. In: Singh S, Sharma HS, De SK (eds) Geomorphology and environment, ACB Publications, Kolkata, pp 204–234

Sharda VN, Dogra P (2013) Assessment of productivity of monetary losses due to water erosion in rainfed crops across different states of India for prioritization and conservation planning. Agric Res 21(4):382–392

Sharda VN, Mandal D, Dogra P (2019) Assessment of cost of soil erosion and energy saving value of soil conservation measuring in India. Indian J Soil Conserv 47(1):1–6

Sharda VN, Mandal D, Ojasvi PR (2013) Identification of soil erosion risk areas for conservation planning in different states of India. J Environ Biol 34:219–226

Sharma RK (2018) Soil loss setbacks to Indian agriculture. Acta Sci Agric 2(6):95–97

Sharma HS (2009) Progress of researches in ravines and gullies geomorphology in India. In: Sharma HS, Kale VS (eds), Geomorphology in India, Prayag Pustak Bhavan, Allahabad, pp 441 – 458

Shit PK, Adhikary PP (2020) Soil disintegration chanracteristics on ephemeral gully collapsing in lateritic belt of West Bengal, India. In: Shit PK, Pourghasemi HR, Bhunia GS (eds) Gully erosion studies from India and surrounding region. Springer Nature, Switzerland, pp 21–33

Shit PK, Bhunia GS, Maiti R (2013) Development of rill networks: an experimental plot scale study. J Water Resour Prot 5:133–141

Shit PK, Maiti R (2012) Mechanism of gully-head retreat: a study at Ganganir Danga, Paschim Medinipur, West Bengal. Ethiop J Environ Stud Manag 5(4):332–342

Shreve RL (1966) Statistical law of stream numbers. J Geol 74(1):17–37

Shreve RL (1967) Infinite topologically random channel networks. J Geol 74(1):178–186

Sidorchuk Y (2020) Assessing the gully potential of a territory (a case study of central Yamal). Geogr Nat Resour 41(2):178–186

Singh S, Agnihotri SN (1987) Rill and gully erosion in the subhumid tropical riverine environment of Teonthar Tahsil, Madhya Pradesh, India. Geogr Ann: Ser A, Phys Geogr 69:227–236

Singh S, Dubey A (2002) Gully erosion and management methods and application: A field manual. National Academic Publishers, New Delhi

Singh LP, Parkash B, Singhvi AK (1998) Evolution of the Lower Gangetic Plain landforms and soils in West Bengal India. CATENA 33(2):75–104

Singh G, Babu R, Narain P, Bhushan LS, Abrol IP (1992) Soil erosion rates in India. J Soil Water Conserv 47(1):97–99

Sinha D, Joshi VU (2012) Application of Universal Soil Loss Equation (USLE) to recently reclaimed badlands along the Adula and Mahalungi Rivers, Pravara Basin, Maharashtra. J Geol Soc India 80:341–350

Sinha R, Singh A, Tandon SK (2020) Fluvial archives of north and northwestern India as recorders of climatic signatures in the Late Quaternary: review and assessment. Curr Sci 119(2):232–243

Smart JS, Wallis JR (1971) Cis and Trans links in natural channel networks. Water Resour Res 7(5):1346–1348
Srivastava RK, Imtiyaz M (2016) Testing of coupled SCS curve number model for estimating runoff and sediment yield for eleven watersheds. J Geol Soc India 88:627–636
Strahler AN (1952) Hypsometric (area-altitude) analysis of erosional topography. Geol Soc Am Bull 63 (11):1117–1142
Strahler AN (1957) Quantitative analysis of watershed geomorphology. Trans Am Geophys Union 38 (6):913–920
Tandon SK, Gibling M, Sinha R, Singh V, Ghazanfari P, Dasgupta A, Jain M, Jain V (2006) Alluvial valleys of the Gangetic Plains, India: causes and timing of incision. In: Dalrymple RW, Leckie DA, Tillman RW (eds) Incised valleys in time and space, SEPM Special Publication No. 85, pp 15–35
Thakkar H, Bhattacharyya S (2006) Reservoir siltation in India: latest studies. SANDRP
Thomas MF (2001) Landscape sensitivity in time and space—an introduction. CATENA 42:83–98
Torri D, Poesen J (2014) A review of topographic threshold conditions for gully head development in different environments. Earth-Sci Rev 130:73–85
Torri D, Poesen J, Borselli L, Knapen A (2006) Channel width—flow discharge relationships for rills and gullies. CATENA 90:76–86
Valentin C, Poesen J, Li Y (2005) Gully erosion: impacts, factors and control. CATENA 63(2–3):132–153
Vandaele K, Poesen J, Govers G, Wesemael B (1996) Geomorphic threshold conditions for ephemeral gully incision. Geomorphology 16(2):161–173
Vandekerckhove L, Poesen J, Wijdnes DO, Nachtergaele J, Kosmas C, Roxo MJ, Figueiredo TD (1998) Thresholds for gully initiation and sedimentation in Mediterranean Europe. Earth Surf Process Landf 25 (11):1201–1220
Vanmaercke M, Poesen J, Radoane M, Govers G, Ocakoglu F, Arabkhedri M (2012) How long should we measure? An exploration of factors controlling the inter-annual variation of catchment sediment yield. J Soils Sediments 12(4):603–619
Vente J, Poesen J (2005) Predicting soil erosion and sediment yield at the basin scale: scale issues and semi-quantitative models. Earth Sci Rev 71:95–125
Wasson RJ (2003) A sediment budget for the Ganga—Brahmaputra catchment. Curr Sci 84(8):1041–1047
Wells RR, Momm HG, Bennett SJ, Gesch KR, Dabney SM, Cruse R, Wilson GV (2016) A measurement method for rill and ephemeral gully erosion assessments. Soil Water Manag Conserv 80:203–214
Wischmeier WH, Smith DD (1978) Predicting rainfall erosion losses: A guide to conservation planning agriculture. Handbook 282, USDA-ARS, USA
Yibeltal M, Tsunekawa A, Haregeweyn N, Adgo E, Meshesha DT, Masunaga T, Tsubo M, Billi P, Ebabu K, Fenta AA, Berihun ML (2019) Morphological characteristics and topographic thresholds of gullies in different agro-ecological environments. Geomorphology 341:15–27
Zaliapin I, Foufoula-Georgiou E, Ghil M (2010) Transport on river networks: a dynamical tree approach. J Geophsyical Res 115:1–15
Zanardo S, Zaliapin I, Foufoula-Georgiou E (2013) Are American river Tokunaga self-similar? New results on fluvial network topology and its climatic dependence. J Geophsyical Res 118:166–183

3 Plan Shape Geomorphology of Alluvial Valley in the Middle-Lower and Deltaic Courses of the Subarnarekha River Basin, India

Subrata Jana and Ashis Kumar Paul

Abstract

The middle-lower and deltaic courses of the Subarnarekha river from Jamsola (upstream) to Chaumukh (mouth) have been considered for the study of plan shape geomorphology partially along the alluvial valley floors between bank margin environment. Geologically and topographically, the depositional environments comprise with Lalgarh formation, Sijua formation, Panskura formation, Basudebpur formation, Daintikri formation, and beach ridge chenier formations of fluvial, fluvio-marine and marine depositional processes with the seaward gradient from Jamsola to Chaumukh. Distinctive sub-environments of the upstream fluvial dominance, ancient delta-fan lobe extension, and areas of sea level fluctuations with lower deltaic beach ridge chenier at downstream section are categorized based on the identified assemblages of landforms in the present study. The geospatial techniques, repeated field observations, Total Station survey and sedimentological analysis of bank margin stratigraphic sections have been considered in the study to explore the spatial diversity of plan shape geomorphology in the different sections of the studied river course. The study reveals that the course of the river sections bears diverse geometry of meander properties with discontinuous straight courses and wider valleys. The section-wise plan shape geomorphological features assemblages of 16 categories in the section of Jamsola–Ragra stretch, 18 categories in Ragra–Dantan stretch, 15 categories in Dantan–Rajghat stretch, and 24 categories in Rajghat–Chaumukh stretch in the form of instream deposition, channel fringe deposition and floodplain deposition. Distinct morphological variations of the three major mid-channel bars at the different channel positions indicate the nature of seasonal hydrodynamics and signatures of catastrophic floods. The layer-wise sedimentological analysis of the younger fill terrace and two mid-channel bars shows the trend of discharge and fluctuating flow regimes in the different flood events of single or multiple years. Identification of palaeo-shorelines, ancient delta-fan lobes and cut and fill terraces along the courses of the river valley highlights the role of dynamic marine, fluvio-marine and fluvial environments in the region.

S. Jana (✉)
Department of Geography, Belda College, Belda, Paschim Medinipur, India

A. K. Paul
Department of Geography and Environment Management, Vidyasagar University, Midnapore, India

P. K. Shit et al. (eds.), *Drainage Basin Dynamics*, Geography of the Physical Environment,
https://doi.org/10.1007/978-3-030-79634-1_3

Keywords

River valley geomorphology · Geological formations · Fluvio-marine depositional environments · Seasonal hydrodynamics · Sedimentological analysis

3.1 Introduction

The distinct geomorphological features within a river valley formed in a suitable position, and their shape and size are modified based on the interacting processes of fluvial hydraulics and sediment loads within a river valley region (Bentham et al. 1993; Ashworth et al. 2000; Sarma 2005; Corenblit et al. 2007; Ventra and Clarke 2018). The existing characteristics of channel geometry also control the dimensions of the geomorphological features (Abrahams 1984; Wharton 1995; Sofia 2020). The diversified river valley geomorphological features form in the longitudinal and transverse section in the different stretches of a river basin (Khan et al. 2018). The types and dimensions of landforms in the hilly or upper catchment areas are far different from the middle and lower catchment areas. Also, the landform types differ according to the lithological features and sedimentary nature (Jana and Paul 2019). In the tropical region, the middle and middle-lower reaches of a river valley are dominated by soft sedimentary deposits (Fryirs and Brierley 2012; Jana 2019). In these reaches, landforms formed under the unidirectional flow, which enhanced during the summer monsoon flood season (Jana and Paul 2018). However, the landforms in the extreme lower reaches or deltaic areas are dominated by silt and clay types of material deposited by the bidirectional tidal flow (Jana and Paul 2018). At the overbank flow regimes, the sediment loaded flood water can reach up to the valley ends. The new or already existed geomorphological features formed or restructured, depending on the flow energy level, the volume of water and sediment and inundation intensities (Gilvear 1999; Fryirs and Brierley 2012). Concurrently, in the instream section, the geomorphic units also evolved according to the erosional and depositional nature of sedimentary environments (Leopold et al. 1992). Therefore, to some extent, landform modifications are related to every flood event. Moreover, anthropogenic activities play a significant role in geomorphic alterations within riparian areas (Best 2019).

The identification and demarcation of different instream geomorphic units are a difficult task for thier ever-dynamic nature in connection with the seasonal fluctuation of the flow regime (Fryirs and Brierley 2012; Roy and Sahu 2018; Sofia 2020). The floodplain geomorphological features are relatively less altered by the flow regime, although, it modified by the anthropogenic activities (Poff et al. 1997; Wu et al. 2008). Despite the dynamic nature and anthropogenic alterations, the distinctive geomorphic units have well preserved in the different extents of the river valley (Bisson et al. 2017). Those landform units can be identified through the analysis of sedimentary environments, terrain diversities, geometric properties and the presence of soil, vegetation and moisture contents in the different landforms. In this concern, field observations and surveying are the prime tasks coupled with satellite image analysis.

In the present study area of the middle-lower and deltaic courses of the Subarnarekha river valley, lots of works have been done regarding the geological and geomorphological study in the different area specific aspects (Niyogi 1975; Bhattacharya and Misra 1984; Paul 2002; Jana and Paul 2014, 2018, 2019, 2020; Jana et al. 2014; Paul and Kamila 2016; Guha and Patel 2017), morphometric analysis and landforms (Ilahi and Dutta 2016), hydrological aspects (Dandapat and Panda 2013; Samanta et al. 2018) and river bank erosion rate and its prediction (Jana 2019). But, most of these works have been done to a discrete extent, which mainly concentrated in the deltaic areas. Although, some studies have been done considering the entire basin area (Ilahi and Dutta 2016; Guha and Patel 2017), these works did not consider the detailed geomorphological features. The entire middle-lower and deltaic courses are yet not considered in a uniform aspect for geomorphological analysis. Therefore, the present study aims to assess the distinct geomorphological features at different spatial extents (along and across the valley) of the middle-lower and deltaic courses of the Subarnarekha basin based on the sedimentological and hydrological analysis coupled with field investigations in the different periods.

3.2 Materials and Methods

3.2.1 Study Area

The present study has been carried out in the middle-lower and deltaic courses of the Subarnarekha river valley, extended from the Jamsola (upstream) to Chaumukh (confluence in the Bay of Bengal) (Fig. 3.1). The study area boundary has been selected based on the extent of lateritic cliffs on both sides of the river valley in the middle-lower course (Fig. 3.1). However, in the deltaplain, about 10 km extents on both sides have been considered as the boundary of the study area. The selected area has been extended between 21°32′34.84″N to 22°19′ 18.48″N and 86°43′21.14″E to 87°28′33.78″E coordinates. The alluvium deposited area has been selected as the study area boundary (Fig. 3.2). However, some patches of laterite and gravel dominated areas have also included within the study area (Fig. 3.2) as it was considered based on the contour values.

Geologically, the selected area was formed during the Late Pleistocene to Late Holocene period comprising the Sijua formation, Panskura formation, Basudebpur formation, Daintikri formation and sand dune and beach formations (Fig. 3.2). The Tertiary gravel bed of Dhalbhum formation and Late Pleistocene laterite surface of Lalgarh formation exists in the extreme boundary extents of the study area (Fig. 3.2) (GSI 1998). The extensive floodplain and valley are observed in the left side (east) of the present river course in comparison with the valley on the right side (west) of the river, particularly within the Jamsola and Rajghat stretch. The present river course is flowing in different directions following the valley slope i.e. west to east (Jamsola to Ragra), north-west to south-east (Ragra to Dantan), north-east to south-west (Dantan to Rajghat), and again it flows from north-west to south-east direction (Rajghat to Chaumukh) before conflux to the Bay of Bengal. Also, the different landform terraces of cut and fill valley and deltaplain (Fig. 3.3) virtually coexists with the distinctive flow directions. The unidirectional river flow is observed up to Baliapal (from upstream) section, which is dominated during the summer monsoonal flood (Jana, 219). However, the downstream section of Baliapal (recent deltaplain) is dominated by the diurnal bidirectional tidal flow with about 2.31 m mean annual range.

3.2.2 Database and Data Processing

In this study, the Geological Quadrangle Map (GQM) has been collected from the open sources of the Geological Survey of India portal (GSI 1998) to comprehend the lithological formations of the area. The 30 m resolution Shuttle Radar Topography Mission (SRTM) images (coordinates: 21/87, 22/86 and 22/87; acquisition date: 11th February 2000) and Landsat 8 OLI image (path/row: 139/45; acquisition date: 27th February 2015) have been collected from the United States Geological Survey (USGS) based data portal (EarthExplorer). All the maps and images have been re-projected and resampled in the Universal Transverse Mercator (UTM) projection concerning the 45 N zone and World Geodetic Survey 1984 (WGS84) datum. The images have also been co-registered (Beuchle et al. 2015) with < 0.5 pixel accuracy of Root Mean Square Error (RMSE). The atmospheric noise effect has been rectified using the Fast line-of-sight Atmospheric Analysis of Spectral Hypercubes (FLAASH) model followed by Jia et al. (2014). The landuse and landcover (LULC) classification have been done considering the corrected Landsat image using the maximum likelihood and support vector machine (SVM) classifiers models in the ENVI 5.1 software (Foody et al. 1992; Jia et al. 2011; Pal and Foody 2012). The diversified soil, vegetation and water bodies (moist areas)

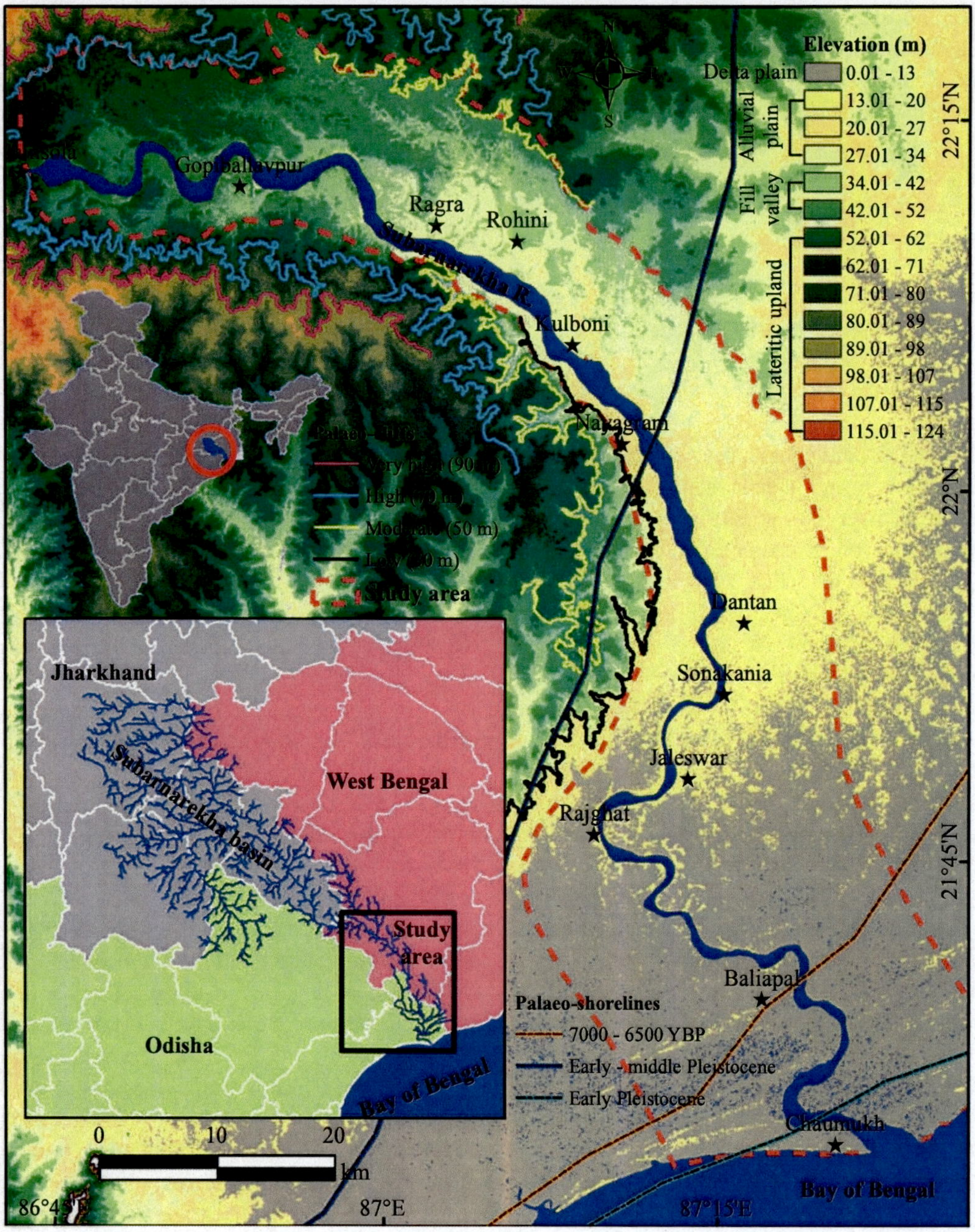

Fig. 3.1 Regional settings of the study area within the fill valley terrace in the middle-lower course and deltaplain of the Subarnarekha river basin

have been identified based on the classified image (Bishop et al. 2012). The elevation data have randomly been taken from ~10,000 locations of different geomorphic units using the DGPS (Differential Global Positioning Systems). Considering the earth as an ellipsoid surface coupled with the WGS84 datum and Earth Gravity Model 1996 (EGM96) the DGPS is

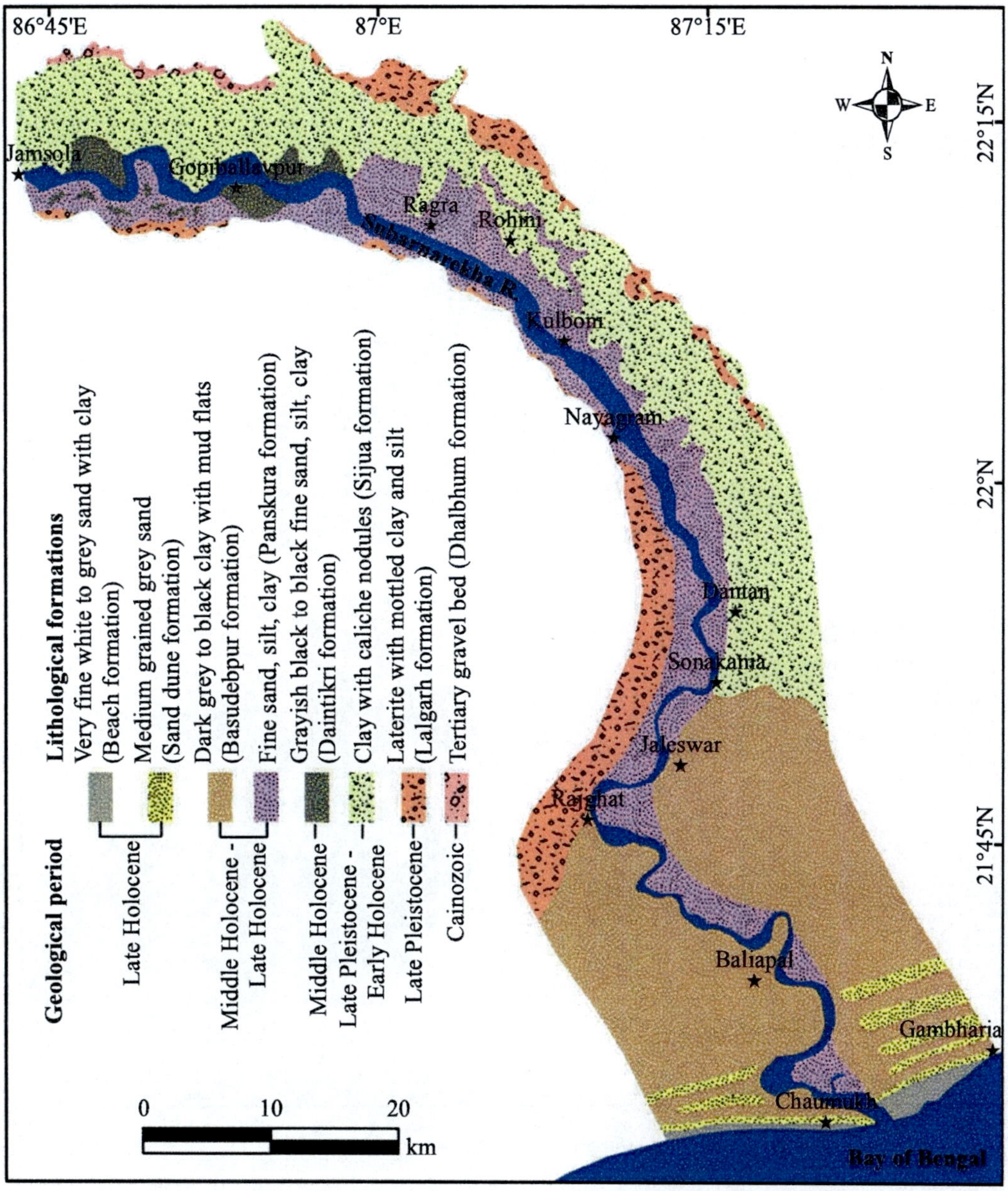

Fig. 3.2 Geological setting of the study area exhibits age-wise different lithological formations

performing. Therefore, the extracted DGPS-based data (elevation) converted into equipotential geoidal surface corresponding to the mean sea level (MSL) as a local vertical datum (Pavlis et al. 2012; Patel et al. 2016). Moreover, the Total Station (TS) survey was conducted during February–April 2015 in the distinct geomorphic units of the river bed, mid-channel bar, natural levees (older and younger), palaeo-courses and oxbow lake areas. Elevation data of ~1500 points were taken from each site at ~10 m of spatial interval (depending on the elevation differences). The SRTM images have been merged (mosaic) and resampled into 10 m resolution after sub-pixel analysis (Mokarrama and Hojati 2018). The vertical accuracy of the resampled

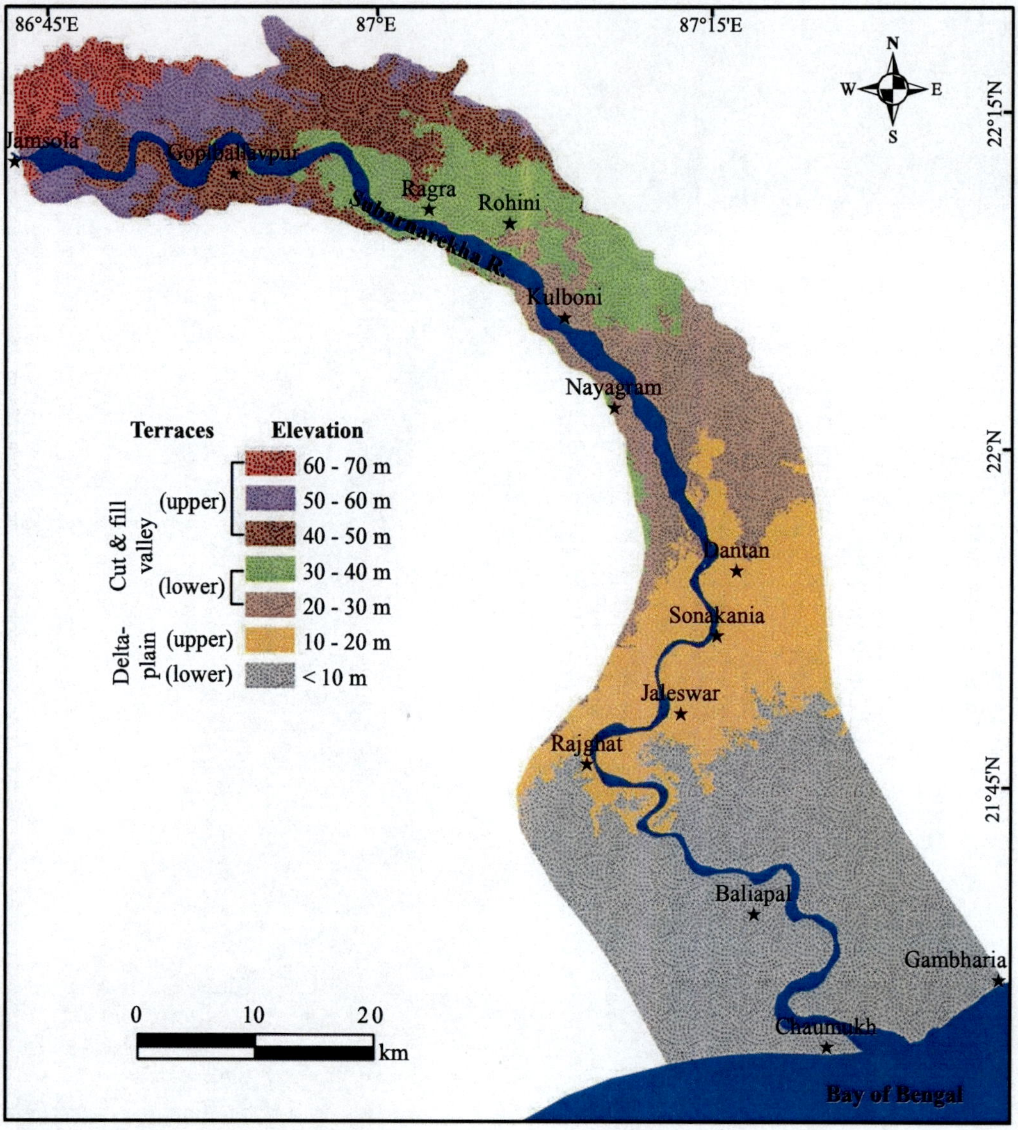

Fig. 3.3 Diversified elevation zones and landform terraces of the study area

SRTM data has been validated with DGPS-based elevations, which reveals the ±0.15 m RMSE. Moreover, the DGPS and TS survey-based elevation data have been converted into raster data (with 10 m resolution) employing the Inverse Distance Weighting (IDW) interpolation method (Patel et al. 2016), which has been superimposed on the resampled SRTM data. Finally, the 10 m spatial resolution raster Digital Elevation Model (DEM) has been primed from the assembled dataset. The multi-temporal Google Earth images have also been used for the extraction and validation of different geomorphic landforms.

3.2.3 Extraction of Landform Terraces and Morphological Features

The different landform terraces (cut and fill valley and deltaplain) have been demarcated based on the extracted contours (10 m interval) from the finally primed high-resolution (10 m) DEM (Fig. 3.3). Only the extended contour lines have been selected for the demarcation and mapping of different terraces. The fragmented and patchy contours zones have been ignored for the distinctive terrace mapping. The upper and lower terraces of cut and fill valley and deltaplain have been discriminated based on the elevation differences. Altogether 44 types of morphological features have been identified applying the DEM-based elevation differences, LULC classification-based soil, vegetation and water (moist) patches, Google Earth images, nature of sediment deposits and field observations. The textural variations of deposited sediments, erosional and depositional signatures, inundation level and periods have been also considered during morphological features identification. The plan shape morphological features of the entire study area have been segmented in four different zones considering the diversified river flow directions coupled with the slope and elevation differences of the terraces.

3.2.4 Geometrical Analysis

The geometrical properties i.e. channel width (w) at bankfull stage and lean-phase, depth of channel (d), width-depth ratio (w/d), meander length (ℓ), meander height (h),radius of curvature (r) of meander and meander arc-length (R_c) have been estimated followed by the Williams (1986), Islam and Guchhait (2017), and Jana (2019) for the sectional river stretches of Jamsola–Ragra, Ragra–Dantan, Dantan–Rajghat, and Rajghat–Chaumukh sections as well as for the entire studied river stretch. The channel width at lean-phase has been estimated based on the Landsat image of 2015. During the extreme flood events (bank full periods) the cloud free image was not freely available. Therefore, the channel width at bankfull stage was estimated from the on field survey during July–September, 2015 in the different sectional reaches of the river. The other meander properties have been estimated from the 2015 image using the ArcGIS 10.1 software after digitizing the respective banklines and channel middle positions (Jana 2019).

3.2.5 Sedimentological Analysis

Sediment samples have been collected from the apparent lithostratigraphic units of the different landforms in the upper and lower river courses. On 15th March 2015, twelve sediment samples have been collected from the equivalent layers of the excavated profiles in the scrub dominated younger fill terrace (S1) at Dharmapur and older mid-channel bar (S2) at Nayabasan. Moreover, six different sediment samples have been collected (on 26th September 2015) from each of the layers of the mid-channel bar (S3) at Saherbazar near Jaleswar. All three profiles have been excavated up to the exposed layer corresponding to the river bed. The collected sediment sample has been processed and analyzed to discriminate the textural variation of the particles using sieving and pipette methods. The percentage distributions of sediment grain-size have been further analyzed employing the GRADISTAT statistical programme (Blott and Pye 2001) to understand the nature of fluvial environments during sediment deposition in the respective layers.

3.3 Results and Discussion

3.3.1 Fluvio-Marine Environments and Landform Terraces

The cut and fill valley terraces (Fig. 3.3) have been formed and modified with due effects from the fluvio-marine environments during the Early Pleistocene to Holocene period (Fig. 3.2). Initially, the marine process was in action around the lower cut and fill valley terrace (Fig. 3.3) when the shoreline was parallel to the laterite surface during the Early–Middle Pleistocene period ((Fig. 3.1) (Banerjee and Sen 1987, 1988; Jana and Paul 2020). During this period, the laterite cliffs were formed with the relatively fluctuating sea level and associated land erosion by sea waves (Niyogi 1975). The tide-water entered into the future interior areas of the river valley and enhanced the retreating rate of laterite layers progressively towards the land. Moreover, the fluvial action was more active during the Late Pleistocene ice melting in the Bengal Basin area (Ghosh and Guchhait 2020). Therefore, the combined erosional effects of the fluvio-marine processes were responsible for the formation of the extensive valley on both sides of the present river course. However, the four different cliffs at various elevations (Fig. 3.1) have been modified by the headward erosion along the tributaries on both sides of the valley in the Holocene period. The marine regression phase was started and the shoreline was aligned at the Baliapal section during 7000–6500 years before present (YBP) (Banerjee and Sen 1987, 1988; Jana and Paul 2020) almost parallel to the present shoreline (Fig. 3.1). Therefore, the Nayagram–Baliapal stretch was dominated by the fluvio-marine transitional environment during the middle Pleistocene to 6500 YBP (Jana and Paul 2020). Moreover, the existence of five successive chenier dune ridges and swales between the Contai and Digha-Talsari coastal stretch reveals the continuous regression phase with a limited period of stillstand phases during 6,000–5,000 YBP, 4,700–4,000 YBP, 3,500–3,000 YBP, 2,500–1,100 YBP, and 600–500 YBP (Goswami 1997, 1999; Jana and Paul 2020).

The extended valley has been filled up by the river carried sediments drained from the upper catchment areas and the nearby laterite uplands during the Late Pleistocene to middle Holocene periods (Jana and Paul 2019, 2020). These cut and fill valley terraces (Fig. 3.3) have remained under the Sijua formation, Panskura formation and Daintikri formations composed with the fine sand, silt and clay type of materials (Fig. 3.2). The river was flowing nearer the laterite surface on the left side of the valley in the past. Recently, the river is flowing through the relatively meandering path within the upper terrace (40–70 m elevation) maintaining the gentle surface gradient. The eroded materials from the retreated lateritic cliffs have been deposited in this area, which is responsible for the higher elevation. The river flows in the straight course within the lower terrace (20–40 m elevation), and the relatively higher gradient is perceived the lower terrace in compared to the upper terrace. The deposited materials in the lower terrace have been redistributed and somehow transported towards the downstream section under the dominance of the fluvial environment. These sediments were deposited (as a delta-fan) in the upper deltaplain during the accelerated rate of marine regression at the later phase of the Middle Pleistocene period (Jana and Paul 2020). Therefore, among the two cut and fill valley terraces, the lower terrace is more extended than the upper terrace, although, the higher elevation differences are experienced in the upper terrace area (Fig. 3.3). Initially, both deltaplain was formed as the submerged delta-fan lobes when the corresponding shorelines remained in the landward extents. However, the submerged delta-fans were emerged due to continuous sediment deposition in the shallow marine environments coupled with the marine regression effects since the middle Pleistocene period (Paul 2002). The initially deposited deltaic sediment of the upper deltaplain has been overlain with the sedimentary deposits of the Sijua and Panskura formations, whereas, most of the lower deltaplain area has remained under the Basudebpur formation (Figs. 3.2 and 3.3). The dominant meandering channel pattern was formed with due effects from the marine regression and short-term stillstand phases within the deltaplain (Paul 2002). Recently, the entire upper deltaplain is controlled by the unidirectional flow, which is dominating during the summer monsoon period, whereas, the tide-water (marine environment) still dominated up to the middle section of the lower deltaplain (Baliapal).

3.3.2 Geometrical Diversity of River Course

Geometrical properties of the river course have been estimated (Jana 2019) considering the four different river stretches and the overall river course within the study area (Table 3.1). In the Jamsola–Ragra section, the 1032 m average channel width in the bankfull stage is reduced up to 265 m in the lean-phase of active channel flow. The instream geometrical properties are also changed depending on such seasonal fluctuation of river discharge. This meandering course is characterized by the sinuosity index (SI) of 1.37 and the average width/depth ratio (w/d) of 122. The geometrical properties of meander i.e. average length (ℓ), height (h), radius of curvature (r) and arc-length (R_c) have resulted as 7513 m, 2898 m, 1624 m and 10,166 m, respectively. The river course become straightens within the Ragra–Dantan section with resultant SI of 1.07. Therefore, the meander geometrical properties have not been estimated in this section. The estimated average channel width (1512 m) in the bankfull stage is greater than the upper stretch. However, the width of the active channel flow (270 m) in the lean-phase is almost similar to the upper section. The highest average w/d (231) has been estimated in this river stretch among the four zones. At the downstream section (Dantan–Rajghat), again the river course becomes meandering with SI of 1.34. The channel width is decreased up to 598 m in the bankfull stage, and the active channel flow is observed only within 200 m width in the lean-phase. In this section, the lowest w/d (100) is observed among the four stretches. The average estimated meander geometrical properties of ℓ (7207 m), h (3240), r (1355 m) and R_c (9955 m) are somehow minimized in comparison to the Jamsola–Ragra stretch. The prominent meandering channel pattern (SI = 1.54) is observed in the deltaic course in-between the Rajghat–Chaumukh stretch. The average channel width is also wider than the preceding stretch (Dantan–Rajghat), which is ~879 m during the peak monsoonal discharge coupled with the extreme high-tide condition. During the lean-phase of monsoon and low-tide conditions, the active flow is concentrated within 500 m width. In this fluctuating hydrodynamic condition, the average w/d is recorded as 208. The average values of meander geometrical properties are estimated as ℓ (5539 m), h (2410 m), r (1055 m) and R_c (8136 m).

3.3.3 Plan Shape Geomorphology and Depositional Environments

The river valley geomorphological features have been classified into three parts i.e. in the instream, at the fringe of river course and within the floodplain. Moreover, the regional diversities in the plan shape geomorphological features have been broadly categorized in four distinct zones of upper and lower terraces in the cut and fill valley and deltaplain terraces. The depositional environments have been assessed by the repeated field observations of the distinct geomorphic units in different seasons coupled with sedimentological characteristics. Within the selected study area, 22 types of geomorphic units with their subdivisions have been demarcated and analyzed considering the process of formation and present status (Table 3.2).

3.3.3.1 Geomorphological Features in the Upper Cut and Fill Valley Terrace

The sixteen major geomorphological features have been demarcated within the upper cut and fill valley terrace extended from Jamsola to Ragra stretch (Fig. 3.4). Four major coupled with thirteen micro-level diversified geomorphological features have been observed in the instream section. Seven major geomorphic units have been observed at the fringe of the river course, whereas, five features have remained in the floodplain areas (Fig. 3.4). The micro-level diversities of the mid-channel bars and fill terraces have been identified within this stretch.

The active channel flow, recent deposits of sand bodies in the channel bed and mid-channel bar, seasonal and matured mid-channel bars have been observed in the instream position (Fig. 3.4).

Table 3.1 Geometrical properties of river course in the different stretches of the Subarnarekha river

Geometrical properties	Jamsola–Ragra			Ragra–Dantan			Dantan–Rajghat			Rajghat–Chaumukh			Overall		
	Max	Min	Avg	Max	Min	Avg	Max	Min	Avg	Max	Min	Avg	Max	Min	Avg
Channel width (w) at bankfull stage (m)	2384	160	1032	2087	870	1512	1051	260	598	2424	308	879	2424	160	1006
Channel width at lean-phase (m)	450	68	265	450	152	270	272	141	200	2424	175	500	2424	68	317
Depth (d) of channel (m)	15.54	5.18	8.84	12.80	3.35	7.24	9.14	3.05	5.79	7.92	3.05	5.12	15.54	3.05	6.75
Width-depth ratio (w/d)	222	58	122	405	144	231	144	63	100	795	61	208	795	58	170
Meander length (m)	10,679	5197	7513	–	–	–	9732	5857	7207	8423	3717	5539	10,679	3717	6532
Meander height (m)	3404	2008	2898	–	–	–	5481	1469	3240	3610	840	2410	5481	840	2760
Radius of curvature of meander (m)	4046	589	1624	–	–	–	2455	569	1355	2671	202	1055	4046	202	1296
Meander arc length (m)	12,116	8426	10,166	–	–	–	14,159	6730	9955	12,953	4357	8137	14,159	4357	9183

Note The '–' in the Ragra–Dantan stretch reveals that the meander geometrical properties did not estimated in the relatively straight river course

Table 3.2 Different morphological features of the study areas

Positions	Morphological features	Formation process	Field description
Instream	Active channel flow	Channel flow path throughout the year	Seasonal changes in flow pattern and it depends on the discharge, sediment load and sediment distribution
	Channel bed sand bodies (recent deposits)	Long-term flood deposits	Exists in same position and or change their position, shape and size depending on the consecutive flood nature
	Mid-channel bar (seasonal)	Resent fluvial deposits	Position, shape and size are changed depending on the seasonal fluctuation of hydraulic and sedimentation nature
	Mid-channel bar (stable)	Long-term flood deposits	Erosive marginal part, and nearly stable and vegetated surface platform
	Tidal shoal	Tidal deposits at turbidity maximum zone	Elongated shaped, grass and mangrove dominated
Fringe of river course	Incipient crevasse channel	Development of a narrow channel after breaching the bar during extreme flood	Narrow channel bifurcates the mid-channel bar deposits
	Chute flow	Secondary flow during high flood events	Sluggish palaeo-flow path and only active flow observed during extreme flood event
	Chute bar	Degradation of mid-channel bar by storm flood	Degraded and almost abolished
	Dead slough	Shifting of active flow path from the bank margin to channel middle position	Sluggish or ponding flow path at the margin of river bed
	Sand splays	Sand deposition during storm flood	Exists in the same location and utilized for cultivable land after removal of sand from the top layer
	Fill terrace (younger)	Sediment deposits (recent) during flood at the position of earlier cut terrace	Exists in the same position (channel margin) and or changing its position, shape and size depending on sedimentation nature
	Fill terrace (older)	Sediment deposits (earlier) during flood at the position of earlier cut terrace of active flow path	Exists at the channel margin position
In the floodplain	Backswamp	Diversion of natural river flow after embankment construction	Depression areas at the palaeo-flow path of river
	Marshy land	Natural depression areas of the floodplain, mainly form in the swale topographic condition	Initially remained as marshy land and habitat for natural aquatic species and recently altered into aquaculture farm
	Oxbow lake	Isolated parts of palaeo-meander course formed after meander neck-cut	
	Natural levee (younger)	Sedimentation over marginal areas of recent river course during high flood discharge	Almost entirely occupied by dwellers for settlement and agricultural purpose
	Natural levee (older)		

(continued)

Table 3.2 (continued)

Positions	Morphological features	Formation process	Field description
		Sediment deposition in the marginal areas of the palaeo-river course during peak flood discharge	
	Meander scrolls (palaeo-channels)	Imprint of the migration of palaeo-river courses	Mostly converted into the agricultural land and fisheries by human activities
	Floodplain	Overbank sediment deposition in the extensive low-lying areas on the both sides of the river course	Intensively used for agricultural activities and human settlement
	Dune ridge	Sand deposits over the beach ridge surface by Aeolian process	Parallel elevated dunes occupied by settlement with orchards and some natural vegetation
	Swale	Tidal mud deposits at the interdune depression areas	Exists as natural wetlands and mostly altered into cultivated areas
	Mudflat	Tidal flat deposits over the shallow marine buffer zone	Dominated by the mangroves and salt marshes

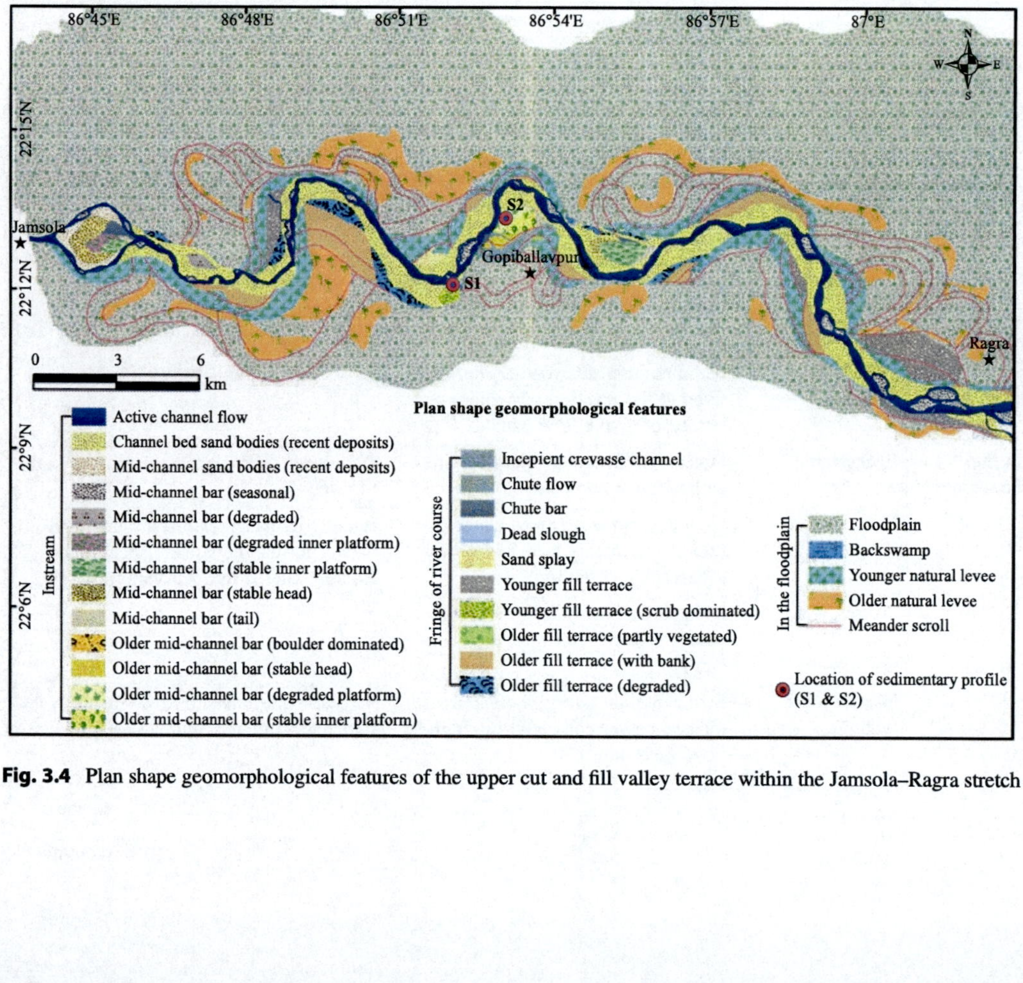

Fig. 3.4 Plan shape geomorphological features of the upper cut and fill valley terrace within the Jamsola–Ragra stretch

Three mid-channel bars have been observed within this river stretch. Among these, one is located in the channel middle position (at Dipapal), and the other two in the channel fringes at Nayabasan (right bank side) and Hatipata (left bank side). Initially, the channel fringe mid-channel bars had remained in the channel middle positions. But, due to the channel shifting on the opposite direction (Jana 2019), these bars have remained in the channel fringe positions. However, these two bars have been surrounded by the river flow during the summer monsoon flood events. The channel bed sand bodies have been deposited on both sides of the channel active flow (during lean-phase). The spatial extents of the sand sheets have been changed depending on the meandering nature of the flow path, hydraulic behaviour of flow during flood events and river bed elevation differences. The seasonal mid-channel bars have been observed in the middle positions of the active flow. These bars have been noticed in the lean-phase of monsoon with due effects from the low magnitude river flow regime. The seasonal mid-channel bars have remained in pseudo nature with their ever-changing positions and extents.

At the fringe of the river course, the incipient crevasse channel, chute flow and bars, dead slough, sand splay, younger and older fill terraces have been found in the Jamsola–Ragra stretch (Fig. 3.4). The incipient crevasse channel has been formed in the Nayabasan bar due to sediment breaching during the extreme flood event (Table 3.2). The chute flow has also been observed in the extreme right bank position of the Nayabasan bar. Initially, this chute flow path remained as the active channel flow path. However, this flow path became gradually inactive with the leftward channel shifting (Jana 2019). The chute bars have been observed in the four different positions, three in the marginal positions of mid-channel bars and one in the margin of the scrub dominated younger fill terrace. The dead slough has formed due to the flow path shifting from the bank margin to the channel middle position and the remnant flow path converted into the sluggish flow path. Such types of micro-geomorphic units have been observed at the five different positions in Jamsola–Ragra stretch (Fig. 3.4). The sand splay has been found in the left bank position, just on the opposite side of scrub dominated younger fill terrace (Fig. 3.4), which has been formed due to the sand deposition at the overbank flow regime during the storm flood events. The younger fill terraces have been formed with recent sediment deposits during the flood events. However, the positions of the fill terraces were remained as the cut terrace in the near past, which were also remained as the active flow path of the earlier river course. The older fill terraces have been formed in the same process likewise the formation process of the younger fill terraces. Only the difference is that the older fill terraces are older and remained far away from the recent river course compared to the younger fill terraces. In this stretch, three types of older fill terraces have been categorized depending on the vegetation significance and degradation level (Fig. 3.4).

In the floodplain areas, the backswamp, younger and older natural levees, meander scrolls coupled with floodplain have been observed in the different sections away from the river course (Fig. 3.4). The backswamp exists in the backside of the bridge protective embankment structure (left bank) at Gopiballavpur (Jana 2019). The younger levees have been formed due to sediment deposition over the recent channel margin areas. The older levees have situated corresponding to the meander scrolls, which were formed due to the sedimentation in the marginal areas of the palaeo-courses. The overall positions of the meander scrolls in the Jamsola–Ragra stretch reveals that the river course has been migrated on both sides of the recent course. However, the positions of palaeo-cliffs, existing meander scrolls and extents of floodplain in this stretch divulge that the river course has shifted at a large extent towards the right in compared to the leftward shifting.

3.3.3.2 Geomorphological Features in the Lower Cut and Fill Valley Terrace

In the lower cut and fill valley terrace of Ragra–Dantan stretch, similar types of geomorphic features have been observed as mentioned in the Jamsola–Ragra stretch, however, the micro-level diversities have resulted in their forms and characteristics (Fig. 3.5). Two outstanding mid-channel bars have been observed near Rohini and Ragra at the confluence positions of Dulung river with the mainstream of Subarnarekha river (Fig. 3.5). The positional extents of the meander scrolls at the confluence region divulge that this area remained as a depression like area in the past. Another mid-channel bar has been situated at the downstream section of Kulboni and near Nayagram. The mid-channel bar located near Rohini is enough elevated and stable, which is well known as Kodopal. The diversified micro-geomorphic units of chute flow with chute bars, dead slough and bar tails have been observed in the Kodopal. The active river flow has dispersed along the wide channel bed in the relatively straight river course. Therefore, a large number of seasonal mid-channel bars are present that river stretch compared to the other stretches. The extended younger fill terrace on the right side of the recent course reveals that the river is migrated towards the left at most of the positions. However, the position of palaeo-cliffs (Fig. 3.1), meander scrolls and associated older natural levees and extensions of floodplains (Fig. 3.5) demonstrate that the river was flowing with a meandering channel pattern in the past and gradually straightened its course by the rightward shifting. The marshy lands have been formed at the positions of palaeo-courses.

3.3.3.3 Geomorphological Features in the Deltaplain

The meandering course of the upper deltaplain is extended within Dantan–Rajghat stretch. This section is now converted into the ancient deltaplain (Jana and Paul 2019). The minimum numbers of geomorphic units have been observed in this stretch (Fig. 3.6) compared to the other stretches. The younger fill terraces have remained at the concave meander bends, whereas, the younger natural levees have been extensively situated almost in both sides of the river course. The positions of the meander scrolls reveal that the palaeo-course was relatively straight, which gradually follow the meandering pattern. However, the most complex channel pattern has been observed in the lower deltaplain of the Rajghat–Chaumukh stretch (Fig. 3.7). This river stretch has been associated with the unidirectional and bidirectional flow nature respectively in the upstream and downstream of Baliapal sections. Though, this entire river stretch was tide-dominated during 7000–6500 YBP (Paul 2002; Jana and Paul 2020). Almost similar types of geomorphological features have remained in this stretch (Fig. 3.7) as observed in the earlier stretches (Figs. 3.4, 3.5, and 3.6). However, the mature and immature tidal shoals, oxbow lakes, dune ridge, swale and mudflat are the additional geomorphic features in the lower deltaplain (Fig. 3.7). The tidal shoals have been formed at the turbidity maximum zones in the extremely tide-dominated lower course (Paul and Kamila 2016). The lower elevated immature tidal shoals have been submerged and emerged depending on the tidal range, whereas, the mature shoals have remained emerged and are significantly covered by saltmarsh and mangroves. The linear array of dune ridges have remained coupled with inter-dune swales within the chenier deltaplain. The elongated mudflat has been formed in the modern barrier coast of the Subarnarekha deltaplain (Jana et al. 2014; Jana and Paul 2019). The existence of oxbow lakes in the palaeo-courses reveals that the river course was active in the near past (Jana and Paul 2020). The younger levees and fill terraces have dominated within this entire river stretch. The older natural levees have been mainly found in the upper section (Rajghat–Baliapal), whereas, the older fill terraces in the lower section of Baliapal. The relative positions of the natural levees and fill terraces (younger and older), oxbow lakes and meander scrolls reveal that the river course has been shifted on both sides in the different sections. However, the river was shifted at a larger extent on the leftward (within the Rajghat–

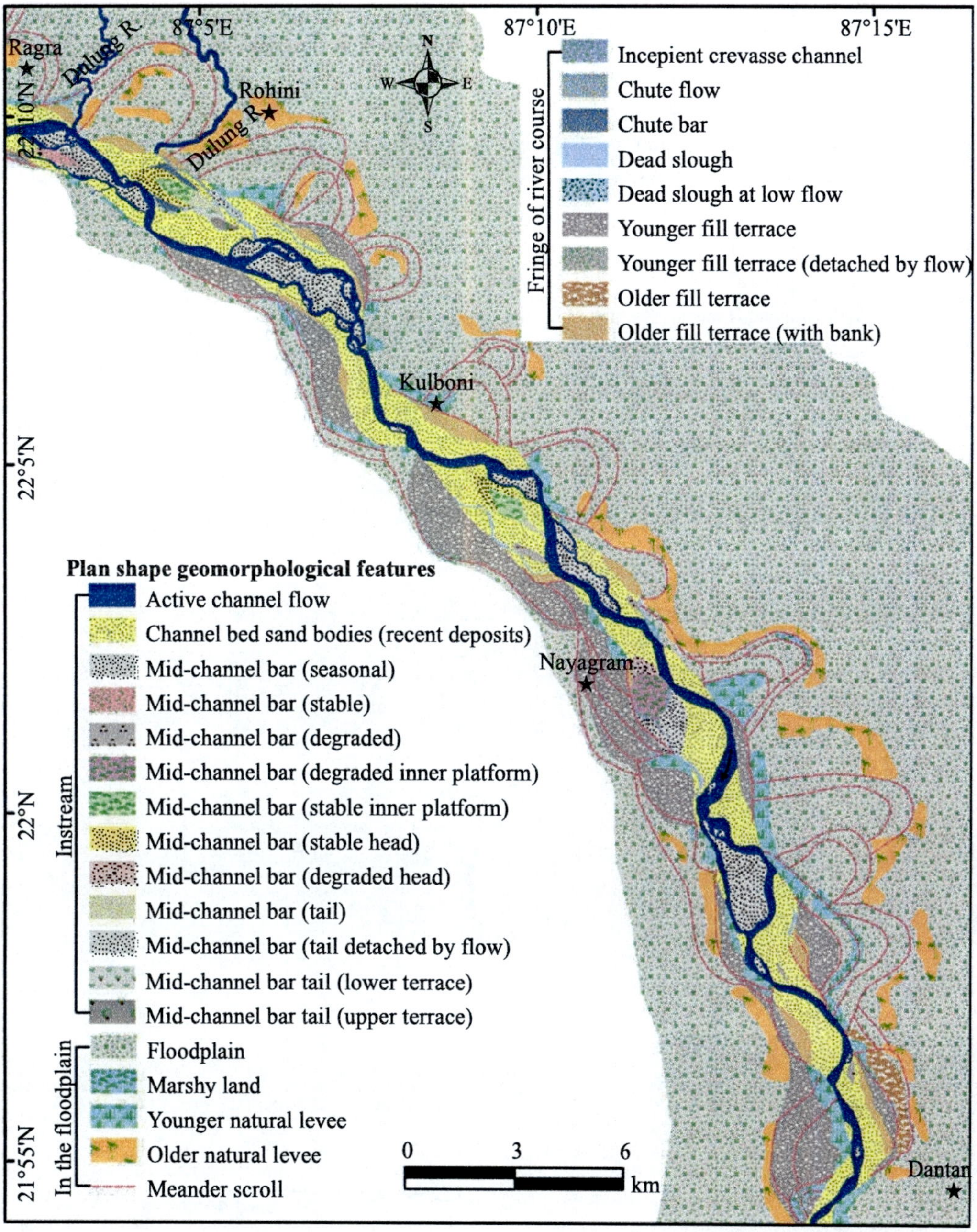

Fig. 3.5 Plan shape geomorphological features of the lower cut and fill valley terrace within the Ragra–Dantan stretch

Baliapal section), both sides at a similar trend (near Baliapal) and rightward (near upstream of Chaumukh) depending on the hydrodynamic nature coupled with the river geometry. Also, the pattern and positions of dune ridges on both sides of the river course (straight and curved ridge cutting) reveals that initially, the river was flowing in a straight course at the downstream of Baliapal, whereas, it was continuously migrating rightward and ensured the meandering flow path

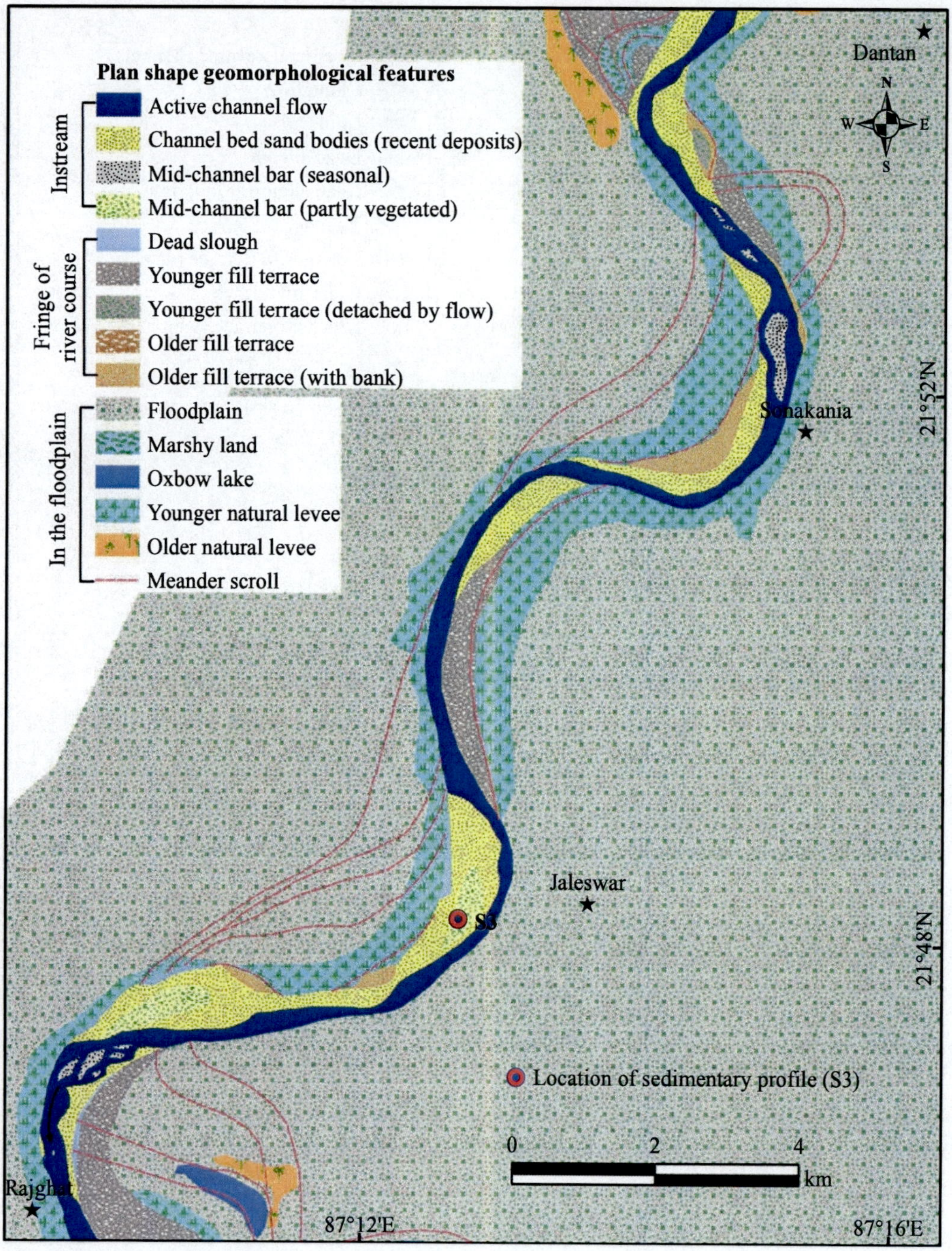

Fig. 3.6 Plan shape geomorphological features of the upper deltaplain within the Dantan–Rajghat stretch

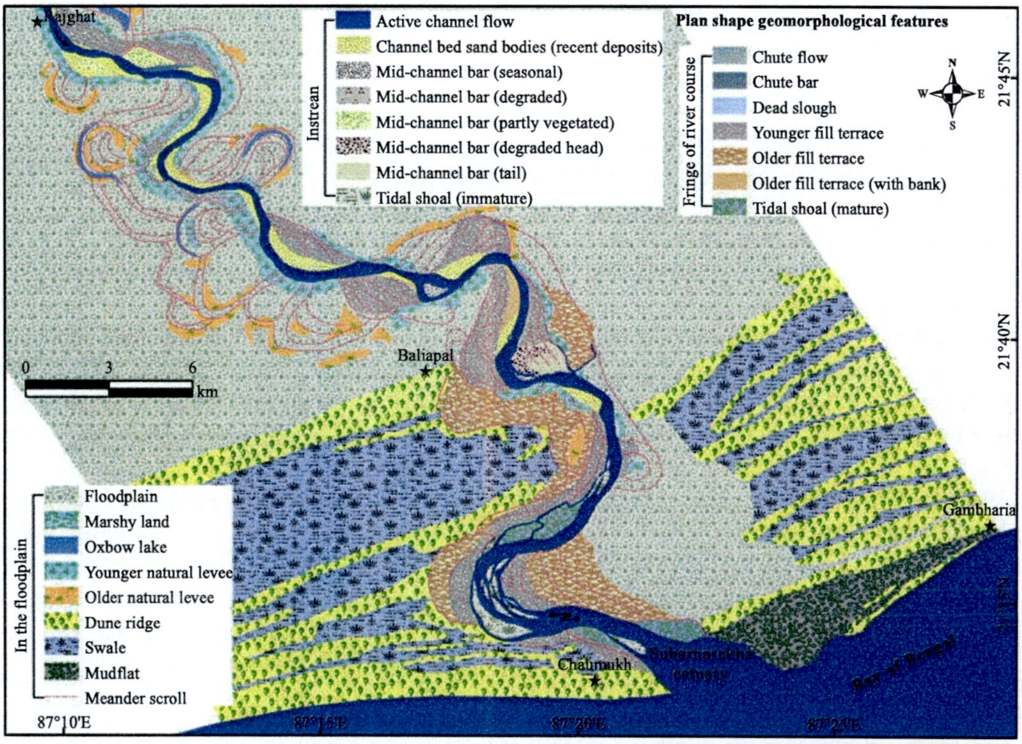

Fig. 3.7 Plan shape geomorphological features of the lower deltaplain within the Rajghat–Chaumukh stretch

in the extreme right bank position. However, the existence of fill terraces in both banks coupled with tidal shoals within the present river course reveals that the river width has been shrunk through maintaining the hydrodynamic adjustments associated with the fluctuating sea level.

3.3.4 Morphology of Mid-Channel bar

Just after the Jamsola gorge section, the channel width has increased from 160 m to 2384 m within 2 km river stretch (Figs. 3.4 and 3.8a). The largest (2.38 km^2) mid-channel bar (surrounded by the mid-channel sand bodies) has been formed at Dipapal with due effects from the abrupt diminution of hydraulic energy in the extensive river course. However, the entire bar is not in a stable form and it has been categorized in five different platforms (bar head, chute bar, bar tail, degraded inner bar platform and stable bar platform) depending on the elevation differences and surface expressions of degradation level (Fig. 3.8a). This bar is about 8.5 m elevated from the surrounding river bed. The river flow is bifurcated and it again converged in a single flow in this section. Another significant mid-channel bar (older) has been observed at Nayabasan (right bank side) covering an area of 2.13 km^2 (Figs. 3.4, and 3.8b). The maximum height of this bar is about 6.75 m in respect to the local river bed. The distinctive geomorphological features of stable inner bar platform, degraded bar platform, stable bar head, chute bar, boulder dominated bar head, bar tail coupled with incipient crevasse channel, dead slough and chute flow path have been found in this section (Fig. 3.8b). The chute flow path initially remained as the main flow path of the active channel. However, recently, flood water can enter within this path only during extreme flood events. During the high magnitude flood events, the river carried boulders have deposited in the marginal areas of the bar. The chute bar has been formed due to erosion of the mid-channel bar. The inner bar platform has remained in stable condition covered by scattered vegetations, whereas, the outer bar platform has been gradually degraded with due effects from the intensive sand mining from the marginal river bed (Jana 2019). The other mid-channel bar has been situated at Kodopal, which is covered about 1.90 km^2 area and the upper bar platform remains in 7 m height from the local river bed (Fig. 3.8c). The stable bar head and inner bar platform, bar tail, degraded bar tail (upper and lower terrace), chute bar and dead slough (at low flow) are the micro-level geomorphic features observed in the mid-channel bar section of the Kodopal (Fig. 3.8c). This bar has been formed with due effects from the combined river discharge and sediment load in the straight and wide river course followed by the relatively curved and narrow course in the downstream section. The voluminous load (water and sediment) might not able to easily drain to the downstream river stretch. Therefore, the lag deposit was responsible for the formation of the mid-channel bar at that position.

3.4 Sedimentary Stratigraphy and Depositional Environments

The sedimentary depositional environments of a landform can be evaluated through the analysis of sedimentary characteristics (Jana and Paul 2020). In this concern, the lithostratigraphic characteristics of three different landforms have been assessed considering the layer-wise diversities of depth (thickness), material types, mean grain-size, mode, sorting, skewness and kurtosis. The sedimentary profile has been excavated up to the 3.58 m.bgl (below ground level) in the channel margin younger fill terrace at Dharmapur

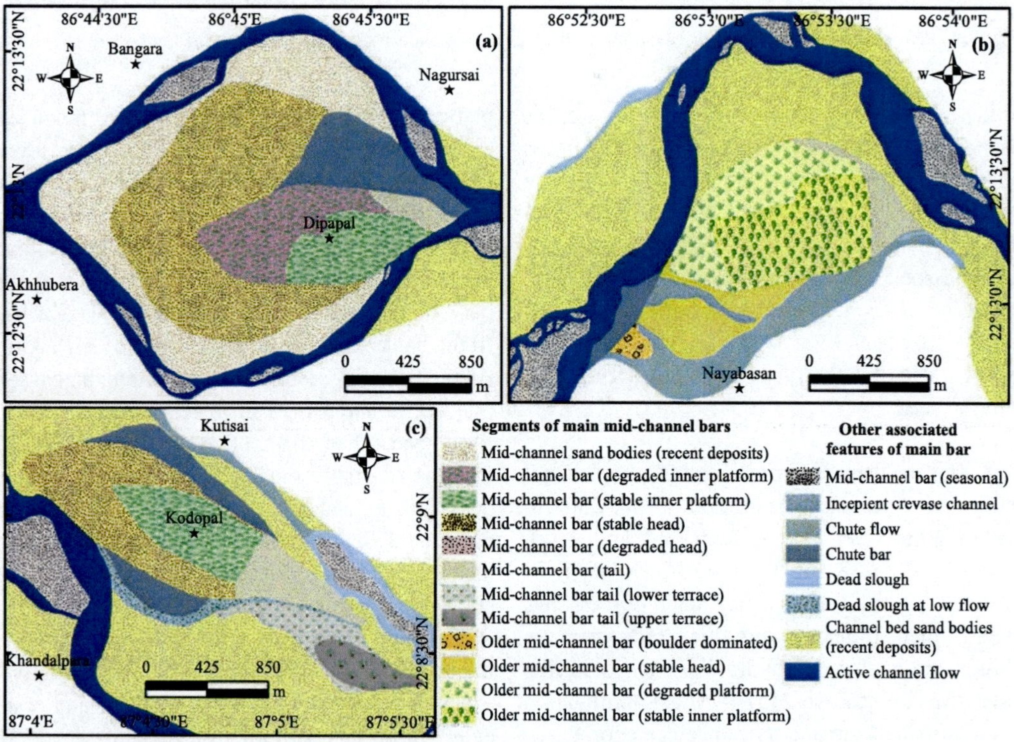

Fig. 3.8 Morphological diversities of the different mid-channel bars situated (**a**) at Dipapal, in the downstream section of Jamsola, (**b**) at Nayabasan near Gopiballavpur, and (**c**) at Kodopal near Rohini

(Fig. 3.4, Table 3.3). The sedimentary profile of the older mid-channel bar has been excavated up to 1.93 m.bgl at Nayabasan (Fig. 3.4 and Table 3.4), whereas, the sedimentary characteristic has been recorded from the 6.11 m thick exposed profile of mid-channel bar at Saherbazar (Fig. 3.6, and Table 3.5). The lithostratigraphic characteristics reveal that the different natures of fluvial environments were effective during the sediment deposition in the distinctive layers (Jana and Paul 2020). Within the three profiles, the material types coupled with the mean grain-size varies from the very coarse sand (with gravels) to clay types, which indicates the very high to very low energy environments. Moreover, the trends of discharge (water and sediment) and energy level coupled with the fluctuating flow regimes have been elucidated through the nature of mode, sorting, skewness and kurtosis (Jana and Paul 2020). The unimodal type of sedimentary characteristics indicates the steady flow of a single flood event, whereas, bimodal nature is characterized by two or more flood events or can be a repetitive fluctuating energy level in a single flood event. The well sorted sedimentary nature reveals the mixed energy environment, whereas, sedimentary nature tends toward well unsorted is deposited by the steady fluvial energy (high or low) environment. Moreover, the higher mean grain-size with well unsorted (poorly sorted) sediment demonstrates the high energy environment, while, the well unsorted with lower mean grain-size exhibits the low energy environment during deposition. The sedimentary deposits resulted with coarse (positive) skewed indicates the enhanced high energy environment followed by continuous receding energy level. The fine (negative) skewness shows a gradual increase in energy level, which rapidly reduced after achieve the peak level, whereas, the symmetrical skewed reveals a similar trend in the rise and fall of flow regime. The leptokurtic type of sedimentary nature conveys the high energy peak discharge, which becomes mesokurtic to platykurtic nature with reducing discharge and energy level during sediment deposition.

The 3.58 m thick younger fill terrace has been composed of twelve distinct sedimentary layers coupled with diversified sedimentological properties (Table 3.3). Based on the response of the local people it is clear that this fill terrace has been formed after the flood event of 1971, which reveals the fill terrace is 45 years old (as in 2015). The material types of these layers have been composed of different mean grain-size material of sands and silts. Very coarse silt (60 μm) and coarse silt (28 μm) has been found respectively in the 2nd and 3rd layers those were deposited in the low energy and high flow regime as a lag deposit. Such type of depositional environment is supported by the nature of mode (unimodal), sorting (moderately sorted), skewness (symmetrical–coarse) and kurtosis (very leptokurtic–leptokurtic) (Table 3.3). The 6th, 8th, 10th and 11th layers have been resulted with bimodal and well—moderately well sorted sedimentary nature, which indicates that these layers were deposited either in the two or more flood events or by the repetitive fluctuating energy level of a single flood event. The layer-wise depositional environment has been assessed in supports of the nature of skewness and kurtosis. In the 6th layer, sediments were deposited with the effects of moderate energy level which was enhanced up to the high level at the end of the flood event. The 8th layer was deposited initially in the increasing energy level, and it was rapidly reducing after achieved the peak discharge and energy level. The 10th layer was deposited with the fluctuating trend of flow regime dominated by high energy peak discharge. The fluctuating flow regime with reducing discharge and energy level was acting during the deposition of sediments in the 11th layer.

In the excavated profile of older mid-channel bar, twelve distinct sedimentary layers have been characterized by very coarse silt (61 μm)–very coarse sand (1512 μm), although, 1.52–2.75 cm diameter gravels have been observed in 12th layer (Table 3.4). The gravels dominated by very coarse sand (12th layer) was deposited under the very high energy steady flow which was diminishing after achieved the peak regime. The 11th

Table 3.3 Lithostratigraphic characteristics of sedimentary profile of younger fill terrace at Dharmapur

Layers	Depth (m.bgl)	Material types	Mean grain-size (μm)	Mode	Sorting	Skewness	Kurtosis
1st (top)	0.00–0.14	Fine sand	202	Unimodal	Well sorted	Coarse	Mesokurtic
2nd	0.14–0.19	Very coarse silt	60	Unimodal	Moderately sorted	Symmetrical	Very Leptokurtic
3rd	0.19–0.30	Fine sand	214	Unimodal	Well sorted	Coarse	Leptokurtic
4th	0.30–0.39	Coarse silt	28	Unimodal	Moderately sorted	Coarse	Leptokurtic
5th	0.39–0.61	Fine sand	206	Unimodal	Moderately well sorted	Coarse	Mesokurtic
6th	0.61–0.88	Fine–medium sand	248	Bimodal	Well sorted	Coarse	Mesokurtic
7th	0.88–1.88	Medium sand	265	Unimodal	Moderately well sorted	Coarse	Platykurtic
8th	1.88–2.00	Medium–fine sand	323	Bimodal	Moderately well sorted	Fine	Platykurtic
9th	2.00–2.28	Medium sand	460	Unimodal	Moderately sorted	Symmetrical	Leptokurtic
10th	2.28–2.63	Medium–coarse sand	481	Bimodal	Well sorted	Symmetrical	Leptokurtic
11th	2.63–3.43	Fine–medium sand	290	Bimodal	Moderately well sorted	Symmetrical	Platykurtic
12th (bottom)	3.43–3.58 (up to exposed layer)	Medium sand	454	Unimodal	Moderately sorted	Coarse	Platykurtic

Table 3.4 Lithostratigraphic characteristics of sedimentary profile of older mid-channel bar at Nayabasan

Layers	Depth (m.bgl)	Material types	Mean grain-size (μm)	Mode	Sorting	Skewness	Kurtosis
1st (top)	0.00–0.10	Fine sand	201	Unimodal	Moderately well sorted	Symmetrical	Platykurtic
2nd	0.10–0.21	Fine–medium sand	242	Bimodal	Moderately sorted	Symmetrical	Leptokurtic
3rd	0.21–0.30	Fine sand	178	Unimodal	Moderately sorted	Coarse	Mesokurtic
4th	0.30–0.37	Fine–medium sand	248	Bimodal	Moderately sorted	Symmetrical	Mesokurtic
5th	0.37–0.75	Medium sand	405	Unimodal	Moderately sorted	Symmetrical	Mesokurtic
6th	0.75–0.87	Very coarse silt	61	Unimodal	Moderately sorted	Coarse	Very Leptokurtic
7th	0.87–1.05	Fine sand	243	Unimodal	Moderately well sorted	Symmetrical	Platykurtic
8th	1.05–1.20	Very fine sand–very coarse silt	68	Bimodal	Poorly sorted	Coarse	Leptokurtic
9th	1.20–1.25	Coarse sand	645	Unimodal	Poorly sorted	Coarse	Mesokurtic
10th	1.25–1.36	Medium sand	451	Bimodal	Moderately sorted	Fine	Very Leptokurtic
11th	1.36–1.63	Fine–medium sand	245	Bimodal	Moderately sorted	Symmetrical	Mesokurtic
12th (bottom)	1.63–1.93 (up to exposed layer)	Very coarse sand with gravels[a]	1512	Unimodal	Poorly sorted	Coarse	Leptokurtic

[a]Average diameter of gravels varies within 1.52–2.75 cm

Table 3.5 Lithostratigraphic characteristics of sedimentary profile of mid-channel bar at Saherbazar

Layers	Depth (m.bgl)	Material types	Mean grain-size (μm)	Mode	Sorting	Skewness	Kurtosis
1st (top)	0.00–1.10	Silt–very fine sand	31	Unimodal	Moderately sorted	Coarse	Mesokurtic
2nd	1.10–2.30	Medium–fine sand	245	Bimodal	Moderately well sorted	Symmetrical	Very Leptokurtic
3rd	2.30–3.80	Silt	15	Unimodal	Poorly sorted	Fine skewed	Leptokurtic
4th	3.80–4.75	Fine sand	134	Unimodal	Moderately sorted	Coarse	Leptokurtic
5th	4.75–5.61	Silt–clay	9	Unimodal	Well sorted	Symmetrical	Mesokurtic
6th (bottom)	5.61–6.11 (up to exposed layer)	Medium–fine sand	254	Unimodal	Moderately well sorted	Coarse	Mesokurtic

layer was deposited with due effects of reducing energy and flow regime, which again enhanced during the deposition of the 10th layer. Also, the 9th layer might be deposited at the same flood event corresponding to the sedimentation in 10th layer, although, during sediment deposition in the 9th layer, the energy level gradually reduced after achieving the peak level. The river carried silt and very fine sand-size materials during the first flood event (Jana and Paul 2019). Therefore, the 8th layer was deposited in a different flood event in the successive year, which was characterized by the voluminous discharge coupled with the high regime and moderate energy level. The 7th layer was also deposited at the end of this flood event associated with a steady and moderate flow energy environment. The 6th to top (1st) layers were deposited in a different flood year with fluctuating energy and flow regimes.

The sedimentary profile of mid-channel bar (at Saherbazar) has been situated in the wide (910 m) river course, which has swiftly reduced to 350 m at Militaribazar bridge and 250 m at Rajghat bridge only within 2.5 km and 7.5 kmdownstream section, respectively. In the downstream section (older deltaplain) such kind of channel dimension is responsible for clay–medium sand types of material deposition in the six diversified sedimentary layers within 6.11 m elevated bar (compared to local river bed) (Table 3.5). The medium–fine sand (254 µm) was deposited at the bottom layer under the moderately steady energy condition which was reduced at the end of the flood event. The silt–clay type materials in the 4th layer (0.86 m thick) were deposited with voluminous sediment loaded low energy steady flow energy environment. In the 4th layer, fine sands were deposited in a separate flood event characterized by moderate energy steady enhancing followed by continuous receding energy environment. The silt was deposited in the 3rd layer under the dominance of low energy and firmly reducing flow regime. The medium–fine sand-size materials (in 2nd layer) were deposited in two different flood events under moderate–high energy fluctuating flow regimes. The top layer (silt–very fine sand) was formed with due effects from the mixed energy fluctuating flow regime of a single flood event. The top layer was deposited during 1978 and after that flood level was unable to inundate the stable platform of the bar. About 1.0–1.5 m diameter tree logs have been observed over the bar surface, which assists to predict the age of the bar. Local people are also confirmed that the bar has been formed about 25 years ago (as in 2015).

3.5 Conclusion

The present study area of the middle-lower and deltaic courses of the Subarnarekha river represent a significant sediment sink. A large amount of sand size sediments is derived from the upper catchment area dominated by weathered granites and gneissic terrain of the Chotanagpur plateau. The depositional environments of the study area represent high fluvial influx in the three sections of the channel reaches from Jamsola to Ragra, Ragra to Dantan, and Dantan to Rajghat. However, the assemblages of landforms, geometry of the channel reach and sediment characteristics of the downstream section indicate the role of concomitant sea level fluctuations in the modification of sedimentary depositional environment. The plan shape geomorphology, properties of channel geometry, sediment texture and stratigraphic sections of the river valley alluviums, morphology of the mid-channel bars, and palaeo-shorelines of the deltaic section reveal the presence of very distinct depositional sub-environments along the present course of the Subarnarekha river from Jamsola to Chaumukh. The study also highlights the role of energy level fluctuations of the river flow since the Late Pleistocene to Early Holocene period, and continuous adjustment of the channel course since Early–Late Holocene period with the base level changes resulted from marine transgression, regression and stillstand phases. The presence of valley cuts, fluvial and marine terraces also indicate the role of neotectonics in the modifications of bank margin environment of the river valley.

Acknowledgements The authors are grateful to the United States Geological Survey (USGS) for making available the satellite images, and also thankful to the Government of India for making available the Geological Quadrangle Map on Open Government Data (OGD) platform free of charge. The corresponding author would like to acknowledge the University Grants Commission, New Delhi, India, for financial support as Junior Research Fellowship [Award Letter No.: F.15-6(DEC.2014)/2015 (NET), UGC Ref. No. 3070/(NET-DEC.2014)] to carry out the research work presented in this paper.

References

Abrahams AD (1984) Channel networks: a geomorphological perspective. Water Resour Res 20(2):161–188

Ashworth PJ, Best JL, Roden JE, Bristow CS, Klaassen GJ (2000) Morphological evolution and dynamics of a large, sand braid-bar, Jamuna River. Bangladesh. Sedimentology. 47(3):533–555

Banerjee M, Sen PK (1987) Palaeobiology in understanding the changes of sea-level and coastline in Bengal Basin during Holocene Period. Indian J Earth Sci 14:307–320

Bannerjee M, Sen PK (1988) Paleobiology and environment of deposition of Holocene sediments of the Bengal Basin, India. In: Paleoenvironment of East Asia from the mid-Tertiary: Proceedings of the Second Conference. Hong Kong Centre of Asian Studies, University of Hong Kong, 703–731

Bentham PA, Talling PJ, Burbank DW (1993) Braided stream and flood-plain deposition in a rapidly aggrading basin: the Escanilla Formation, Spanish Pyrenees. In: Best JL, Bristow CS (eds) Braided Rivers. Geol Soc Spec Publ 75(1):177–94. https://doi.org/10.1144/GSL.SP.1993.075.01.11

Best J (2019) Anthropogenic stresses on the world's big rivers. Nat Geosci 12(1):7–21

Beuchle R, Grecchi RC, Shimabukuro YE, Seliger R, Eva HD, Sano E, Achard F (2015) Land cover changes in the Brazilian Cerrado and Caatinga biomes from 1990 to 2010 based on a systematic remote sensing sampling approach. Appl Geogr 58:116–127

Bhattacharya A, Misra AK (1984) Geological remote sensing in parts of Subarnarekha—Baitarani basin. Eastern India. J Indian Soc Remote 12(1):19–28

Bishop MP, James LA, Shroder JF Jr, Walsh SJ (2012) Geospatial technologies and digital geomorphological mapping: concepts, issues and research. Geomorphology 137(1):5–26

Bisson PA, Montgomery DR, Buffington JM (2017) Valley segments, stream reaches, and channel units. In: Hauer FR, Lamberti GA (eds) Methods in Stream Ecology. Elsevier, San Francisco, pp 21–47

Blott SJ, Pye K (2001) GRADISTAT: a grain size distribution and statistics package for the analysis of unconsolidated sediments. Earth Surf Proc Land 26 (11):1237–1248

Corenblit D, Tabacchi E, Steiger J, Gurnell AM (2007) Reciprocal interactions and adjustments between fluvial landforms and vegetation dynamics in river corridors: a review of complementary approaches. Earth-Sci Rev 84(1–2):56–86

Dandapat K, Panda GK (2013) Drainage and floods in the Subarnarekha Basin in Paschim Medinipur, West Bengal, India–a study in applied geomorphology. Int J Sci Res 4:791–797

Foody GM, Campbell NA, Trodd NM, Wood TF (1992) Derivation and applications of probabilistic measures of class membership from the maximum-likelihood classification. Photogramm Eng Rem S 58(9):1335–1341

Fryirs KA, Brierley GJ (2012) Geomorphic analysis of river systems: an approach to reading the landscape. John Wiley & Sons

Geological Survey of India (1998) Geological Quadrangle Map, Geological Survey of India, Government of India, available on https://www.gsi.gov.in/webcenter/portal/OCBIS/pageMAPS/pageMapsSeries?_adf.ctrl-state=wd51swrgt_5&_afrLoop=27454911959312140!, retrieved on 6th December, 2015.

Ghosh S, Guchhait SK (2020) Laterites of the Bengal Basin: Characterization. Springer, Geochronology and Evolution

Gilvear DJ (1999) Fluvial geomorphology and river engineering: future roles utilizing a fluvial hydrosystems framework. Geomorphology 31(1–4):229–245

Goswami AB (1997) A morphostratigraphic hydrologic and hydrochemical appraisal. Reprint of 8th National symposium on Hydrology, Jadavpur University, Calcutta, Medinipur coastal belt, WB, pp 30–40.

Goswami AB (1999) Quaternary mapping concept constraints, aims and approaches with special reference to Bengal Basin. Workshop manual on coastal quaternary. Bengal Basin, Bose inst., Calcutta, 4–9(1):1–20.

Guha S, Patel PP (2017) Evidence of topographic disequilibrium in the Subarnarekha River Basin, India: A digital elevation model based analysis. J Earth Sys Sci 126(7):106

Ilahi RA, Dutta S (2016) Quantification and Mapping of Morphometric Parameters of Subarnarekha River Basin in Eastern India using Geo-Spatial Techniques. Indian Cartog 16:184–197

Islam A, Guchhait SK (2017) Analysing the influence of Farakka Barrage Project on channel dynamics and meander geometry of Bhagirathi river of West Bengal. India. Arab J Geosci 10(11):245

Jana S (2019) An automated approach in estimation and prediction of riverbank shifting for flood-prone middle-lower course of the Subarnarekha River. Int J River Basin Manag, India. https://doi.org/10.1080/15715124.2019.1695259

Jana S, Paul AK (2014) Morphodynamics of the meandering river: A study along the Subarnarekha river of

Gopiballavpur section, West Bengal, India. Int J Geol Earth Env Sci 4(3):219–230

Jana S, Paul AK (2018) Genetical Classification of Deltaic and Non Deltaic Sequences of Landforms of Subarnarekha Middle Course and Lower Course Sections in Odisha and Parts of West Bengal with Application of Geospatial Technology. J Coast Sci 5 (1):16–26

Jana S, Paul AK (2019) Assessment of morphogenetic sedimentary depositional environments of different morphological surfaces of middle-lower and deltaic courses of Subarnarekha River. J Coast Sci 6(1):1–11

Jana S, Paul AK (2020) Chronological evolution of the channel functional units in association with palaeo-hydrogeomorphological environment in the ancient delta fan of Subarnarekha basin. India. Environ Earth Sci 9:331. https://doi.org/10.1007/s12665-020-09093-1

Jana S, Paul AK, Islam SM (2014) Morphodynamics of Barrier Spits and Tidal Inlets of Subarnarekha Delta: a study at Talsari-Subarnapur spit, Odisha, India. Indian J Geo Env 13:23–32

Jia K, Wei X, Gu X, Yao Y, Xie X, Li B (2014) Land cover classification using Landsat 8 operational land imager data in Beijing. China. Geocarto Int 29 (8):941–951

Jia K, Wu B, Tian Y, Zeng Y, Li Q (2011) Vegetation classification method with biochemical composition estimated from remote sensing data. Int J Remote Sens 32(24):9307–9325

Khan A, Rao LA, Yunus AP, Govil H (2018) Characterization of channel planform features and sinuosity indices in parts of Yamuna River flood plain using remote sensing and GIS techniques. Arab J Geosci 11 (17):525

Leopold LB, Wolman MG, Miller JP (1992) Fluvial processes in geomorphology, 2nd edn. Dover Publishers, New York, p 522

Mokarrama M, Hojati M (2018) Landform classification using a sub-pixel spatial attraction model to increase spatial resolution of digital elevation model (DEM). Egypt J Remote Sens Space Sci 21(1):111–120

Niyogi D (1975) Quaternary Geology of the coastal plain of West Bengal and Orissa. Indian J Earth Sci 2 (1):51–61

Pal M, Foody GM (2012) Evaluation of SVM, RVM and SMLR for accurate image classification with limited ground data. IEEE J Sel Topics Appl Earth Obs Remote Sens 5(5):1344–1355

Patel A, Katiyar SK, Prasad V (2016) Performances evaluation of different open source DEM using Differential Global Positioning System (DGPS). Egypt J Remote Sens Space Sci 19(1):7–16

Paul AK (2002) Coastal Geomorphology and Environment: Sundarban Coastal Plain, Kanthi Coastal Plain. ACB publications, Kolkata, Subarnarekha Delta Plain

Paul AK, Kamila A (2016) Coastal Mud Banks of Subarnarekha Delta with Special Reference to Degradation and Accretion under Physical Processes. Indian Cartog 16:13–24

Pavlis NK, Holmes SA, Kenyon SC, Factor JK (2012) The development and evaluation of the Earth Gravitational Model 2008 (EGM2008). J Geophys Res-Sol Ea 117:B044406. https://doi.org/10.1029/2011JB008916

Poff NL, Allan JD, Bain MB, Karr JR, Prestegaard KL, Richter BD, Sparks RE, Stromberg JC (1997) The natural flow regime. Bioscience 47(11):769–784

Roy S, Sahu AS (2018) Road-stream crossing an in-stream intervention to alter channel morphology of headwater streams: case study. Int J River Basin Manag 16(1):1–9

Samanta RK, Bhunia GS, Shit PK, Pourghasemi HR (2018) Flood susceptibility mapping using geospatial frequency ratio technique: a case study of Subarnarekha River Basin. India. Model Earth Syst Environ 4 (1):395–408

Sarma JN (2005) Fluvial process and morphology of the Brahmaputra River in Assam. India. Geomorphology. 70(3–4):226–256

Sofia G (2020) Combining geomorphometry, feature extraction techniques and Earth-surface processes research: The way forward. Geomorphology. https://doi.org/10.1016/j.geomorph.2020.107055

Ventra D, Clarke LE (2018) Geology and geomorphology of alluvial and fluvial fans: current progress and research perspectives. Geol Soc Spec Publ 440(1):1–21

Wharton G (1995) The channel-geometry method: guidelines and applications. Earth Surf Proc Land 20 (7):649–660

Williams GP (1986) River meanders and channel size. J Hydrol 88(1–2):147–164

Wu B, Xia J, Fu X, Zhang Y, Wang G (2008) Effect of altered flow regime on bankfull area of the Lower Yellow River. China. Earth Surf Proc Land 33 (10):1585–1601

4 Quantitative Assessment of Channel Planform Dynamics and Meander Bend Evolution of the Ramganga River, Ganga Basin, India

Rameswar Mukherjee

Abstract

The present study explores the use of channel planform and meander bend parameters quantitatively analyse the spatio-temporal variations of the river channel morphology. The study was undertaken on ~ 37 km reach of the Ramganga river over the period of 91 years (1923–2014) based on topographical map (1923) and satellite images (1972, 1981, 1993, 2003, 2014). All the map data were put into GIS in order to evaluatethespatio-temporal changes of the chosen parameters. For assessing channel planform dynamics, widely recognised parameters viz. active channel width; active channel total channel width ratio; channel length; channel sinuosity; active channel area; channel belt area; active channel area channel belt area ratio; sandbar number and types; sandbar area active channel area ratio was utilized. For examining the meander bend evolution, bend radius, width, bend curvature channel width ratio, amplitude, wavelength, length of the bend and sinuosity was considered. The study revealed that all the planform and bend parameters underwent significant modification over the assessed periods. It expressed that the channel was laterally unstable. Within decadal scale it also showed remarkable variations. The analysis showed that the channel narrowing phenomenon was prevalent throughout the assessed period. The tendency of channel braiding increased substantially. The most of the meander bends of the Ramganga river in this study reach did not get enough time to grow. The cut off and lateral migration of the channel largely hampered the bend growth for the 60% of the total assessed bends. Moreover, few high-curvature meander bends were persisted in each assessed period due to some local resistance.

Keywords

Channel planform · Meander bend · Centreline migration · Bank line migration · Western Gangetic Plain

4.1 Introduction

The natural rivers maintain the state of equilibrium with its existing channel form through space and time (Leopold and Wolman 1957). This condition is achieved by a river when a balance between water and sediment discharge exists with available energy. This equilibrium may break due to change among the factors that enforced on the channel morphology (independent variables) with those that are adjusted with enforced condition (dependent variables)

R. Mukherjee (✉)
Department of Geography, Samsi College, Samsi
Malda, West Bengal, India

P. K. Shit et al. (eds.), *Drainage Basin Dynamics*, Geography of the Physical Environment,
https://doi.org/10.1007/978-3-030-79634-1_4

(Schumm 1981; Hogan and Luzi 2010; Ghimire 2020). Whenever this condition arises, it drives to channel instability resulting in marked changes in the planform morphology. The spatio-temporal variations of channel planform occur due to adjustment of the internal geometry of the channel viz. channel width, depth, slope and meander form (Knighton 1984). The adjustment mechanism includes the channel erosion and deposition, change in slope and movement of the channel bed materials (Schumm 1981; Morisawa 1985; Nanson and Knighton 1996; Montgomery and Buffington 1997). Hooke (1977) identified two types of channel planform adjustment. The first occurs where the channel is in the state of dynamic equilibrium, translates downstream but maintains the channel planform. But in dynamic non-equilibrium state, the changes in various channel parameters and hydraulic geometry attributes drive major changes in the channel planform.

Meandering planform is a major subject research in fluvial geomorphology because of its picturesque form, ubiquitous distribution of form, its spatio-temporal dynamics and its practical consequences (Hooke 1977, 2003, 2007). The knowledge of meander is very important for understanding the natural dynamics of fluvial systems, also for planning, engineering constructions, conservation and restoration of the channel (Hooke and Yorke 2010). Meandering rivers in unconfined alluvial reaches are highly mobile. Although, the rate or scale of change in meandering are varying widely across the world, wherever the rate of change in meander morphology and its downstream migration is higher in propensity, it causes huge practical concern for the geomorphologists, river engineer and planners.

For understanding the pattern of planform dynamics and also to predict the future change, one has to consider the past histories of the river (Winterbottom 2000). The availability of the aerial photographs, satellite images provide a historical perspective for studying present-day fluvial processes as well to predict the pattern of channel planform change (Petts 1989; Trimble and Cooke 1991; Gurnell et al. 1994; Gurnell 1997). It provides ample visual information on channel planform. But these data are available only for around the last 60 years. Although, few topographic maps, survey maps, charts exist for the last 300 years in the world, which can be useful for further detail understanding of river dynamics. But these maps or charts are available at wide range of intervals (mostly >50 years). Hence, sequential change in decadal or annual range cannot be attributed.

The channel planform study involves quantification of planform parameter and assessment of the spatio-temporal variation. Various studies also address the role of determining factors in planform dynamics (Anderson and Calver 1980; Hooke and Redmond 1992). Within a short time period, the variations of discharge and stream power are considered as potent causal factors for channel planform modification (Lewin 1983; Hickin and Nanson 1984).

Most of the present-day channels have been attributed dramatic planform variation due to climate change (Ashmore and Church 2001; Feng et al. 2011; Chang 2008; Kiss and Blanka 2012) and anthropogenic disturbances like land use change (Karwan et al. 2001; Ghimire and Higaki 2015), construction of dam (Williams and Wolman 1984; Surian 1999; Grant et al., 2003; Vörösmarty et al. 2003; Yang et al. 2014), sand and gravel mining (Wishart et al. 2008).

The channel planform of the rivers of the Gangetic Plain is highly dynamic. The substantial changes in the planform occur by adjusting channel width, deposition of bar, bank erosion in response to the varying discharge and sediment load. A limited number of in-depth study was conducted in the Western Gangetic Plain rivers (e.g. Sinha et al. 2005; Roy 2009; Bawa et al. 2014; Khan et al. 2018; Yunus et al. 2019; Agnihotri et al. 2020) examining the changes in the channel planform. Hence, in the present paper, an attempt has been made to evaluate channel planform dynamics and evolution of the meander bends over the period of 91 years on the selected stretch of the lower Ramganga river, flowing in the Western Gangetic Plain. In this region, the Ramganga river is flowing in unconfined reach thus it is free to laterally adjust its channel size and morphology due to variable sediment flux and water discharge. It causes severe damages to the land, people, properties, local infrastructures and effectually hinders the growth of this region. Therefore, it is necessary to understand the nature of planform dynamics and plausible causes of that which would be effective for the river planning and management. The objectives of this study are (i) to quantitatively assess the dynamics of various chosen parameters of the channel planform (ii) to evaluate meander bend bend evolution of the river (iii) to explore the causative factors of channel planform dynamics and mechanisms of meander bend evolution.

4.2 Study Area

The Ramganga river is the first major tributary of Ganga river that joins in the Indo-Gangetic plain. The total catchment area of the Ramganga basin is 30635.1 km^2. The Ramganga river flows through the bedrock and alluvial reaches, originates from the Lesser Himalaya over the Dudhatoli Crystalline Formation at an elevation of 2926 m ASL near Gairsain of Chamoli District, Uttarakhand (Mukherjee et al. 2017). The total length of this river is ~649.11 km, out of which ~167.91 km is in the Himalaya and ~481.2 km in the Western Gangetic Plain. A dam was constructed on Ramganga river along the foothill of Himalaya at Kalagarh, Uttarakhand. It had been functioning properly since 1974. The present study reach is loacted in Western Gangetic Plain locally known as *Rohilkhand Plain* (Fig. 4.1). Here, the Ramganga river is notorious for frequent migration of channel. The presence of meander scar and series of swale and ridges demonstrate horizontal instability of the meander bends of the river. Besides, a wide number of abandoned channels, chute channels, palaeochannels provide ample shreds of evidence for frequent migration of the Ramganga river. The channel pattern alternates between single-thread meandering to partly braided types. The river gradient is relatively higher than the major rivers of Eastern Gangetic Plain. The river is also characterized by higher stream power (Sinha et al. 2005).

According to Koppen's climatic classification, the present study reach falls under Cwg types of climate. It is characterized by dry winter. More than 80% of rainfall happens during monsoon season (June to September). This region receives around 160–180 cm rainfall annually. In summer, the temperature rises more than 40 °C, while in winter, it goes below 10 °C. The geological formation of the Rohilkhand Plain comprises of Varanasi older alluvium, Ramganga terrace alluvium and Ramganga recent alluvium (Khan and Rawat 1992). The Ramganga recent alluvium is formed by the river-borne sediments of recent origin lying adjacent to the present course of the Ramganga river (Fig. 4.2). This unit is largely reworked by the recurring floods. The Ramganga terrace alluvium is associated with older floodplain and upland terrace surface. As it is located well above the present channel floor (>6 m), so it is only inundated during episodic mega flood events. Here, the concentration of clayey-silt soil along with calcium-rich *kankar* is high. Varanasi older alluvium is found in upland terrace surface and sandy alluvial ridges are locally known as *Bhur*. The *Bhur* is formed over the abandoned courses of the Ramganga river (Shrivastava et al. 2000).

The fluvial regime of the Ramganga river is largely controlled by the monsoon climate. In the monsoon season, a large amount of precipitation is received in the Ramganga watershed that caused an exceptionally higher amount of water discharge. Apart from water discharge season, no such significant variation in the discharge is witnessed.

The monthly average discharge data (2003–2012) of the Ramganga river for Dabri gauging station, Shahjahanpur, Uttar Pradesh has been displayed in Fig. 4.3. The discharge rises from July and peaks in September and then starts to decline from October. High monthly discharge is recorded from July to October. November to June is designated as the period of low monthly discharge. The months of August and September register as an exceptionally higher amount of

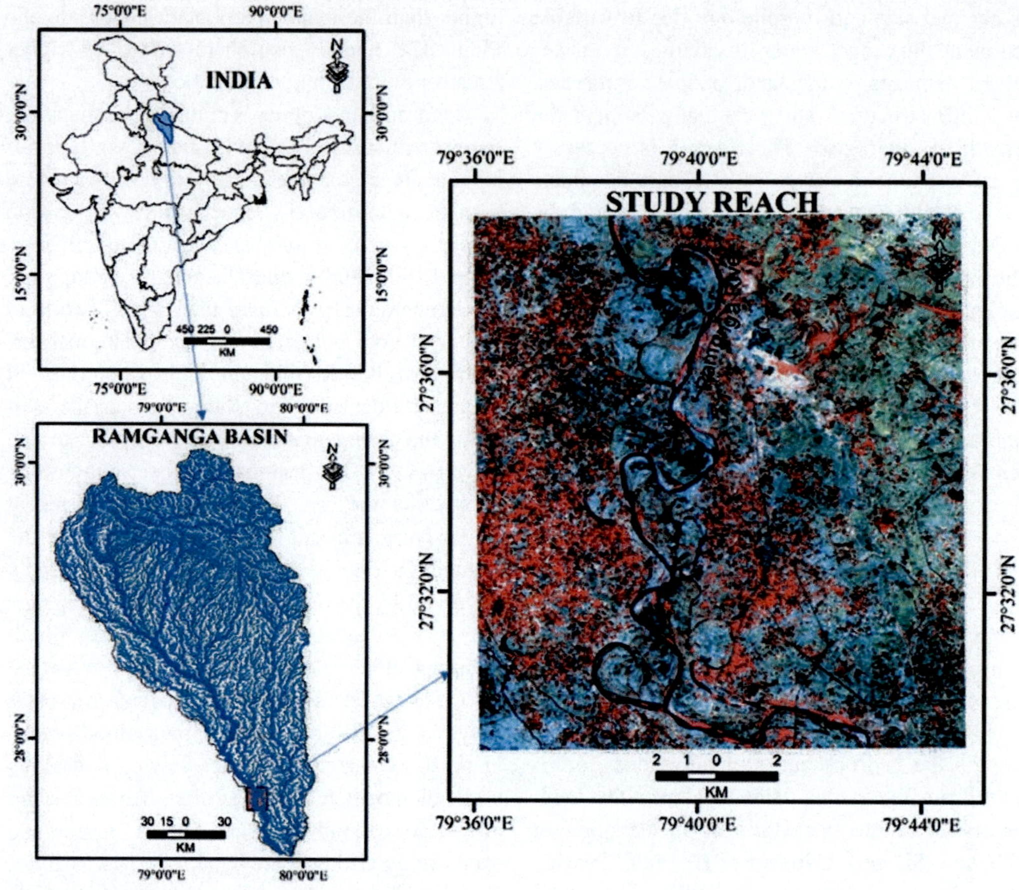

Fig. 4.1 Study area

water discharge. The bank full discharge of the Ramganga is documented as 665 m^3/s. The mean annual discharge is estimated as 3050 m^3/s. The river bears 5.09 mt sediment load annually (Roy and Sinha 2007).

Locally, the soil of the study reach can be classified into three categories: *Khadar*; *Bhangar*; and *Bhur*. The *Khadar* soil occupies active floodplain zone. The texture of the soil is sandy to siltyloam. The *Bhangar* soil is found in the older flood plain, upland terrace surface and interfluve zone. The soil profile is quite mature and composed of sandy loam to clayey-silt texture. The *Bhur* is formed by the deposition of large amount of sand, highly localized along the Ramganga river tract.

4.3 Database and Methodology

4.3.1 Data Acquisition

Georeferenced and precisely orthorectified Landsat data were downloaded from Landsat look viewer website (Table 4.1). Landsat data were georeferenced with the following parameters:

Projection Type: Universal Transverse Mercator (UTM).

Spheroid Name: WGS 84.

Datum Name: WGS 84.

The study area falls under UTM Zone 44. Whole study reach has been covered in single

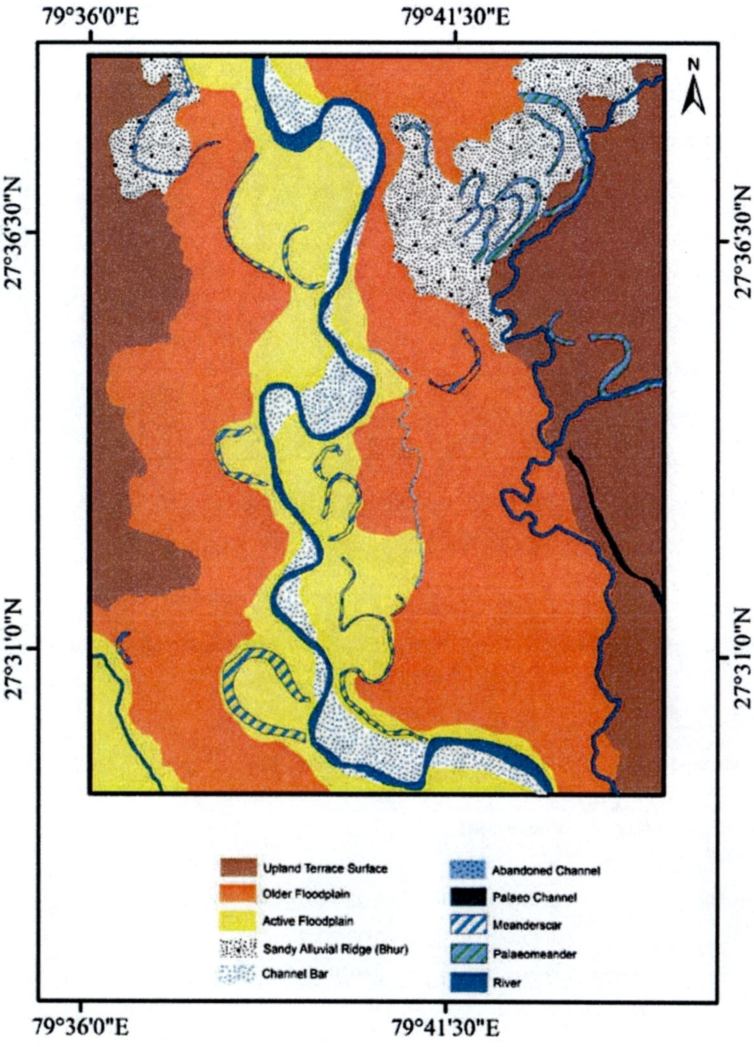

Fig. 4.2 Geomorphological map of the study reach

scene of Landsat data, whose path and row are 145/41 (for 2014, 2003 and 1993 Images) and 155/41 (for 1972 and 1981 images).

4.3.2 Image Processing

The Landsat-MSS images of 1972 and 1981 were rescaled into 30 m, by using resample operation in ERDAS-IMAGINE 2014 Software to maintain uniformity in spatial resolution among all the satellite images used for the present study. The Topographical Sheet (NG 44–2) published by US Army Map Service was geo-referenced by following the same parameters as used for Landsat Satellite Images.

4.3.3 Channel Planform Parameters

Overall, nine variables were used for characterizing planform dynamics of the Ramganga river: Active channel width; active channel total channel width ratio; length; channel sinuosity; active channel area; channel belt area; active channel area channel belt area ratio; sandbar

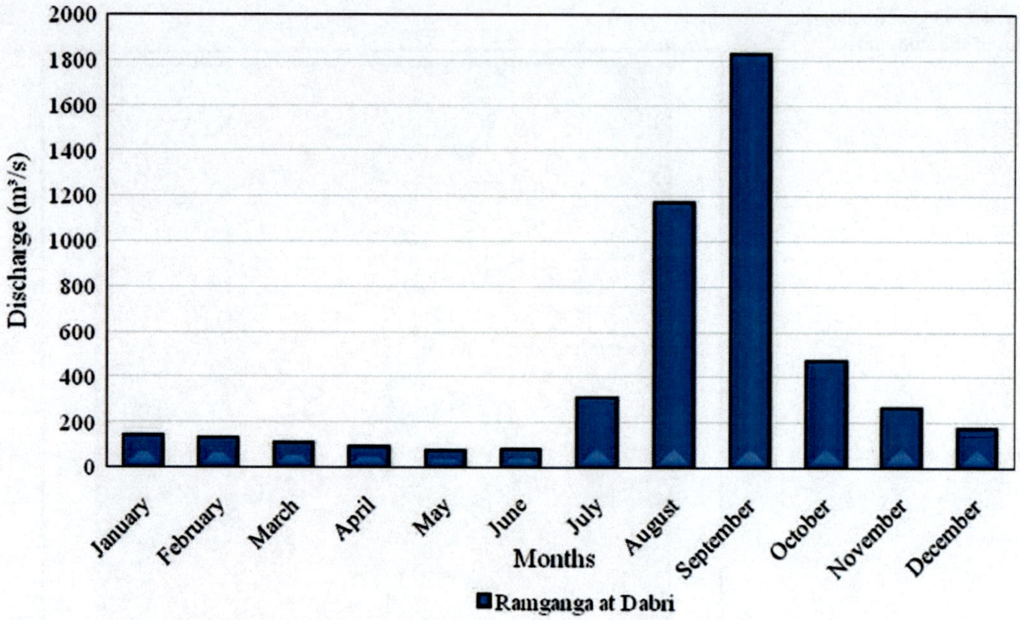

Fig. 4.3 Average Monthly Water Discharge (2003–2012) of the Ramganga river at Dabri Gauging Station

Table 4.1 Data used in the study

Data type	Satellite-sensor/Map No.	Date (mm. dd. year)	Spatial resolution (m) or scale	Source
Remote sensing Images	Landsat-MSS	2/19/1972	79	USGS
	Landsat-MSS	1/15/1981	79	USGS
	Landsat-TM	2/22/1993	30	USGS
	Landsat-ETM	2/27/2003	30	USGS
	Landsat 8-OLI	2/9/2014	30	USGS
Topographical Maps	NG 44-2	1922–1923 (Surveyed) 1955 (Published)	1:2,50,000	Series U502, US Army Map Service

number and types; sandbar area active channel area ratio. For showing lateral migration of the channel, the channel centreline migration and bank line migration were assessed.

For determining the active channel width, the total extent of wetted channel from a particular point was considered. For the images of Landsat series, the MNDWI was applied for extracting channel area by using the following formula:

$$\mathrm{MNDWI} = (\rho_{\mathrm{Green}} - \rho_{\mathrm{SWIR}})/(\rho_{\mathrm{Green}} + \rho_{\mathrm{SWIR}}) \quad (4.1)$$

where ρ_{Green}; ρ_{SWIR} are the reflectance of green and shortwave infrared bands, respectively.

Then, the active channel width had been measured through transects lying at an interval of 1 km. The active channel total channel width ratio (ACTCWR) had been calculated transect wise by dividing active and total channel width. The total channel width was calculated by measuring orthogonal distance between both sides of the banks along the transects. The total length of the channel centreline was taken into consideration for determining the channel length. The sinuosity of the Ramganga channel was calculated in Arc GIS 10.1 by using the following equation:

$$CS = CL/VL \quad (4.2)$$

where CS = channel sinuosity

CL = distance between two points along the channel centreline

VL = straight line distance between two points.

The sinuosity of the channel was calculated by dividing total channel length of the study reach to the straight line distance of both the ends of the channel. To show the spatial variation of CS within the reach, equispaced section lines were drawn at an interval of 3 km.

The braiding index was calculated by using the formula of Brice (1964):

$$\text{Braiding Index (BI)} = \frac{\text{Total length of the bar}}{\text{Length of the reach}} \quad (4.3)$$

The active channel area (ACA) was calculated from Landsat images by extracting water pixels through MNDWI method. After that, these pixels were converted to polygon, and ACA was measured by using calculate geometry command in Arc GIS. For 1923 top sheets, the channel area was digitized manually. The channel belt area was marked digitally by observing the extent of river channel dynamics within a valley. In this study, the channel belt area had been demarcated by tracing abandoned channels, meander scars and tracts of palaeochannels. The Centreline was drawn by joining the midpoint of active channel width lying at an interval of 50 m from each other. After the superimposition of the centrelines of the successive years, the lateral deviations between them were marked by polygons. After that, the area and perimeter of the polygons were calculated using following formula of Micheli and Larsen (2011):

$$\text{Mean lateral change} = \text{Area/Half Perimeter} \quad (4.4)$$

4.3.4 Delineation of Bank Lines

The frequent shifting of the Ramganga river created difficulties in the bank line delineation. Although numerous automated bank line extraction methods are available, but those did not provide reliable results for frequently migrating channels. The problem became more critical along the region where the bank accretion predominated. Because spectral reflectance of riverine tracts near the bank of those channels is quite similar. Hence, field knowledge of the expert was necessary to delineate the banks. In the present study, manual digitization was done after conducting several fieldworks. For digitizing bank lines from the topographical map, the task became quite easy because all the banks were marked on the topo sheets.

4.3.5 Determining Bank Line Migration

To trace out the bank line migration at spatio-temporal scale, 37 transects were drawn at an interval of 500 m. All the transects were drawn parallel to each other. The shifting distance of the bank line was marked from each transect. The westernmost point of each transect was taken as a reference point for determining the bank line migration.

4.3.6 Meander Bend Morphology

For evaluating meander bend morphology, bend radius, width, bend curvature channel width width ratio, amplitude, wavelength, length of the bend and sinuosity were considered (Fig. 4.4).

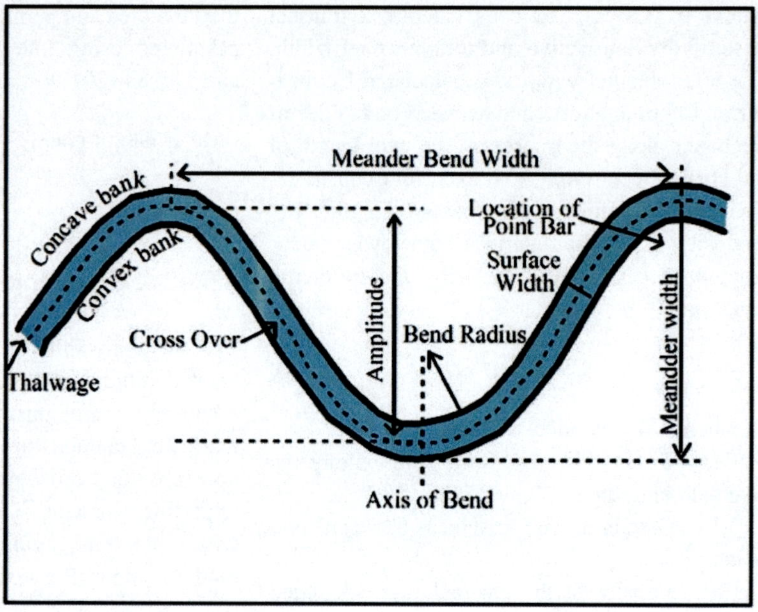

Fig. 4.4 Standardized parameters of meander morphology

For extracting meander bend parameters, the channel centreline and bank lines were digitized, and each meander bend were distinguished. Temporal as well as spatial changes were marked for each meander bend. The morphological changes of the bends were identified based on Hooke's models (1984,1995, 2003) (Fig. 4.5).

For analysing the spatio-temporal variations of the meander bends, the morphometric parameters of meander bend viz. radius, width, bend curvature channel width ratio, amplitude, length, wavelength and sinuosity were taken into consideration. The actual, maximum, mean, percentage change, as well as coefficient of variation (CV) of each bend parameter were calculated to show the pattern of spatial and temporal change.

4.4 Result

The sequential changes of the the channel planform of the Ramganga river can be observed in Fig. 4.6. Over the assessed period, increasing tendency of braiding was noticed. However, the channel also contained several well-developed meander bends that can be observed in each of the assessed period. Here, the river designated the characteristics of the partly braided channel.

4.4.1 Active Channel Width (ACW)

The active channel of a river refers to the portion of a river whereby the consistent flow of water occurs. The highest mean ACW (331.9 m) was observed in 1923, but it decreased considerably in the subsequent periods. The maximum reduction in ACW was noted in 1981–1993 period (−38.37%). Over the assessed periods, the coefficient of variation (CV) of ACW showed a scarce difference, suggesting a lower rate of spatial variation.

4.4.2 Active Channel Total Channel Width Ratio (ACTCWR)

It is a very important parameter for characterizing a channel, specifically to recognize whether the channel is aggrading or degrading or transport-dominated. The lower value of this ratio indicates lack of discharge within a channel creating lower channel competence. Therefore, the channel

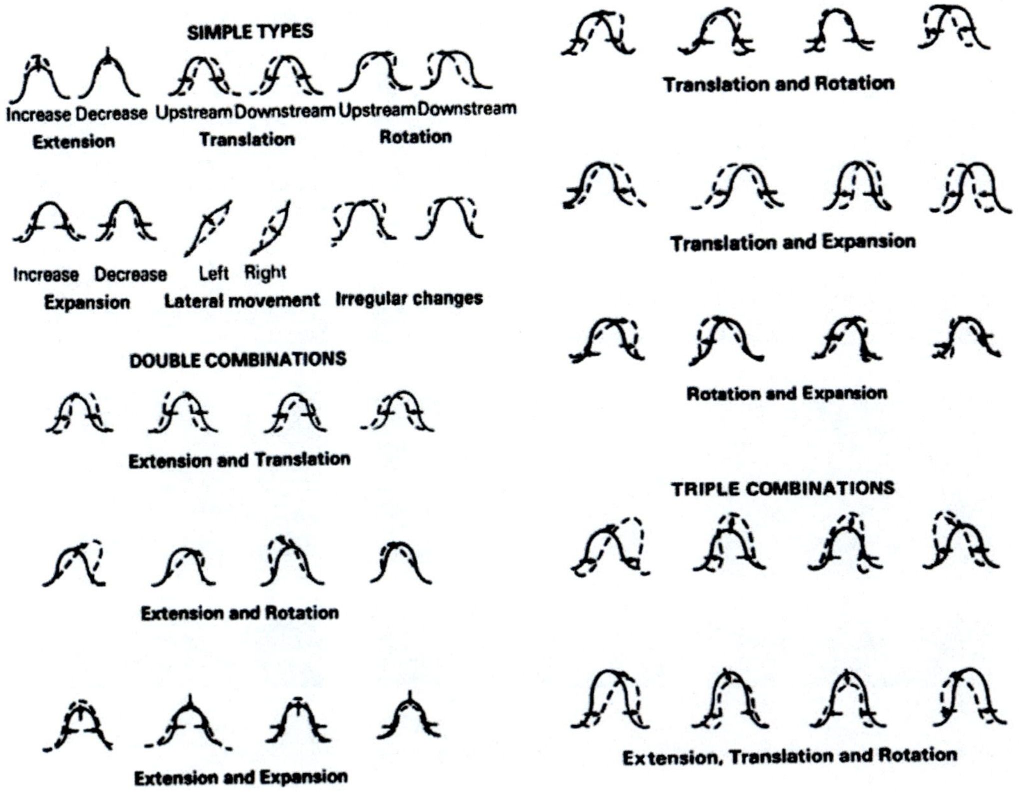

Fig. 4.5 Modes of meander change based on Hooke (1984)

aggradation can be observed within the channel. The higher value (close to 1) suggests that the river utilizes its whole channel area for maintaining the flow. This situation may arise during bankfull stage of the river or in a transport-dominated channel.

The mean value of ACTCWR showed some notable decrease in 1923–1972 (Fig. 4.7). However, in the succeeding periods, no specific changes in ACTCWR were experienced. On the other hand, the spatial variation of this ratio became more prominent over the studied period. The CV of the ACTCWR ranged from 0.51 (1993) to 0.424 (2003).

4.4.3 Channel Length (CL)

The channel length (CL) refers to the longitudinal distance of the channel between any two points. The channel length varies temporally as well as spatially in the downstream direction. It increases with the increasing curvature of the channel. In contrast, when the river straightens its course through cut off, the length of the channel declines. The avulsion of the river course in the alluvial reach also causes significant variation in the channel length.

In the present study, the highest channel length was observed in 1923 (40.97 km) but it showed remarkable decrease in the subsequent periods (Fig. 4.7a). In 1923, the Ramganga channel exhibited a highly meandering course. Therefore, the channel length became maximum than other assessed periods. In 1972, the channel was straightened by cut off, although several new meander bends were also evolved herewith resulting 10.01 km reduction in the CL. During 1923–2014, it decreased by 5.8 km at the rate of 61.37 m/year. Within the decadal time scale,

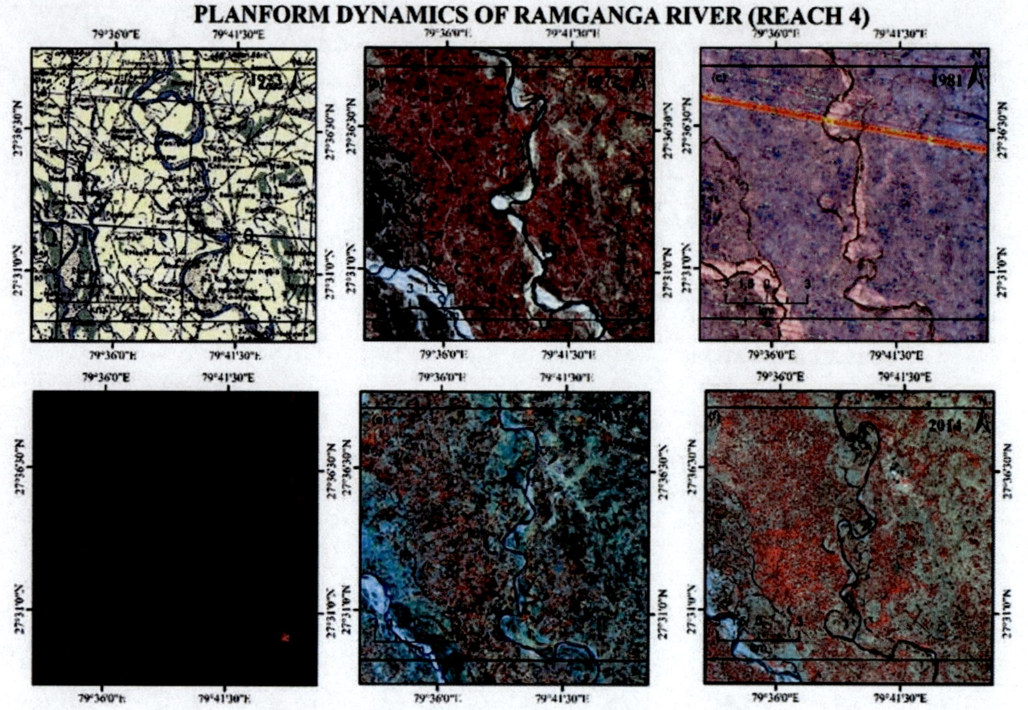

Fig. 4.6 Sequential changes of Channel Planform of the Ramganga River

significant reduction in the length of the course was observed in 1993–2003 (2.988 km), whereas substantial increase in CL (3.72 km) was noted in 2003–2014 as a consequence of complex evolution of the meander bend. The CL decreased sharply (−3.19 km) during 1981–1993 while a remarkable increase in CL (6.73 km) was registered in 1993–2014.

4.4.4 Channel Sinuosity (CS)

The natural river has an inherent tendency to flow in a curved path. The Ramganga river here exhibits a meandering course. The channel sinuosity of this reach remained remarkably higher in 1923 (2.26). Throughout the period of observation, it reduced by -15.32%. The channel sinuosity drastically subdued (−23.91%) in 1923–1972. The similar trend was continued up to 1981–1993. Since 1993, a considerable increase in CS was experienced. As a result of that, the CS reached to 1.92 in 2014 (Fig. 4.7b).

Along all the sections, the fluctuations in CS were convincingly consistent. The coefficient of variation was maximum for Sect. 4.6 (0.42), followed by Sect. 4.2 (0.41). It was minimal for Sect. 4.5 (0.22).

4.4.5 Braiding Index (BI)

This index is used to designate the channel pattern. It is a very useful parameter to recognize the nature of braiding of a channel. The significant increase in BI was observed between 1923 and 1972 (51.63%). In this period, the river transformed from partly braided to the moderately

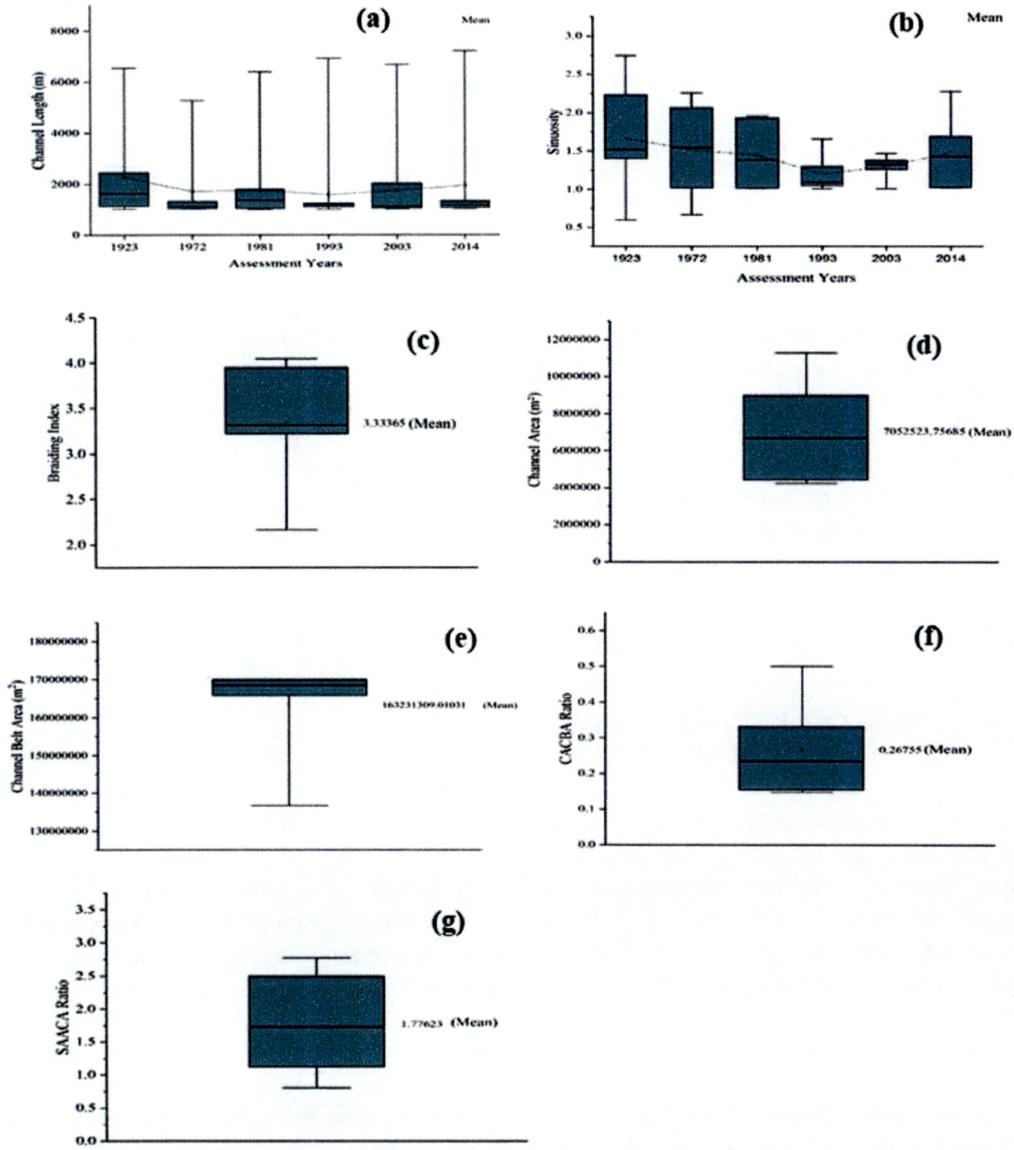

Fig. 4.7 Depicts dynamics of Channel planform parameters: (**a**) channel length; (**b**) sinuosity; (**c**) braiding index; (**d**) channel area; (**e**) channel belt area; (**f**) CACBA ratio; (**g**) SAACA ratio

braided. During 1972–1993, it also exhibited the attributes of moderately braided streams. However, in 2003–2014, the BI value indicated higher braiding intensity of the channel. The BI depicted a remarkable increase (82.45%) during the period of 1923–2014 (Fig. 4.7c).

4.4.6 Active Channel Area (ACA)

It refers to the total area of a channel that comes under the region of consistent water flow. Over the assessed periods, the active channel area contracted by −62.51%. Since 1923 to 1993, the

ACA contracted persistently, however in 1993–2003, a little increment was noted (20.53%). Although in the subsequent period, it decreased significantly (Fig. 4.7d). The most notable change in CBA was witnessed in 1981–1993 (−45.12%).

4.4.7 Channel Belt Area (CBA)

It is denoted by the area of overall lateral extension of the river channel within a valley floor. Since 1923, the Channel Belt Area had been increasing (Fig. 4.7e). It indicated lateral instability of the channel. In 1923, the CBA was 13, 66, 91,466.79 m^2 but it increased up to 16, 99, 74,902.46 m^2 in 2014. Throughout assessed periods, the CBA increased by 24.35%. However, decadal variations of the CBA were not so prominent.

4.4.8 Active Channel Area-Channel Belt Ratio (ACACBR)

The ACACBR expresses the proportion of total channel belt area inundated by continuous stream flow which discourages the active presence of the vegetation. Since 1923–2014 period, the ACACBR decreased by −70.49% (Fig. 4.7f). Not only this, the variation of ACACBR was also sharp. The highest value of this ratio was noted in 1923 (0.50), while it was lowest in 2014 (0.15). This observation revealed that the channel narrowing phenomenon had been increased throughout the assessed periods. The most remarkable temporal variations of this ratio were observed in 1981–1993 (−45.69%) followed by 1923–1972 (−33.40%). The only positive change was noted in 1993–2003 (20.07%).

4.4.9 Sandbar Number and Types

In all the evaluated periods, significant variation in the sand bar area was noticed. During 1923–1972, the area of the sandbar increased by 49.22%. On the other hand, in 1972–1981, it decreased by -33.50%. In 1923, the scroll bars only occupied around 58.55% to the total sand bar area (Table 4.2). In 1972, one large-sized concave bar was formed which accounted for 22.47% of the total sandbar area. Although the number of sandbar remained same in 1923 and 1972, the sand bar area increased. During 1981–1993, the total sand bar area increased by 35.94%. In 1993, three large-sized scroll bars were formed incorporating 43.66% of total area. Around 54.30% area was covered by point and sidebars. The number of sand bar increased from eighteen (1981) to thirty-one (1993). Several new mid-channel bars were formed that aggravated the total number of the sand bar in 1993.

The number of sand bar increased again in 2003, although the area of the sand bar decreased by -14.95%. The point bar and scroll bar occupied around 78.32% of the total sand bar area. In 2014, five scroll bars formed covering 45.61% of the total sand bar area.

4.4.10 Sandbar Area Active Channel Area Ratio (SAACAR)

It is defined as the ratio between total sandbar area and total channel area. It expresses the channel contraction or expansion as well as the pattern of channel aggradation. Along this reach, the channel area became higher than the sandbar area only in 1923 (0.81). The highest value of this ratio was observed in 1993 (2.77) (Fig. 4.7 g). From this period, a dramatic increase in the sandbar area was observed. As a result of that, degree of braiding in the channel increased manifolds.

4.4.11 Channel Centreline Migration

In this reach, the maximum annual centreline migration was noted in 1972–1981 with the rate of 42.96 m/year. It was followed by 1993–2203 (31.94 m/year). The considerably lower rate of centreline migration (~12.0 m/year) was registered in 1923–1972 and 1972–2014. Nonetheless,

Table 4.2 Sandbar number and area

Sand bar types	Sand bar number and area in assessment years											
	1923		1972		1981		1993		2003		2014	
	Number	Area (%)	Number	Area (%)	Number	Area (%)	Number	Area (%)	Number	Area (%)	Number	Area (%)
Point bar	6	19.69	3	25.36	5	41.33	7	30.08	6	38.71	7	36.25
Side bar	3	18.18	3	14.47	4	25.50	5	24.22	8	17.57	8	13.03
Scroll bar	1	58.55	5	36.83	3	23.71	3	43.66	4	39.61	5	45.61
Mid-channel bar	3	1.45	3	0.88	2	1.63	12	0.77	11	1.62	6	0.52
Concave bar	1	0.76	1	22.47	2	3.95	0	0	3	1.48	3	3.17
Longitudinal bar	1	1.37	0	0	2	3.87	4	1.27	6	1.01	6	1.41

on the decadal scale, the annual rate of migration was appreciably higher. During 1982–1993 and 2003–2014, the channel centreline migration remained ~25.0 m/year (Fig. 4.8).

4.4.12 Bank Line Migration

During 1923–1972, the trend of eastward migration of both of the banks was quite evident (Figs. 4.9 and 4.10). The average eastward shifting of the left and right bank was recorded as 28.41 m/year and 34.26 m/year, respectively. In the same period, the average annual westward migration of the right bank also became higher than the left bank (1.51 times). In 1972–1981, the average annual westward migration rate increased momentously than the preceding period. In the same period, the average annual westward migration of the right and left bank accounted for 122.94 m/year and 143.96 m/year, respectively. The maximum limit of the westward migration recorded as 3,020.42 m for the right and 3,808.056 m for the left bank. The eastward extension of the right bank was also pronounced (139.94 m/year) but for the left bank it was quite lower.

The maximum migration of the right and the left bank was registered 5,314.51 m and 1,582.05 m, respectively. In the period of 1981–1993, the rate of westward bank line migration reduced significantly, and the rate was also declined for eastward migration of the right bank. In this period, the average rate of eastward migration was higher (2.46 times) as opposed to the westward migration of the right bank. In the left bank, the pattern of migration along both sides was quite similar. In 1993–2003, the westward migration of the right bank showed quite significant increase.

At the same time, the average eastward migration of the left bank showed remarkable

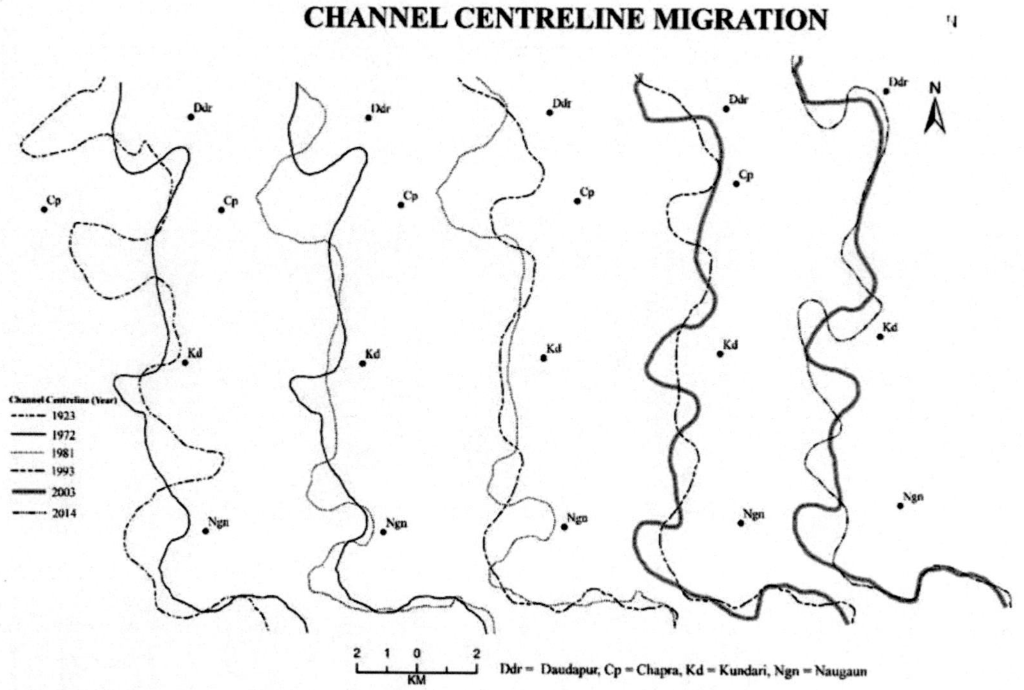

Fig. 4.8 Channel centreline migration

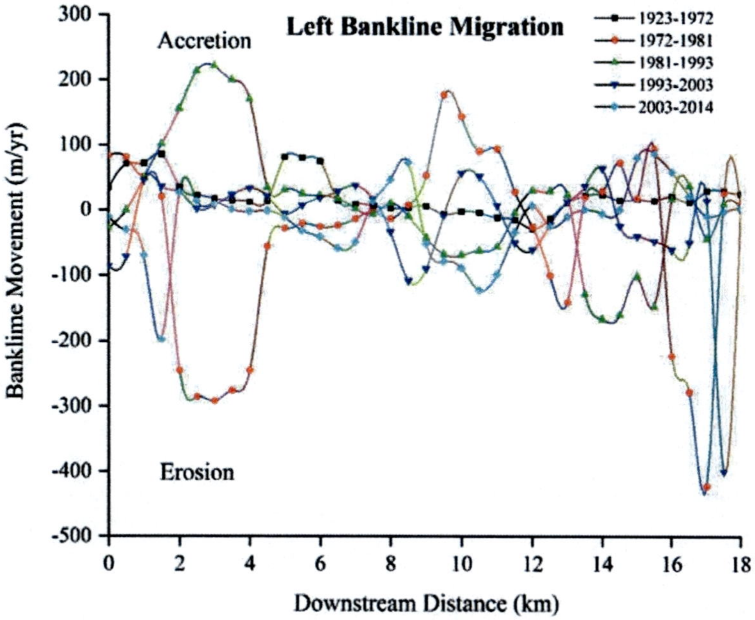

Fig. 4.9 Left Bankline migration

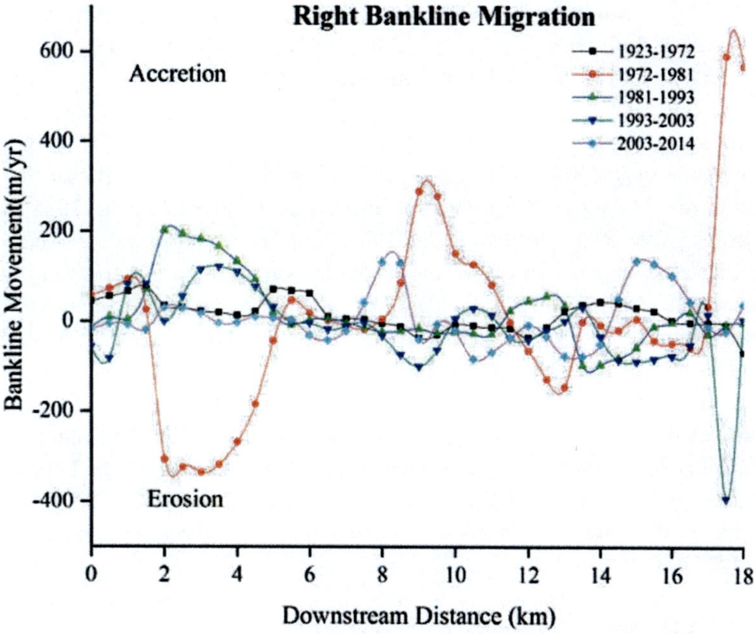

Fig. 4.10 Right Bankline migration

decrease than the previous decade. In 2003–2014, the westward migration of both the banks increased momentously. Although considerable increase in the eastward migration was evident in both of the banks.

4.4.13 Meander Bend Dynamics

Along this reach, ten meander bends had been evolved (Fig. 4.11). The quantitative assessment of the meander bend parameters was discussed spatio-temporally.

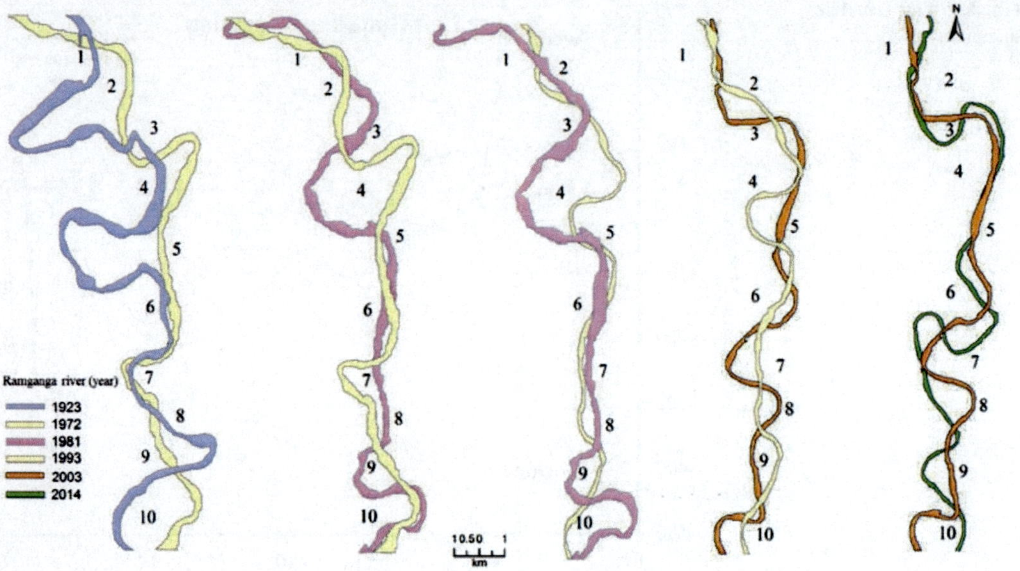

Fig. 4.11 Meander bend evolution

4.4.13.1 Meander Bend Radius (Rmb)

The mean Rmb ranged from 915.72 m (1993) to 539.52 m (2014). The value of Rmb decreased once the meander bend underwent progressive growth. It indicated the formation of high-curvature bends. The mean Rmb increased from 1923 to 1972 suggesting reduction in bend sinuosity. The Rmb decreased in 1981, but a remarkable increase in the mean Rmb was observed in 1993 owing to bend cut off and development of new meander bends. In the subsequent period, there was a rapid reduction in the mean Rmb (Fig. 4.12a). Among the ten meander bends, the coefficient of temporal variations of Rmb was more significantly observed in bend 8–10 (CV ~0.672), whereas a little amount of change was witnessed in bend 3, 6 and 7 (CV ~0.26).

4.4.13.2 Meander Bend Width (Wmb)

Over the assessed period, the Wmb did not show any remarkable spatio-temporal variations (Fig. 4.12b). The mean Wmb ranged from 287.30 m (1972) to 161.97 m (1993). The considerable increase in the CV was registered in 1981 (0.411) and 2003 (0.383). Moderate range of CV (~0.38) was observed on all the assessed meander bends.

4.4.13.3 Meander Bend Curvature Channel Width Ratio (Rcw)

From 1923 to 1981, the mean Rcw was considerably lower (~2.68). In 1993, there was a rapid increase in this ratio from 2.68 in 1981 to 4.76 in 1993. It decreased in the succeeding period, however, it increased to 4.17 in 2003–2014 (Fig. 4.12c). The highest range of CV was observed in 1993 (0.725), followed by 2003 (0.641), in other periods it varied ~0.44. Among all the bends, the Rcw varied from 4.69 (bend 8) to 2.50 (bend 5). The CV registered highest for bend 8 (0.979), followed by bend 4 (0.836). In all other bends, moderate to higher coefficient of variations were found.

4.4.13.4 Length of Meander Bend (Lmb)

The length of the meander bends had been modified in all of the assessed periods because of frequent bend migration or intensification of the bend growth. The length of the meander bend

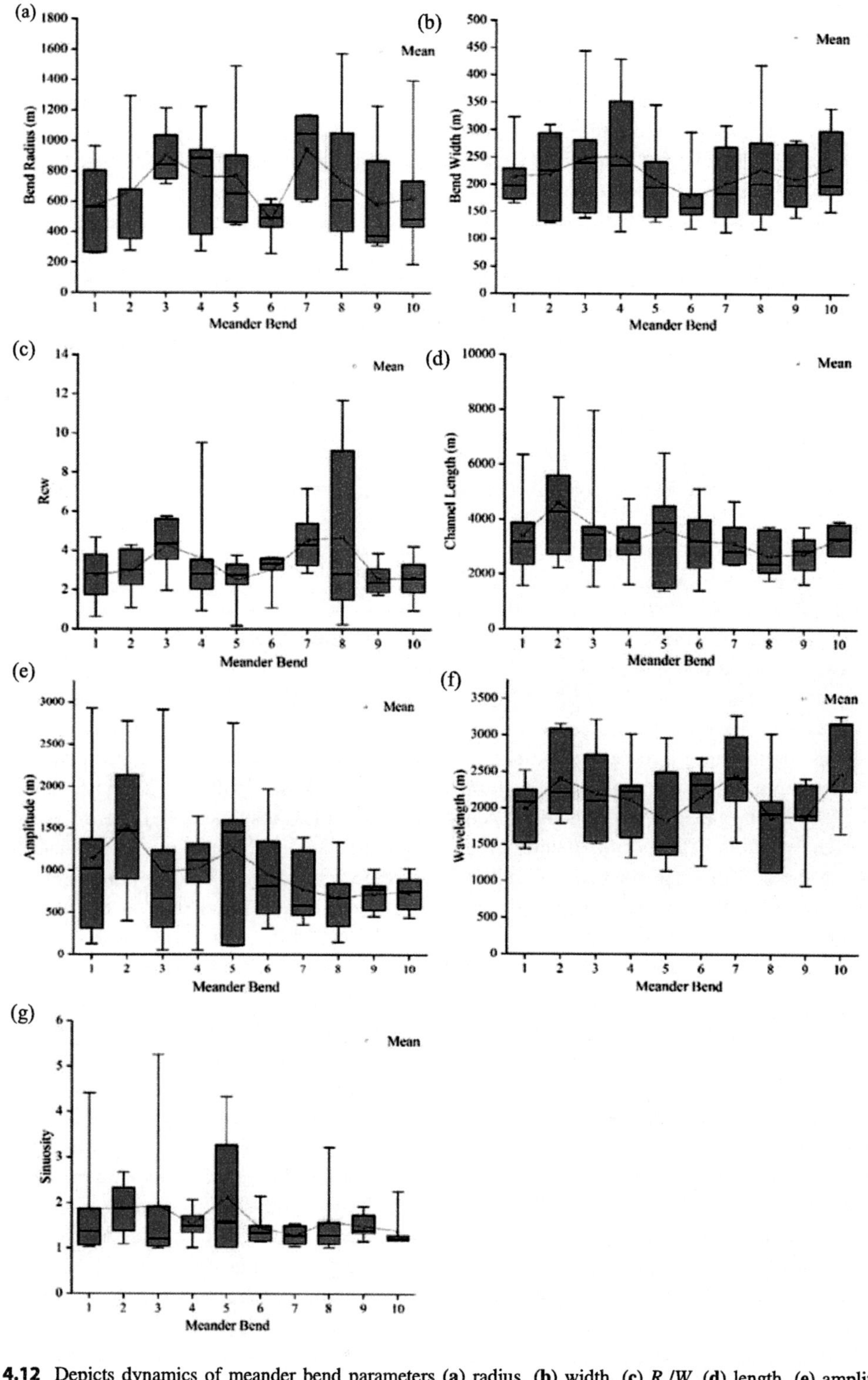

Fig. 4.12 Depicts dynamics of meander bend parameters (**a**) radius, (**b**) width, (**c**) R_c/W, (**d**) length, (**e**) amplitude, (**f**) wavelength, (**g**) sinuosity

registered highest during 1923 (4930.86 m) as most of them appeared as high-curvature bend while it remained lowest in 1981 (2223.19 m) due to straightening of the channel course by cut off and lower bend growth. There was a significant increase (36. 79%) in the Lmb in 1993–2003.

The similar trend continued in the subsequent periods. Among the ten meander bends, the mean Lmb ranged from 4594.43 m (Bend 2) to 2666.51 m (Bend 8) (Fig. 4.12d). The coefficient of variations among all the meander bends registered moderate (~0.413) except for bend 10 (0.18).

4.4.13.5 Meander Bend Amplitude (Amb)

The mean Amb ranged from 1714.12 m (1923) to 454.41 m (1981). From 1923 to 1981, the mean Amb decreased remarkably (Fig. 4.12e). However, in the succeeding period, it started to increase and became 1106.05 m in 2014. The drastic changes in the mean Amb were specifically observed in all of the studied periods. The major changes happened during 1972–1981 and 1923–1981. The lateral expansion of the meander bends aggravated the bend's amplitude. Thereafter, these bends were subjected to cut off. Among the ten meander bends, the mean Amb value ranged from 1527.78 m (bend 2) to 674.49 m (bend 8). Except for bends 9 and 10, very high coefficient of variation was registered (~0.71).

4.4.13.6 Wavelength of Meander Bend (Wmb)

The average Wmb was recorded highest in 1972 (2,426.18 m), while it was lowest in 1981 (1,691.50 m). In 1972–1981 period, an abrupt reduction in the mean Wmb (−30.3%) was observed. Since 1993, appreciably higher amount of mean Wmb was noted (Fig. 4.12f). The CV of Wmb registered higher value in each evaluated period except for 1993 (0.193). Among ten meander bends, the mean Wmb ranged from 2477.57 m (bend 10) to 1820.83 m (bend 5). The coefficient of variations remained moderate. The highest range of variation was observed in bend 5 (0.402).

4.4.13.7 Meander Bend Sinuosity (Smb)

The highest mean value of Smb was recorded in 1923 (2.33) as the entire channel remained highly sinuous that time. From 1923 to 1993, the sinuosity of meander bend decreased considerably. After 1993, the meander bend sinuosity started to increase and it became 1.89 in 2014 (Fig. 4.12g). The mean value of Smb of the ten meander bends depicted that the sinuosity of the meander bends had been constantly changing with time and most of them were heralded with higher sinuosity. Over the assessed period, the meander bend 5 achieved the highest amount of mean sinuosity (2.13), whereas the bend 7 had lower value (1.29). The coefficient of variations among the ten meander bends represented major ranges of variability for the bends 3 (0.855), 1 and 5 (0.698 and 0.639).

4.5 Discussion

The major rivers located in the Indo-Gangetic Plain show remarkable variation in channel morphology, migration and erosion rate which can be witnessed within a short span of time (usually in less than a decade). Hence, a detailed explanation of chronology and sequence of development of the planform parameters of the highly dynamic alluvial river of this region can be very useful in order to designate the mode of river adjustment in response to varying water discharge and sediment load. In the present study, a detailed account of decadal and mid-term scale variation of channel planform and meander bend morphology was evaluated. A careful investigation was made to show the spatial variation of the planform and meander bend parameters for each assessed period. The actual, maximum, mean, per cent change and coefficient of variations were calculated to quantify the chosen parameters and also to express the nature of their dynamicity.

For multi-temporal analysis of planform dynamics, channel migration, bank erosion and accretion assessments, most of the studies have used the images of Landsat series of orbital sensors: MSS, TM, ETM+, Landsat-OLI8 (Kalliola et al. 1992; Peixoto et al. 2009; Baki and Gan 2012). In addition to that, topographical maps, aerial photographs, navigational charts have also been used (Mukherjee et al. 2017; Yunus et al. 2019). Among all the available satellite images, the Landsat data is still considered as the most convenient and economical resources for the aforesaid studies (Peixoto et al. 2009). In the present analysis, all the orthorectified Landsat images with zero per cent of cloud cover were downloaded in winter season. Because during this season, the discharge remains more or less constant over the years. Hence, it provided better results for temporal analysis.

4.5.1 Dynamics of Channel Planform Parameters

After examining the active channel width and active channel area, it was found that the channel narrowing phenomenon has been aggravated widely over the assessed period. It indicates downstream impact of channel impoundment (Grams and Schmidt 2005). Although its rate can be observed with more prominence in the upstream reaches from the present study reach in the western Gangetic plain where lesser number of tributaries meet to the Ramganga river. Here, more than 90% of the major tributary channels contribute water to the Ramganga river, hence it helps to reduce the variability of the water discharge.

Although the active channel narrowing phenomenon was widely observed over the assessed period, the spatial variation of it in downstream direction was widely experienced. The ACTCW ratio remained lower in all of the assessed period. It depicted that the flow did not cover most of the part of the channel. The rivers in the Indo-Gangetic plain with lower ACTCWR value suggests higher probability of river bed and bank erosion due to periodic wetting and drying that loosen river bed and bank materials. When the bank full discharge or channel overflow occur, the stream erodes those loosened particles and deposits in the downstream direction. In each assessed period, significant variations of the channel length was noticed. It expressed the horizontal instability of the channel. The sinuosity of the channel varied remarkably within 91 years of the study period. Although the sinuosity decreased from 1923 to 1981, but it showed a considerable increase from 1993. During 1923, the Ramganga channel exhibited truly meandering channel, however, in due course of time, the intensity of braiding had been increased remarkably. The braiding index reflected that since 1993, the tendency of braiding of the Ramganga river increased substantially.

4.5.2 Changes in Channel Pattern/Growth of Bars and Islands

As the tendency of braiding in Ramganga river had been increased, the size and area of the sandbar increased remarkably. Here, the river flows with heavy sediment load through high-curvature meander bend. As a consequence, the transported sediments mostly dropped in the convex side of the bank which eventually formed point bar. In the present study reach, large mid-channel bar also came into existence. As the meander bend grew in the downstream direction by the extension process, the point bar formed in the convex side of the bend. According to Mertes et al. (1995, 1996), point bars are extremely unstable and migrate laterally when intercepted by secondary channels. Consequently, the migration of point bars leads to the formation of scroll bar. During flood, the Ramganga river often creates small chute and join the parent channel on the opposite side of the bend. With the progression of time, several scroll sloughs have been produced which eventually form new scroll bars. Throughout the assessed periods, the concentration of the scroll bars was widely discerned. Most of them were larger in size and occupied major areas of the total sand bar area. Due to progressive shifting, partial and chute cut off of the channel, the area of the sand bar modified effectively.

The SAACAR ratio gives a detailed account of the proportion of the deposited sediments within a river channel. The variation in this ratio is determined by the several conditions: if the total wetted channel area increases at the expanse of sand bar area, the value of this ratio will be below 1. On the contrary, if the active channel area decreases and the sandbar area increases, SAACAR value will be more than 1. Throughout the study period, this ratio became more than 1 only in 1923. Hence, it can be said that in all other periods, the area of the sandbar remained higher than the active channel area.

4.5.3 Trend of Channel Migration

At decadal scale, the incidence of channel centreline migration was significantly higher, however, in mid-term scale (for the period of 1923–1972 and 1972–2014), considerably lower amount of migration was noticed. After multi-temporal analysis of the pattern of channel shifting, it had become evident that the channel mostly oscillated either in east or westward direction. It was observed that at mid-term scale, the channel drifted its position towards the older channel, hence a little amount of centreline migration was noticed. However, in decadal scale, the rate of shifting towards the previous channel was quite lower which caused greater shifting of the channel centreline.

4.5.4 Pattern of Bank Line Migration

The sequential analysis of the bank line migration revealed that along the left bank, the eastward erosion was prevalent in most of the assessed periods. During 1923–1972, this phenomenon was observed along 32 transects. The highest rate of eastward migration of the channel was registered in 1981–1993 (65.95 m/year). The westward migration was associated with accretion predominated over erosion in 2003–2014 (24 transects) and 1972–1981 (19 transects). For the right bank, the westward migration predominated only in 1923–1972 (23 transects). In other assessed periods, westward migration of the bank became significantly higher which was associated with bank erosion. The highest annual rate of the migration was registered in 1972–1981 (122.94 m/year).

4.5.5 Role of Flood in Channel Planform Change and Channel Migration

The lower reach of the Ramganga river is located in western Gangetic plain. The rivers flowing in this region are characterized by higher stream power and carry lower sediment load than the river of the Eastern Gangetic Plain (Sinha et al. 2005). The rivers of the western Gangetic plain including Ramganga are incisional in nature. During monsoon, heavy water discharge with lower bedload, higher degree of slope causes very high stream power which causes severe erosion of the river bank. In the study reach, the medium to high magnitude flood had been experienced in decadal time period. It caused frequent shifting of the channel and and meander bends.

4.5.6 Role of Dam on Channel Planform

The Kalagarh dam was constructed on the Ramganga at Himalayan foothill region for irrigation and power supply. The dam has been operating since 1974. In the present study, toposheet in 1923 and Landsat image of 1972 were available for pre-dam period and rest of the images for the post-dam. In the post-dam period, the study was conducted by maintaining decadal scale (almost). But, it can be asserted from the present investigation that the Kalagarh dam did not make any hindrances in the rate of dynamicity of the channel in terms of its morphology and horizontal position, meanwhile, it drives to increase the rate.

4.5.7 Evolution of Meander Bend Parameters

Over the assessed periods, 61.67% of the meander bend radius was below 750 m. Only for two times, the Rmb appeared below 250 m. More than 1000 m Rmb was experienced in 21.67% of the evaluated bends. It was experienced that the bends below 750 m of Rmb were associated with progressive growth of bends which eventually transformed into symmetric bends.

The significant reduction in mean bend width was witnessed from 1923–2014. In 1923 and 1972, more than 70% of the bends were observed with >250 m of Wmb. However, since 1981, the bend width drastically reduced. In 2014, 80% of the bend width was experienced with less than 150 m. It expressed considerable decrease in discharge within the assessed period.

For about 56% of the assessed bends, the Rcw value ranged from 2 to 4 and 78.33% of them remained below 4 Rcw value. The present study found that the most significant cut off occurred in the bends with the range of 2–4 (Rcw).

The amplitude of the bend reflects the progressive growth of the meander bends in the downstream direction. Around 2/3 of the bends registered more than 500 m bend amplitude. Around 56.67% of the bend amplitude existed within the range of 500–1500 m. Only nine bends showed more than 1500 m bend amplitude. Out of which, several bends were developed in 1923–1972 and the rest of them in 2014.

The length of the bends also did not remain constant over the assessed period. It expressed unstable nature of the channel. Around 65% of the bend existed within 2000–4000 m of channel length. More than 4000 m channel was observed for 13 bends during 1923–1972 and 2003–2014.

For around 78.63% of the meander bends, the wavelength was lying within the range of 1500–2500 m. Only 6% of them were below 1500 m of length. It was observed that where the bend was considerably lower, the chance of chute or neck cut off was greater. On the contrary, partial cut off or channel avulsion predominated in the higher bend wavelength.

The less sinuous (CS 1.00–1.25) to sinuous bends (CS 1.25–1.50) were predominantly observed from 1972 to 1993. Since 2003, the tendency of meandering had been increased. The eight bends became highly sinuous (CS > 1.5) in 2014. Over the assessed periods, extremely high meandering bends were noticed in 1923 (2) and 1972 (1). The changes in the bend sinuosity in a short interval expresses the dynamic nature of the channel. The meanders did not get enough time to grow up. With increasing sinuosity, the bends became more vulnerable to cut off. Moreover, few bends evolved as high-curvature meander bend, most possibly due to some local resistance imposed by bank materials or by the progressive growth of the point bars.

4.5.8 Mechanism of Change in Meander Bend Morphology

In the present reach, a significant number of the compound meander bends consistently persisted in all of the evaluated periods. In 1923, six bends (out of ten bends) exhibited the characteristics of compound bends. In 1972, all the meander bends (except bend 9) became simple types. Among them, three bends became simple asymmetric whereas three bends became straight. During 1923–1972, the cut off and lateral migration of the channel was observed in the bends 1 to 9 causing asymmetry of the bends, and the translation process had operated in the bend 10 (Table 4.3). In 1981, growth of compound bend had been aggravated than the preceding periods. Among the compound bends, the majority of them are characterized by symmetry types of bend suggesting progressive growth of the bends. The channel had become straight in the bends 3–5. During 1972–1981, 1–7 bends became highly unstable owing to lateral shifting of the channel and formation of new meander bends. In the bends 8 and 9, the extension process, meanwhile in the bend 10, the rotation process was actively involved in meander bend evolution. In 1993, the number of simple bends increased than the previous periods, and the compound bends were

Table 4.3 Changes in in Meander Bend morphology and the mechanisms involved in the meander bend evolution

Bend	1923 shape	1972 shape	1923–1972 shape change	1981 shape	1971–1981 shape change	1993 shape	1981–1993 shape change	2003 shape	1993–2003 shape change	2014 shape	2003–2014 shape change
1	SS	SA	Cut off and new bend development	CS	Cut off and new bend development	SS	Cut off and new bend development	CS	New loop	SA	Rotation and extension
2	SA	SS	Confined migration	CS	Cut off and new bend development	SS	Cut off and new bend development	SS	Cut off and new bend development	CA	Translation and rotation
3	CS	S	Cut off	S	Lateral Migration	SS	Cut off and new bend development	S	Cut off	CA	Rotation
4	CA	SA	Lateral Migration	S	Lateral Migration	SS	Extension	SA	Triple combination	CS	Triple combination
5	SA	SS	Confined migration	S	Cut off	SA	Lateral migration	SS	New loop	CA	Complex change
6	CS	SA	Cut off	SA	Cut off	SS	Lateral migration	SS	Complex change	SA	Cut off
7	CS	SS	Lateral Migration and new bend formation	CS	New Bend formation	SA	Cut off and new bend development	SA	Cut off and new bend development	CS	New loop
8	CA	SA	Cut off	CS	Extension (Increase)	SA	Cut off and new bend development	SA	Confined migration	SA	Cut off and new bend development
9	CA	CA	Cut off	CS	Extension (Increase)	CA	Lobing	SA	New loop	CA	Translation
10	SA	SA	Translation	CA	Rotation	CA	Lobing	CA	Rotation	CA	Translation

only observed in the bends 9 and 10. Except for the bends 4, 9 and 10, the morphological changes of the meander bends occurred in all other bends due to cut off, lateral migration and development of new bends. In the bends 9 and 10, lobing process gave rise to compound asymmetric bends. The simple types of meander bends also predominated in 2003. During 1993–2003, cut off and new loop formations were witnessed in most of the bends. The complex change and triple combination processes had operated in the bends 6 and 4, respectively, whereas the bend 10 experienced the rotation. In 2003–2014, the meander bend migration was relatively constrained compared to all other evaluated periods. During this period, the compound bend development was prevalent than the previous periods. The seven bends were characterized by compound bends. In this period, the translation and rotation processes were actively involved in meander morphological change. The single process operated in the bends 3 (rotation), 7 (new loop), 9 and 10 (translation). The triple combination and complex mechanisms were responsible for the morphological change in the bends 4 and 5, respectively.

4.6 Conclusion

The purpose of this study was to quantify the dynamics of channel planform and meander bend metrics ~37 km reach of the Ramganga river by using topographical map and LANDSAT images acquired between 1923 and 2014. The study was integrated with GIS that provide an objective evaluation of spatio-temporal changes of various attributes of channel planform and meander bend.

In the study reach, the Ramganga river is very dynamic in nature. The most of the parameters chosen for the present study underwent remarkable variations. Over the assessed periods, the active channel width, ACTCWR, channel length, channel area, ACACBR, SAACAR were reduced dramatically. On the contrary, the channel belt area, sand bar number and sand bar area showed significant increase. After examining the temporal trend of all planform parameters, it could be mentioned that over the assessed periods the active channel narrowing phenomenon had been aggravated substantially. On the other hand, the channel aggradation and braiding tendency of the channel increased remarkably.

Most of the meander bends of the study reach never remained stable within the assessed periods. The bend morphology changed substantially in response to constant river adjustment with the available stream discharge and sediment load. Due to progressive shifting of the channel, the meander bends did not get sufficient time to evolve as high-curvature bends. Although, few bends got favourable condition to grow, most probably by the local resilience of the fluvial adjustment. Over the assessed period, it was observed that considerably greater number of high-curvature bends were developed in 1923, 2003 and 2014 periods. However, since 1972 to 1993, cut off and local channel avulsion created hindrance for the bend growth. The increasing stability of highly sinuous bends were found since 2003. The cut off and lateral migration processes were responsible for the changes in the meander bend morphology for about 60% of the total assessed bends. Around 12% of the bends were associated with the further complex evolution of the bends.

References

Agnihotri AK, Ohri A, Mishra S (2020) Channel planform dynamics of lower Ramganga River, Ganga Basin, GIS and remote sensing analyses. Geocarto Int 35(9):934–953

Anderson MG, Calver A (1980) Channel plan changes following large floods. In: Cullingford RA, Davidson DA, Lewin J (eds) Timescales in geomorphology. Wiley, Chichester, pp 43–52

Ashmore P, Church M (2001) The impacts of climate change on rivers and river processes in Canada. Geological Survey of Canada Bulletin, Ontario: Ottawa

Baki ABM, Gan TY (2012) Riverbank migration and island dynamics of the braided Jamuna River of the Ganges-Brahmaputra basin using multi-temporal Landsat images. Quat Int 263:148–161

Bawa N, Jain V, Shekhar S, Kumar N, Jyani V (2014) Controls on morphological variability and role of stream power distribution pattern, Yamuna River, western India. Geomorphology 227:60–72

Brice, J.C.(1964). *Channel patterns and terraces of the Loup rivers in Nebraska.* United States Geological Survey Professional Paper, P 0422-D: pp. D1–D41.

Chang HH (2008) River morphology and river channel changes. Trans. Tianjin Univ., 254–262

Feng X, Zhang G, Yin X (2011) Hydrological responses to climate change in Nenjiang River basin Northeastern China. Water Resour Manag 25:677–689

Ghimire M (2020) Basin characteristics, river morphology, and process in the Chure-Terai landscape: a case study of the Bakrahariver, East Nepal. Geogr J Nepal 13:107–142

Ghimire S, Higaki D (2015) Dynamic river morphology due to land use change and erosion mitigation measures in a degrading catchment in the Siwalik Hills Nepal. Int J River Basin Manag 13(1):27–39. https://doi.org/10.1080/15715124.2014.963860

Grams PE, Schmidt JC (2005) Equilibrium of indeterminate? Where sediment budgets fail: sediment mass balance and adjustment of channel form, Green River downstream from Flaming Gorge Dam Utah and Colorado. Geomorphology 71(1):156–181

Grant GE, Schmidt JC, Lewis SL (2003) A geological framework for interpreting downstream effects of dams on rivers. In: O'Connor JE, Grant GE (eds) A Peculiar River: geology, geomorphology, and hydrology of the Deschutes River, Oregon. American Geophysical Union, Washington DC, pp 203–219

Gurnell AM, Downward SR, Jones R (1994) Channel planform change on the River Dee meanders, 1976–1992. Regul Riv 9:187–204

Gurnell AM (1997) Channel change on the River Dee meanders, 1946–1992, from the analysis of air photographs. Regul Rivers Res Manag 12:13–26

Hickin EJ, Nanson GC (1984) Lateral migration rates of river bends. J Hydraul Eng-ASCE 110:1557–1567

Hogan DL, Luzi DS (2010) Channel geomorphology: fluvial forms, processes, and forest management effects. In: Compendium of forest hydrology and geomorphology in British Columbia, vol 1, pp 331–372

Hooke J (1984) Changes in river meanders—a review of techniques and results of analyses. Prog Phys Geogr 8:473–508

Hooke J (2003) River meander behaviour and instability: a framework for analysis. Trans Inst Br Geogr 28:238–253

Hooke JM (1977) The distribution and nature of changes in river channel patterns: the example of Devon. In: Gregory KJ (ed) River Channel Changes. Wiley UK, Chichester, pp 206–220

Hooke JM (1980) Magnitude and distribution of rates of river bank erosion. Earth Surf Proc Land 5:143–157

Hooke JM (1995) Processes of channel planform change on meandering channels in the UK. In Gurnell A, Petts GE (eds) Changing river channels. Chichester, Wiley, pp 87–116

Hooke JM (2007) Complexity, self-organization and variation in behaviour in meandering rivers. Geomorphology 91:236–258

Hooke JM, Redmond CE (1992) Causes and nature of river planform change. In: Billi P, Hey RD, Thorne CR, Tacconi P (eds) Dynamics of Gravel-bed Rivers. Wiley, Chichester, pp 557–571

Hooke JM, Yorke L (2010) Rates, distributions and mechanisms of change in meander morphology over decadal timescales: River Dane UK. Earth Surf Process Landforms 35(13):1601–1614

Kalliola R, Salo J, Puhakka M, Rajasilta M, Häme T, Neller RJ (1992) Upper Amazon channel migration: Implications for vegetation disturbance and succession using bitemporal Landsat MSS images. Naturwissenschaften 79:75–79

Karwan DL, Allan JD, Bergen KM (2001) Changing near-stream land use and river channel morphology in the Venezuelan Andes. J Am Water Resour Assoc 37 (6):1579–1587

Khan A, Rao L, Yunus AP, Govil H (2018) Characterization of channel planform features and sinuosity indices in parts of Yamuna River flood plain using remote sensing and GIS techniques. Arab J Geosci 11:525

Khan AU, Rawat BP (1992) Quaternary geology and geomorphology of a part of Ganga basin in parts of Bareilly, Badaun, Shahjahanpur and Pilibhit district, Uttar Pradesh. Hyderabad: G.S.I.

Kiss T, Blanka V (2012) River channel response to climate and human induced hydrological changes: case study on the meandering Hernad River, Hungery. Geomorphology 175–176:115–125

Knighton D (1984) Fluvial forms and processes: a new perspective. Arnold, London

Leopold, L.B., Wolman, M.G. (1957). River Channel Patterns: Braided, Meandering and Straight. US Geological Survey Professional Paper 282-B, Washington DC, 39–85

Lewin J (1983) Changes of channel patterns and floodplains. In: Gregory KJ (ed) Background to paleohydrology. Wiley, Chichester, pp 303–319

Mertes LAK, Daniel DL, Melack JM, Nelson B, Martinelli A, Forsberg BR (1995) Spatial patterns of hydrology, geomorphology, and vegetation on the floodplain of the Amazon River in Brazil from a remote sensing perspective. Geomorphology 13:215–232

Mertes LAK, Dunne T, Martinelli LA (1996) Channel-floodplain geomorphology along the Solimões-Amazon River Brazil. Geol Soc Am Bull 108 (9):1089–1107

Micheli ER, Larsen EW (2011) River channel cutoff dynamics, Sacramento River, California, USA. River Res Applic 27:328–344

Montgomery DR, Buffington JM (1997) Channel-reach morphology in mountain drainage basins. Geol Soc Am Bull 109:596–611

Morisawa M (1985) Rivers. Forms and processes. Longman, London

Mukherjee R, Bilas R, Biswas SS, Pal R (2017) Bank erosion and accretion dynamics explored by GIS techniques in lower Ramganga river, Uttar Pradesh. India. Spat Inf Res 25(1):23–38. https://doi.org/10.1007/s41324-016-0074-2
Nanson GC, Knighton A (1996) Anabranching Rivers: their cause. Character and classification. Earth Surf Process Landf 21:217–239
Peixoto JMA, Nelson BW, Wittmann F (2009) Spatial and temporal dynamics of river channel migration in central Amazonian white-water floodplains by remote sensing techniques. Remote Sens Environ 113:2258–2266
Petts GE (1989) Historical analysis of fluvial hydrosystems. In: Petts GE, Muller H, Roux AL (eds) Historical change of large Alluvial Rivers: Western Europe. Wiley, Chichester, pp 1–18
Roy NG, Sinha R (2007) Understanding confluence dynamics in the alluvial Ganga-Ramganga valley, India: An integrated approach using geomorphology and hydrology. Geomorphology 92(3–4):182–197
Roy N (2009) Morphodynamics and late quaternary valley-interfluve statigraphy of the western Ganga Plains: Rhythmic sequences and monsoon forcing. Unpublished PhD thesis. Dept. of Civil Engineering, IIT Kanpur
Schumm SA (1981) Evolution and response of the fluvial system, sedimentologic implications. Water Resources Publications, Lit-tleton, CO.
Sinha R, Jain V, Babu GP, Ghosh S (2005) Geomorphic characterization and diversity of the fluvial systems of the Gangetic Plains. Geomorphology 70(3–4):207–225
Srivastava P, Shukla UK, Mishra P, Sharma M, Sharma S, Singh IB (2000) Luminescence chronology and facies development of Bhur sands in the interfluve region of central Ganga Plain, India. Curr Sci 78:498–503
Surian N (1999) Channel changes due to river regulation: the case of The Piave River, Italy. Earth Surf Proc Land 24:1135–1151
Trimble SW, Cooke RU (1991) Historical sources for geomorphological research in the United States. Profess Geogr 43:212–228
Vörösmarty CJ, Meybeck M, Fekete B, Sharma K, Green P, Syvitski JPM (2003) Anthropogenic sediment retention: major global impact from registered river impoundments. Glob Planet Chang 39:169–190
Williams GP, Wolman MG (1984) Downstream effects of dams on Alluvial Rivers. US Geological Survey Professional Paper 1286, Washington DC, pp 1–83
Winterbottom SJ (2000) Medium and short-term channel planform changes on the River Tay and Tummel, Scotland. Geomorphology 34:195–208
Wishart D, Warburton J, Bracken L (2008) Gravel extraction and planform change in a wandering gravel-bed river: The River Wear, Northern England. Geomorphology 94 (1–2):131–152
Yang SL, Milliman JD, Deng B, Zhang XY, Luo XX (2014) Downstream sedimentary and geomorphic impacts of the Three Gorges Dam on the Yangtze River. Earth-Sci Rev 138:469–486
Yunus AP, Dou Jie D, Armugha K, Sravanthi N, Rao LAK, Hao C (2019) Channel migration characteristics of the Yamuna River from 1954 to 2015 in the vicinity of Agra, India: a case study using remote sensing and GIS. Int J River Basin Manag. https://doi.org/10.1080/15715124.2019.1566238

5 Changes of Flow Regime in Response to River Interventions in the Barakar River, India

Sumantra Sarathi Biswas and Padmini Pani

Abstract

The flow regime of a river is an essential aspect of river hydrology. The changes of flow regime impact on peak flood discharge, sedimentation process, seasonal flow pattern, morphology of the rivers. The Barakar River, the main tributaryriver of the Damodar River in the Chhotanagpur plateau and plateau fringe region of India, has a significant flow character that is altered by human interventions. The discharge data of the Barakar River are available from Maithon and Panchet hydraulic stations. Mann–Kendall test depictsa decreasing trend of discharge. However, rainfall has an increasing trend at Maithon, thereby the relationship between discharge and rainfall is very insignificant. In the seasonal variation, a rising trend of water discharge in the pre-monsoon and the post-monsoon season is significant. Conversely, in the monsoonal season, there is an indication of a decrease in the flow over time. The uneven release of water from the dams may be the probable cause for the reduction in monsoon flow discharge. At the Maithon station, a tendency of late shifting of hydrograph indicates that the peak flow season has shifted towards the right side over time. The late release of water from the dams has caused the late shifting of the hydrograph. The annual maximum discharge for both Tilaiya and Maithon tends to decrease across the period of 1980–2013. Nevertheless, the annual minimum discharge of the Maithon shows a tendency to increase. The increase in water discharge depicts the increase in water availability during dry months of the year. The overall trend of the water flow discharge has decreased from 1980 to 2013 at both the gauging stations. The changes inflow regimes are the results of human interventions through the modification of water discharge from the dams.

Keywords

Dams · Hydrograph · Human Interventions · Discharge · Mann-Kendall Test

S. Sarathi Biswas (✉) · P. Pani
Centre for the Study of Regional Development, Jawaharlal Nehru University, New Delhi 110067, India

S. Sarathi Biswas
Department of Geography, Sukumar Sengupta Mahavidyalaya, Keshpur, West Bengal 721150, India

5.1 Introduction

The hydrology of a river is primarily controlled by rainfall and ice or glacier melting. In the tropical monsoon region, the seasonal rainfall

P. K. Shit et al. (eds.), *Drainage Basin Dynamics*, Geography of the Physical Environment,
https://doi.org/10.1007/978-3-030-79634-1_5

mostly regulates the river hydrology in the monsoon season. The uncertainty, variability and seasonally of the rainfall are natural; thereby, fluctuation of river hydrology is frequent. However, the climatic variability and human interferences accelerate the fluctuation and changes in river hydrology globally (Arnell 1998; Middelkoop et al. 2001; Christensen et al. 2004; Steele-Dunne et al. 2008; Boyer et al. 2010; Grillakis et al. 2011; Panda et al. 2013; Xu et al. 2013; Biswas 2014). Streamflow is the primary indicator to evaluate the impact of extrinsic factors on the river flow regime. The increase in the streamflow, base flow and water discharge is correlated with the increase of precipitation and decrease of potential evapotranspiration (PET) (Tomer and Schilling 2009; Coe et al. 2011). The reduction of annual rainfall enhances the annual mean PET and thereby, reduction in the annual mean runoff and discharge (Xu et al. 2013). Rainfall variability induces the shift of peak flow and hydrograph (Sharma and Shakya 2006; Moran-Tejeda et al. 2011; Panda et al. 2013) and it leads to scarcity and uncertainty of water availability (Gosain et al. 2006). On the contrary, hydrograph shifting, reduction of water availability and peak flow are very significant as an effect of human interventions on the river system (Cheng et al. 2019; Gierszewski et al. 2020; Uday Kumar and Jayakumar 2020).

The human impacts on river mainly include the changes through engineering construction, sand mining and land-use changes (Knighton 1998; Biswas 2014; Gibling 2018). Kiss and Blanka (2012) have analysed the hydro-morphological changesas a result of climate change, along with the human impacts of the Hernád River. They have used the long-term changes in the stages and discharge hydrographs. Similarly, Xu et al. (2007) have studied impacts of climate and anthropogenic activities on water resources and sediment by using the trend of precipitation and runoff with the aid of non-parametric test like the Mann–Kendall Test. They found that over 50 years, the constructions of over 50,000 dams results in small variations of monthly and yearly water discharges of the Yangtze River. Several methods like Soil and Water Assessment Tool (SWAT), Regional Climate Model (RCM), Indicators of Hydrologic Alteration (IHA), Range of Variability (RVA), Flow Health (FH) have been adopted to assess the hydrological alterations due to human interventions (Mittal et al. 2014; Gain and Giupponi 2014; Uday Kumar and Jayakumar 2018, 2020; Borgohain et al. 2019; Gierszewski et al. 2020). The decrease of streamflow is significant as a result of the human interferences through dam construction (Magilligan and Nislow 2005; Graf 2006; Milliman et al. 2008; Biswas and Pani (2021). Flow augmentation and flow variability such as reduction of high flow during monsoon season and enhancement of low flow are the joint impacts of dam construction (Mittal et al. 2014; Pal 2016; Borgohain et al. 2019). However, the changes in climate-related phenomena such as river flow, air temperature are also the result of human induced changes (Barnett et al. 2008).

The study on the rivers of peninsula India is more significant than the extra-peninsular river in context to the discharge variability as most of the rivers of the peninsula region are non-perrenial. So change of flow characters is very significant. The studied Barakar River is a peninsular river, and it is mostly disturbed by human interventions. The changes in the flow regime should be concerned for the people of the Barakar river basin. Hence, the study aims to identify the changes in the flow regime in the Barakar River as an effect of human interventions.

5.2 Barakar River: Geomorphic and Climatic Settings

Originating from Padma village, the Barakar River runs as a major tributary of the Damodar River which is the part of the Ganga River system in eastern India (Fig. 1). The length and catchment area of the Barakar River is ~256 km and ~6159 km^2, respectively. Several tributaries flow out into the Barakar River from the different directions of the Chotanagpur Plateau. The land use and land cover changes of the river basin alter the morphology and hydrology of the river significantly (Biswas and Pani 2016b). The

Barakar River has a steady flow in its lower section. Two dams characterise it at Tilaiya and Maithon in the upper and lower section of the river, respectively (Biswas 2014). The river has hydraulic gauging stations at Tilaiya and Maithon. In the upper section, the undulating physiography of the Chotanagpur Plateau is discernible, and it is very active in denudation that forms several narrow valleys. The general gradient of the Barakar River basin is toward the east and south-east. A wide, shallow channel along with a steep-sided, narrow valley of the river is marked over the flat plateau and plateau fringe region. The long profiles are disrupted by the number of rapids (Singh 1971). In the upper section, the river is very potential to erode. It is characterised by topographic features of rugged banded gneisses (Sen and Prasad 2002). The middle section of the basins distinguished by the large size of interfluves that influence on the increase of water volume and capacity of erosion (Prasad and Gupta 1982). In the lower section of the river basin, the river flows through several joints, and these are characterised by steep-sided granitic and gneisses walls with the presence of rugged blades of gneiss (Sen and Prasad 2002). The river has a significant character of mixed bedrock and alluvium nature (Biswas and Pani 2016a). In the upper regime, sinuous bedrock with the incised valley is significant. Conversely, in the lower regime, the river has a braiding character. Paleochannel study reveals that the river flowed by 235–415 m width, 3.7–5.5 m deep and over a slope of 0.00035 (Khan 1987). Thus the rapid formation, knick point development, channel incision is the results of long-term rejuvenation processes in Triassic to the Quaternary period (Biswas and Pani 2016a; Biswas et al. 2019).

The monsoon climate character prevails over the Barakar River basin. There are three distinct seasons in Jharkhand as well as in the Barakar River basin (Biswas 2014). The most pleasant weather of the year is cold weather that extends from November to February. Rainfall occurs mainly from July to September, and it accounts for more than 90% of the total rainfall of the year. In the lower portion of the basin, rainfall is high than in the north-western side of the upper portion of the Barakar River basin (Fig. 5.2). The lower section of the catchment area experienced more rainfall in the monsoon season and the hot summer season (April–May) through the thunderstorms that are locally known as *Kalbaishaki* (Nor-Wester).

5.3 Materials and Methods

The change of hydrology of the Barakar River is one of the main concerns for the livelihood of people of the catchment area. The rainfall and discharge data have been used over 50 years to assess the hydrological changes of this river system. The discharge data of two hydraulic stations have been collected from the Hydraulic Data Division, Damodar Valley Corporation (DVC), Maithon. The daily discharge data of Maithon and Tilaiya station is available from 1980 to 2013. However, the yearly discharge data of Maithon station is available from 1945 to 2013. The daily rainfall data of Dhanbad and Asansol stations have been collected from the Indian Meteorological Department (IMD), New Delhi. The rainfall data is available from 1950 to 2000 at Dhanbad and Asansol and from 1987 to 1996 at Tilaiya. Moreover, daily rainfall data of six rainfall gauging stations (Barhi, Tilaiya, Barki Saria, Palganj, Nandadih, Maithon) of the Barakar River Basin is collected from Hydraulic Data Division, DVC, Maithon (Fig. 5.1).

To understand the effect of rainfall on discharge, the non-parametric test, i.e. the Mann–Kendall test is applied (Smith 2000; Salmi et al. 2002; Kampata et al. 2008; Zhang et al. 2009; Xu et al. 2010; Tabari and Talaee 2011). To avoid the problem of data skewness, the non-parametric test has been adopted rather than using a parametric test (Smith 2000). The Mann–Kendall test is suitable where the increasing or decreasing trend is monotonic without the presence of seasonal or other cycles. It is based on the rank series of progressive and retrograde rows of random variables (Salmi et al. 2002; Zhang et al. 2009). The Mann–Kendall trend has been analysed using the rainfall and discharge

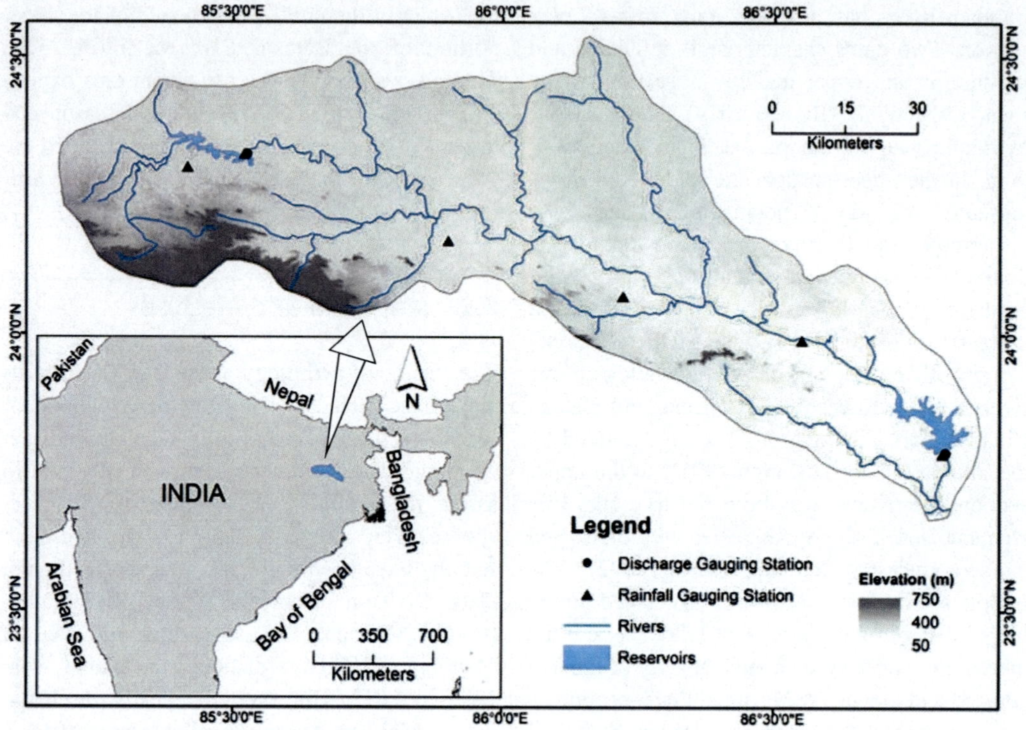

Fig. 5.1 Location of the Barakar River Basin. The discharge and rainfall gauging stations of the study area with elevation of the catchment area.

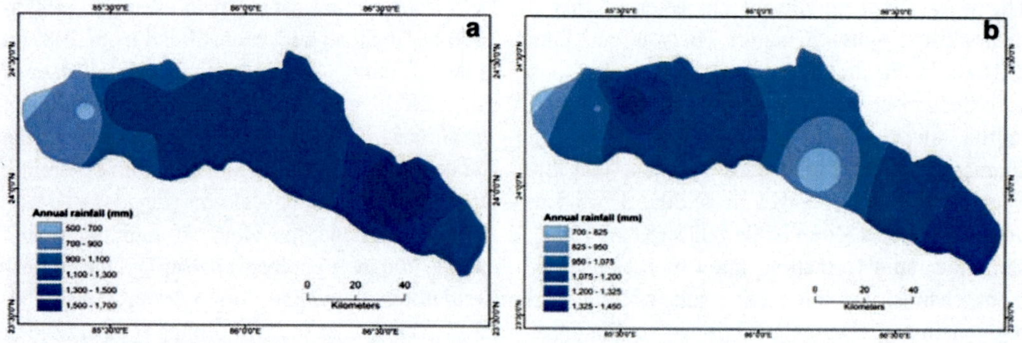

Fig. 5.2 Annual precipitation of the Barakar River Basin over 20 years. **a** Precipitation before 2000 (1990–2000) and **b** after 2000 (2000–2010).

data of the Maithon hydraulic station. The trend analysis is not performed for the Tilaiya as there was a limitation in the availability of the long-time annual rainfall data at the Tilaiya.

Mann–Kendall statistics (S) are calculated as follows:

$$S = \sum_{k=1}^{n-1} \sum_{j=k+1}^{n} \operatorname{sgn}(x_j - x_k) \quad (5.1)$$

where S = Mann–Kendall statistics and the ranked series $sgn(x_j - x_k)$ is generated from the time series data of rainfall and precipitation and comparing each value x_j with the subsequent values x_k with applying the following criterion:

$$\operatorname{sgn}(x_j - x_k) = \begin{Bmatrix} 1 if x_j - x_k > 0 \\ 0 if x_j - x_k = 0 \\ -1 if x_j - x_k < 0 \end{Bmatrix} \quad (5.2)$$

The test is based on the rank of data that have been used. There is not any tied data that have been observed. The variation of S has been generated by using the following formula:

$$Var(S) = \frac{n(n+1)(2n+5)}{18} \quad (5.3)$$

The values of $Var(S)$ and S have been used to compute the Z statistics

$$Z = \begin{Bmatrix} \frac{S-1}{\sqrt{Var(S)}} S > 1 \\ 0 S = 0 \\ \frac{S+1}{\sqrt{Var(S)}} S < 1 \end{Bmatrix} \quad (5.4)$$

A positive value of Z is the indication of an increasing trend, and a negative value is the indication of a decreasing trend.

Seasonal variation of flow has been discussed in three main seasons (pre-monsoon, monsoon and post-monsoon) of the study area to explain the temporal variation of the discharge in the different seasons or the seasonal flow variation on the Barakar River. Month-wise plotting of discharge at two stations (Maithon and Tilaiya) reveals the spatial variation of the discharge on the Barakar River. The hydrograph at two stations have been produced to explain the availability of water in the different months of the year based on temporal discharge data at an interval of 10^1 years (Sharma and Shakya 2006). The trends of the maximum (Q_{max}) and minimum (Q_{min}) discharge of both the stations reveal the temporal trend of peak flood discharge and the water availability in the dry season, respectively. The discharge has been analysed using maximum, minimum annual discharge to show the peak discharge and lowest discharge, respectively. Moreover, the relationship between Qmax/Qmin and Qmax/Qmean has been established to express the relationship between discharge variability and high peak flow flood regime of the Barakar River, respectively (Latrubesse et al. 2005). The trend of annual water flow has been produced to explain the overall discharge or water available at two hydraulic stations. A graph is established to show the relationship between rainfall and discharge and also the similar occurrences of peak flow and high rainfall over 40 years (1950–1990) at Maithon. The discharge data of the Maithon station is available in the pre-dam period. Thus, the effects of the Maithon dam are assessed over the pre- (1945–1957) and post- (1958–2013) dam period as the Maithon dam has been constructed in 1957 (Biswas 2014). Due to the limitation of the long-term discharge data in the pre-dam period, the change in recurrence interval in pre- and post-dam periods could not be calculated. The decadal change and seasonal change in the rainfall have been assessed using yearly and monthly rainfall of the six rainfall gauging stations, respectively (Moran-Tejeda et al. 2011).

Furthermore, the relationship between rainfall and discharge with scatter diagram plotting has been established at three stations Tilaiya, Dhanbad and Asansol.The student t-test is applied to test the relationship between rainfall and discharge. A test of significance of r is used to presume whether the correlation of coefficient of rainfall and discharge is zero or not. It is tested with the null hypothesis (H_0) of only rainfall controls the water discharge against the alternative hypothesis of (H_1). The student t distribution is calculated using the formula (Mahmood 1998):

$$t = r\sqrt{\frac{n-2}{1-r^2}} \quad (5.5)$$

where r = correlation coefficient, n = number of observations.

5.4 Results and Discussions

5.4.1 Temporal Trend of Rainfall and Discharge Using the Mann–Kendall Test

In the analysis of the Mann–Kendall test, the rainfall at Maithon has an increasing trend over the year 1970–2005 (Table 5.1). However, the discharge shows a decreasing trend across the period 1970–2012. The changes in the rainfall and discharge are not significant as the value of Z statistics is very low compared with the critical value of Z-statistics at 5 and 10% significance level. During the period 1970–2005, the annual rainfall has increased. However, it is insignificant as there is only an increase of 8.11 percent. On the contrary, from 1970 to 2012, the annual water discharge has been decreased by about 4.56%. From the Mann–Kendall test, it can be said that rainfall and discharge are not significant at 0.05 levels (Two-tailed). It indicates that the rainfall does not solely affect the discharge. The probable cause of insignificant relation is the construction of the Maithon dam, where the maximum upper channel flow of the upper part of the basin is consumed.

5.4.2 Variability in Hydrological Characteristics of the Barakar River

5.4.2.1 Temporal Change in Seasonal Flow Discharge

In the tropical monsoon rivers, the river hydrology is solely dependent on the monsoon precipitation of the river basin. Similarly, monsoon rainfall is the only source of the precipitation of the Barakar River. The rainfall occurs primarily from June to October in which is the monsoon season. Three prominent hydrological seasons are dominant in the Barakar basin, such as the Pre-monsoon season (March to May), Monsoon season (June to October), and Post-monsoon season (November to February). The discharge data are available at Maithon and Tilaiya, where two concrete dams have been constructed. There is an increasing trend of water discharge in the pre- and post-monsoon season. However, the trend is not of much significance as the R^2 value is 0.1 and 0.9, respectively (Fig. 5.3a, c). In this season, the discharge is not fed by the precipitation. It is supplied by the groundwater and the outflow from the Tilaiya dam in the upper stream regime. On the contrary, in the monsoon season, an indication of a decrease in the flow is discernible (Fig. 5.3b).

In the monsoon season, the release of water from the dams depends on the amount and the intensity of rainfall in the upstream reaches and the storage capacity of the dam reservoir. Dams release excess water when it approaches the storage capacity. Thus, intense rainfall in the upstream reaches is the only source of water availability in the downstream reaches of the dams. The amount of discharge is high at the Maithon than at the Tilaiya (Fig. 5.3d). High discharge at the Maithon is the result of the water flow contribution from most of the tributaries of the basin, and it is located just 12 km above the Barakar confluence with the Damodar River.

5.4.2.2 Late Shifting of the Hydrograph at the Maithon Hydraulic Station

Hydrographs at Maithon and Tilaiya indicate the availability of water in the different months of the

Table 5.1 Mann–Kendall trend analysis of the Barakar River Basin

Hydraulic station	Rainfall				Discharge			
	Year	Change (%)	Z statistics	Confidence (%)	Year	Change (%)	Z statistics	Confidence (%)
Maithon	1970–2005	8.11	0.026	NS	1970–2012	−4.56	−0.023	NS

NS—Not significant

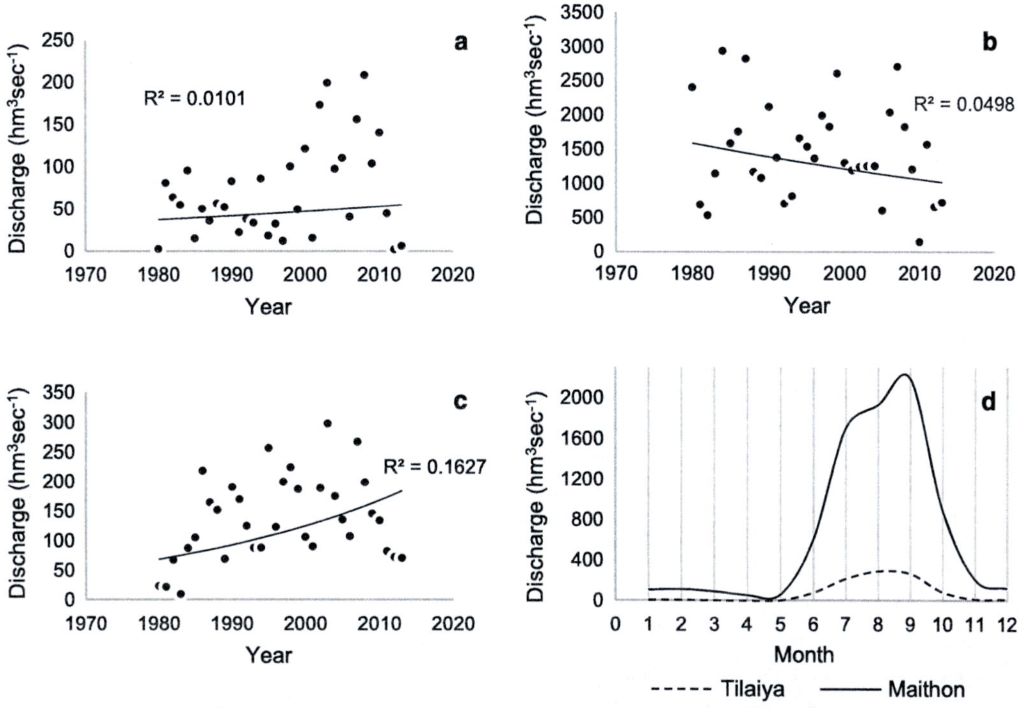

Fig. 5.3 Trend of flow discharge. **a** Pre-monsoon, **b** Monsoon, **c** Post-monsoon and **d** flow discharge in a year at the Maithon and Tilaiya station

year at these hydrological stations. The late shifting tendency of the hydrograph at the Maithon station is very significant (Fig. 5.4a). The peak flow season has shifted towards the right side over time. In 1980–89, the peak flow was from July to August, but after that from 1990 to 2013, the peak flow is in September which is a one and half month delay in the peak flow discharge. The late shifting of the rainfall period is the main factor for the shifting of the hydrograph. A significant observation is the initiation of two peaks after the period of 1980–89. In the hydrograph at the Tilaiya station, this scenario is different (Fig. 5.4b). The hydrograph is more or less constant except for one period. The shifting of the hydrograph is only noticed during 2000–09. During this period, the peakflow has shifted to September from August. Nevertheless, for the rest of the period, the peak flow occurs in August. In between 2010 and 2013, low peak discharge indicates the unevenness of the precipitation and the low water availability from the upstream section.

The uneven release of water from the dams along with the late arrival of the monsoon has caused the late shifting of the hydrograph. The late shifting of hydrograph can cause the flood in the downstream section in September and October (e.g. late arrival of monsoon creates the cloud burst in October 2013, and flood had occurred in the downstream of its mainstream, and it was accelerated by the sudden release of the water from the Maithon and Panchet dam). From April to May, both the gauging stations faced dry season along with the rest of the months of the years as hardly any type of precipitation occurs.

5.4.2.3 The Trend of Maximum (Q_{max}) and Minimum (Q_{min}) Flow Discharge at Tilaiya and Maithon Hydraulic Station:

The annual maximum discharge(Q_{max}) for both Tilaiya and Maithon tends to decrease across the period of 1980–2013, but the trend at the two stations is different. The decreasing tendency of

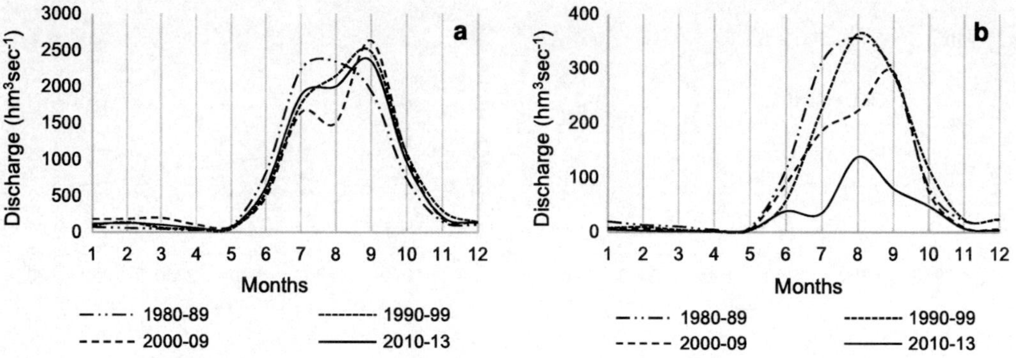

Fig. 5.4 Late shifting hydrograph of Maithon (**a**) and fluctuating hydrograph of Tilaiya (**b**) hydraulic station.

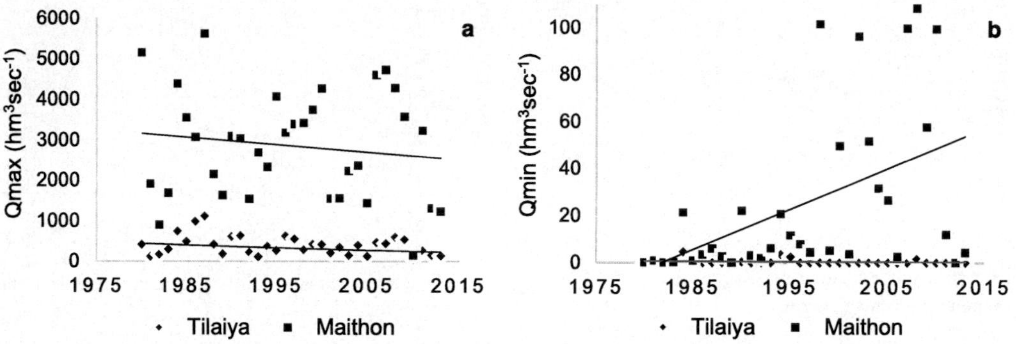

Fig. 5.5 Maximum (Q_{max}) (**a**) and minimum discharge (Q_{min}) (**b**) of Maithon and Tilaiya hydraulic gauging station.

maximum discharge (Q_{max}) at Maithon represents the decrease in the magnitude of a flood (Fig. 5.5a). On the contrary, the annual Q_{min} of the Maithon shows a tendency to increase, but the Q_{min} of the Tilaiya is near to zero in most of the year. As a result, the trend is not significant, and it lies on the baseline (Fig. 5.5b).

The increase in water discharge in the non-monsoon season depicts the increase in water availability in the driest season of the year (Sharma and Shakya 2006). It indicates the modification of the flow regime caused by the construction of the dams. Due to the dam closure, the reservoir storage is increased in the non-monsoonal season and peak discharge in the monsoon season is reduced (Biswas and Pani 2021). Thus, the frequencies of high floods tend to be reduced, while the low magnitude flood is slightly increased.

The overall trend of the water flow discharge has decreased from 1980 to 2013 at both gauging stations (Fig. 5.6). Both of the stations indicate a trend of decrease in the discharge of water, but the trend is a little high at Tilaiya than Maithon. Some occurrences of the peakflow at the Maithon is above 1200 $hm^3\ s^{-1}$ for the years 1984, 1987 and 2007, simultaneously, the Tilaiya shows peakflow in the same years. However, the amount of flow discharge that meagres at the Tilaiya dam was experienced with only above 200 $hm^3\ s^{-1}$. The position of the Tilaya at upper regime favours the low peak discharge.

Since the Tilaiya dam is located near the source of the river, the magnitude of interventions is lower than the Maithon dam. The peak flow of discharge is related to the high intensity and amount of precipitation and local cloudburst in the Chotanagpur plateau region. It is also

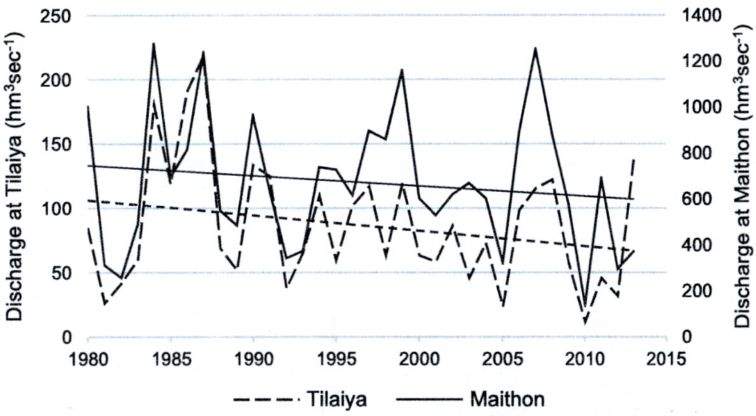

Fig. 5.6 Decreasing trend of discharge both at Maithon and Tilaiya hydraulic station. Darker and dotted ones represent discharge at Maithon and Tilaiya respectively.

related to the peak flow discharge at the Maithon dam. It is noteworthy that the water release from the Maithon Dam result floods in the downstream regime of its mainstem river, the Damodar River.

5.4.3 Relationship Between Discharge and Rainfall of Two Stations on the Barakar River Basin

In the analysis of the relationship between rainfall and discharge of two gauging stations, a different scenario is recognisable. Significantly, the peak discharge correlates with the high amount of rainfall, or the peak discharge coincides with the high rainfall in Maithon (Fig. 5.7). In the case of Tilaiya, the rainfall is significantly correlated ($r = 0.69$) to the discharge (Fig. 5.8c). For Tilaiya, the t-test is not significant at the 99% significance level, but it is significant at the 95% significance level. From the significance level test of the relationship between rainfall and discharge at Tilaiya, it can be said that the rainfall is controlling the discharge. Similarly, the relationship between the rainfall of two stations (Dhanbad and Asansol) and the flow discharge at the Maithon depicts the positive but weak relationship ($r = 0.29$) over an extended period (Fig. 5.8a, b). At Maithon, the peak flood discharge in the different years (1959, 1963, 1978 and 1980) simultaneously coincides with the peak combined rainfall of Dhanbad and Asansol. However, the relationship is not very significant ($r = 0.40$ and 0.22) Asansol and Dhanbad. The *t*-test is very significant at 99, 95 and 90% significance levels. Thus, the two variables are independent in this observation and from a large number of observations also. In the studied section of the river basin, precipitation has the least influence on the water discharge of the river. Such type of result was established in the previous study at the middle course of the Damodar River (the mainstream of the Barakar River) by Biswas (2014). But, a strong positive relationship between precipitation and water discharge has been observed in the lower section of the Damodar River by Bhattacharyya (2011). Hence, from the significance level test of the relationship between rainfall and discharge at Maithon, it is concluded that other factors control the discharge rather than rainfall only. Thus, the weak relationship may be the result of discharge modification by the dams, other anthropogenic alterations (sand mining, increase of urbanisation) and the late arrival of the monsoon (Biswas 2014). Thus, these factors can be treated as responsible for recent hydrological changes for the Barakar River.

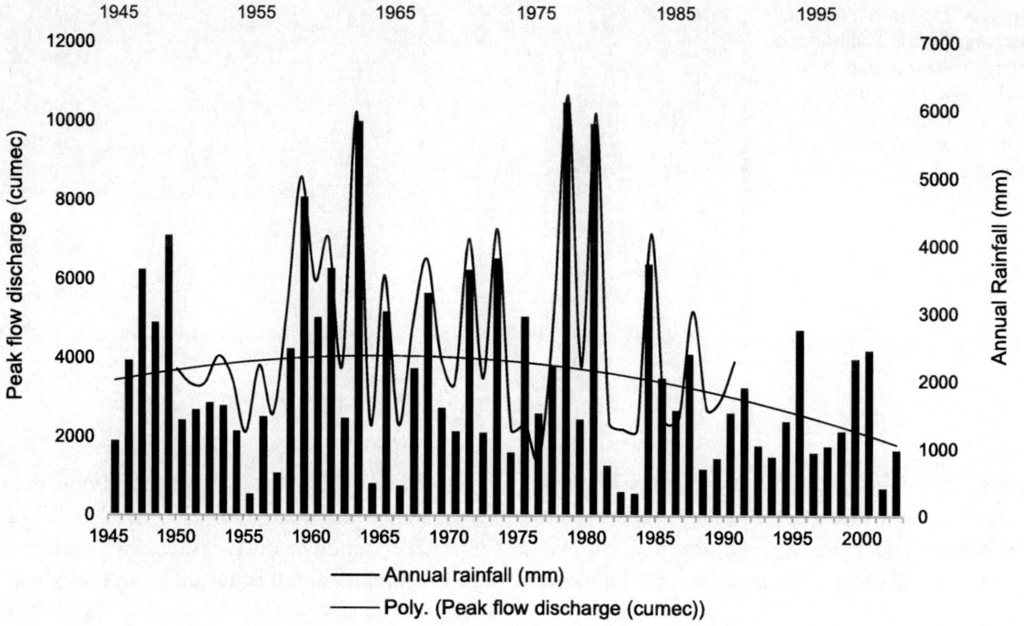

Fig. 5.7 Temporal coincide of peak flow discharge and high amount of rainfall depicts the positive relation between rainfall and discharge at Maithon station.

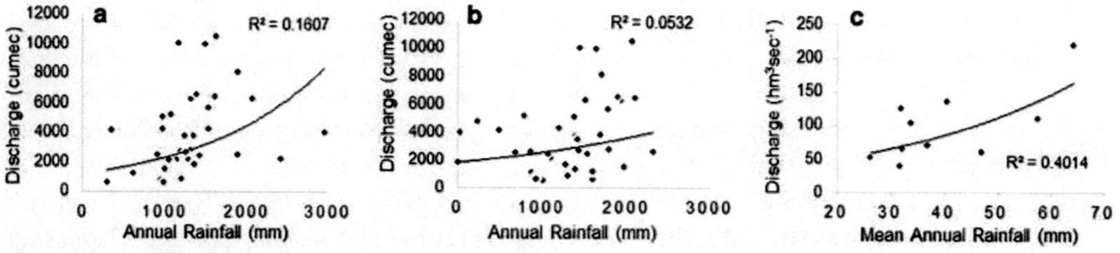

Fig. 5.8 The relationship between annual rainfall and discharge at **a** Asansol, **b** Dhanbad and **c** Tilaiya station. The discharge of Maithon is considered to show relationship in case of Asansol and Dhanbad.

5.5 Conclusion

The discharge variability on the Barakar River is a basin-scale phenomenon. The seasonal variability, water availability, changes in the rainfall-discharge relationship, shifting of the hydrograph, changes of flood pattern are significant indications of the human interventions in the river hydrological system. The decreasing trend of discharge in the monsoon season and the increasing trend of minimum discharge are very discernible in context to the changes in hydrological characters. Weak relation in rainfall and discharge also indicates the influences of dams on the flow discharge. Moreover, the late shifting of the hydrograph and reduction of water availability in the monsoon season reflect in agricultural practices of the basin. The people of this basin have to depend on irrigation for agricultural production. The increase in water availability in the dry season somehow meets the needs of agriculture. Hence, the analysis of hydrological change is indispensable for the morphology of the river as well as the livelihood of people of the Barakar River basin.

Acknowledgements The first author would like to express his acknowledgement to the University Grant Commission (UGC), New Delhi, India, for the financial support of the research work. This paper is a part of M. Phil. Dissertation (Unpublished) of the first author.

References

Arnell N (1998) Climate change and water resources in Britain. Clim Change 39(1):83–110. https://doi.org/10.1023/a:1005339412565

Barnett T, Pierce D, Hidalgo H, Bonfils C, Santer B, Das T, … Dettinger M (2008) Human-induced changes in the hydrology of the western United States. Science 319(5866):1080–1083.https://doi.org/10.1126/science.1152538

Bhattacharyya K (2011) The lower Damodar River, India: understanding the human role in changing fluvial environment. Springer, New York

Biswas SS (2014) Influences of changes in Land use/Land cover and precipitation variability on hydrology and morphology of middle course of Damodar River in East India. Int J Earth Sci Eng 7(1):326–338

Biswas SS, Pani P (2021) Changes in the hydrological regime and channel morphology as the effects of dams and bridges in the Barakar River, India. Environ Earth Sci 80(5). https://doi.org/10.1007/s12665-021-09490-0

Biswas SS, Pani P (2016a) Characteristics of a mixed bedrock-alluvial channel in a plateau and plateau fringe region: a study on the Barakar River of the Chotanagpur Plateau India. Environ Process 3(4):981–999. https://doi.org/10.1007/s40710-016-0190-y

Biswas SS, Pani P (2016b) Fluvial systems responses to landuse landcover change: a study on lower section of Barakar River Basin, India. J Indian Geomorphol 4:29–46

Biswas SS, Pal R, Pani P (2019) Application of remote sensing and GIS in understanding channel confluence morphology of Barakar River in Western most fringe of lower Ganga Basin. In Das BC, Ghosh S, Islam A (eds), Quaternary geomorphology in India—case studies from the lower Ganges basin, 1st ed. Springer International Publishing, pp 139–153. https://doi.org/10.1007/978-3-319-90427-6_8

Borgohain PL, Phukan S, Ahuja DR (2019) Downstream channel changes and the likely impacts of flow augmentation by a hydropower project in River Dikrong, India. Int J River Basin Manag 17(1):25–35

Boyer C, Chaumont D, Chartier I, Roy AG (2010) Impact of climate change on the hydrology of St. Lawrence tributaries. J Hydrol 384(1–2):65–83. https://doi.org/10.1016/j.jhydrol.2010.01.011

Cheng J, Xu L, Fan H, Jiang J (2019) Changes in the flow regimes associated with climate change and human activities in the Yangtze River. River Res Appl 1–13. https://doi.org/10.1002/rra.3518

Christensen NS, Christensen NS, Wood AW, Wood AW, Voisin N, Voisin N, … Palmer RN (2004) The effects of climate change on the hydrology and water resources of the Colorado River Basin. Climatic Change 62:337–363.https://doi.org/10.1023/B:CLIM.0000013684.13621.1f

Coe MT, Latrubesse EM, Ferreira ME, Amsler ML (2011) The effects of deforestation and climate variability on the streamflow of the Araguaia River, Brazil. Biogeochemistry 105(1):119–131. https://doi.org/10.1007/s10533-011-9582-2

Latrubesse EM, Coe, MT, Ferreira ME, Amsler ML (2011) The effects of deforestation and climate variability on the streamflow of the Araguaia River, Brazil. Biogeochemistry 105(1):119–131https://doi.org/10.1007/s10533-011-9582-2

Gain AK, Giupponi C (2014) Impact of the Farakka dam on thresholds of the hydrologic: flow regime in the lower Ganges River Basin (Bangladesh). Water (Switzerland) 6(8):2501–2518. https://doi.org/10.3390/w6082501

Gosain AK, Rao S, Basuray D (2006) Climate change impact assessment on hydrology of Indian river basins. Curr Sci 90(3):346–353. Retrieved from http://www.ias.ac.in/currsci/feb102006/346.pdf

Gibling MR (2018) River systems and the Anthropocene: a late Pleistocene and Holocene timeline for human influence. Quaternary 1(3):21. https://doi.org/10.3390/quat1030021

Gierszewski PJ, Habel M, Szmańda J, Luc M (2020) Evaluating effects of dam operation on flow regimes and riverbed adaptation to those changes. Sci Total Environ 710:136202. https://doi.org/10.1016/j.scitotenv.2019.136202

Graf WL (2006) Downstream hydrologic and geomorphic effects of large dams on American rivers. Geomorphology 79(3–4):336–360. https://doi.org/10.1016/j.geomorph.2006.06.022

Grillakis MG, Koutroulis AG, Tsanis IK (2011) Climate change impact on the hydrology of Spencer Creek watershed in Southern Ontario Canada. J Hydrol 409 (1–2):1–19. https://doi.org/10.1016/j.jhydrol.2011.06.018

Kampata JM, Parida BP, Moalafhi DB (2008) Trend analysis of rainfall in the headstreams of the Zambezi River Basin in Zambia. Phys Chem Earth 33(8–13):621–625. https://doi.org/10.1016/j.pce.2008.06.012

Khan ZA (1987) Paleodrainage and Paleochannel Morphology of a Barakar River (Early Permian) in the Rajmahal Gondwana Basin, Bihar, India. Palaeogeogr Palaeoclimatol Palaeoecol 58:235–247

Knighton D (1998) Fluvial forms and processes a new perspective (1st ed). Routledge, New York

Kiss T, Blanka V (2012) Geomorphology River channel response to climate- and human-induced hydrological changes: case study on the meandering Hernád River, Hungary. Geomorphology 175–176:115–125. https://doi.org/10.1016/j.geomorph.2012.07.003

Latrubesse EM, Stevaux JC, Sinha R (2005) Tropical rivers. Geomorphology, 70(3–4 SPEC. ISS.):187–206. https://doi.org/10.1016/j.geomorph.2005.02.005

Mahmood A (1999) Statistical methods in geographical studies (1st ed). Rajesh Publications, New Delhi

Magilligan FJ, Nislow KH (2005) Changes in hydrologic regime by dams. Geomorphology 71(1–2):61–78. https://doi.org/10.1016/j.geomorph.2004.08.017

Middelkoop H, Daamen K, Gellens D, Grabs W, Kwadijk JCJ, Lang H, … Wilke K (2001) Impact of climate change on hydrological regimes and water resource management in the Rhine River Basin. Climatic Change 105:128.https://doi.org/10.1023/a:1010784727448

Milliman JD, Farnsworth KL, Jones PD, Xu KH, Smith LC (2008) Climatic and anthropogenic factors affecting river discharge to the global ocean, 1951–2000. Glob Planet Chang 62:187–194. https://doi.org/10.1016/j.gloplacha.2008.03.001

Mittal N, Mishra A, Singh R, Bhave AG, Van Der Valk M (2014) Flow regime alteration due to anthropogenic and climatic changes in the Kangsabati River India. Ecohydrol Hydrobiol 14(3):182–191. https://doi.org/10.1016/j.ecohyd.2014.06.002

Moran-Tejeda E, Lopez-Moreno JI, Ceballos-Barbancho A, Vicente-Serrano SM (2011) River regimes and recent hydrological changes in the Duero basin (Spain). J Hydrol 404(3–4):241–258. https://doi.org/10.1016/j.jhydrol.2011.04.034

Pal S (2016) Impact of Massanjore dam on hydrogeomorphological modification of Mayurakshi River, Eastern India. Environ Dev Sustain 18(3):921–944. https://doi.org/10.1007/s10668-015-9679-1

Panda DK, Kumar A, Ghosh S, Mohanty RK (2013) Streamflow trends in the Mahanadi river basin (India): linkages to tropical climate variability. J Hydrol 495:135–149. https://doi.org/10.1016/j.jhydrol.2013.04.054

Salmi T, Määttä A, Anttila P, Ruoho-Airola T, Amnell T (2002) Detecting trends of annual values of atmospheric pollutants by the Mann-Kendall test and Sen's slope estimates—the Excel template application MAKESENS. Finnish Meteorological Institute

Sen PK, Prasad N (2002) An Introduction to the Geomorphology of India. Allied Publishers Pvt. Ltd., New Delhi

Sharma RH, Shakya NM (2006) Hydrological changes and its impact on water resources of Bagmati watershed Nepal. J Hydrol 327(3–4):315–322. https://doi.org/10.1016/j.jhydrol.2005.11.051

Singh RL (1971) India: a regional geography. National Geographical Society of India, Varanasi

Smith LC (2000) Trends in Russian Arctic river-ice formation and breakup, 1917 to 1994. Phys Geogr 21 (1):46–56. https://doi.org/10.1080/02723646.2000.10642698

Steele-Dunne S, Lynch P, McGrath R, Semmler T, Wang S, Hanafin J, Nolan P (2008) The impacts of climate change on hydrology in Ireland. J Hydrol 356 (1–2):28–45. https://doi.org/10.1016/j.jhydrol.2008.03.025

Tabari H, Talaee PH (2011) Temporal variability of precipitation over Iran: 1966–2005. J Hydrol 396(3–4):313–320. https://doi.org/10.1016/j.jhydrol.2010.11.034

Tomer MD, Schilling KE (2009) A simple approach to distinguish land-use and climate-change effects on watershed hydrology. J Hydrol 376(1–2):24–33. https://doi.org/10.1016/j.jhydrol.2009.07.029

Uday Kumar A, Jayakumar KV (2018) Assessment of hydrological alteration and environmental flow requirements for Srisailam dam on Krishna River, India. Water Policy (August), 1176–1190. https://doi.org/10.2166/wp.2018.203

Uday Kumar A, Jayakumar KV (2020) Hydrological alterations due to anthropogenic activities in Krishna River Basin India. Ecol Ind 108:105663. https://doi.org/10.1016/j.ecolind.2019.105663

Xu K, Milliman JD, Xu H (2010) Temporal trend of precipitation and runoff in major Chinese Rivers since 1951. Glob Planet Chang 73(3–4):219–232. https://doi.org/10.1016/j.gloplacha.2010.07.002

Xu K, Millima JD, Yang Z, Xu H (2007) Climatic and anthropogenic impacts on water and sediment discharges from the Yangtze River (Changjiang), 1950–2005. Large Rivers: Geomorphol Manag, 609–626. https://doi.org/10.1002/9780470723722.ch29

Xu YP, Zhang X, Ran Q, Tian Y (2013) Impact of climate change on hydrology of upper reaches of Qiantang River Basin, East China. J Hydrol 483:51–60. https://doi.org/10.1016/j.jhydrol.2013.01.004

Zhang Q, Xu C, Zhang Z, Chen YD, Liu C (2009) Spatial and temporal variability of precipitation over China, 1951–2005. Theor Appl Climatol 95(1–2):53–68. https://doi.org/10.1007/s00704-007-0375-4

Prasad N, Gupta SP (1982) Streamline surface of the barakar basin-a perspective of landscape evolution. In: Sharma HS (ed) Perspective in geomorphology (vol 4) Essay on Indian geomorphology. Concept Publishing Company, New Delhi, pp 103–108

6 Landscape Characterization using Geomorphometric Parameters for a Small Sub-Humid River Basin of the Chota Nagpur Plateau, Eastern India

Jayesh Mukherjee and Priyank Pravin Patel

Abstract

River responses to endogenic perturbations and climatic attributes lead to the morphological development of basins, the quantification of whose terrain facets to decipher the ambient process-response mechanisms has long been an important aspect of geomorphological studies. The availability of newer, higher resolution datasets, however, entail that past exercises in this domain be looked at anew, in terms of the greater sensitivity and diversity of information that can now be extracted/collated. Here, we combine classical and modern morphometric methods to examine the terrain characteristics of the Kharkai River Basin in eastern India, with this basin chosen for its diversity of landforms and human activities. While traditional methods have been based on eliciting terrain information from topographical maps, we use a higher resolution Digital Elevation Model (DEM) to extract such attributes. The stream network was derived using flow-routing and flow accumulation algorithms and a net of 1 km × 1 km grids was overlain on the DEM to compute various morphometric parameters. These were combined using Principal Component Analysis to demarcate distinct physiographic/landscape entities, in conjunction with the corresponding lithological and soil attributes of the area. Multi-temporal land cover and land use layers extracted from Landsat datasets were overlain on the extracted terrain units to estimate changes in the same across different landscape types. The demarcated terrain units strongly correlated with the lithology, as expected, and this also controlled local slope and drainage development. More rugged locales had greater vegetation cover but were also threatened by deforestation due to agricultural expansion and mining.

Keywords

Geospatial techniques · Terrain analysis · Geomorphometry · Kharkai River · Map overlays · Image processing

J. Mukherjee · P. P. Patel (✉)
Department of Geography, Presidency University, 86/1, College Street, Kolkata, West Bengal 700073, India
e-mail: priyank.geog@presiuniv.ac.in

J. Mukherjee
Centre for the Study of Regional Development, School of Social Sciences, Jawaharlal Nehru University, New Delhi 110067, India

6.1 Introduction

The term 'Morphometry' was initially adopted by De Martonne (1934), and 'morphometric methods' encapsulate the numerical characterization of different landform/landscape attributes/elements as enumerated from digital elevation datasets or

P. K. Shit et al. (eds.), *Drainage Basin Dynamics*, Geography of the Physical Environment,
https://doi.org/10.1007/978-3-030-79634-1_6

from topographical maps. Morphometry, then is widely considered as the mathematical measurement of the Earth's surface together with the shape, dimension and distribution of its constituent landforms (Clarke 1996; Agarwal 1998; Obi Reddy et al. 2002; Vaidya et al. 2013; Singh et al. 2014; Asfaw et al. 2019). This provides the basis for 'quantitative geomorphology' and as such can be viewed as the means of creating a '*census handbook of the landform*' with a purpose to provide an empirical description of its inherent characteristics without delving into or trying to formulate a hypothesis of its origin. The main aspects of the landform features which provide the differentiating characteristics and which can be studied from a topographical map are (Hammond 1954)—the area (surface arrangement), altitude, relief and volume (horizontal dimension) and the slope (deviation of the surface from the horizontal). All these individual elements, along with the overall different aspects of the drainage basin within which they are situated, are investigated in morphometric examinations and terrain analysis.

The eventual aim of these analyses is *terrain characterization leading to terrain modelling* (Patel and Sarkar 2010), not only to decipher just the general surface attributes but also garnered towards any type of geo-applications. Investigating such indicators helps reveal the geo-hydro morphological functioning of drainage basins (Horton 1945; Evans 1972, 1984), which encapsulates ambient factors like climate, topography (Strahler 1952, 1964; Chorley et al. 1984), structure (Shreve 1969; Merritts and Vincent 1989; Oguchi 1997), tectonics and its geomorphology (Mueller 1968; Ohmori 1993; Cox 1994; Burrough and McDonnell 1998; Hurtrez et al. 1999). Hence, such attributes are crucial factors in gauging landscape evolution (Goodbred 2003; Das et al. 2018a) and profoundly impact the basin network (Zhang 2005; Das et al. 2018b) developed within a watershed. The combined assessment of these characteristics successfully mirror the overall denudational evolution of a locale (Patel et al. 2021). Thus, any landscape is readily explainable with the help of morphometry, in terms of architectural unevenness and watershed characteristics, that together with the underlying hydro-geological processes have been operative in creating the present landforms (Horton 1932; Smith 1958; Miller 1953; Soni 2017; Gizachew and Berhan 2018; Asfaw et al. 2019), and in enabling soil erosion and other surficial processes (Khare et al. 2014; Asfaw et al. 2019).

Thus *geomorphometry* intertwines principles from mathematics, computer science and earth system analysis (or simple morphometry). This field has been revolutionized by geocomputational advances over the last three decades and the greater development and availability of coarser and higher resolution digital elevation models (DEMs) (Maune 2001). Morphometric exercises are now largely undertaken through various geographic information systems (GIS) or dedicated software suites. Newer remotely sensed datasets and relevant methods have simplified terrain and hydro-morphometric analyses of drainage basins from DEMs (Jensen 1991; Wise 2000; Aparna et al. 2015; Patel 2013; Patel et al. 2016; Gutema et al. 2017; Gizachew and Berhan 2018; Kabite and Gessesse 2018; Afsaw et al. 2019). Harinath and Raghu (2013) have further explained how in situ methods of landscape and terrain evaluation are laborious, time consuming and capital intensive, while geomorphometric analysis provides far easier evaluative measures of drainage basins.

However, continuously representative surfaces are hard to characterize using traditional measures and newer various parameters have been framed to enumerate terrain facets that older measures may fail to describe. Two such methods are the 'terrain fabric' (Guth 2001) and the surface 'openness' (Yokoyama 2002) parameters. Terrain fabric characterization is the probability that a surface is organized as linear ridges rather than as isotropic topography, while openness reveals the dominance of exposure over enclosure of a site. Yet many computed variables are similar in nature, e.g. the hypsometric integral and elevation skewness parameters are quite similar (Pike 2001). Currently, DEM analysis could be best termed as 'a modern, analytical, cartographic approach to represent the bare-earth topography by the computer manipulation of terrain heights' (Tobler 2000).

Landscape interlinkages form the basic foundation of the economy in most developing countries, encapsulated by its water and land attributes (Chattopadhyay and Carpenter 1990). Surface drainage characteristics have been quantified and mapped worldwide, mostly employing traditional methods (e.g. Horton 1945; Morrisawa 1959; Langbein and Leopold 1964; Strahler 1954, 1957, 1964). Such analyses have aimed to assess drainage properties by measuring various channel network aspects, e.g. stream ordering, watershed area and perimeter extents, drainage length, density and frequency, bifurcation and texture ratios (Kumar et al. 2000), to mention but a few. For such quantitative characterization of a watershed, measurements of linear/areal aspects, stream gradients and valley slopes are required (Kumar et al. 2000; Nag and Chakraborty 2003; Krishnamurthy 1996; Vijith and Satheesh 2006). These analyses help reveal the ambient drainage character and its interactions with the local topography (Obi Reddy et al. 2002, 2004; Sutradhar 2020).

The underlying geology, exogenic and endogenic processes and major climatic alterations profoundly influence drainage network development (Ghosh 2016), the genesis and morphology of landforms, soil properties and the alterations in present-day land use and land cover due to both natural and anthropogenic stimuli, causing degradation (Subhramanyan 1981; Javed et al. 2009, 2011; Altaf et al. 2014; Islam and Barman 2020). Morphometric aspects are thus used in various catchment-based assessments like terrain analysis (Patel and Sarkar 2010), delineation of geomorphological features, quantitative geomorphology and watershed prioritization (Sarkar and Patel 2011, 2012).

The novelty of this research is its combining of traditional and the modern morphometric techniques to bridge the remaining gaps found across the literatures that either use the newer or the more conventional morphometric techniques to explain river basin landscapes. This work underlines the multi-temporal alterations observed along different topographic classes occurring over various morphogenetic regions using statistical measure like factor score, which helps to decipher a new comparative way of investigating and comprehending the river basin landscape. This in turn will also help to identify those areas in urgent need of attention in terms of effectuating proper land and water preservation exercises.

The major objectives of the present research thus include extraction and exploration of the structural properties of the Kharkai River basin, underlining its terrain aspects using conventional grid-based morphometry and more recent GIS-based automated algorithms. The second objective is to understand the relation between the geomorphometric parameters using statistical measures like Principal Component Analysis (PCA) to delineate distinct terrain units present within the basin, highlighting their erodibility and validating the results obtained from geomorphometric evaluation. The final objective is to bring out the multi-temporal changes occurring within the river basin by detecting the land use and land cover (LULC) alterations across the past thirty years, which would help pinpoint areas that need attention for implementing land and water conservation practices.

6.2 Materials and Methods

6.2.1 Database

The methodology followed in this study requires diverse datasets from four main sources—the remotely sensed SRTM DEM (Shuttle Radar Topographic Mission Digital Elevation Model) tiles and the LANDSAT imageries of the study area from the Earth Explorer repository of the USGS (United States Geological Survey) (https://earthexplorer.usgs.gov/); Survey of India topographical maps for corroboration of river basin outline and DEM-extracted stream network; and finally geology, lithology and geomorphology maps from the Geological Survey of India quadrangles. Table 6.1 lists the data types and their respective sources in detail.

Table 6.1 Enumerated list of parameters in the present study

Sl. No	Parameters	Data type	Data source	Scale/resolution	Time period
1.	Elevation, Terrain & Drainage	Raster	SRTM DEM (USGS), version 3.0	30 m	September, 2014
2.	Topography & Kharkai River Channel	Line and polygon coverage	Survey of India Topographical Maps, Sheet Numbers—73F/5, F/6, F/7, F/9, F/10, F/11, F/12, F/13, F/14, F/15, F/16; 73 J/2, J/3, J/4, J/7, J/8; 73 K/1 and K/5	1:50,000	2006
3.	Geology & Lithology	Polygon coverage	Geological Survey of India Quadrangles	1:250,000	2006
4.	Geomorphology	Polygon coverage	Geological Survey of India Quadrangles	1:250,000	2006
5.	Land use & Land cover	Raster	Landsat TM 4–5 C-1 Level-1 and Landsat 8 OLI/TIRS C1 Level 1; Paths/Rows: 140–44 & 140–45	30 m	December, 1990 & April, 2020

6.2.2 Data Analysis

The analysis undertaken was to primarily extract the terrain information on various GIS platforms like MapInfo Professional GIS and ArcMap 10.3, while other allied mapping and statistical software like Whitebox GAT-3.4, SAGA GIS 7.5 and IBM SPSS-23 were used for creating the geo-database. Terrain aspects and drainage networks were demarcated and extracted and then the attribute data pertaining to the stream and basin parameters were attached accordingly to generate the required database for producing the maps. Three-dimensional (3-D) surfaces of relevant geomorphic parameters of the basin were subsequently generated and the respective isopleth maps prepared. Map overlays were done to show spatial relations between different attributes. The empirical relationships derived between the various morphometric variables—liner, areal, relief attributes enabled terrain classification for demarcating distinct physiographic units within the basin using PCA-based factor analysis and hierarchical clustering. These were validated and correlated with the existent LULC types and patterns.

6.2.3 Extraction of Morphometric Parameters

Following conventional practice, the entire river basin was covered by grids of 1 km × 1 km dimension for extraction of the morphometric attributes for each grid and this starts with the visual analysis of the DEM surface to distinguish various relief features (Patel 2012). This was followed by the correlation of the ascertained relief features with the basin geology by draping the litho-cover map over the DEM surface and the generation of contours at 10 m interval. The extraction of stream networks using the D8 method of flow routing and flow accumulation methods then followed, with this eliciting a more denser river network than is usually obtainable from topographical maps (Patel and Sarkar 2009; Das et al. 2016). The grid-wise extraction of various morphometric parameters like *maximum grid elevation, mean grid elevation, relative relief, average slope, dissection index, stream frequency, hypsometric integral surrogate, drainage density and texture and ruggedness index* was done subsequently. Through interpolation of the extracted parameter values from the centroid of each morphometric grid using triangulation technique- inverse distance weighted (IDW), the respective isoline maps were generated. Computation of numbers of grids and amount of basin area falling under the respective iso-zones in each isoline map generated above led to the preparation of frequency or percentage area histograms on this basis. The other terrain morphometric parameters such as *terrain surface texture, terrain surface convexity, topographical wetness index and topographical position index* were calculated based on automated algorithms which are already coded in a GIS environment and can thus be extracted directly from the input DEM tiles. Brief explanations of the enumerated parameters are as follows:

- ***Mean elevation***: The average altitude within each grid is recorded on the basis of enumeration of all the DEM pixels falling within it.
- ***Relative relief***: Considered as the 'amplitude of available relief' or 'local relief', it is the elevation difference between the lowest and highest points in an areal unit (Smith 1935).
- ***Average slope***: Slopes (the angular terrain inclinations between the ridge crest and valley bottom) result from the combination of the local geological structure, absolute/relative relief, vegetation cover, climate, drainage and degree of dissection. It is a vector quantity and Wentworth (1930) put forward a method for the calculation of the average slope of an area (in degrees). In this method, however, the counting of contour crossings does not strictly take into account the nature of the gradient since the same contour may cut a grid on numerous occasions and thus a higher slope value will be obtained which is not actually the case. Furthermore, the sensitivity of the derived slope value is dependent on the contour interval and the smaller the contour interval, the better minor gradient changes can be enumerated.
- ***Dissection index***: It is the ratio of the relative relief and the highest elevation in a grid, and indicates the dissection/magnitude of a terrain. 'It takes into account the dynamic potential state of the area as well, i.e., the ratio between relative relief (relief energy) and the perpendicular distance from the erosion base' (Miller 1953; Nir 1957). The classification of the dissection index values is as follows (modified from Kumar and Pandey 1982): less dissected (below 0.1), medium dissected (0.1–0.3), much dissected (0.3–0.6), highly dissected (above 0.6). The values for the dissection index so derived vary between 0 (when the entire depth of altitude is dissected) and 1 (where minimum altitude is equal to the maximum altitude—no dissection has taken place). It reveals gently rolling uplands and dissections in mountainous areas particularly (Pal 1972). This index can be indicative of the erosional cycle stage, with old, mature and young stages related to dissection values.
- ***Hypsometry surrogate***: The grid-wise Hypsometric Integral (HI) was computed across the basin surface as proposed by Pike and Wilson (1971).
- ***Terrain Surface Texture***: Terrain textures consider relief (Z factor) and spacing (X, Y factors), which represent measures of spatial intricacy per unit area, incorporating the drainage density and slope curvature. This measure highlights the 'fine-versus coarse expression of topographic spacing', i.e. the 'grain' (Ayalew and Yamagishi 2004; Iwahashi and Pike 2007), and was earlier referred to as 'frequency of valleys and ridges' or 'roughness' (Iwahashi 1994; Iwahashi and Kamiya 1995; Iwahashi et al. 2001).
- ***Terrain Surface Convexity***: For automatic classification of a high gradient topography, slope gradient and surface texture of the topography play a combinational and fundamental role, but are inadequate to classify low relief features, for instance, segregating older river terraces from the younger ones. So in order to better demarcate these, the local convexity or positive surface curvature was utilized by Iwahashi and Pike (2007). It is commonly seen that low surface convexity conforms to broad valleys and mountain foot slopes, while higher values are typically associated features like alluvial fans or terraces.

- ***Topographical Wetness Index***: The TWI parameter measures the relief effect on the generation of runoff (O'Loughlin 1986) and thereby approximates surface saturation zones (Beven and Kirkby 1979; Barling et al. 1994). Similar TWI value zones are likely to behave in a similar hydrological manner when precipitation occurs, if the other ambient conditions match (Qin et al. 2011). The TWI depends on the algorithms used to compute the upslope contributing area and the slope gradient (Qin et al. 2009).
- ***Topographical Position Index***: The TPI parameter (Guisan et al. 1999) is an automated algorithm meant for measuring topographic slope positions and enable landform classification. It has been used in geology (Mora-Vallejo et al. 2008; Deumlich et al. 2010; Illés et al. 2011), geomorphology (Tagil and Jenness 2008; Liu et al. 2009; McGarigal et al. 2009; De Reu et al. 2013), hydrology (Lesschen et al. 2007; Francés and Lubczynski 2011; Liu et al. 2011) and many more to such allied branches and quantifies the elevation difference of a central point and the mean elevation within a predetermined radius from it (Gallant and Wilson 2000; Weiss 2000, 2001). Positive TPI values are discerned for central point that are located above their surroundings, and negative values indicate lower topographic positions. The TPI range thus depends on the predetermined radius and elevation difference (e.g. Grohmann and Riccomini 2009). Higher predetermined radius values usually denote major landforms while lower values correspond to minor valleys and ridges (De Reu et al. 2013).
- ***Stream frequency***: It denotes the total number of streams flowing per unit area (Horton 1945) and is closely related with the drainage density, and the various parameters that influence drainage density. It is indicative of the nature of runoff in an area, giving insights into the overland flow length and channel spacing. To compute it, all drainage lines were clipped using each morphometric grid and the number of channel segments lying within each grid were counted. The sum of the length of each of these drainage lines is later used while computing the grid-wise drainage density.
- ***Drainage density***: Horton (1945) enumerated this as the ratio of the sum of all stream lengths within the basin to its total area. As a principal component of the landscape that could be investigated using maps and aerial photographs, drainage density has been the subject of numerous studies with the underlying belief being that if the numerous independent variables that control drainage density could be quantitatively related to it, the results would be of great academic interest and practical value (Schumm 1956; Lin and Oguchi 2004). Thus, it is an important variable that has been related to climate change (Rodgriguez-Iturbe and Escobar 1982; Moglen et al. 1998), slope failure (Oguchi 1997), hillslope processes (Tucker 1998), stream flow (Dingman 1978; Carlston 1966), flood peaks (Pallard et al. 2009; Ogden et al. 2011), mean annual discharge and sediment yields (Branson and Owen 1970; Wasson 1994; Binger et al. 1997; Biswas et al. 1999). It has also been used to detect variations of rock types and structure by photo-geologists to document the stage of erosional evolution of a drainage system and in land reclamation studies.
- ***Drainage texture***: Postulated by Smith (1950), it is obtained by multiplying the stream frequency and drainage density parameters, and is indicative of the mesh of drainage lines that have developed over a surface. Thus indirectly, it also points towards the lithological, structural, pedological and climatic set up of an area. The scale of drainage texture after Pal (1972) is coarse (<4), intermediate (4–10), fine (10–50) and ultrafine (>50).
- ***Ruggedness index***: It is a parameter used to describe how rugged the terrain is (Horton 1945), with terrain ruggedness being a feature of areas having high relief variations as a result of dissection sub-aerial denudational processes. The more dissected the topography, the more rugged it appears, being a combined expression of relief, texture and slope steepness.

6.3 Study Area

6.3.1 River Course and Basin Physiography

The Kharkai River flows due NNE (north north-east) and then NNW (north north-west) for a distance of 63 km from its source in the Bamanghati–Kolhan upland area through the Singhbhum plain, before the river takes a sharp turn due east within the Achaean terrain of southern Singhbhum that includes areas of Jharkhand and Odisha (Fig. 6.1). The Kharkai follows a roughly east north-east (ENE) course till it confluences with the Subarnarekha, its trunk stream, at Adityapur near Jamshedpur in Saraikela-Kharsawan district of Jharkhand (Fig. 6.1). The river basin covers approximately 6255.75 sq.km. and has a perimeter of about 774.05 km.

The Kharkai, with a width of 15 to 30 m, has a meandering course and a steep gradient for the first 40 km of its course and a relatively steeper gradient between the 350 and 225 m contours, situated mainly within the iron ore series of rocks. Part of the gentle gradient section is found in the NNW course and the rest is in the eastern course on either side of the abrupt right-angled turn in the Kharkai River beyond Chaibasa. The upper reaches are marked by meandering index (mean) of 1.3 with a mean gradient 1:30 and 1:350 for the southern and northern segments, respectively.

This Kharkai Basin mostly covers the south-western part of Subarnarekha basin, and geomorphologically it consists of a several distinct planation surfaces at varying altitudes (Fig. 6.2).

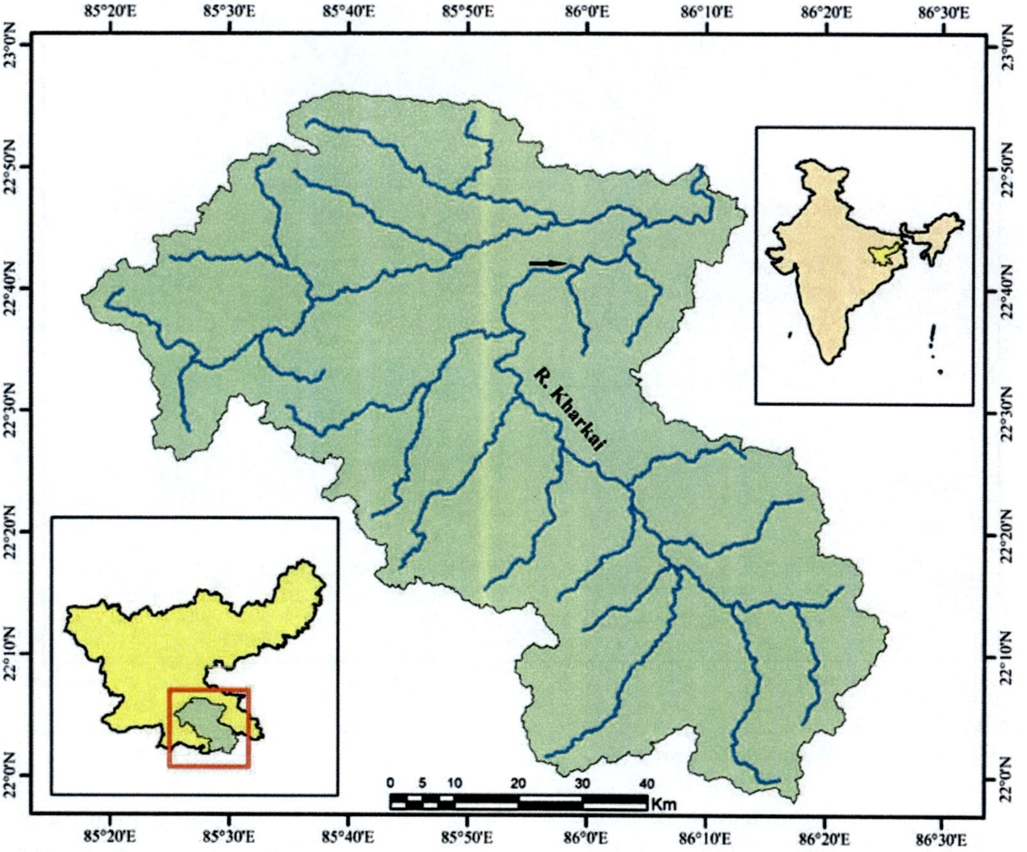

Fig. 6.1 Location of the Kharkai River Basin

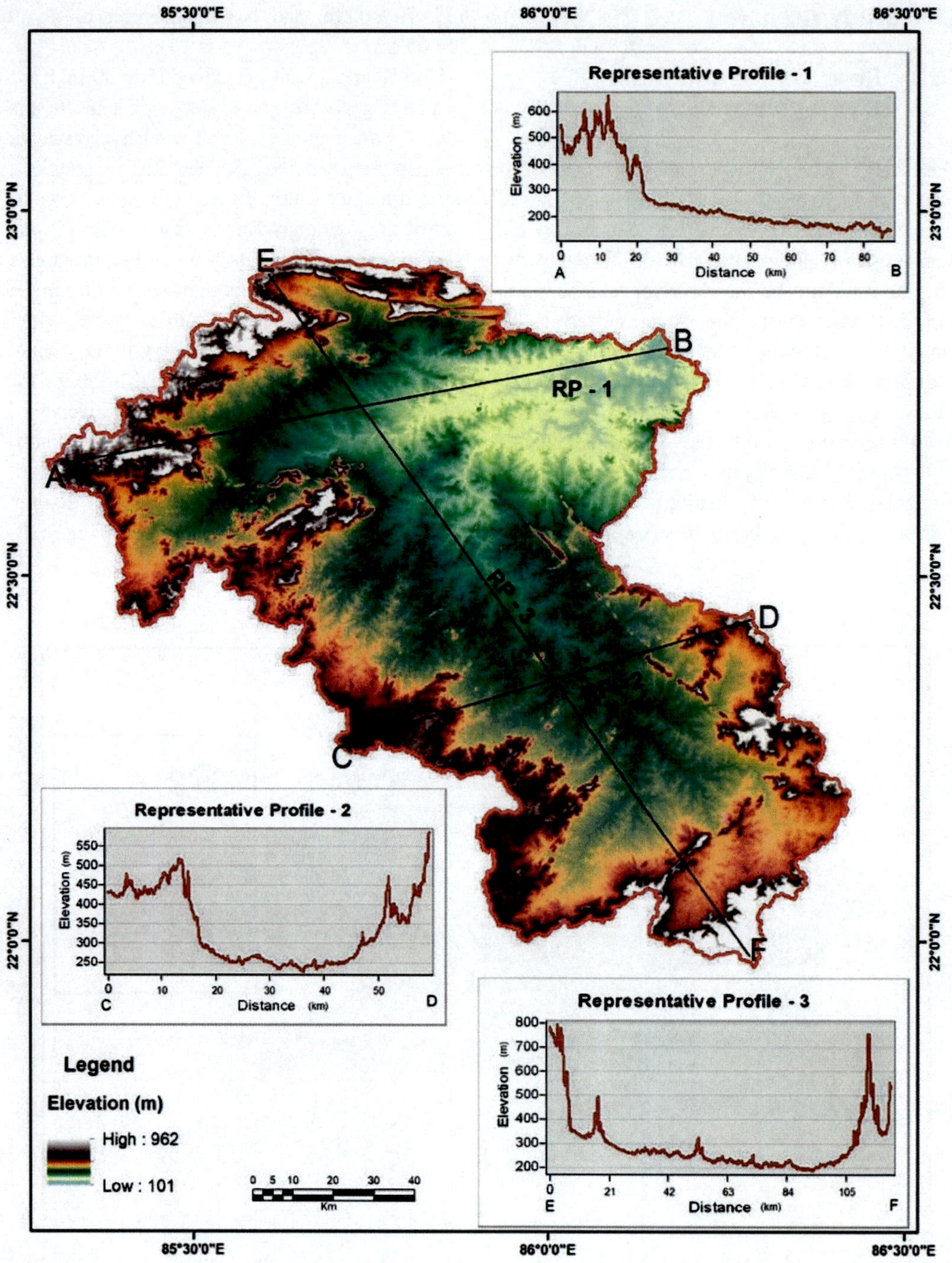

Fig. 6.2 Digital elevation model of the Kharkai River Basin, showing the elevation distribution and representative surface profiles with major planation surfaces

The elevation is between 960 and 130 m, respectively, with the altitude range being 830 m. The highest surface is along the southern rim built up by the Bamanghati–Kolhan upland and the northern Porahat hill formations. This rises to over 600 m. above sea level, within which the river's meandering index varies from 1.2 to 1.38 and mean gradient is 1:30. This ridgeline actually forms the water divide in between the Koel (South), Baitarani and Subarnarekha drainage systems. The second surface within which the river has a meandering index from 1.2 to 1.38 and mean gradient less than 1:350, lies below an average altitude of 600 m. It is an extensive belt stretching in an irregular manner, thus fencing the basin in its north, west and southern part in a striking way. This planation surface rises to above 300 m and corresponds mostly to the iron ore series and Singhbhum granite formations (Mukhopadhyay 1980).

6.3.2 Climatic Attributes and Soil Cover

This river basin has a tropical monsoon climatic regime with alternate dry and wet seasons (Singh 1975). The hot weather commences in March and temperatures rise sharply till May, when the mean monthly temperature is between 29 and 32 °C. During the summer southwest monsoon (June–October), the rainfall received is between 100 and 150 cms, which is almost 90% of the total. Since, the Kharkai is seasonal, its discharge is highest during July–August, with a strongly leptokurtic hydrograph. January has the lowest temperatures (mean of around 16.4 °C).

The soil cover varies according to the parent rock and mostly contains high ferric oxide and bauxite, which tinges them red. A mixture of lateritic, black and red soils is predominantly found over the area. These soils range from laterite and lateritic soils on the high plateau surfaces in the north along Dalbhum and in the south around Simlipal (Singh 1975). Yellow grey loams and black and brown soils are found within valley floors or in predominantly lowland areas along the trunk stream of Kharkai and its major tributaries. The tertiary soils contain gravel and grit with high alluvial content at Adityapur where the Kharkai meets the Subarnarekha. Loams of reddish yellow variety are found here with marked lateritic formations due to intense leaching by pluvial events, which are eroded and reworked by the initial stream orders that dissect the river basin. Admixture of kaolin, siliceous matter and potash makes the soils here wet and sticky, retaining moisture for a long time but which become hard and friable when dry.

6.3.3 Basin Lithology

The area is principally underlain by Precambrian igneous and metamorphic rocks (Fig. 6.3a). Some sedimentary rocks are found along the courses of the main channels and there are some laterite patches in the area. The majority of the exposed rocks are granitoids, which are often weathered and fractured and have gneissic banding. Besides these, amphibolite and metabasics intrusions occur within the granitoid mass. The basin also contains a great variety of sheared and foliated formations since it is situated along and close to some major shear zones and fault lines that have created great contortions in the surface topography, forming twisted linear ridges (Mahadevan 2002). The sedimentaries usually occupy higher elevations, overlying the basics and metamorphics below. The sheared rocks are of many varieties and show different stages of prograding metamorphism, with the presence of micas, phyllites, schists, migmatites and gneiss.

The entire basin area may broadly be divided into two broad geological provinces though even within these two major divisions, there are the occasional patches of rocks of a differing character (Fig. 6.3b). This first group comprises Achaeans of granite and gneiss and such meta-igneous rocks, in the central and southern part of the basin. The second group is an amalgamation of sheared, heavily contorted and foliated rocks belonging to the schist and phyllite group (centering around large formations of quartzites) that are cut and intruded into by many formations of amphiboles and pyroxenes in the northern to central portion of the basin. This group shows a wider variety of lithological variations than the first one that is largely uniform, except where intruded into by amphiboles and epidiorites in the extreme south, along linear narrow ridges. All formations are intruded by pegmatites and quartz veins in lineament swarms.

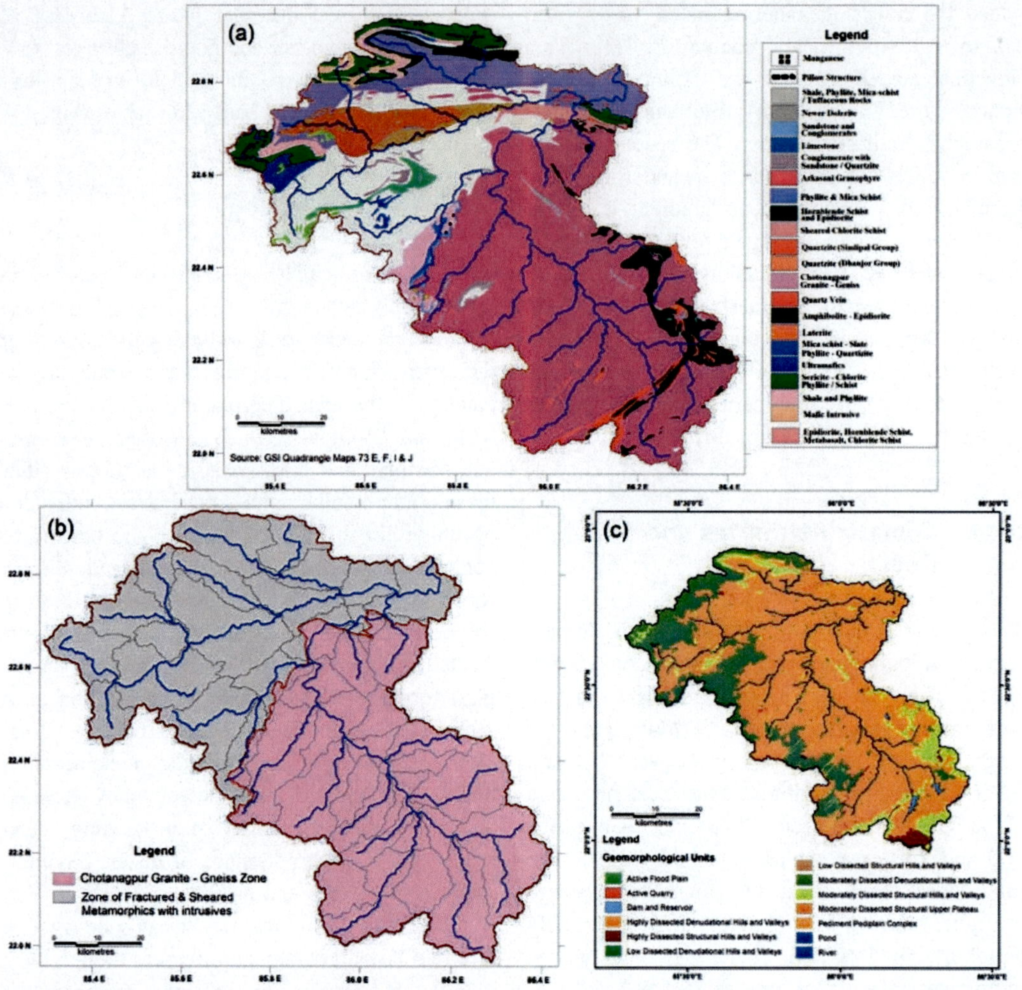

Fig. 6.3 Landscape attributes- **a** Detailed lithological units, **b** Broad lithological groups, and **c** Geomorphological units

6.3.4 Basin Geomorphology

The plateau landscape is heavily eroded by the drainage network of the Kharkai and its major tributaries. The pediment-pediplain complex comprises almost 70% of the basin landscape (Fig. 6.3c). Next is the moderately dissected denudational hills and valleys along the edges of the river basin which account for 16.5% of the landscape while moderately dissected structural hills and valleys comprise the higher elevations and form 8.5% and are the source regions of most streams. Only 1% of the basin geomorphic constituents is comprised of highly dissected structural hills and the smaller geomorphic units are very localized, contributing less than 1% of the landscape.

6.4 Results and Discussions

6.4.1 Enumerated Morphometric Parameters

Morphometric parameters consider variations in terrain attributes (shape/shape/alignment) and there are several parameters that reveal the varying aspects of the spatial geometry of the

Table 6.2 Morphometric parameters used for terrain analysis of the Kharkai River Basin

Parameter type	Morphometry base	Morphometric parameters	Enumeration	References
Terrain	Grid	Mean elevation	Average elevation of the river basin	
		Relative relief	Maximum Elevation—Minimum Elevation	Smith (1935)
		Average slope	$\tan^{-1}$ ((N * CI)/636.6) (N = Average no. of contour crossings in each grid, CI = Contour Interval and 636.6 is a constant)	Wentworth (1930)
		Dissection index	Relative Relief/Absolute Relief	Miller (1953) and Nir (1957)
		Surrogate of hypsometry	(Mean Elevation—Minimum Elevation)/ Relative Relief	Pike and Wilson (1971)
	Automated algorithm	Terrain surface texture	Spatial intricacy per unit area (i.e. drainage density and changes in sign of slope aspect or curvature)	Iwahashi and Pike (2007)
		Terrain surface convexity	Segregation of high and low relief features	
		Topographical wetness index	log [Flow Accumulation/tan (Slope in degree)]	Beven and Kirkby (1979)
		Topographical position index	Measuring topographic slope positions and to automate landform classifications	Guisan et al. (1999)
Drainage	Grid	Stream frequency	No. of Stream segments flowing through a grid/Grid Area	Horton (1945)
		Drainage density	Total Length of Streams of all order/Grid Area	
		Drainage texture	Drainage Density × Stream Frequency	Smith (1950)
Terrain & drainage		Ruggedness index	(Relative Relief × Drainage Density)/1000	Horton (1945)

landscape, each with their own benefits and weaknesses. The quantitative interpretation of the Kharkai River Basin using morphometric techniques provides a standardized scale of measurement and comparison of its various aspects. The statistical and graphical methods deal mainly with the relationship of the basin area to each surface attribute and their cartographical representation portrays the total character of the landscape. The type of morphometry, their method of extraction, individual formulae and description used to discuss each of them are presented in detail (Table 6.2).

The highest mean elevation zones are found along the extreme northern, north-western, south-eastern and eastern edges of the basin where it ranges between 750 and 800 m (Fig. 6.4a). The lowest values of 50–150 m abound around the confluence zone of the two major streams in the north-eastern section. Mostly, the mean elevation ranges between 150 and 200 m and comprises the entire central zone of the basin. Numerous small streams dissect the higher elevation zones as they descend to the lower heights. The highest relative relief zones are observed along the northern, north-eastern, north-western, south-eastern and southern fringes of the river basin, ranging between 400 and 500 m, while rest of the basin is highly dissected by numerous small streams for which the relative relief is low, ranging between 50 and 150 m.

In the Kharkai River Basin, the highest average slope values are obviously found in the higher elevation zones, i.e. along the northern and southern periphery, with values between 24° and 30° (Fig. 6.4b). On the whole, the greatest part of the basin has slope values ranging between 6° and 12°. Almost the entire basin area is moderately dissected (values 0–0.2°) by the Kharkai, its main tributaries and numerous smaller streams, except a few more affected patches in the southern, northern and south-eastern fringes where values are around 0.5° –0.9°, due to the higher ambient slope and elevation (Fig. 6.5a). Higher HI values denote a greater volume of material still to be eroded from within that basin grid area while lower values point to more eroded tracts (Fig. 6.5b). Higher dissection has caused most of the basin landscape to have relatively lower HI values between 0.2 and 0.4, while higher values are present along the water divides to the south-west, south and north, where they are between 0.6 and 0.8. The higher HI values in a landscape that is otherwise geologically quite old are possible indicators of an ambient topographic disequilibrium (e.g. Guha and Patel 2017).

The terrain surface texture (TST) shows that the central portion of the basin, through which the major rivers drain, have the least values, i.e. below 8, which means that the texture of the terrain surface is flat and therefore this corresponds to the pediplain-pediment complex (Fig. 6.6a). The foot slope areas of the dissected denudational hills are represented in lighter brown shades, with a value range of 8–16, and show a higher degree of texture roughness, while the highest range of > 32 is for the highland areas having remnants of few structural hills that are dissected by first order streams. The terrain surface convexity (TSC) map (Fig. 6.6b) shows that the higher elevation structural and denudational hills are having a clear convex slope with a value of more than 60, and stretching out till the foot slopes or pediments. Followed by this, the continuation of this value range of 40–60 clearly shows the undulating slopes of the pediment areas within the river basin of the Kharkai. Furthermore, the lowest range values of the below 20 group is the actual pediplan complex where the trunk streams of Kharkai and its major tributaries are continuously eroding the landscape in due achievement of gradation. The TWI map (Fig. 6.6c) shows that the Kharkai basin has negative values, i.e. below –10 in the higher elevation areas implying that these zones have higher runoff and least surface saturation. Contrastingly, the primary drainage area has higher values ranging from – 1 to + 10, indicating how surface saturation along the stream courses increase and this rises to > 10 in those tracts that are adjacent to the central drainage network of the Kharkai and its tributaries.

In the Kharkai basin, the TPI values predominantly range between –5 and + 5, implying that the topography is mostly dissected by denudational process, i.e. the major streams

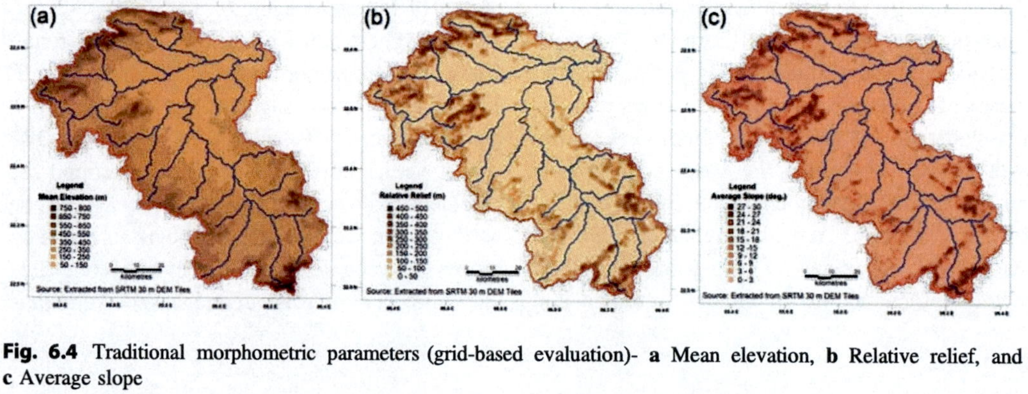

Fig. 6.4 Traditional morphometric parameters (grid-based evaluation)- **a** Mean elevation, **b** Relative relief, and **c** Average slope

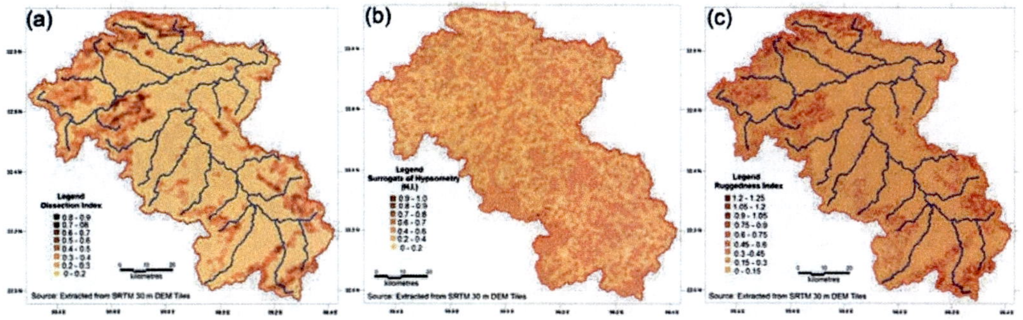

Fig. 6.5 Traditional morphometric parameters (grid-based evaluation)- **a** Dissection Index, **b** Hypsometry, and **c** Ruggedness index

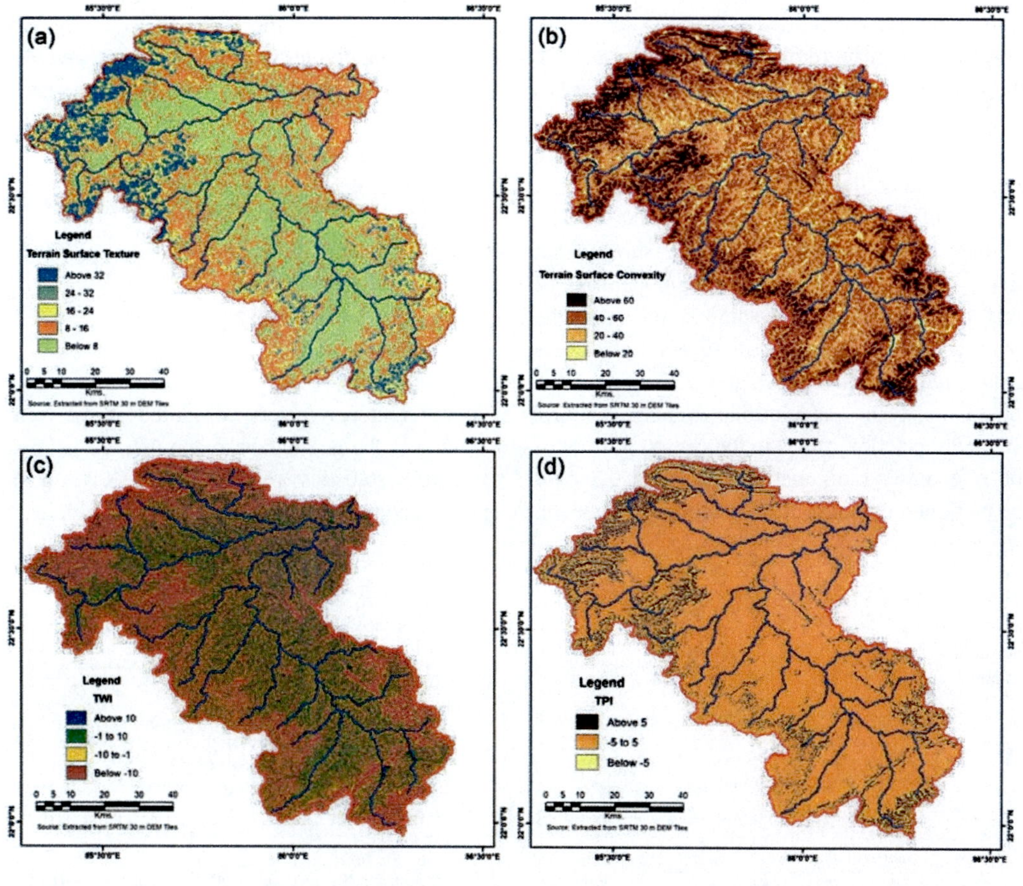

Fig. 6.6 New topographic variables (DEM pixel-based evaluation)- **a** Terrain surface texture, **b** Terrain surface convexity, **c** Topographic wetness index, and **d** Topographic position index

(Fig. 6.6d). The higher altitude areas show values > 5, especially in the structural and denudational hill complex around the fringes of the river basin. Stream frequencies in the Kharkai basin range between 10 and 20 stream segments per sq. km. (Fig. 6.7a), except for some patches in the south-central, north-eastern and north-western parts, where the stream frequency is between

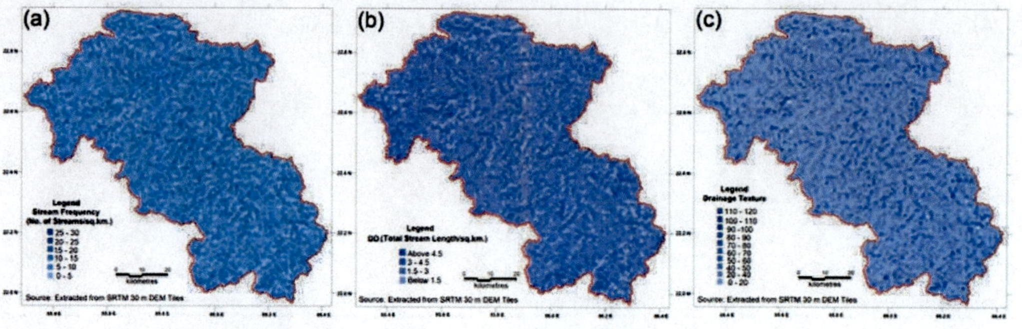

Fig. 6.7 Drainage parameters- **a** Stream frequency, **b** Drainage density, and **c** Drainage texture

25 and 30, along the higher water divides that are drained by numerous smaller streams. The drainage density map of the Kharkai River basin (Fig. 6.7b) reveals that almost the entire area has a drainage density of 3.0 to 4.5 km/sq.km, with few patches ranging above 4.5 km/sq.km, along the principal confluence zones of the major trunk streams of the Kharkai and its tributaries. The drainage texture map (Fig. 6.7c) shows that drainage texture values lie between 40 and 70 for most of the basin, except some zones along the main stream course and its tributaries where values range between 90 and 120. Thus, the region mostly has an ultrafine drainage texture. Ruggedness index values in the central portion of the river basin range between 0.0 and 0.3 over the pediplain complex, with this being an uneven plain or almost flat landscape drained by the major tributaries and the Kharkai itself (Fig. 6.5 c). The water divides exhibit greater ruggedness due to the presence of moderately dissected structural hills and valleys carved out by the numerous streams issuing forth therein.

6.4.2 Statistical Analysis of enumerated Morphometric Parameters

The different parameters computed above were arranged in the geo-database and their respective descriptive statistics were computed (Table 6.3). Their computed correlation coefficients

Table 6.3 Descriptive statistics for the enumerated morphometric variables

Measures	RR	AS	DI	RI	SF	DD	DT	HI
Mean	73.70	5.89	0.17	0.14	8.61	2.30	23.32	0.38
Median	32.01	3.82	0.12	0.07	8.50	2.35	19.75	0.39
Mode	16.99	3.15	0.07	0.00	9.00	0.00	0.00	0.44
Standard deviation	78.51	4.92	0.12	0.16	4.12	1.03	17.86	0.11
Sample variance	6164.18	24.25	0.01	0.03	16.99	1.07	319.07	0.01
Kurtosis	2.19	1.57	0.30	6.39	−0.06	−0.31	1.23	−0.16
Skewness	1.65	1.49	1.11	2.28	0.18	−0.18	1.05	−0.11
Range	439.01	28.11	0.61	1.30	29.00	5.43	118.28	0.71
Minimum	6.00	0.00	0.02	0.00	0.00	0.00	0.00	0.02
Maximum	445.01	28.11	0.63	1.30	29.00	5.43	118.28	0.73
Count	6616	6616	6616	6616	6616	6616	6616	6616

Note RR: Relative Relief; AS: Average Slope; DI: Dissection Index; RI: Ruggedness Index; SF: Stream Frequency; DD: Drainage Density; DT: Drainage Texture; HI: Hypsometric Integral

Table 6.4 Correlation coefficients among the enumerated morphometric parameters

	RR	AS	DI	RI	SF	DD	DT	HI
RR	1.00							
AS	0.86	1.00						
DI	0.93	0.83	1.00					
RI	0.72	0.76	0.73	1.00				
SF	−0.29	−0.14	−0.27	0.15	1.00			
DD	−0.33	−0.17	−0.30	0.20	0.83	1.00		
DT	−0.28	−0.18	−0.27	0.14	0.93	0.90	1.00	
HI	0.05	0.03	−0.09	−0.22	−0.35	−0.40	−0.37	1.00

Note RR: Relative Relief; AS: Average Slope; DI: Dissection Index; RI: Ruggedness Index; SF: Stream Frequency; DD: Drainage Density; DT: Drainage Texture; HI: Hypsometric Integral

(Table 6.4) reveal their interlinkages. Obviously, the topographic variables (RR, AS, DI and RI) and the drainage variables (SF, DD and DT) are strongly and positively correlated among themselves. Mostly, the drainage and topographic parameters are inversely related with each other, pointing to the paucity of drainage development in the higher elevations within the basin and the concentration of streamlines along the lower valley floors.

6.4.3 Factor Analysis and delineation of Morphometric Regions

While a correlation matrix states the relationship type between pairs of variables, factor analysis employs a 'factor loadings' matrix, that distinguishes the various 'basic' or 'abstract' variables, expressing the relationship level between these and the original variables and provides a simplified data matrix, i.e. the 'factor score' (or weightings) matrix. Beginning the analysis without any preconceived ideas as to the relative importance of the variables, it is initially reasonably presumed that each variable contributes the same amount of information to the study, and the lengths of the corresponding vectors arc standardized. To ensure that all the vectors have the same origin, the mean of each variable is made equal to zero. Using such standardized vectors, the cosine of the angle separating any two vectors equals the coefficient of correlation between the corresponding variables.

Data reduction from the original eleven variables to a few significant factors was obtained by extracting and rotating the factor-loadings matrix, and the corresponding relationships were noted in terms of the percentage contributed to the variance of each variable by each rotated factor (Table 6.5). Four factors were significant, as they explained >80% of the total variation. The initial loadings on Factor-1 seemed to align closely with the variances. Factor One scores were thus mapped across the Kharkai Basin (Fig. 6.8) to derive regions which are the best representative amalgamation of the combinations of the different morphometric variables. The prepared maps show good visual correlation with the different terrain features and using this a number of terrain units were demarcated (Table 6.6). The proportion of basin area falling under each unit was also enumerated. It shows that the Kharkai Basin is mostly comprised of an undulating, at times broken, plain surface with local higher tracts. The unevenness of topographic classes is greater in the northern and western parts of the basin over the zone of fractured and sheared high grade metamorphics than over the Chota Nagpur granite-gneiss complex that forms the rest of the region. These maps of Factor One Scores describe the various types landforms developed from intensive and continuous riverine erosion. The central part of the basin has values between –2 and 0, indicative of gently sloping undulating plains. The northern fringes show summital convexities and some

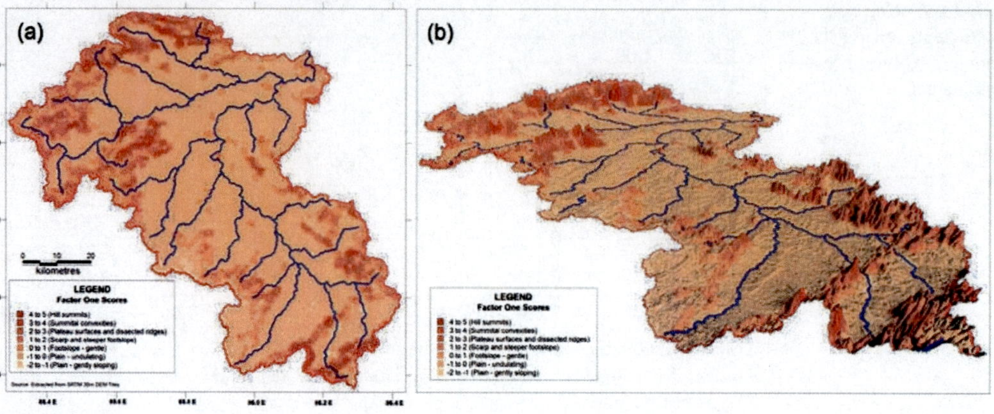

Fig. 6.8 Factor One score-based landscape unit delineation in the Kharkai River Basin- **a** Planform view, and **b** 3-D perspective

hilly peaks, the north-western and west-central zones show dissected ridges and plateau surfaces with some intermittent heights. In the south, the range of topographic variation is from 0 to 3, i.e. mostly gentle foot slopes to dissected ridges and plateau surfaces.

6.4.4 Relating changing LULC attributes with the Terrain Units

Modification of the natural landscape by human activities leaves profound impacts and changing LULC patterns are a major component of many current ecological concerns, being recognized as key drivers of environmental change (Turner et al. 1993; Iqbal and Sajjad 2014). A comparative analysis of LULC units within the Kharkai basin was thus done to elicit the major changes within the Kharkai river basin from 1990 to 2020 (Fig. 6.9). There are six major LULC classes (Table 6.7), which are typically fragmented and conform to the usual types seen in the Chota Nagpur Plateau region (e.g. Chatterjee and Patel 2016; Sarkar and Patel 2016) and marked changes have occurred in each. Water bodies (which includes ponds and lakes) were seen to shrink with increasing prevalence of agriculture and forestry. Vegetation cover has depleted by over 246.99 km^2 and fallow land have declined by 243.44 km^2 from 1990 to 2020, with both of these contributing to more areal coverage under agriculture, which has seen a sharp rise of 494.38 km^2. Settlement coverage and those of mines and quarries have risen markedly during this time period, particularly the huge urban expansion of Jamshedpur and Adityapur, at the mouth of the river in the north-eastern part of the basin. While the rivers have also shown a minimal decline in its share of LULC units by 15.15 km^2, this may be due to image classification issues as well as from the continuous encroachment onto the river bed as a result of sand mining and farming on the principal stream and its tributaries.

The interrelations among the lithology, major geomorphic units, LULC with the factor-one score based topographic units and the automated algorithm-based geomorphometric parameters needs to be delved into further to understand the landscape fabric (Table 6.6). Higher TST values can be well related to the higher topographic classes discerned from factor one scores (like the hill summits and summital convexities) as the maximum and mean values express the high amount of ruggedness with greater values. The contrary is observed for the gentle undulating slopes in the pediment and pediplain complex, consisting of LULC units like agricultural lands, water bodies, settlements and fallow lands. The higher TSC values demarcating the hill tops and convex slopes of the moderately to highly

Table 6.5 Explanation of total variance

Component	Initial Eigen values					Extraction Sums of Squared Loadings			Rotation Sums of Squared Loadings		
	Total	% of Variance	Cumulative %		Total Variance	% of %	Cumulative	Total Variance	Total	% of Variance	Cumulative %
1	5.574	50.677	50.677	5.574	50.677	50.677	4.328	39.346	4.328	39.346	39.346
2	2.862	26.015	76.692	2.862	26.015	76.692	2.987	27.158	2.987	27.158	66.504
3	1.209	10.995	87.687	1.209	10.995	87.687	1.977	17.971	1.977	17.971	84.475
4	0.665	6.045	93.732	0.665	6.045	93.732	1.018	9.257	1.018	9.257	93.732
5	0.274	2.489	96.221								
6	0.181	1.646	97.867								
7	0.132	1.196	99.063								
8	0.055	0.503	99.566								
9	0.034	0.307	99.873								
10	0.012	0.108	99.982								
11	0.002	0.018	100								

Extraction Method: Principal Component Analysis with Varimax Rotation using Kaiser Normalization

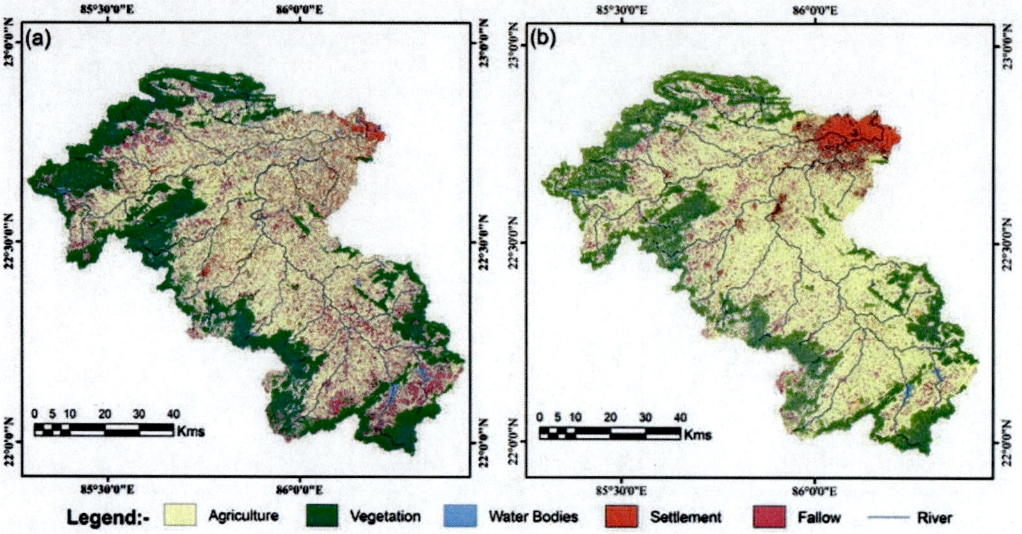

Fig. 6.9 LULC attributes in the Kharkai River Basin for the years- **a** 1990, and **b** 2020

dissected denudational hills and valleys predominantly have natural vegetation cover (forests). The other extreme of the TSC value range shows that lower mean values represent gentle or undulating slopes of the pediplain complex where the convexity is almost nil or near to zero. In case of TWI values, higher elevation areas from hill top summits to footslopes, highlight tracts where the amount of runoff is higher. The pediplain zone having the major rivers, settlements and agricultural lands are represented by low TWI scores and are underlain either by the Chota Nagpur gneiss-granite complex or the fractured and sheared metamorphic lithology that contains major intrusions. Lastly, the TPI values indicate the position of the topographic classes and is one of the best parameters to validate the morphometric regions derived based on the factor one scores. Extremely low TWI scores represent the central section of the river basin that is drained by major streams and is therefore ideal for practicing agriculture due to presence of major water bodies and available fallow land cover in the pediplain complex. The higher TWI values corroborate quite perfectly with the factor one scores in showing higher elevations (scarplands, summital convexities and the hill summits majorly) that are moderate to highly dissected.

As is obvious, the nature of the terrain has an influence on the kind of landscape modification undertaken within the basin and the different terrain units demarcated on the basis of the morphometric attributes. Such modifications of the ambient geomorphic diversity of the Chota Nagpur region and its adjoining tracts through anthropogenic activities are quite common (Patel and Mondal 2019; Patel et al. 2020), while this also has a marked impact on the streamside ecology and degree of riparian naturalness and vegetation cover (Saha et al. 2020) along the principal river corridors (cf. Banerji and Patel 2019). The smaller hills and ridges and hill summits have been largely levelled in a series of broad terraces to enable cultivation (Fig. 6.10a) while the higher dissected plateau tracts have undergone marked clearing in their original vegetation cover (Fig. 6.10b). In the piedmont-pediplain zone and the footslopes of the ridges,

Table 6.6 Different topographic classes or morphometric regions delineated using Factor One scores

Factor score one range	Topographic class	Area (sq.km)	Proportion of Basin Area (%)	Main lithological attributes	Major landform assemblages	Dominant LULC Class	Terrain Surface Texture			Terrain Surface Convexity			Topographical Wetness Index			Topographical Position Index		
							Max	Min	Mean	Max	Min	Mean	Max	Min	Mean	Max	Min	Mean
−2 to −1	Plain—gently sloping, almost level	92.09	1.47	Chotanagpur granite-gneiss and fractured and sheared metamorphics with intrusives	Pediplain complex and active floodplain	Agriculture, water bodies, rivers, settlements and fallow land	29.37	0.00	4.63	55.43	0.00	36.46	6.54	−18.52	−13.62	3.98	−3.06	0.01
−1 to 0	Plain—undulating and dissected	4100.28	65.35	Chotanagpur granite-gneiss, fractured and sheared metamorphics, mafic intrusives, laterite and conglomerate/quartzite	Pediplain complex	Agriculture, water bodies, rivers, settlements and fallow land	32.79	0.00	3.85	60.49	0.00	37.31	13.52	−19.59	−13.53	7.44	−5.36	−0.00
0 to 1	Footslope—gentle	1073.10	17.10	Chotanagpur granite-gneiss, phyllite, mica schist and fractured and sheared metamorphics with intrusives	Pediment-pediplain complex	Agriculture, settlements and rivers	37.10	0.17	10.88	61.43	1.9405	41.41	10.11	−19.63	−15.20	10.53	−6.72	−0.03
1 to 2	Scarp and steeper footslope	674.58	10.75	Chotanagpur granite-gneiss, hornblende, schist and epidiorite and ultramafic instrusives	Pediment complex and low dissected structural/denudational hills	Vegetation cover and rivers	40.17	0.41	15.05	63.15	8.791	43.60	8.88	−20.04	−16.04	11.93	−7.58	−0.00
2 to 3	Plateau surfaces and dissected ridges	259.65	4.14	Chotanagpur granite-gneiss, amphibolite-epidiorite and limestone	Moderate to low dissected denudational/structural hills and valleys	Vegetation cover and rivers	37.30	1.16	15.98	65.09	10.072	45.25	4.55	−20.48	−16.43	15.95	−9.75	0.04
3 to 4	Summital convexities	66.53	1.06	Chotanagpur granite-gneiss, amphibolite-epidiorite, serite-chlorite, phyllite/ schist and fractured and sheared metamorphics with intrusives	Moderate to low dissected denudational/structural hills and valleys	Vegetation cover	35.81	2.56	16.06	63.92	27.582	46.20	6.33	−20.49	−16.57	11.90	−15.86	0.07
4 to 5	Hill summits	7.71	0.12	Chotanagpur granite-gneiss, amphibolite-epidiorite and fractured and sheared metamorphics with intrusives	High to Moderate dissected denudational/structural hills	Vegetation cover	32.59	2.17	15.15	63.74	27.992	46.05	0.80	−20.82	−16.71	14.20	−11.70	−0.01

Table 6.7 Changes in LULC observed within the Kharkai River Basin from 1990 to 2020

Sl. no	LULC class	Years		Change matrix 2020 −1990	Interpretation
		2020	1990		
		Area (sq. km.)			
1	Water bodies	18.61	60.31	−41.70	Major shrinkage and drying up of water bodies
2	Vegetation	1547.62	1794.61	−246.99	Depletion of vegetation cover
3	River	97.44	112.59	−15.15	Decrease of riverine drainage due to drying up of streams and encroachment of river beds for agriculture
4	Settlement	357.79	304.89	+52.90	Massive urban expansion—Jamshedpur and Adityapur
5	Fallow	254.37	497.81	−243.44	Major decline in fallow land coverage
6	Agriculture	3979.92	3485.54	+494.38	Major rise of area under agriculture

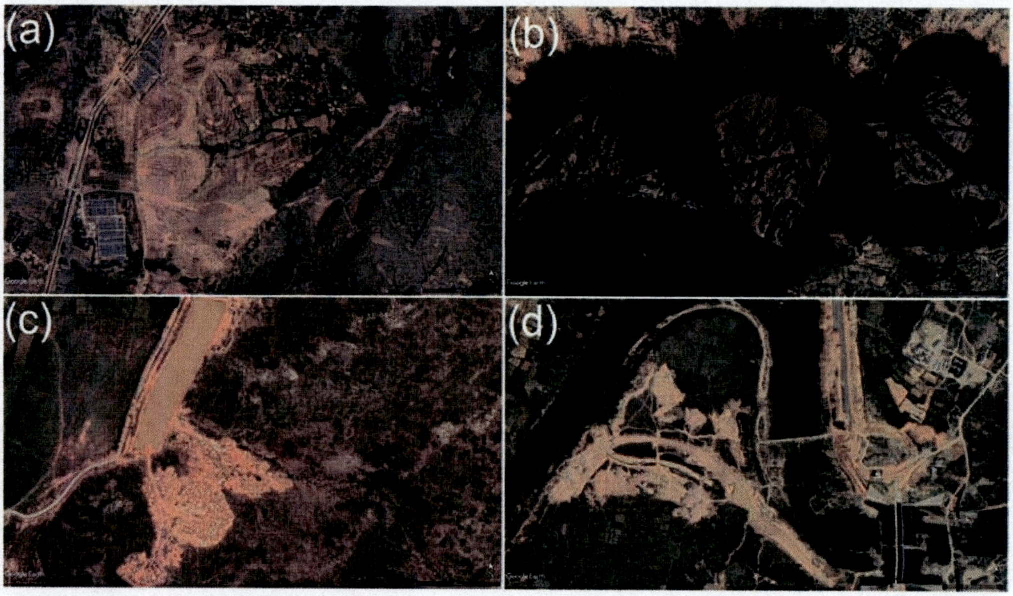

Fig. 6.10 Google Earth screenshots of LULC changes in the Kharkai Basin- **a** Planation and deforestation of the small hills for agriculture, **b** Deforestation occurring in plateau surface and dissected ridges, **c** Hollowing of pediplain-piedmont and foot slope areas for local water harvesting, and **d** Excavations along gently undulating surfaces and floodplains for irrigation and transport infrastructure. *Note* All images are aligned to north vertically upwards

local surface water harvesting measures have formed landforms of excavation and accumulation (Szabo 2006), with ponds dug up and mounds of earth piled around (Fig. 6.10c). Along the gently sloping floodplains and undulating surfaces by the major stream courses, more extensive earth-moving measures have ensued to create transport and irrigation facilities (Fig. 6.10d). Further assessment of the soil loss occurring from the basin as a result of these LULC changes can be discerned by using methods like the Universal Soil Loss Equation (Majhi et al. 2021). Further assessment of the soil loss occurring from the basin as a result of these LULC changes can be discerned by using methods like the Universal Soil Loss Equation (Majhi et al. 2021).

6.5 Conclusion

Landscape characterization has been performed in this study to develop an understanding about how different geomorphometric parameters are derived based on conventional grid system and from recently developed automated algorithms in a geospatial environment. The central section of the Kharkai River basin was found to be heavily eroded by the principal tributaries and by the Kharkai itself, while the edges of the river basin are composed of high to moderately dissected denudational ridges and some structural hills and valleys. The various morphometric parameters were statistically correlated with each another to delineate distinct terrain units using PCA. Strong positive correlations were obtained among the drainage and topographic variables while inverse relations were elicited between the drainage and topographic parameters. An abundance of drainage development in the higher elevations and clustering of major tributaries along the lower valley floors was discerned. The correlation matrix was further used to develop major physiographic zones using factor one scores to decipher the most pertinent topographic classes to describe the landscape of this river catchment. Two major lithological formations, i.e. the Chota Nagpur granite-gneiss complex in the south and the fractured and sheared metamorphics containing intrusives in the northern half of the river basin were corroborated with the present-day LULC units. The respective areal coverage under settlements and agriculture has seen a major rise in the past thirty years while there has been a considerable decline in the natural vegetation cover (forests), fallow land and water bodies. Some evidences from Google Earth imagery were collected across the basin to showcase the major alterations occurring under the various topographic classes. The need for planning and implementation of land and water conservation schemes can hence be practiced for the most anthropogenically modified and erodible portions to protect the landscape from further degradation.

Acknowledgements JM and PPP would like to acknowledge the Department of Geography, Presidency University for providing with the necessary infrastructure to conduct this research work. JM would also like to express his sincere gratitude to various experts from regional, state and international conferences where this paper was presented for giving their constructive views and helpful suggestions which proved to be immense helpful for improving certain parts of this research paper. Ms. Anuva Chowdhury, Coordinator, Policy and Impact, Partners in Prosperity, New Delhi, is also gratefully acknowledged for her kind help with certain maps and fruitful discussions to improve this research paper further.

References

Agarwal CS (1998) Study of drainage pattern through aerial data in Naugarh area of Varanasi district, U. P. J Indian Soc Rein Sens 24(4):169–175

Altaf S, Meraj G, Romshoo SA (2014) Morphometry and land cover based multi-criteria analysis for assessing the soil erosion susceptibility of the western Himalayan watershed. Environ Monit Assess 186 (12):8391–8412

Aparna P, Nigee K, Shimna P, Drissia TK (2015) Quantitative analysis of geomorphology and flow pattern analysis of Muvattupuzha River basin using geographic information system. J Aquatic Procedia 4:609–616

Asfaw D, Workineh G (2019) Quantitative analysis of morphometry on Ribb and Gumara watersheds: Implications for soil and water conservation. Int Soil Water Conserv Res 7(2):150–157. https://doi.org/10.1016/j.iswcr.2019.02.003i

Ayalew L, Yamagishi H (2004) Slope failures in the Blue Nile basin, as seen from landscape evolution perspective. Geomorphology 57:95–116

Banerji D, Patel PP (2019) Morphological aspects of the bakreshwar river corridor, West Bengal, India. In: Das B, Ghosh S, Islam A (eds) Advances in micro geomorphology of lower ganga basin - Part I: fluvial geomorphology, Springer International Publishing, Cham, pp 155–189. https://doi.org/10.1007/978-3-319-90427-6_9

Barling RD, Moore ID, Grayson RB (1994) A quasi-dynamic wetness index for characterizing the spatial distribution of zones of surface saturation and soil water content. Water Resour Res 30:1029–1044

Beven KJ, Kirkby MJ (1979) A physically based, variable contributing area model of basin hydrology/Un modèle à base physique de zone d'appel variable de l'hydrologie du bassin versant. Hydrol Sci J 24(1):43–69

Bingner RL, Garbrecht J, Arnold JG, Srinivasan R (1997) Effect of watershed subdivision on simulation runoff and fine sediment yield. Transact ASAE 40(5):1329–1335

Biswas S, Sudhakar S, Desai VR (1999) Prioritisation of sub watersheds based on morphometric analysis of drainage basin: a remote sensing and GIS approach. J Indian Soc Remote Sens 27(3):155

Branson FA, Owen JB (1970) Plant cover, runoff, and sediment yield relationships on Mancos Shale in western Colorado. Water Resour Res 6(3):783–790

Burrough PA, McDonnell RA (1998) Principles of geographical information systems. Oxford University Press Inc., New York

Carlston CW (1966) The effect of climate on drainage density and streamflow. Hydrol Sci J 11(3):62–69. https://doi.org/10.1080/02626666609493481

Chatterjee S, Patel PP (2016) Quantifying landscape structure and ecological risk analysis in Subarnarekha sub-watershed, Ranchi. In: Mondol DK (ed) Application of geospatial technology for sustainable development. University of North Bengal, India, North Bengal University Press, Raja Rammohunpur, pp 54–76

Chattopadhyay S, Carpenter RA (1990) A theoretical framework for sustainable development planning in the context of a river Basin. Unpublished Report, East-West Centre, Hawaii, 327–335

Chorley RJ, Schumm SA, Sugden DE (1984) Geomorphology. Methuen, London

Clarke JI (1996) Morphometry from Maps. Essays in geomorphology, Elsevier Publications, New York, 235–274

Cox RT (1994) Analysis of drainage-basin symmetry as a rapid technique to identify areas of possible quaternary tilt-block tectonics: an example from the Mississippi embayment. Geol Soc Am Bull 106:571–581

Das S, Pardeshi SD, Kulkarni PP, Doke A (2018) Extraction of lineaments from different azimuth angles using geospatial techniques: a case study of Pravara basin, Maharashtra, India. Arab J Geosci 11(8):160. https://doi.org/10.1007/s12517-018-3522-6

Das S, Pardeshi SD (2018) Morphometric analysis of Vaitarna and Ulhas river basins, Maharashtra, India: using geospatial techniques. Appl Water Sci 8(6):158. https://doi.org/10.1007/s13201-018-0801-z

Das S, Patel PP, Sengupta S (2016) Evaluation of different digital elevation models for analyzing drainage morphometric parameters in a mountainous terrain: a case study of the Supin—upper tons Basin, Indian Himalayas. SpringerPlus **5**(1544). DOI: https://doi.org/10.1186/s40064-016-3207-0

De Martonne E (1934) Problemes Morphologiques du Bresil Tropical Atlantique. Annales De Geographie. 49:16–27

De Reu J, Bourgeois J, Bats M, Zwertvaegher A, Gelorini V, De Smedt P, ... Van Meirvenne M (2013) Application of the topographic position index to heterogeneous landscapes. Geomorphology 186:39–49. https://doi.org/10.1016/j.geomorph.2012.12.015

Deumlich D, Schmidt R, Sommer M (2010) A multiscale soil–landform relationship in the glacial-drift area based on digital terrain analysis and soil attributes. J Plant Nutr Soil Sci 173:843–851

Dingman SL (1978) Drainage density and streamflow: a closer look. Water Resour Res 14(6):1183–1187

Evans IS (1972) General geomorphometry, derivatives of altitude, and descriptive statistics. In Chorley RJ (ed) Spatial analysis in geomorphology. Harper and Row, New York, pp. 17–90

Evans IS (1984) Correlation structures and factor analysis in the investigation of data dimensionality: statistical properties of the Wessex land surface, England. In Proceedings of the Int. Symposium on Spatial Data Handling, Zurich., v 1. Geographisches Institut, Universitat Zurich-Irchel. pp 98–116

Francés AP, Lubczynski MW (2011) Topsoil thickness prediction at the catchment scale by integration of invasive sampling, surface geophysics, remote sensing and statistical modeling. J Hydrol 405:31–47

Gallant JC, Wilson JP (2000) Primary topographic attributes. In: Wilson JP, Gallant JC (eds) Terrain analysis: principles and applications. Wiley, New York, pp 51–85

Ghosh KG (2016) Spatial heterogeneity of catchment morphology and channel responses: a study of Bakreshwar River Basin, West Bengal. J Ind Geomorphol 4:47–64

Gizachew K, Berhan G (2018) Hydro-geomorphological characterization of Dhidhessa River Basin, Ethiopia. Int Soil Water Conserv Res 6:175–183. https://doi.org/10.1016/j.iswcr.2018.02.003

Goodbred SL Jr (2003) Response of the Ganges dispersal system to climate change: a source-to-sink view science the last interstade. Sediment Geol 162:83–104. https://doi.org/10.1016/S0037-0738(03)00217-3

Grohmann CH, Riccomini C (2009) Comparison of roving-window and search-window techniques for characterising landscape morphometry. Comput Geosci 35:2164–2169

Guha S, Patel PP (2017) Evidence of topographic disequilibrium in the Subarnarekha River Basin, India: a digital elevation model based analysis. J Earth Syst Sci 126:106. https://doi.org/10.1007/s12040-017-0884-1

Guisan A, Weiss SB, Weiss AD (1999) GLM versus CCA spatial modeling of plant species distribution. Plant Ecol 143:107–122

Gutema D, Kassa T, Sifan A, Koriche (2017) Morphometric analysis to identify erosion prone areas on the upper blue Nile using GIS: case study of Didessa and Jema sub-basin, Ethiopia. Intern Res J Eng Technol 04 (08)

Guth PL (2001) Quantifying terrain fabric in digital elevation models. Environ Legacy Military Operat Geol Soc Am Rev Eng Geol 14:13–25

Hammond EH (1954) A geomorphic study of the Cape Region of Baja California (Vol. 10). University of California Press

Harinath V, Raghu V (2013) Morphometric analysis using Arc GIS techniques a case study of Dharurvagu, south eastern part of Kurnool district, Andhra Pradesh, India. Int J Sci Res 2(1):2319–7064

Horton RE (1932) Drainage basin characteristics. EOS Trans Am Geophys Union 13(1):350–361

Horton RE (1945) Erosional development of streams and their drainage basins; hydrophysical approach to quantitative morphology. Geol Soc Am Bull 56 (3):275–370

Hurtrez JE, Sol C, Lucazeau F (1999) Effect of drainage area on hypsometry from an analysis of small-scale drainage basins in the Siwalik hills (central Nepal). Earth Surf Process Landform 24:799–808

Illés G, Kovács G, Heil B (2011) Comparing and evaluating digital soil mapping methods in a Hungarian forest reserve. Can J Soil Sci 91:615–626

Iqbal M, Sajjad H (2014) Watershed prioritization using morphometric and land use/land cover parameters of Dudhganga Catchment Kashmir Valley India using spatial technology. J Geophy Remote Sens 3:12–23. https://doi:https://doi.org/10.4172/2169-0049.1000115

Islam A, Barman SD (2020) Drainage basin morphometry and evaluating its role on flood-inducing capacity of tributary basins of Mayurakshi River. India. https://doi.org/10.1007/s42452-020-2839-4

Iwahashi J, Pike RJ (2007) Automated classifications of topography from DEMs by an unsupervised nested-means algorithm and a three-part geometric signature. Geomorphology 86:409–440

Iwahashi J (1994) Development of landform classification using the digital elevation model. Ann Disaster Prev Res Inst, Kyoto Univ 37(B-1):141–156 (in Japanese with English abstract and illustrations)

Iwahashi J, Kamiya I (1995) Landform classification using digital elevation model by the skills of image processing—mainly using the Digital National Land Information. Geoinformatics 6(2):97–108 (in Japanese with English abstract)

Iwahashi J, Watanabe S, Furuya T (2001) Landform analysis of slope movements using DEM in Higashikubiki area. Japan. Comput. Geotech. 27:851–865

Javed A, Khanday MY, Ahmed R (2009) Prioritization of sub-watersheds based on morphometric and land use analysis using remote sensing and GIS techniques. J Indian Soc Remote Sens 37(2):261. https://doi.org/10.1007/s12594-009-0079-8

Javed A, Khanday MY, Rais S (2011) Watershed prioritization using morphometric and land use/land cover parameters: a remote sensing and GIS based approach. J Geol Soc India 78(1):63. https://doi.org/10.1007/s12594-011-0068-6

Jensen SK (1991) Applications of hydrologic information automatically extracted from digital elevation models. Hydrol Proc 5(1):31–44

Kabite G, Gessesse B (2018) Hydro-geomorphological characterization of Dhidhessa River basin, Ethiopia. Int Soil Water Conser Res 6(2):175–183. https://doi.org/10.1016/j.iswcr.2018.02.003

Khare D, Mondal A, Mishra PK, Kundu S, Meena PK (2014) Morphometric analysis for prioritization using remote sensing and GIS techniques in a Hilly catchment in the state of Uttarakhand, India. Indian J Sci Technol 7(10):1650–1662

Krishnamurthy J, Srinivas G, Jayaraman V, Chandrasekhar MG (1996) Influence of rock types and structures in the development of drainage networks in typical hardrock terrain. ITC J 3–4:252–259

Kumar A, Pandey RN (1982) Quantitative Analysis of Relief of the Hazaribagh Plateau Region. Perspect Geomorphol 1:235

Kumar R, Kumar S, Lohani AK, Nema RK, Singh RD (2000) Evaluation of geomorphological characteristics of a catchment using GIS. Gis India 9(3):13–17

Langbein WB, Leopold LB (1964) Quasi-equilibrium states in channel morphology. Am J Sci 262(6):782–794

Leopold LB, Wolman MG, Miller JP (1995) Fluvial processes in geomorphology. Courier Corporation

Lesschen JP, Kok K, Verburg PH, Cammeraat LH (2007) Identification of vulnerable areas for gully erosion under different scenarios of land abandonment in southeast Spain. CATENA 71:110–121

Lin Z, Oguchi T (2004) Drainage density, slope angle, and relative basin position in Japanese bare lands from high-resolution DEMs. Geomorphology 63(3–4):159–173

Liu H, Bu R, Liu J, Leng W, Hu Y, Yang L, Liu H (2011) Predicting the wetland distributions under climate warming in the Great Xing'an Mountains, northeastern China. Ecol Res 26:605–613

Liu M, Hu Y, Chang Y, He X, Zhang W (2009) Land use and land cover change analysis and prediction in the upper reaches of the Minjiang River, China. Environ Manage 43:899–907

Mahadevan TM (2002) Text book series 14: Geology of Bihar and Jharkhand; Geol. Soc. India, Bangalore
Majhi A, Shaw R, Mallick K, Patel PP (2021) Towards improved USLE-based soil erosion modelling in India: a review of prevalent pitfalls and implementation of exemplar methods. Earth-Sci Rev 221:103786. https://doi.org/10.1016/j.earscirev.2021.103786
Maune DF (2001) Digital elevation model technologies and applications: the DEM user's manual (Bethesda, MD: American Society for Photogrammetry and Remote Sensing)
McGarigal K, Tagil S, Cushman S (2009) Surface metrics: an alternative to patch metrics for the quantification of landscape structure. Landscape Ecol 24:433–450
Merritts D, Vincent KR (1989) Geomorphic response of coastal streams to low, intermediate, and high rates of uplift, Mendocino junction region, northern California. Geol Soc Am Bull 101:1373–1388
Miller VC (1953) Quantitative geomorphic study of drainage basin characteristics in the Clinch Mountain area, Virginia and Tennessee. Technical report (Columbia University. Department of Geology); no. 3, 189–200
Moglen GE, Eltahir EA, Bras RL (1998) On the sensitivity of drainage density to climate change. Water Resour Res 34(4):855–862
Mora-Vallejo A, Claessens L, Stoorvogel J, Heuvelink GBM (2008) Small scale digital soil mapping in southeastern Kenya. CATENA 76:44–53
Morisawa ME (1959) Relation of morphometric properties to runoff in the Little Mill Creek. Department of Geology-Columbia University, Ohio
Mueller JE (1968) An introduction to the hydraulic and topographic sinuosity indexes. Ann Assoc Am Geogr 58(2):371–385
Mukhopadhyay SC (1980) Geomorphology of the Subarnarekha Basin: the Chota Nagpur Plateau. University of Burdwan, Eastern India
Nag SK, Chakraborty S (2003) Influence of rock types and structures in the development of drainage network in hard rock area. J Indian Soc Remote Sens 31(1):25–35
Nir D (1957) Maps of dissection index of terrain. Proceedings of the Yorkshire Geological Society, 29
Obi Reddy GP, Maji AK, Gajbhiye KS (2004) Drainage morphometry and its influence on landform characteristics in a basaltic terrain, Central India–a remote sensing and GIS approach. Int J Appl Earth Obs Geoinf 6(1):1–16
Obi Reddy GE, Maji AK, Gajbhiye KS (2002) GIS for morphometric analysis of drainage basins. Gislndia 11 (4):9–14
Ogden FL, Raj Pradhan N, Downer CW, Zahner JA (2011) Relative importance of impervious area, drainage density, width function, and subsurface storm drainage on flood runoff from an urbanized catchment. Water Res Res **47**(12)
Oguchi T (1997) Drainage density and relative relief in humid steep mountains with frequent slope failure. Earth Surface Proc Landforms J British Geomorphol Group 22(2):107–120
Ohmori H (1993) Changes in the hypsometric curve through mountain building resulting from concurrent tectonics and denudation. Geomorphology 8:263–277
O'Loughlin EM (1986) Prediction of surface saturation zones in natural catchments by topographic analysis. Water Resour Res 22(5):794–804
Pal SK (1972) A classification of morphometric methods of analysis: an appraisal. Geog Rev India 34(1):61–84
Pallard B, Castellarin A, Montanari A (2009) A look at the links between drainage density and flood statistics. Hydrol Earth Syst Sci 13(7):1019
Patel A, Katiyar KS, Prasad V (2016) Performances evaluation of different open source DEM using Differential Global positioning system (DGPS). Egyptian J Remote Sens Space Sci 19(1):7–16
Patel PP (2012) An exploratory geomorphological analysis using modern techniques for sustainable development of the Dulung river basin. Unpublished PhD Thesis, University of Calcutta, Kolkata. https://shodhganga.inflibnet.ac.in/handle/10603/156681
Patel PP (2013) GIS techniques for landscape analysis—case study of the Chel River Basin, West Bengal. Proceedings of State Level Seminar on Geographical Methods in the Appraisal of Landscape, held at Dept. of Geography, Dum Dum Motijheel Mahavidyalaya, Kolkata, on 20th March, 2012, pp 1–14.
Patel PP, Sarkar A (2009) Application of SRTM data in evaluating the morphometric attributes: a case study of the Dulung River Basin. Pract Geograp 13(2):249–265
Patel PP, Sarkar A (2010) Terrain characterization using SRTM data. J Indian Soc Remote Sens 38(1):11–24. DOI: https://doi.org/10.1007/s12524-010-0008-8
Patel PP, Mondal S (2019) Terrain—landuse relation in garbeta-I block, paschim medinipur district, West Bengal. In: Mukherjee S (ed) Importance and Utilities of GIS, Avenel Press, Burdwan, pp 82–101
Patel PP, Mondal S, Prasad R (2020) Modifications of the geomorphic diversity by anthropogenic interventions in the silabati river basin. In: Das BC, Ghosh S, Islam A, Roy S (eds) Anthropogeomorphology of bhagirathi-hooghly river system in India. Routledge, pp 331–356
Patel PP, Dasgupta R, Chanda S, Mondal S (2021) An investigation into longitudinal forms of gullies within the "Grand Canyon" of Bengal Eastern India. Transactions in GIS. https://doi.org/10.1111/tgis.12828
Pike RJ (2001) Topographic fragments of geomorphometry, GIS, and DEMs. In DEMS and Geomorphology, Geographic Information Systems Association (Japan) Special Publication. 5th International Conference on Geomorphology, Chuo University: Tokyo, Japan (Vol. 1, pp. 34–35)
Pike RJ, Wilson SE (1971) Elevation-relief ratio, hypsometric integral, and geomorphic area-altitude analysis. Geol Soc Am Bull 82(4):1079–1084

Qin CZ, Zhu AX, Pei T, Li BL, Scholten T, Behrens T, Zhou CH (2011) An approach to computing topographic wetness index based on maximum downslope gradient. Precision Agric 12(1):32–43. https://doi.org/10.1007/s11119-009-9152-y

Riebsame WE, Meyer WB, Turner BL (1994) Modeling land use and cover as part of global environmental change. Clim Change 28(1–2):45–64. https://doi.org/10.1007/BF01094100

Rodriguez-Iturbe I, Escobar LA (1982) The dependence of drainage density on climate and geomorphology. Hydrol Sci J 27(2):129–137

Saha D, Das D, Dasgupta R, Patel PP (2020) Application of ecological and aesthetic parameters for riparian quality assessment of a small tropical river in eastern India. Ecol Ind 117:106627. https://doi.org/10.1016/j.ecolind.2020.106627

Sarkar A, Patel PP (2011) Topographic analysis of the Dulung R. Basin. Indian J Spatial Sci II(1, Article 2)

Sarkar A, Patel PP (2012) Terrain classification of the Dulung drainage basin. Indian J Spatial Sci III(1, Article 6)

Sarkar A, Patel PP (2016) Land use—terrain correlations in the piedmont tract of eastern India: a case study of the Dulung river basin. In: Santra A, Mitra S (eds) Handbook of research on remote sensing applications in earth and environmental studies, IGI Global, USA, pp 147–193. DOI: https://doi.org/10.4018/978-1-5225-1814-3.ch008

Schumm SA (1956) Evolution of drainage systems and slopes in badlands at Perth Amboy, New Jersey. Geol Soc Am Bull 67(5):597–646

Shreve RW (1969) Stream lengths and basin areas in topologically random channel networks. J Geol 77:397–414

Singh RL (1975) Regional geography of India. National Geographical Society of India (Banaras Hindu University. Dept. of Geography. Varanasi. India), 654–655

Singh P, Gupta A, Singh M (2014) Hydrological inferences from watershed analysis for water resource management using remote sensing and GIS techniques. Egypt J Remote Sens Space Sci 17:111–121

Smith KG (1958) Erosional processes and landforms in Badlands national monument South Dakota. Bull Geol Soc 69:975–1008

Smith GH (1935) The relative relief of Ohio. Geogr Rev 25(2):272–284

Smith KG (1950) Standards for grading texture of erosional topography. Am J Sci 248(9):655–668

Soni S (2017) Assessment of morphometric characteristics of Chakrar watershed in Madhya Pradesh India using geospatial technique. Appl Water Sci 7:2089–2102. https://doi.org/10.1007/s13201-016-0395-2

Strahler AN (1952) Hypsometric (area-altitude) analysis of erosional topography. Geol Soc Am Bull 63 (11):1117–1142

Strahler AN (1954) Statistical analysis in geomorphic research. J Geol 62(1):1–25

Strahler AN (1957) Quantitative analysis of watershed geomorphology. EOS Trans Am Geophys Union 38 (6):913–920

Strahler AN (1964) Part II. Quantitative geomorphology of drainage basins and channel networks. Handbook of Applied Hydrology: McGraw-Hill, New York, 4–39

Subramanyan V (1981) Geomorphology of the Deccan volcanic province. Memoir-Geol Soc India 3:101–116

Sutradhar H (2020) Assessment of drainage morphometry and watersheds prioritization of Siddheswari River Basin, eastern India. J Indian Soc Remote Sens 1–18. https://doi.org/10.1007/s12524-020-01108-5

Szabo J (2006) Anthropogenic geomorphology: subject and system. In: Szabo J, David L, Loczy D (eds) Anthropogenic geomorphology: a guide to man-made landforms. Springer, Dordrecht, pp 3–12

Tagil S, Jenness J (2008) GIS-based automated landform classification and topographic, landcover and geologic attributes of landforms around the Yazoren Polje, Turkey. J Appl Sci 8:910–921

Tobler W (2000) The development of analytical cartography: a personal note. Cartogr Geogr Inf Sci 27 (3):189–194

Tucker GE, Bras RL (1998) Hillslope processes, drainage density, and landscape morphology. Water Resour Res 34(10):2751–2764

Turner B, Moss RH, Skole DL (1993) Relating land use and global land-cover change. IGBP Report # 24/HDP Report #5. Stockholm

Vaidya N, Kuniyal JC, Chauhan R (2013) Morphometric analysis using Geographic Information System (GIS) for sustainable development of hydropower projects in the lower Satluj river catchment in Himachal Pradesh, India. Int J Geomat Geosci 3 (3):464–473

Vijith H, Satheesh R (2006) GIS based morphometric analysis of two major upland sub-watersheds of Meenachil river in Kerala. J Indian Soc. Remote Sens 34(2):181

Wasson RJ (1994) Annual and decadal variation of sediment yield in Australia, and some global comparisons. IAHS Publications-Series of Proceedings and Reports-Intern Assoc Hydrological Sciences 224:269–280

Weiss AD (2000) Topographic position and landforms analysis. Poster http://www.jennessent.com/downloads/tpi-poster-tnc_18x22.pdf

Weiss AD (2001) Topographic position and landforms analysis. Poster Presentation, ESRI Users Conference, San Diego, CA

Wentworth CK (1930) A simplified method of determining the average slope of land surfaces. Am J Sci 117:184–194

Wilson JP, Gallant JC (2000) Primary topographic attributes. In: Wilson JP, Gallant JC (eds) Terrain analysis: principles and applications. John Wiley and Sons, pp 51–85

Wise S (2000) Assessing the quality for hydrological applications of digital elevation models derived from contours. Hydrol Process 14(11–12):1909–1929

Yokoyama H, Watanabe Y, Shimizu Y, Bousmar D, Zech Y (2002, September) Numerical simulation of sandbars using 2-D shallow water equation under unsteady flow. In River flow 2002. Proceedings of the international conference on fluvial hydraulics. Louvain-La-Neuve, Belgium (pp. 4–6)

Zhang Y (2005) Global tectonics and climatic control of mean elevation of continents and Phanerozoic sea level change. Earth Planet Sci Lett 237:524–531. https://doi.org/10.1016/j.epsl.2005.07.015

7 River Raidak-I Migration Dynamics Within Himalayan Foreland Basin Applying Quaternary Sedimentological Bank Facies and Geospatial Techniques

Supriya Ghosh and Biswajit Bera

Abstract

River channel migration is a significant phenomenon in the alluvial tract of Himalayan foreland basin in India. Natural processes such as rapid sedimentation due to low channel gradient, high discharge during monsoon and non-cohesive bank materials primarily control the channel migration pattern and erosion-accretion dynamics of the river system in quaternary geological sites of sub-Himalayan alluvial floodplain of West Bengal. Additionally, human interference in the fluvial setting also regulates the channel dynamics in this area. This study deals with lateral channel migration rates and erosion-accretion dynamics of different study reaches of river Raidak-I within the sub-Himalayan alluvial tract of West Bengal, India. The study of geospatial techniques and quaternary sedimentary bank facies hasbeen used to assess the reach based channel migration rates and erosion-accretion dynamics. Channel planform shows ameandering pattern recording sinuosity index 2.14 in 2019. Variation of channel width (mean channel width of entire reach was 139.03 m in 1972, 146.84 m in 1979, 88.58 m in 1996, 107.76 m in 2009 and 93.94 m in 2019), channel sinuosity (2.42 in 1972 and 2.14 in 2019) and the mean of all radius of curvature along the meander loops (237.5 m in 1972 and 181.25 m in 2019) reflects the channel instability in different study reaches over 47 year spanning of study period. The study reveals that average rate of net migration along the left bank is higher in reach-2 (60.02 m/y) than that of reach-3 (33.12 m/y) and reach-1(3.73 m/y) while on right bank, it is comparatively higher in reach-1(96.38 m/y) than reach-2 (68.92 m/y) and reach-3 (25.81 m/y). Therefore, this scientific study would be helpful for policy making purposes during long term flood management.

Keywords

Channel migration · Erosion-accretion dynamics · Quaternary · Foreland · Bank facies · Channel sinuosityXE "channel sinuosity

7.1 Introduction

The nature and characteristics of river channel migration or avulsion on floodplain of an alluvial river is an interesting field of research in contemporary fluvial geomorphology. In the last few decades, many tropical populous countries of the world have incessantly occupied floodplain and different developmental activities have been executed along both sides of the rivers.

S. Ghosh · B. Bera (✉)
Department of Geography, Sidho-Kanho-Birsha University, Ranchi Road, Purulia, West Bengal 723104, India

P. K. Shit et al. (eds.), *Drainage Basin Dynamics*, Geography of the Physical Environment,
https://doi.org/10.1007/978-3-030-79634-1_7

Subsequently, landform building mechanism on fast changing landscape or more specifically Himalayan foredeep basin region has been partially disturbed. In the recent years, people have been facing devastating fluvio-hydrological hazard or disaster due to large scale anthropogenic stresses on river channel or dynamic floodplains. In case of alluvial channels, the responses have been registered in different magnitudes at different geomorphic scale on quaternary geological sites (Heitmuller 2014; Sinha and Ghosh 2012). The river channel migration simply means the change of river channel geometry across the cross section of the channel (Gregory 1977). Generally, migration of an alluvial channel occurs on floodplain due to the dual processes of bank erosion and accretion (Charlton 2008). River bank erosion is regarded as the most important factor for lateral channel migration. Besides, channel bed aggradation is also accountable factor to influence channel migration as sediment deposition in the form of channel bar which accelerates the rate of bank erosion particularly along the concave river bank side. Therefore, combined processes of aggradation and degradation within river channel bring channel instability which ultimately leads to channel migration (Ahmed and Das 2018; Ahmed et al. 2018). The magnitude of bank erosion is primarily dependent on nature of bank materials and its strength against high flood water (Knighton 1984). During monsoonal flood, river bank experiences bank full discharge and water molecules enter in between the particles (especially for sandy river bank). When water level goes down, bank material with water flow starts rapidly at the stage of liquefaction. The water pressure stress exceeds the material strength of river bank and natural river bank failure occurs automatically (Thorne and Tovey 1981; Thorne 1982; Bera et al. 2019b). However, channel migration process is carried out by the complex interaction among flow, sediment mobility and bank materials (Duan et al. 2001). In space–time context, lateral migration of channel is a mechanism of natural as well as anthropogenic agents (Wallick et al. 2007; Bera et al. 2019b). The river metamorphosis, flow disturbance, confinement of sediments etc. have been accelerated in the tropical and sub-tropical rivers of the world due to execution of different developmental projects or processes of channelization (construction of culverts, bridge, sluice gate and concrete embankments) within river basin. Linear structure mostly changes the geomorphic action of water resulting into floodplain modification (Szabo et al. 2010). Besides, cross-sectional interventions on river channel alters the channel morphology by modifying the discharge volume (Islam and Guchhait 2020). The longitudinal and horizontal engineering structures sometime obstruct river channel migration but occasionally these structures help to make new spill channels or directly channel avulsion within floodplain (Bera et al. 2019b). The regional or micro scale landscape modifications, shape and form of flood plain alterations exist due to lateral river channel migration (Thakur et al. 2012) and sometimes, floodplain human habitations have been extremely threatened for devastating bank erosion as well as switch over of meandering channels. In case of lower Ganga plain on quaternary geological sites of India, thousands of hectares of agricultural land along with settlements have been engulfed by the river course of Ganga due to lateral shifting or intersection of meanders (Bera et al. 2019b). More than 200 sq. km. land has been eroded from the left bank of Ganga in Malda district and that is about 356 sq. km. in Murshidabad district (Rudra 2011). Moreover, 195 villages have been affected due to bank erosion and channel migration along the bank of river Bhagirathi in three district namely Murshidabad, Nadia and Barddhaman in Gangetic West Bengal (Islam and Guchhait 2017). In the last two decades, the spatio-temporal confluence and off-take dynamics, sediment yield and channel bar dynamics have been analyzed applying remote sensing and geographical technique (Ophra et al. 2018). In the recent years, various research studies have been conducted by applying geospatial techniques to evaluate channel course modifications, degree of anthropogenic stress, cost benefit analysis and quantification of river bank erosion (Bera et al. 2019b). Similarly, geospatial studies such as channel migration, estimation of river bank erosion and accretion through overlay analysis (using historical maps) have been successfully done in different parts of the world for various channel types and patterns (Downward et al. 1994; Gurnell 1997). Since historical era, many authentic research works have been documented on channel migration and mechanism of different types of channel avulsion on quaternary floodplain geological sites (Baki and Gan 2012; Dury 1977; Knight 1975). Sarma et al. (2007) have carried out a scientific study on the lateral channel migration of Burhi Dihing river using remote sensing data from 1934 to 2004 and they have focused on regional geological factors which are principal responsible factors for river channel migration on fast changing landscape. Thorne et al. (1993) made a study on mechanism and rate of riverbank migration of the river Jamuna using Remote Sensing data and old maps. The applied fluvio-geomorphological study had been conducted on magnitude of accretion and erosion of the river Yellow in the year 2006 applying Landsat imageries (Chu et al. 2006) and also studied channel dynamics of river Brahmaputra (Takagi et al. 2007). The digital composite maps require very much to understand not only the rate of shifting but also integrated regional planning in the modern context.

The river Raidak, a right bank tributary of river Sankosh, originates from the Bhutan Himalaya and after flowing through the Himalayan terrain it enters into the plain of Alipurduar, West Bengal. After debouching from the Himalayas, river Raidak is bifurcated into two branches like Raidak I and Raidak II. The river Raidak-I takes meandering channel pattern beyond Himalayan Frontal Thrust (HFT) towards the downstream and largely flowing on the Himalayan foreland basin. This foreland basin is mainly interlacing drainage system and fast changing landscape due to active tectonic setup of the Himalayas and dynamic landscape. The Raidak-I river is characterized by enormous discharge(average peak discharge is 385.27 cumec at station Chepani and 13,240 cumec at Bhutanghat)with huge sediment loads(mean annual sediment load is 1.89 million m^3) particularly during rainy season, speedy channel bed aggradation, bank erosion and bank line change (Subba 2014). In the last century, river Raidak-I had been changed its course due to natural as well as anthropogenic stresses. Natural factors include flood, heavy rainfall, and silt deposition while anthropogenic factors like construction of bridge, railway, road, dams and deforestation are responsible for the modifications of the downstream reaches of river Raidak (Subba 2014). Here, the authors have considered the river Raidak-I which is well-known for its notorious channel dynamics, the vulnerabilities through bank erosion, floods and bank line shifting. This work is very unique because any research work on lateral channel migration and erosion-accretion dynamics have yet not been conducted on river Raidak-I and previously, no one had tried to find out the nature of lateral channel migration trends of Raidak-I flowing over the sub-Himalayan Alipurduar and Cooch Behar district of West Bengal. Thus the main objective of the scientific paper is (i) to assess the nature of channel migration and (ii) to measure the erosion and deposition rate along the middle and lower reaches of river Raidak-I. The present research has been tried to analyze the nature of historical channel modifications in the Sub-Himalayan foreland basin within the territory of West Bengal for a time span of 47 years (1972 to 2019) based on remotely sensed data, GIS mapping and field based experiments. This study focuses on new dimension for short and long term mechanism of channel migration of river Raidak-I as there is no as such past studies conducted on river Raidak-I. Thus, the results of scientific study will help directly for sustainable integrated floodplain management exclusively for the dynamic landscape at the proximity of Himalayas.

7.2 Study Area

The study area falls under the tectonically active Himalayan foreland basin and it is bounded within $26^{0}12$/N to $26^{0}27$/N latitudes and $89^{0}36$/E to $89^{0}48$/E longitudes. The river Raidak, a right bank tributary of river Sankosh, originates from the Mt. Akungphu of the Bhutan Himalaya at an elevation of 6400 m. It flows through the three countries namely Bhutan, India and Bangladesh. In the mountain tract of Bhutan, the river has many tributaries and it is locally known as the Wong Chu, Paro Chu and Ha Chu. After flowing through the hilly tracts, it debouches into the plains of Alipurduar district at Bhutan Ghat and then flows through Cooch Behar district in the state of West Bengal in India. At Tiyabari in Alipurduar district, the river is bifurcated into Raidak I and Raidak II. The western branch or Raidak-I joins with river Torsha near Balabhut village in Tufanganj-I block of Coochbehar district and eastern branch or Raidak-II falls into river Sankosh at Bainaguri in Baxirhat block of Coochbehar. Before joining with the river Sankosh, Raidak-II drains about 349 km^2 of Indian territory. The catchment area of Raidak-I is 760 km^2 within the Indian territory (421 sq.km in Alipurduar district and 339 sq.km in Coochbehar district). The total length of the river Raidak-I in the Indian part is 90 km (48 km in Alipurduar district and 42 km in Coochbehar district). Gadadhar and Dhaksi are the main tributaries of Raidak-I.

Physiographically, the study area stands on Sub-Himalayan alluvial fan zone having altitude ranging between less than 50 and 100 m. There is no as such slope variations have been noticed. Geologically, the area is covered with Holocene newer sediments (Bandyopadhyay et al. 2014). Very fine sand, silt and clay and in some places light gray silt loams in varying proportions are found along the middle and lower courses of Raidak-I and its floodplain. The floodplain of the upper course of the river is mainly covered with Pleistocene older alluvium (Fig. 7.1).

Sandstone and shale have been exposed in the source region of theriver Raidak-I (Fig. 7.2).

The Sub-Himalayan foreland basin is one of the most tectonically active foreland basins in the world. It has been emerged due to the obliquity of Indian plate owing to the huge crustal load of the gradually rising Himalaya (Molnar 1984; Lyon-Caen and Molnar 1985; Duroy et al. 1989) and the river system has been directly influenced by active tectonic movements of the Himalayas (Holbrook and Schumm 1999; Schumm et al. 2002). As the region comes under tropical monsoon climatic zone, this region is received maximum rainfall in monsoon season (from June to September). In monsoon season, the river is characterized by high discharge and high velocity (maximum average velocity 1.8 m/sec). The average peak discharge is 385.27 cumec at station Chepani (Table 7.2). Silt content is also high in rainy season (1.3gm/lt) while it becomes low in non-monsoon season (0.02gm/lt). The lower courses of riverRaidak-I have low to moderate sloping plain which helps huge accumulation of sediments along the channel path especially during monsoon. The river bank materials along with floodplain sediments move towards river Raidak-I by surface and sub-surface runoff during monsoon and catastrophic flood. The present study around 54.4 km river course of the Raidak-I (in downstream from the confluence point of river Dhaksi and Raidak-I near Uttar Chikliguri up to Balabhut in Tufanganj-I block of Coochbehar district) has been considered and about 54.4 km length of this river has been subdivided into 3 reaches for detailed fluvio-geomorphological analysis (Fig. 7.5a).

7.3 Materials and Methods

7.3.1 Data Acquisition

In order to find out the nature and amount of bank line shifting, satellite data have been obtained from different sources. Few research works have been done applying the given techniques in the terai-dooars region and abroad (Lawler et al. 1997; Gogoi and Goswami 2014; Sarkar et al. 2012; Robinson 2013; Nath et al. 2013). In order to identify 47 years channel migration trend, five satellite imageries of the year 1972, 1979, 1996, 2009 and 2019 have been collected from United States Geological Survey (USGS) website. LANDSAT 1–5 MSS of 1972 and 1979 (Path/Row: 148/42; Spatial Resolution: 80 m), LANDSAT-4/5 TM of 1996 (Path/Row: 138/41, 138/42; Spatial Resolution: 30 m),

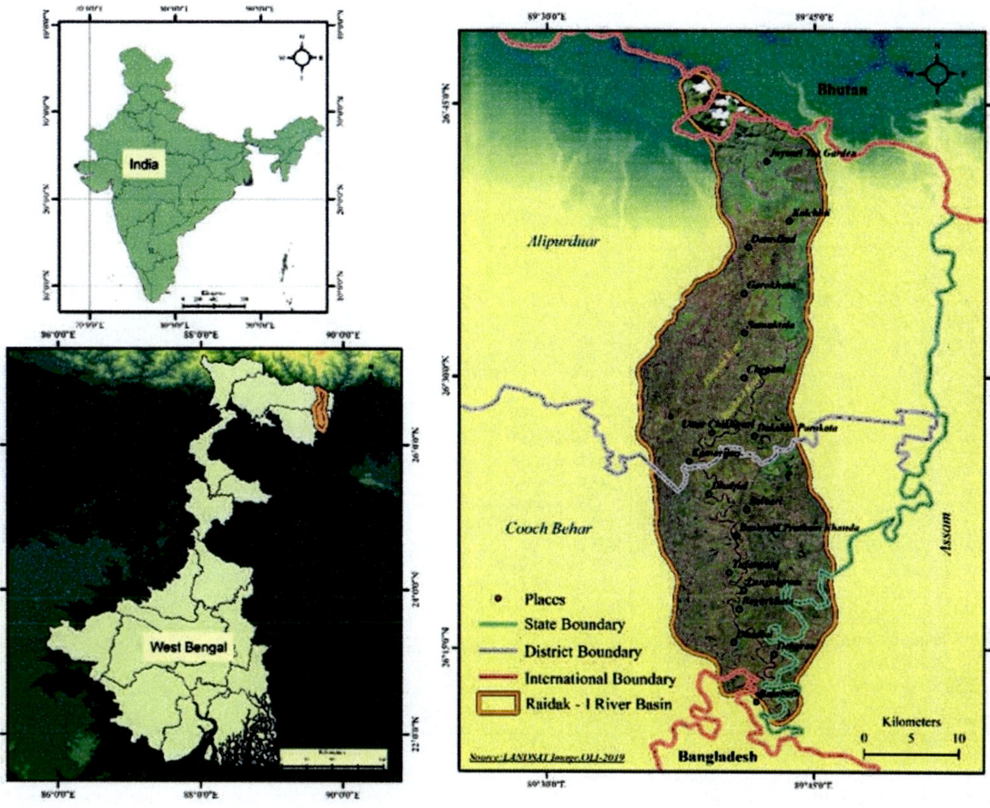

Fig. 7.1 Location map of the study area

LANDSAT-7ETM + of 2009 (Path/Row: 138/41, 139/42; Spatial Resolution: 30 m) and LANDSAT 8OLI/TIRS of 2019 (Path/Row: 138/41, 139/42; Spatial Resolution: 30 m) satellite images have been collected from USGS Earth Explorer. To maintain the accuracy of the work, the post monsoonal data or satellite imageries have been used. The data regarding discharge and velocity have been collected from the Central Water Commission, Govt. of India (CWC, Govt. of India), Jalpaiguri. To quantify river bank shifting along with channel migration, GIS techniques are applied which is very much significant and suitable tool for computation of river channel dynamics (Winterbottom 2000). Arc GIS 10.2 and ERDAS imagine 2014 softwares have been used to prepare the digital composite maps. The data sets are imported and geo-referenced in Arc GIS 10.2 software applying the following steps- Projection type: Universal Transverse Mercator (UTM),Spheroid name: WGS 84, Datum: WGS 84, UTM Zone: 45 North. In order to bring all spatial data in a same scale, resampling and geometric correction have been done using ERDAS imagine 2014 software. After georeferencing, digitization of bank lines has been done from the orthorectified satellite images using digitizing tool of Arc GIS 10.2 Software. Vector data base is prepared from the Raster data in order to know different measurements like channel width, the rate of lateral migration of bank lines and channel length.

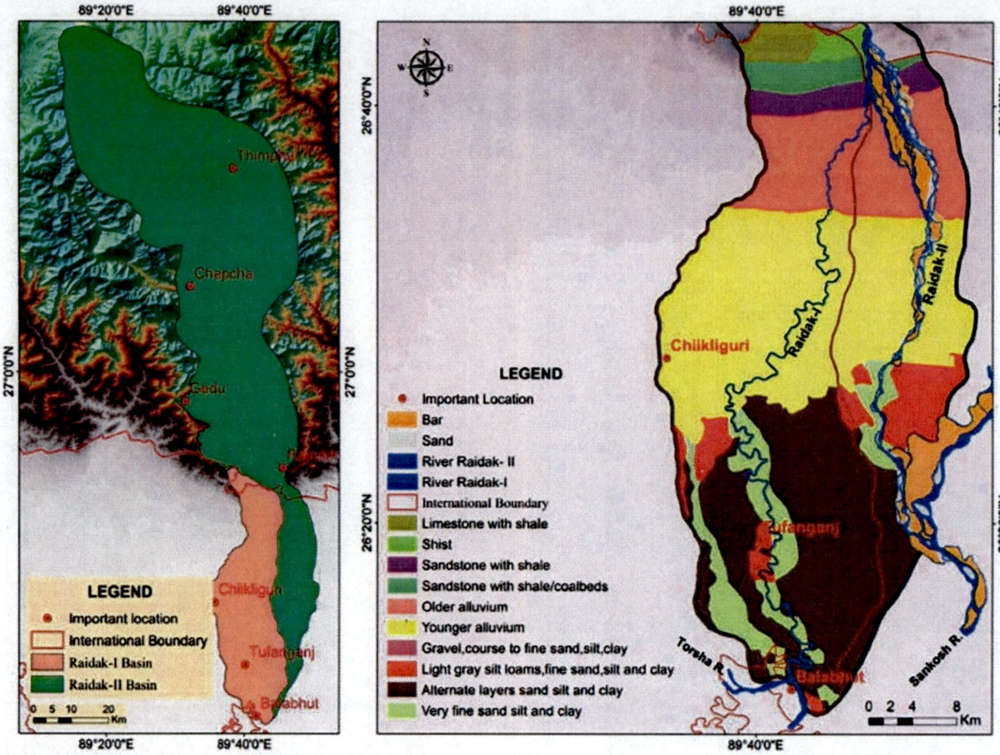

Fig. 7.2 Geological map of the study area

7.3.2 Bank Line Change Detection and Measurement of Lateral Channel Migration

For bank line change detection, 54.4 km river course has been divided into 3 reaches and total 26 cross sections have been considered at different locations across the course of river Raidak-I. Reachwise bank line change detection method has been used by different researchers (Das et al.2014; Sarkar et al. 2012) over the decades. Different fluvio-hydrological parameters (reach length, sinuosity, channel type, channel pattern, dominant channel bed materials etc.) have been taken to find out the nature of channel dynamics and associated problems related with river health(Table 7.1) as fluvial hydrodynamics largely control the formation of different channel bed features (Bera et al. 2019a) as well as channel geometry.

Computed by the Authors from field study and 2019 LANDSAT 8OLI/TIRS image.

Channel geometry along with channel bed configuration and river hydraulics study have been done to detect bank line change detection. The channel sinuosity study had been specially conducted to understand the nature of channel planform and its change. For bank line shifting analysis, both the left and right banks are delineated from satellite images of the year 1972, 1979, 1996, 2009 and 2019 respectively and these bank lines are digitized using Arc GIS 10.2 software. Then the bank lines are overlaid to measure the distance among the bank lines of different years at each cross section of the river course. In order to know the short-term channel migration, two consecutive images are used i.e. 1972–1979, 1979–1996, 1996–2009 and 2009–2019. On the other hand, long-term channel migration has been measured in respect of ear-

Table 7.1 Details of different parameters of three reaches of the river Raidak-I, 2019

Parameters	Reach-1	Reach-2	Reach-3
Reach extension	From downstream of the confluence point of river Dhaksi and Raidak-I near Uttar Chikliguri up to Bashraja Dwitia Khanda	Bashraja Dwitia Khanda to Begarkhata	Begarkhata to confluence of Raidak-I with Torsa at Balabhut: 17 km
Reach length (in km)	16.5	20.9	17
Channel type	Alluvial	Alluvial	Alluvial
Sinuosity	2.48	1.97	1.99
Channel pattern	Meandering	Meandering	Meandering
Dominant channel bed materials	Sands & silts	Sand, silts & clay	Sand, silts & clay
Total catchment area (km^2)	760 km^2 (within the Indian territory, from Bhutanghat to confluence)		
Avg. peak discharge (Cumec)	385.27 (at chepani station)		
Average annual Rainfall (mm)	3049.8 mm in Cooch Behar district and 3720.9 mm in Jalpaiguri district (IMD)		

Table 7.2 Reach wise short-term erosion and accretion rates of river Raidak-I

Year	Short-term riverbank migration											
	Reachwise Right bank shifting average rate (m/y)						Reachwise Left bank shifting average rate (m/y)					
	Erosion			Accretion			Erosion			Accretion		
	R-1	R-2	R-3	R-1	R-2	R-3	R-1	R-2	R-3	R-1	R-2	R-3
1972–1979	67.37	258.47	167.52	440.76	490.80	8.81	244.63	447.87	21.97	379.49	260.78	147.92
1979–1996	0.00	2.54	59.77	76.99	4.58	45.56	35.62	2.97	21.66	9.36	3.80	72.37
1996–2009	0.15	233.11	30.73	32.53	239.05	48.47	32.14	251.18	47.70	15.20	219.58	33.02
2009–2019	1.36	149.18	18.92	20.01	299.31	9.81	12.02	281.03	8.34	1.83	169.54	8.73

R-1: Reach-1; R-2: Reach-2 and R-3: Reach-3

liest image of 1972 and other images, namely, 1972–1996, 1972–2009 and 1972–2019. Migration distance has been measured by previously used methods of Leopold (1973), Gurnell et al. (1994) and Giardino and Lee (2011). The formula is used for measuring the migration distance as suggested by Giardino and Lee, 2011 which is as follows:

$$D_m = T_1 - T_2 \quad (7.1)$$

where, D_m is the migration distance and T_1 and T_2 are the survey years of channel migration.

Furthermore, the annual rate of channel migration has been measured by using the following formula applied by Wallick et al. (2006):

$$Rm = Dm/T \quad (7.2)$$

where, Rm is the annual rate of channel migration, Dm is the migration distance between two particular year of study and T is the time interval between the years.

In addition, bank erosion and accretion rates at each cross section are computed from the overlaid bank lines which are digitized from different images where negative value is considered as erosion and positive value as accretion. The channel width is measured at each cross section on the basis of distance between two bank lines of channel in a particular year.

Channel sinuosity has been measured for each reaches following the equation given by Schumm (1963).

$$SI = OL/EL \quad (7.3)$$

where, SI is Sinuosity Index, OL is the observed (actual) path of a stream and EL is the expected straight path of a stream. In order to explore meander geometry, radius of curvature has been computed at different meander bends of river Raidak-I as it is related to meander migration.

The Radius of curvature (Rc) is measured by applying the following universal formula previously used by several authors (Bag et al. 2019).

$$Rc = \sqrt{(A/\Pi)} \text{ or, } P/2\Pi \quad (7.4)$$

where, Area (A) = Πr^2, Perimeter (P) = 2 Πr.

7.3.3 Bank Facies Study

Bank facies study was conducted in order to unveil the lithological characteristics as well as composition of river bank during the month of November, 2019. Facies may be classified into two types such as lithofacies and biofacies. A lithofacies can be defined on the basis of its unique features, sedimentological composition and structure. Each lithofacies reveals unique depositional characteristics (Miall). A quaternarysedimentary bank facies of 7.7 m length is arranged before detailed study along the right bank of river Raidak-I near village Chikliguri Dwitia Khanda in Tufanganj-I block of Cooch Behar district. Another 4.2 mbank facies is also conducted along the left bank of the river near Sauerkhata village in Tufanganj-II block of Cooch Behar district. Both of the facies were previously exposed due to the continuous bank failure. Soil samples have been collected from the different micro layers of the bank facies sites and these samples are tested in the laboratory to obtain the nature of bank materials. Sedlog 3.1 software is used to draw the bank facies profiles and to acquire the data about the different micro layers withinthe bank facies.

7.3.4 Statistical Techniques

Descriptive statistics has been used in this study to measure the variability of data regarding channel migration. Mean, standard deviationand coefficient of variation have been used to evaluate the channel migration of Raidak-I river (Mandal et al. 2017). Higher the value of coefficient of variation reflects irregular pattern of channel migration or high dynamicity of channel migration patterns in nature.

7.4 Results

7.4.1 Channel Sinuosity

Channel sinuosity is an important fluviogeomorphic parameter to analyze the channel pattern. It refers to the deviation of channel path from its original signature. Sinuosity Index (SI) is used to measure the changes of river Raidak-I dividing into three important segments from the successive satellite images of 1972,

1979, 1996, 2009 and 2019. The lateral migration rate has been calculated applying such important fluvial indicator (Richard et al. 2005). In 1972, all the three reaches are characterized by high sinuosity index value (2.43 in reach-1, 2.28 in reach-2 and 2.57 in reach-3) which indicates meandering channel pattern (Singh 2007)but in 1979, Sinuosity values were decreased in all the three segments (Fig. 7.3a) in comparison to previous year (1972) due to decrease in channel length. In 1996, the Sinuosity Index has been measured in three reaches ranging from 1.80 to 2.42 which are indicating meandering pattern (2.42 in reach-1, 1.80 in reach-2 and 1.84 in reach-3). Between the years 1996 and 2009, sinuosity values increase to all the three reaches (Fig. 7.3a) while it has been again decreased in between 2009 and 2019 in reach-1 and reach-2 except the reach-3 (Fig. 7.3a). It exhibits shortening of reach-1 and reach-2 within the time frame (2009–2019). Moreover, the overall Sinuosity value of entire reach was 2.42 in 1972 whereas it was decreased to 2.14 in 2019. However, Since 1972 to 2019, the overall result shows decreasing trend of Sinuosity Index which is indicating the fact that the river has shorten its course by 13.34 km between downstream of the confluence point of river Dhaksi and Raidak-I near Uttar Chikliguri up to confluence of Raidak-I with Torsa at Balabhut. The presence of several cut-offs and palaeo channels on the floodplain of river Raidak-I validate that these geomorphic features are the part of river Raidak-I and subsequent detachment or abandonment of these features from the main course have made the channel length shorter.

7.4.2 Radius of Curvature (Rc)

Radius of curvature (Rc) is the radius of a circle drawn through the apex of the bend and the two crossover points (Singh 2007). In the channel geometry study, radius of curvature is measured at various meander bends of river Raidak-I. The mean of all radius of curvature measured along the meander loops of entire reaches was 237.5 m in 1972 but it is found to be decreased by 181.25 m in 2019. In 1972, highest radius of curvature was 375 m and that was found at meander bend 3 in reach-1 near Paschim Chikliguri but it decreased by 143.75 m from 1972 (375 m) to 2019 (231.25 m). Between 1972 and 2019, numbers of meander loops have been increased within reach-1 (Fig. 7.3b) and due to tremendous spatial migration of meander bends, most of the bends have been dislocated resulting into formation of new mender bends within this segments of river course. The study also shows that at meander bend 4 near Tufanganj in reach-2, the radius of curvature was 175 m in 1972 while it is found to be increased by 12.50 m from 1972 (175 m) to 2019 (187.50 m) and it was situated at meander bend 7 of reach-2 in the year 2019 (Fig. 7.3c). This meander bend has migrated westwards direction and a new cut-off has been formed. Meanwhile, it was also the part of main river course before 1979. The Radius of curvature (Rc) near Dwiparpar bend (within reach-3) was 250 min 1972 and it has decreased to 187.5 m in 2019. Therefore, the variation of mean radius of curvature (Rc) in different meander bends of river Raidak-I imply the high rate of meander migration and continuous alterations of meander geometry in the entire study period.

7.4.3 Channel Width

The changes of channel width are an important indicator of river channel migration. In order to know the erosion-accretion dynamics, (variation of channel width and nature of channel migration) 26 cross-sections have been drawn at different locations across the entire study course of river Raidak-I. Within the reach-1, maximum channel width (146.51 m) was found at cross section ‘a-b’ in 1972 while highest channel width (103.95 m) was measured at cross sections ‘g-h’ in 2019. In 1972, the lowest channel width within reach-1 has been recorded at cross section ‘c-d’ (106.52 m) and that was at cross section ‘a-b’ (74.52 m) in 2019. Between 1972 and 1979, within reach-1, channel widening has been documented at cross sections ‘c-d’, ‘g-h’, ‘i-j’, ‘k-l’,

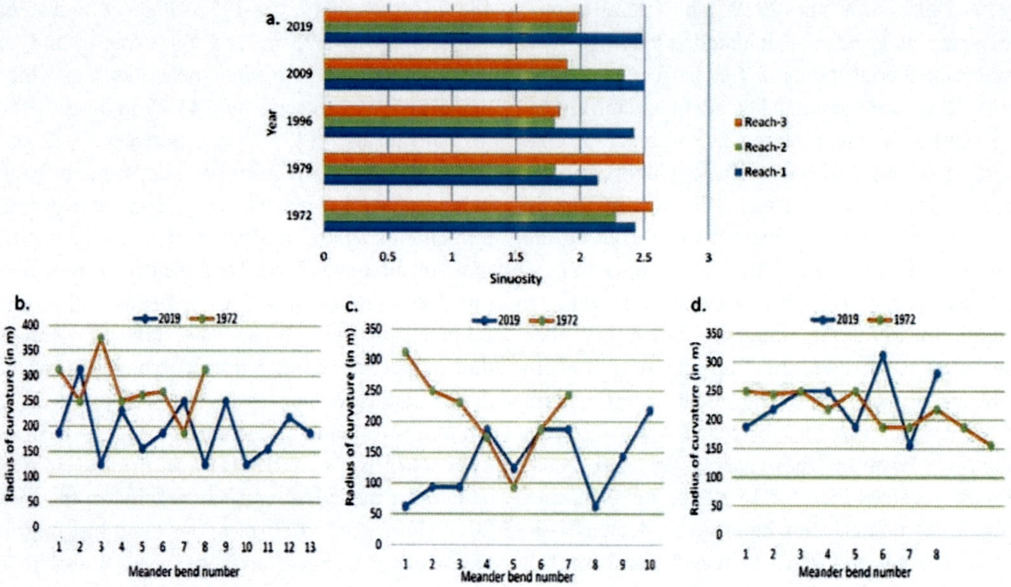

Fig. 7.3 **a** Variation of sinuosity in different reaches of river Raidak-I (1972–2019); Variation of radius of curvature in different meander bends of river Raidak-I during 1972 and 2019 in **b** reach-1, **c** reach-2 and **d** reach-3

'm–n', 'o-p' and 'q-r' except cross sections 'a-b' and 'e–f' but channel narrowing is also found at all the cross sections during the study period 1979 to 1996. The channel widening was found across all the cross-section except 'm–n' during 1996 to 2009 study period. Between 2009 and 2019, channel widening was also documented at cross sections 'c-d', 'g-h' and 'm–n' within reach-1 while channel narrowing was found at cross sections 'a-b', 'e–f', 'i-j', 'k-l', 'o-p' and 'q-r'. Over 47 years of study period (1972–2019), variations of channel width have also been found at different cross sections within reach-2 and reach-3. Mean channel width of entire reach was measured as 139.03 m in 1972, 146.84 m in 1979, 88.58 m in 1996, 107.76 m in 2009 and 93.94 m in 2019. The processes of (Fig. 7.4b and c) channel widening and narrowing are primarily due to intensive bank erosion and deposition within the flow path of river Raidak-I in Himalayan foreland basin. However, the result also indicates the lateral channel migration at different points of river Raidak-I.

7.4.4 Short-Term Riverbank Erosion and Accretion Trend

To inspect the micro level channel dynamics along with floodplain modifications, the whole study reaches (54.4 km) have been subdivided into three reaches; reach-1 (from downstream of the confluence point of river Dhaksi and Raidak-I near Uttar Chikliguri up to Bashraja Dwitia Khanda: 16.5 km), reach-2 (Bashraja Dwitia Khanda to Begarkhata: 20.9 km) and reach-3 (Begarkhata to confluence of Raidak-I with Torsa at Balabhut: 17 km). The magnitude and attitude of the erosion and accretion for both short-term and long term measurement were carried out along the left and right bank of the active channel (Figs. 7.5 and 7.6). For short-term analysis, satellite images of two consecutive years have been taken and the rate of erosion and accretion in details are measured simultaneously (Table 7.2). Reach wise short-term erosion and accretion dynamics along the right and left banks are presented in Figs. 7.7, 7.8 and 7.9.

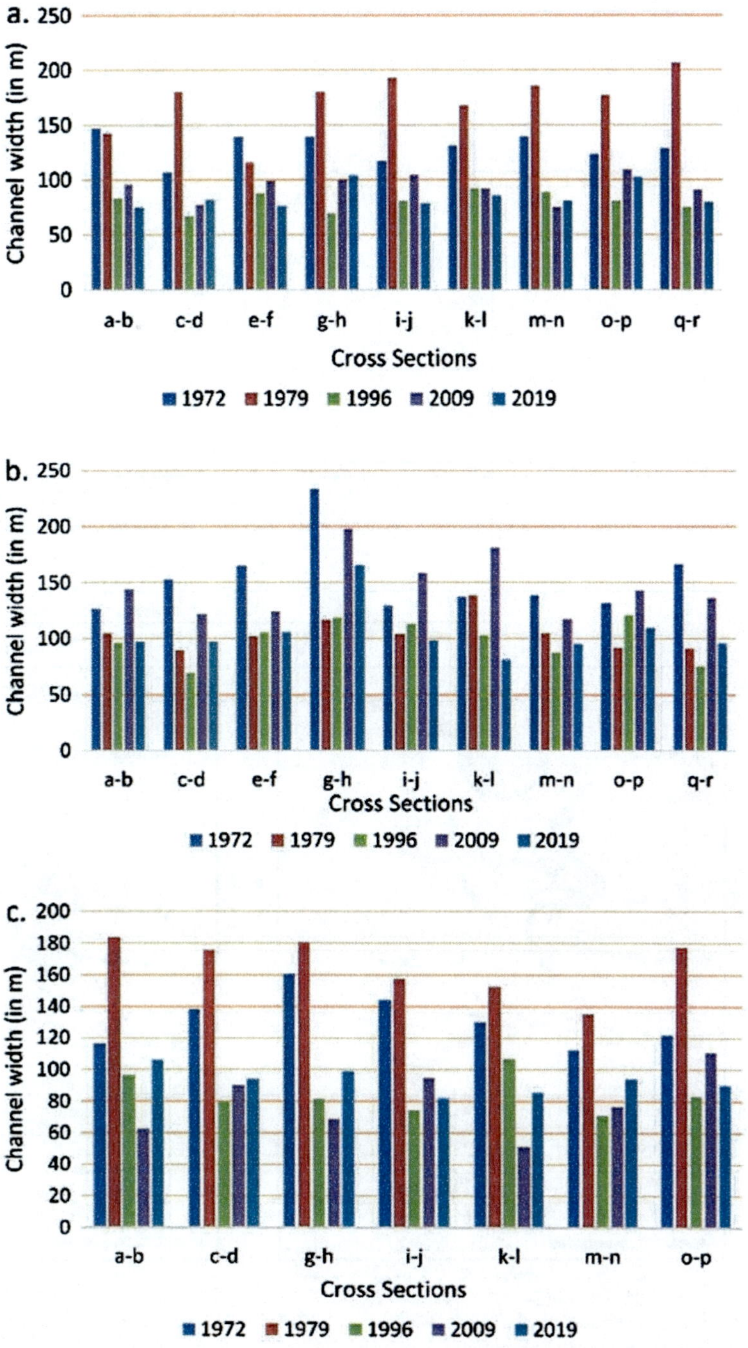

Fig. 7.4 Temporal variation of channel width in different cross sections across the channel Raidak-I (1972–2019); **a** reach-1, **b** reach-2, **c** reach-3

Between the years 1972 and 1979, average erosion rate along the right bank was 493.36 m/y while accretion rate was 940.37 m/y. Along the left bank, mean erosion rate was 714.47 m/y and accretion rate was 788.19 m/y within the whole studied reach of Raidak-I (Table 7.3). The seven years (1972 to 1979) study showed that the maximum erosion had been occurred within the

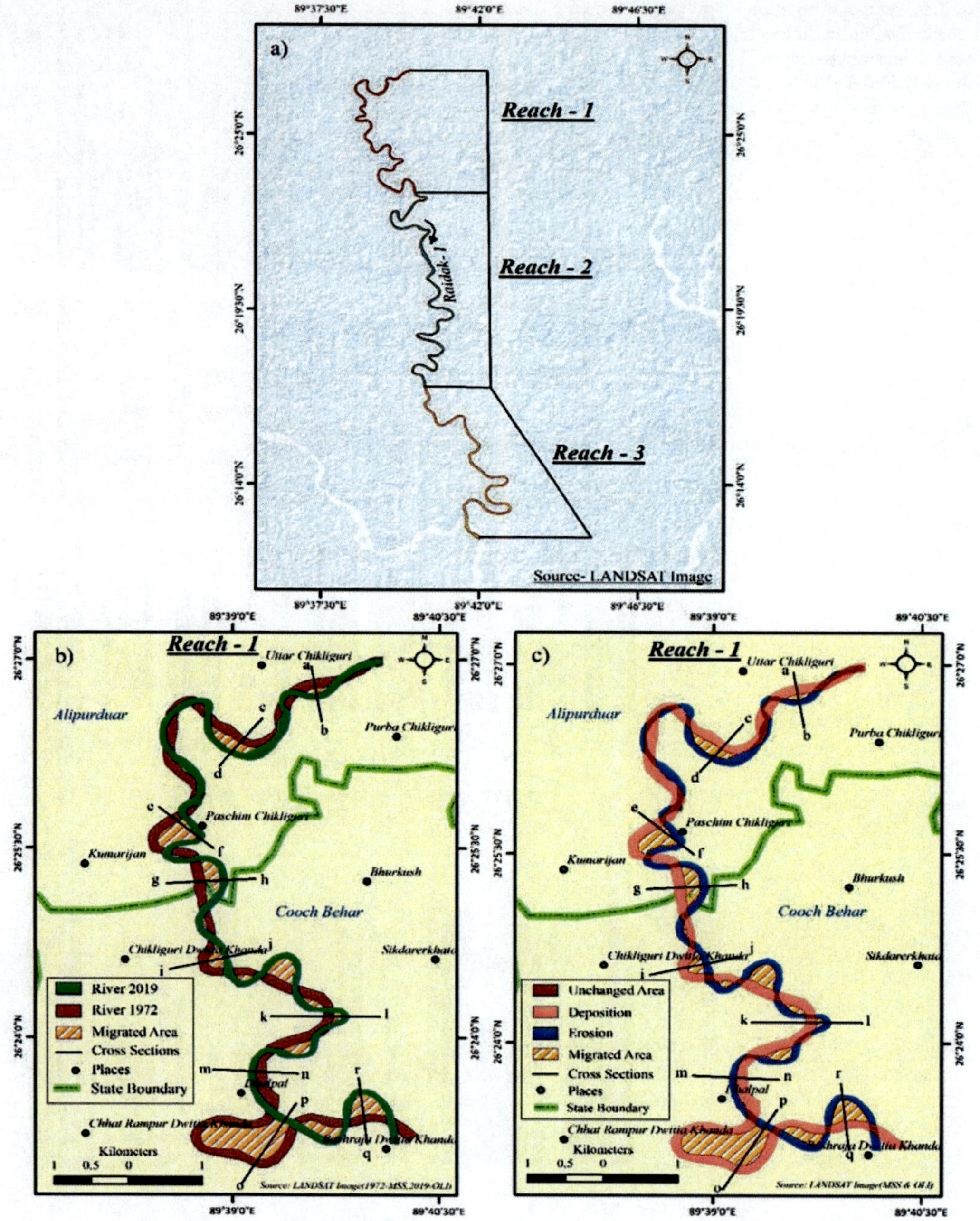

Fig. 7.5 **a** Different reach of Raidak-I river (2019); (Reach-1: near Uttar Chikliguri to Basraja Dwitia Khanda, Reach-2: Basraja Dwitia Khanda to Begarkhata, Reach-3: Begarkhata to Balabhut); **b** Reach-1 of Raidak-I river during 1972 and 2019 and **c** channel migration scenario in reach-1 from 1972 to 2019

reach-2 in an average rate of 258.47 m/y along the right bank whereas on the left bank side it was 447.87 m/y. In that period, maximum accretion rate had been recorded around 490.80 m/y along the right bank within the reach-2. On the other hand, the left bank side

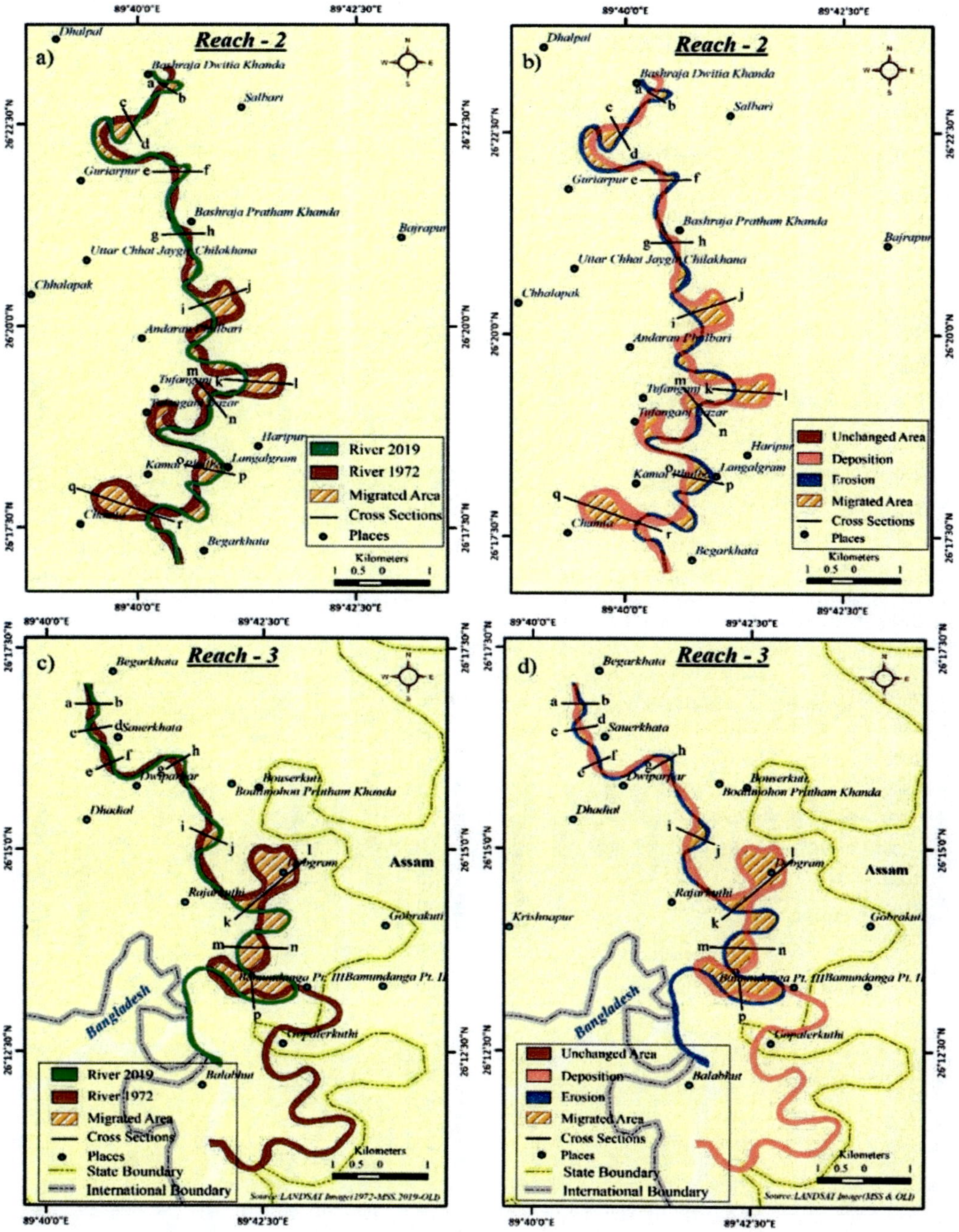

Fig. 7.6 **a** Reach-2 of Raidak-I river during 1972 and 2019; **b** channel migration scenario in reach-2 from 1972 to 2019; **c** Reach-3 of Raidak-I river during 1972 and 2019 and **d** channel migration scenario in reach-3 from 1972 to 2019

average accretion rate was about 379.49 m/y (maximum value) within reach-1. The reach wise erosion-accretion trend shows that the average accretion rate is greater within reach-1 and 2 than the average erosion rate on the right bank of channel while the average erosion rate in reach-3

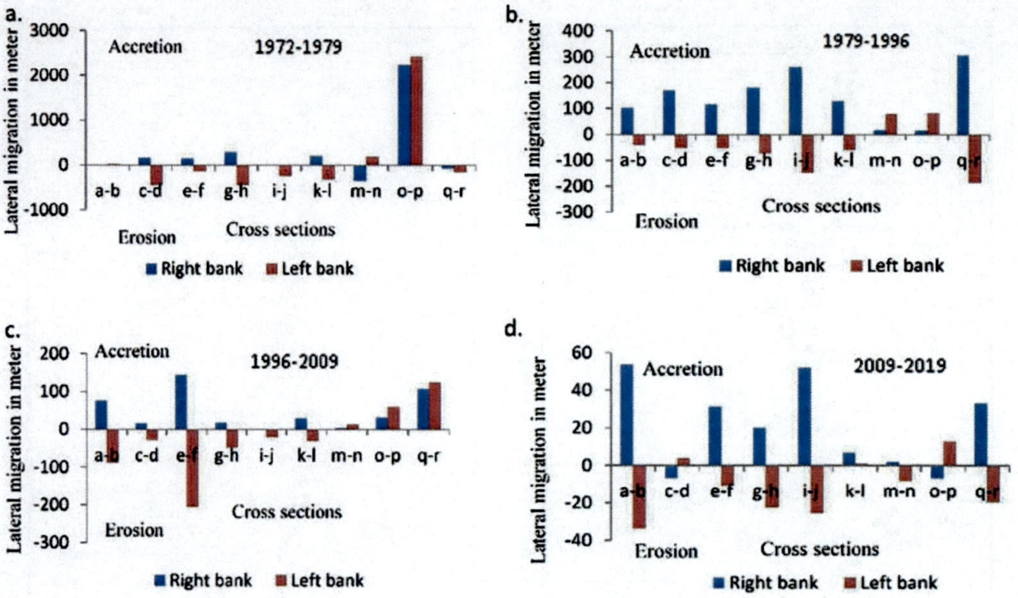

Fig. 7.7 Bank wise erosion and accretion scenario in reach-1 of river Raidak-I during different study period **a** 1972–1979 **b** 1979–1996 **c** 1996–2009 and **d** 2009–2019

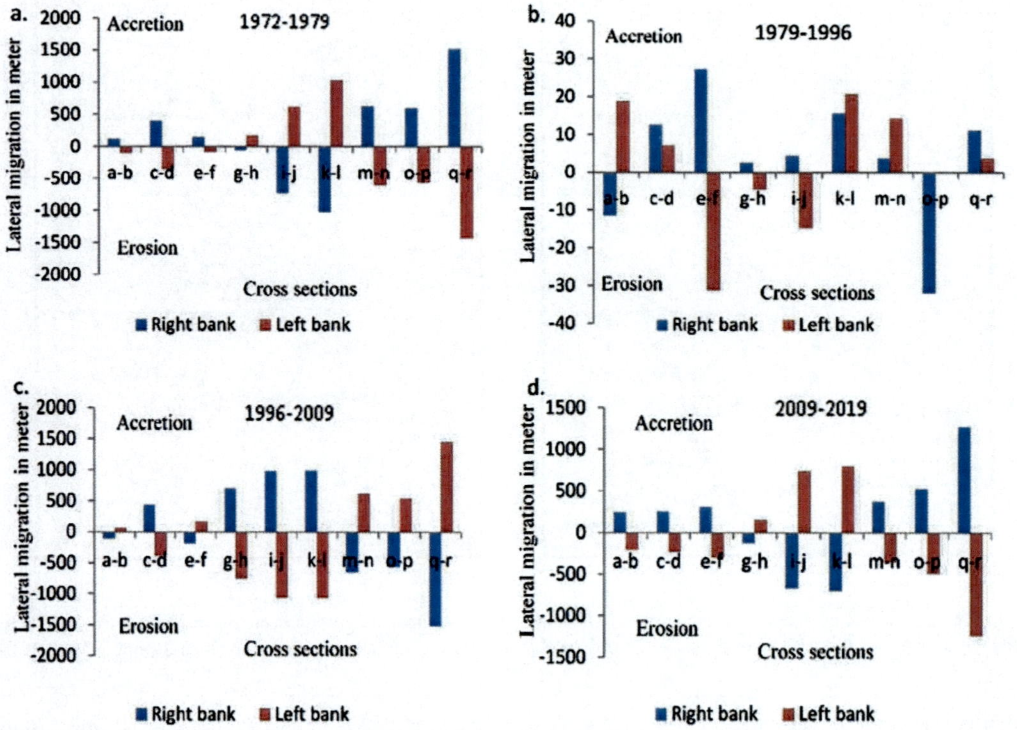

Fig. 7.8 Bank wise erosion and accretion scenario in reach-2 of river Raidak-I during different study period **a** 1972–1979 **b** 1979–1996 **c** 1996–2009 and **d** 2009–2019

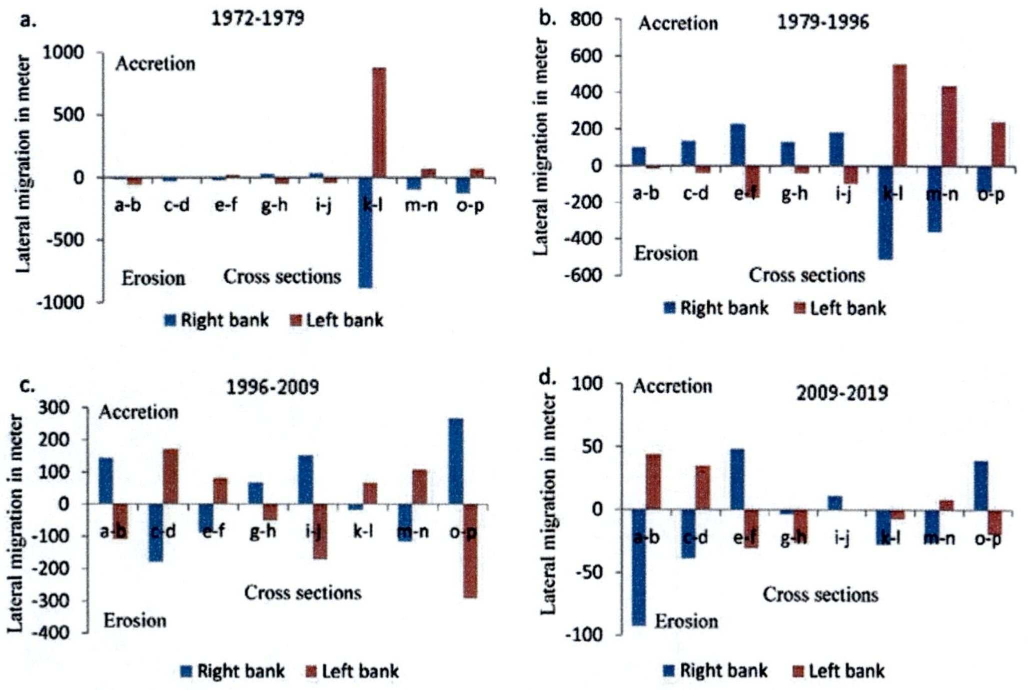

Fig. 7.9 Bank wise erosion and accretion scenario in reach-3 of river Raidak-I during different study period **a** 1972–1979 **b** 1979–1996 **c** 1996–2009 and **d** 2009–2019

exceeds the accretion rate. Accretion rate was higher than the erosion rate in both reach-1 and 3 along the left bank of the channel whereas erosion rate exceeded the accretion rate in reach-2 (Table 7.2). Therefore, it is clearly stated that the right bank side (within reach-3) was more vulnerable than reach-1 and 2 while the left bank side (in reach-2) was more prone to bank erosion than the other two reaches.

In time period of 1979 to 1996, the erosion rate along the right bank side of the river Raidak-I had been reduced to 62.31 m/y in comparison to last 7 years period (1972 to 1979). The erosion rate was 493.36 m/y and accretion rate also found to be decreased from 940.37 m/y to 127.13 m/y along the right bank side of river Raidak-I. Along the left bank side of the channel, both erosion and accretion rate were decreased from 714.47 m/y to 60.25 m/y and 788.19 m/y to 85.53 m/y respectively in between two time period of 1972 to 1979 and 1979 to 1996 (Table 7.3). Between 1979 and 1996, erosion rate was maximum (59.77 m/y) at reach-3 on the right bank while highest accretion rate was also measured at reach-1 (76.99 m/y). Along the left bank side, erosion and accretion rate was recorded maximum within reach-1 and 3 respectively. From 1979 to 1996 in the study area, both along left and right bank side, average accretion rate were higher than the average erosion rate (Table 7.2).

Within the time frame 1996 to 2009, both annual rate of erosion and accretion were amplified from the previous 18 years period. This trend has been observed on both right bank and left bank side of the river. Average erosion and accretion rate along the right bank side have been augmented to 263.99 m/y and 320.05 m/y from the previous studied years and rate of 62.31 m/y and 127.13 m/y respectively while along the left bank the rate of increase of average erosion and accretion have been measured around 331.02 m/y and 267.8 m/y with respect to the previous period about 60.25 m/y and 85.53 m/y respectively (Table 7.3). Furthermore, the reach wise study highlights that along the

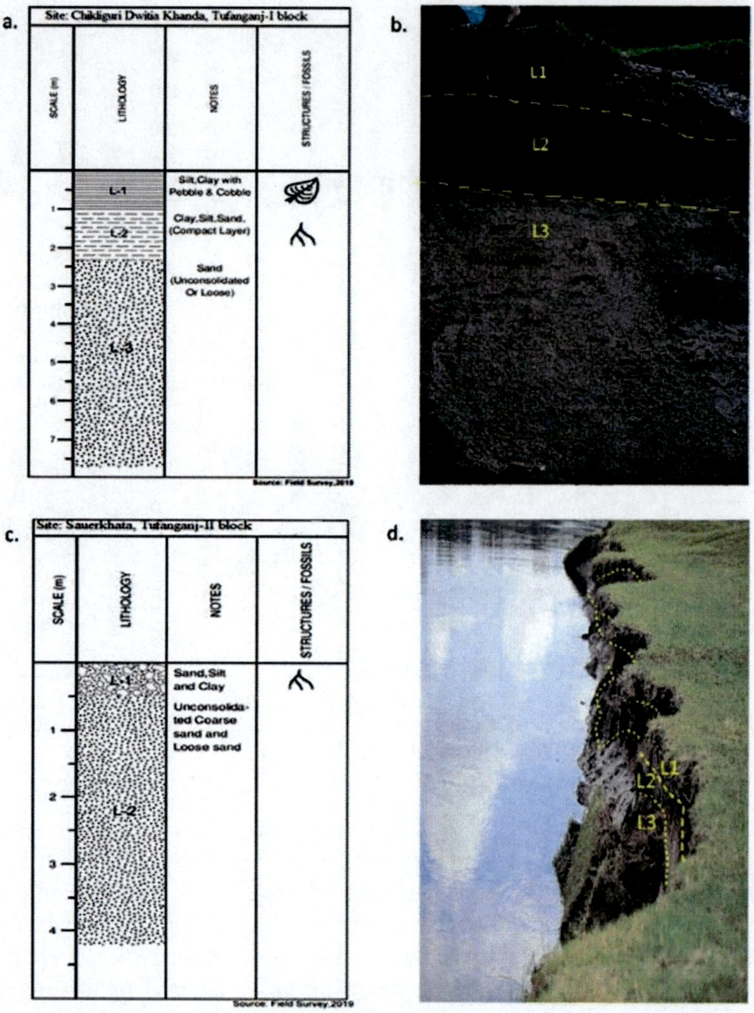

Fig. 7.10 **a** Quaternary sedimentary bank facies along the right bank of river Raidak-I near village Chikliguri Dwitia Khanda in Tufanganj-I block of Cooch Behar district; **b** field photography of exposed right bank composed of mixed materials; **c** Quaternary sedimentary bank facies along the left bank of river Raidak-I near village Sauerkhata in Tufanganj-II block of Cooch Behar district and **d** field photography of exposed vulnerable left bank

Table 7.3 Short-term erosion and accretion rates of entire study reach of Raidak-I river

Year	Right bank shifting average rate (m/y)		Left bank shifting average rate (m/y)	
	Erosion	Accretion	Erosion	Accretion
1972–1979	493.36	940.37	714.47	788.19
1979–1996	62.31	127.13	60.25	85.53
1996–2009	263.99	320.05	331.02	267.8
2009–2019	169.46	329.13	301.39	180.1

right bank side, maximum erosion and deposition had been occurred in reach-2 with an average rate of 233.11 m/y and 239.05 m/y (during the time span of 1996 to 2009) respectively. Along the left bank side, the similar scenario was also found due to highest average rate of erosion (251.18 m/y) and deposition (219.58 m/y) within reach-2 (Table 7.2). Therefore, reach-2 was more dynamic in compared with reach-1 and 3 and left bank was more prone to erosion than right bank side for the time period (1996 to 2009).

Between 2009 and 2019, the average erosion rate along the right bank again decreased to 169.46 m/y from previous 13 years period and average rate was around 263.99 m/y while average accretion rate has been increased slightly from 320.05 m/y (1996 to 2009) to 329.13 m/y (2009 to 2019). The study further showed that the average bank erosion rate along the left bank again decreased to 301.39 m/y from the earlier rate of 331.02 m/y (1996 to 2009) and accretion rate is found to be decreased from 267.8 m/y (1996 to 2009) to 180.10 m/y (2009 to 2019). However, the channel migration analysis shows an interesting fact that during this period, both on left and right bank, erosion and accretion rate was highest in reach-2 (Table 7.2). It indicates reach-2 was more dynamic in nature than the other two reaches of river Raidak-I during the time span of 2009 and 2019.

The Coefficient of variation (CV) of both the river banks along with different reaches show the ranges between 0.34 and 1.94 which indicate moderate to high variability for both erosion and accretion mechanism (Table 7.4). The result also demonstrates that, along the both river banks, the CV value of reach-1 for both erosion and accretion rate are very high which specifies the rate of channel migration is highly unstable or less consistency. However, it is also proved that there was wide variation of erosion and accretion rate along the river banks of Raidak-I over the last 47 years. The average rate of short-term migration for both banks reveals a dynamic nature of erosion and accretion mechanism which influences the channel migration of river Raidak-I.

R-1: Reach-1; R-2: Reach-2 and R-3: Reach-3.

7.4.5 Long-Term Riverbank Erosion and Accretion Trend

Like short-term erosion-accretion rates, reach wise long-term average bank erosion and accretion rates have also been measured along the three reaches of river Raidak-I (Table 7.5). In long-term trend, the erosion and accretion rate varies significantly due to several factors acting independently or together upon the mechanism of bank erosion and accretion.

Over 24 years study (1972 to 1996), the average bank erosion rate along the right bank was highest in reach-2 (77.18 m/y) and average accretion rate was also highest in reach-1 (183.08 m/y) while on the left bank, highest erosion and accretion took place in reach-2 (132.73 m/y) and reach-1 (117.31 m/y) respectively (Table 7.5). Between the years 1972 and 2009, maximum.

erosion and accretion rate had been recorded in reach-2 (131.97 m/y and 178.94 m/y) along the right bank of the channel and left bank was also experienced the similar incident as highest erosion and accretion rate measured in reach-2 with an average rate of 174.34 m/y and 128.23 m/y respectively. Over 47 years of study period (1972 to 2019), maximum erosion and accretion rate was also recorded in reach-2 in case of both right and left banks (Table 7.5). So the result exhibits that the reach-2 was more dynamic in nature than reach-1 and reach-3 during both the time span of 1972 to 2009 and 1972 to 2019. However, the long-term study shows that along the right bank side, the overall mean erosion rate of three studied reaches is 67.53 while it is 90.45 on left bank of the river. So, mean erosion rate in long-term perspective is significantly less than overall short-term mean erosion rate for right and left bank (82.42 and 117.25). Similarly, the mean accretion rate along the right bank is 117.51 whereas in left bank it is 92.63. These are also less than short-term mean accretion rate for both the banks (143.05 and 110.13).

The study exhibits that in reach-1, significant bank erosion took place in Paschim Chikliguri and Uttar Chikliguri village of Alipurduar-II block, Chikliguri Dwitia Khanda and Dhalpal village of Tufanganj-I block and Bashraja Dwitia Khanda village of Tufanganj-II block.Reach-2 experienced considerable erosion near village Guriarpur and Andaran Phulbari under

Table 7.4 Descriptive statistics of lateral channel migration of Raidak-I river during the entire study period of 1972–2019

	Short-term riverbank migration											
	Right bank shifting average rate (m/y)						Left bank shifting average rate (m/y)					
	Erosion			Accretion			Erosion			Accretion		
	R-1	R-2	R-3	R-1	R-2	R-3	R-1	R-2	R-3	R-1	R-2	R-3
Mean	17.22	160.82	69.23	142.57	258.43	28.16	81.10	245.75	24.91	101.47	163.42	65.51
SD	33.43	115.39	67.73	200.28	200.40	9.51	109.51	183.54	16.46	185.42	112.76	60.87
CV	1.94	0.71	0.97	1.40	0.77	0.34	1.35	0.74	0.66	1.82	0.69	0.92

Table 7.5 Reach wise long-term erosion and accretion rates of river Raidak-I

Year	Long-term riverbank migration											
	Right bank shifting average rate (m/y)						Left bank shifting average rate (m/y)					
	Erosion			Accretion			Erosion			Accretion		
	R-1	R-2	R-3	R-1	R-2	R-3	R-1	R-2	R-3	R-1	R-2	R-3
1972–1996	19.64	77.18	91.19	183.08	146.39	34.84	96.58	132.73	21.75	117.31	78.75	94.4
1972–2009	12.79	131.97	69.95	130.19	178.94	39.62	73.93	174.34	30.86	81.43	128.23	72.83
1972–2019	10.36	135.63	59.09	106.74	204.55	33.28	60.76	197.04	26.07	64.49	137.02	59.19

Tufanganj-I block, Basraja Pratham Khanda and Langalgram under Tufanganj-II block and near Tufanganj Municipality and Kamat Phulbari of Tufanganj-I block. Dwiparpar, Rajarkuthi and Balabhut of Tufanganj-I block and Begarkhata and Sauerkhata of Tufanganj-II block are most vulnerable location of bank erosion in lowermost reach-3 (Table 7.7).

Over 47 years of study period, there was a wide variation of erosion and accretion rate along both the banks of river Raidak-I because the river channel is not following the general geomorphological principles for short-term or long-term period. If discharge and sediment load remains unchanged, the bank erosion rate would be decreased with increasing channel width and vice versa. It indicates the greater variation of discharge and flow velocity with gradual advances of time. Subsequently, the complex interaction among erosion, deposition and transportation are being happened on the dynamic channel path of Raidak-I river within the Himalayan foreland basin.

7.4.6 Short-Term Channel Migration

Channel migration occurs through bank line shifting. In order to obtain a detail picture, bankline migration rates both on right bank and left bank have been measured at 26 cross-sections over the three study reaches of river Raidak-I for 47 years of study period (1972–2019).

1972 to 1979.

Between 1972 and 1979, in reach-I, average net migration rate on right bank was 373.39 m/y and that was about 134.86 m/y on left bank while in reach-2 from Bashraja Dwitia Khanda to Begarkhata(20.9 km) in cooch Behar district, average net migration were 232.33 m/y and 187.09 m/y respectively. In reach-3 from

Begarkhata to Balabhut (17 km), average net bank line migration was 158.71 m/y and 125.95 m/y on right bank and left bank respectively (Table 7.6).

1979 to 1996.

The average net migration rate of right bank was comparatively higher in reach-1(76.99 m/y) than reach-2 (2.04 m/y) and reach-3 (14.21 m/y) whereas in left bank, it is found to be higher in reach-3 (50.71 m/y) than reach-1(26.26 m/y) and reach-2 (0.83 m/y). Along the left bank, averagenet bank line migration rate was relatively higher in reach-3(50.71 m/y) than reach-1 (26.26 m/y) and reach-2(0.83 m/y) (Table 7.6).

1996 to 2009.

Between 1996 and 2009, it is found that in reach-1, the average rate of net migration on right bank is 32.39 m/y while left bank average net migration rate is 16.94 m/y. In reach-2, right bank exibits very low net migration rate (5.94 m/y) whereas left bank shows comparatively higher rate of net bank line migration (31.6) than right bank. Average net migration rate is almost uniform both on left and right bank in reach-3 i.e. 14.68 m/y and 17.74 m/y respectively (Table 7.6).

2009 to 2019.

During 2009–2019, average net right bank migration rate was 18.65 m/y in reach-1 whereas it was 10.19 m/y on left bank. In reach-2, both on right and left bank, high rates of bank line migration (150.13 m/y and 111.49 m/y) have been predominant compared to the other reaches. Reach-3 shows very low rate of net migration in both the banks (9.11 m/y on right bank and 0.39 m/y on left bank) of the river. However, the study unfolds that at most of the cross-sections in reach-1 and reach-3, channel migration rate was very low. Besides, it has been found that the average rate of bank line migration was high at maximum cross-sections in reach-2 (Table 7.6).

7.4.7 Long-Term Channel Migration

Over 47 years (1972–2019), the average rate of net migration along the right bank was higher in reach-1 (96.38 m/y) compared to reach-2 (68.92 m/y) and reach-3 (25.81 m/y). In left bank, net migration rate was very low in reach-1 (3.73 m/y) while in reach-2 and reach-3 it was about 60.02 m/y and 33.12 m/y respectively (Table 7.6).

In reach-1, at several places across thecross-sections channel migrationhave been registered during the entire study period (1972–2019). Substantial channel migrationhas taken place near Uttar Chikliguri, Paschim Chikliguri, Chikliguri Dwitia Khanda, Dhalpal and Bashraja Dwitia Khanda due to inward and outward migration of both the river bank. South-easterly, south-westerly, north-easterly, easterly and northerly migration has recorded in this stretch (Fig. 7.5).

In reach-2, longterm channel migrationhas also been measured at different location of river Raidak-I. Near Bashraja Dwitia Khanda, channel has migrated in south-easterly direction across the cross section a-b due to high rate of erosion (6.55 m/y) occurred on left bank than accretion (1.81 m/y). At cross section c-d,1.8 km downstreamof Bashraja Dwitia Khanda, south-easterly shifting of channel is also found during the entire study period of 1972–2019. Near Guriarpur,

Table 7.6 Reach wise net migration rates along the right and left bank of river Raidak-I

Year	Right bank net migration rate (m/y)			Left bank net migration rate (m/y)		
	Reach-1	Reach-2	Reach-3	Reach-1	Reach-2	Reach-3
1972–1979	373.39	232.33	158.71	134.86	187.09	125.95
1979–1996	76.99	2.04	14.21	26.26	0.83	50.71
1996–2009	32.39	5.94	17.74	16.94	31.6	14.68
2009–2019	18.65	150.13	9.11	10.19	111.49	0.39
1972–2019	96.38	68.92	25.81	3.73	60.02	33.12

Table 7.7 Volume of bank erosion in different location along the river Raidak-I (1972–2019)

Sl. No	Location of bank erosion	G. P/Municipality/Census Town	Block	Area of erosion (in hectares)	Duration (year) (1972–2019)
	Reach-1				
1	Uttar Chikliguri	Bhatibari	Alipurduar-II	21.10	47
2	Paschim Chikliguri	Parokata	Alipurduar-II	19.83	47
3	Chikliguri Dwitia Khanda	Dhalpal-I	Tufanganj-I	49.05	47
4	Dhalpal	Dhalpal-I	Tufanganj-I	31.10	47
5	Bashraja Dwitia Khanda	Salbari-I	Tufanganj-II	30.42	47
	Reach-2				
6	Guriarpur	Dhalpal-II	Tufanganj-I	39.08	47
7	Bashraja Pratham Khanda	Salbari-I	Tufanganj-II	34.00	47
8	Andaran Phulbari (P)	Andaran Fulbari-II	Tufanganj-I	32.25	47
9	Tufanganj	Municipality	Tufanganj-I	33.90	47
10	Langalgram	Barokodali-I	Tufanganj-II	29.60	47
11	Kamat Phulbari (P)	Census Town	Tufanganj-I	9.53	47
	Reach-3				
12	Begarkhata	Barokodali-II	Tufanganj-II	27.35	47
13	Sauerkhata	Barokodali-II	Tuafanganj-II	29.25	47
14	Dwiparpar	Nakkati Gacchi	Tufanganj-I	21.11	47
15	Rajarkuthi	Nakkati Gacchi	Tufanganj-I	29.21	47
16	Balabhut	Balabhut	Tufanganj-I	48.18	47

there has been a easterly shift of river Raidak-I along the cross section e–f owing to the outward migration of left bank and inward migration of right bank. Cross section o-p shows that the channel has migrated about 350 m in easterly direction near Langalgram due to the high rate of bank erosion (22.45 m/y) along the left bank of the river (Fig.7.6). In the lower most segment of the study reaches, shifting nature of the channel can be observed at different location of the river. The study shows that channel has migrated to the east at cross section a-b about 1 km downstreamof Begarkhata due to the higher average erosion rates (3.82 m/y) than the average accretion rates (0.94 m/y) on left bank. Cross section c-dshows channel migrationhas taken place towards west just opposite of Sauerkhata because of the outward migration of right bank and continuous lateral erosion at an average rate of 5.27 m/y over the accretion rate of 2.84 m/y. It is observed from the Fig.7.6that, near Dwiparpar, the channel shifted eastward across the cross section g-h. The cross section i-j shows that channel moved eastward just 1.6 km downstream of Dwiparpar due to lateral shifting of river banks to the eastern direction. From cross section m–n, it is found, that, the channel moved about 660 m westward few meter upstream of Bamundanga

Pt. III resulted from the continuous lateral erosion of right bank (12.80 m/y) at this location (Fig.7.6).

7.4.8 Quaternary Sedimentary Bank Facies

Sedimentary bank facies study was conducted to reveal the bank lithology and structure during post-monsoon period. Three distinct layers have been identified on right bank facies near Chikliguri Dwitia Khanda in Tufanganj-I block (Fig. 7.5c). Approximately 7.7 m sedimentary bank facies has three micro layers i.e. L-1: 1 m (Silt, clay with pebble and cobble), L-2: 1.3 m (clay, silt and sand) and L-3: 0.5.4 m (sand) (Fig. 7.10a and b). Laminated litho-structure(Sh) is found at the uppermost layer L-1 with existence of plant residues and activities of small living organisms. Horizontal and diagonal alignment of sand, silt and clay is prominent in the middle layer L-2 with activities of plant and animals. The lower most layer L-3 is most vulnerable as it contains unconsolidated sands which is non-cohesive in nature. Another 4.2 msedimentary bank facies identified on left bank near Sauerkhata village in Tufanganj-II block (Fig. 7.6d) having two distinct micro layers namely L-1: 0.5 m (sand, silt and clay) and L-2: 3.7 m (unconsolidated sand) (Fig. 7.10c and d). The upper layer L-1 has laminated sedimentary structure(Sh) while the lower layer is not integrated owing to the existence of loose or unconsolidated sands in maximum proportion.

Bank erosion process is regulated by several factors such as channel pattern, channel geometry, channelgradient, discharge, velocity, bank materials, presence of riparian vegetation etc. Though, bank erosion is mainly regulated by nature of bank materials and river discharge (Youdeowei 1997). Secondly, nature of channel flow and velocity also has a great influence on bank erosion process (Majumdar and Das (Pan) 2014). However; bank facies study show that both the river banks have non-cohesive bank materials at their base or lower layer contains unconsolidated sands which are responsible for bank erosion due to low resistive power.

Minimum 20 cm s^{-1} flow velocity is enough for erosion and transportation of fine sand from the bank. The high velocity with turbulence during monsoon month erode the non-cohesive bank materials and enhance the bank erosion and lateral shifting of channel (Guchhait et al. 2016). However, selective entrainment of non-cohesive bank materials during high flow weakens the overall structure which leads to bank failure.

7.5 Discussion

From short and long-term analysis of channel migration, it has been found that river bank erosion along with annual migration rate is very high at few pockets within quaternary geological site. There are specific factors which have been practically responsible for high degree of river bank erosion as well as channel shifting within alluvium flood plain of Alipurduar and Coochbehar plain or beyond the piedmont zone of Bhutan Himalaya. This river system receives intense and prolonged rainfall during rainy season (Annual rainfall varies between 2400–3500 mm) which causes high discharge (According to central water commission data average peak discharge in monsoonal month is 385.27 cumec) and high velocity (avg. velocity in monsoon is 1.67 m/sec) during monsoon season resulting into severe bank erosion and lateral channel migration in many locations along the river course. However, the area under study is characterized by seasonal floods which cause bank erosion and often damage the embankments at many pockets of the river Raidak-I. In addition with these, some anthropogenic factors accelerate the process of bank erosion. The channel depth has been decreasing since post-independence period due to rapid rate of sedimentation in the channel bed. Subsequently, channel bed does not accommodate huge volume of water with sediment during rainy season resulting into occurrence of devastating floods and severe bank erosion in the study area. In the last three or four decades, the physical landscape has been completely modified due to unscientific landuse practices (Chamling and Bera 2020a). The Himalayan foreland basin is basically fast changing landscape with interlacing drainage system. Since British Era, Colonial rulers wanted to execute western river training mechanism without properly knowing the hydro-geomorphological setup of this area. After independence, we borrowed the western philosophy with techniques and constructed horizontal as well as longitudinal hard engineering structures along the natural river channels and across the rivers. The above mentioned Himalayan mighty river when enters in this region, they debouch and (due to sudden drop of channel hydraulic gradient) multiple channels with bars and islands have been emerged within the river courses. Simultaneously, due to the construction of concrete embankments or river jacketing, the excess water with sediment overtops and many new spill channels have been originated with in the floodplain. Anthropogenic stresses in the river system are being increased in different parts of the world and as a result people face multiple fluvio-hydrological hazards and disaster. Meanwhile, people who are particularly residing in tropical and sub-tropical region of the world, they are facing as well as inviting such type of hazards. Deforestation in the upper catchments of the river and excessive dolomite mining in Bhutan hills are the main cause of sedimentation in the channel bed (Chamling and Bera 2020b). Besides, construction of roads, culverts and railway bridges across the river Raidak-I are another triggering factors for high rate of bank erosion on both the banks. On the other hand, bank erosion rates in some locations has been very low or some portions remain unchanged after construction of embankments at different points along the river course although embankments may not be the permanent solution of bank erosion as it gives immediate flood protection. Therefore, an in-depth study is required for future management of the river which will incorporate a proper hydrological knowledge about the study area.

7.6 Conclusion

From the geospatial and field based sedimentary bank facies study, it is clearly stated that river Raidak-I is very much unstable within quaternary geological site of Himalayan foreland basin. The sinuosity value indicates meandering channel pattern for individual reaches as well as temporal variation of SI value over 47 years study period. The decreasing trend of overall SI value of entire reaches since 1972 (2.42) to 2019 (2.14) unfolds shortening of river course by 13.34 km between downstream of the confluence point of river Dhaksi and Raidak-I near Uttar Chikliguri up to confluence of Raidak-I with Torsa at Balabhut. The study reveals that, over 47 years of study period (1972 to 2019), average rate of net migration measured along the left bank is higher in reach-2 (60.02 m/y) and reach-3 (33.12 m/y) than the reach-1 (3.73 m/y) while on right bank, it is comparatively higher in reach-1(96.38 m/y) than reach-2 (68.92 m/y) and reach-3 (25.81 m/y). Reach-wise short-term analysis of erosion and accretion trend unveils that on the right bank, reach-3 was more vulnerable than reach-1 and reach-2 and that on the left bank, reach-2 was more prone to bank erosion than the other two reaches of the study site probably because of non-cohesive bank materials. The study also exhibits reach-2 was more dynamic than reach-1 and reach-3 during the study period 1996–2009 and 2009–2019. Reach wise erosion-accretion dynamics shows that reach-2 was more dynamic in nature than reach-1 and reach-3 during both the time span of 1972–2009 and 1972–2019. However, the study revealed that in reach-1, considerable bank erosion has occurred in Paschim Chikliguri and Uttar Chikliguri village of Alipurduar-II block, Chikliguri Dwitia Khanda and Dhalpal village of Tufanganj-I block and Bashraja Dwitia Khanda village of Tufanganj-II block. Reach-2 experienced significant erosion near village Guriarpur and Andaran Phulbari under Tufanganj-I block, Basraja Pratham Khanda and Langalgram under Tufanganj-II block and near Tufanganj Municipality and Kamat Phulbari of Tufanganj-I block.

Dwiparpar, Rajarkuthi and Balabhut of Tufanganj-I block and Begarkhata and Sauerkhata of Tufanganj-II block are most vulnerable location of bank erosion and lateral channel migration in lowermost reache-3. The study area is a quaternary geomorphological sites and being a part of tectonically sensitive Himalayan foreland basin controls the fluvial system of the area. In general, natural factors are sole responsible for channel dynamics. Every year during monsoon excessive sedimentation in the channel bed of river Raidak-I resulting into numerous shoal and bar formation. Dynamic nature of shoal and mid channel or point bars causes thalweg line shifting and consequent channel shifting by lateral bank erosion. Furthermore, anthropogenic activities like construction of Dam, railway bridges and roads across the river, sand and gravel mining and embankment building exaggerated the channel dynamics and related hazards in the study area. Hectares of hectares agricultural lands have been engulfed in the river bed due to bank erosion and associated lateral migration of river Raidak-I. Many people became homeless and they faced a great misery. Therefore, this scientific study would be helpful for policy making purposes during long term floodplain management.

Acknowledgements The authors are acknowledged to the Geological Survey of India for providing geological map of the study area. Authors are also highly thankful to the USGS for supplying satellite images.

Declare of interest statement
The authors declare that they have no conflict of interest.

References

Ahmed I, Das N (2018) Sedimentation induced depositional lands of the Gumti River of Tripura and its land use pattern. In Mal S, Singh RB, Huggel C (eds) Climate change, extreme events and disaster risk reduction, Germany: Springer, pp 135–145. https://doi.org/10.1007/978-3-319-56469-2_9

Ahmed I, Das N, Debnath J (2018) An assessment toprioritise the critical erosion-prone sub-watersheds for soil con-servation in the Gumti basin of Tripura, North-East India. Environ Monit Assess 189:1–15. https://doi.org/10.1007/s10661-017-6315-6

Bag R, Mondal I, Bandyopadhyay J (2019) Assessing the oscillation of channel geometry and meander migration cardinality of Bhagirathi River, West Bengal, India. J Geograp Sci 29(4):613–634. https://doi.org/10.1007/s11442-019-1618-z

Baki ABM, Gan TY (2012) Riverbank migration and island dynamics of the braided Jamuna River of the Ganges-Brahmaputra basin using multi-temporal Landsat images. Quatern Int 263:148–161. https://doi.org/10.1016/j.quaint.2012.03.016

Bandyopadhyay S, Kar NS, Das S, Sen J (2014) River systems and water resources of West Bengal: a review. Geol Soc India Spec Publ 3:63–84. https://doi.org/10.17491/cgsi%2F2014%2F62893

Bera B, Bhattacharjee S, Ghosh A, Ghosh S, Chamling M (2019a) Dynamic of channel potholes on Precambrian geological sites of Chhota Nagpur plateau, Indian peninsula: applying fluvio-hydrological and geospatial techniques. SN Appl Sci 1(5):494. https://doi.org/10.1007/s42452-019-0516-2

Bera B, Bhattacharjee S, Roy C (2019b) Estimating stream piracy in the lower Ganga plain of the Quaternary geological site in West Bengal, India applying sedimentological bank facies, log and geospatial techniques. Curr Sci 117(4):662–671. https://dx.doi.org/https://doi.org/10.18520/cs/v117/i4/662-671

Chamling M, Bera B (2020a) Spatio-temporal patterns of Land Use/Land Cover change in the Bhutan-Bengal foothill region between 1987 and 2019: Study towards Geospatial applications and Policy Making. Earth Syst Environ 4(7). https://doi.org/10.1007/s41748-020-00150-0

Chamling M, Bera B (2020b) Likelihood of elephant death risk applying kernel density estimation model along the railway track within biodiversity hotspot of Bhutan-Bengal Himalayan Foothill. Model Earth Syst Environ. https://doi.org/10.1007/s40808-020-00849-z

Charlton CO (2008) Fundamentals of fluvial geomorphology. Routledge, London and New York, p 97

Chu ZX, Sun XG, Zhai SK, Xu KH (2006) Changing pattern of accretion/erosion of the modern Yellow River (Huanghe) subaerial delta, china: based on remote sensing images. Mar Geol 227:13–30. https://doi.org/10.1016/j.margeo.2005.11.013

Das S, Adak K, Samanta K (2014) Hydrodynamic changes of river course of part of Bhagirathi-Hugli in Nadia district—a Geoinformatics appraisal. Int J Geomat Geosci 5(2):284–299. https://scinapse.io/papers/2288103395

Downward SR, Gurnell AM, Brookes A (1994) A methodology for quantifying river channel planform change using GIS. IAHS Publ Series Proc Reports-Intern Assoc Hydrol Sci 224:449–456

Duan JG, Wang SSY, Jia Y (2001) The application of the enhanced CCHE2D Model to study the alluvial channel migration processes. J Hydraul Res 39 (5):469. https://doi.org/10.1080/00221686.2001.9628272

Duroy V, Farah A, Lillie RJ (1989) Subsurface densities and lithospheric flexure of the Himalayan foreland in Pakistan. In: Malinconico Jr., LL, Lillie RJ (eds) Tectonics of Western Himalaya, Geol. Soc. Am. Spec. Pap 232:217–236. https://doi.org/10.1130/SPE232-p217

Dury GH (1977) Underfit streams: retrospect and prospect. In River Channel changes. Edited by K.J. Gregory (Chichester: Wiley Interscience), pp 281–293

Giardino JR, Lee AA (2011) Brazos rates of channel migration on the River. Texas Water Development Board, pp 1–41. https://www.twdb.texas.gov/publications/reports/contracted_reports/doc/0904830898_Brazos.pdf

Gogoi C, Goswami DC (2014) A study on channel migration of the Subansiri River in Assam using remote sensing and GIS technology. Curr Sci 106 (8):1113–1120. https://www.jstor.org/stable/24102308

Gregory KJ (1977) The context of river channel changes. In River Channel Changes, edited by K.J.Gregory. Chichester: Wiley Interscience, pp 1–12

Guchhait SK, Islam A, Ghosh S, Das BC, Maji NK (2016) Role of hydrological regime and floodplain sediments in channel instability of the Bhagirathi River, Ganga-Brahmaputra Delta, India. Phy Geog 37 (6):476–510. https://www.tandfonline.com/doi/abs/https://doi.org/10.1080/02723646.2016.1230986

Gurnell AM, Downward SR, Jones R (1994) Channel planform change on the River Dee meanders, 1876–1992. Regul River: Res Appl 9:187–204. https://doi.org/10.1002/rrr.3450090402

Gurnell AM (1997) Channel change on the River Dee meanders, 1946–1992, from the analysis of air photographs. River Res Appl 13(1):13–26. https://doi.org/10.1002/(SICI)1099-1646(199701)13:1%3C13::AID-RRR420%3E3.0.CO;2-W

Heitmuller FT (2014) Channel adjustments to historical disturbances along the lower Brazos and Sabine Rivers, south-central USA. Geomorphology 204:382–398. https://doi.org/10.1016/j.geomorph.2013.08.020

Holbrook J, Schumm SA (1999) Geomorphic and sedimentary response of rivers to tectonic deformation: a brief review and critique of a tool for recognizing subtle epeirogenic deformation in modern and ancient settings. Tectonophysics 305:287–306. https://doi.org/10.1016/S0040-1951(99)00011-6

Islam A, Guchhait SK (2017) Search for social justice for the victims of erosion hazard along the banks of river Bhagirathi by hydraulic control: a case study of West Bengal, India. Environ Devel Sustain 19(2):433–459. https://link.springer.com/article/https://doi.org/10.1007/s10668-015-9739-6

Islam A, Guchhait SK (2020) Characterizing cross-sectional morphology and channel inefficiency of lower Bhagirathi River, India, in post-Farakka barrage condition. Nataral Hazards. https://doi.org/10.1007/s11069-020-04156-9

Knight MJ (1975) Recent crevassing of the Erap River, New Guinea. Aust Geogr Stud 13:77–82. https://doi.org/10.1111/j.1467-8470.1975.tb00068.x

Knighton D (1984) Fluvial forms and processes. Edward Arnold, London, p 218

Lawler DM, Couperthwaite J, Bull LJ, Harris NM (1997) Bank erosion events and processes in the upper Severn basin. Hydrol Earth Syst Sci 1(3):523–534. https://hal.archives-ouvertes.fr/hal-00304421/

Leopold LB (1973) River channel change with time: an example. GSA Bull 84:1845–1860. https://doi.org/10.1130/0016-7606(1973)84%3C1845:RCCWTA%3E2.0.CO;2

Lyon-Caen H, Molnar P (1985) Gravity anomalies, flexure of the Indian plate, and the structure, support, and evolution of Himalaya and Ganga Basin. Tectonics 4:513–538. https://doi.org/10.1029/TC004i006p00513

Majumdar S, Das (Pan) N (2014) Spatio-temporal shift of right bank of the Gumti River, Amarpur town, Tripura and its impact. In Singh M et al. (eds) Landscape Ecolgy and Water Management: Proceedings of IGU Rohtak Conference, 2, Advances in Geographical and Environmental Sciences. https://link.springer.com/chapter/https://doi.org/10.1007/978-4-431-54871-3_16

Mandal AC, Patra P, Majumder R, Ghosh DK, Bhunia GS (2017) Evaluating meander shifting dynamics (1977–2017) of the Bhagirathi river course in Murshidabad District, West Bengal, India. Spat Inf Res 26(1):33–45. https://doi.org/10.1007/s41324-017-0153-z

Miall AD (1990) Principles of Sedimentary Basin analysis. Springer-Verlag, New York

Molnar P (1984) Structure and tectonics of the Himalaya: Constraints and implications of Geophysical data. Annual Rev Earth Planet Sci 12:489–518. https://doi.org/10.1146/annurev.ea.12.050184.002421

Nath B, Sultana NN, Paul A (2013) Trends analysis of river bank erosion at Chandpur, Bangladesh: a remote sensing and GIS approach. Int J Geomat Geosci 3 (3):454–463. http://www.indianjournals.com/ijor.aspx?target=ijor:ijggs&volume=3&issue=3&article=006

Ophra SJ, Begum S, Islam R, Islam MN (2018) Assessment of bank erosion and channel shifting of Padma River in Bangladesh using RS and GIS techniques. Spat Inf Res 26(6):599–605. https://doi.org/10.1007/s41324-018-0202-2

Richard GA, Julien PY, Baird DC (2005) Statistical analysis of lateral migration of the RioGrande, NewMexico. Geomorphology 71:139–155. https://doi.org/10.1016/j.geomorph.2004.07.013

Robinson BA (2013) Recent (circa 1998 to 2011) channel-migration rates of selected streams in Indiana. U.S. Geological Survey, Scientific Investigation Report pp 2013–5168, 46. https://pubs.usgs.gov/sir/2013/5168/

Rudra K (2011) The encroaching Ganga and social conflict: the case of West Bengal, India. https://www.

indiawaterportal.org/articles/encroaching-ganga-and-social-conflicts-case-west-bengal

Sarkar A, Garg RD, Sharma N (2012) RS-GIS based assessment of river dynamics of Brahmaputra River in India. J Water Resour Prot 4:63–72. https://doi.org/10.4236/jwarp.2012.42008

Sarma JN, Borah D, Goswami U (2007) Change of river channel and Bank erosion of the Burhi Dihing River (Assam), assessed using remote sensing data and GIS. J Indian Soc Remote Sens 35(1):94–100. https://doi.org/10.1007/BF02991837

Schumm SA (1963) Sinuosity of alluvial rivers on the Great Plains. Bull Geol Soc Am 74:1089–1100

Schumm SA, Dumont JF, Holbrook JM (2002) Active tectonics and Alluvial rivers. Cambridge University Press

Singh S (2007) Geomorphology. Allahabad: Prayag Pustak Bhawan

Sinha R, Ghosh S (2012) Understanding dynamics of largerivers aided by satellite remote sensing: a case study from LowerGanga plains, India. Geocarto Int 27 (3):207–219. https://doi.org/10.1080/10106049.2011.620180

Subba S (2014) The changes of the lower courses of river Raidak-Sankosh interfluves and impact on their surrounding environment. Geo-Analyst, ISSN pp 2249–2909. http://gswb.in/wp-content/uploads/2015/04/Saroja-Subba.pdf

Szabo J, David L, Loczy D (2006) Anthropogenic geomorphology: a guide to man-made landforms. Springer, Dordrecht Heidelberg London New York

Takagi T, Oguchi T, Matsumoto J, Grossman MJ, Sarker MH, Matin MA (2007) Channel braiding and stability of the Brahmaputra River, Bangladesh, since 1967: GIS and remote sensing analyses. Geomorphology 85:294–305. https://doi.org/10.1016/j.geomorph.2006.03.028

Thakur PK, Laha C, Aggarwal SP (2012) River bank erosion hazard study of river Ganga, upstream of Farakka barrage using remote sensing and GIS. Nat Hazards 61:967–987. https://link.springer.com/article/10.1007%252Fs11069-011-9944-z

Thorne CR (1982) Processes and mechanisms of river bank erosion. In Hey RD, Bathurst JC, Thorne CR (eds) Gravel-Bed Rivers. Wiley, Chichester, England, pp 227–271. https://notts.rl.talis.com/items/3EEAD2B9-BCF5-6966-8449-B3FA139DA17B.html

Thorne CR, Russell APG, Alam MK (1993) Planform pattern and channel evolution of the Brahmaputra River, Bangladesh. Geol Soc London, Spec Public 75:257–276. https://doi.org/10.1144/GSL.SP.1993.075.01.16

Thorne CR, Tovey NK (1981) Stability of composite river banks. Earth Surf Proc Land 6:469–484. https://doi.org/10.1002/esp.3290060507

Wallick JR, Grant GE, Lancaster ST, Bolte JP, Denlinger RP (2007) Patterns and controls on historical channel change in the Willamette River, Oregon, USA. In. Gupta A (ed) Largerivers: geomorphology and management. John Wiley & Sons Ltd, pp 492–516. https://www.fs.usda.gov/treesearch/pubs/33121

Wallick JR, Lancaster ST, Bolte JP (2006) Determination of bank erodibility for natural and anthropogenic bank materials using a model of lateral migration and observed erosion along the Willamette River, Oregon, USA. River Res Appl 22:631–649. https://doi.org/10.1002/rra.925

Winterbottom SJ (2000) Medium and short-term channel planform changes on the rivers Tay and Tummel, Scotland. Geomorphology 34:195–208. https://doi.org/10.1016/S0169-555X(00)00007-6

Youdeowei PO (1997) Bank collapse and erosion at the upper reaches of the Ekole creek in the Niger delta area of Nigeria. Bull Int Assoc Bull Eng Geol Environ 55(1):167–172. https://doi.org/10.1007/BF02635419

8 Spatio-Temporal Variation of Morphological Characteristics in Bhagirathi River—Case Study in Murshidabad District, West Bengal (India)

Anukul Chandra Mandal and Gouri Sankar Bhunia

Abstract

In order to understand the periodical evolution pattern of the river system in the Murshidabad district, multi-temporal Landsat satellite data (1990–2017) were used based on river plan sequential maps. The Landsat satellite's standard false color composite (SFCC) image was used to identify geomorphic features of the study area. Normalized Difference Water Index was used for automatic extraction river course based on the threshold value. During the period 1990 and 2017, the erosion and accretion zones were identified using the overlay function in QGIS software. Major geomorphic characteristics including active river channel/main river course, tributary river channel, the seasonal stretch of waterways, meander scar, paleochannel, shallow depression, deep depression, ox-bow lake, and wet sand were delineated in the study area. The maximum deposition was observed in the Beldanga block along the Bhagirathi River course. This information may help to investigate the morphological features of the prevailing flood program and to demarcate sensitive areas, the planning and management plan for the river.

Keywords

Landsat · Bhagirathi River · Morphological characteristics · Fluvial landforms

8.1 Introduction

River basins are the fundamental units of the river landscape, and many studies have focused on their geometric physiognomies, including stream network's relationships and quantifiable hydrological parameters (Huggett and Cheesman 2002). Water resources management and assessment in the river basin are extremely important methodological criteria. Fluvial basins cover a discrete morphological area and are especially relevant to the area's pattern of streams and geomorphology (Strahler 1957). Geology, elevation, and climate are shown to be essential components of complex watershed ecosystems (Mesa 2006). The drainage patterns reflect the river dynamics of the variables. In different topographical and climatic conditions, the connection between morphometric parameters differs from one riverbank to another. Understanding these relationships would allow us to identify the dominant variables of the river basin

A. C. Mandal (✉) · G. S. Bhunia
Department of Geography, Seacom Skills University, Kendradangal, Birbhum, Bolpur, West Bengal 731236, India

P. K. Shit et al. (eds.), *Drainage Basin Dynamics*, Geography of the Physical Environment,
https://doi.org/10.1007/978-3-030-79634-1_8

(Samal et al. 2015; Rai et al. 2018). Thus, the geomorphology of the river basin and its channel network is essentially linked.

The natural and inevitable phenomenon of rivers is erosion by the banks of the river. Erosion on the banks of the river is a natural and dynamic process. It consists of washing the materials from the river banks. Bank erosion occurs when the size of water flowing exceeds the solidity of materials on the river banks (Ghosh and Sahu 2018; Das et al. 2014). Three moribund Bengal delta rivers, the Bhagirathi River, the Mathabhanga River, and the Jalangi River, came to be known during the British Colonial period as Nadia Rivers (Majumder 1978; Biswas 2001). River Hooghly is called the combined flow of Jalangi and Bhagirathi. Soil cutting by brick-fields from banks and beds represents one of the human activities that most affect the river. In the West Bengal area of Maldah, enormous losses in the Ganga River bank are reported because of erosion (Laha and Bandyopadhyay 2013). The illegality of this practice leads to bank erosion, loss of property and death, which leads to a low river and rapid deterioration (Das 2014).

The land type processes, soil-physical characteristics, and erosions are important elements that can best be assessed by comparing their morphometrical parameters (Dar et al. 2013) with their effect on the LULC of the basin. The basic characteristics of the watershed are calculated. Remote Sensing (RS) and GIS technologies are now efficiently being used to produce substantially accurate maps of multiple watershed lines (Ozdemir and Bird 2009). RS and GIS are cost-effective tools to evaluate the hydrological impacts of spatial–temporal river course changes. It provides valuable information to support the correct planning and decision-making capacity in Franklin in 2001 and provides valuable, multi-temporal satellite data for analyzing changes in river flow processes (Rai et al. 2018). Hence, the present chapter describes about the periodical evolution pattern of the river system in the Murshidabad district, multi-temporal Landsat satellite data (1990–2017) were used based on river plan sequential maps.

8.2 Study Area

A hundred years ago, the largest flow of the Ganges River was Bhagirathi, basically along the modern river of Hooghly. Geologists claim that there might be two large channels in the state of Bengal, to the west of the Madhupur forest, that is to say Ganges was flowing through central Bengals and Teesta, through southern Bengal before it was diverted to the Padma to the east (Fig. 8.1). The Bhagirathi River divided the district of Murshidabad into two broad geographical districts almost equal in area, according to the district statistics handbook for Murshidabad (2010–2011). On the west side of the Bhagirathi River lies the western tract or the Rarh district. It is an estate of the Sub-Vindhyan area of laterite clay with a nodular glut. It's somewhat undulate. The soil is greyish, reddish and has a high lime and iron oxide content (Census of India 2011). The Bagri region is the eastern section of the Bhagirathi River. It consists generally of alluvial deposits of Gangetic. The region of Bagri is very fertile due to fresh silt deposition nearly every year (Census of India 2011). The slope is found to be zero to five degrees in almost all areas of study. The area has highly warm summers and moisture all year long. Rainfall from June to September is caused by the south-west monsoon. Murshidabad's economy depends mainly on agriculture. The Murshidabad (2010 to 2011) District Statistical Handbook states that the majority of the population are engaged in farming activities and that a small share of industrial activities is involved.

8.3 Materials and Methods

8.3.1 Data Used

For the assessment of river bank vulnerability of Bhagirathi River channel, LANDSAT-5 (Thematic Mapper, DoP- 21/11/1990), LANDSAT -7 Enhanced Thematic Mapper + (ETM$^+$, DoP-26/10/2001), LANDSAT-8 Optical Land Imager (OLI, DoP -07/11/2017) having a Path/Row—

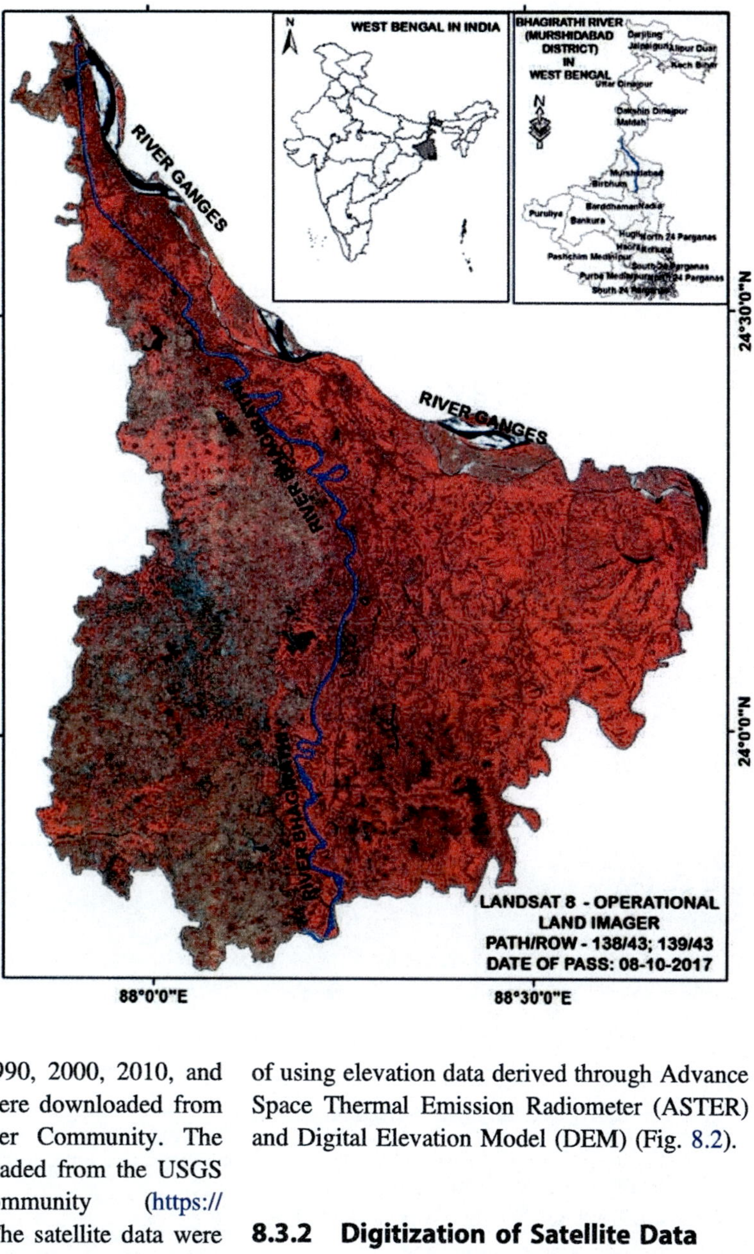

Fig. 8.1 Location map of the Bhagirathi River course in Murshidabad district

138/043 in the year of 1990, 2000, 2010, and 2017. The satellite data were downloaded from the USGS Earth Explorer Community. The satellite data were downloaded from the USGS Earth Explorer Community (https://earthexplorer.usgs.gov/). The satellite data were chosen for the month of October and November because of the maximum discharge in the study area. The acquired images were rectified in Universal Transverse Mercator (UTM) Projection System and World Geodetic Survey (WGS) 84 datum and North 45 zone. All the data were resampled into 30 m spatial resolution. The cross profiles have been drawn at 1 km interval of using elevation data derived through Advance Space Thermal Emission Radiometer (ASTER) and Digital Elevation Model (DEM) (Fig. 8.2).

8.3.2 Digitization of Satellite Data

The Normalized Difference Water Index (NDWI) was used to delineate the river water course at different study period in Murshidabad district. The NDWI is calculated using Shortwave Infrared (SWIR) and Near-Infrared (NIR) spectrum of Landsat data based on the following equation (Gao 1996), as follows:

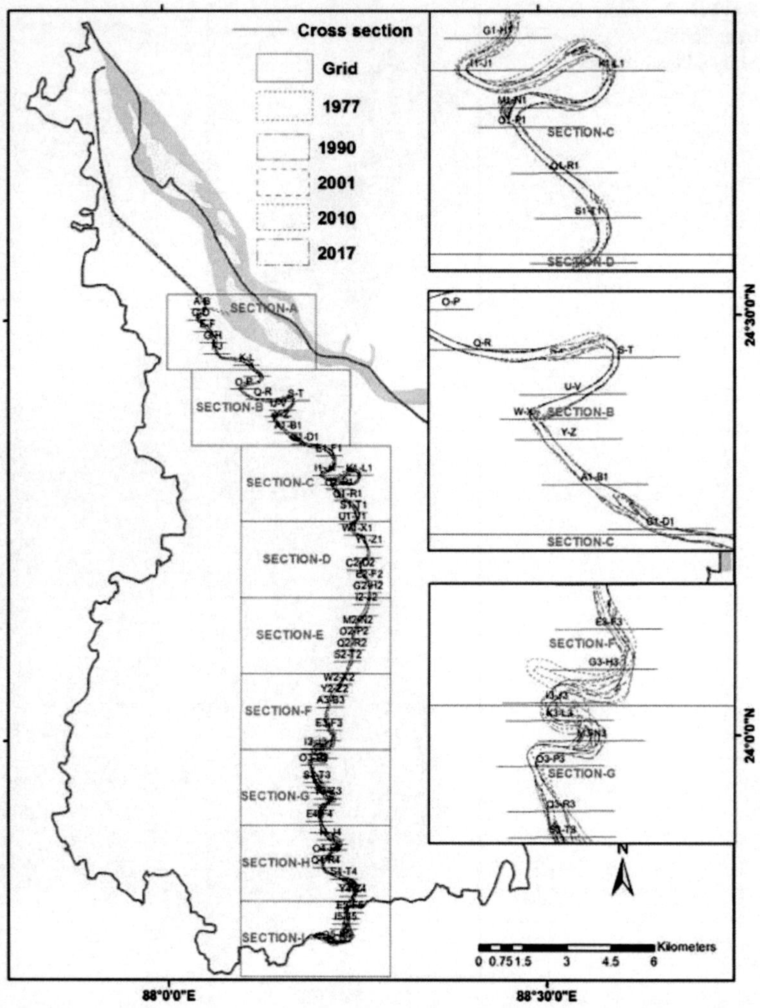

Fig. 8.2 Section and cross section along the Bhagirathi River course in Murshidabad district

$$NDWI = \frac{Shortwave\ infrared\ -\ Near\ in frared}{Shortwave\ infrared\ +\ Near\ infrared}$$

The result from the NDWI picture helps to identify the bank line of the river. The shallow water channels were replicated as a river part. The bank lines of the Bhagirathi River were restricted by the visual clarity and digital image processing techniques for the years 1990, 2000, 2010, and 2017. The highlighted bank lines were used to create a midpoint streak channel (e.g., the middle point between both channel margins) via on screen digitization in QGIS Software v2.4.

8.3.3 Cross-Sectional Analysis

The entire riverside of Bhagirathi was alienated to 99 strips with a distance of around 1 km in the Murshidabad district (from Roshanpur from Barokulberia) (Fig. 8.3). At the edge of each strip the reference cross sections were strained. Through 11 cross sections are grouped together as a region with numbers upstream to downstream of the river. Such grids were used for the study of expected parameters. A number starting with the upper reach of ‘A’ and the lower reach of the river in the district of Murshidabad are assigned to each grid index feature.

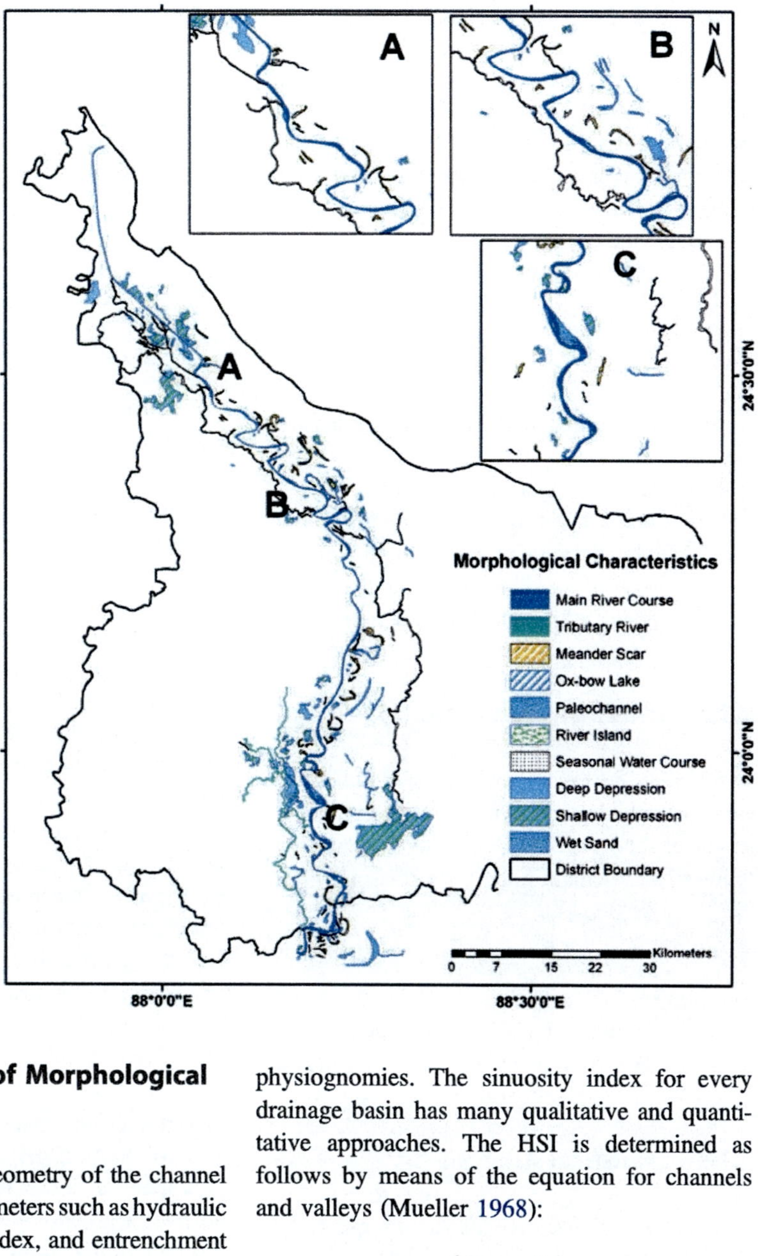

Fig. 8.3 Morphological characteristics of the Bhagirathi River planform for the year of 1990

8.3.4 Identification of Morphological Characteristics

In its sinuous shape, the geometry of the channel instigates. The various parameters such as hydraulic sinuosity index, braiding index, and entrenchment ratio were calculated for each section.

8.3.5 Hydraulic Sinuosity Index (HSI)

Sinuosity studies help to identify the drainage basin topographical and hydrological physiognomies. The sinuosity index for every drainage basin has many qualitative and quantitative approaches. The HSI is determined as follows by means of the equation for channels and valleys (Mueller 1968):

$$HSI = \frac{(CI - VI)}{(CI - 1)} \times 100$$

CL = Channel Length (The length of the channel in the stream)

VI = Valley index, i.e., is the ratio between channel length and air length

8.3.6 Braiding Index (BI)

The gradient index shows the number of channels enabled by the bars at the base (Brice 1964). Braided bars are the middle of the channel walls, where the flux is small and the river flows along the path of lower strength and lower height. In this study, the braiding index determined using CWPRS (2017), Pune:

$$BI = \frac{B_m - B_r}{B_m}$$

where B_m is the gap of each twin bar between the outer bars

B_r is the interiors of the bars width.

The value of braiding index varies between 0 and 1. The braiding index value ranges from 0 to 1. So because of its subjective nature, this definition is quite useful than the Brice index.

8.3.7 Entrenchment Ratio (ER)

The ER is the ratio between the width of the surge vulnerable to flooding and the maximum length of the river. The following equation was followed in calculating the ER (Rosgen 1996):

$$ER = \frac{W_{fpa}}{W_{bkf}}$$

W_{fpa} = Width of the flood prone areas

W_{bkf} = Width of the channel at bank full stage

8.3.8 Statistical Analysis

The statistics descriptive of the morphometric characteristics of the channel (mean, standard deviation, kurtosis, skewness) were analyzed. In Microsoft Excel sheet all of the statistical analysis was carried out. The study was performed at a confidence interval of 95%.

8.4 Results

Geomorphology of the river covenants with the modification of river planform and cross sections because of the deposition and river erosion. For example, the natural cut-off developments along the river Bhagirathi alter the morphometric pattern of the channel. The strong aggravation of the river's course often raises flood rates, which contribute to the overturning of banks and reservoirs during comparatively small incidences of flooding.

The Bhagirathi River course has been characterized by different types of geomorphological features at different times (Fig. 8.3). The most significant geomorphological features were marked by an active channel, tributary river, palaeochannel, meandering scar, low- and deep-lying depression, ox-bow lake, islands of the river, wet sand, and seasonal watercourses. The distribution of geomorphic elements in the study area for 1990, 2000, 2010, and 2017 are illustrated in Figs. 8.3, 8.4, 8.5, and 8.6, respectively.

In the middle of the Bhagirathi River most of the meander scar is seen; however, in 1990 the density of meander scar was highest. The lake of the ox-bow was mostly located in the mid and lower part of the river Bhagirathi. In the lower part (right side of the channel) and the upper part (left side), there are several small and shallow depression. In the lower course of the Bhagirathi River, many wet sand deposits and a small river island were found. On the right side of the river, some seasonal streams were found while the bottom left of the river course was defined by the river tributary spatial allocation. In addition, at the upper and lower reaches, the flux pattern of the river was observed. In the Bhagirathi River meander route (upper and lower section of Murshidabad district) river chute, point bar, splay bar, and scrollbars are found too.

Routine turbulent, rainy season overflows of waters throughout and around the course of the

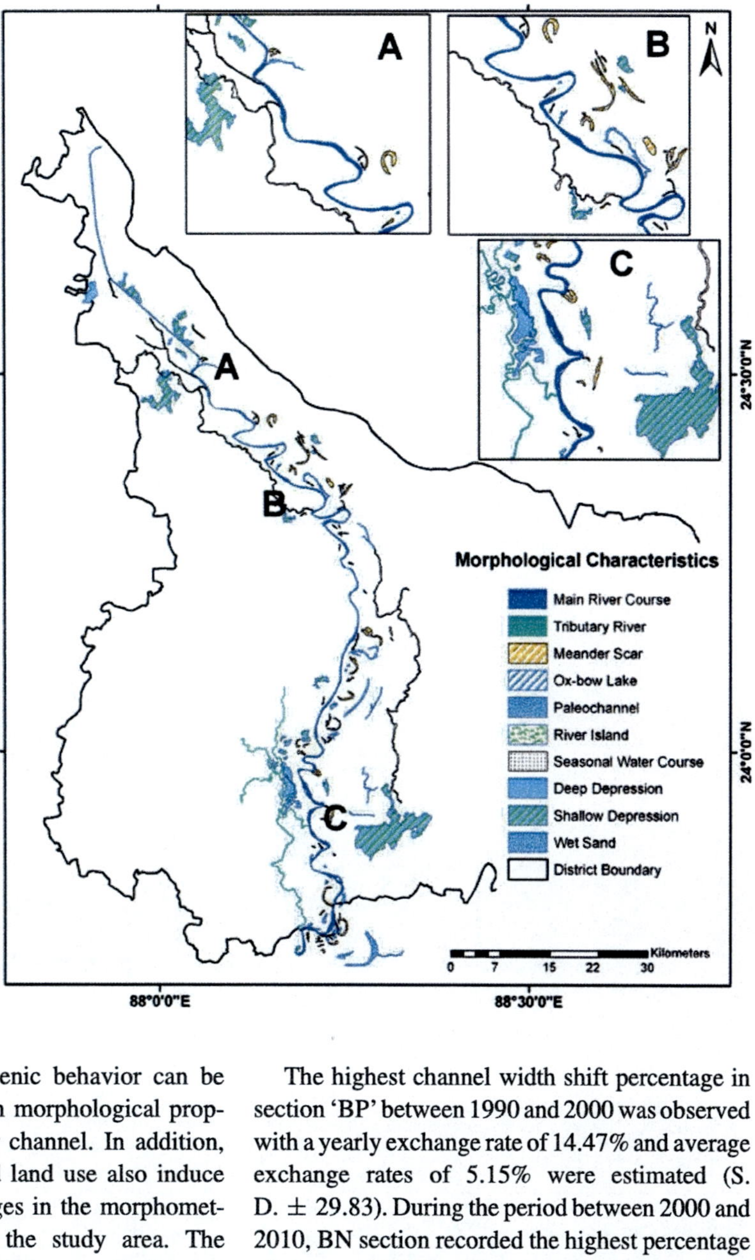

Fig. 8.4 Morphological characteristics of the Bhagirathi River planform for the year of 2000

river and other anthropogenic behavior can be due to spatial variability in morphological properties throughout the river channel. In addition, changes in urban area and land use also induce spatial and temporal changes in the morphometric characteristics within the study area. The average value of channel width was calculated as 0.52 km (S.D. ± 0.42) and 0.47 km (S. D. ± 0.33) in 1990 and 2017, respectively. In 1990, 'CT' cross section recorded the average width of the fluvial channel and the lower width at 'AQ'. 'AA' cross section was found to have an average range of 1.73 km in 2017.

The highest channel width shift percentage in section 'BP' between 1990 and 2000 was observed with a yearly exchange rate of 14.47% and average exchange rates of 5.15% were estimated (S. D. ± 29.83). During the period between 2000 and 2010, BN section recorded the highest percentage of change (375%) with an annual change rate of 37.50%. The 'CA' sector was the highest change percentage between 2010 and 2017 (52.24%) and this section has an average change rate of 3.66%. The 'CA' sector was the highest change percentage between 2010 and 2017 (52.24%) and this section has an average change rate of 3.66%.

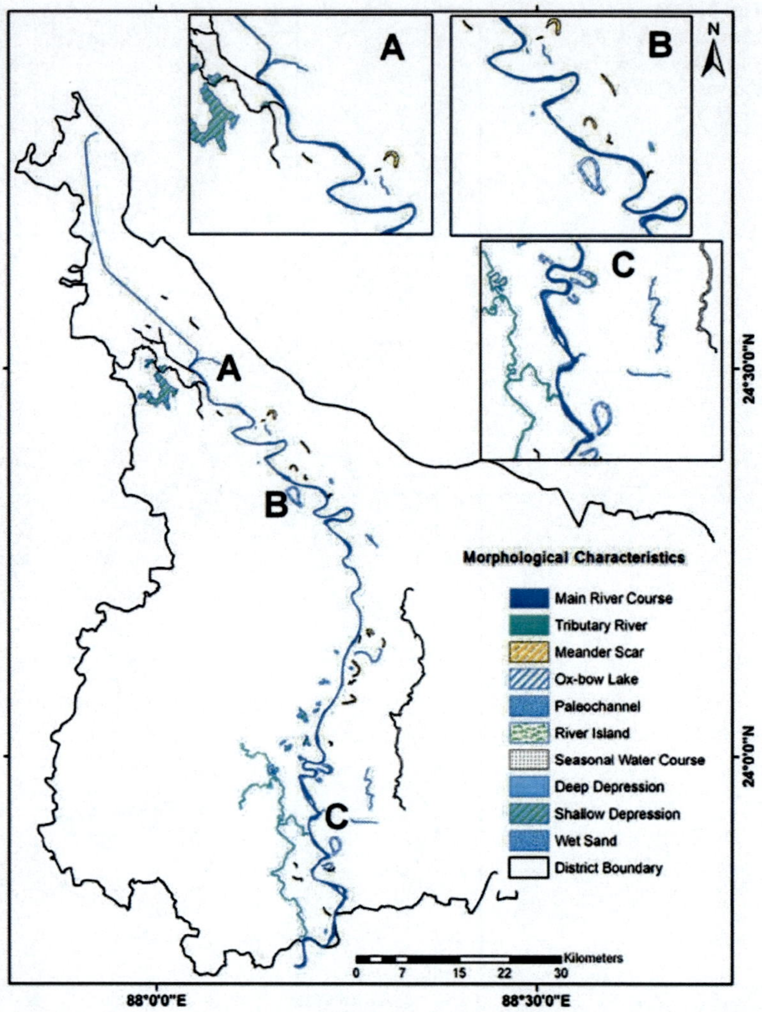

Fig. 8.5 Morphological characteristics of the Bhagirathi River planform for the year of 2010

8.4.1 Morphometric Pattern at Different Time Intervals

In determining the expansion process and the scheduling factors of IS, the Hydraulic Sinuosity Index (HSI) and Topography Sinuosity Index (TSI) are an important morphometric predictor. The higher HSI values indicated that rivers have formed their floodplains, while the lesser HSI value showed rivers crossing hills.

As far as TSI is concerned, in 1990, 2000, 2010, and 2017, the mean value of TSI was calculated as 27.61, 30.45, 27.68, and 28.29. Section-A estimated the average TSI value in 1990 and then it remained nearly constant until 2017 (Table 8.1). The variability of the TSI value is very small and almost zero in Section-B, Section-D, and Section-E. The supremacy of hydraulic activity in this segment can be due to this. In Section-C, the TSI value was not consistent throughout the entire study period. In Section-F, TSI's pattern between 1990 and 2000 decreased (Table 8.2), the high topography gradient represented and then the trend grew. Section-G showed a smaller topographical gradient in terms of an increase in TSI during the period 1990 to 2010 (Table 5.6). In fact, the higher TSI value suggested that the preliminary surface was loophole. As a consequence, TSI's

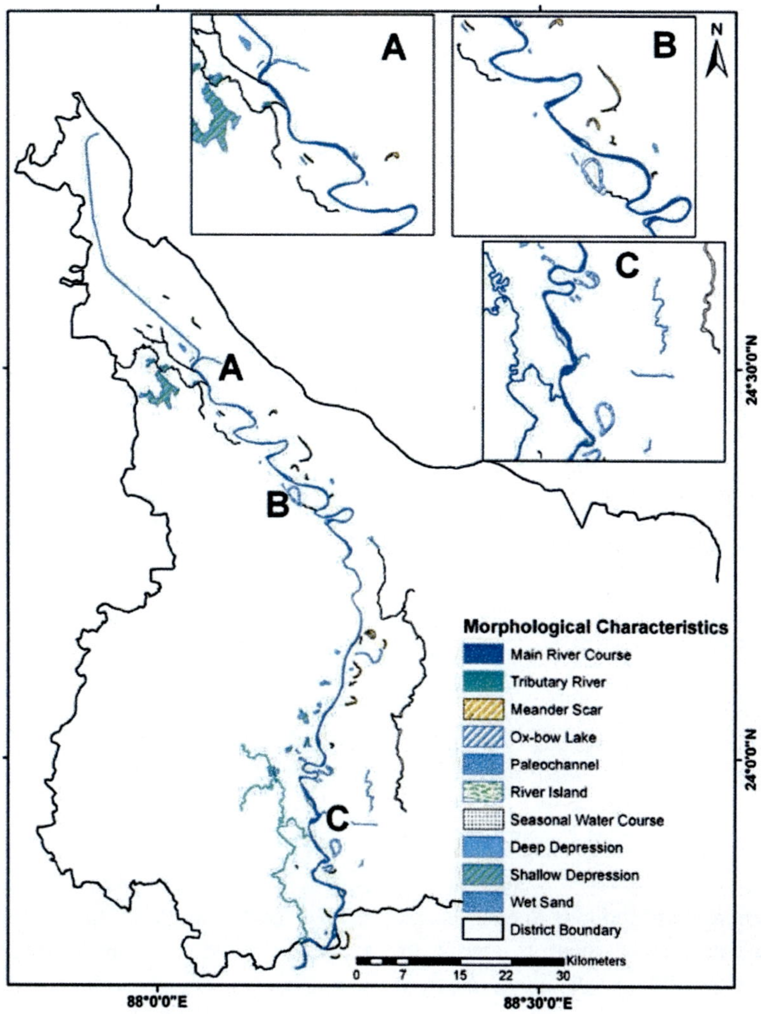

Fig. 8.6 Morphological characteristics of the Bhagirathi River planform for the year of 2017

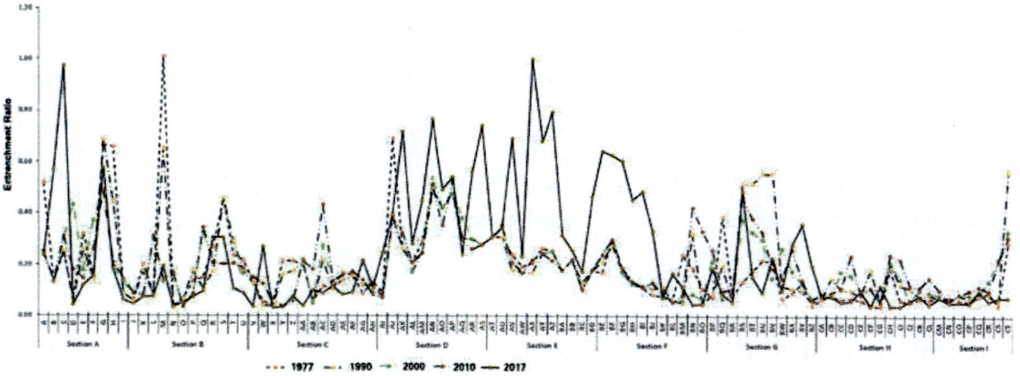

Fig. 8.7 Entrenchment ratio of the river Bhagirathi during the period between 1990 and 2017

Table 8.1 Calculation of Topographic Sinuosity Index (TSI) of Bhagirathi River during the period between 1990 and 2017

Section	1990	2000	2010	2017
A	17.24	18.68	17.58	17.95
B	46.93	48.69	49.08	47.07
C	50.21	50.24	48.57	49.57
D	15.81	18.07	16.11	15.48
E	2.21	2.25	2.30	2.44
F	16.03	17.27	17.46	20.77
G	30.84	38.14	39.01	36.90
H	37.80	36.18	28.74	30.63
I	31.45	44.49	30.24	33.77

Table 8.2 Calculation of Hydraulic Sinuosity Index (HSI) of Bhagirathi River during the period between 1990 and 2017

Section	1990 (km^2)	2000 (km^2)	2010 (km^2)	2017 (km^2)
A	2.71	1.95	2.07	1.47
B	2.28	3.96	3.64	3.91
C	0.52	2.16	1.90	0.47
D	1.53	0.92	1.30	1.32
E	23.68	4.35	6.52	8.33
F	4.31	2.74	2.47	4.26
G	1.72	6.22	29.70	10.55
H	22.81	18.49	0.65	11.14
I	5.60	57.06	20.94	16.49

value for the entire study period in Section-H and Section-I was highly inconsistent as a result of the recurring floods and currents along the river course.

HSI estimation specifies the hydraulic meander sinuosity factor (Sharma et al. 2018). The HSI of the channel was also measured in different sections (Table 8.2). In most of the river at different times, the hydraulic sinuosity index has not been high. The mean HSI value for the years 1990, 2000, 2010, and 2017 was 6.90, 10.39, 7.71, and 6.04. The negativity of the HSI shows that in floodplains the river has developed. The Section-A, Section-B, Section-C, Section-D, and Section-F were recorded very low value of HSI for the entire study period. In the bottom section (Section-G to Section-I), the HSI value was different. Section-I reported the highest value of HSI in 2000 (57.06%), with the lowest in the same section in 1990 to 1990. In addition, in Section-H, in 2010 the lower value was reported compared with in 1990 (22.81%), 2000 (18.49%), and 2017 (11.14%).

The equally defined 9 sections of the study area as shown in Table 8.3 observed variation of the Braiding Index (BI) from 1990 and 2017. It was observed for all 9 parts that the radical differences among the BI were small at various times. The BI value ranged from 0 to 1. The river with less braided bars and the river with the lower river value specified the extremely braided river zone.

In Section-A, it was observed that the value of BI varied between 0 and 0.07. Values were the highest and slowly declined in 2017 to 0.003 except in 2010 (Table 8.3). This situation indicates the previously high load of sediments in the channel, which slowly decreased during the study period and varied.

The BI value for 1990 (0.023), 2010 (0.022), and 2017 (0.037) is estimated in Section-B. For

Table 8.3 Calculation of Braiding Index (BI) of Bhagirathi River during the period between 1990 and 2017

Section	1990	2000	2010	2017
A	0.016	0.011	0.000	0.003
B	0.023	0.000	0.022	0.037
C	0.027	0.012	0.073	0.055
D	0.000	0.000	0.000	0.000
E	0.004	0.000	0.000	0.000
F	0.000	0.000	0.180	0.007
G	0.446	0.107	0.341	0.213
H	0.074	0.087	0.217	0.234
I	0.061	0.072	0.051	0.189

the years 1990 and 2000, the estimated value in this section was almost zero. Section-C's BI variation was gradually changing, with the highest value recorded in 2010. Nevertheless, for the whole study period the BI for Section-D and Section-E was almost zero.

Section-G showed the highest BI value and it changed year after year constantly. This section recorded the highest value (0.446) in 1990. Nevertheless, the BI value gradually increased between 1990 and 2017 in Section-H and Section-I (Fig. 8.3).

The entrenchment ratio was calculated between 1990 and 2017 at 99 locations along the river (Fig. 8.4). At every 1 km radius, the fixation ratio was determined. Section A, Section-B, Section-D, and Section-G have been found to be the highest entrenchment ratio. Section-D was determined for the maximum value of the entrenchment ratio and Section-H for the whole of the study period (19977–2017) was the lowest value. The total deviation of the entrenchment ratio at Section-E station was calculated overall, and the average deviation of the entrenchment ratio at station M was calculated in Section-B.

The average entrenchment ratio was 0.19 km (±0.15) in 1990. The highest entrenchment ratio at G (Section-A) has been measured as 0.69 km and the lowest value at N (Section-B), at 0.03 km. Kurtosis was found to be 1.61 km which is larger than 0.19 km as normal. The distribution is therefore platykurtic.

In 2000, AN (Section-D) recorded the highest value of the entrenchment ratio and Y (Section-C) recorded the lowest value of the entrenchment ratio. The average value of the ER was measured at 0.19 km in the study area with a standard deviation of ± 0.113.

The average ER in the study area by 2010 was 0.20 km (S.D. ± 0.11). This was calculated in 2010. At the 'AP' Section-E station, the highest calculated value of the study site was 0.527 km. In addition, at 'O' station (0.035 km), the lowest measured value of the ER.

By 2017, the highest estimated ER value was estimated in Section-A (1.011 km) and the lowest ER value in X (0.030 km) in Section-C (1.011 km) was estimated. In addition, in the study area average and standard deviation of ER was calculated, respectively, as 0.24 and ±0.23.

8.5 Discussion

Channel migration and erosion are multi-faceted acts instigated by an interplay of many aspects which include emancipation of the river, the river currents, riverbank geology, and geomorphology (Debnath et al. 2017). The satellite imagery sometimes does nothing to assess the morphology of baselines on the Bhagirathi River systems without a proper resolution. The remotely sensed image can therefore be responsible for better results in the demarcation of the river parameters morpho-dynamics by assimilating cross-sectional data. Between 1990 and 2017, the activities of the fluvial Bhagirathi were studied for 28 years. There are several fluvially geomorphic landscapes sprinkling the alluvial plains in and around the Bhagirathi Rivers. In order to effectively renovate fluvio-geomorphological methods and to chart the paleochannel and dry river course, the satellite information has been used. Most of the river geomorphic characteristics are characterized by meanders, which have many cut-off cycles. Certain scars are soothed by natural vegetation, and different are recycled for farming. Several of the abandoned meander hoops on the land plantation concurrent with the point bars in the left-over channel have been identified.

In assessing the development stage and the regulation of the sinuosity variables, the HSI and TSI are useful morphometric devices. The higher value of HSI revealed that these rivers have developed their flood plains. The lower HSI value suggested that the valleys were restricted and they indicated implicitly that the sinuosity of the valleys did not belong to the initial denudation period but was replenished and the topography function was thus enhanced. Greater irregularity of the initial surface is indicated in the higher TSI values. In Soman (2002) any flood plain river, narrow canal, laterite mound canal, ox-bow lakes, TSI > HIS indicate imposed ripeness, has been suggested. Most of the BI observations were shown near to '0' which indicates river in this section is much straighter during this specified period. The role of HSI increases and the role of TSI decreases as the water advances in the erosion process. Consequently, role of hydraulic factors become prominent in the lower reaches. It is clear that TSI decreases as height decreases and HIS increases subsequently.

The results also showed the significant changes in BI for the whole period of study in Section-G and Section-H, which could be attributed to influences in this zone that sediments the larger along the course of the river. Khan et al. (2018) suggested to be close to high-torture shoal, while the stretch of the river connects with youth and neotectonic events in the area. The Bhagirathi River course is therefore assumed to be characterized by small SSIs and is mature or mid-stage between the old and young erosion cycle. A widespread use of historic maps, satellite images, and geospatial technologies has documented historical channel flow (Das et al. 2013). Due to water overflowing into the river, the deep canals smoking through and down the banks risk significant losses of precious land, transportation, and nearby towns and cities in monsoon. The Bhagirathi River is usually silted throughout its flow. The sediment components appear to remove gravitational energy from the floor of the river but can be still disturbed in stormy flow by rising waves that exaggerate the gravitational power.

8.6 Conclusion

GIS and remote sensing techniques are very inventive and effective in determining erosion levels and movement of river channels, without the use of field measurement at the sites. There have been numerous interesting fluviomorphic features formed in the study field, such as bills, ox-bow lakes, cut-offs, paleochannels, etc., during the years period (1990–2017). The HIS, TSI, and BI are useful morphometric tools for demarcating the parameters of the Bhagirathi River by assimilating cross-sectional data. The value of BI in most of the cross section is near to zero indicating that the river course is straight nature. The maximum entrenchment ratio is observed in Section-D and Section-E. The TSI value is inconsistent for the entire river course indicated the dynamic nature of topographic condition that may be attributed to recurrent flood in the study area. The Bhagirathi is still very dynamic in nature and causes severe bank erosion in some particular sector within the study region. Bank erosion sites are the concave banks of the river channel where the steepening of the banks induces extreme bank erosion by undercutting. Some type of shift in fluvial geomorphology oscillates the specific place's age-old economic and occupational structure. People's life in the region is extremely unpredictable and unstable. Some steps taken by the government are the building of spurs and embankments, boulder fencing, etc. However, long-term initiatives such as floodplain zoning and restoration with sufficient economic help, growing people's understanding and preparedness for hazards, eradicating social and psychological hazards, etc., will be very beneficial for erosion victims.

References

Biswas KR (2001) Rivers of Bengal, Vol. I, Government of West Bengal, pp. xviii, xxix, 87, plate-18 and 19.

Brice N (1964) Fundamentals of very low frequency emission generation mechanisms. J Geoph Res, 69 (21):4515–4522. https://doi.org/10.1029/jz069i021p04515

Das BC (2014) Impact of in-bed and on-bank soil cutting by brickfields on moribund deltaic rivers: a study of NadiaRiver in West Bengal. NEHU J. XII(2):101–111

Das TK, Halder SK, Gupta ID, Sen S (2014) Riverbank erosion induced human displacement andits consequences. Living Rev. Landscape Res http://dx.doi: https://doi.org/10.12942/lrlr-2014-3

Dar RA, Chandra R, Romshoo SA (2013) Morphotectonic and lithostratigraphic analysis of intermontane Karewa basin of Kashmir Himalayas, India. J Mt Sci 10(1):731–741

Debnath J, Das (Pan) N, Ahmed I, Bhowmik M (2017) Channel migration and its impact on land use/land coverusing RS and GIS: a study on Khowai River of Tripura, North-East India. Egypt J Remote Sens Space Sci20(2):197–210. https://doi.org/10.1016/j.ejrs.2017.01.009

Franklin SE (2001) Remote sensing for sustainable forest management. Lewis Publishers, Boca Raton, p 407

Gao B (1996) NDWI—A normalized difference water index for remote sensing of vegetation liquid water from space. Rem Sen Environ 58(3):257–266. https://doi.org/10.1016/s0034-4257(96)00067-3

Ghosh D, Sahu AS (2018) Problem of river bank failure and the condition of the erosion victims: a casestudy in Dhulian, West Bengal, India. Reg Sci Inquiry 10 (2):205–214

Huggett R, Cheesman J (2002) Topography and the environment. Prentice Hall, Pearson Education, Upper Saddle River, p. 274

Khan A, Rao LAK, Yunus AP. et al. (2018) Characterization of channel planform features and sinuosity indices in parts of Yamuna River flood plain using remote sensing and GIS techniques. Arab J Geosci 11:525. https://doi.org/10.1007/s12517-018-3876-9

Laha C, Bandyopadhyay S (2013) Analysis of the changing morphometry of River Ganga, shift monitoring and vulnerability analysis using space-bornetechniques: a statistical approach. Int J Sci Res Publ 3(7):1–10

Majumder D (1978) West Bengal District Gazetters Nadia, Government of West Bengal, pp. 5,7,16.

Mesa LM (2006) Morphometric analysis of a subtropical Andean basin (Tucuman, Argentina). Environ Geol 50 (8):1235–1242

Mueller JE (1968) An Introduction to the Hydraulic and Topographic Sinuosity Indexes 1. Annals of theAssociation of American Geographers, 58:371–385. https://doi.org/10.1111/j.1467-8306.1968.tb00650.x

Ozdemir H, Bird D (2009) Evaluation of morphometric parameters of drainage networks derived from topographic maps and DEM in point of floods. Environ Geol 56:1405–1415

Rai PK, Chandel RS, Mishra VN, Singh P (2018) Hydrological inferences through morphometric analysis of lower Kosi river basin of India for water resource management based on remote sensing data. Appl Water Sci 8:15. https://doi.org/10.1007/s13201-018-0660-7

Rosgen D (1996) Applied River Morphology. Wildland Hydrology, Pagosa Springs.

Samal DR, Gedam SS, Nagarajan R (2015) GIS based drainage morphometry and its influence on hydrology in parts of Western Ghats region, Maharashtra, India. Geocarto Int 30(7):755–778

Sharma A, Singh P, Rai PK (2018) Morphotectonic analysis of Sheer Khadd River basin using geo-spatial tools. Spat Inf Res 26(4):405–414. https://doi.org/10.1007/s41324-018-0185-z

Singh P, Gupta A, Singh M (2014) Hydrological inferences from watershed analysis for water resource management using remote sensing and GIS techniques. Egypt J Remote Sens Space Sci 17:111–121

Soman K (2002) Geology of Kerala, 2nd edn. Geological Society of India, Bangalore, pp 179–204

Strahler AN (1957) Quantitative analysis of watershed geomorphology. Trans Am Geophys Union 38:913–920

9 Sedimentation and Shifting of Lower Mundeswari and Rupnarayan River, West Bengal, India

Souvik Das and Subodh Chandra Pal

Abstract

The two important embranchments of the River of Damodar and Dwarkeswar are Mundeswari and Rupnarayan, respectively, which have joined with each other at Marokhana panchyet, Hooghly district, West Bengal, India. In this study area, Hooghly (Pansuli village), Howrah (Kaijuri village), and Midnapur (Vatora village), these three districts are separately located at conjunction point of lower Rupnarayan (main channel) and lower Mundeswari. It's a fact that, due to continuous sedimentation, the lower part of those rivers has deteriorated heavily. So, various soil samples have been collected from riverside surroundings and analyzed sediment sizes by sieving techniques to explain the causes, mechanism, dynamicity, and present situation of sedimentation. This area is located in the Monsoon region. So, during the rainy season, the water flow or stream power becomes very high due to heavy rainfall, but during the summer or late monsoon season, it is comparatively low due to a lack of sufficient rain, which is the main cause of massive sedimentation. In comparison to the Mundeswari River, the River Rupnarayan has a higher energy flow of water, depth, stream power, discharge, slope, bank-full index, and so on which are measured by different instruments such as Echo-sounder and Current meter, Dumpy level, and so on. River shifting is another important incident in this area. From very past to present, several secondary data shows a shifting trend of these two rivers which is a threat for that area. By methods of sedimentation analysis and GIS techniques, it has been noticed that the shifting trend of the two rivers is uni-directional. By sedimentation analysis and Landsat imagery classification, it is showed that from the year 1979–1999, the rate of shifting is 6.5mt and from the year 1999–2019, the rate of shifting is 4.75 mt. In the rainy season, devastating floods and bank erosion are very common and dangerous incidents every year. It is so affrighting for local peoples and their resources. So, it is clear that these two rivers are the gift of that place, denunciation also.

Keywords

Sedimentation · Embranchment · Rupnarayan · Mundeswari · Shifting trends · Deteriorated · GIS · Staff reading

S. Das · S. C. Pal (✉)
Department of Geography, The University of Burdwan, Burdwan, West Bengal, India

P. K. Shit et al. (eds.), *Drainage Basin Dynamics*, Geography of the Physical Environment,
https://doi.org/10.1007/978-3-030-79634-1_9

9.1 Introduction

Damodar is an important part of Bhagirathi-Hooghly Rivers, in West Bengal, India. The western part of the Bengal Basin is controlled by the floodplain of this river. This western part is associated with mainly Ajoy–Damodar interfluves "Rarh Plain". Toward southeast region, this river has created a mature fan-Delta. This basin area is one kind of funnel-shaped around 23,370.98 km^2. It covers a large area in the states of Jharkhand (73.7%) and West Bengal (26.3%). Damodar River is Monsoon dominated and flood-prone also. Various geomorphic Phenomena like erosion, transportation, and deposition occur in this basin. The Summer and Rainy Monsoon seasons are very effective on Morphometry and Lithology of this channel. Three main physiographic characteristics are dominant factors in the whole river basin (1) Upper and (2) Middle Basin is covered by Archean to Gondowana formation, generally a high topographic region (Asansol, Durgapur, Bardhaman districts) (3) Lower Basin is below the Durgapur–Asansol–Bardhaman–Arambag region mainly Champadanga and surroundings, which have older alluvium and generally lower relief (Maity and Maiti 2016).

Dwarkeswar is a long (200.5 km long) and another most important river at the western part of West Bengal and have important transitional characteristics of flash flood, Monsoon flood, and tidal flood in various parts of the basin area. It has originated from an extended part of Chota Nagpur plateau (86°31′E to 87°02′E) in Purulia District and ended in West Midnapur (22°42′30′′ N to 23°31′N) of West Bengal.

Our planet is made of different resources like water, land, and vegetations and various geomorphologic features like mountains, plateau, plains, rivers, etc. People all are simultaneously bounded in a web. All the features are connected to each other with a chain and all the systems on the earth's surface are running continuously in a cyclic order. Damodar river basin is the first multipurpose river valley, planning of India. Several resources of river Damodar have been developed in a unified manner. Damodar gets water by integrated method from Panchet Reservoir, controlled by DVC. The functional work of the Damodar river system follows the dynamic equilibrium between the fluvial and anthropogenic processes. River Rupnarayan is the part of Dwarkeswar which is deteriorated by sedimentation. By historical studies, it was found out that 26.57 million sediments submerged at the lower part of Rupnarayan in the last 25 years. From various reports, a conclusion can be drawn that due to as frequent floods and year-to-yearsea-level changes, there is a huge amount of sediment deposition is going on in this area (Bianchi and Allison 2009). Actually, in most of the rivers globally, bank erosion by human activity has simultaneously increased in present by the huge amount of sediment transportation (Syvitski et al. 2005; Azarang et al. 2017; Malik et al. 2017). In the condition of the upper, middle, and lower part of a river, rapid change can be found out because of numerous dams construction, soil erosion, or water-soil conservation in various places (Yang et al. 2006; Ekeleme and Agunwamba 2018). Various kinds of organic matters like Mn, Fe, Cu, Zn, Cd, and carbonate content (CO3-2) are very important for contaminants in sediments of the dam to assess the extent of environmental pollution (Ghrefat and Yusuf 2006). At the lower part of the channel, various kinds of sediments are heavily loaded with zinc, manganese, copper, nickel, phosphorus and cadmium, phosphorus and zinc, and lead (Loska and wiechula 2003), and rapid sedimentation is not related to salinity (Mondal and Bandhyapadhya 2016; Malik and Pal 2020a). In this Ganga lower region areas, rivers Mundeswari and Rupnarayan are the main facts. Sedimentation, discharge, the direction of flow, velocity, carrying capacity, etc. are so important for agricultural activity and irrigation activity and for controlling the settlement pattern in that region. For controlling river system as well as hydrological, geomorphological, and ecological functioning, the fine-grained sediments are an important essential matter. The quantity and quality of fine-grained sediment within the river is a natural and essential component for controlling river morphological systems. Its effects on construction and management of reservoirs, and due to lack of proper construction and management in different countries, heavily contaminated sediment in these rivers (Owens et al. 2005). Along the course of the estuary, there have various metallic and non-metallic elements. These elements make a variation and can be attributed to a different rate of discharges from industrial, agricultural, aqua-cultural as well as domestic sewage (Sarkar Saha et al. 2004). Actually, in the coastal area of the river, various kinds of metal can be found out in the following order: $Fe > Mn > Zn > Pb > Cu$ as an increased pattern from the estuary to the coastal area (Venkatramanan et al. 2014). The river gradient, width, stream power, and various climate-induced hydrological changes controlled lithology, grain size, and channel morphology which have a long-term stability of the total landscape (Ghosh and Guchhait 2014). It is a fact that Grain-size analysis by sediment sample collection from different areas at different banks shows the variation of grain size from 1 mm to 1/256 mm range and among these variations, clay portion was observed throughout basin area and at the distance from channel silt and rest sand type of sediment and higher grain size shows. For more than 50 years, the flood is the most devastating incident of this region. With the monsoon flow regulation, the sinuosity index is gradually increasing at downstream areas and topographic factors are totally dominated by hydraulic conditions. Actually, the hydrological change influenced the lithological control, grain size, and channel morphology, which have the long-term stability of the total landscape (Ghosh and Guchhait 2014; Chakrabortty et al. 2018; Malik and Pal 2019). Near Marokhana panchyet in the Arambagh subdivision, Mundeswari and Rupnarayan have connected to each other. Different features are so much important to analyze the overall concept about that area and these are exceptional from the various places because it is a lower joining area of two different rivers; therefore, the present survey has studied about changing status of channel and surrounding area, different features like meander, natural levee, floodplain, badland, etc. So, this paper is mainly highlighting the sedimentation and shifting trends of the channel in different courses of the river by several parameters with special features and characteristics.

9.2 Study Area

The study area is Panseuli, Maroklana block, Arambag subdivision, and Latitude and Longitude are from (22°36′9.81″N - 87°50′42.55E) to (22°35′19.10″N - 87°51′24.49″E) and (22°35′49.13″N - 87°51′7.99″E) to (22°36′1.63″N - 87°51′29.94E). It is the conjunction point of three districts, Hooghly, Howrah, and West Midnapur. In this region, three individual villages belong (Fig. 9.1). The village Pansuli is at Hooghly district, the village Kaijuri is at Howrah district, and the village Vatora is at West Midnapur district. In these areas, two rivers, tributaries of Mundeswari and the main channel of Rupnaraaayan, have assimilated. Actually, the mainstream is Rupnarayan and it is the lower bank of its main flow. The tributary of Mundeswari has detached from the main channel at Natibpur (22°39′45.63″N - 87°53′00.21E) and at the Pansuli region (Study area), it has joined with Rupnarayan main channel (22°35′49.13″N - 87°51′7.99″E). After joining with its tributary Mundeswari, the main channel Rupnarayan has flowed 20.05 km and again at Geonkhali region (22°31′41.89″N - 87°53′16.48″E), it has joined with Mundeswari main channel and named as Hooghly River. So the area has a clear expression about the nature and characteristics of those rivers individually and how its nature changes after assimilation to each other (Main channel Rupnarayan & tributary of Mundeswari). This area have also a clear expression about various reverie features, which gives a clear-cut idea about the present status and future planning for that surroundings.

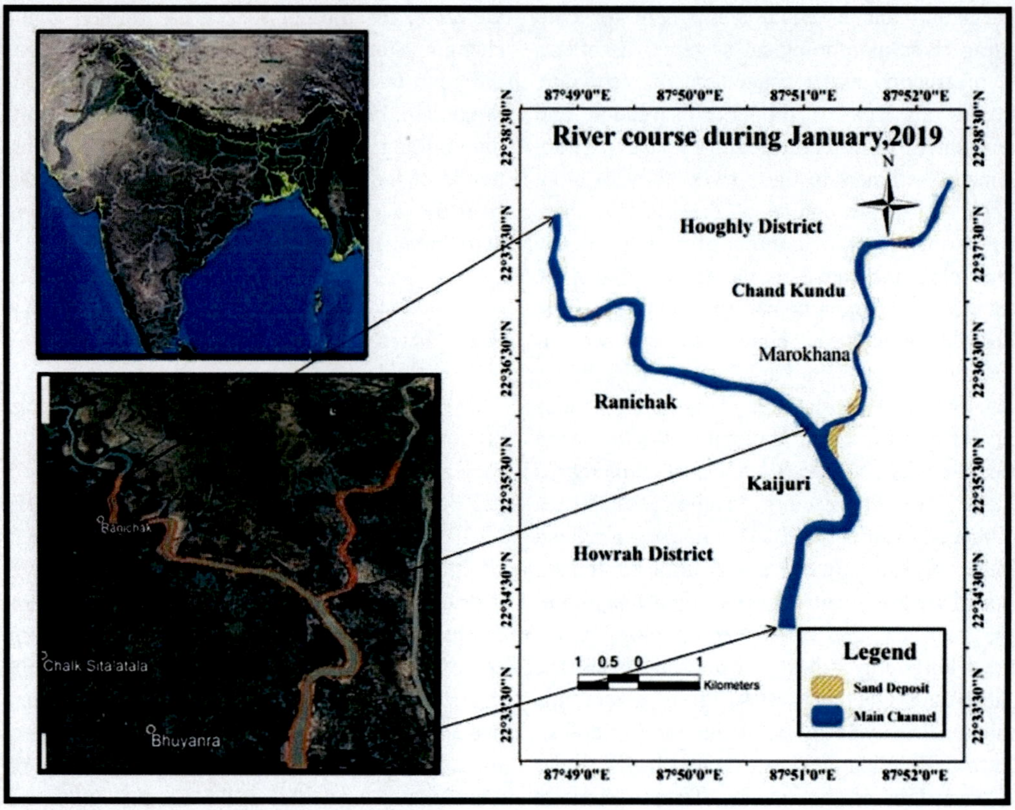

Fig. 9.1 Location map of the study area

9.3 Database and Methodology

Several methodologies have been followed to describe and depict different characteristics of two individual rivers and their joining area also. Various types of fluvial features such as velocity, stream power, discharge, carrying capacity, slope analysis, bank-full index, width, and cross profile (primary data) measure different kinds of instruments at the time of field survey. Various approaches have been adopted for the study and confirmation of primary and secondary data, such as sedimentological analysis, image classification, and overlay analysis. Sedimentation, Grain-size variation and shifting trends have been demonstrated by sediment sample analysis with laboratory sieving techniques (Fig. 9.2) using GPS and Google Maps, area identification, location analysis, Latitude–Longitude consideration, Height measurement, and Direction consideration. Google Earth Pro, Satellite data, SRTM DEM, and Topographical map are also very useful for Basin Identification, Location or study area consideration, River shifting trend analysis, channel cross-sectional analysis, different channel analysis, etc. Various GIS data are also important for the analysis of sedimentation, grain-size variation, or particle size.

Different kinds of instrumental methods and techniques have been followed to generalize and evaluate the entire study region. Theodolite, Tape, Staff, Dumpy level, Current meter, and Variable rod are needed mainly for the measurement of height, depth, distance, location, and so many characteristics (Fig. 9.3).

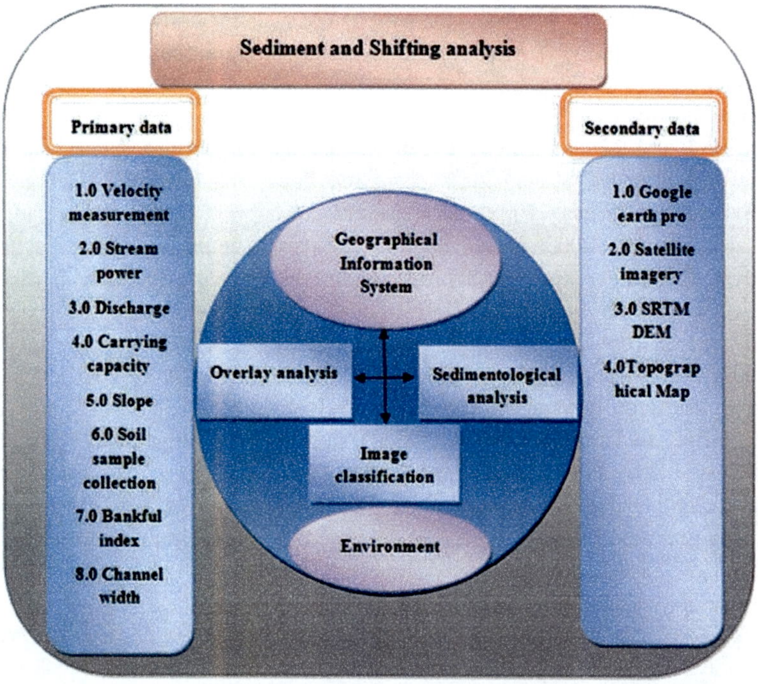

Fig. 9.2 Methodology flow chart

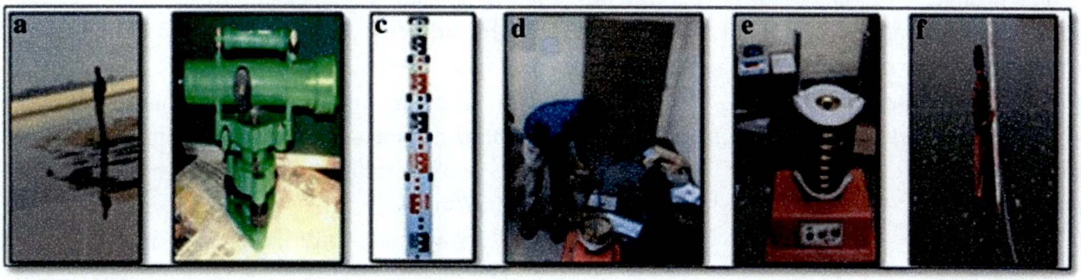

Fig. 9.3 Field survey and sedimentation analysis

9.3.1 River Velocity

The quality of water and various kinds of activities in the river are controlled by the power of velocity, mainly on the residence time of water and thus on high and the low water (K. Schulze et al. 2005). Various kinds of methods are very important for measurements velocity and to assess the quality and detect abnormalities (Goring and Nikora 2002). The duration of time and the length of the macro-turbulent flow is affected by average flow velocity, where ejections and the low-speed wedges also decrease with flow velocity as they have a negative relation to each other, definitely relation for the low-speed wedges is $L = 2.4Y$ that's a slope value within the range (Roy et al. 2004). The current meter is an important instrument for measuring river velocity. $V = D/A$ (Where: V is velocity; D is river discharge, and A means total area). In the study area, it is noticed that the velocity of the river Mundeswari is high rather than the river Rupnarayan.

9.3.2 River Discharge

Discharge is so important for the measurement of stream power and this computation depends on the application of the analysis with the choice of peak flow (Phillips and Slattery 2007), bank-full flow (Fonstad and Marcus 2003), mean annual flow (Finnegan et al. 2005; Larsen and Montgomery 2012), and mean annual maximum flow, and this flow occurs with a particular interval like 2-year, 10-year, or 100-year flow (Phillips and Slattery 2007; Phillips and Desloges 2014). To analyze river discharge, several models, methods, and techniques are so important like divided channel method (DCM), single-channel method (SCM), the coherence method (COHM), and the exchange discharge method (EDM), which are so important for the analysis of several characteristics of the river also (Sahu et al. 2011). Actually, the fluctuating flows follow the period of discharge for the cause of the erosion of separation and reattachment deposits, and this eroded part of sand transports to the main channel (Patel and Joshi 2017).

For measurement of the amount of discharge of two individual river systems and how it is being different after assimilation, individually *area* (*sq mt*) and velocity (*mt3/minute*) have been measured.

$$D = A * V.$$

Here D is the discharge, A is the area, and V is the velocity. In the case of the lower Damodar and Dwarkeswar rivers, a huge volume of water flow during the Monsoon between June and August used to be fury in the upper valley region and it overflows and flooded in lower valley regions.

9.3.3 Carrying Capacity

At the time of investigation of science and morph dynamic analysis of the channel, Carrying capacity is very important to trust entirely. This capacity considered a river as an individual organism in a particular region. This capacity and qualitative–quantitative characteristics of water are significant indexes for the flow of civilization and economic activity (Yang et al. 2006). For sustainable development of a region or planning for a region, the carrying capacity is most important. After proposed the concept of carrying capacity by Burgess and Park in 1920, several researchers applied this concept in different fields. This concept dominated in ecology science, resources science, demography, environmental carrying capacity, water capacity, river carrying capacity, and so on (Yang et al. 2006; Duan et al. 2010). Carrying capacity can reduce or lose its power due to degradation and pollution such that it does not affect the sustainability of the region. As the velocity and rate of discharge of river Mundeswari are low, similarly the carrying capacity is also very low.

9.3.4 Slope Analysis

Actually with a hydraulic geometry relation, the mean annual flood, valley gradient, channel width, and other various criteria can definitely be predicted (kleinhans 2010). In the basin of a river, the amount of precipitation and mean temperature are so important for drainage intensity, slope, gradient, and discharge (Malik and Pal 2020b).

Slope = tan^ − 1(h/d) or; tan^− 10.005.

The slope of the two rivers (Rupnarayan and Mundeswari lower region) is comparably similar; together, they flow from the North to South direction.

9.3.5 Bank-Full Index

The hydraulic geometry framework (Leopold and Maddock 1953) has given a solution with the basis of different kinds of criteria as like river width, depth, and velocity to depict power-law relationships with discharge.

$$d = cQ^f \;\&\; W = aQ^b$$

This relationship is very applicable for stable, single-channel rivers, and other various kinds of related environments also (Smith and Pavelsky 2008). Bank-full discharge is the discharge which fills the channel to the floodplain level. Actually, the effective discharge and bank-full discharge are so equal. Bank-full Depth is the average depth measured at Bank-full Discharge. Bank-full Discharge can also be calculated by bank-full width area. And the bank-full Width is the Channel width of Bank-full Discharge. The average bank-full discharge of river Rupnarayan is relatively high than river Mundeswari.

9.3.6 Stream Power

Stream power is very important for stream discharge, stream slope, and the weight of water. It has a direct relationship with sediment transport and this relationship analyzes the instantaneous rate of transportation of sediments and investigated various kinds of geomorphic questions across a range of time and spatial scales in river cross section which is so important for the depiction of river long profiles also (Kirby and Whipple 2001). Stream power is given by the equation: Stream power Ω = pgqs (where Ω is the stream power, ρ is the density of water (1000 kg/m3), g is the acceleration due to gravity (9.8 m/s2), Q is discharge (m3/s), and S is the channel slope). So, clearly express that the stream power of river Rupnarayan is medium and the stream power of the river Mundeswari is very low, but the stream power of mainstream Rupnarayan is growing after joint with Mundeswari.

9.3.7 Size of Grain and Particles

The mode of river transportation and grain size is so important for the mechanization of natural bunds and Reverie features. It also affects different kinds of physical processes and environmental and morphological activities related to the deltaic depositional system (Orton and Reading 1993). Mean, sorting, and skewness, these various statistical forms are so important for grain-size analysis which is used to speculate the sediment transport direction and trend (McLaren and Bowles 1985). The grain-size parameter method is suggested by Folk and Word. The mean grain size can be expressed by various kinds of Equations. But Folk and Word method is very important, which has given the formula with mean deviation, standard deviation, and Skewness. The mean deviation:

$M = \phi 16 + \phi 50 + \phi 843$. The standard deviation σ is expressed by Equation: $\sigma = \phi 84 + \phi 164 + \phi 95 - \phi 56.6$. The coefficient of skewness (Skew) is expressed by Equation:

$$SK = \phi 16 + \phi 84 - 2\phi 502\,(\phi 84 - \phi 16) + \phi 5 + \phi 95 - 2\phi 502(\phi 95 - \phi 5).$$

9.3.8 Channel Shifting

Different kinds of methods and techniques are so important for measuring channel migration with different time scales. Actually, various kinds of evidence like Sediment basis evidence, botanical evidence, and historical sources are very important for the analyses of the river channel shifting over a long time scale. The plan metric resurveying is utilized over intermediate time scales and over short time scale, terrestrial photogrammetric and erosion pins are utilized (Lawler, D. M. (1993) "The measurement of river bank erosion and lateral channel change: A review"). Channel shifting is a geomorphic process. By sedimentation analysis, it has been noticed that river Rupnarayan is shifting from Koijuri village toward Pansuli village, and on the other hand, the river Mundeswari is shifting from the village toward Pansuli (Fig. 9.4).

9.3.9 Sedimentation Analysis

Various kinds of effective evidence like Land use changing pattern, soil and water conservation, land clearance, mining activity, various kinds of sediment control programmers, and climate change records; all of these give an important idea about the stream of sediment sensitively

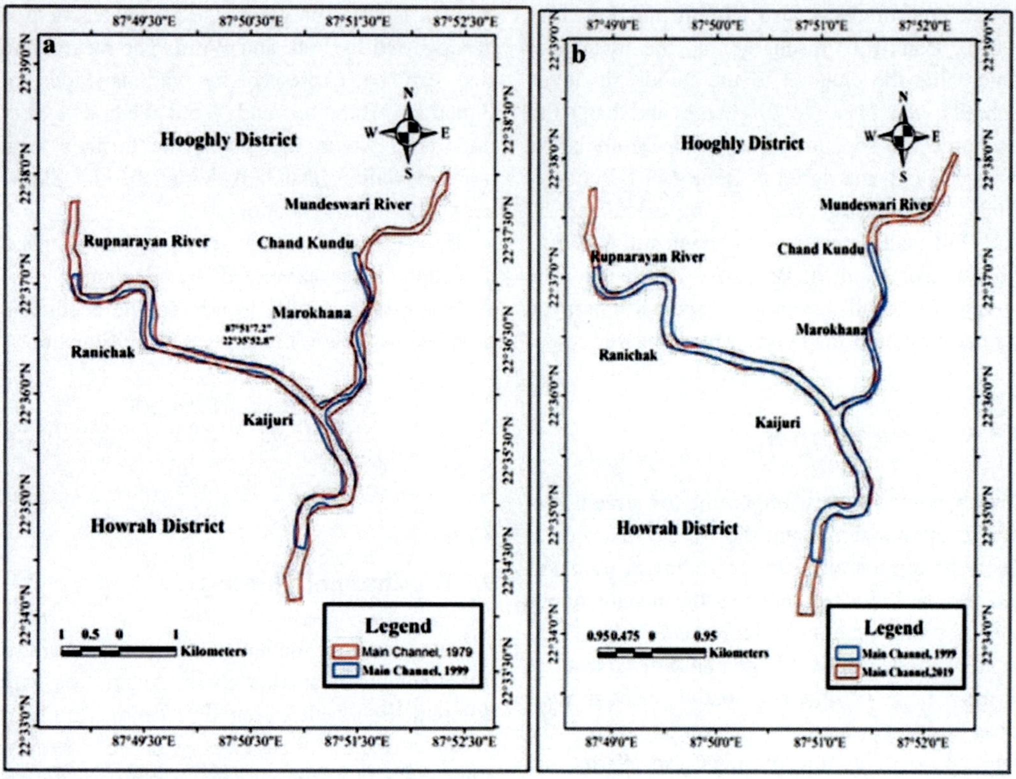

Fig. 9.4 Channel shifting

(Walling and Fang 2003). Actually, sedimentation analyses have been done by a collection of several samples of sediments with mineral composition from two banks of the lower reach of the Mundeswari and Rupnarayan. There have several processes and methods for the collection and analysis of sediment samples, which is important to understand the causes and mechanisms of sedimentation (Fig. 9.5a and b).

9.3.9.1 River Channel Cross Profile

The river changes its shape, size, width, and several geomorphic features from the source point to end point in a particular way. So, the cross and long profile characteristics are also Changeable with the change of time (Ghosh and Mistri 2012). The study area is a lower basin of two different embranchments, and the cross profile has also been drawn there to depict several characteristics and features (Fig. 9.6).

9.4 Results

River channel cross-sectional profiles have been drawn on basis of satellite imagery data 2019. Showing these profiles, it can be interpreted that the bed of River Rupnarayan is very rough due to heavy stream power and discharge. And in the cross profile of the river Mundeswari, it has an equal pattern due to low stream power and erosion. But it is interesting that after the joining of two rivers, the heavy flow of water or velocity, heavy discharge, and heavy stream power convert the channel smooth and flat also (Table 9.1).

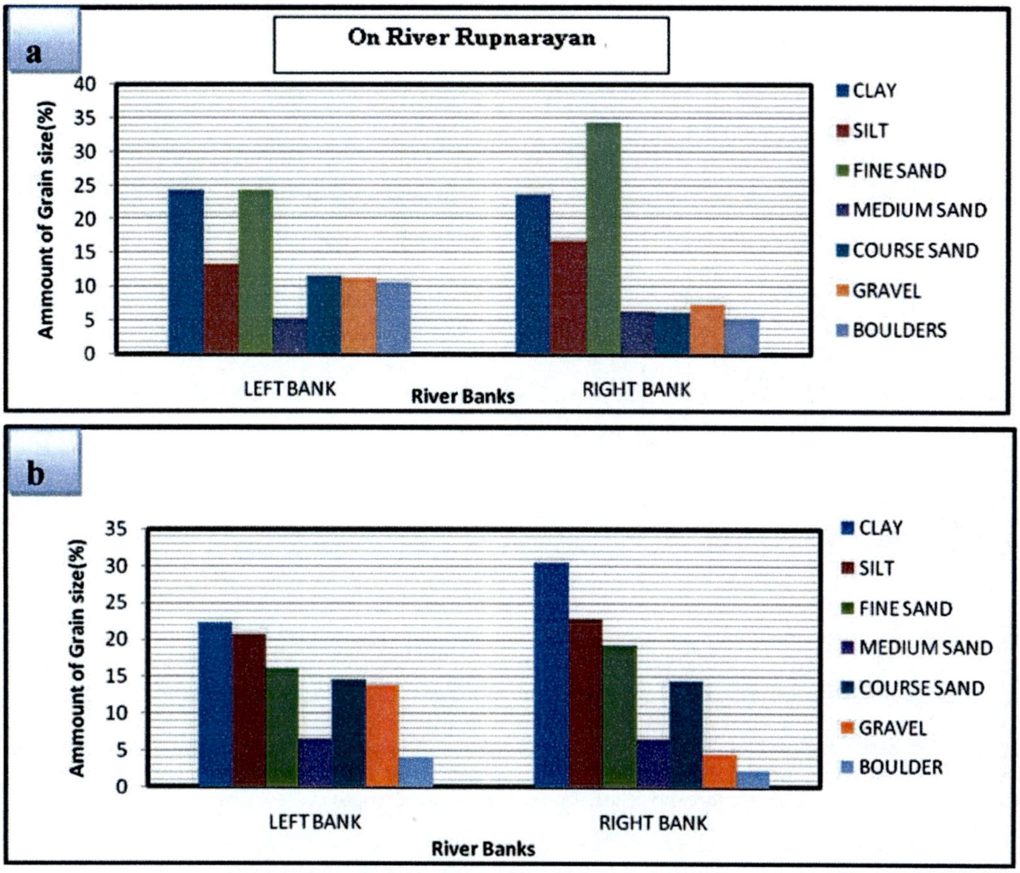

Fig. 9.5 **a**, **b** Grain size of River Rupnarayan and Mundeswari River

9.4.1 Different Kinds of Criteria on River Rupnarayan and Mundeswari

On basis of these diagrams, it is clearly noticed that these various kinds of natural criteria are occurring in different scales in different zones. Actually, in the main flow of Rupnarayan, the water level is very high. So, on the basis or relation with water flow power or stream power, different kinds of natural criteria like discharge, velocity, erosion–deposition, and carrying capacity have been measured (Fig. 9.7). As a result, there has a positive relation between Stream power and others criteria. In the study area, the channel Mundeswari has detached from the main channel in the Natibpur region. Generally, in the main channel Rupnarayan, the scale of natural criteria (discharge, velocity, stream power, carrying capacity) is relatively high than the detachment Channel Mundeswari (Because the natural flow of water is very low), normally the velocity and discharge are also very low. After joining these two rivers, the slope of the channel becomes high due to continuous bed erosion. On the adjoining area and onwards the stream power, velocity and discharge are relatively high than single-channel flow.

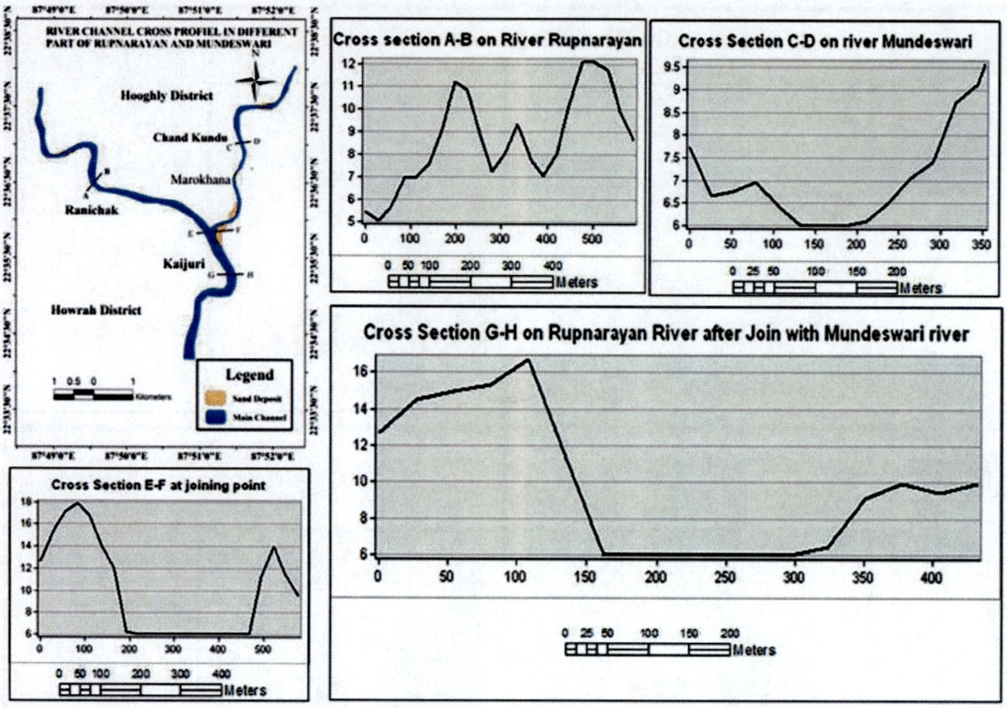

Fig. 9.6 Different Cross profiles of River Rupnarayan and Mundeswari (Cross section A–B on Rupnarayan River, Cross section C–D on Mundeswari River, Cross section E–F on joining point of two Rivers, and Cross section G–H on Rupnarayan River after joining with River Mundeswari)

Table 9.1 Comparisons between both the rivers on the basis of their selected criteria

Rivers/criteria	Velocity (*mt/ minute*)	Stream power	Discharge (cubic mt / second)	Carrying capacity (*mt^2/s.*)	Slope (In *degree*)	Bankful-Index (*mt^2*)	Width (*mt.*)
Mundeswari	1.50	1,389,811.50	58.1175	0.65	0.2865	112.5	90
Rupnarayan	3.20	163,176.51	495.00	2.75	0.2865	450	180

9.5 Discussion

The river channel course is dynamic because it is continuously changeable for effective different environmental factors. During 1979, the main channel Rupnarayan was very strong in carrying capacity and bank-full index but the river Mundeswari was relatively low. There is no sand deposition in the main channel Rupnarayan but in Mundeswari. During 1999, Rupnarayan's activity was relatively low compared to 1979. In the Mundeswari channel, the amount of sand deposition and the mid-channel bar is comparatively high than in 1979. In the current year 2019, discharge, velocity, erosion–deposition, and carrying capacity, all these criteria are very active in the Rupnarayan channel, so there is no

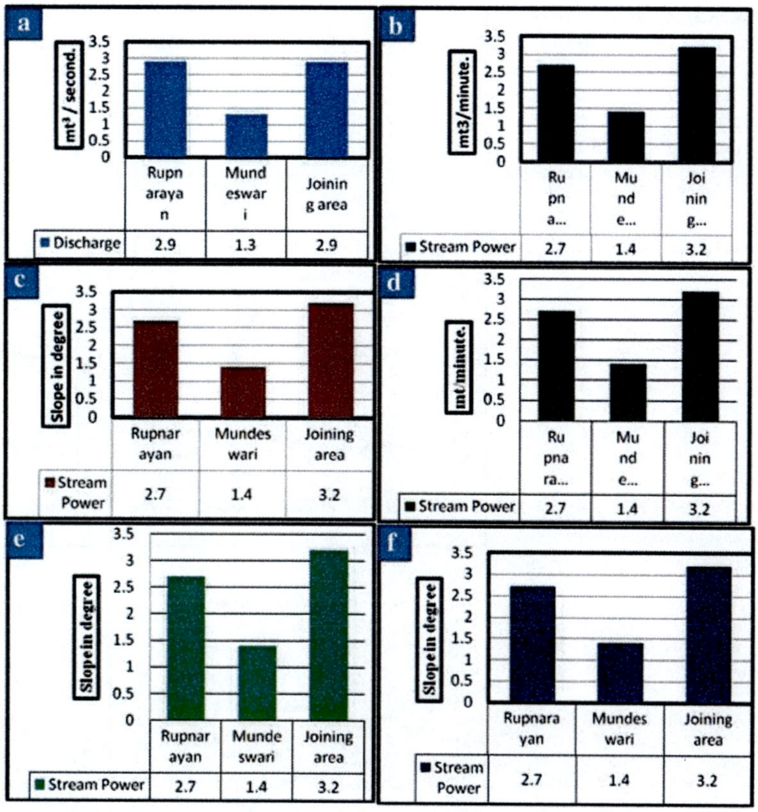

Fig. 9.7 Discharge (**a**), Velocity (**b**), Channel slope (**c**), Carrying capacity (**d**), Stream power (**e**), Bank-full index (**f**) of River Rupnarayan and Mundeswari River

mid-channel bar. And the heavy power of the stream causes erosion due to the sedimentation of the lower area. But, it's not so strong in Mundeswari (study area), since this channel is the main flow detachment, so there are too many mid-and side-channel bars. From 1979 to 2019, another significant event was the shifting of the river. From 1799 to 1999, the average displacement of the Rupnarayan River was 8.3 mt and the average displacement of Mundeswari was 11.25 mt. (Fig. 9.8).

Sedimentation analysis is one kind of method of particle or grain-size analysis. Actually, in the study area, from the left and right side banks of river Mundeswari, soil samples have been collected from A1points to A10 points (Fig. 9.10). On river Rupnarayan, B1 to B9 cross section wise samples and after the joining region C1to C12 cross section wise samples have been collected with a particular interval (Table 9.2 and Fig. 9.9). These sediment samples have been analyzed by laboratory procedure. By this

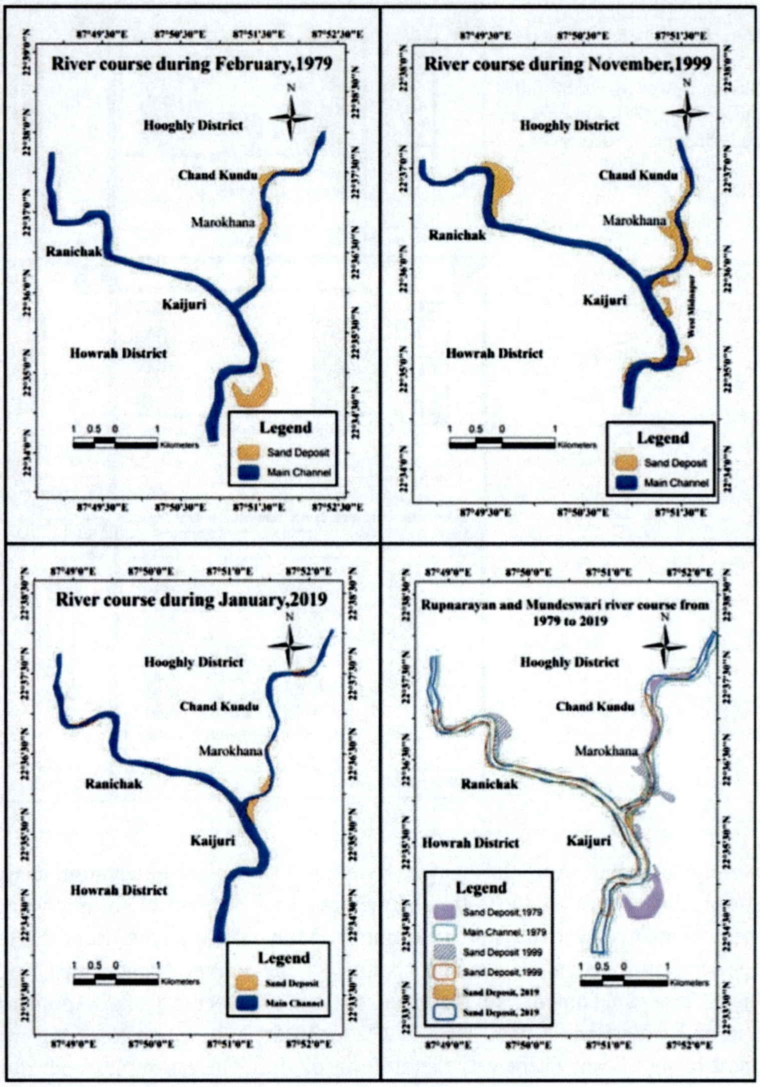

Fig. 9.8 Position and shifting trend of different River courses (**a** River course during 1979; **b** River course during 1999, **c** River course during 2019, and **d** All the course of River in one frame)

analysis, the grain-size variation of sediments in different banks can be identified. It also helps to measure the erosion and depositional area besides banks. It also gives a minor idea about channel shifting trends from the past year to the present. This sediment sample collection is also important for the analysis of seasonal change of bank-full cross-sectional area. So, it gives an overall clear idea about the channel and related information.

Table 9.2 Samples location for sedimentation analysis [Latitude and longitude of study areas a cross section on river Rupnarayan (A1–A20) and Mundeswari (C1 to C20)]

Points	Latitude	Longitude
A1	87°50′45.96″	22°36′59.76″
A2	87°50′48.84″	22°37′00.84″
A3	87°51′08.28″	22°37′00.84″
A4	87°51′20.52″	22°37′00.84″
A5	84°61′31.32″	22°37′00.84″
A6	87°51′37.08″	22°37′02.28″
A7	87°51′37.08″	22°37′01.56″
A8	87°52′04.80″	22°37′01.92″
A9	87°16′48.00″	22°37′01.56″
A10	87°52′35.04″	22°37.48.00″
B1	87°50′54.96″	22°36′27.36″
B2	87°50′32.64″	22°36′27.00″
B3	87°50′22.92″	22°36′25.56″
B4	87°50′11.04″	22°36′25.56″
B5	87°50′00.24″	22°36′26.28″
B6	87°48′46.56″	22°36′21.24″
B7	87°49′30.72″	22°36′16.56″
B8	87°49′13.8″	22°36′11.52″
B9	87°48′54.00″	22°36′25.56″
B10	87°48′38.16″	22°35′58.56″
C1	87°52′19.56″	22°35′38.40″
C2	87°52′05.88″	22°35′38.04″
C3	87°51′53.28″	22°35′36.96″
C4	87°51′40.32″	22°35′35.52″
C5	87°51′28.08″	27°35′33.72″
C6	87°51′21.96″	22°35′33.72″
C7	87°51′10.08″	22°35′33.36″
C8	87°50′59.28″	22°35′33.36″
C9	87°50′46.32″	22°35′31.92″
C10	87°50′46.32″	22°35′33.00″
C11	87°49′57.00″	22°35′30.84″
C12	87°50′03.12″	22°35′30.48″

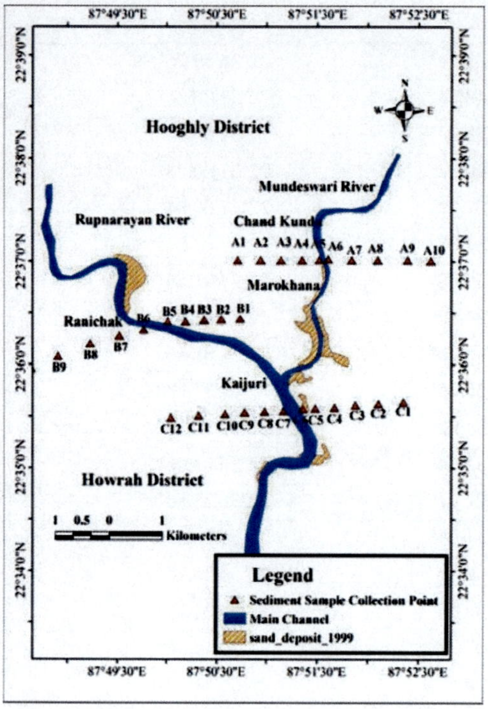

Fig. 9.9 Position of sediment sample along both the Rivers

9.6 Conclusion

This selected study area, which includes the monsoon climate, is prone to catastrophic floods due to heavy rainfall during the monsoon season and the influx of water from several nearby rivers into the area. So, a devastating flood in lower Damodar and Darkeswar is a normal annual occurrence in the consequence of high Discharge. Actually, an Analysis of the sinuosity index of this region shows that the river in this region is located in a meander pattern as well as much funnel-shaped to look at it from an external point of view. Exterior imagery of the area reveals the discontinuous natural embankment on the left bank of the Mundeswari River and the unstable floodplain over a large area. River Terrence, tides and ebbs, badlands, and areas are also some special characteristics of that area. There is also a "Doab region" between two rivers. The Doab region of that area is named as Pansuli village. This village is a very floodprone region and the river bank erosion is very high and the shifting trends of two embranchment, toward this village. So, this village is very vulnerable.

To identify erosion and depositional activities, the relations among dynamic geometric phenomenon, stratification, and discharge are much essential and the Channel profile's characteristics can be focused by sediment structure, color, grain size, mapping, etc. So, from the primary (field) survey and secondary survey, these arbitraries express that, in the lower Rupnarayan and Mundeswari, channel morphology and morphological criteria as like the size of grain and particles, stream power, and so on are present in different rates in two different Rivers. Thus, a particular perspective of the study area selected by the external image of the overall region and the overall primary data collection and analysis as well as the different types of secondary data collection can be explained.

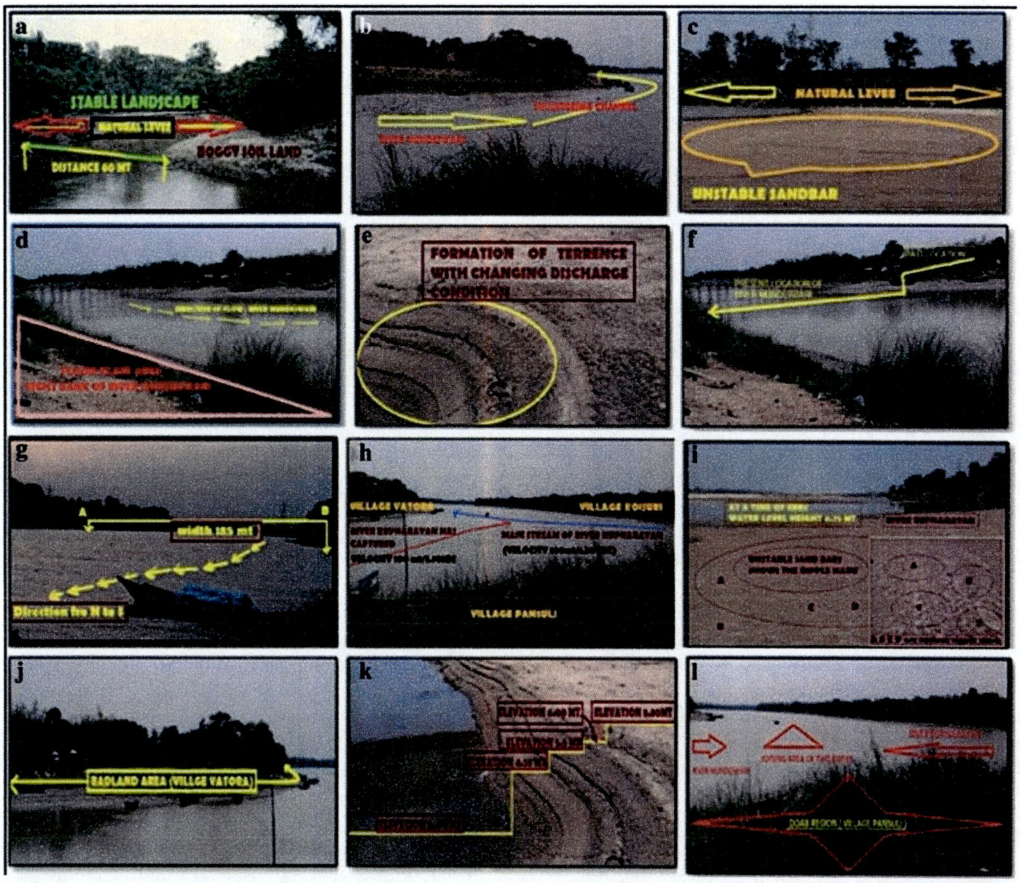

Fig. 9.10 Different riverine features

References

Acharyya SK, Shah BA (2007) Arsenic-contaminated groundwater from parts of Damodar fan-delta and west of Bhagirathi River, West Bengal, India: influence of fluvial geomorphology and Quaternary morphostratigraphy. Environ Geol 52(3):489–501. DOI https://doi.org/10.1007/s00254-006-0482-z

Adams TB, Sinclair DC, West AR (2002) Giant barrier layer capacitance effects in CaCu3Ti4O12 ceramics. Adv Mater 14(18):1321–1323. https://doi.org/10.1002/1521-4095(20020916)14:18%3C1321::AID-ADMA1321%3E3.0.CO;2-P

Akahashi M, Gonzalez CA, Chanson H (2006) Self-aeration and turbulence in a stepped channel: influence of cavity surface roughness. Int J Multi Flow 32 (12):1370–1385. https://doi.org/10.1016/j.ijmultiphaseflow.2006.07.001

Amos CB, Burbank DW (2007) Channel width response to differential uplift. J Geophy Res Earth Surf 112 (F2). https://doi.org/10.1029/2006JB004291

Amos CB, Burbank DW (2007) Channel width response to differential uplift. J Geophy Res Earth Surface 112 (F2). https://doi.org/10.1029/2006JB004291

Andreadis KM, Clark EA, Lettenmaier DP, Alsdorf DE (2007) Prospects for river discharge and depth estimation through assimilation of swath-altimetry into a raster-based hydrodynamics model. Geophys Res Lett 34(10). https://doi.org/10.1029/2007GL029721

Angillieri MYE (2008) Morphometric analysis of Colangüil river basin and flash flood hazard, San Juan, Argentina. Environ Geol 55(1):107–111. DOI: https://doi.org/10.1080/10106049.2013.868042

Aricò C, Nasello C, Tucciarelli T (2009) Using unsteady-state water level data to estimate channel roughness and discharge hydrograph. Adv Water Res 32 (8):1223–1240. DOI: https://doi.org/10.1016/j.advwatres.2009.05.001

Azarang F, Jafari G, Karami M, Bejestan MS (2017) Protecting river environment through proper management of material mining by matrix method (Case Study of A'la River in Iran). Civil Eng J 3(12). DOI: https://doi.org/10.28991/cej-030959

Bera S, Mistry B (2014) Flood in the lower Damodar Basin and channel morphology: a case study at the bifurcation zone into Damodar and Mundeswari river, West Bengal. Intern J Geol Earth Environ Sci. ISSN, 2277–2081. http://www.cibtech.org/jgee.htm

Best JL, Ashworth PJ, Bristow CS, Roden J (2003) Three-dimensional sedimentary architecture of a large, mid-channel sand braid bar, Jamuna River, Bangladesh. J Sedimen Res 73(4):516–530. https://doi.org/10.1306/010603730516

Bianchi TS, Allison MA (2009) Large-river delta-front estuaries as natural "recorders" of global environmental change. Proce National Acad Sci 106(20):8085–8092. DOI: https://doi.org/10.1073/pnas.0812878106

Blum MD, Roberts HH (2009) Drowning of the Mississippi Delta due to insufficient sediment supply and global sea-level rise. Nature Geosci 2(7):488. DOI: https://doi.org/10.1038/NGEO553

Bridge JS, Tye RS (2000) Interpreting the dimensions of ancient fluvial channel bars, channels, and channel belts from wireline-logs and cores. AAPG Bull 84 (8):1205–1228. https://doi.org/10.1306/A9673C84-1738-11D7-8645000102C1865D

Bridge JS, Tye RS (2000) Interpreting the dimensions of ancient fluvial channel bars, channels, and channel belts from wireline-logs and cores. AAPG Bull 84 (8):1205-1228. https://doi.org/10.1306/A9673C84-1738-11D7-8645000102C1865D

Bridge JS, Smith ND, Trent F, Gabel SL, Bernstein P (1986) Sedimentology and morphology of a low-sinuosity river: Calamus River, Nebraska Sand Hills. Sedimentology, 33(6):851–870. https://doi.org/10.1111/j.1365-3091.1986.tb00987.x

Chabukswar S, Maji D, Manoj CR, Anil KG, Rao VR, Crupi F, ... & Collaert N (2010) Implications of fin width scaling on variability and reliability of high-k metal gate FinFETs. Microelec Eng 87(10):1963–1967. https://doi.org/10.1016/j.mee.2009.12.013

Chakrabortty R, Pal SC, Islam S, Das B, Malik S (2018) Subsurface lithofacies and its geomorphic significance: a case study of Teesta River, West Bengal, India. Int J Environ Sci Nat Res 10(2):01–013

Coleman JM (1969) Brahmaputra river: channel processes and sedimentation. Sediment Geol 3(2–3):129–239. https://doi.org/10.1016/0037-0738(69)90010-4

Das BC (2015) Modeling of most efficient channel form: a quantitative approach. Model Earth Syst Environ 1 (3):15. https://link.springer.com/article/https://doi.org/10.1007/s40808-015-0013-6

Dhali MK, Sahana M (2017) Spatial variation in fluvial hydraulics with major bed erosion zone: a study of Kharsoti river of India in the post monsoon period. Arab J Geosci 10(20):451. DOI:https://doi.org/10.1007/s12517-017-3205-8

Dhali MK, Biswas M. MCA on mechanism of river bed potholes growth: a study of middle Subarnarekha River basin, South East Asia. Environ Devel Sustain 1–25. https://link.springer.com/article/https://doi.org/10.1007/s10668-017-0069-8

Doble R, Brunner P, McCallum J, Cook PG (2012) An analysis of river bank slope and unsaturated flow effects on bank storage. Groundwater 50(1):77–86. DOI:https://doi.org/10.1111/j.1745-6584.2011.00821.x

Duan CQ, Liu CM, Chen XN, Liu WH, Zheng HX (2010) Preliminary research on regional water resources carrying capacityconception and method. Acta Geographica Sinica 65(1):82–90

Ekeleme AC, Agunwamba JC (2018) Experimental determination of dispersion coefficient in soil. Em Sci J 2(4):213–218. DOI: http://dx.doi.org/https://doi.org/10.28991/esj-2018-01145

Finnegan NJ, Roe G, Montgomery DR, Hallet B (2005) Controls on the channel width of rivers: Implications for modeling fluvial incision of bedrock. Geology 33 (3):229–232. http://gis.ess.washington.edu/grg/publications/pdfs/Noahpaper.pdf

Fonstad M, Marcus WA (2003) Selforganized criticality in riverbank systems. Annals Associate AmGeo 93 (2):281–296. https://doi.org/10.1111/1467-8306.9302002

Gaeuman D, Schmidt JC, Wilcock PR (2005) Complex channel responses to changes in stream flow and sediment supply on the lower Duchesne River, Utah. Geomorphology 64(3–4):185–206. https://doi.org/10.1016/j.geomorph.2004.06.007

Gellis AC, Cheama A, Lalio SM (2001) Developing a geomorphic approach for ranking watersheds for rehabilitation, Zuni Indian Reservation, New Mexico. Geomorphology 37(1–2):105–134. https://doi.org/10.1016/S0169-555X(00)00065-9

Ghosh PK, Bandyopadhyay S, Jana NC, Mukhopadhyay R (2016) Sand quarrying activities in an alluvial reach of Damodar River, Eastern India: towards a geomorphic assessment. Int J River Basin Manag 14(4):477–489. https://doi.org/10.1080/15715124.2016.1209509

Ghosh S, Mistri B (2012) Hydrogeomorphic significance of sinuosity index in relation to river instability: a case study ofDamodar River, West Bengal, India. Int J Adv Earth Sci 1(2):49–57

Ghosh S, Guchhait SK (2014) Hydrogeomorphic variability due to dam constructions and emerging problems: a case study of Damodar River, West Bengal, India. Environ Develop Sustain 16(3):769–796. DOI: https://doi.org/10.1007/s10668-013-9494-5

Ghrefat H, Yusuf N (2006) Assessing Mn, Fe, Cu, Zn, and Cd pollution in bottom sediments of Wadi Al-Arab Dam, Jordan. Chemosphere 65(11):2114–2121.

Gleason CJ, Smith LC (2014) Toward global mapping of river discharge using satellite images and at-many-stations hydraulic geometry. Proce National Acad Sci 111(13):4788–4791. https://www.pnas.org/content/111/13/4788.short

Goring DG, Nikora VI (2002) Despiking acoustic doppler velocimeter data. J Hydraulic Eng 128(1):117–126. https://doi.org/10.1061/(ASCE)0733-9429

Gupta H, Chakrapani GJ (2005) Temporal and spatial variations in water flow and sediment load in Narmada River Basin, India: natural and man-made

factors. Environ Geol 48(4–5):579–589. http://www.iisc.ernet.in/~currsci/mar102007/679.pdf
Kale VS, Hire PS (2004) Effectiveness of monsoon floods on the Tapi River, India: role of channel geometry and hydrologic regime. Geomorphology 57(3–4):275–291. https://doi.org/10.1016/S0169-555X(03)00107-7
Kale VS, Baker VR, Mishra S (1996) Multi-channel patterns of bedrock rivers: an example from the central Narmada basin, India. Catena, 26(1-2):85-98. https://s3.amazonaws.com/academia.edu.documents/2799980/kale_baker_mishra1996.pdf?response-content-disposition=inline%3B%20filename%3DKale_V._S._V._R._Baker_and_S._Mishra
Kleinhans MG (2010) Sorting out river channel patterns. Progress Phys Geog: Earth Environ 34(3):287–326. https://doi.org/10.1177/0309133310365300
Kleinhans MG, van den Berg JH (2011) River channel and bar patterns explained and predicted by an empirical and a physics-based method. Earth Surface Proc Landf 36(6):721–738. https://doi.org/10.1002/esp.2090
Kou HS, Lee JJ, Chen CW (2008) Optimum thermal performance of microchannel heat sink by adjusting channel width and height. Intern Commun Heat Mass Transfer 35(5):577–582. https://doi.org/10.1016/j.icheatmasstransfer.2007.12.002
Lai AC, Byrne MA, Goddard AJ (2001) Aerosol deposition in turbulent channel flow on a regular array of three-dimensional roughness elements. J Aerosol Sci 32(1):121–137. DOI: https://doi.org/10.1016/S0021-8502(00)00051-3
Latrubesse EM (2008) Patterns of anabranching channels: The ultimate end-member adjustment of mega rivers. Geomorphology 101(1–2):130–145. https://doi.org/10.1016/j.geomorph.2008.05.035
Larsen IJ, Montgomery DR (2012) Landslide erosion coupled to tectonics and river incision. Nat Geosci 5 (7):468–473.https://doi.org/10.1038/ngeo1479
Lemme M, Mollenhauer T, Henschel W, Wahlbrink T, Heuser M, Baus M, Kurz H (2003) Influence of channel width on n-and p-type nano-wire-MOSFETs on silicon on insulator substrate. Microelectron Eng 67:810–817. https://doi.org/10.1016/S0167-9317(03)00191-6
Leopold LB, Maddock T (1953) The hydraulic geometry of stream channels and some physiographic implications, vol 252, USGovernment Printing Office
Liu B, Lu YW, Jin GR, Zhao Y, Wang XL, Zhu QS, Wang ZG (2010) Surface roughness scattering in two dimensional electron gas channel. Appl Phys Lett 97 (26):262111. https://doi.org/10.1063/1.3532967
Loska K, Wiechuła D (2003) Application of principal component analysis for the estimation of source of heavy metal contamination in surface sediments from the Rybnik Reservoir. Chemosphere 51(8):723–733. DOI: https://doi.org/10.1016/S0045-6535(03)00187-5
Ludwig W, Probst JL, Kempe S (1996) Predicting the oceanic input of organic carbon by continental erosion. Glob Biogeochem Cycles 10(1):23–41. https://doi.org/10.1029/95GB02925
Maity SK, Maiti R (2016) Understanding the sediment sources from mineral composition at the lower reach of Rupnarayan RiverWest Bengal India – XRD-based analysis. Geo Res J 9(12):91–103. https://doi.org/10.1016/j.grj.2016.09.004
Malik S, Pal SC (2020a) Is the topography playing a dual role in controlling downstream channel morphology of a monsoon dominated Dwarkeswar River, Eastern India? HydroResearch. https://doi.org/10.1016/j.hydres.2020.04.002
Malik S, Pal SC (2020b) Application of 2D numerical simulation for rating curve development and inundation area mapping: acase study of monsoon dominated Dwarkeswar river. Int J River Basin Manage 1–11. https://doi.org/10.1080/15715124.2020.1738447
Malik S, Pal SC, Das B, Das B (2017) Chute cut-off processes along a small alluvial channel: a case study of Sangra Khal, sub-tributary of Gour Nadi, West Bengal, India. Model Earth Syst Environ 3(1):17. https://doi.org/10.1007/s40808-017-0307-y
Malik S, Pal SC (2019) Is the construction of Groynes accelerating the degradation of channel morphology and paved the way for human encroachment in The Bengal Basin? Adv Space Res 64(8):1549–1576. https://doi.org/10.1016/j.asr.2019.07.024
Martinez J, Martinez RV, Garcia R (2008) Silicon nanowire transistors with a channel width of 4 nm fabricated by atomic force microscope nanolithography. Nano Letters 8(11):3636–3639. https://doi.org/10.1021/nl801599k
McLaren P, Bowles D (1985) The effects of sediment transport on grain-size distributions. J Sedimen Res 55 (4). DOI: https://doi.org/10.1306/212F86FC-2B24-11D7-8648000102C1865D
Meade RH, Dunne T, Richey JE, Santos UDM, Salati E (1985) Storage and remobilization of suspended sediment in the lower Amazon River of Brazil. Science 228(4698):488–490. DOI:https://doi.org/10.1126/science.228.4698.488
Mondal I, Bandyopadhyay J (2016) Physicochemical Analysis of Ichamati river and estimation of soil parameters using geospatial technology. J Inst Eng (India): Series E 97(2):151–158. DOI:https://doi.org/10.1007/s40034-016-0086-4
Moore ID, Burch GJ (1986) Physical basis of the length-slope factor in the universal soil loss equation 1. Soil Sci Soc Am J 50(5):1294–1298. doi:https://doi.org/10.2136/sssaj1986.03615995005000050042x
Nagano Y, Hattori H, Houra T (2004) DNS of velocity and thermal fields in turbulent channel flow with transverse-rib roughness. Int J Heat Fluid Flow 25 (3):393–403. https://hal.archives-ouvertes.fr/hal-01321586/document
Nohara D, Kitoh A, Hosaka M, Oki T (2006) Impact of climate change on river discharge projected by multimodel ensemble. J Hydromet 7(5):1076–1089. https://doi.org/10.1175/JHM531.1
Orton GJ, Reading HG (1993) Variability of deltaic processes in terms of sediment supply, with particular emphasis on grain size. Sedimentology 40(3):475–

512. DOI: https://doi.org/10.1111/j.1365-3091.1993.tb01347.x

Owens PN, Batalla RJ, Collins AJ, Gomez B, Hicks DM, Horowitz AJ, ... & Petticrew EL (2005) Fine-grained sediment in river systems: environmental significance and management issues. River Res Applic 21(7):693–717. https://doi.org/10.1002/rra.878

Patel AB, Joshi GS (2017) Modeling of rainfall-runoff correlations using artificial neural network-A case study of Dharoi Watershed of a Sabarmati river basin, India. Civil Eng J 3(2):78–87. DOI: https://doi.org/10.28991/cej-2017-00000074

Patra S, Mitra P, Pradhan SK (2011) Preparation of nanodimensional CdS by chemical dipping technique and their characterization. Mater Res 14(1):17–20. DOI: https://doi.org/10.1590/S1516-14392011005000015

Peizhen Z, Molnar P, Downs WR (2001) Increased sedimentation rates and grain sizes 2–4 Myr ago due to the influence of climate change on erosion rates. Nature 410(6831):891. DOI:https://doi.org/10.1038/35073504

Perron JT, Royden L (2013) An integral approach to bedrock river profile analysis. Earth Surface Proc Landf 38(6):570–576. DOI: https://doi.org/10.1002/esp.3302

Peterson BJ, Holmes RM, McClelland JW, Vörösmarty CJ, Lammers RB, Shiklomanov AI, ... Rahmstorf, S (2002) Increasing river discharge to the Arctic Ocean. Science 298(5601):2171–2173. DOI: https://doi.org/10.1126/science.1077445

Phillips JD, Slattery MC (2007) Downstream trends in discharge slope and stream power in a lower coastal plain river. JHydrol 334(1–2):290–303. https://doi.org/10.1016/j.jhydrol.2006.10.018

Phillips RTJ, Desloges JR (2014) Glacially conditioned specific stream powers in low-relief river catchments of the southernLaurentian Great Lakes. Geomorphol 20(6):271–287. https://doi.org/10.1016/j.geomorph.2013.09.030

Prasannakumar V, Shiny R, Geetha N, Vijith H (2011) Spatial prediction of soil erosion risk by remote sensing, GIS and RUSLE approach: a case study of Siruvani river watershed in Attapady valley, Kerala, India. Environ Earth Sci 64(4):965–972
https://www.researchgate.net/profile/Vijith_H/publication/225394075

Purkait B (2002) Patterns of grain-size distribution in some point bars of the Usri River, India. J Sediment Res 72(3):367–375. https://doi.org/10.1306/091001720367

Rai UN, Tripathi RD, Vajpayee P, Jha V, Ali MB (2002) Bioaccumulation of toxic metals (Cr, Cd, Pb and Cu) by seeds of Euryale ferox Salisb. (Makhana). Chemosphere 46(2):267–272. DOI:https://doi.org/10.1016/s0045-6535(01)00087-x

Rawool AS, Mitra SK, Kandlikar SG (2006) Numerical simulation of flow through microchannels with designed roughness. Microfl Nanofluidics 2(3):215–221. DOI:https://doi.org/10.1007/s10404-005-0064-5

Rehana S, Mujumdar PP (2011) River water quality response under hypothetical climate change scenarios in Tunga-Bhadra river, India. Hydrol Proc 25 (22):3373–3386. http://citeseerx.ist.psu.edu/viewdoc/download?doi=10.1.1.702.3902&rep=rep1&type=pdf

Roy AG, Thomas B, Helene L, Alisttair D (2004) Size, shape and dynamics of large-scale turbulent flow structures in a gravelbedriver. J Fluid Mech 500:1–27

Roy S, Mistri B (2013) Estimation of peak flood discharge for an ungauged river: a case study of the Kunur River, West Bengal. Geog J 2013. https://doi.org/10.1155/2013/214140

Sahu M, Khatua KK, Mahapatra SS (2011) A neural network approach for prediction of discharge in straight compound open channel flow. Flow Measure Instrumen 22(5):438–446. https://doi.org/10.1016/j.flowmeasinst.2011.06.009

Sarkar SK, Frančišković-Bilinski S, Bhattacharya A, Saha M, Bilinski H (2004) Levels of elements in the surficial estuarine sediments of the Hugli River, northeast India and their environmental implications. Environ Intern 30(8):1089–1098. DOI: https://doi.org/10.1016/j.envint.2004.06.005

Savini J, Bodhaine GL (1971) Analysis of current-meter data at Columbia River gaging stations, Washington and Oregon (No. 1869-F). US Govt. Print. Off. [for sale by the Supt. of Docs.]. https://doi.org/10.3133/wsp1869F

Schulze K, Hunger M, Döll P (2005) Simulating river flow velocity on global scale. Adv Geosci 5:133–136. https://hal.archives-ouvertes.fr/hal-00296854/document

Shepherd RG (1989) Correlations of permeability and grain size. Groundwater 27(5):633–638. https://doi.org/10.1111/j.1745-6584.1989.tb00476.x

Shi Z, Li YH (2003) Experimental studies of mean velocity profiles in vegetated river flow. J Shanghai Jiaotong Univer Chinese Edition 37(8):1254–1260. http://en.cnki.com.cn/Article_en/CJFDTotal-SHJT200308027.htm

Squires KD, Simonin O (2006) LES–DPS of the effect of wall roughness on dispersed-phase transport in particle-laden turbulent channel flow. Int J Heat Fluid Flow 27(4), 619–626. https://asu.pure.elsevier.com/en/publications/les-dps-of-the-effect-of-wall-roughness-on-dispersed-phase-transp-2

Smith LC, Pavelsky TM (2008) Estimation of river discharge propagation speed and hydraulic geometry from space: Lena RiverSiberia. Water Res Res 44(3). https://doi.org/10.1029/2007WR006133

Syvitski JP, Vörösmarty CJ, Kettner AJ, Green P (2005) Impact of humans on the flux of terrestrial sediment to the global coastal ocean. Science 308(5720):376–380. DOI: https://doi.org/10.1126/science.1109454

Venkatramanan S, Ramkumar IT, Anithamary S, Vasudevan (2014) Heavy metal distribution in surface sediments of the Tirumalairajan river estuary and the surrounding coastal area east coast of India. Arabian J Geosci 7(1):123–130.https://doi.org/10.1007/s12517-012-0734-z

Walling DE, Fang D (2003) Recent trends in the suspended sediment loads of the world's rivers. Glob Planet Change 39(1–2):111–126. DOI: https://doi.org/10.1016/S0921-8181(03)00020-1

Wu W, Shields FD, Bennett SJ, Wang SS (2005) A depth-averaged two-dimensional model for flow, sediment transport, and bed topography in curved channels with riparian vegetation. Water Resourc Res 41(3). https://doi.org/10.1029/2004WR003730

Yang ZS, Wang HJ, Saito Y, Milliman JD, Xu K, Qiao S, Shi G (2006) Dam impacts on the Changjiang (Yangtze) River sediment discharge to the sea: The past 55 years and after the Three Gorges Dam. Water Resourc Res 42(4). https://doi.org/10.1029/2005WR003970

Yorke TH, Oberg KA (2002) Measuring river velocity and discharge with acoustic Doppler profilers. Flow Measur Instrum 13(5–6):191–195. file:///C:/Users/DELL/Desktop/downloads/MeasuringrivervelocityanddischargewithacousticDopplerProfilers.pdf

Zhang C, Chen Y, Shi M (2010) Effects of roughness elements on laminar flow and heat transfer in microchannels. Chem Eng Proc Process Intensif 49 (11):1188–1192. DOI: https://doi.org/10.1016/j.cep.2010.08.022

Role of Controlling Factors in the Development of Drainage Around Rajmahal Hills, Jharkhand and West Bengal

10

Shuvasish Karmokar, Senjuti Nandy, and Manasi De

Abstract

Development of any drainage network on a landscape is regulated by its surface terrain, generated landforms through evolution processes, geomorphic and climatic conditions, underlying geological, lithological, tectonic and structural control, soil, vegetation cover, etc. Quantification of drainage morphometric characteristics highlights the drainage development and evolution of a region. In this present paper, the hydrological, topographical, geological, climatological characteristics and vegetation cover are evaluated through the analysis of DEM, satellite images, geological map and rainfall data in GIS environment using the different morphometric parameters. Various statistical techniques assist the evolution of drainage characteristics and help to understand the contribution of different controlling factors on the development of drainage networks of Rajmahal hills and its surroundings. The Rajmahal Hills and its surrounding is characterized by unique lithology, varied topography, different vegetation covers with a perceptible climatic difference. Thus, diverse drainage conditions provide an excellent tract to examine the effect of topography, geology, rainfall and vegetation cover on the development of drainage characteristics of a region. The result of different morphometric parameters such as geology, rainfall, vegetation cover reveals that the landscape of the study area is going through the mature stage of landform evolution. Analyses of different morphometric parameters divulge that there is high geological and geomorphological control on the development of drainage characteristics in the study area. Hence, it is concluded that this type of fluvio-hydrological study can give a detailed overview of landscape development and hydrological behaviour and definitely assists for integrated drainage basin management.

Keywords

Drainage network · Landscape · Lineament · Rajmahal hill · Chhotanagpur fringe region (CFR)

10.1 Introduction

Drainage is the most important fundamental units in terms of geometric characteristic of the fluvial landscape (Altın and Altın 2011). Development of drainage network is regulated by surface

S. Karmokar (✉) · M. De
Department of Geography, Lady Brabourne College, Kolkata, India

S. Nandy
Department of Geography, Visva-Bharati, Santiniketan, India

P. K. Shit et al. (eds.), *Drainage Basin Dynamics*, Geography of the Physical Environment,
https://doi.org/10.1007/978-3-030-79634-1_10

topography, processes of landform evolution, geomorphic and climatic conditions, underlying geological, lithological, tectonic and structural control, soil, vegetation cover, etc. (Strahler 1956). Thus, stream network systems reflect the relation between surface topographical features and subsurface geology along with detailed hydrogeomorphological characteristics of the land surface. Various characteristics of drainage are determined and primarily controlled by three determinants, i.e. topography, climate and geology (Frissell et al. 1986; Nag and Chakraborty 2003; Mesa 2006) as they are the key regulators of running water ecosystems functioning at the basin scale (Lotspeich and Platts 1982; Frissell et al. 1986). Hence, spatial variation of drainage characteristics reflects the influence of these three determinants (Sujatha et al. 2015). The topography in terms of relief and slope regulates the development of drainage, whereas geology controls the pattern of drainage. In most of the cases, topography and geology (especially structures) are inseparable in nature and operated together to determine the nature of the drainage network. Two relief parameters, i.e. relative relief and slope of the terrain influence the development of drainage in a region. Relative relief determines the hillslope morphology and influences the hillslope processes which in turn influenced the drainage density. Schumm (1956) noted a positive relationship between drainage density and relief ratio. Oguchi (1997) found a negative correlation between drainage density and relative relief in the Japanese mountains. The drainage characteristics of a region also depend on the geomorphic stage of landscape development. Climate also controls the development of drainage in a region (Chorley 1957; Melton 1957, 1958; Strahler 1964; Soukup 2006; Garcia-Castellanos 2006; Thomas et al. 2010; Altın and Altın 2011; Sangireddy et al. 2016; Sojka 2020). Rodríguez-Iturbe and Escobar (1982) highlighted on the dependence of drainage density on climate and geomorphology and outlined that the drainage density cannot be analysed on the basis of climate alone and for that, it is necessary to consider the response characteristics of the basin in any spatial or climatic study of drainage densities. Tucker and Slingerland (1997) studied about drainage basin evolution responses to climate change, in which they show that increase in runoff intensity will lead to a rapid expansion of the channel network with increasing sediment supply and aggradation along with the main network, followed by down cutting and made a conclusion that there is a positive relationship with drainage to the direction of climate change.

The geomorphic development of landscape and its drainage characteristics could be understood by computing and analysing different morphometric parameters (Strahler 1956). Morphometry is the measurement of geometric characteristics of the outward configuration of landforms and the drainage network associated with the landscape. The landforms and drainage networks acquire different geometric characteristics with respect to underlying geology, processes governing the landscapes and climatic condition. Hence, there is a distinct relationship that exists between geometric properties of landforms and drainage networks with their controlling factors and the process shaping them (Karmokar 2017). This relationship is established through the works of several geologists, geomorphologists and hydrologists (Strahler 1952; Schumm 1963; Dury 1951; Morisawa 1957, 1958; Gregory and Walling 1973). Thus, quantification of drainage morphometric characteristics with reference to these controlling factors could highlight the drainage development and evolution of a region.

Since the introduction of Horton's law of morphometry, several attempts have been made to study the landscape evolution, hazard assessment and watershed management by computing the morphometric parameters (Smith 1950; Miller 1953; Starhler 1957, 1964; Schumm 1963; Dury 1951; Morisawa 1957, 1958; Gregory and Walling 1973; Mesa 2006; Nag 1998; Nag and Chakraborty 2003; Mahamaya 2007; Thomas et al. 2010; Zaidi 2011; Magesh et al. 2012a, b; John Wilson et al. 2012; Altaf et al. 2013; Vandana 2013; Nandy 2016; Karmokar 2017; Mahala 2019; Karmokar and De 2020). Most of the morphometric studies are based on drainage basins with the aim of landscape development, hazard studies or watershed management. But there are very few studies that focus on the role of different controlling factors in the development of drainage characteristics of a region. In the present study, we have tried to investigate the contribution of topography, geology, rainfall, and vegetation on the development of drainage characteristics of Rajmahal Hills and its surroundings. Two main objectives were taken to fulfil the aim of the present research: (1) decipher the drainage characteristics of the study area using various morphometric parameters and (2) evaluate the role of different controlling factors on the development of drainage characteristics of the area.

The Rajmahal Hills and its surrounding areas characterized by various lithology, varied topography, variety in vegetation cover with a perceptible climatic difference and diverse drainage condition thus, provide an excellent tract to examine the effect of topography, geology, climate and vegetation on the development of drainage characteristics of this region. The lithology of the study area varies from the Archean metamorphic rocks in the west to the Quaternary alluvial deposits in the east with igneous rocks in the central part. The western part of the study area is plateau fringe of Chhotanagpur plateau, the central portion is characterized by the hills, which gradually descends to piedmont and plain to the east. The western part of the study area is characterized by semi-arid climatic condition with intense monsoonal rainfall. Rainfall decreases from the west to the east and south-east. The western and central part of the study area are endowed with very well-developed drainage network while the eastern part is suffering from drainage congestion.

In this paper, we have tried to explore the contribution of different controlling factors on the development of drainage surrounding the Rajmahal Hills by applying the Remote Sensing data and GIS tools powered by different statistical techniques coupled with field investigation. This type of drainage analysis can give a detailed overview of the hydrological behaviour of the study area, and the results generated from this analysis can also be used for various types of hydrological studies and management. This study also demonstrates the usefulness of morphometric techniques and Remote Sensing and GIS to highlight the drainage characteristics of an area and to evaluate the contribution of different controlling factors in the development of drainage characteristics of a particular region.

10.2 Study Area

Rajmahal hill is located between Chhotanagpur Plateau of Jharkhand and Gangetic Delta of West Bengal, including the districts, namely, Bhagalpur, Godda, Dumka and Pakur, of Jharkhand and Birbhum and Murshidabad of West Bengal (Fig. 10.1a). It is bounded by Ganga Basin in the north, Greater Gangetic Delta in the east, Chhotanagpur Plateau in the west and Rarh Bengal in the south. The study area lies within 23°57′52.64″ N.–25°16′56.73″ N. and 87°8′ 26.78″ E.–88°8′44.16″ E. The total geographical area under study is about 9787 km^2. The general elevation of the Rajmahal Hill varies from 17 to 587 m (Fig. 10.1b). Higher elevations are observed in the middle part of the study area, where the main Rajmahal Hill is situated and gradually decreased towards the south-eastern and north-western periphery of the study area. Several streams originated from this part of the area and flowed in different directions. Dominating rivers of the Rajmahal Hill include Bansloi, Baghmari, Brahmani, Gumani, Pagla, Tarini, Mayurakshi, etc. (Fig. 10.1c). There is a moderate type of precipitation in the area, varying from 134 to 181 cm, which is gradually increasing towards the west (Fig. 10.1d). Geologically, the study area is a mosaic of Chhotanagpur Gneissic Complex, Gondwana Super Group, Eastern Ghat Super Group and Quaternary Super Group where all three types of rocks, namely, Igneous, Sedimentary and Metamorphic are noted with major dominance of later one (Fig. 10.2). Igneous rock is concentrated in the Rajmahal Trap and areas of Rhyolite flow, whereas sedimentary rock acquires the south-eastern section which is dominated by the Gangetic Alluvial Region.

10.3 Database and Methodology

The drainage of a region is controlled by geomorphic and climatic conditions, underlying geological, lithological, tectonic and structural control, etc. Thus, to understand the drainage characteristics of Rajmahal Hills and its surroundings, we consider the topography, drainage, geology, structure and climate in a holistic manner. Figure 10.3 illustrates our assessment methodology with data used from the sources tabulated in Table 10.1.

In this work, Sentinel 2A/B and Landsat 8 OLI satellite images, geological maps (Rajmahal-Bansloi, Godda, Pakur, Dumka, Birbhum, Murshidabad and southern parts of Bihar), climatic data (Rainfall) and Digital Surface Model (DSM) along with acquired data from field survey have been used for comparative analysis to detect the control over a heterogeneous topography. First of all, we computed the important morphometric parameters, such as drainage density, stream frequency, constant of channel maintenance, length of overland flow (Horton 1945; Strahler 1952), infiltration number (Faniran 1968) and stream length gradient index (Hack 1973). To explore the geomorphic stage of drainage development, we have fit mathematical functions to the main rivers and their principal tributaries (Snow and Slingerland 1987; Rădoane et al. 2002; Lee and Tsai 2009). To understand how topography controls the development of drainage in surroundings of Rajmahal Hill, we calculated important relief morphometric characteristics, such as absolute relief, relative relief, dissection index and average slope and explored their correlation with drainage morphometric parameters (Table 10.2). We adopted grid-wise approach and used 1sq. km grid to calculate different morphometric parameters. To understand how far the drainage characteristics of the area is controlled by the underlying geology, we computed the mean and standard deviation of each drainage morphometric parameters for separate geological formation/lithologic units. We also extracted lineaments from Landsat 8 OLI satellite image in order to examine the structural control on the drainage network. A stream length gradient index map followed by Hack (1973) was prepared to evaluate the control of lithology/tectonics on the development of longitudinal profile of the streams. To understand how far the drainage characteristics of the study area is controlled by the rainfall, first of all, we performed IDW interpolation using annual average rainfall of 16 grid points falling within

Table.10.1 Data used in the study

Data	Source	Year of acquisition	Resolution/scale	Application
ALOS PALSAR DEM	NASA Earth Data (https://search.asf.alaska.edu/#/)	2008	12.5 m spatial resolution	Quantitative relief information used for computation of geomorphometric indices
Sentinel 2A	European Space Agency (https://scihub.copernicus.eu/dhus/#/home)	2016	10 m spatial resolution	Drainage networkdigitization, geological mapping
Landsat 8 OLI	United States Geological Survey (https://earthexplorer.usgs.gov/)	2016	30 m spatial resolution	Lineamentmapping, NDVI
Geological Map	Geological Survey of India, Ghose et al. (2017)	-	1:2,50,000	Lithological units and tectonic structures
Rainfall Data	Indian Meteorological Department, Pune	1989–2018	0.25-degree grid	Rainfall zonation map

Table.10.2 Morphometric parameters and index used in the study

Morphometric parameters	Definition/formula	Reference	Significance
Stream order (o)	1 + 1 = 2 2 + 1 = 2 2 + 2 = 3 and so on	Horton (1945) Strahler (1952)	Helpful to compute other linear drainage morphometric parameters
Stream number (n)	Total number of stream segments in each order	Horton (1945) Strahler (1952)	It indicates the erosional status and development stage of a given watershed
Mean stream length	Total lengths of all stream of particular order/	Horton (1945) Strahler (1952)	Indicates the effect of slope and topography on stream
Stream length ratio	$R_b = \frac{L_u}{L_u - 1}$	Horton (1945) Strahler (1952)	It indicates the geomorphic stage of landscape development
Bifurcation ratio (R_b)	$R_b = \frac{n+1}{n}$	Horton (1945) Strahler (1952)	Highlights the effect of topography, geology, structure and climate on the development of streams
Stream Length-Gradient Index (SL)	$SL = (\Delta H/\Delta L) * L$	Hack (1973)	Anomalously high values of SL Index highlight the effect of active tectonics on river profiles
Best fit curve to the river longitudinal profile	The linear function $y = ax + b$ (1) The exponential function $y = ae^{bx}$ (2) The power function $y = ax^b$ (3) The logarithmic function $y = a\ ln\ x + b$ (4)	Snow and Slingerland (1987), Rădoane et al. (2002) and Lee and Tsai (2009)	Indicates the stages of geomorphic development of river longitudinal profiles
Drainage Density (Dd)	Total length of streams within one sq. km grid	Horton (1945)	Dd is an indicator of landscape dissection. It could also be used to infer the lithology, climate and vegetation of an area
Drainage Frequency (Fd)	Total number of stream segments within a one sq. km grid	Horton (1945)	It could be used to infer the lithology, climate and vegetation of an area
Constant of Channel Maintenance (CCM)	CCM = 1/*Dd*	Schumm (1956)	Express the role of lithology on the development of stream network
Length of Overland flow (Lo)	Lo = 1/2*Dd*	Horton (1945)	Influence the development of stream
Infiltration Number (In)	In = Dd*Fd	Faniran (1968)	Indicates landscape dissection by fluvial erosion
Relative relief (Rr)	Maximum elevation-Minimum elevation	Schumm (1956)	Relative relief determines thehillslope morphologyand influences the hillslope processes which in turn influenced the drainage development
Hypsometric Integral (HI)	$HI = \frac{(H_{mean} - H_{min})}{H_{max} - H_{min}}$	Strahler (1952) Schumm (1956) Pike and Wilson (1997)	Indicates the stage of landscape development and fluvial dissection

(continued)

Table.10.2 (continued)

Morphometric parameters	Definition/formula	Reference	Significance
Average Slope (As)	$S = \frac{Lc * Ic * 100}{A * 43560}$	NRCS	Influence the development of drainage

and surrounding the study area to prepare the rainfall zonation map. We computed mean and standard deviation of each drainage morphometric parameters for separate rainfall zone. Apart from the spatial relationship, we also examine the topographic control on drainage at basin scale through linear approaches such as stream number, mean stream length, stream length ratio, bifurcation ratio and SL Index. To do so, first of all, we extracted basins and sub-basins by using the Arc Hydro tool and computed the important drainage morphometric parameters at the basin/sub-basin level. A flow chart of the methodology adopted in the study is presented in Fig. 10.3.

10.4 Result and Discussion

To decipher the drainage characteristics of the study area, we have considered some important linear and areal drainage morphometric parameters, such as stream number, mean stream length, stream length ratio, bifurcation ratio, longitudinal profiles, drainage density, stream frequency, drainage intensity, infiltration number, constant of channel maintenance and length of overland flow.

10.4.1 Drainage Characteristics

10.4.1.1 Linear Aspects

Stream Number

Horton's (1945) *law of stream number* states that the number of stream segments of each order forms an inverse geometric sequence with the order number. Actually, it is the total number of stream segments involved in each order. The physiographic and structural condition of any region controls the variation of stream numbers with the order and size of tributary basins. It also indicates the erosional status and development stage of a given watershed (Debelo et al. 2017). Table 10.3 reveals that the maximum frequency is noted in first-order stream segments with total 8715 stream segments in Rajmahal Hill region. All the drainage basins of the study area have a decreasing tendency of stream number with increasing order that show a linear relationship, with a small deviation from a straight line (Fig. 10.4). Notable deviation from the linear trend is observed in case of Brahmani river, most probably due to the effect of tectonic upliftment (Chakraborty 1985). Among 11,461 segments of the major basins of Rajmahal Hill, 76.04% (8715) is 1st order, 18.26% (2093) is 2nd order, 4.26% (488) is 3rd order, 1.04% (119) is 4th order, 0.29% (33) is 5th order, 0.1% (11) is 6th order and 0.02% is 7th order (Table 10.3). Presence of hard metamorphic and igneous rock, low water percolation, steep slope in the main Rajmahal Hill create a favourable condition to form a large number of 1st order segments in this part, which gradually decreases with increasing order and gentle slope in the north-west and south-east section of the area. The consistent decrease of stream number in the study area with orders is an indicator of the presence of erosional landform throughout the area (Mahala 2019).

Mean Stream Length

From the point of hydrology, stream length directly deals with surface runoff. Mean Stream Length is the ratio between total stream length of any order of a basin and the number of stream segments of that particular order. It shows the characteristic size of drainage network components and its contributing basin surface (Strahler 1964). Different streams of the Rajmahal Hill

Table.10.3 Order wise stream number, mean stream length, bifurcation ratio and stream length ratio

Watersheds	Stream Number							Mean Stream Length							Stream Length Ratio						Bifurcation Ratio					
	I	II	III	IV	V	VI	VII	I	II	III	IV	V	VI	VII	I-II	II-III	III-IV	IV-V	V-VI	VI-VII	I-II	II-III	III-IV	IV-V	V-VI	VI-VII
Bansloi	1303	314	74	20	6	1	-	0.73	1.14	3.02	7.15	6.58	114.99	-	1.55	2.65	2.37	0.92	17.48	-	4.15	4.24	3.70	3.33	6.00	-
Baghmari	493	123	32	8	2	1	-	0.91	1.35	3.48	8.19	12.91	42.89	-	1.49	2.57	2.35	1.58	3.32	-	4.01	3.84	4.00	4.00	2.00	-
Brahmani	1752	416	93	17	5	2	1	0.62	0.89	2.74	7.86	11.13	43.21	101.03	1.42	3.09	2.86	1.42	3.88	2.34	4.21	4.47	5.47	3.40	2.50	2.00
Mayurakshi	1019	246	54	12	3	1	-	0.68	1.21	2.90	7.89	32.63	101.31	-	1.77	2.39	2.72	4.14	3.11	-	4.14	4.56	4.50	4.00	3.00	-
Gumani	1848	483	113	32	7	3	1	0.70	1.06	2.88	6.19	5.91	28.33	60.04	1.51	2.72	2.15	0.95	4.79	2.12	3.83	4.27	3.53	4.57	2.33	3.00
Pagla	287	64	13	3	1	-	-	1.13	2.57	4.15	9.01	81.28		-	2.28	1.62	2.17	9.02		-	4.48	4.92	4.33	3.00		-
Dholia	417	89	22	7	3	1	-	0.71	1.03	3.02	13.81	24.02	18.87	-	1.45	2.94	4.57	1.74	0.79	-	4.69	4.05	3.14	2.33	3.00	-
Domjola	190	44	10	3	1	-	-	0.66	1.68	3.79	7.54	15.69		-	2.53	2.26	1.99	2.08		-	4.32	4.40	3.33	3.00		-
Jalhara	643	149	38	7	3	1	-	0.42	0.65	1.62	7.27	4.43	15.87	-	1.54	2.48	4.49	0.61	3.58	-	4.32	3.92	5.43	2.33	3.00	-
Dhulia	763	165	39	10	2	1	-	0.51	1.09	2.86	11.06	29.53	3.07	-	2.12	2.62	3.87	2.67	0.10	-	4.62	4.23	3.90	5.00	2.00	-

Table.10.4 Fitting of best fit mathematical functions to the main rivers and their principal tributaries of the study area

Rivers	Functions	Correlation coefficient (r)	Determination of coefficient (R^2)	Best fit functions	Geomorphic stage of evolution
Gumani	Linear	0.759	0.5757	Logarithmic	Graded
	Exponential	0.959	0.9198		
	Logarithmic	0.977	0.9542		
	Power	0.960	0.9214		
Morai	Linear	0.802	0.6433	Logarithmic	Graded
	Exponential	0.924	0.8543		
	Logarithmic	0.985	0.9704		
	Power	0.982	0.9637		
Baghmari	Linear	0.708	0.5006	Power	Transitional stage between Dynamic equilibrium and Graded
	Exponential	0.921	0.8488		
	Logarithmic	0.960	0.9219		
	Power	0.983	0.967		
Bansloi	Linear	0.832	0.6915	Logarithmic	Graded
	Exponential	0.990	0.9804		
	Logarithmic	0.993	0.9854		
	Power	0.917	0.8411		
Bansloi sub	Linear	0.775	0.6009	Logarithmic	Graded
	Exponential	0.906	0.8205		
	Logarithmic	0.948	0.8987		
	Power	0.937	0.8782		
Pagla	Linear	0.836	0.6983	Logarithmic	Graded
	Exponential	0.962	0.9258		
	Logarithmic	0.992	0.9843		
	Power	0.959	0.9201		
Brahmani	Linear	0.911	0.8295	Exponential	Dynamic equilibrium
	Exponential	0.989	0.9781		
	Logarithmic	0.973	0.9472		
	Power	0.900	0.8092		
Tripati	Linear	0.843	0.7099	Logarithmic	Graded
	Exponential	0.969	0.9398		
	Logarithmic	0.994	0.9878		
	Power	0.921	0.8488		
Mayurakshi	Linear	0.896	0.8029	Exponential	Dynamic equilibrium
	Exponential	0.988	0.9764		
	Logarithmic	0.984	0.9683		
	Power	0.921	0.8487		
Tarini	Linear	0.734	0.5394	Power	
	Exponential	0.915	0.8374		

(continued)

Table.10.4 (continued)

Rivers	Functions	Correlation coefficient (r)	Determination of coefficient (R^2)	Best fit functions	Geomorphic stage of evolution
	Power	0.983 0.960	0.9662 0.9214		Transitional stage between Dynamic equilibrium and Graded

Table.10.5 Correlation between relief and drainage parameters

Variables	Ar	Rr	Mr	Hi	As	Dd	Sf	CCM	Lo	In
Ar	1	0.883	0.983	0.432	0.102	0.479	0.613	0.212	0.212	0.565
Rr	0.883	1	0.807	0.308	0.092	0.417	0.548	0.230	0.230	0.497
Mr	0.983	0.807	1	0.513	0.100	0.484	0.618	0.201	0.201	0.571
Hi	0.432	0.308	0.513	1	0.032	0.156	0.286	0.092	0.092	0.231
Ds	0.287	0.659	0.146	-0.096	0.030	0.033	0.090	0.132	0.132	0.062
As	0.102	0.092	0.100	0.032	1	0.131	0.120	0.115	0.115	0.128
Dd	0.479	0.417	0.484	0.156	0.131	1	0.898	0.208	0.208	0.972
Df	0.613	0.548	0.618	0.286	0.120	0.898	1	0.310	0.310	0.969
CCM	0.212	0.230	0.201	0.092	0.115	0.208	ty	1	1.000	0.236
Lo	0.212	0.230	0.201	0.092	0.115	0.208	0.310	1.000	1	0.236
Di	0.555	0.517	0.555	0.316	0.113	0.530	0.777	0.677	0.677	0.657
In	0.565	0.497	0.571	0.231	0.128	0.972	0.969	0.236	0.236	1

Where, Ar = Absolute relief, Rr = Relative relief, Mr = Mean relief, Hi = Hypsometric integral, Sf = Stream frequency, CCM = Constant of channel maintenance, Lo = Length of over land flow, In = Infiltration number

experience a range of Lμ like Bansloi varies from 0.73 to 114.99 km; Baghmari varies from 0.91 to 42.89 km; Brahmani varies from 0.62 to 101.03 km; Mayurakshi varies from 0.68 to 101.31 km; Gumani varies from 0.7 to 60.04 km; Pagla varies from 1.12 to 81.28 km, etc., and in total, Lμ varies from 0.42 km in 1st order of Jalhara Basin stream to 114.99 km in 6th order of Bansloi River (Table 10.3). The result shows that Horton's *law of mean stream length* (Lμ of a given order is greater than that of the lower and less than that of its next higher-order) is followed in the study region. The greater sinuosity of the stream segments in a lower order than that of higher-order due to variation in slope and topography seems to be the main cause behind the increase of mean stream length with order.

Stream Length Ratio (RL)

The 'length ratio' is the ratio of the mean length of stream segments of one order to mean length of segments of next lower order, which tends to be constant throughout the successive orders of a watershed (Horton 1945). Horton's (1945) *law of stream length* states that the mean lengths of stream segments of each of the successive orders of a basin tend to approximate a direct geometric sequence in which the first term is the average length of segments of the first order (Strahler 1964). The stream length ratio plays a vital role in surface flow discharge and erosional stage of the basin (Sreedevi et al. 2004). From the study area, it reveals that there is both intra- and inter-level variation in RL among all the major streams of the area (Table 10.3) which help to determine the age of the basin (Rudraiah et al. 2008). The

Table.10.6 Geological formation/lithology wise average value of different drainage parameters

Formations/lithological group	Area (sq. km)	Dd	Df	CCM	Lo	In
Alternating layers of Sand, Silt and Clay	301.00	0.57	0.10	1.74	0.87	0.06
Barakar Formation	301.15	1.47	1.57	0.68	0.34	2.30
Belhar Formation	217.86	1.00	0.64	1.00	0.50	0.64
Chhotanagpur Gneissic Complex	545.27	2.35	2.96	0.42	0.21	6.97
Chhotanagpur Granite Gneissic Complex	757.02	2.32	2.73	0.43	0.22	6.32
Diara Formation	74.63	1.08	0.55	0.93	0.46	0.59
Dubrajpur Formation	53.16	1.54	3.05	0.65	0.32	4.69
Eastern Ghat Super Group	36.96	1.96	2.19	0.51	0.25	4.31
Ganga-Kosi Formation	405.35	0.54	0.22	1.85	0.92	0.12
Granite Gneiss with enclaves of Metamorphites	110.40	1.78	1.92	0.56	0.28	3.42
Jamui Formation	2657.68	0.68	0.27	1.46	0.73	0.19
Laterite and Lateritic Soil	832.52	0.78	0.52	1.29	0.64	0.40
Metamorphics of Chhotanagpur Mica Belt	62.93	1.64	2.02	0.61	0.30	3.31
Rajmahal Trap	3347.18	1.63	1.42	0.61	0.31	2.31
Rhyolite Flow	39.56	1.42	1.14	0.71	0.35	1.61
Talcher Formation	21.28	1.59	1.88	0.63	0.32	2.98

Table.10.7 Rainfall zone wise summary statistics of drainage density, drainage frequency and infiltration number

Rainfall zones (cm)	Area (sq. km)	Drainage density				Drainage frequency				Infiltration number			
		Min	Max	Mean	Std	Min	Max	Mean	Std	Min	Max	Mean	Std
< 140	2393.46	0.00	5.39	0.62	0.75	0.00	17.99	0.95	1.66	0.00	97.02	1.97	5.66
140–150	3233.62	0.00	5.84	1.39	0.79	0.00	22.99	2.55	1.87	0.00	126.93	5.46	6.24
150–160	2601.38	0.00	7.25	1.54	0.91	0.00	21.99	3.08	2.22	0.00	120.76	7.35	7.84
170–180	1253.88	0.00	6.15	1.61	0.89	0.00	15.00	3.33	2.22	0.00	70.65	7.92	7.46
> 180	300.70	0.00	5.14	1.85	0.76	0.00	11.99	3.82	1.82	0.00	43.29	9.05	6.45

intra-level change from one order to another suggests late youth stage of geomorphic development (Singh and Singh 1997). In total, the highest value (17.48) is found in 6th order of Bansloi River, whereas the lowest value is enlisted as 0.1 in 6th order of Dhulia River (Table 10.3), which indicates difference in prevailing geomorphic stage of the rivers.

Bifurcation Ratio (Rb)

We have used bifurcation ratio to explore the impact of different factors, such as topography, geology, structure and climate on the development of streams. It is expressed as the ratio of the number of streams of any given order to the number of streams in the next higher-order in a drainage basin (Schumm 1956). Due to variations in watershed geometry, the bifurcation ratio will not be precisely the same from one order to the next but will tend to be constant throughout the series (Strahler 1964). Uniform rock type, similar underlying geology, lithology, structure and stage of development, the same type of physiographic division and climatic condition of an area help to characterize more or less constant values of bifurcation ratio from one

Table.10.8 NDVI zone wise average values of drainage density, drainage frequency and infiltration number

NDVI (value)	Area (sq. km)	Vegetational characteristics	Geomorphic province	Dd (km/sq. km)	Df (no./sq.km)	In
-0.62–0.2	259.11	Aquatic vegetation, seasonal wetlands, swamps, marshes	Floodplain	1.19	0.35	0.42
0.21–0.26	942.83	Agricultural land	Depositional alluvial plain	0.42	0.21	0.09
0.27–0.37	2856.65	Agricultural land, grass land and shrubs	Depositional piedmont zone	1.02	1.83	1.87
0.38–0.49	2607.09	Agricultural land, grass land, deciduous open forest and shrubs	Plateau fringe	1.61	2.97	4.78
0.50–1	3123.10	Deciduous dense forest	Hills	1.54	2.21	3.40

order to next. Low Rb value indicates minimum structural disturbances, relatively circular basin and less distorted drainage patterns and vice versa (Strahler 1964). The highest value of Rb (5.47) is achieved by high variation in number of stream segments between two successive orders due to discrimination in juvenile topography with greater slope of higher order and comparatively mature topography with moderate slope. Maximum values of high Rb is noted in the parts of metamorphosed rock of middle-western side Badlands, which is prevailed by various sub-basins of the major streams. The Rb values of lower order streams for all the basins indicate the homogeneity of rocks in the study area. But anomalously higher values of Rb for different order in the major basins of the study area certainly indicate the inhomogeneity of the rock surface and structural disturbance.

Stream Length-Gradient Index

Normally, undisturbed rivers (both tectonically and lithologically) typically develop a smoothly changing, concave, graded/near graded longitudinal profile. Hence, variations in the lithology of the riverbed, rock resistance or tectonic activity may depart the river gradient from this ideal smooth shape and create an anomaly. Whenever such disturbances ceased, rivers are rapidly approached to a gradient profile (Snow and Slingerland 1987). Thus, instability in river profiles may be interpreted as a response to ongoing tectonism (Sarp et al. 2011), deviations in lithological compositions and rock resistance. It is defined as the product of channel slope of the reach and the distance of the midpoint of the reach from the divide in a watershed (Hack 1973). According to Hack 1973, the SL Index is expressed by:

$$SL = (\Delta H/\Delta L) * L$$

where, ΔH = the difference in elevation between the ends of the reach, ΔL = the length of the reach, L = the channel length from the divide to the midpoint of the channel reach including the bends in the stream. The value of this index increases significantly in the places of river flow over the active uplifted zones and over relatively hard rocks, whereas it has quite lesser value when flows parallel to structures (such as valleys made by strike-slip faulting) and over relatively soft rocks (Keller and Pinter 2002).

In this present work, to identify the anomalies along with the longitudinal profile of the streams, and their correlation with lithology or structure of the area, the spatial distribution map of the SL index is generated. The spatial distribution map of the SL index is obtained by interpolating the calculated midpoint values along the major rivers and their principal tributaries using the IDW method. The SL values of the area vary spatially from 0.1595474 to 1393.39 (Fig. 10.5a). In the middle part of the Rajmahal Trap, the projection of basaltic scarp over the almost peneplain surface of the Archean metamorphosed province creates a substantial difference between topography. The presence of the Saithiya-Bramhani Fault in this region also helps to separate the two different provinces and marks as a clear breakpoint between two elevated zones of igneous basalt and Archean metamorphic. Hence, rivers emerging from this scarp and flowing over this region have obtained a high SL value. High SL value along the Bansloi River is obtained due to prominent lithological difference (Fig. 10.7a and b) and thereby deep incision as several parts of the area have washed out the main basaltic layer and exposed shale, sandstone, etc., of the lower layer. Though Chakraborty (1985) has argued about the presence of tectonic activity in the Mayurakhi-Bramhani basin to generate the lithological difference. High SL Value has also been found in the extreme north-east part of the study area, where Motijhorna waterfall is situated (Fig. 10.7c). The steep scarp, which is formed due to the lithological difference between the basaltic layer of Rajmahal Trap and the alluvial flood plain of Ganga-Koshi formation, seems to be the prime factor to experience a high SL value in this part. The high SL value of the central Rajmahal Hill gradually decreases radically due to the only existence of a monotonous flood plain in both north-western and south-eastern sections which in turn creates a favourable ground to form the profiles of rivers near to equilibrium.

The Best-Fit Functions to River Longitudinal Profiles

In order to determine the geomorphic stage of evolution of river course, we have fitted the mathematical model for longitudinal profile of major rivers and their principal tributaries following Rădoane et al. (2002) and Lee and Tsai (2009). According to this model, there are four stages of geomorphic evolution of a river course determined by the four mathematical functions.

The linear function

$$y = ax + b \quad (10.1)$$

The exponential function

$$y = ae^{bx} \quad (10.2)$$

The power function

$$y = ax^{b} \quad (10.3)$$

The logarithmic function

$$y = a\,In\,x + b \quad (10.4)$$

According to this model, when the grain size of bedload is greater than river transport capacity, the longitudinal curve of the channel fit to the linear function. Long profiles better fit to exponential functions, when rivers achieve a dynamic equilibrium between erosion and deposition. When the river reaches to a balance between erosion and resistance and channel grain size decrease downstream, i.e. river attended graded profile as postulated by Hack (1973), the resulting longitudinal profile of the river better fits to a logarithmic function. Rivers characterized by high profile concavity and downstream increase of discharge and sediment load better fit to a power function. The geomorphic evolution sequence of rivers should be linear → exponential → power → logarithmic (Rădoane et al. 2002; Lee and Tsai 2009). The curve with the highest R^2 is considered as best fit curve to the normalized longitudinal profile.

We have considered 10 rivers, including 3 main rivers and 7 principal tributaries. The result is presented in Table 10.4. Considering the statistical point of view, it could be said that all four mathematical equations show a higher degree of fit. The very higher correlation values, generally > 0.7 approve the fact that we are right to choose these four mathematical curve functions to show the form of the longitudinal profiles of the rivers in the study area (Rădoane et al. 2002).

The best fit to logarithm for Gumani, Morai, Bansloi sub, Pagla and Tripati rivers indicate that they have reached to the graded stage (Hack 1973). Brahmani and Mayurakshi rivers are fitted to the exponential functions, thus they are currently in the stage of dynamic equilibrium. Only Tarini and Baghmari rivers better fit to a power function. The landscape of this region is very old (Ghose et al. 2017). The graded condition of the rivers is in accordance with the stage of landscape development in these areas. The dynamic equilibrium of Brahmani and Mayurakshi rivers is most probably due to the tectonic upliftment and supply of huge sediment and water by their tributaries. The best fit to the power function for the rivers Tarini and Baghmari is due to very higher curvature at their upstream which is a characteristic of a transitional stage between dynamic equilibrium and graded.

Drainage Density (Dd)

Drainage density is the ratio of total channel-segment lengths cumulated for all orders within a basin to the basin area (Strahler 1964). The concept of drainage density was first introduced by Horton (1932) and subsequently been followed by Smith (1950). It is a measure of the degree of fluvial dissection and thus, it reflects both the tendency of the drainage basin to generate surface runoff and the erodibility of the surface materials (Chorley 1969). It is also influenced by a number of factors like topography, lithology, climate, soil, and vegetation (Nag 1998; Mesa 2006). In the study area, higher Dd values (>2.3) are found over the metamorphosed Archean landmass, to the west of Rajmahal hills (Fig. 10.5b). Jointing system of Archean terrain (Chakraborty 1985) and the development of gullies along these weak zones produces numerous streams. The torrential monsoonal rainfall after the prolonged semi-arid condition in the pre-monsoon (summer) creates heavy surface runoff by the lower order streams (generally 4th or lower order) over the basaltic terrain of Rajmahal hills, the underlying shale and sandstone beds have been exposed in many places (Fig. 10.7a and b). Exposure of these soft bed rocks has paved the path of further erosion and thus enhances the development of drainage networks. All these factors contribute to producing higher Dd over the Rajmahal hills. The western rim of Rajmahal Hills is also characterized by higher Dd. Many first-order streams originate from the scarp of Rajmahal and join with the Tarini, Bansloi, Brahmani and Baghmari rivers. The central part of Rajmahal hills is associated with moderate to highly moderate drainage density due to higher slope, low permeability of the terrain and higher vegetation cover. Moderate value of Dd is the characteristics of the piedmont regions of the study area. Higher porosity of the colluvial soil of piedmont is mainly responsible for the lower drainage density than the Hilly and metamorphosed areas. But inspite of porous soils, this zone has higher drainage density than the alluvial plain, only because of coalsence of several tributaries emerging from the Rajmahal hills. Very low *Dd* is found over the depositional alluvial plain due to very high infiltration capacity of the underlying subsurface geology, very low relief and very gentle average slope. Due to the absence of prominent surface slope, drainage is not well developed and thus this zone suffers from drainage congestion. Several swampy areas, marshes, lakes and anastomosing channels are developed in this zone (Fig. 10.7d). It seems topography and its underlying geology control the drainage density of the area.

Stream Frequency (Sf)

According to Horton (1945), stream frequency (Fs) is defined as the ratio of the total number of stream segments of all the orders per unit area. Stream frequency of an area is controlled by the permeability and infiltration capacity of the rock, rainfall and relief (Strahler 1956; John Wilson et al. 2012). A combination of high rainfall and greater slope increase stream frequency, whereas high infiltration capacity and low relief decrease the stream frequency (Bali et al. 2011; Mahala 2019).

In the case of the study area, the values of Fs ranges between 0 and 21 (Fig. 10.5c), which have been categorized into six classes. Higher stream frequency occurs over the metamorphosed Archean rocks and around the northern and southern part of Rajmahal hills. The torrential monsoonal rainfall over Archean metamorphosed terrain characterized with jointing rock system and sparse vegetation, after a prolonged semi-arid condition in the pre-monsoon (summer) creates heavy surface runoff and thus accelerates soil erosion. This factor also enhances the gully erosion in this part and creates numerous streams. Rocky terrain, rigid underlying rocks, high relief conditions, steep slope, high amount of rainfall and low infiltration capacity of the rock are the main reasons behind the higher stream frequency over the Rajmahal hills. Intense and prolonged erosion of Basaltic upper bed and exposure of underlying sandstone and shale beds enhance the erosion and facilitate further drainage network development. Except for northern part and some residual hills in the southern part of Rajmahal hills, most of the portion of Basaltic Rajmahal hills and the piedmont to the east of Rajmahal hills are characterized by moderate to highly moderate drainage frequency. Decrease of terrain slope and absence of high relief are the main reasons behind the slightly lower values of Fs in the central part of Rajmahal hills than its northern and southern portion. On the other hand, despite its porous and highly permeable colluvial soils, piedmont zone has slightly higher Fs than the depositional alluvial plain due to convergence of much lower order stream emerging from the eastern scarp of Rajmahal hills. Very lower Fs values are observed over the depositional alluvial plain in the north-west and east of Rajmahal hills. The presence of high permeable subsurface material, very low relief and very gentle slope are the prime causes behind low Fs value. Although some patches of moderate drainage frequency are found over the Gangetic depositional plain. Very low slope of the terrain, stream flows in meandering and anabranching patterns, are the main cause of slightly higher stream frequency over the porous alluvial soils.

Constant of Channel Maintenance (CCM)

Constant of channel maintenance is defined as the inverse of drainage density (Schumm 1956). Thus, it is characterized by the drainage area and required to maintain a unit length of channel. Actually, the number of sq. ft. (grid) of a watershed surface required to sustain one liner feet of channel is shown by CCM (John Wilson et al. 2012). Low value of CCM suggests weak or low resistance soils, sparse vegetation and mountain terrain (Shulits 1968). In the Rajmahal Hill, most of the areas are influenced by low to moderate CCM (0.25–0.85) which indicates mature stage of landscape development in area (Fig. 10d) (Thomas et al. 2010). High structural control (areas of Gumani, Bansloi and Baghmari Basin), low permeability, presence of residual hills, steep to very steep slope and high surface runoff seem as a regulatory factor to achieve very low to low CCM (seldom moderate) over the maximum area. High to very high CCM is observed over ridge top in Rajmahal hills and over the depositional alluvial plain to the north-west and east and south-east of the Rajmahal hills. Very higher values of CCM over the alluvial plain is mainly attributed to high infiltration capacity of the soil.

Length of Overland Flow (Lo)

Length of Overland Flow is one of the most important independent variables which affects both the hydrological and physiographic development of a drainage basin (Horton 1932). Length of overland flow is defined as the length of the flow path, projected to the horizontal, of non-channel flow from a point on the drainage divide to a point on the adjacent stream channel (Horton 1945). According to Chorley (1969), it is the mean horizontal length of the flow path from the divide to the stream in a first-order basin and is a measure of stream spacing and degree of dissection is approximately one-half the reciprocal of the drainage density. Horton (1945) also had taken it to be roughly equal to half the reciprocal of the drainage density as at an average, it is about half the distance between the stream channels. Basically, average slope of the channel is inversely related to it whereas the length of sheet flow to a large degree is quite synonymous with it. In this study, length of overland flow varies from < 0.23 to 4.9 within five categories. Most of the areas of metamorphic rock in the western side and the middle part of Rajmahal Hill have the shortest flow path with low surface runoff, steep slope and high relief fragmentation whereas north-western side and parts of south-east with high Lo value imply the longest flow path with high surface runoff and gentle slope over the plain surface.

Drainage Texture

Drainage texture is an important morphometric parameter to highlight the landscape dissection by the fluvial process. Drainage texture of terrain is affected by its relief, lithology, soil, vegetation cover and geomorphic stage of landscape development (Smith 1950). Horton (1945) opined that drainage texture incorporates both the drainage frequency and drainage density. Smith's (1950) method of drainage texture is the most used method by the geomorphologists and geologists, but this method is not suitable for grid-based morphometric calculation. Thus, we adopted Faniran's (1968) infiltration number index to quantify the landscape dissection in the present study. The values of infiltration number denote the infiltration characteristics of any watershed. It is the product of drainage density and stream frequency (Faniran 1968). Infiltration number is directly related to drainage intensity. Thus, higher the infiltration number, higher will be stream frequency, drainage density as well as drainage intensity and vice versa. Where Faniran (1968) used it to show the infiltration capacity of the basins, we have used it to highlight the dissection of terrain by the effect of the fluvial process (Horton 1945) as well as to infer the impact of controlling factors on drainage texture (Smith 1950). The low value of infiltration number indicates that there is a limited effect of drainage density and stream frequency in the process of surface denudation and vice versa. Moderate to very high values of In is observed over the metamorphosed rocks of Archean and over the dissected Basaltic terrain of Rajmahal hills (Fig. 10.6a). These areas show higher dissection of fluvial erosion. The piedmont zone to the east of Rajmahal hills is associated with low to moderate In. The higher infiltration of the colluvial soil of piedmont produce low stream frequency and thereby produce low In. Although the piedmont zone has higher value of In compared to depositional alluvial plain due the coalesce of many lower-order streams emerging from the eastern scarp of Rajmahal hills. Very low values of infiltration number observed in the north-west and east and south-east of Rajmahal hills is a characteristic feature of depositional plains, where fluvial deposition is the main geomorphic process and far more pronounced than the erosion.

10.4.2 Topographic Characteristics

We consider three important relief morphometric parameters, i.e. relative relief, hypsometric index and slope to highlight the landscape development and to understand the topographical control over the drainage characteristics of the study area.

10.4.2.1 Relative Relief (Rr)

Relative relief termed as 'local relief' or 'relative energy' or 'amplitude of relief', is the elevation difference of the highest and lowest point of a unit area. It represents the actual variation of altitude in a unit area with respect to its base level. It indicates the general steepness of an area (Vandana 2013). Relative relief determines the hillslope morphology and influences the hillslope processes which in turn influenced the drainage density. For example, Schumm (1956) noted a positive relationship between drainage density and relief ratio. On the other hand, Oguchi (1997) found a negative correlation between drainage density and relative relief in the Japanese mountains. The relative relief of Rajmahal hills and its western metamorphosed province is the result of a process of pronounced peneplanation (Chakraborty 1985) while the eastern part associated with depositional process. Analysis reveals that the derived values of relative relief in Rajmahal Hill varies from 0.041 to 310 m, which are classified in the six categories (Fig. 10.6b). The piedmont zone between the Rajmahal Hills and Gangetic alluvial plain is characterized with moderate (16–30 m) to highly moderate (31–60 m) values of relative relief. Due to prolonged erosion of the Archean landmass of Granitic-Gneiss complex of Chotonagpur Lava plateau in the western part of Rajmahal hills an almost featureless plain is created. But due to deep incision of lower order tributaries (4th order or less) of Bansloi, Brahmani and Mayurakshi rivers, narrow and deep valleys are created (Fig. 10.7e), and together with interfluves and residual (Fig. 10.7f) hills, these valleys cause moderate (16–30 m) to highly moderate relief (31–60 m) over the peneplain surface. Gully erosion is the main geomorphic process of erosion in plateau fringe areas (Fig. 10.7g). The entire Rajmahal hills are characterized with highly moderate (31–60 m) to very high (121–310 m) categories of relative relief. High (61–120 m) categories of RR are mainly associated with scattered residual hillocks of Rajmahal hills. The striking feature of relative relief of the study area is the occurrence of very high categories of RR (121–310 m) in the form of finger-like extension, especially in the northern part of Rajmahal hills (Fig. 10.7h). Bhattacharji (2012) assumed that these finger-like extensions of ridges were the water divides of adjacent lake basins. She argued that these finger-like extensions of ridges are actually the remnants of these water divides. Projections of these spur on the present-day valley floor create great vertical difference which results in very high RR. On the other hand, very high RR in the southern part of the Rajmahal hills is caused by the deep fluvial erosion of lower-order streams. Fluvial erosion, gully erosion and mass wasting are the main geomorphic process of erosion (Fig. 10.7i). Very low categories of RR (<7.5 m) are found over the Gangetic plain due to its very low relief. This area is characterized by swamps, marshes and anastomosing rivers (Fig. 10.7d).

10.4.2.2 Average Slope (as)

According to Strahler (1956), slope is the property of inclination of the ground surface or of a profile line from the horizontal—*i.e.* an angular measure, or dip. There are various factors like geologic, tectonic, geomorphic and climatic that govern the slope of a region. Wenthworth in the year 1930 compared erodibility with average slope and made a conclusion that if all other parameters are kept constant, then more the percentage of slopes leads to more erosion. Among the various methods of average slope, here the slope method of NRCS is followed.

$$S = \frac{Lc * Ic * 100}{A * 43560}$$

where, S = slope (in degree), Lc = length of contour (in feet), Ic = contour inter (in feet), A = area (in acres).

The distribution of average slopes is classified into six zones (Fig. 10.7c). The low slope categories mainly confined in the depositional alluvial plain, to the eastern side of Rajmahal hills. It is the combined floodplain of Gumani,

Bagheri, Bansloi, Pagla, Brahmani, Mayuraksi and Bhagirathi rivers. This zone is characterized by marshes, swamps and anastomosing river (Fig. 10.7d) and inundates more or less every year. The gentle slope of the alluvial plain is highly modified by the settlement patches. The north-western part of the study area, to the north-west of Rajmahal hills is also a depositional plain and characterized with very low slope. The piedmont zone in between Rajmahal hills and depositional Gangetic plain is characterized with moderate to high slopes. The metamorphosed province to the west of Rajmahal hills also characterized with moderate to high slopes. Basaltic province of Rajmahal hills is characterized with high to very high slopes. Due to prolonged erosion, the southern part of Rajmahal hills eroded away and thus have very low slope.

10.4.2.3 Hypsometric Index (Hi)

Hypsometric index indicates the stage of landscape development (Strahler 1956), which in turn may reflect the development of drainage of a region. In the early stage of development, the landscape is characterized by a fewer number of streams and low drainage density. As landscape approaches to maturity, numerous streams develop and thus increases stream frequency and drainage density (Davis 1899; Horton 1945). Most of the portion of the study area is characterized by high to very high values of Hi. With reference to threshold suggested by Strahler (1956), most of the portion of Rajmahal Hills and the metamorphic province to the west of the Rajmahal Hills presently are going through the mature stage of landscape development. Higher stream frequency and higher drainage density of these regions are in accordance with the stage of landscape development. On the other hand, the plain region to the east of Rajmahal Hills is characterized with very low to low Hi values due to its very low relief. Some higher patches of Hi values spotted over the plain region is due to artificial relief caused by infrastructure and settlement.

10.4.3 Geology

Geology is widely accepted as a prime factor in the development of drainage characteristics of a region (Schumm 1956; Strahler 1956). Lithologic control of drainage density is well understood. In general, lower values of drainage density occurs over the highly resistant rocks or permeable lithology. On the other hand, high drainage density occurs over the impermeable subsurface lithology characterized with sparse vegetation cover and high relief (Schumm 1956; Gregory and Walling 1973; Krsihnamurthy et al. 1996).

The Rajmahal Hill is a part of Chhotanagpur Gneissic Complex, which forms the basement of the Gondwana Supergroup (Ghose et al. 2017). It is characterized by 18 geological formations on the basis of the major geotectonic unit (Fig. 10.2). All these formations are composed of igneous, metamorphic as well as sedimentary rock types. Among them, the Rajmahal Trap is the largest one followed by Jamui formation (Table 10.3). The central part, characterized by Rajmahal Trap and Rhyolite Flow, is mainly formed with core igneous rocks and developed in the phase of the early eruption of Rajmahal (Ghose et al. 2017). Due to direct relation with intrusion, the main Rajmahal Trap is rather complex in nature as it is characterized by Flows of Basalt with Inter-Trappean Sediments and Dolerite Dykes (Probodh 2018). The Jamui formation of the south-eastern and eastern section of the study area, which is developed during the time-bound between Middle Pleistocene to Early Holocene, is made up with older alluvium in general and khaki green clay with brown silts and highly oxidized sands, those are impregnated with caliche nodules (GSI). Alternative layering of oxidized sand, silt and clay of older flood plain deposits of Ganga-Koshi Formation are the prime characteristics of the rest part of the eastern and south-eastern section, which is the part of Birbhum and Murshidabad districts of West Bengal. Metamorphic rock is concentrated in the whole western and north-western section, where Chhotanagpur Granitic Gneissic Complex, Chhotanagpur Gneissic Complex, Dubrajpur Formation, Barakar Formation, Talcher Formation, Metamorphics of Chhotanagpur Mica, etc., belt is found (Fig. 10.2). Over large parts, the metamorphism reached up to the upper amphibolite facies and locally granulites facies (Lalnunmawia 2011). The most abundant rock types within the famous Chhotanagpur Granitic Gneissic Complex (CGGC) are granitic gneisses and migmatites (Lalnunmawia 2011) along with layers and enclaves of high Mg–Al granulite and two-pyroxene granulites (Karmakar et al. 2011). Dubrajpur Formation of Lower Gondwana system, which dates back to Upper Triassic period (Probodh 2018), is comprised of Ferruginous sandstones, shales and conglomerates (Raja Rao 1953; Shrivastava and Shah 1966). The Talcher Formation, which is the basal member of the lower Gondwana Formations (Sengupta 1988), is of Carboniferous period and composed of Coal Seams with Sandstones, Clays and Boulder Beds boundaries (Probodh 2018). Barakar formation overlies the Talchir formation with maximum thickness of 152 m (Sengupta 1988; Bhattacharyya 1996) and dates back to Permian Age (Probodh 2018). The lithology of this formation is composed of a sequence of course to medium sandstone with intercalations of lenticular carbonaceous shale containing plant fossils (Bhattacharyya 1996).

To examine how far the drainage network of the study area is controlled by the structure, we have considered lineaments azimuth. Lineaments are linear topographic features of regional extent that may reflect crustal structure and/or is an expression of underlying structural features (Morrison 1985). Such lineaments are the topographic expression of faults, joints, valleys, dykes, etc., and may be associated with regional tectonic activities. According to Clark and Wilson (1994), lineaments are the representative of geological and/or geomorphological phenomena. Thus, lineaments guide the course of the river (Ribolini and Spagnolo 2008), hence in order to assess the effects of lineaments on drainage network, we extracted lineaments and assessed their orientation and compared the orientation of lineaments with the orientation of different orders of stream separately. Lineaments are extracted from bands 7 of Landsat8 with spatial resolution of 30 m using the LINE module in PCI Geomatica 2015. The automatically generated lineaments were visually inspected and lineament that matched artificial features such as canals, road, railway lines, etc., were removed. For determination of dominant trend, a length-weighted rose diagram was constructed as suggested by Ricchetti and Palombella (2007) and Aubry et al. (2012) using Arc GIS 10.3 and Rock Works 15 software.

The whole region is characterized by numerous lineaments. The density of lineaments is greater in the central part of Rajmahal Hill over metamorphic rock, due to presence of numerous ridges and valleys. Higher lineament density over the basaltic terrain of Rajmahal hills is mainly attributed to the geomorphic evolution of igneous landforms (Bhattacherji 2012). The Jointing system of the Archean rocks in the western part of the study area produces numerous lineaments (Chakraborty 1985). The lineament density decreases over the depositional alluvial plain due to depositional monotonous topography. The petal of the rose diagram shows that lineaments in the study area oriented in all direction with preferential W-E and N-S direction (Fig. 10.8a). We have also considered order-wise stream orientation (Fig. 10.8b) in order to compare them with lineament orientation.

10.4.4 Vegetation

Drainage characteristics especially, drainage density, drainage frequency, constant of channel maintenance, length of overland flow is controlled by vegetation cover (Horton 1945; Schumm 1956; Strahler 1956). To explore the impact of vegetation on the drainage characteristics of the study area, we have calculated Normalized Difference Vegetation Index (NDVI) (Rouse et al. 1974) using the Landsat 8 OLI satellite image. To calculate the NDVI, we have used the following formula

$$NDVI = \frac{(NIR - Red)}{(NIR + Red)} \qquad (10.5)$$

We have classified the resulting NDVI raster into five categories (Fig. 10.8c). The NDVI values < 0.20 shows the vegetation growth in waterbodies or the areas having a very higher moisture content in the soil. The areas showing higher moisture content in the NDVI map actually are the seasonal wetlands, bogs, swamps, marshes, etc. The plain zone devoted to intensive agricultural practice shows lower values of NDVI (0.21–0.26). Despite the semi-arid condition, the western metamorphosed province also exhibits higher NDVI values (0.38–0.49) than the alluvial zones due to its sparse deciduous forest cover and intensive agricultural practices (Fig. 10.7j and k). The Rajmahal basaltic province shows the highest NDVI values (0.50–1) in the study area as the entire Rajmahal hills is covered by dense deciduous forest.

10.4.5 Controls on Drainage

10.4.5.1 Topography

To explore the control of topography over the drainage of the study area, the correlation between relief and drainage parameters are estimated. Besides, relative relief, average slope and hypsometric integral, we have also considered absolute relief and mean relief to decipher the effect of topography on the development of drainage characteristics. To perform the correlation, at first, we checked the nature of the distribution of data. The result shows that no single variable is normally distributed. Thus, we adopted Spearman's rank correlation method, which is developed to examine the correlation of non-parametric data. To perform Spearman's rank correlation, at first, we rank the different morphometric (relief and drainage both) parameters in ascending order, i.e. greater the value higher the rank. Considering the ranks of different morphometric parameters, we performed the Spearman's rank correlation. The result shows that a significant positive correlation exists between different drainage and relief parameters. From the correlation matrix, it is apparent that Dd, Df, Di and In are moderately controlled by Ar, Rr and Mr. Although, higher Df and higher Dd are observed over the high Hi value zones, there is no statistically significant correlation found between Hi and Dd or Hi and Df. Correlation matrix also revealed that As, an important relief parameter, has no significant impact on any drainage parameters. On the other hand, CCM and Lo have no significant relationship with any relief parameters. It is evident from the result that the drainage of the study area is moderately controlled by the relief parameters. That means there exist other controlling factors that determine the drainage characteristics of the area.

10.4.5.2 Geology

Geology in terms of lithology and structure broadly control the origin and evolution of landscape as well as the drainage flowing over it (Thornbury 1969). Hence, we consider geology as a factor responsible for variation in drainage characteristics throughout the study area. We evaluate different lithological units and lineaments separately.

10.4.5.3 Lithology

Lithology has a definite role in facilitating or hindering the development of drainage in a region. A visual comparison of drainage density, drainage frequency, CCM, Lo and In with lithological spatial distribution of the study area reveals that the spatial contrast of Dd, Df, CCM, Lo and In corresponds to the location of geological contrast. A statistical comparison (Table 10.5) of mean values of parameters among different geological formation also support the fact. Highest drainage density with a mean value of greater than 2.3 km/sq. km is observed over the metamorphosed terrain of Chhotanagpur plateau fringe areas. The major formation of the study area is Rajmahal basaltic flow, which covers about 35% area of the study area. Mean Dd, Df, CCM, Lo and In over the Rajmahal basaltic province are 1.63 km/sq. km, 1.42/sq. km, 0.61 sq. km/km, 0.31 km and 2.31, respectively. In terms of areal coverage, Jamui formation is the second-largest geological formation of the study area. The Jamui formation is a sedimentary formation built up by the deposition of sediment transported by the rivers from the Chhotanagpur plateau fringe areas. Due to higher infiltration rate of the sedimentary composition, the mean drainage frequency (0.27/sq. km) and drainage density (0.68 km/sq. km) are very low, thus drainage texture (0.19). On average, 1.46 sq. km area is required to maintain 1 km channel length over this formation which is second highest after Ganga-Kosi alluvial formation within the study area.

The drainage pattern of the study area is also controlled by geology. The metamorphosed province is characterized by dendritic drainage pattern with high drainage density. The upstream drainage network of Bansloi, Brahmani and Mayurakshi spread over the Chhotanagpur Gneissic Complex and Chhotanagpur Granite Gneissic Complex exhibit dendritic drainage pattern. The basaltic terrain of Rajmahal hills is also characterized with dendritic drainage pattern. Trellis drainage pattern developed over the Gangetic alluvium due to the presence of soft unconsolidated alluvium and is more related to sudden flattening of topography (Madabhusi 1991). Above analysis concluded that the drainage characteristics of the study area are largely determined by the surface lithology.

10.4.5.4 Drainage Network and Lineament in Relation to the Structure

Lineaments guide the course of the river, hence to assess the effect of lineaments on drainage network, we have extracted lineaments and assessed their orientation and compared the orientation of lineaments with the orientation of different order streams separately. The rose diagrams of different stream order of drainage network show that 1st order to 4th order streams of the study area joins the higher-order streams from all directions, though preferential flow in N-S and E-W (Fig. 10.8b). The 5th order streams have similar trends as lower-order streams while 6th and 7th order streams have W-E dominant trend. The linkage between lineaments and the stream courses helps to understand the relationship between tectonics and stream orientation (Capolongo et al. 2005). A comparison of lineaments and stream channel network shows that some parts of the drainage network are likely to be controlled by tectonic features. The abrupt change in the direction of a tributary of Gumani river from N-S to W-E and sudden bends in the course of Bansloi river from NNW-S to W-E and remarkably straight channels in some parts of Brahmani, Bansloi and Torai rivers are some of the examples of tectonic controls over the drainage in the study area (Fig. 10.7l). It could be inferred from the analysis of drainage and lineament orientation that there is a very positive relationship between lineament and drainage, which proves that the formation of drainage is fully guided by lineament. The Rajmahal basaltic and metamorphosed sections have experienced both high lineament and maximum stream number due to the presence of joints in the Archean rock system (Chakraborty 1985), topography incorporated with successive ridges and valleys. Most of the lineaments in the study area are related to the geomorphological evolution of landforms (Bhattacharji 2012), lithological contrast and zone of lithological weakness (Chakraborty 1985).

From the analysis, it could be concluded that the drainage of the alluvial plain and the metamorphosed province is controlled by the lithology, while the drainage over basaltic terrain of Rajmahal hills is mainly regulated by the structure and topography (Madabhusi 1991).

10.4.6 Climate

Although the study area is not so large, the climate produces perceptible variations in the study area. The western part of the study area is characterized by semi-arid climatic condition with intense monsoonal rainfall. The amount of precipitation in the study area ranges between 134 and 181 cm, which is gradually increasing from east to west (Fig. 10.1d). The torrential monsoonal rainfall over Archean metamorphosed terrain characterized with jointing rock system and sparse vegetation, after the prolonged semi-arid condition in the pre-monsoon (summer) creates heavy surface runoff and thus accelerates soil erosion. The higher amount of precipitation over the metamorphosed province and basaltic province certainly helps in greater erosion which causes higher drainage density. Statistical comparison has also been performed between different drainage parameters with different rainfall zones to show the actual condition. Rainfall zone wise calculated summary statistics of Dd, Df and In presented in the Table 10.7 shows that Dd increases with the increase in rainfall. The average value of Df also increases with the increase of rainfall but the higher value of SD suggests a greater variation of Df within the rainfall zones itself. But greater intra variability of infiltration number, i.e. drainage texture as indicated by SD suggests the climate is not the sole element to control the drainage characteristics.

10.4.7 Vegetation

To investigate the impact of vegetation on the drainage characteristic, we have computed NDVI using Landsat 8 OLI and categorized the resulting NDVI map into five classes (Fig. 10.8c). We have calculated average value of Dd, Df and In for each NDVI zone. The result presented in Table 10.8 shows that the lowest value of NDVI (Fig. 10.8c) is associated with aquatic vegetation, seasonal wetlands, swamps, bogs and marshy areas of the floodplain zone. This zone is characterized with very low drainage frequency although slightly higher drainage density mainly attributed to the very flat terrain and meandering stream course. Very low NDVI zone mostly indicates the agricultural lands over the depositional alluvial plain. This zone has lowest drainage frequency and drainage density thus very coarse drainage texture. Due to the absence of pronounced surface slope, drainage is not developed well. Moderate NDVI zone is mainly associated with the piedmont zone of the study area and characterized by sparse vegetation, shrubs and intensive agriculture. This zone has a moderate drainage density with slightly higher stream frequency. Drainage characteristics of this zone is basically determined by its terrain characteristics and surface lithology. Comparatively higher values of NDVI is observed over the fringe area of Chhotanagpur plateau. This zone is sparsely covered by deciduous trees and intensive agriculture is practised in the gullies. Due to these reasons, the NDVI value is higher than the floodplain, alluvial plain and piedmont zone. Drainage frequency and stream density are highest over this zone due to sparse vegetation cover, weaker lithology and intense monsoonal rainfall. The very high NDVI zone is spread over the entire Rajmahal hills. The zone is covered by dense deciduous forest. Despite high rainfall and impermeable lithology, steep slope, and high relief, drainage density and stream frequency are slightly lower due to dense forest cover. Thus, the above analysis and visual comparison among drainage density, drainage frequency and NDVI indicate that the vegetation influences the drainage characteristics where the vegetation cover is higher.

10.5 Conclusion

In the present study, we have applied Remote sensing and GIS coupled with field investigation to highlight the drainage characteristics and tried to explore the contribution of topography, geology, climate, and vegetation in the development of drainage characteristics of Rajmahal Hills and surroundings.

Computed morphometric parameters highlight the drainage characteristics and indicate the stage of landscape development. The moderate values of HI, Rr and fit to the logarithmic functions of the long profile for most of the rivers suggest that the landscape of the area is under the mature stage of geomorphic development. Higher drainage density, higher stream frequency, moderate relative relief, moderate average slopes, and lower values of dissection index are all in accordance with the mature stage of landscape development for the metamorphosed province and basaltic terrain of Rajmahal Hills. The correlation analysis between drainage and relief parameters shows that drainage characteristics of the study area are moderately controlled by the topographic characteristics of the study area. Computed statistics of different drainage parameters for separate lithological units or geological formations suggests that drainage density, stream frequency and drainage pattern of the study area are broadly controlled by the underlying lithology. Comparison between lineaments and drainage azimuth for different order streams indicates that the drainage orientation is controlled by the topographic lineaments and by the lineaments associated with the zone of lithological weakness. Stream length gradient index map of the study area highlights that some parts of the drainage network is also controlled by the tectonic activities. Analysis of NDVI suggests that drainage characteristics of the study area are to some extent influenced by vegetation characteristics. Analysis suggests that among the four-

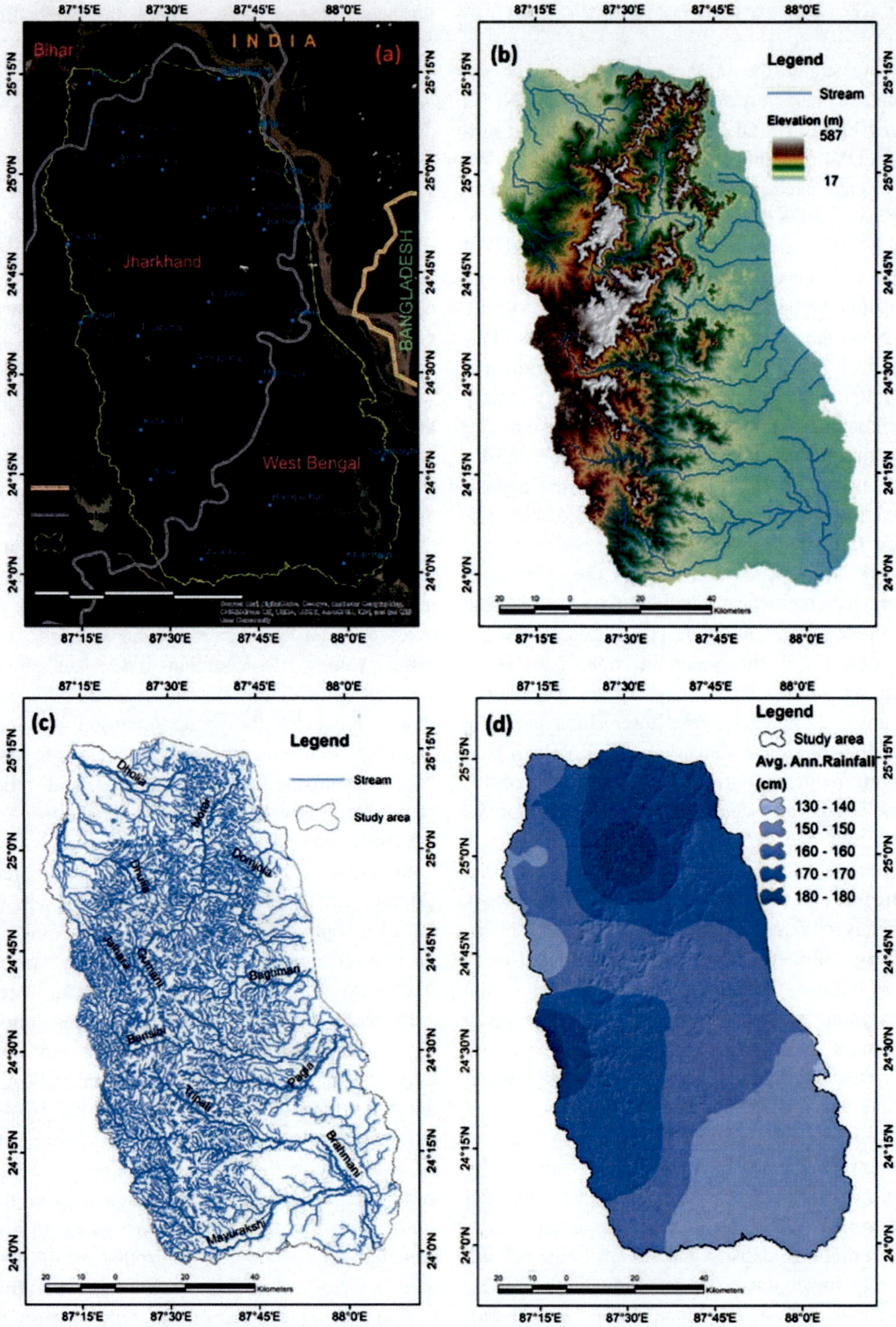

Fig. 10.1 **a.** location map of the study area, **b.** elevation map of the study area, **c.** drainage map of the study area digitized from Sentinel 2 B 10 m satellite images, **d.** rainfall map prepared using the IMD, Govt of India data

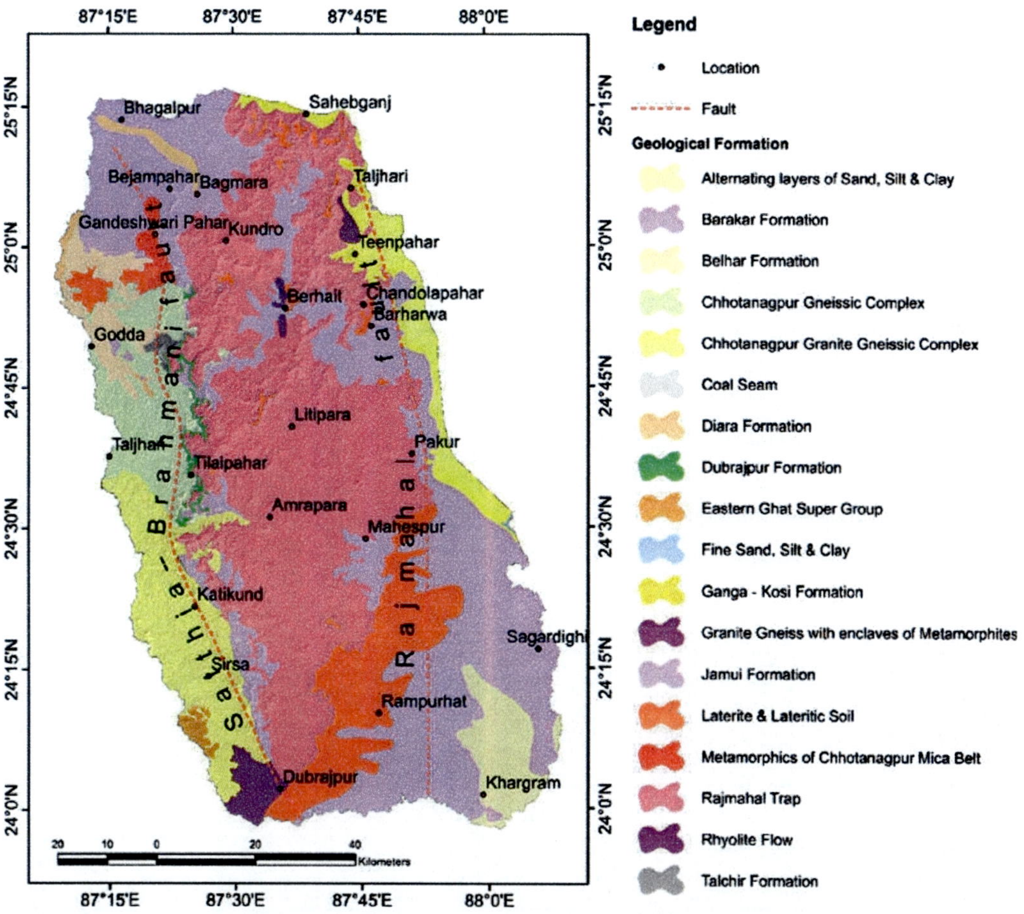

Fig. 10.2 Geological map of the study area (after GSI 2005; Ghose et al. 2017)

dominant controller of the drainage, here relief (topography) and geology play a vital role to determine the hydrological behaviour, whereas the presence of climatic control is very limited and less prominent to develop drainage characteristics. From the integrated results of the study, it can be concluded that it has immense utility from hydrological behaviour in response to landform evaluation and natural resource management. The spatial variability of selected morphometric parameters helps to understand present terrain characteristics along with hydrology. Thus, it is hoped that the detailed study of these geoprocessing techniques will assist the planner and decision-makers for various types of management through sustainable development in the Rajmahal region.

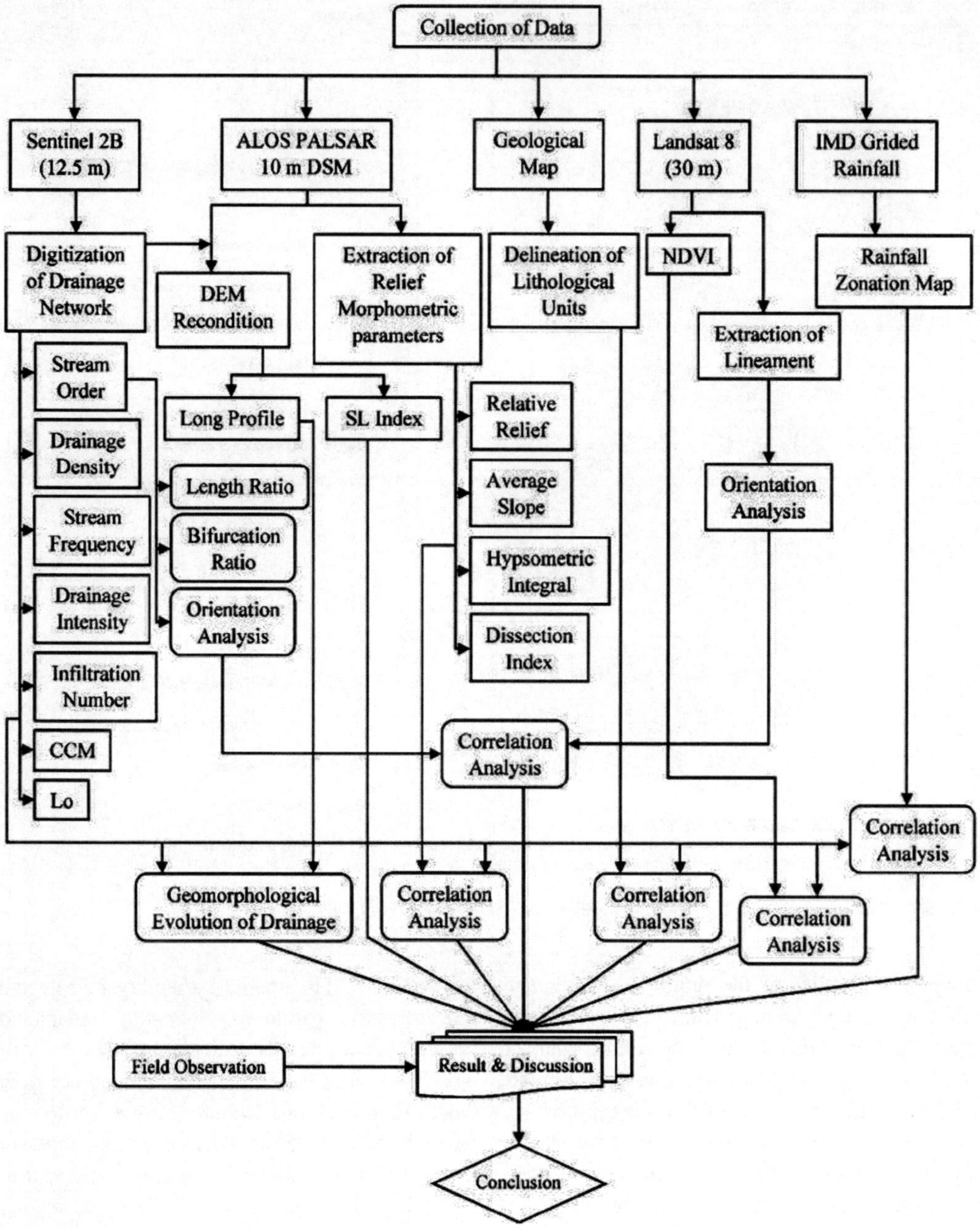

Fig. 10.3 Flowchart of the methodology

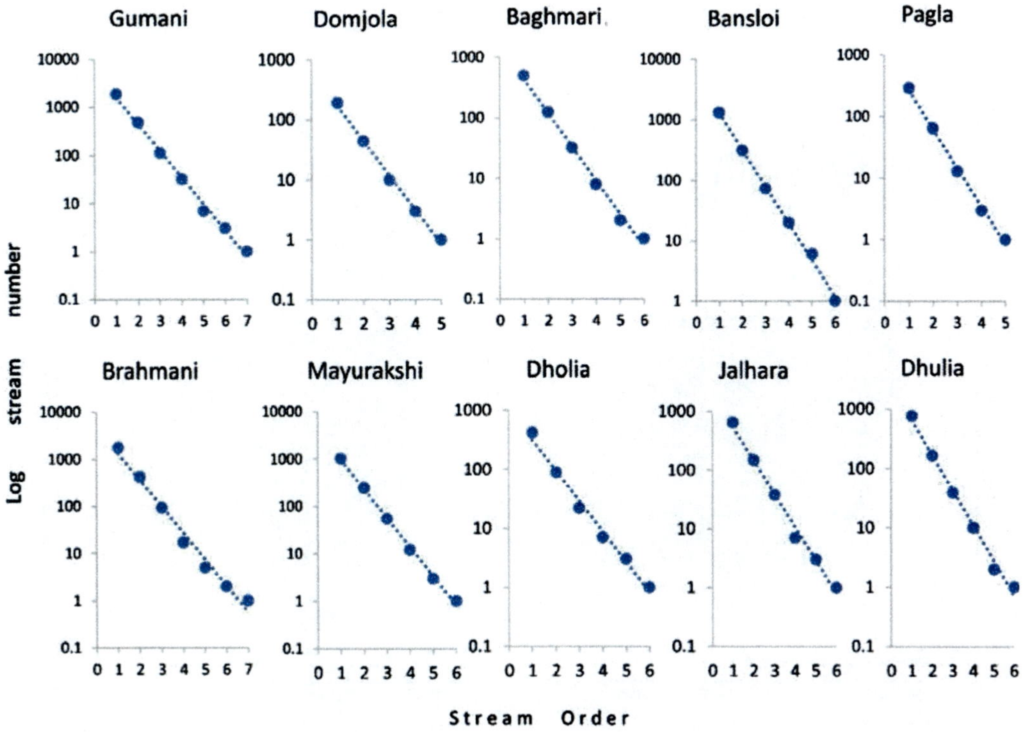

Fig. 10.4 Linear regression between stream order and number of streams

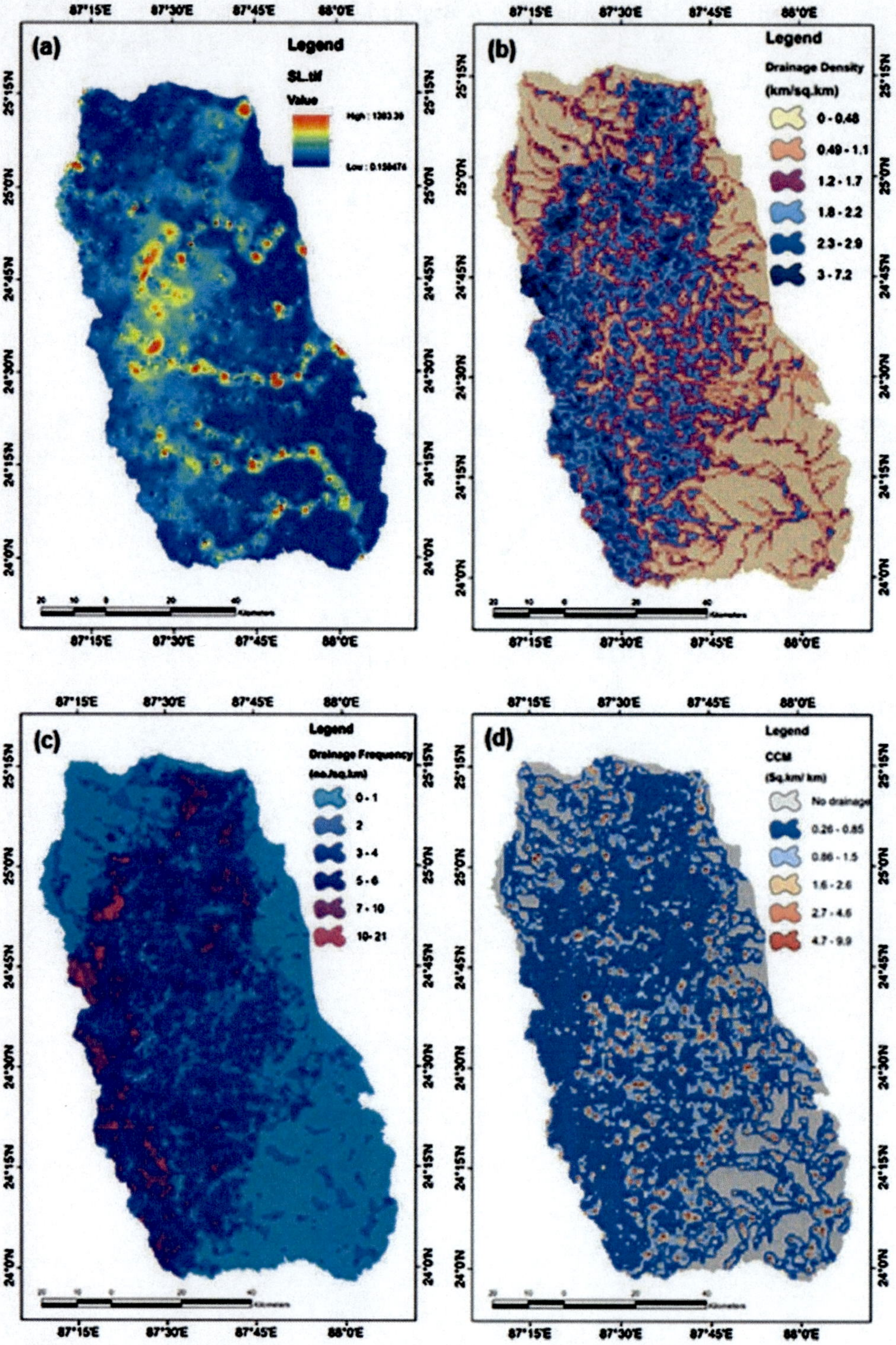

Fig. 10.5 **a.** SL index map after Hack (1973), **b.** Drainage density, **c.** Drainage frequency, **d.** Constant of channel maintenance

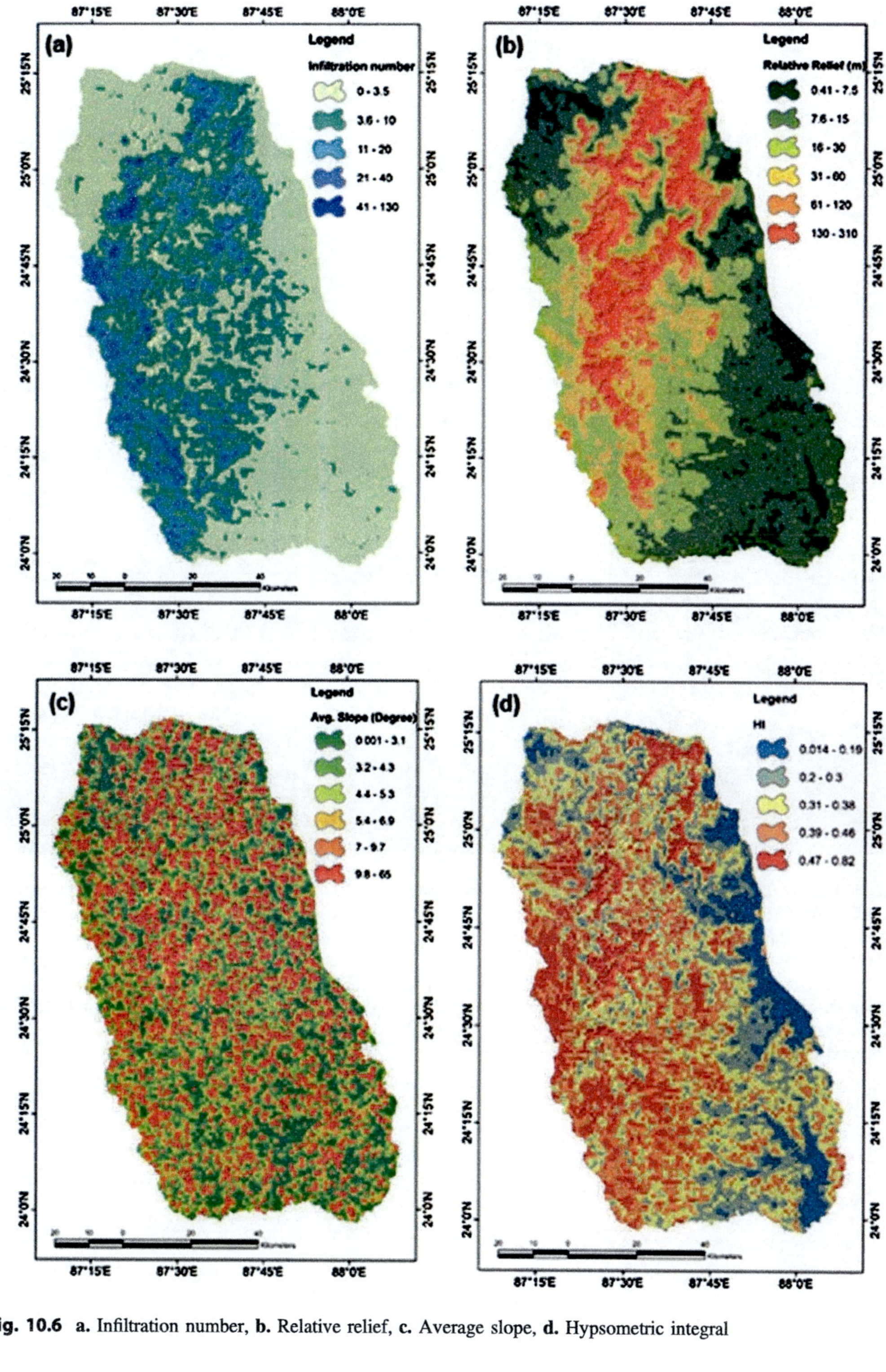

Fig. 10.6 **a.** Infiltration number, **b.** Relative relief, **c.** Average slope, **d.** Hypsometric integral

Fig. 10.7 **a** & **b**. underlying sandstone is exposed in Bansloi river due to intensive fluvial erosion, **c**. Moti jharna, **d**. swampy area developed in alluvial plain in Pagla basin, **e**. deep and narrow channel formed in plateau fringe areas over the metamorphosed terrain in Kathikund, Dumka District, Jharkhand **f**. residual hills in plateau fringe areas, **g**. extensive gully erosion over the metamorphosed terrain of plateau fringe areas, **h**. finger-like spur in the northern part of Rajmahal hills, **i**. gully erosion and mass wasting on the eastern scarp of Rajmahal hills, near Tin Pahar, Sahebganj District of Jharkhand, **j**. sparse deciduous vegetation cover in the western part of Rajmahal hills, **k**. extensive cultivation in gullies, **l**. Brahmani river guided by Saithia-Brahmani fault in Kathikund

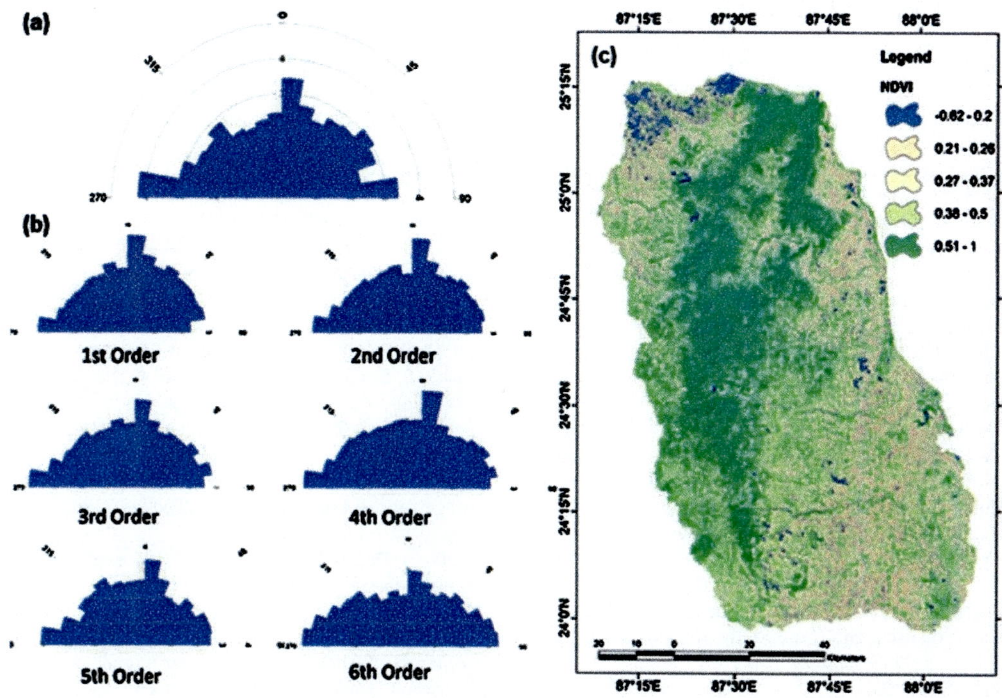

Fig. 10.8 **a**. rose diagram showing the orientation of lineaments over the study area, **b**. stream order wise orientation. **c**. NDVI

References

Altaf F, Meraj G, Romshoo SA (2013) Morphometric analysis to infer hydrological behaviour of Lidder Watershed, Western Himalaya, India. Geography J 1–14. https://doi.org/dx.doi.org/10.1155/2013/178021

Altın TB, Altın BN (2011) Drainage morphometry and its influence on landforms in volcanic terrain, Central Anatolia, Turkey. Proc Soc Behav Sci 19:732–740. https://doi.org/10.1016/j.sbspro.2011.05.192

Aubry T, Luis L, Dimuccio LA (2012) Nature vs. Culture: present-day spatial distribution and preservation of open-air rock art in the Côa and Douro River Valleys (Portugal). J Archaeol Sci 39(4):848–866. https://doi.org/10.1016/j.jas.2011.10.011

Babault J, Teixell A, Struth L, Driessche JV, Arboleya ML, Tesón E (2013) Shortening, structural relief and drainage evolution in inverted rifts: insights from the Atlas Mountains, the Eastern Cordillera of Colombia and the Pyrenees. In Nemčok M, Mora A, Cosgrove JW (eds) *Thick-skin-Dominated Orogens: From Initial Inversion to Full Accretion* (Special Publications ed. Vol. 377, pp. 141–158). London: Geological Society of London. doi:https://doi.org/10.1144/SP377.14

Bahrami S (2013) Analyzing the drainage system anomaly of Zagros basins: implications for active tectonics. Tectonophysics 608:914–928. https://doi.org/10.1016/j.tecto.2013.07.026

Bali R, Agarwal KK, Ali SN, Rastogi SK, Krishna K (2011) Drainage morphometry of Himalayan Glacio-fluvial basin, India: hydrologic and neotectonic implications. Environ Earth Sci 66(4):1163–1174. https://doi.org/10.1007/s12665-011-1324-1

Bhattacharyya M (1996) Rajmahal Hills: an enquiry into some problems of structural geomorphology with particular reference to the Gumani Basin. Ph.D. Thesis, Calcutta University, Department of Geography, Calcutta

Bhattarji M (2012) Topographic features of the Gumani River Basin, Rajmahal volcanic province, Jharkhand: an example of exhumed topography or Paleotopographic control? J Indian Geom 1:45–54

Bourne JA, Twidale CR (2011) Neglected and cryptostructural effects in drainage development. Cadernos Do Laboratorio Xeolóxico De Laxe 36:41–60

Bretis B, Bartl N, Grasemann B (2011) Lateral fold growth and linkage in the Zagros fold and thrust belt (Kurdistan, NE Iraq). Basin Res 23(6):615–630. https://doi.org/10.1111/j.1365-2117.2011.00506.x

Capolongo D, Cecaro G, Giano SI, Lazzari M, Schiattarella M (2005) Structural control on drainage network of the south-western side of the Agri River upper valley (Southern Apennines, Italy). Geogr Fis Din Quat 28:169–180

Chakraborty B (1985) A geomorphological analysis of the Mayurakshi river basin. http://hdl.handle.net/10603/68600

Chen Y-C, Sung Q, Cheng K-Y (2003) Along-strike variations of morphotectonic features in the Western Foothills of Taiwan: tectonic implications based on stream-gradient and hypsometric analysis. Geomorphology 56(1):109–137. https://doi.org/10.1016/S0169-555X(03)00059-X

Chitra C, Alaguraja P, Ganeshkumari K, Yuvaraj D, Manivel M (2011) Watershed characteristics of Kundah sub basin using Remote Sensing and GIS techniques. Int J Geomat Geosci 2(1):311–335

Chorley RJ (1957) Climate and morphometry. J Geol 65 (6):628–638

Chorley RJ (1969) Introduction to physical hydrology. Methuen and Co., Ltd., Suffolk

Clark CD, Wilson C (1994) Spatial analysis of lineaments. Comput Geosci 20(7–8):1237–1258. https://doi.org/10.1016/0098-3004(94)90073-6

Davis WM (1899) The geographical cycle. Geogr J 14 (5):481–504

Debelo G, Tadele K, Koriche SA (2017) Morphometric analysis to identify erosion Prone areas on the upper blue Nile using GIS (Case Study of Didessa and Jema Sub-Basin, Ethiopia). Int Res J Eng Technol (IRJET) 4(8):1773–1784

Dury GH (1951) Quantitative measurement of available relief and depth of dissection. Geol Mag 88(5):339–343. https://doi.org/10.1017/S0016756800069776

Faniran A (1968) The index of drainage intensity - a provisional new drainage factor. Aust J Sci 31:328–330

Fenta AA, Yasuda H, Shimizu K, Haregeweyn N, Woldearegay K (2017) Quantitative analysis and implications of drainage morphometry of the Agula watershed in the semi-arid northern Ethiopia. Appl Water Sci 7:3825–3840. https://doi.org/10.1007/s13201-017-0534-4

Frissell CA, Liss WJ, Warren CE, Hurley MD (1986) A hierarchical framework for stream habitat classification: viewing streams in a watershed context. Environ Manage 10(2):199–214. https://doi.org/10.1007/BF01867358

Garcia-Castellanos D (2006) Long-term evolution of tectonic lakes: climatic controls on the development of internally drained basins. In Willett SD, Hovius N, Brandon MT, Fisher DM (eds) Tectonics, climate, and landscape evolution (Vol. Special Paper 398, pp. 283–294). Geological Society of America. doi:https://doi.org/10.1130/2006.2398(17)

Ghose NC, Chatterjee N, Windley BF (2017) Subaqueous early eruptive phase of the late Aptian Rajmahal volcanism, India: evidence from volcaniclastic rocks, bentonite, black shales, and oolite. Geosci Front 8 (4):809–822. https://doi.org/10.1016/j.gsf.2016.06.007

Gregory KJ, Walling DE (1973) Drainage basin form and process: a geomorphological approach. Edward Arnold, London

Hack JT (1973) Stream-profile analysis and stream-gradient index. J Res U.S. Geol Survey 1(4):421–429

Horton RE (1932) Drainage basin characteristics. Trans Am Geophys Union 13:350–361

Horton RE (1945) Erosional development of streams and their drainage basins: Hydrophysical approach to quantitative morphology. Geol Soc Am Bull 56 (3):275–370. https://doi.org/10.1130/0016-7606(1945)56[275:EDOSAT]2.0.CO;2

John Wilson JS, Chandrasekar N, Magesh NS (2012) Morphometric Analysis of Major Sub-Watersheds in Aiyar & Karai Pottanar Basin, Central Tamil Nadu, India Using Remote Sensing & GIS Techniques. Bonf Int J Ind Eng Manag Sci 2(Special Issue 1):8–15

Karmakar S, Bose S, Basu Sarbadhikari A, Das K (2011) Evolution of granulite enclaves and associated gneisses from Purulia, Chhotanagpur Granite Gneiss Complex, India: Evidence for 990–940 Ma tectonothermal event(s) at the eastern India cratonic fringe zone. J Asian Earth Sci 48(1):69–88. https://doi.org/10.1016/j.jseaes.2010.12.006

Karmokar S, De M (2020) Flash flood risk assessment for drainage basins in the Himalayan foreland of Jalpaiguri and Darjeeling Districts. West Bengal. https://doi.org/10.1007/s40808-020-00807-9

Karmokar S (2017) Balasan Basin: morphology and landscape evolution- A morphometric analyis in seminar proceed. ILEE, 256–266

Keller EA, Pinter N (2002) Active Tectonics: Earthquakes, Uplift, and Landscape, 2nd edn. Prentice Hall, New Jersey

Kleidorfer M, Mikovits C, Jasper-Tönnies A, Huttenlau M, Einfalt T, Rauch W (2014) Impact of a changing environment on drainage system performance. Proc Eng 70:943–950. https://doi.org/10.1016/j.proeng.2014.02.105

Krishnamurthy J, Kumar NV, Jayaraman V, Manivel M (1996) An approach to demarcate ground water potential zones through remote sensing and geographic information system. Int J Remote Sens 17 (10):1867–1884. https://doi.org/10.1080/01431169608948744

Lalnunmawia J (2011) Petrochemistry of granulites in the vicinity of dumka town jharkhand. Ph.D. Thesis, Mizoram University, Department of Geology, Aizawl

Langbein WB (1947) Topographic characteristics of drainage Basins. Washington DC: United States Government Printing Office

Lee C-S, Tsai LL (2009) A quantitative analysis for geomorphic indices of longitudinal river profile: a case study of the Choushui River, Central Taiwan. Environ Earth Sci 59:1549–1558. https://doi.org/10.1007/s12665-009-0140-3

Leopold LB, Langbein WB (1962) The concept of entropy in landscape evolution. United States Government Printing Office, Washington, DC

Lotspeich FB, Platts WS (1982) An integrated land-aquatic classification system. Am J Fish Manag 2 (2):138–149. https://doi.org/10.1577/1548-8659(1982)2%3c138:AILCS%3e2.0.CO;2

Madabhusi S (1991) Drainage, lithology and structure in the north-western part of Rajmahal Hills, Bihar, India. Indian J Geomorphol (edited by HS Sharma)

Magesh NS, Jitheshlal KV, Chandrasekar N, Jini KV (2013) Geographical information system-based morphometric analysis of Bharathapuzha river basin, Kerala, India. Appl Water Sci 3:467–477. https://doi.org/10.1007/s13201-013-0095-0

Magesh NS, Chandrasekar N, Kaliraj S (2012a) A GIS based automated extraction tool for the analysis of basin morphometry. Bonfring Int J Ind Eng Manag Sci 2(1):32–35

Magesh NS, Jitheshlal KV, Chandrasekar N, Jini KV (2012b) GIS based morphometric evaluation of Chimmini and Mupily watersheds, parts of Western Ghats, Thrissur District, Kerala, India. Earth Sci Inform 5(2):111–121. https://doi.org/10.1007/s12145-012-0101-3

Mahala A (2019) The significance of morphometric analysis to understand the hydrological and morphological characteristics in two different morpho-climatic settings. Appl Water Sci 10:1–16. https://doi.org/10.1007/s13201-019-1118-2

Mahamaya M (2007) Morphometric analysis of the Periyar River, Kerala, India. The Geographer 54:1–16

Manjoro M (2015) Structural control of fluvial drainage in the western domain of the Cape Fold Belt, South Africa. J Afr Earth Sc 101:350–359. https://doi.org/10.1016/j.jafrearsci.2014.10.001

Melton MA (1957) An analysis of the relations among elements of climate, surface properties and geomorphology, Project NR 389–042. Columbia University, Department of Geology. Office of Naval Research, New York

Melton MA (1958a) Correlation structure of morphometric properties of drainage systems and their controlling agents. J Geol 66(4):442–460

Melton MA (1958b) Geometric properties of mature drainage systems and their representation in an E4 phase space. J Geol 66(1):35–54. https://doi.org/10.1086/626481

Mesa LM (2006) Morphometric analysis of a subtropical Andean Basin (Tucumán, Argentina). Environ Geol 50(8):1235–1242. https://doi.org/10.1007/s00254-006-0297-y

Miller VC (1953) A quantitative geomorphic study of drainage basin characteristics in Clinch mountain area, Virginia and Tennesse. Technical Report:3, Project NR 389–042, Columbia University, Department of Geology, New York

Morisawa ME (1957) Accuracy of determination of stream length from topographic maps. Trans Am Geophys Union 38:86–88

Morisawa ME (1958) Measurement of drainage basin outline form. J Geol 66(5):587–591

Morrison WG (1985) A dictionary of geology (Paperback ed.). CBS Publication & Distributors, New Delhi, 506 p

Nag SK (1998) Morphometric analysis using remote sensing techniques in the Chaka subbasin Purulia district, West Bengal. J Indian Soc Remote Sens 26(1–2):69–76. https://doi.org/10.1007/BF03007341

Nag SK, Chakraborty S (2003) Influences of rock types and structures in the development of drainage network in hard rock area. J Indian Soc Remote Sens 31(1):25–35

Nanda AM, Ahmed P, Kanth TA, Hajam RA (2014) Morphometric analysis of sandran drainage basin (J & K) using Geo–Spatial Technology. Earth Sci India 7 (II):55–66

Nandy S (2016) Morphometric analysis as a component of hydrological characteristics of Himalayan upper Pindar Basin, Uttarakhand. Geo-Analyst 6(1):25–41

Oguchi T (1997) Drainage density and relative relief in humid steep mountains with frequent slope failure. Earth Surface Proc Land 22(2):107–120. https://doi.org/10.1002/(SICI)1096-9837(199702)22:2<107::AID-ESP680>3.0.CO;2-U

Patton PC (1988) Drainage basin morphometry and floods. In: Baker V, Kochel RC, Patton PC (eds) Flood geomorphology. Wiley, New York, pp 51–65

Pike RJ, Wilson SE (1997) Elevation-relief ratio, hypsometric integral, and geomorphic area-altitude analysis. Bull Geo Soc Am 82(4):1079–1084. https://doi.org/10.1130/0016-7606(1971)82[1079:ERHIAG]2.0.CO;2

Probodh MK (2018) Geological study of Rajmahal trap area and its resource base:a regional perspective. IOSR J Appl Geol Geophys 6(2):23–27

Rădoane M, Rădoane N, Dumitriu D (2002) Geomorphological evolution of longitudinal river profiles in the Carpathians. Geomorphology 50(4):293–306. https://doi.org/10.1016/S0169-555X(02)00194-0

Raj R (2007) Strike slip faulting inferred from offsetting of drainages: Lower Narmada basin, western India. J Earth Syst Sci 116(5):413–421. https://doi.org/10.1007/s12040-007-0040-4

Raj R (2012) Active tectonics of NE Gujarat (India) by morphometric and morphostructural studies of Vatrak River basin. J Asian Earth Sci 50:66–78. https://doi.org/10.1016/j.jseaes.2012.01.010

Raja Rao CS (1953) Detailed mapping of parts of northern portions of Rajmahal hills and adjoining areas, Santhal Parganas district Bihar. Geological Survey of India, Government of India, Calcutta

Ribolini A, Spagnolo M (2008) Drainage network geometry versus tectonics in the Argentera Massif (French–Italian Alps). Geomorphology 93(3–4):253–266. https://doi.org/10.1016/j.geomorph.2007.02.016

Ricchetti E, Palombella M (2007) Production of geological lineament map of Southern Italy using Landsat 7 ETM+ imagery. Bollettino della Societa Geologica Italiana 126(3):567–572

Rodríguez-Iturbe I, Escobar LA (1982) The dependence of drainage density on climate and geomorphology. Hydrol Sci J 27(2):129–137. https://doi.org/10.1080/02626668209491095

Rouse JW, Haas RH, Schell JA, Deering DW (1974) Monitoring vegetation systems in the Great Plains with ERTS. Goddard Space Flight Center 3rd ERTS Symposium SP–351. 1, Section A, pp. 309–317. NASA

Rudraiah M, Govindaiah S, Vittala SS (2008) Morphometry using remote sensing and GIS techniques in the sub-basins of Kagna river basin, Gulburga district, Karnataka, India. J Indian Soc Remote Sens 36 (4):351–360. https://doi.org/10.1007/s12524-008-0035-x

Sangireddy H, Carothers RA, Stark CP, Passalacqua P (2016) Controls of climate, topography, vegetation, and lithology on drainage density extracted from high resolution topography data. J Hydrol 537:271–282. https://doi.org/10.1016/j.jhydrol.2016.02.051

Sarp G, Geçen R, Toprak V, Duzgun S (2011) Morphotectonic properties of yenicaga basin area in Turkey. The GEOSS Era: towards operational environmental monitoring. In: 34th International Symposium on Remote Sensing of Environment, Sydney

Schumm SA (1956) Evolution of drainage systems and slopes in Badlands at Perth Amboy, New Jersey. Bull Geol Soc Am 67(5):597–646. https://doi.org/10.1130/0016-7606(1956)67[597:EODSAS]2.0.CO;2

Schumm SA (1963) Sinuosity of Alluvial Rivers on the great plains. Geo Soc Am Bull 74(9):1089–1100. https://doi.org/10.1130/0016-7606(1963)74[1089:SOAROT]2.0.CO;2

Sengupta S (1988) U per gondwana stratigraphy and palaeobotany of Rajmahal Hills, Bihar, India (Palaeontologia Indica), vol 98. Geological Survey of India, Government of India, India

Sharma HS, Padmaja G (1982) Quantitative fluvial characteristics of streams of the Mej basin (Rajasthan). In: Sharma HS (ed) Perspectives in geomorphology: quantitative fluvial geomorphology, vol 2. Concept Publishing Co, New Delhi, pp 143–190

Shimano Y (1992) Characteristics of the stream network composition drainage basins in the Japanese Islands. Environ Geol Water Sci 20:5–14. https://doi.org/10.1007/BF01736105

Shrivastava RN, Shah SC (1966) Ginko (Ginkgoites) digitata, Brong. from the Rajmahal hills, Santhal Parganas (Bihar). Geological Survey of India, Government of India

Shulits S (1968) Quantitative formulation of stream and watershed morphology. Bull Intern Assoc Sci Hydrol 3:201–207

Singh S, Dubey A (1994) Geoenvironmental planning of watersheds in India. Chugh Publications, Allahabad

Singh S, Singh MC (1997) Morphometric analysis of kanhar river Basin. Nat Geograp J India 43(1):31–43

Smith KG (1950) Standards for grading textures of erosional topography. Am J Sci 248:655–668

Snow RS, Slingerland RL (1987) Mathematical modeling of graded river profiles. J Geol 15–33. https://doi.org/10.1086/629104

Sojka M, Kozłowski M, Kęsicka B, Wróżyński R, Stasik R, Napierała M, Liberacki D et al. (2020) The effect of climate change on controlled drainage effectiveness in the context of groundwater dynamics, surface, and drainage outflows. Central-western poland case study. Agronomy 10(5):1–21. https://doi.org/10.3390/agronomy10050625

Soukup M (2006) Drainage systems and their water management function with regard to probable climatic and hydrological changes. Soil Water Res 1(1):32–38

Sreedevi PD, Subrahmanyam K, Ahmed S (2004) The significance of morphometric analysis for obtaining groundwater potential zones in a structurally controlled terrain. Environ Geol 47(3):412–420. https://doi.org/10.1007/s00254-004-1166-1

Strahler AN (1952) Dynamic basis of geomorphology. Geo Soc Am Bull 63:923–938

Strahler AN (1956) Quantitative slope analysis. Geol Soc Am Bull 67:571–596

Strahler AN (1957) Quantitative analysis of watershed geomorphology. Transac Am Geophys Union 8 (6):913–920

Strahler AN (1964) Quantitative geomorphology of drainage basin and channel networks. In: Chow VT (ed) Handbook of applied hydrology. McGraw Hill Book Company, New York, pp 4–39–4–76

Sujatha ER, Selvakumar R, Rajasimman UA, Victor RG (2015) Morphometric analysis of subwatershed in parts of Western Ghats, South India using ASTER DEM. Geomat Nat Haz Risk 1–16. https://doi.org/10.1080/19475705.2013.845114

Thomas J, Joseph S, Thrivikramaji KP (2010) Morphometric aspects of a small tropical mountain river system, the southern Western Ghats, India. Intern J Dig Earth 3(2):135–156. https://doi.org/10.1080/17538940903464370

Thornbury WD (1969) Principles of geomorphology, 2nd edn. New Age International Pub

Tucker GE, Slingerland R (1997) Drainage basin responses to climate change. Water Resour Res 33 (8):2031–2047

Twidale CR (2004) River patterns and their meaning. Earth Sci Rev 67(3–4):159–218. https://doi.org/10.1016/j.earscirev.2004.03.001

Vandana M (2013) Morphometric analysis and watershed prioritisation: a case study of Kabani river basin, Wayanad district, Kerala, India. Indian J Geo-Marine Sci 42(2):211–222

Walker F, Allen MB (2012) Offset rivers, drainage spacing and the record of strike-slip faulting: The Kuh Banan Fault, Iran. Tectonophysics 530–531:251–263. https://doi.org/10.1016/j.tecto.2012.01.001

Wentworth CK (1930) A simplified method of determining the average slope of land surfaces. Am J Sci Series (5th Ser) 20(117):184–194

Wikipedia (2017, November 23) Wikipedia, the free encyclopedia. Retrieved December 20, 2017, from en.wikipedia.org. https://en.wikipedia.org/wiki/Altitudinal_zonation

Yadav KP (1988) Application of morphometry in geomorphic analysis of the Brahmani basin. HYPERLINK http://hdl.handle.net/10603/69226

Zaidi FK (2011) Drainage basin morphometry for identifying zones for artificial recharge: a case study from the Galas river basin. J Geol Soc India 77:160–166

Zelilidis A (2000) Drainage evolution in a rifted basin, Corinth graben, Greece. Geomorphology 35(1–2):69–85. https://doi.org/10.1016/S0169-555X(00)00023-4

Zovoili E, Konstantinidi E, Koukouvelas IK (2004) Tectonic geomorphology of escarpments: the cases of kompotades and nea anchialos faults. Bull Geol Soc Greece 36(4):1716–1725. doi:https://doi.org/10.12681/bgsg.16579

11 Analyzing Morphometric Attributes of Kopai River Basin of West Bengal, India, Using Geospatial Technology

Biraj Kanti Mondal, Indrasish Mukherjee, and Sanchita Saha

Abstract

Drainage basin's morphometry refers to the measurement of different geomorphic and hydrologic attributes of a river basin. The geospatial technology has an efficient contribution in geographical and geological studies as it includes Remote Sensing (RS) and Geographical Information System (GIS); therefore, it has been adopted in the present study for the identification of different geomorphic attributes and their existence in Kopai river basin of West Bengal, India. The inter-relation between different morphometric parameters within each major morphometric attribute along with various dominant controlling factors of landform development was taken into consideration for the geospatial analysis and find out the stage of landform evolution of Kopai river basin. The various morphometric parameters have been correlated with each other and morphometric characteristics have been analyzed by explaining the relief, slope and drainage attribute and geometry of basin to understand their underlying relationship and control over the basin hydro-geomorphology. To synthesize the upshot of the drainage basin, the knowledge of geoinformatics was applied by using spatial data obtained from ASTER GDEM. These data have been analyzed through Arc Map (10.3), Ms Excel and IBM SPSS (23) Statistics software and the application of geospatial technology was carried out minutely to explore the spatial and geomorphic features of Kopai river basin. It is evident that such kind of morphometric analysis is very much effective in determining the landform features and the mechanism behind its development, hence the morphometric parameters have been discussed with respect to linear, relief and aerial aspects in the present study. The current study revealed that Kopai river basin has fifth order drainage networks (according to Strahler's method) with the dendritic drainage pattern. Moreover, the obtained values of different morphometric parameters indicate that the basin produces direct runoff from west to east and it is elongated in shape which eventually controls basin's geomorphic and hydrological response.

Keywords

Morphometric attribute · Kopai river basin · Hydro-geomorphology · Geospatial · Geoinformatics · DEM · Statistical techniques

B. K. Mondal (✉) · I. Mukherjee · S. Saha
Department of Geography, Netaji Subhas Open University, Kolkata, India

P. K. Shit et al. (eds.), *Drainage Basin Dynamics*, Geography of the Physical Environment,
https://doi.org/10.1007/978-3-030-79634-1_11

11.1 Introduction

The quantification and numerical analysis of the configuration of the earth's surface; the shape and dimensions of its landforms are termed as morphometry (Clarke 1973). The current study considersa drainage basinas it is the basic hydrological unit (Strahler 1952). River basin contains a distinctive morphological characteristic along with drainage pattern and geomorphology (Doornkamp and CuChlaine 1971; Strahler 1952; Singh and Singh 1997; Nag 1998; Grohmann 2004; Grohmann et al. 2007; Pareta and Pareta 2011; Magesh and Chandrasekar 2012; Magesh et al. 2013; Withanage et al. 2014; Sarita 2015; Vikhe and Patil 2016). The morphometry of any river basin is related to its hydrological and geomorphic processes (Kessali 1941; Abrahams 1984; Frissel et al. 1986; Sen 1993; Morisawa 1985; Agrawal 1998; Nagand Chakraborty 2003; Esper 2008; Singh and Awasthi 2011; Mohd and Bhat 2013); therefore the responses of runoff, floods and draughts, river sedimentation, flowing pattern and its changing nature have been taken to understand the dimensions of the landforms (Grade 2005; Leopold and Maddock 1953; Gardiner 1995; Sharma et al. 2014). The various morphometric attributes and parameters are often acting as the chief and a dominant controlling factor of landform development and it is considered to evaluate them for the prediction of the flood occurrences. Therefore, the current effort is designed to inspect the geospatial analysis of Kopai river basin with the application of different geomorphic attributes. Moreover, it also investigates to find out the dominant controlling factor for landform development of Kopai river basin.

The strength of the current study lies in its geomorphic analysis which has been designed through understanding the relief, slope, drainage attribute and basin geometry, whereas the morphometry was investigated through relief attribute: absolute relief, relative relief, relief variability index dissection index, etc.; slope attribute: average slope, slope variability index, etc.; drainage attribute: stream order, drainage density, etc. All these attributes and parameters were performed with the help of Arc GIS software and establishing the relationship between diverse morphometric parameters with the help of IBM SPSS Statistics (23) and MS Excel. Thus, the morphometric analysis of Kopai river basin assists to understand the geo-hydrologic characteristics and flood potentiality based on morphometric features. Henceforth, the significance of such studies inculcates in the truth that it will provide database and knowledge about morphometric and hydrological properties which can be used in strategic planning, decision making and definitely it facilitates future management of Kopai river basin.

The prime objectives of the current study are (i) to quantify the morphometry which includes relief, slope, drainage aspect, basin geometry and (ii) to establish the nature of relationships between different morphometric attributes and parameters.

11.2 Study Area

The current study has been designed for Kopai river basin, which mostly belongs to West Bengal and a little part of Jharkhand state of India. Kopai, as the main tributary of Mayurakshi river, rises in the Jamtara district of Jharkhand and flows toward south-east and joins with Bakreswar river in Labpur block of Birbhum district then flows toward east and joins with Mayurakshi river in Talgram village of Murshidabad district of West Bengal. Kopai river basin extends from 23° 37′ 55″ N to 23° 56′ 45″ N latitude and 87° 16′ 53″ E to 88° 4′ 42″ E longitude with 1714.099 sq. km. area.

11.3 Materials and Methods

The current effort has been completed by using secondary data; mostly spatial data obtained from ***ASTER GDEM*** (Global Digital Elevation Model). Delineation of Kopai river basin and preparation of drainage map is based on ASTER GDEM with the help of Arc GIS 10.3 software. Various morphometric parameters namely relative relief, average slope, dissection index, relief variability index, slope variability index, mean drainage density, mean ruggedness index have been calculated by using ***Inverse Distance Weighted (IDW)*** method for multivariate interpolation through dividing the study area into some rectangular grids having 29.45 sq. km. area each through grids which have been considered for making maps. Actual relief, actual slope, actual drainage density, actual ruggedness index have been calculated by direct analysis of elevation and stream raster data in Arc GIS (10.3) software and the results were represented by maps. To analyze the pattern of relationship among actual relief, actual slope, actual drainage density and actual ruggedness index some pixels have been selected through simple random sampling method in Arc GIS software. Relationship of different morphometric parameters was represented in MS Excel through ***Pearson's correlations*** and dominant controlling factors for landform development and complex relationship among all of them calculated in IBM ***SPSS Statistics*** (23) through ***Principal Component Analysis*** (PCA). At the final stage, some graphical representations have been carried out for better understanding and some primary observations have also been prepared to comprehend the role of Kopai river in the surrounding environment, human life and livelihood. The attributes, parameters and techniques have been used for geomorphic analysis (Table 11.1).

11.4 Result and Discussion

11.4.1 Quantitative Study of Different Geomorphic Attributes

11.4.1.1 Relief Attributes

Relief attributes of a drainage basin has three-dimensional features involving area, volume and vertical dimension of landform, therefore different morphometric methods have been applied to analyze for understanding Kopai river basin relief features.

Absolute Relief

Absolute relief has been calculated by classifying the Digital Elevation Model (DEM) of Kopai river basin (Fig. 11.1). The highest relief is identified in the north western part of the basin and it gradually decreases toward the north east corner of the basin area. The north western part of the basin belongs to the plateau region whereas the north eastern part is under the Ganges delta.

Relative Relief

Relative relief is the difference between the maximum and minimum elevation (Schumm 1956). It is the essential parameter to determine the denudation features of the drainage basin (Hadley and Schumm 1961). A small division of north western part of the basin shows high relative relief, whereas the eastern part confirms the lowest relative relief (Fig. 11.2). Thus, it is evident that the variation of denudational characteristics with the increasing relative relief and higher stream gradient decreases the time of concentration of runoff.

Average Relief

Average relief determines the grid-specific mean relief and therefore the average relief map of Kopai river basin has been prepared. The average relief is gradually decreasing toward east probably due to different tectonic activity, dissimilar formation and diverse erosional and depositional process of the river basin (Fig. 11.3).

Relief Variability Index

It refers to grid-specific variability of relief and the relief map shows that Kopai river basin has less relief variability. Some parts of middle and western section of the map show comparatively high relief variability due to erosional process of the river especially in the plateau region.

Dissection Index

It is the comparative ratio of relative relief to the absolute relief. It is an important morphometric parameter for the understanding of magnitude of the dissection of terrain. Dissection index is also

Table 11.1 Attributes, parameters and techniques used for geomorphic analysis

Geomorphic attributes	Measurement parameters	Formula	References
Relief	Absolute relief	Height above mean sea level	
	Relative relief	Maximum relief–minimum relief	Strahler (1952)
	Average relief	Sum of absolute relief of each pixels/no. of observations or pixels	
	Dissection index	Relative relief/maximum relief	
	Relief variability index	Standard deviation of absolute relief × 100/average relief	
Slope	Actual slope	Slope of a particular pixel area	Wentworth (1930)
	Average slope	Sum of actual slope of a pixel area/no. of pixels	
	Slope variability index	Standard deviation of actual slope of each pixels × 100/average slope	
	Aspect	Direction of slope to geographical north	
Drainage	Stream order	Hierarchical rank	Strahler (1964)
	Stream number	Total number of streams in particular stream order	
	Stream length	Total length of streams in particular stream order	
	Bifurcation ratio	Number of streams of one order / number of streams next higher order	Schumm (1956)
	Actual Drainage density	Length of a stream/sum of the area of pixels through which the stream flows	
	Mean drainage density	Sum of actual drainage density of a particular grid/number of pixels within a particular grid	
	Actual ruggedness index	Ruggedness index of a particular pixel	
	Mean ruggedness index	Sum of actual ruggedness index of a particular grid/number of pixels within a particular grid	
	Drainage texture	Total number of stream/basin perimeter	Horton (1945)
Basin geometry	Circulatory ratio	4π × basin area/square of basin perimeter	Miller (1953)
	Elongation ratio	Diameter of a circle having area as same as basin/maximum length of basin	Schumm (1956)
	Form factor	Basin area/square of maximum basin length	Horton (1932)

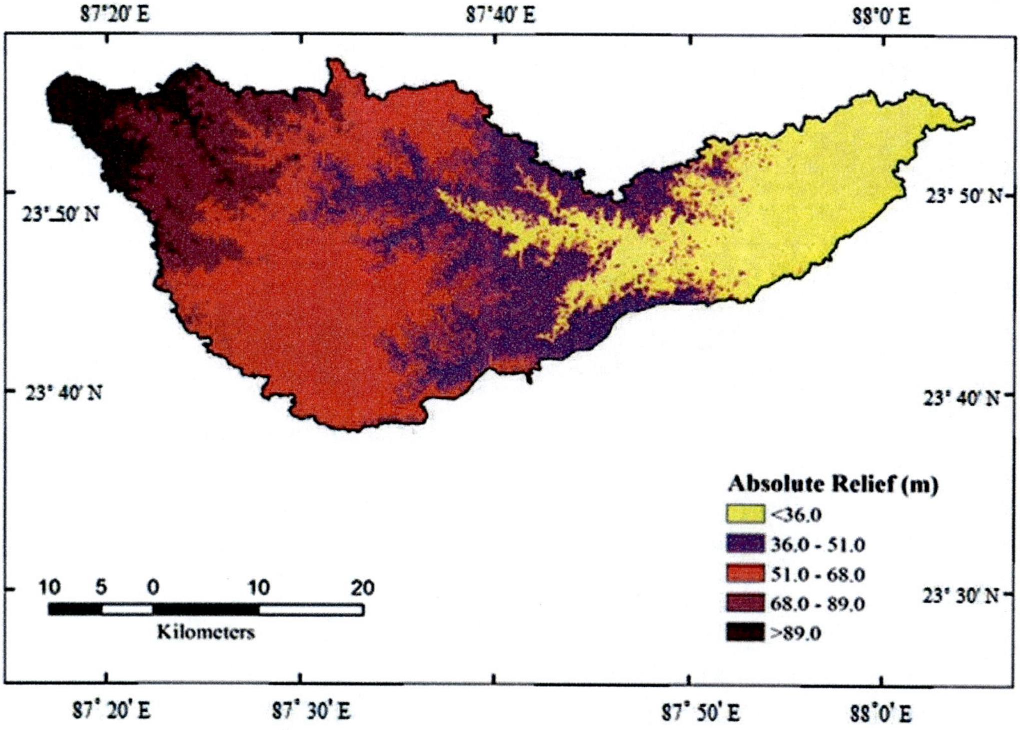

Fig. 11.1 Absolute relief map of Kopai river basin

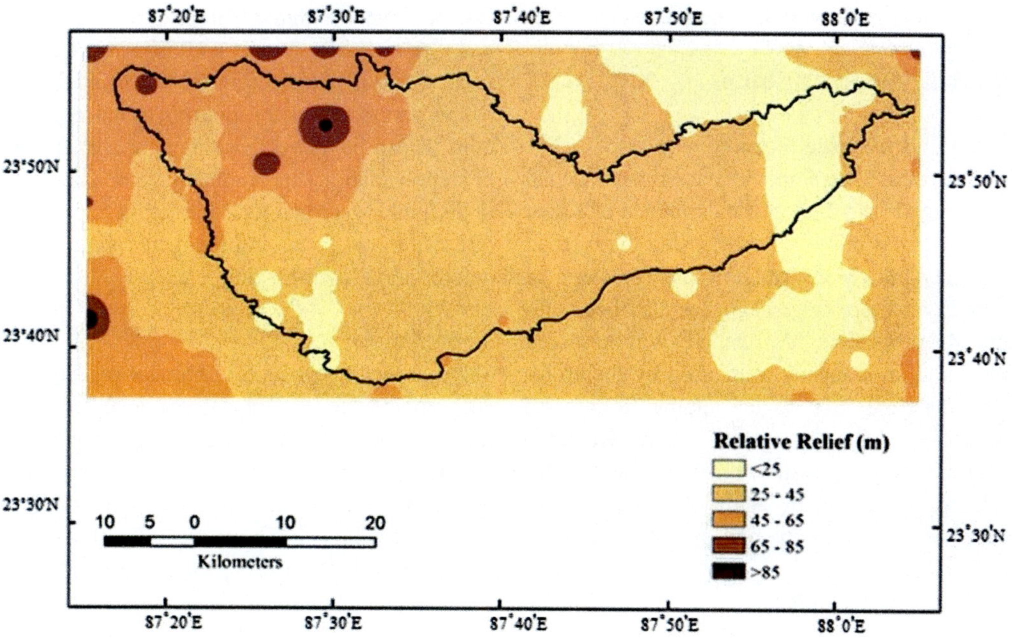

Fig. 11.2 Relative relief map of Kopai river basin

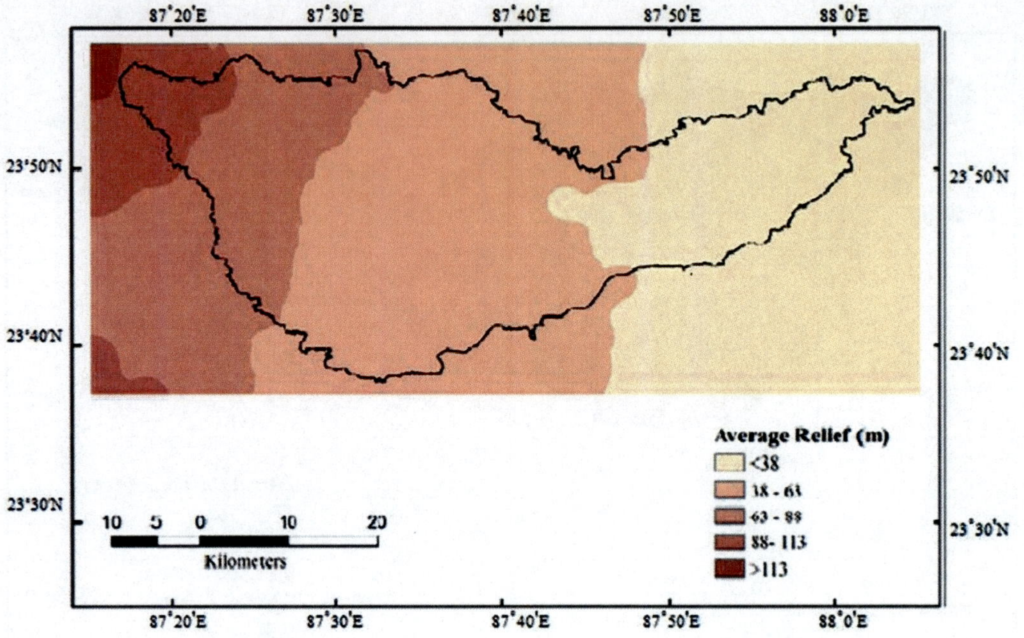

Fig. 11.3 Average relief map of Kopai river basin

useful to explain the stages of the cycle of erosion. The calculated value of dissection index of Kopai river basin shows that most part of the basin largely belongs in the matured stage.

11.4.1.2 Slope Attributes

Actual and Average Slope

To determine the slope characteristics, actual slope (Fig. 11.4) of the particular pixels of Kopai river basin has been calculated using the method of Wentworth. Moreover, the average slope of Kopai river basin (Fig. 11.5) has also been calculated by Wentworth's method. It is observed that the basin is mostly influenced by the erodibility therefore it is large slopping from north western to south eastern direction. Geologic upliftment, rock structure, erosional rate of fluvial process and valley deposition is the major causes of slope variation of Kopai river basin.

Slope Variability Index

Slope variability index (Fig. 11.6) shows the variability and stability of slope in each grid of the Kopai basin. As the basin is belonged to lower plateau and plain land region, consequently the region has less slope variability. Comparatively higher slope variability is observed in the eastern and middle part and lower value is perceived in the west middle part of the region. From the field visit and observations, it can be stated that this particular pattern of slope variability has probably come about due to dissimilar geologic rock formation, unlike rate of erosion and some anthropogenic gesture like deforestation, over-grazing, etc.

Slope Aspect

Slope aspect refers to the direction of slope to geographical north. This aspect analysis is an important parameter as it can easily influence the direction of river or stream and thereafter, it is taken into consideration to identify Kopai river basin morphometry. The aspect map (Fig. 11.7) affirms that the northern part of the basin confirms south direction of slope which compelled most of the streams of this region to flow toward south.

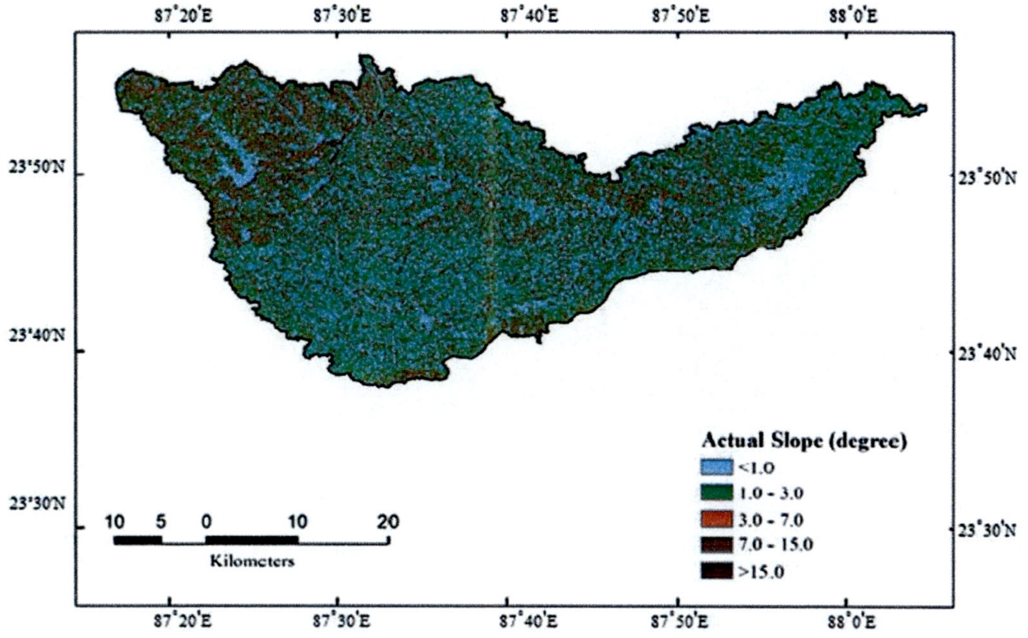

Fig. 11.4 Actual slope map of Kopai river basin

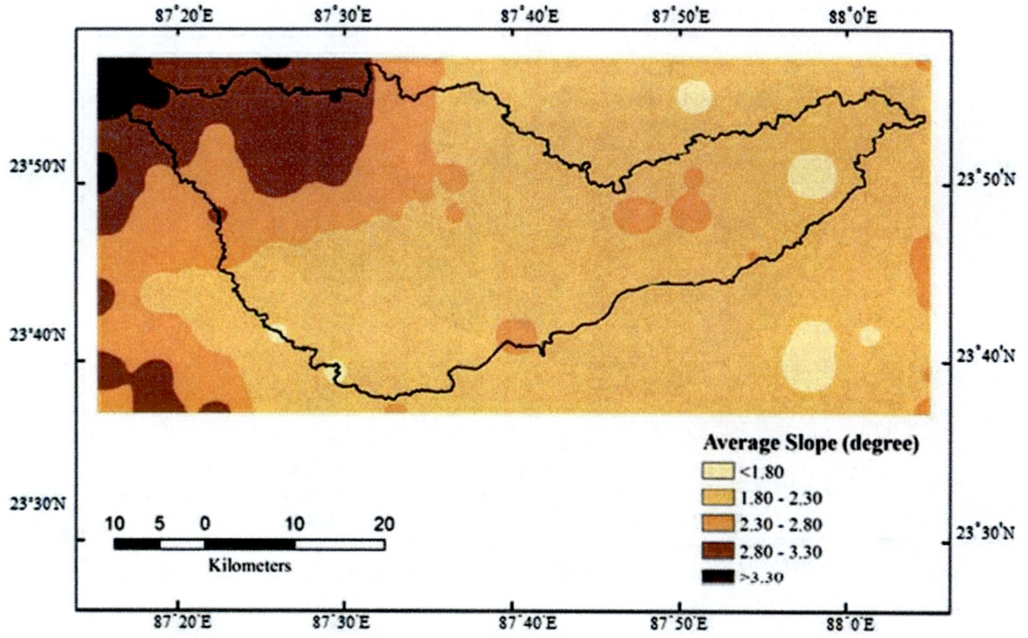

Fig. 11.5 Average slope map of Kopai river basin

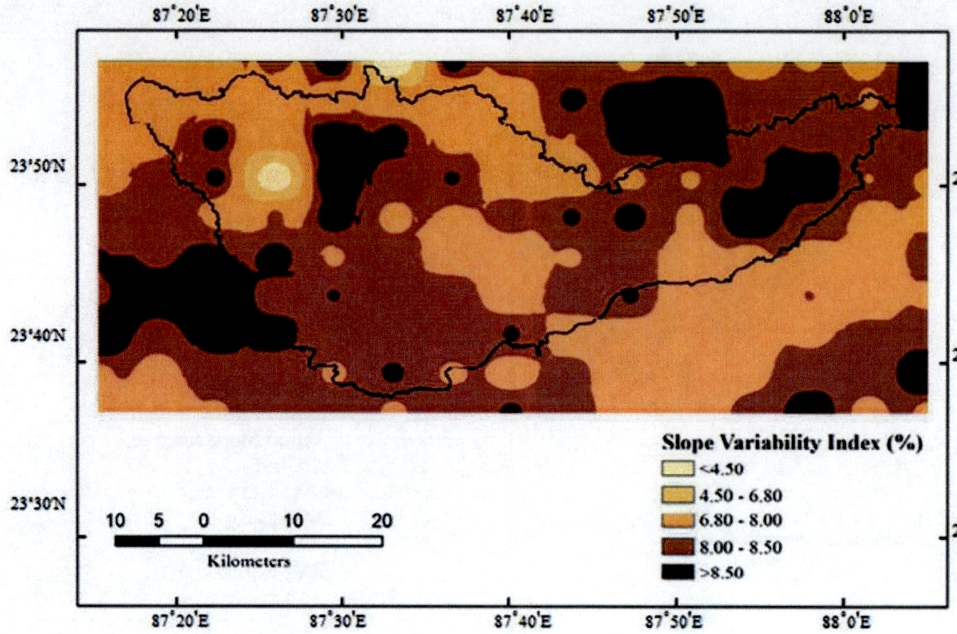

Fig. 11.6 Slope variability index map of Kopai river basin

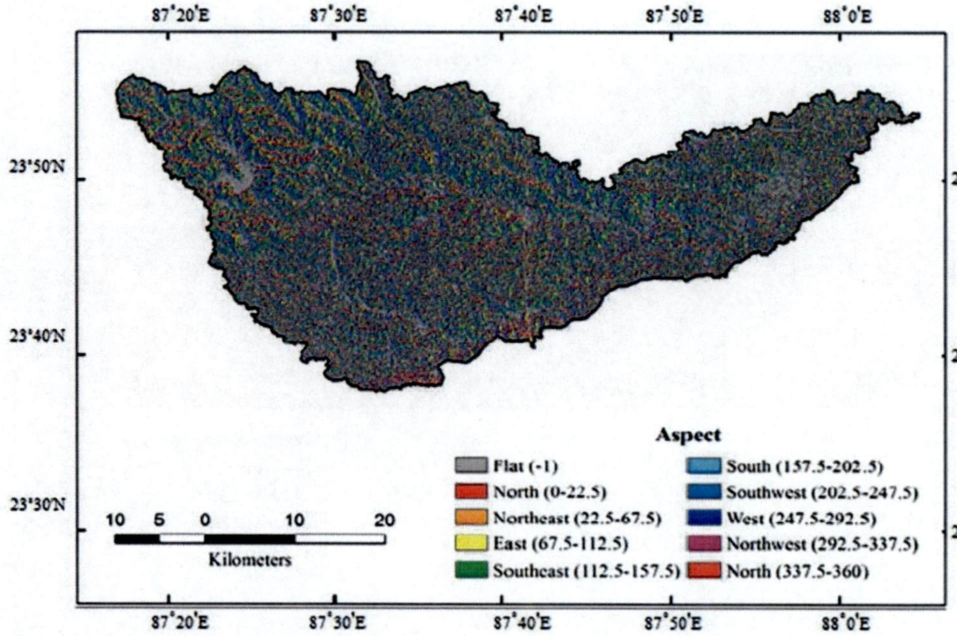

Fig. 11.7 Aspect map of Kopai river basin

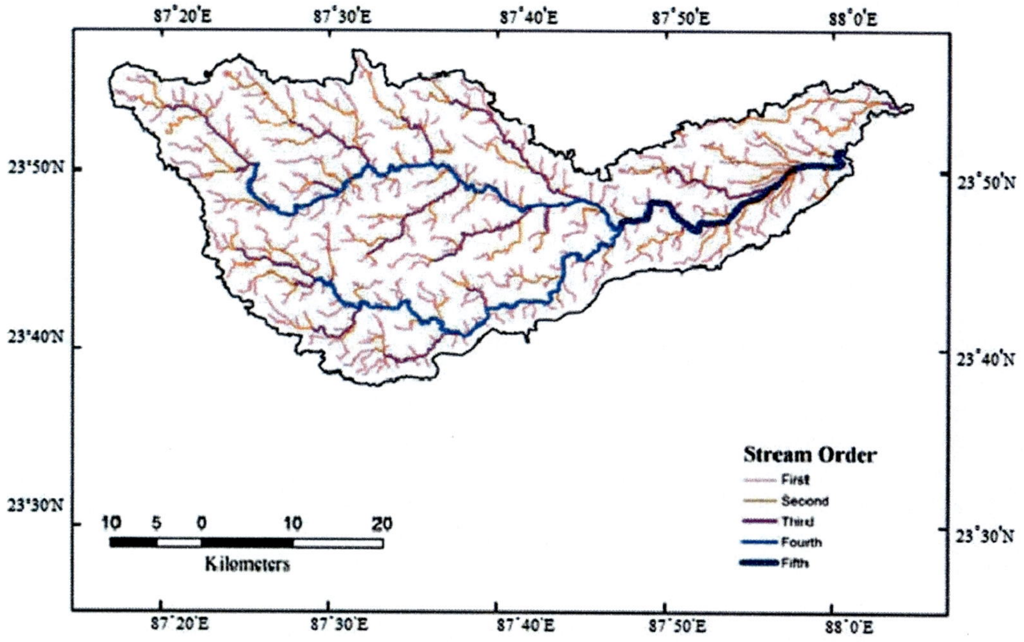

Fig. 11.8 Stream ordermapof Kopai river basin

11.4.1.3 Drainage Attributes

Stream Order

Stream order refers to the determination of the hierarchical position of streams within a drainage basin. To find out the stream order, Strahler's method is applied in the present study and it is established that Kopai river basin has fifth order stream networks (Fig. 11.8).

Stream Number

The total number of streams of each order in the basin is called stream number and as per the calculation, Kopai river basin has a total of 938 stream segments of which 472 are 1st order, 225 are 2nd order, 119 are 3rd order, 95 are 4th order and 27 are 5th order streams. This huge figure of different streams has a tremendous influence in shaping the morphometry and geo-hydrological personality of the basin area.

Stream Length

The total stream length of Kopai river basin has been calculated and categorized according to the stream orders. Total stream length and mean stream length of Kopai river basin are 1423.0812 km and 1.5171 km, respectively. In particular, first order stream has larger length with 758.452 km, followed by 349.473 km 2nd order, 165.861 km 3rd order,113.850 km 4th order and 35.4452 km 5th order stream length. These small streams exert a critical influence on downstream portions of the drainage network of the study basin.

Bifurcation Ratio

It is the ratio between the number of streams of one order and the number of streams of the next higher order (Horton 1945; Strahler 1952). Thus the bifurcation ratio is an index of relief and dissection and is often considered as a measure of the level of the upshot of drainage system, which has a strong control above the runoff (Chorley 1969; Chorley et al. 1957; Mesa 2006). Therefore, the bifurcation ratio of Kopai river basin has been calculated and the value ranges from 1.25 to 3.51 with mean bifurcation ratio 2.19. Moreover, the highest bifurcation ratio is found in between 4th and 5th order stream and lowest bifurcation ratio is found in between 3rd and 4th order stream. The study also reveals that Kopai river basin is principally near elongated in shape which indicates moderate probability of flood considering its circular shape in the upper basin and lower basin area is elongated shape.

There is a good connection between stream order with its number and stream order with its length, both of which are studied (Fig. 11.9) here to establish the significance and influence of the morphometry of Kopai river basin. Both the graphs show positive relationship which reflecting a dissimilarity between slope and topography in variant stream orders. Hence, it symbolizes an essential control on water discharge and erosional activities which assist to understand the denudational features of Kopai river basin.

Drainage Density

Drainage density is the ratio of total length of all streams and area of the basin and it has a very significant role on surface runoff, influencing the intensity of torrential floods. Therefore, in the current study, actual drainage density (Fig. 11.10) and grid specific drainage density (mean drainage density) (Fig. 11.11) are taken into consideration to explain the drainage basin. The middle and eastern portion of the basin demonstrates high drainage density and some parts of western and northern portion confirm lowest drainage density values due to physical properties of underlying rock, vegetation envelop and variable relief assets.

Ruggedness Index

Ruggedness index indicates surface roughness and unevenness of a region and expansively high ruggedness indicates elevated relief with high stream density (Melton 1965). Therefore, to analyze the morphometry of the study area, actual ruggedness index (Fig. 11.12) and mean ruggedness index (Fig. 11.13) have been calculated and mapped. The present study basin indicates comparatively lower relief, subsequently it denotes less surface roughness in general. In the mean ruggedness index map, the western part of the basin demonstrates comparatively high ruggedness index, which confirms the comparatively high relief and steep slope while the eastern part of the basin point toward low ruggedness index which indicates low relief, depositional surface and gentle slope.

Drainage Texture

Drainage texture depends on causal lithological structure, elevation, infiltration rate and the capacity of any river basin and therefore it explains the quantity of virtual spacing of a drainage network. According to Smith (1950), drainage texture may be of five categories, like very coarse (<2), course (2–4), moderate (4–6), fine (6–8) and very fine (>8). Here, the intended value of the drainage texture of Kopai river basin is 2.781 which indicates course nature of drainage texture.

11.4.1.4 Basin Geometry

Basin geometry is the study of basin shape and size with the help of some variables like circulatory ratio, elongation ratio and form factor, etc. Basin geometry of a basin controls drainage discharge, surface runoff and determines the probability of flood.

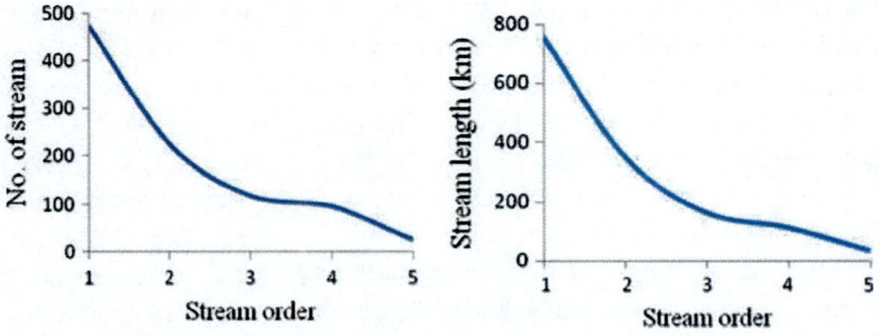

Fig. 11.9 Relationship between stream order and number and stream order and length of Kopai river basin

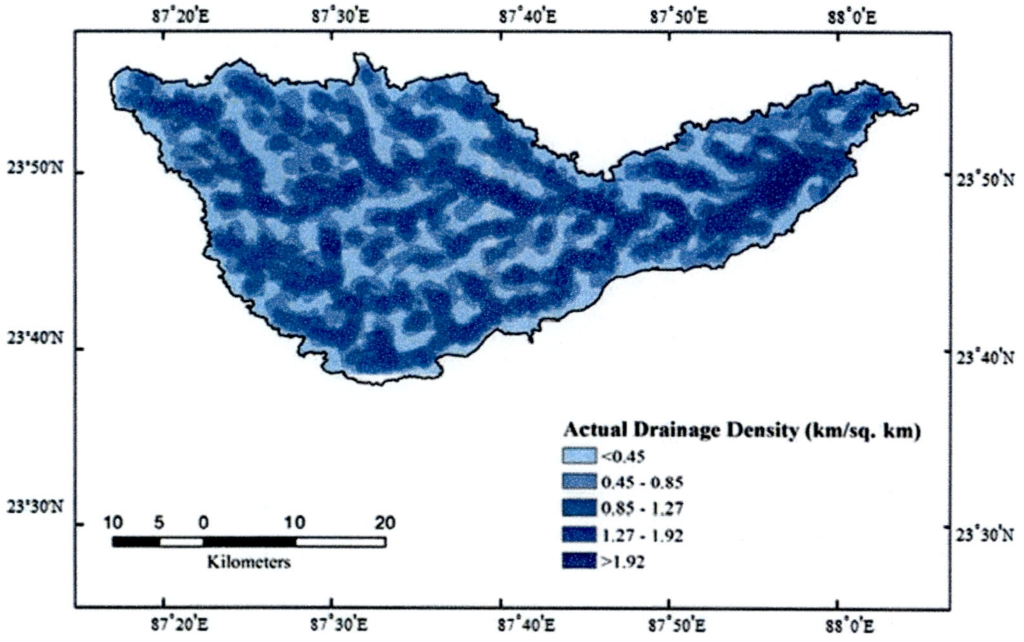

Fig. 11.10 Actual drainage density mapof Kopai river basin

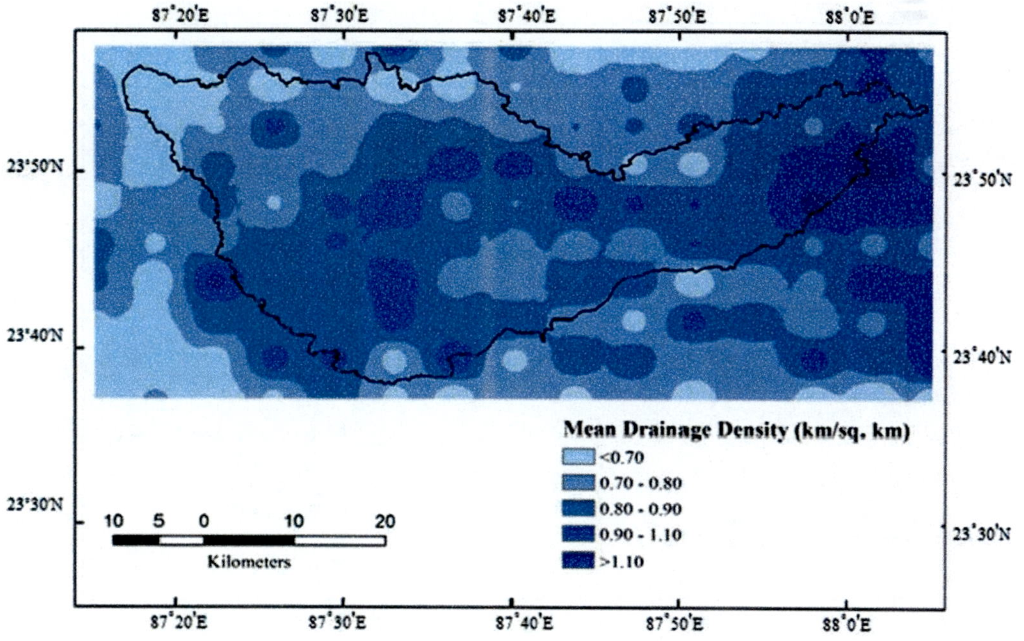

Fig. 11.11 Mean drainage density map of Kopai river basin

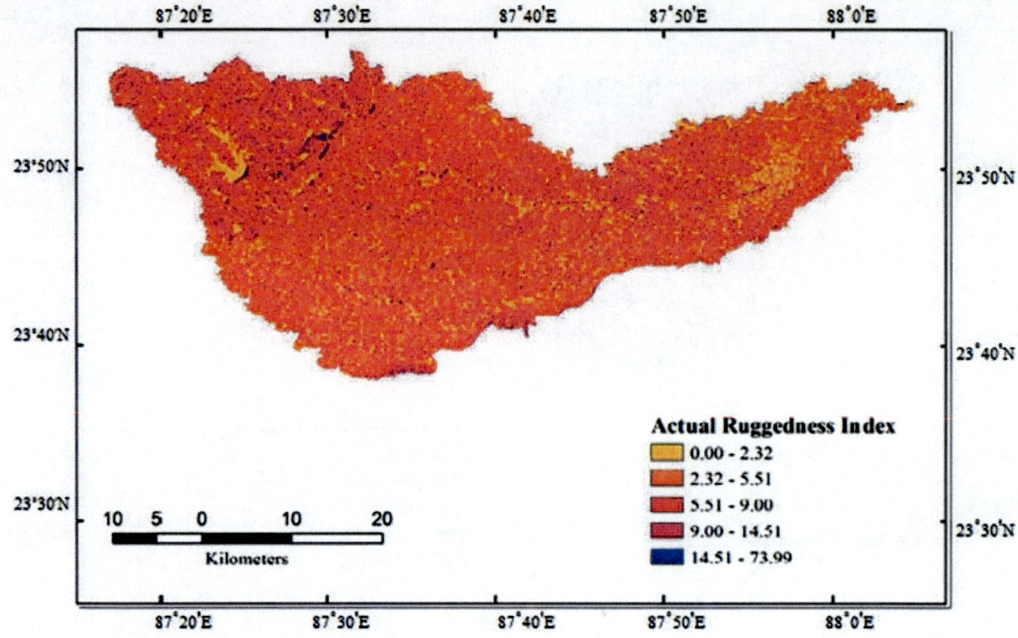

Fig. 11.12 Actual ruggedness index map of Kopai river basin

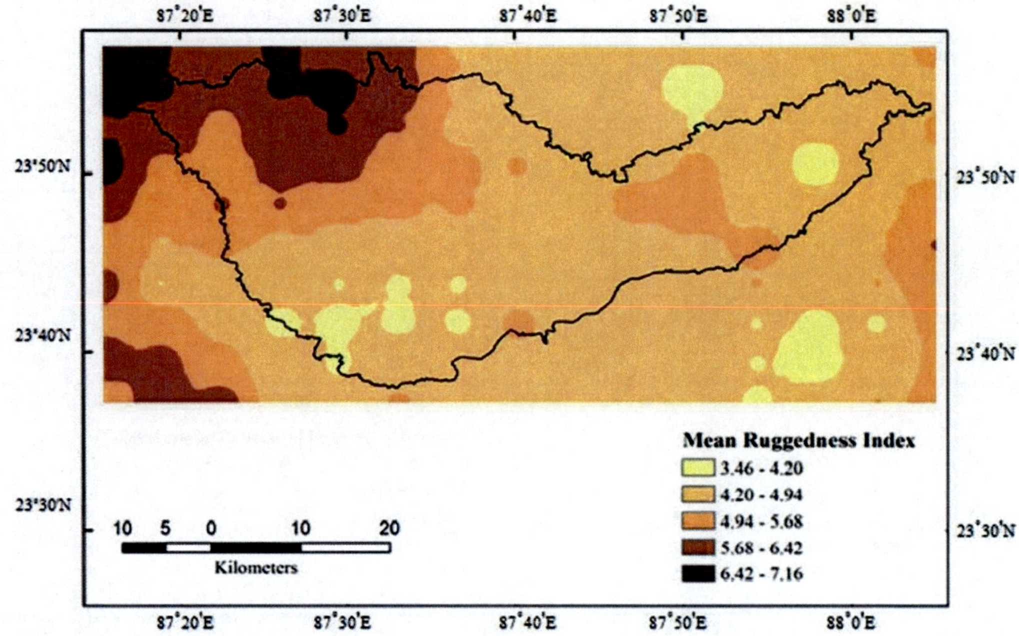

Fig. 11.13 Mean ruggedness index map of Kopai river basin

Circulatory Ratio

It is considered as the ratio of the areas of a circle which has the same boundary as that on the outskirts of the river basin (Miller 1953). Generally, the circulatory value of one indicates the circular shape of the basin and it is related with land use, land cover, elevation properties, slope, length and frequency of the stream of the river basin, etc. The calculated value of circulatory ratio of Kopai river basin is 0.19, which signifies its elongated nature with low discharge of runoff and highly permeable sub-soil conditions.

Elongation Ratio

The ratio between the wideness of the circle and utmost extent of the area of the river basin and is considered as the significant parameter to understand the basin's shape and hydrological characteristics. The deliberated value of the elongation ratio of Kopai river basin is 0.06096, which designates the low relief, moderate slope and moderate tectonic activity.

Form Factor

It is important to analyze the basin's size and shape wherein low form factor indicates elongated and narrow basin and high form factor indicates the broadness and wideness of the basin (Gregory and Walling 1985). Besides, low form factor signifies less amount of rainfall and elongated shape of the basin with low flow for longer duration. The computed value of the form factor of Kopai river basin is 0.2921, which signifies the elongated shape and narrow nature of the basin with less acquisition of rainfall.

11.4.2 Relationship Among Different Geomorphic Attributes

In this section the analysis of inter-relation and intra-relations among three major geomorphic attributes viz. relief, slope and drainage have been analyzed with the help of simple and partial ***correlation*** (r) and ***coefficient of determination*** (r^2). The result of these relations has been categorized differently, like general relationship among relief, slope and drainage parameters separately and tabulated.

11.4.2.1 Intra-relation Among Different Relief, Slope and Drainage Parameters

To understand the pattern of inter-relation, two major types of statistical techniques; simple correlation analysis and partial correlation analysis have been made.

Simple Correlation Analysis

Here Pearson's simple correlation coefficient of determination has been calculated among each morphometric parameter which signifies general nature of inter-relation. The results of these different types of relationships have been explained in Tables 11.2, 11.3 and 11.4.

Partial Correlation Analysis

The inter-relationship among the diverse parameters of relief, slope and drainage have also been analyzed by the partial correlation (Tables 11.5 and 11.6). Partial correlation is a type of correlation in which degree of association between two variables assuming other variables as constant.

The above correlation analysis of Kopai river basin designates that when slope and drainage parameters are constant then average relief and drainage density show strong negative correlation but in normal bivariate correlation it shows exactly the opposite correlation (0.05 approximately). Therefore, it can be stated that in case of relationship between these two variables, it is too much influenced by drainage and slope parameters. Furthermore, same type of announcement can be drawn in case of relative relief and average relief relationship. In such case of the relationship between average slope and slope variability index, partial correlation and normal correlation values are roughly the same which signifies drainage and slope attributes have no significant role on slope parameters.

Table 11.2 General relationship among relief parameters

Relief parameters	*r* value	r^2 value	Remarks
Average relief and relative relief	0.62988	0.3967	Moderately positive relationship r^2 value shows near about 40% relative relief controlled by average relief
Relative relief and relief variability index	0.21214	0.045	Poor positive relationship r^2 value shows insignificant role of relative relief on relief variability index
Average relief and relief variability index	0.4864	0.2366	Moderate positive relationship r^2 value indicates almost 24% relief variability index controlled by average relief
Average relief and dissection index	0.49775	0.2478	Moderate positive relationship r^2 value indicates almost 25% dissection index controlled by average relief
Relative relief and dissection index	0.28815	0.083	Poor positive relationship and near about 08% of dissection index is controlled by relative relief according to r^2 value

Table 11.3 General relationship between slope parameters

Relationship between different slope parameters	*r* value	r^2 value	Remarks
Average slope and slope variability index	0.23103	0.0534	Poor positive correlation and only 05% of relief variability index is controlled by average slope

Table 11.4 General relationship among drainage parameters

Relationship between different drainage parameters	*r* value	r^2 value	Remarks
Stream order and stream number	−0.9259	Not applicable	Very strong negative correlation
Stream order and stream length	−0.9200	Not applicable	Very strong negative correlation

Table 11.5 Partial correlation among relief parameters

Constant variables (slope and drainage attributes)	Variables having partial correlations (relief attributes) → ↓	Average relief	Relative relief	Dissection index	Relief variability index
Average slope, slope variability index, mean drainage density	Average relief	+1.0000	+0.0952	−0.8107	+0.5533
	Relative relief	+0.0952	+1.0000	+0.3873	+0.1705
	Dissection index	−0.8107	+0.3873	+1.0000	−0.4025
	Relief variability index	+0.5533	+0.1705	−0.4025	+1.0000

Table 11.6 Partial correlation between slope parameters

Assumed constant variables (relief and drainage attributes)	Variables having partial correlations (Slope attributes)	Value of partial correlations
Average relief, relative relief, dissection index, relief variability index, mean drainage density	Average slope and Slope variability index	+0.135

Table 11.7 General relationship among different relief, slope and drainage parameters

Relationship between different relief, slope and drainage parameters	r value	r^2 value	Remarks
Relative relief and average slope	0.816092	0.666	Strong positive correlation and near about 67% of average slope is controlled by relative relief
Relative relief and slope variability index	−0.24262	0.0589	Poor negative correlation in which relative relief controls only about 06% of slope variability index
Average relief and drainage density	−0.35702	0.1275	Moderate negative correlation and near about 13% of drainage density is controlled by average relief
Drainage density and relief variability index	−0.23383	0.0547	Poor negative correlation in which drainage density determines only near about 05% of relief variability index
Relative relief and drainage density	−0.24091	0.058	Poor negative correlation in which relative relief controls only about 06% of drainage density
Relief variability index and slope variability index	−0.15383	0.0237	Poor negative correlation and only about 02% of slope variability index is controlled by relief variability index
Drainage density and average slope	−0.3122	0.0975	Moderate negative correlation and near about 10% of drainage density is controlled by average slope
Drainage density and slope variability index	0.05815	0.0034	Poor positive correlation and only less than 01% of slope variability index is controlled by drainage density
Drainage density and dissection index	0.1487	0.0221	Poor positive relationship r^2 value shows no significant role of drainage density on dissection index
Average relief and average slope	0.72889	0.5313	Strong positive relationship and almost 53% average slope value controlled by average relief
Average relief and slope variability index	−0.17149	0.0294	Poor negative relationship and r^2 value indicate no significant role of average relief on slope variability index
Dissection index and slope variability index	−0.02245	0.0005	Insignificant negative correlation and no such controlled in each other
Average slope and dissection index	0.08239	0.0068	Insignificantly poor positive correlation and <01% dissection index value controlled by average slope

Table 11.8 Correlation values obtained through principal component analysis

Controlling variables of Geomorphic properties	Correlations of different components						
	1	2	3	4	5	6	7
Average relief (m)	0.913	−0.235	0.083	0.229	−0.171	−0.076	0.131
Relative relief (m)	0.801	0.507	0.058	0.140	0.149	−0.226	−0.078
Dissection index	−0.228	0.896	0.025	−0.136	0.341	0.040	0.092
Relief variability index (%)	0.525	−0.543	−0.211	0.038	0.614	0.080	−0.005
Average slope (degree)	0.849	0.379	0.129	0.134	−0.163	0.270	−0.041
Slope variability index (%)	−0.344	−0.162	0.863	0.282	0.178	0.008	−0.005
Mean drainage density (km/sq. km)	−0.506	0.156	−0.359	0.768	0.017	0.017	0.001

Table 11.9 Component and explained variance

Variance explained	Components						
	1	2	3	4	5	6	7
Total variance	2.897	1.604	0.946	0.779	0.604	0.138	0.33
Explained variance (%)	41.387	22.909	13.508	11.125	8.626	1.970	0.476
Cumulative explained variance (%)	41.387	64.296	77.803	88.928	97.554	99.524	100.000

11.4.2.2 Inter-relationship of Different Relief, Slope and Drainage Parameters

In order to find out the pattern of inter-relationship, two major statistical techniques; simple correlation analysis and principal component analysis have been used.

Simple Correlation Analysis

Simple correlations have been completed to recognize the general pattern of relationship among different geomorphic variables which have been remarked and tabulated in Table 11.7.

Principal Component Analysis to Identify Major Controlling Variables

Major controlling factor has been identified by using principal component analysis of seven geomorphic variables. Through this method, the entire basin has been divided into seven component and correlations of each component has been determined (Table 11.8) while each component explain certain portions and amount of that portions are illustrated in tabular format (Table 11.9).

The outcome of principal component analysis disclosed the fact that average relief is the most primary controlling factor of morphology for about 41% cases. That means other morphometric factor varies with the change of average relief at most for 41% cases. While dissection index is revealed as the second most key controlling factor which controls the other variables for about 23% cases. Relative relief is the least controlling factor here, which never acts as a primary controlling factor, but it is a significant derivative controlling factor. It shows that relative relief always depends on at least one factor, but its controlling nature is low. The calculated principal component analysis represents complex inter-relations in a simple manner (Fig. 11.14) where the correlations have been graphically represented. The relative relief and average slope have the most strong correlation whereas dissection index and relief variability index have weakest correlation. That means higher range of maximum and minimum relief leads to steep slope in most of the cases.

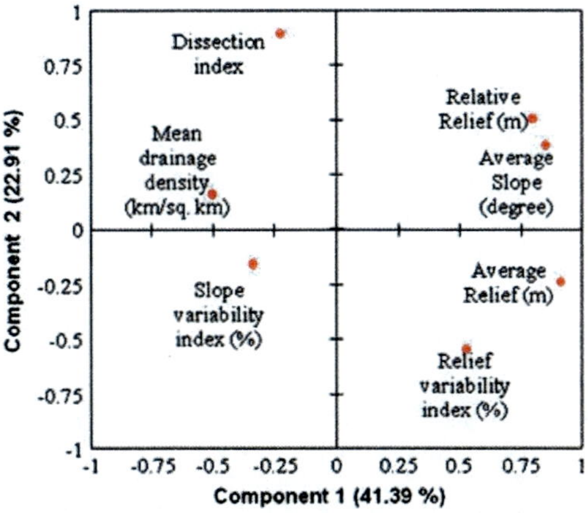

Fig. 11.14 Graphical representation of principal component analysis

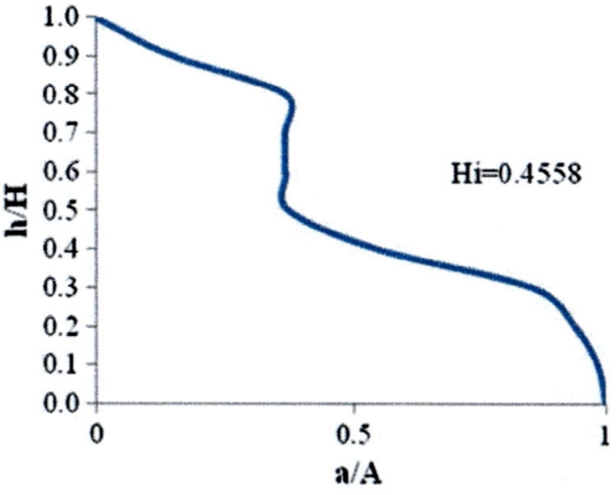

Fig. 11.15 Hypsometric curve ofKopai river basin

Additionally, the dissection index has insignificant impact on relief variability since dissection index depends on relative relief, but relief variability is controlled by average relief.

11.4.3 Identification of Stage of Landform Evolution

The stages of landform can be distinguished through the hypsometric curve and it can be designated by finding the relationship of the distribution of ground surface area with respect to its elevation. The deliberated shape of the curve indicates the sequential changes over the slope of the basin (Giamboni et al. 2005) and it often facilitates to indulge the stages of the cycle of erosion of the river basin (Strahler 1952). Furthermore, the shape of the curve also assists that whether the landform is gone through diverse tectonic activity or not, along with the disruption and or the period of the cycle of erosion. The smoothness of the curve indicates an entire cycle of erosion do not disturb by any activity, but an uneven shape of the curve makes clear of diverse annoyance of the cycle of erosion. Therefore, to identify the stages of landform evolution of Kopai River Basin, hypsometric curve has been constructed and hypsometric integral (Hi) has also been calculated.

The curve is acquired by plotting cumulative relative area in the 'X axis' and cumulative relative height in the 'Y axis'. Here, the relative area is intended by calculating the ratio of the area in picky contours to the sum of the area of the entire basin and relative height is calculated as the ratio of the height of the given contour to the maximum altitude of the basin. The shape of the hypsometric curve of Kopai river basin (Fig. 11.15) reveals that the upper portion is convex type, the middle portion is concave type and the lower portion is again convex. The overall shape of hypsometric curve explicates that the landform of the basin area is under the near mature stage of cycle of erosion but the steep slope of middle portion of hypsometric curve illuminates of the trouble due to some tectonic activities.

Hypsometric integral (Hi) is calculated of Kopai river basin from the hypsometric curve to illustrate the shape of the profile of the river basin. Therefore, the shape of the hypsometric curve and hypsometric integral value provide valuable information regarding identification of the tectonic, climatic and lithological controlling factors and announce the erosional stage of the basin (Rhea 1993; Moglen and Bras 1995; Willgoose and Hancock 1998; Huang and Niemann 2006). The calculated value of the hypsometric integral of Kopai river basin is 0.4558 which indicates the landform is under the mature equilibrium stage.

11.5 Conclusion

The current effort of morphometric analysis of Kopai river basin is intended with the measurement of relief, slope and drainage parameters byusing geospatial techniques and it reveals that the remotely sensed datasets merged with the GIS based tools to execute the morphometric attributes and their influence on landform, especially at the river basin level. The present study has been carried out through the measurement of linear, relief and aerial aspects with diverse sets of parameters with accurate mapping. It is found that Kopai river basin has a fifth order of stream and the entire drainage network is dendritic in nature. Furthermore, the values of form factor, elongated ratio and circularity ratio reveal that the basin is elongated in shape and has a predominance of extremely permeable homogeneous geologic materials. The bifurcation ratio of the river basin indicates the presence of moderate drainage density; highly permeable sub-soil and coarse drainage texture which indicating moderate probability of flood occurrences, soil erosion and debris flow. Furthermore, the shape of hypsometric curve and hypsometric integral value of Kopai river basin indicates that the streams are in mature equilibrium stage. It is also articulated from the above analysis that relief parameters like average relief, dissection index are diversely correlated with other morphometric attributes and thus carry the principal role in making variant morphometric surface. Thenceforth, the diverse geomorphic and hydrologic attributes are minutely observed in the basin morphometry and the inter-relation between different morphometric parameters within each major morphometric attribute of Kopai river basin has been derived and mapped using the geospatial technology and it is really much useful for the management of flood and future planning of the region. The appropriate outcomes from such analysis are found very significant to determine the dominant controlling factors in the evolution and the development of landform of Kopai river basin assisting its maintenance and management. Such kind of morphometric analysis may further be extended to watershed management, flood hazard management, soil erosion, groundwater studies and natural resource management at micro level by using the geospatial techniques as it provides accurate and existing information like in the case of Kopai river basin.

References

Abrahams AD (1984) Channel networks: a geomorphological perspective. Water Resour 20:161–168

Agrawal CS (1998) Study of drainage pattern through aerial data in Navgarh area of Varanasi district, U. P. J Indian Soc Remote Sens 26:169–175

Chorley RJ, Donald Malm EG, Pogorzelski HA (1957) A new standard for estimating drainage basin shape. Am J Sci 255(2):138–141. https://doi.org/10.2475/ajs.255.2.138

Chorley RJ (1969) Introduction to fluvial processes. Methuen and Co Ltd., Routledge, London, Bungay, UK

Clarke JI (1973) Morphometry from maps. Essays in geomorphology. Elsevier Publication Corporation, India, Delhi

Doornkamp JC, CuChlaine AMK (1971) Numerical analysis in geomorphology: an introduction. Published by Edward Arnold, London

Esper AMY (2008) Morphometric analysis of Colanguil River Basin and Flash Flood Hazard, San Juan. Argent Environ Geol 55:107–111. https://doi.org/10.1007/s00254-007-0969-2

Frissel CA, Liss WJ, Warren CE, Hurley MD (1986) A hierarchical framework for stream habitat classification-viewing streams in a watershed context. Environ Manag 10:199–214

Garde RJ (2005) River morphology. New Age International (Pvt) Ltd. Publishers, New Delhi

Gardiner V (1995) Channel networks: progress in the study of spatial and temporal variations of drainage density. In: Gornell A, Petts GE (eds) Change in river channels. Wiley, New York, pp 65–85

Giamboni M, Carretier S, Niviere B, Winter T (2005) Do river profiles record along stream variations of low uplift rate? J Geophys Res 111(F02024). https://doi.org/10.1029/2005JF000419

Gregory KJ, Walling DE (1985) Drainage basin form and process: a geomorphological approach. Published by Edward Arnold, London. https://doi.org/10.1080/02626666809493583

Grohmann CH (2004) Morphometric analysis in geographic information systems: applications of free software GRASS and R. Comput Geosci 30:1055–1067. https://doi.org/10.1016/j.cageo.2004.08.002

Grohmann CH, Riccomini C, Alves FM (2007) SRTM – based morphotectonic analysis of the Pocos de caldas alkaline massif Southeastern Brazil. Comput Geosci 33:10–19. https://doi.org/10.1016/j.cageo.2006.05.002

Hadley R, Schumm S (1961) Sediment sources and drainage basin characteristics in Upper Cheyenne River Basin. US Geological Survey Water-Supply Paper 1531-B, Washington DC, 198

Horton RE (1932) Drainage basin characteristics. Trans Am Geophys Union 13:350–361

Horton RE (1945) Erosional development of stream and their drainage basin—hydrogeological approach to quantitative morphology. Bull Geolog Soc Am 56 (3):275–370

Huang XL, Niemann JO (2006) An evaluation of the geomorphically effective event for fluvial processes over long periods. J Geophys Res 111:1–17

Kessali JE (1941) Concept of the graded river. J Geol 49:561–588

Leopold LB, Maddock T (1953) The hydraulic geometry of stream channels and some physiographic implications. USGS Prof Paper 252:1–57

Magesh NS, Jitheshlal KV, Chandrasekar N, Jini KV (2013) Geographical information system based morphometric analysis of Bharathapuzha River Basin, Kerala, India. Appl Water Sci 3:467–477

Magesh NS, Chandrasekar N (2012) GIS model-based morphometric evaluation of Tamiraparani subbasin, Tirunelveli district, Tamil Nadu India. Arab J Geosci 7 (1):131–141

Melton MA (1965) The geomorphic and paleoclimatic significance of alluvial deposits in Southern Arizona. J Geol 73:1–38. https://doi.org/10.1086/627044

Mesa LM (2006) Morphometric Analysis of a Subtropical Andean basin (Tucuman, Argentina). Environ Geol 50:1235–1242

Miller VC (1953) A quantitative geomorphic study of drainage basin characteristics in the Clinch Mountain area Varginia and Tennessee. J Geol 65(1):112–120

Moglen GE, Bras RL (1995) The importance of spatially heterogeneous erosivity and the cumulative area distribution. Geomorphology 12(3):173–185. Elsevier

Mohd I, Haroon S, Bhat FA (2013) Morphometric analysis of Shaliganga sub-catchment, Kashmir Valley, India using geographical information system. Int J Eng Trends Technol 4(1):10–21

Morisawa M (1985) Rivers-forms and process. Longman Group, London, pp 54–56, pp 70–73

Nag SK (1998) Morphometric analysis using remote sensing techniques in the Chaka sub basin, Purulia District, West Bengal. J Indian Soc Remote Sens 26(1 & 2):69–76

Nag SK, Chakraborty S (2003) Influence of rock types and structures in the development of drainage network in the hard rock area. J Indian Soc Remote Sens 31 (1):25–35

Pareta K, Pareta U (2011) Quantitative morphometric analysis of a watershed of Yamuna Basin, India using ASTER (DEM) data and GIS. Int J Geomat Geosci 2 (1):248–269

Rhea S (1993) Geomorphic observations of rivers in the Oregon Coast Range from a regional reconnaissance perspective. Geomorphology 6(2):135–150. Elsevier

Sarita G (2015) Morphometric analysis of a Shakkar river catchment using RS and GIS. Int J u- and e- Serv Sci Technol 8(2):11–24

Schumm SA (1956) The evolution of drainage system and slopes in Badlands at Perth Amboy New Jersey. Bull Geolog Soc Am 67:214–236

Sen PK (1993) Geomorphological analysis of drainage basins. The University of Burdwan, Burdwan

Sharma SK, Gajbhiye S, Tignath S (2014) Application of principal component analysis in grouping geomorphic parameters of a watershed for hydrological modeling. Appl Water Sci 5:89–96

Singh DS, Awasthi A (2011) Implication of drainage basin parameters of Chhoti, Gandak River, Ganga plain, India. J Geol Soc India 78:370–378

Singh S, Singh MC (1997) Morphometric analysis of Kanhar river basin. Natl Geograph J India 43:31–43

Smith GK (1950) Standards for grading texture of erosional topography. Am J Sci 248:655–668

Strahler AN (1952) Hypsometric (area-attitude) analysis of erosional topography. Geol Soc Am Bull 63:1117–1142

Strahler AN (1964) Quantitative geomorphology of drainage basins and channel networks. In: Chow VT (ed) Handbook of applied hydrology. McGraw Hill Book Company, New York, Section 4–76

Vikhe SD, Patil KA (2016) Morphometric analysis of a basin using remote sensing and GIS—a review. Int J Innov Res Sci Eng Technol 5(5):7029–7034

Wentworth CK (1930) A simplified method of determining the average slope of land surfaces. American J Sci 5–20(117):184–194

Willgoose GR, Hancock G (1998) Revisiting the hypsometric curve as an indicator of form and process in transport limited catchment. Earth Surf Proc Land 23:611–623

Withanage NS, Dayawansa NDK, DeSilva RP (2014) Morphometric analysis of the Gal Oya River basin using spatial data derived from GIS. Trop Agric Res 26(1):175–188

12 Impact Assessment of Check Dam in the Pappiredipatti Watershed (South India) Using LULC and NDVI Signatures

S. Satheeshkumar and S. Venkateswaran

Abstract

Agriculture is an essential sector because of the changing phenomenon of weather conditions, which is further complicated by the interaction of vegetation with the environment. The scarcity of rainfall causes physiognomic changes that can be identified by satellite images with reference to vegetation signature and land use and land cover changes. The rate of plant growth as well as to the amount of growth is indicated by vegetation indices. They are sensitive to the vegetation changes affected by moisture stress. Hence systematically, scarcity can be monitored using techniques than ground-based methods of information collection. The capabilities of geospatial techniques have been used to demarcate effective sites for monitoring the functional characteristics of water conservation structures in the Pappiredipatti watershed. In this study, land use/land cover has been used to delineate the existing sites to water harvesting measures. An increase in groundwater resources of runoff storage structures is proposed in the watershed by constructing check dams, percolation tanks and gabion structures. Water conservation sites are erected by considering spatial distribution of land use/land cover and normalized difference vegetation index (NDVI) information of the watershed. Geospatial techniques have been used to integrate spatial information pertaining to functional characteristics of land cover within buffering distances in the proposed site of watersheds. For detailed changes around the recharge area assessment, an attempt was made over the watershed. Crop growth area was assessed in and around forest, barren land and dense vegetation areas using the temporal normalized difference vegetation index (NDVI) in 1985, 2005 and 2015. They were compared with vegetation area regions, and crop-wise area growing was identified. Based on the results obtained from the study area on vegetation cover, the buffers around the recharge area impact were assessed by the NDVI. A decrease in barren land in 2005 was identified by identifying the NDVI threshold over predominantly sparse vegetation growing regions, and crop-wise area was identified. Furthermore, the water body was estimated from the NDVI data of 2005 and 2015, which have been validated for vegetation assessment. The results of the Landsat data use study indicated a decrease in the dense vegetation area that has led to widespread drought conditions in some parts of the watershed.

S. Satheeshkumar (✉) · S. Venkateswaran
Department of Geology, Periyar University, Salem, India

P. K. Shit et al. (eds.), *Drainage Basin Dynamics*, Geography of the Physical Environment,
https://doi.org/10.1007/978-3-030-79634-1_12

Keywords

Landsat image · QGIS · Land use/Land cover · NDVI · Buffer technique

12.1 Introduction

Land use and land cover changes can easily indicate that the growth of agriculture plays a significant part in the watershed. when allow adequate water to coniferous forest it will grow but deciduous trees intake more to allow into streams whereas evergreen forestland and plains areas use less water than forest (Bosch and Hewlett 1982; Brown et al. 2013) examined river harvests through remotely sensed variations of cropland cover in a great rocky fynbos catchment. Because of the consequences of changes, one of the spectral vegetation indices of NDVI with differences was chosen to have a negative and positive connection with the river crop area. Depending on the environment and climatic conditions, plant life is strongly influenced by the surface water process. Robinson et al. (2003) suggested that there is no effect of a modification in the forest cover area of mountains across N-W Europe, though deforestation directed to a surge in improper flow in more moderate environments of situation (Hornbeck et al. 1993). Land use and land cover modification have a straight influence of water resource processes on the catchment (Tang et al. 2005; Bhaduri et al. 2000; Ott and Uhlenbrook 2004). This impact can be used to dignify the change flow of water in rivers or forecast by a hydrological simulation model to combine the changes of land use and land cover, allowing for assessing the impression of the fluctuation on groundwater recharge and discharge of river sites. The influences of land use change in the context of recharge studies have been deliberated by various researchers, such as Albhaisi et al. 2013; Narjary et al. 2014; Raposo et al. 2013; Van Ty et al. 2012, Sashikkumar et al. 2017, Jinno et al. 2009; Dams et al. 2008; Hosseinimarandi et al. 2014 Leterme and Mallants 2011; Stiefel et al. 2009; Parsa et al. 2016; Ala-aho et al. 2015; Chatterjee et al. 2009; Varni et al. 2013; Rani et al. 2009; Oke et al. 2013. The main objective of this study is to validate and the changes of the impact assessment of land use and land cover variation on groundwater recharge sites in the area of data resulting from Landsat satellite images by using a simulated model (Molusce) for the Pappiredipatti watershed, South India. Pappiredipatti watershed is chosen because it has faced many temporal variations in land use and soil erosion in recent decades. Check dams are one of the barriers on river sites that are making a new environmental cycle of the system behind them and natural hill slope of living plants. It was imagined that recharge would be further augmented due to modification of land cover.

12.2 Study Area

The Pappiredipatti watershed is situated in the study area and is shown in Fig. 12.1. This study area lies between 78°18′ 0″ E and 78° 30′ 0″ E and 11° 48′ 30″ N and 11° 58″ 0′northern latitudes in Tamil Nadu, India. The total geographical area of the watershed is 184.55 sq. km, of which 67,343.75 acres is under cropland. Data on general cropland were collected from field investigations, which may indicate that sugarcane and coconut trees are the major crops in most of the area, followed by paddies.

12.3 Material and Methods

The main data are used for the mapping of LULC and NDVI lineaments that are LANDSAT 7 enhanced thematic mapper plus (ETM+) downloaded from U.S. Geological Survey. The imagery used to comprise subsets from the original scenes of path 143 and row 52 and path 154 and row 52 of 2005 and 2000, respectively. Geometric rectification generated NDVI after applying corresponding sensor calibration coefficients. NDVI images over a period of months are required to obtain cloud-free NDVI images of the

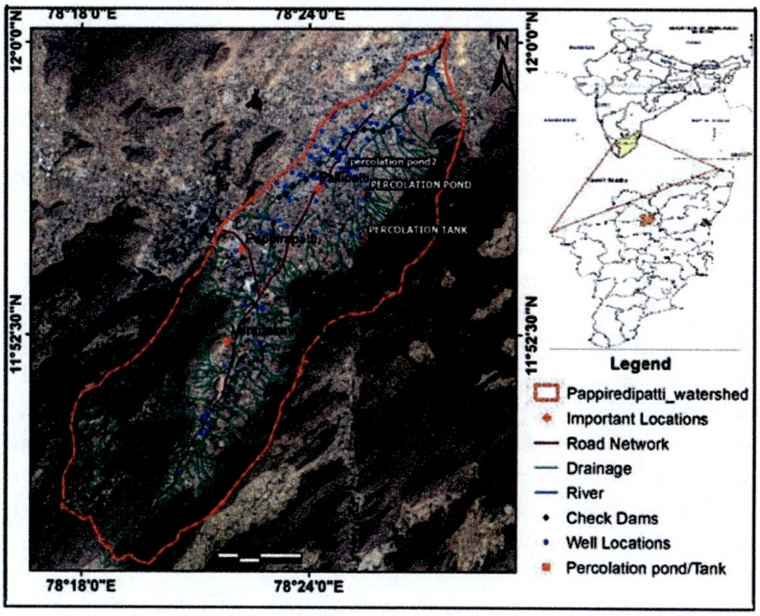

Fig. 12.1 Location of the Pappiredipatti watershed

area, which can represent the total scenario of vegetation conditions in that particular month. The LULC was predicted in the watershed using QGIS software. LU/LC was prepared by using unsupervised classification and NDVI used in Erdas software. Important developments in the field depend on the recharge sites of land use and land cover change. An actual modelling is supported to integrate various parameter, designated by the arrows in Fig. 12.2.

LU/LC prediction is necessary for its accuracy and future development of changes in land patterns in the recharge of groundwater. Using combined LU/LC and surface runoff can manage the potential of vegetation through water conservation structures.

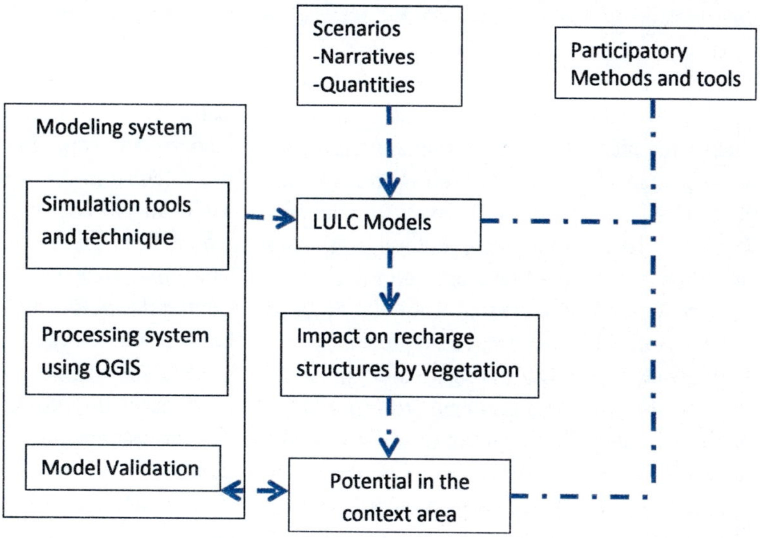

Fig. 12.2 Indication of the potential vegetation in the context area of LU/LC

12.4 Result and Discussion

12.4.1 Land Use and Land Cover Changes

The barren lands have 1.45% in 2005 and 2% in 2015 of the total surface area. The results show that 2005–2015 continuous intense trends of reducing barren land and fallow land but that this trend was relatively less intensive in 2005. The groundwater has the potential to be close to the water conservation structure, whereas natural recharge is low compared to previous recharge because it is far from those structures. Its validity is found in the context of wells.

In this watershed, Land Use and Land Cover 2005 (45,503.3 acres) include Barren land (679.75 acres), Cropland (12,766.6 acres), Deciduous Forest (3544.92 acres), Evergreen Forest (22,288.9 acres), Fallow Land (2878.79 acres), River (353.608 acres), Road network (586.233 acres), and Urban (2404.53 acres). Land Use and Land Cover of 2015 (45,501.661 acres) comprises Barren land (935.171 acres), Cropland(11,431 acres), Deciduous Forest (3445 acres), Evergreen Forest (2383.9), Fallow Land (21,919.1), (River 1058.49 acres), Road network (1609 acres), and Urban (2720 acres) are shown in Fig. 12.3.

An order to assess the real image and the simulated images that have model output to the present land use and land cover by comparing simulated LC/LC representing the 2005 LULC with the present LULC-based molusce model (QGIS) to validate land use and land cover change for prediction. Satellite image classification between 2005 and 2015 showed eight distinct LULC classes such as barren land, cropland, deciduous forest, evergreen, fallow land, river and water bodies, road network and urban.

Agricultural land increased from 20% in 2005 to 38% in 2015 due to the augmentation of groundwater by suitable construction of water conservation structures (buffered 500 m). LU/LC 2015, real and simulated land use by visual inspection analysis indicates that the simulated LU/LC map and actual map have moderately close resemblance. Agriculture land has the finest choice where the simulated area is 15645.4 acres. The simulated LULC map, barren land and agricultural land area are underestimated, but the predicted amount of forestland and urban land is overestimated. The accuracy of yield is in the assessment process by the molusce module using QGIS software. The K values of all wells above 0.9 are shown in a satisfactory level of accuracy. The model validation of land use and land cover indicates that in 2015 optimum changes occurred in the watershed. The simulation map is a respectable promise with the situation of 2015. Therefore, it is concluded that molusce models are suitable for the prediction of future challenges of planning LU/LC for the sustainable development of agricultural land and urban areas.

12.4.2 Validation and Simulations of LU/LC

QGIS is a quick and convenient tool for the analysis of land cover changes. Molusce is toolboxes to partially automate the process to make a model that can anticipate arrival utilize changes between two decades of period for predicting future sustainable development. The raster configuration of land utilizes classifications for the year 2005; raster of land utilizes classifications for the year 2015 and rasters of illustrative factors. Training a model that predicts arrival utilizes changes from 2005 to 2015, as shown in Fig. 12.4. The predicted future land use changes utilize the determined model, current condition of land use and current elements. Display (Molusce) is a calculation that is utilized for forecasting land use changes. The state raster is a one-band raster where every pixel is allotted to the land use class. The input state raster is a one-band raster portraying the past. The output state raster is a one-band raster portraying present that prepares the model. The change guide is a whole number one-band raster that stores data about moves. Classification estimations of the progress guide are mapped coordinated to move classes.

The simulator module performs the land use change evaluation process. Introductory state

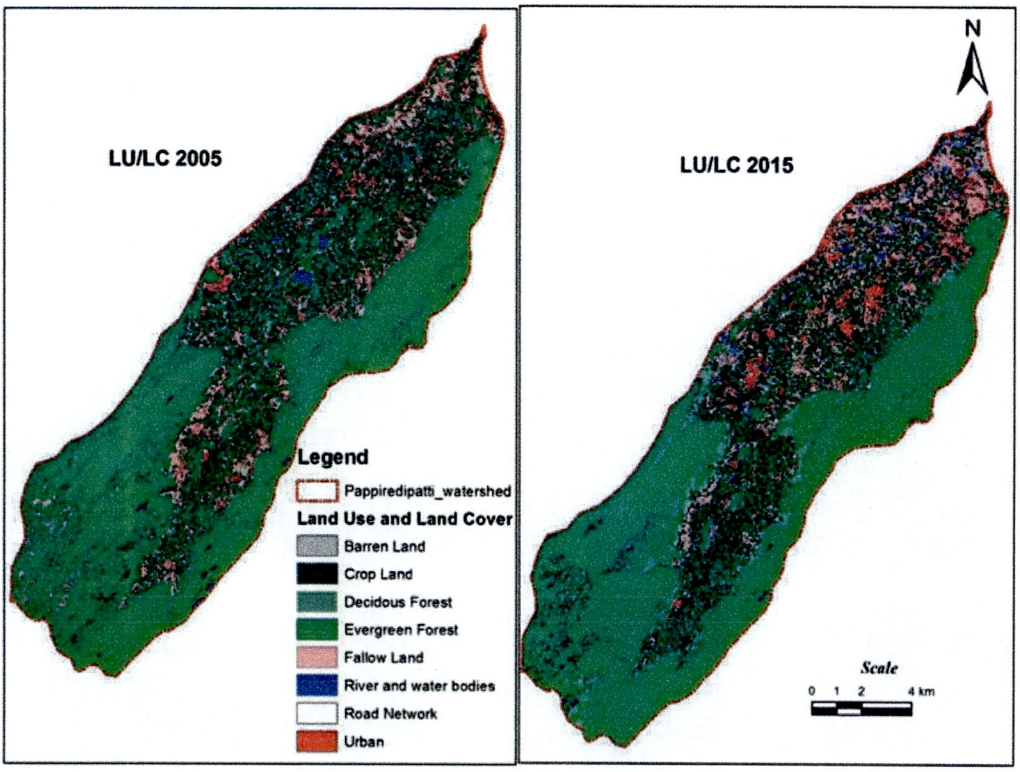

Fig. 12.3 Land use and land cover in the area of interest

Fig. 12.4 Validation of land use and land cover

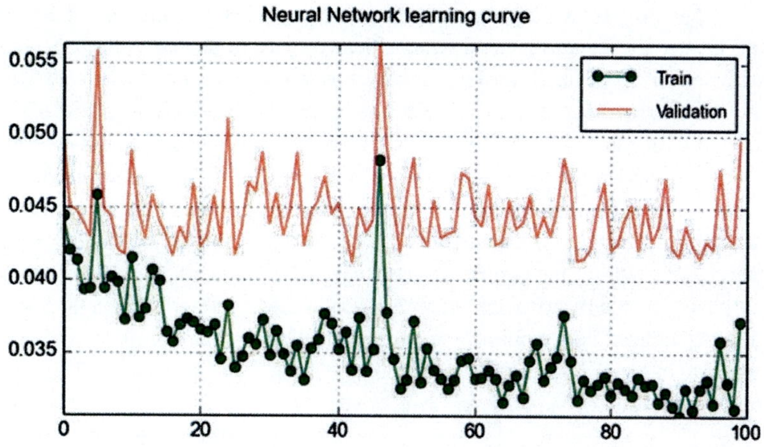

rasters contain data about current land use classes; figure rasters contain data about illustrative factors. The model is an indicator that computes change possibilities in the state of the variables and current land use. The area impact is accomplished if a model uses neighbourhood amid preparing, Simulator considers just broad examples. The test system take probabilities from the framework and figure number of the pixels that must be changed test system, which go to it starting state of raster component. The model sweeps pixels of the rasters and figures move possibilities of each move class.

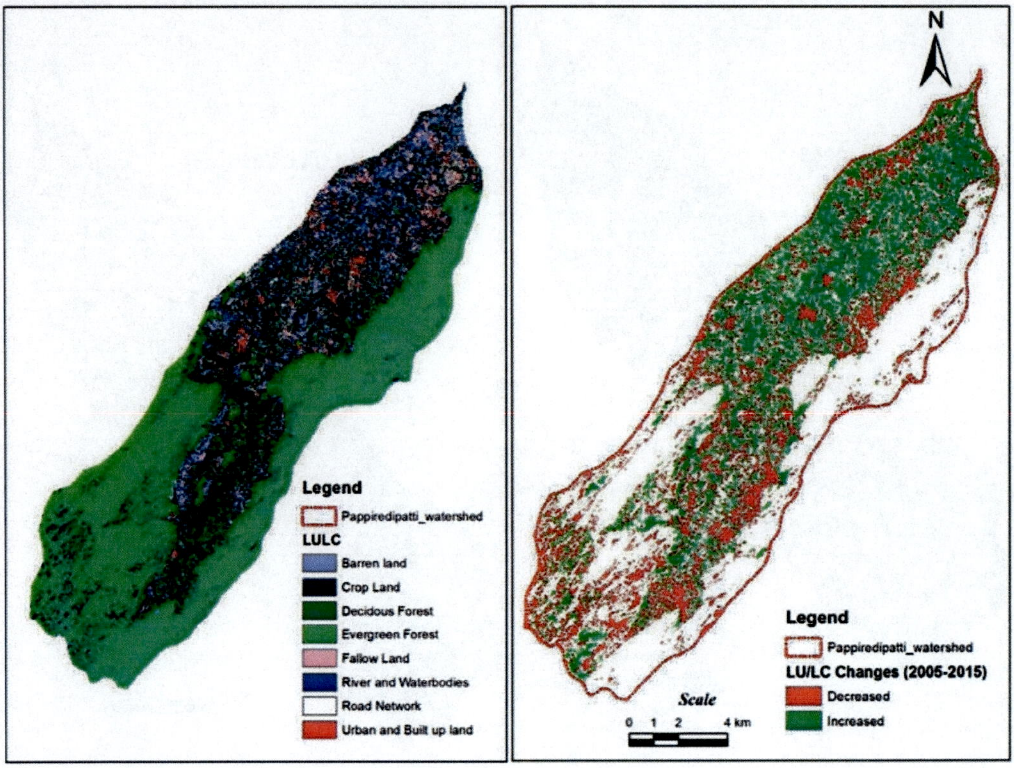

Fig. 12.5 Simulated land use/land cover and changes from 2005 to 2015

As a result, the raster contains the model certainty: the greater the contrast, the simulator develops a raster of the most likely moves, and the pixels of the raster are the move class with the greatest capability of move. This raster is utilized amid the following stage for each class. The simulator seeks in the raster of the most plausible moves a required check of pixels with the best certainty and changes the classification of the pixels. At least two pixels are close, and afterward, an arbitrary decision of the pixel is utilized to emphasize reproduction. The approval module is permitted to check the precision of the recreation of land use and land cover.

12.4.3 Effect of Land Use Change

In the watershed, irrigation has increased by erection of check dam construction. Socioeconomically, it is strongly indicated that impact of crop yield has been improved. By artificial recharge structures presence at suitable site, it has increased water in the surrounding wells on winter season that crop yield increased under cultivation. An impact assessment of check dams around 500 m depending upon recharge structures which is a strong benefit perceived.

The contributions of the different LU/LC classes are for effects on groundwater storing. The consequences of groundwater revive assurance for unmistakably appear from the impact of land use change (Fig. 12.5). The observational model aims to revive the yearly groundwater of the watershed. The commitment of the woodland is arriving in subsurface water revive at locales diminished from 2005 to 2015 after deforestation. The barren land contribution increases due to the proper selection of suitable sites from the 4th order stream. By comparison, the contributions of fallow land are in major change with values of 6.4% in 2005 and 48.17%in 2005 and cropland 28% in 2005 and 25.12% being small rates of change, respectively.

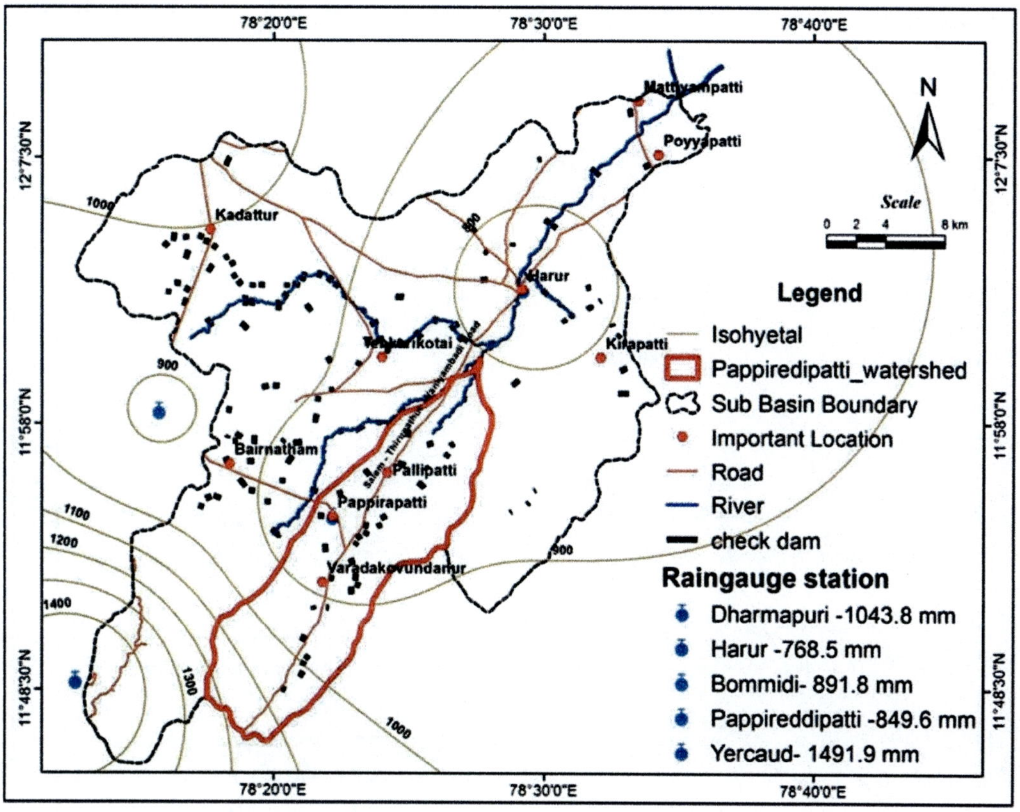

Fig. 12.6 Isohyetal map of the study area

The monthly rainfall data were collected from the Public Work Department (PWD) and converted into average seasonal rainfall. The data have been interpreted for 2000–2015. Taking data, two periods of time have helped to assess recharge sites by comparing low and high rainfall years (Fig. 12.6). The groundwater level has been increased due to the recharge through annual rainfall of 971.24 mm. The average rainfall from 2000 was 990.62 mm higher than the rainfall from 2015 was 971.24. For comparison reasons, Fig. 12.7 presents the average monthly values of temperature and potential evapotranspiration observed from climatological

Fig. 12.7 Potential evapotranspiration with average temperature

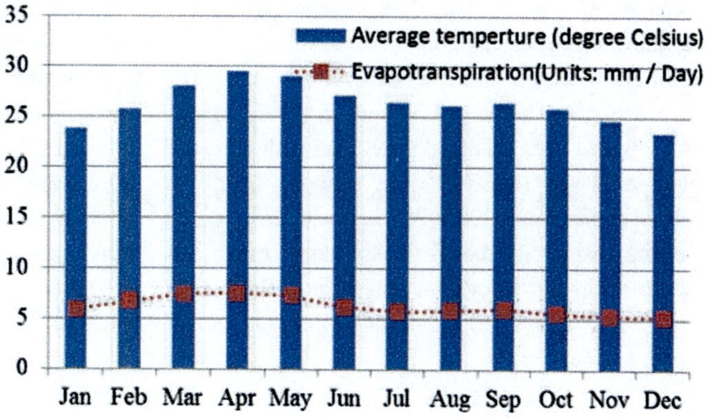

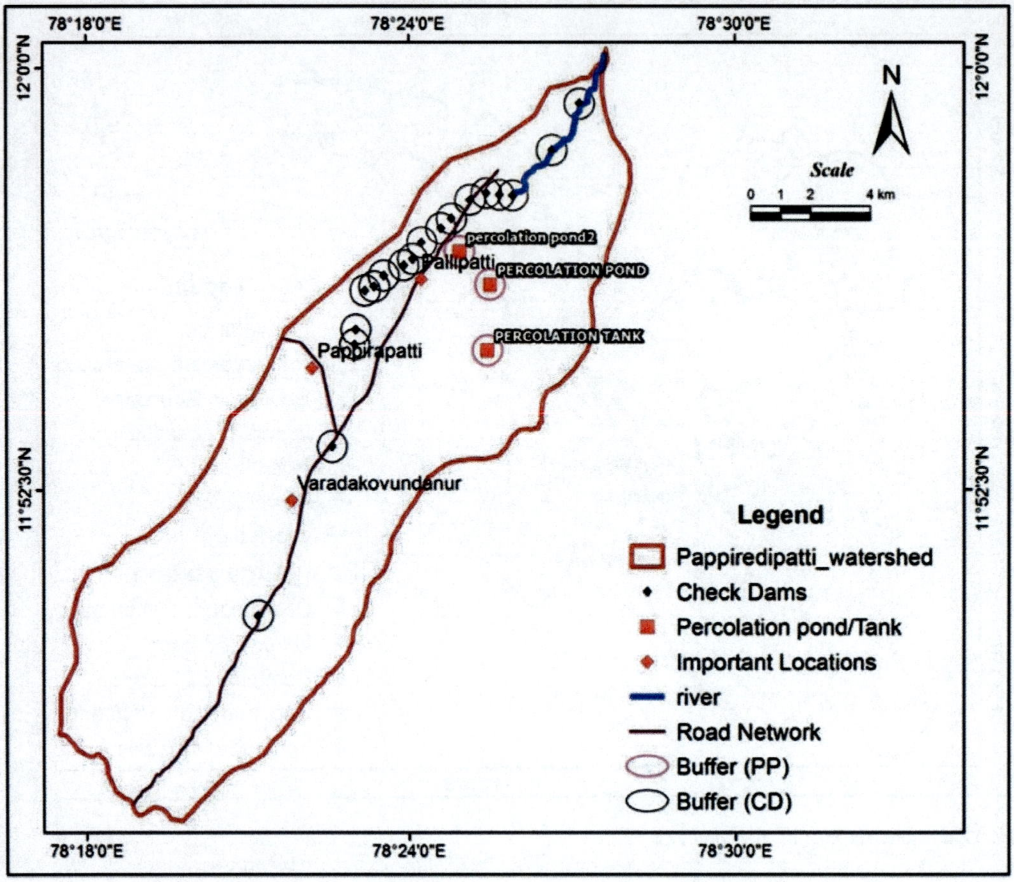

Fig. 12.8 Area of interest buffered in the context of water conservation structures

data. From this comparison, it occurred that the winter temperature was slightly high whereas the potential evapotranspiration also increased. Therefore, it was decided on reasons due to catchment scale, the mean historical data being useful instead of the baseline climatic since the observed differences in and around artificial recharge structures for further recharge calculation.

In addition, the areal scope and ascending of land use type in the watershed and the appropriation and area of these types can influence in suitable water recharge area amount used to delineate the year 2005–2015. Notwithstanding the expansion in the fruitless land zone and reduction in the woodland zone, the estimations of groundwater energize contrasted in 2015. The decrease in groundwater revitalization could be an effect of the circulation of the bush terrains in the ranges of high energizes values; these regions near the seepage arrange in the catchment of the Vaniyar Dam. As the precipitation in the period 2015 was high, changed land use has a significant contribution to expand revitalization in this way; vegetation contributes to the anticipated increment in groundwater energization. From 2005 to 2015 with the land utilize outline that there is a deliberate extreme of revive increment of 18% in late year due to land utilize little transform from 2015, which confirms that the significant land utilize changes.

12.4.4 Proximity of WCS Function by NDVI

NDVI data were generated using the yearly (10-year interval) time composited over the watershed in 1985, 2005 and 2015 by Landsat images. This is presented in the area of interest buffered in the context of Water Conservation Structures (Fig. 12.8) using ArcGIS Software. Buffer techniques can create polygons around input (vegetation indices) features to a specified (500 m) distance. The normalized difference vegetation index (NDVI 1985) is in the study area, as shown in Figs. 12.9 and 12.10. With verification, the temporal variations in NDVI are through the years 1985 NDVI over the region, which indicates low crop conditions, whereas 2015 indicates high crop conditions. Based on the prepared GIS layers of different land uses, the total crop area was extracted from the satellite. The major crops were classified by temporal variation. From the data, a reduction in crop area is observed in 1985 when compared with 2015 because of the prolonged dry spell that occurred in the month of July–August 2015, which is critical for crop sowings. Barren land (0.33–0.10 in the whole area, around check dam −0.07 to 0.10 and percolation pond −0.03 to 0.10), sparse vegetation 0.10–0.50 compared to recharge sites is 0.10–0.43. The area of interest buffered (NDVI 1985) in the context of water conservation structures is more sparse vegetation than dense vegetation. The normalized difference vegetation index (NDVI 2005) in the watershed of the area

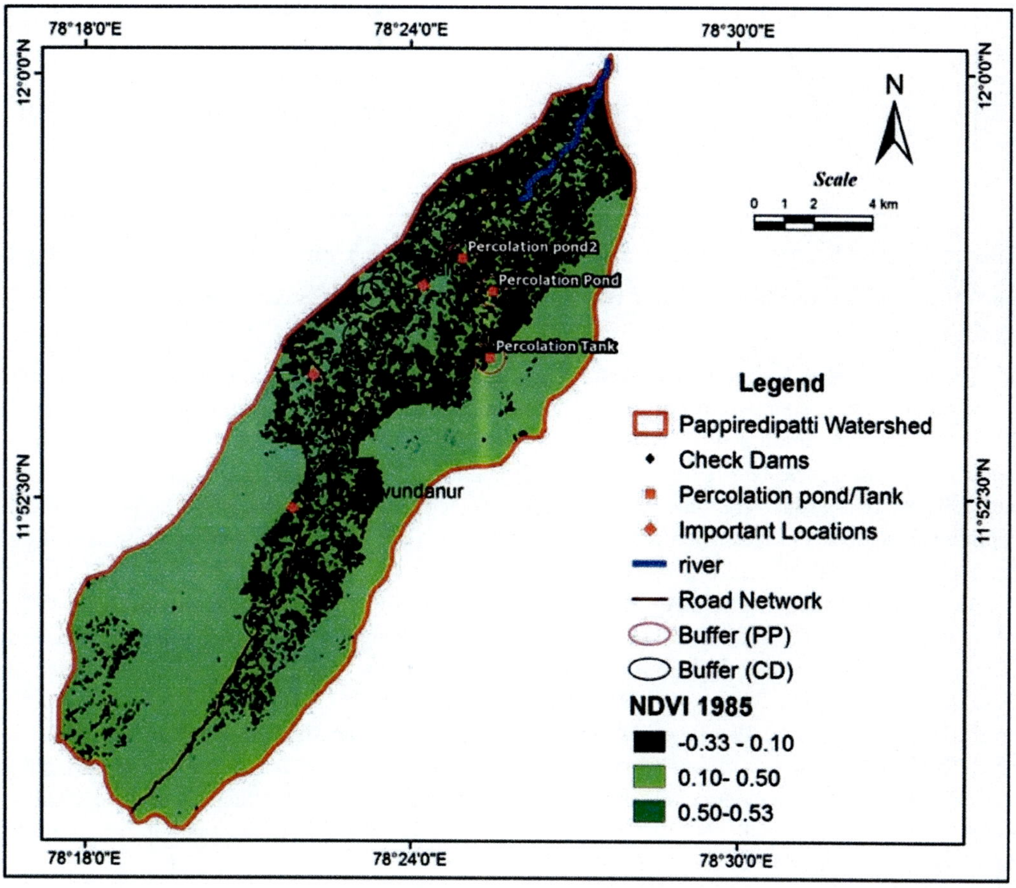

Fig. 12.9 Normalized Difference Vegetation Index (NDVI 1985) in the study area

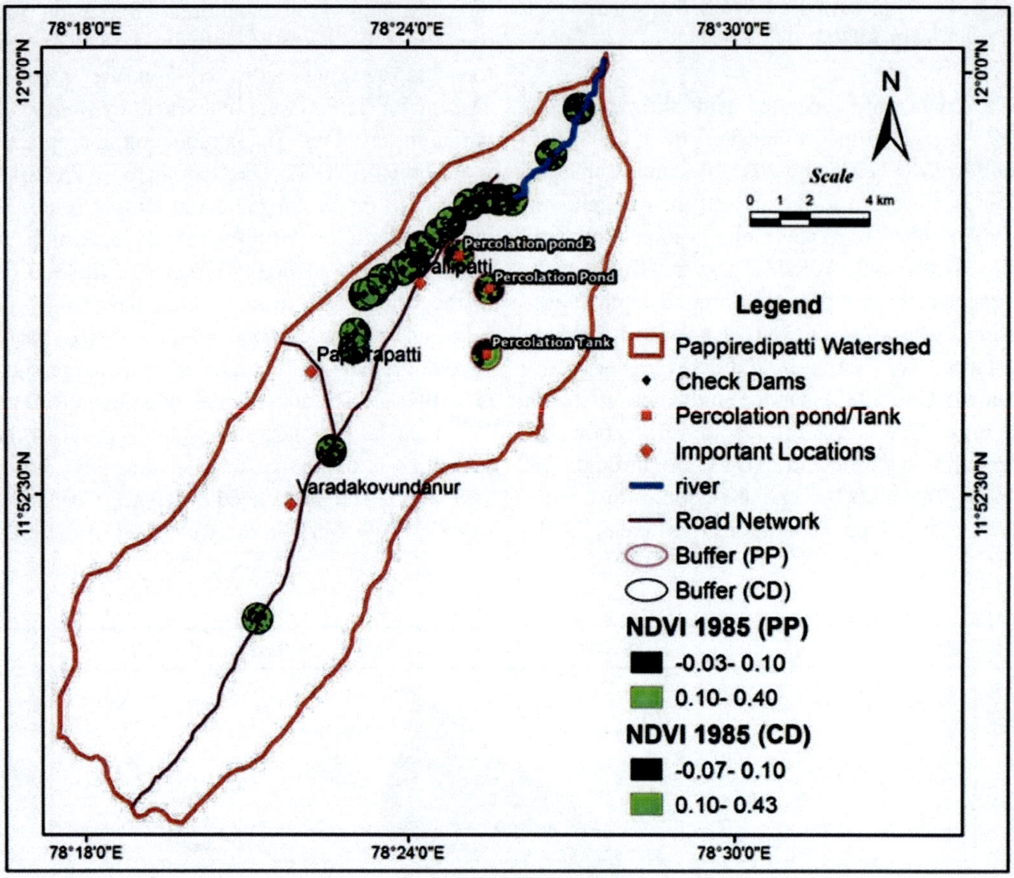

Fig. 12.10 Area of interest buffered (NDVI 1985) in the context of water conservation structures

of interest buffered in the context of water conservation structures is shown in Figs. 12.11 and 12.12. The crop condition in 1985 is assessed by comparing the temporal NDVI data of the 2015 and 2005 kharif seasons. The mandal wise crop condition image for the watershed indicates that most of the mandals are facing the deficiency situation with the indication of lower NDVI 1985 when compared to the NDVI 2015. The normalized difference vegetation index (NDVI 2015) and area of interest buffered (NDVI 2015) in the study area (Fig. 12.13) are shown in Figs. 12.13 and 12.14.

Changes in the vegetation condition are weighted for average with NDVI and water deficit values obtained from the water planning procedure for crop comparison. The results indicated that the paddy-dominated mandals deficit water in July and the NDVI value have fallen 0.37–0.17. Due to continuous deficit values, the NDVI continuously falls until September. It is observed that due to a constant water deficit over Kharif, the NDVI has fallen in August and September even though the NDVI is good in July. In the mandal, due to the high deficit in the month, the NDVI decreased in August and September. A scatter plot (Fig. 12.15) showing 70–100% of the NDVI in the context of water conservation structures indicates that rainfall influence occurred, and 30–70% indicates a natural store of water in the watershed.

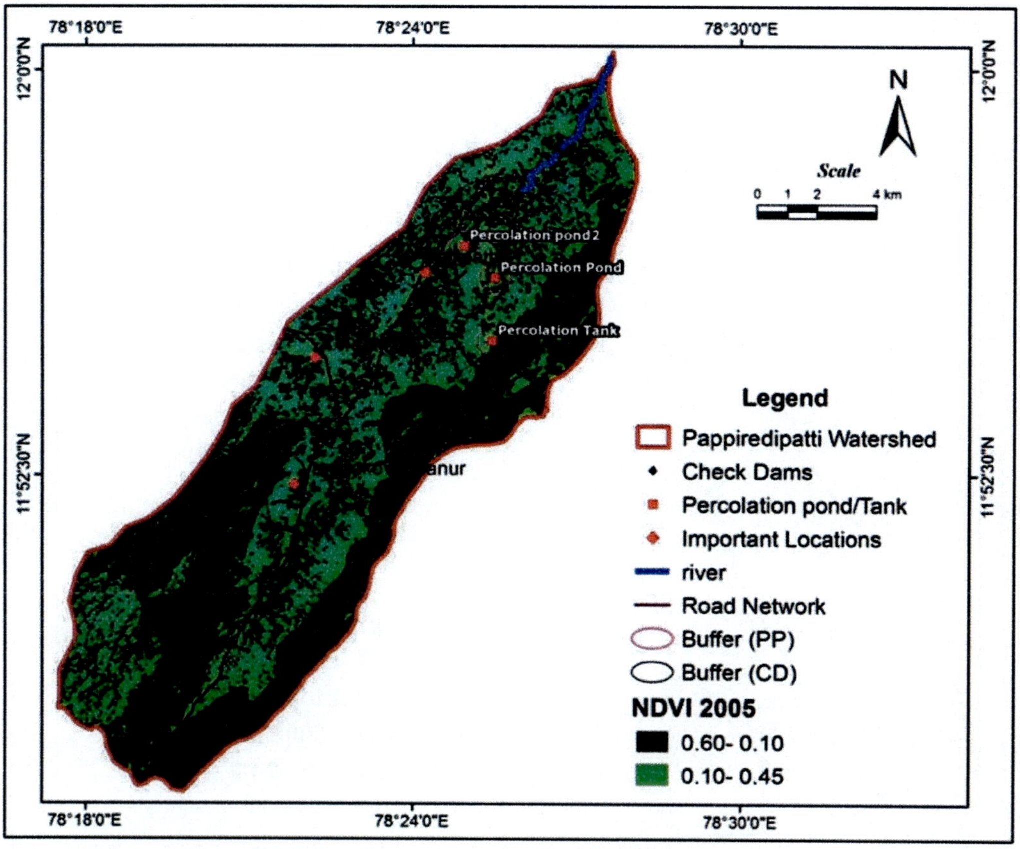

Fig. 12.11 Normalized Difference Vegetation Index (NDVI 2005) in the study area

12.4.5 Land Use and Land Cover Change Analysis

LULC changes from 2005 to 2015 periods, and the area in acres of two satellite images was projected around check dams. ERDAS Imagine was helpful for conversion matrix among land use classes for the two periods to the change analysis before and after construction of the check dam which provided the functional characteristics of the area changes, as shown in Fig. 12.16. Shannon's and Simpson's diversity index (H) are used for changes of each category by comparing pixel values of the pre- and post-intervention classified using images for change analysis.

12.4.5.1 Shannon's Diversity Index (H)

In the present research, Shannon's diversity index (H) species are represented by LU/LC categories based on two components. These are richness of LC/LC categories and abundance of relative proportion. This index is computed using the following equation:

$$H = -\sum (P_i * \log P_i)$$

where

H Shannon indcx of diversity,

i LU/LC category,

Pi probability of i = fi/n.

where,

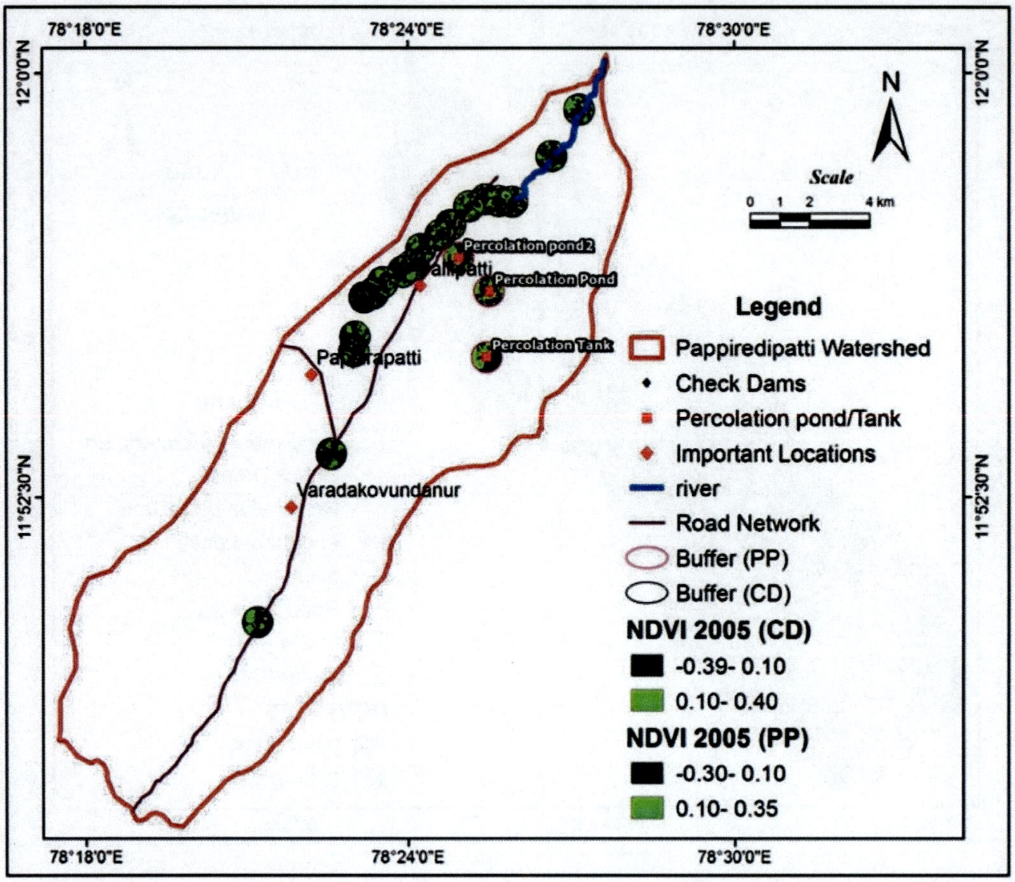

Fig. 12.12 Area of interest buffered (NDVI 2005) in the context of water conservation structures

f_i frequency of species, i.e. patch frequency of different LU/LC categories (i) and n = number of species (LU/LS categories) (equal to 4 in the present study) Log p log of Pi.

In the present study, by the relative abundance of the eight LU/LC categories; for example, in 2005(H), the diversity ranged from 1.45 to 1.29, and after 2015 (H), the diversity ranged from 1.68 to 1.5. The evenness in prior images ranged from 0.75 to 0.62 after 0.80 to 0.55 in the study area. The highest abundance of the individual indicates their functional performance in Table 12.1.

12.4.5.2 Simpson's Diversity Index (D)

Simpson's diversity index (D) measures the diversity of LU/LC and their accountability of spatial variation (Off well Woodland and Wildlife Trust 1998) is expressed by the following equation:

$$D = \sum N\ (n-1)/N\ (N-1)$$

where

n = Selective total number of land use and land cover.

N = Selective total number of all land use and land cover.

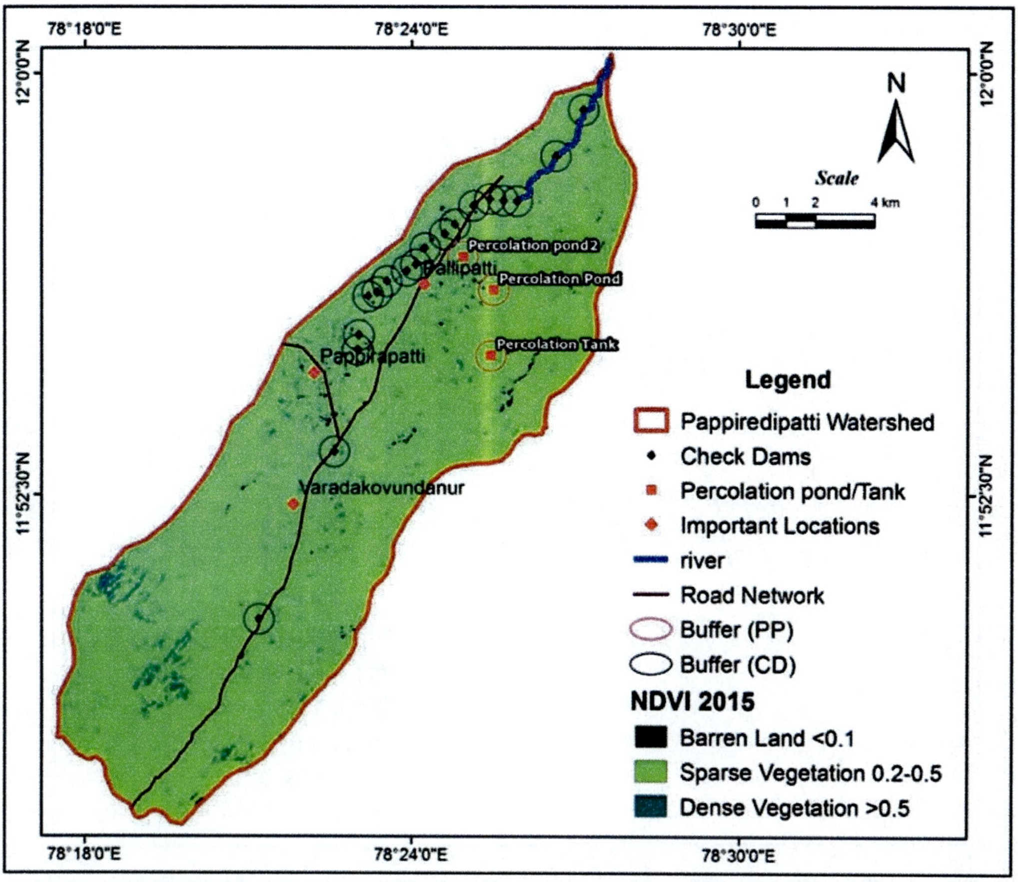

Fig. 12.13 Normalized Difference Vegetation Index (NDVI 2015) in the study area

Simpson's reciprocal index is computed in the present study to study the diversity of landscapes pattern between 2005 and 2015.

In this watershed, Simpson's index represents the range between 0 and 1 selected from a two different Land cover. Simpson's reciprocal index starts with 1 as the lowest possible. Table 12.2 represents one individual dominant land use and land cover. Higher values indicate greater diversity.

12.4.6 Field Photographs of the Inspected Area

Siltation in this area is characterized by dry and heavy monsoon rainfall and high erosion because the high sediment load occurs at the check dam site which originates from the hillside of flows near the check dam carrying a high fine-sediment load, as shown in Fig. 12.7. However, the sediment carriers carried by sediments which are

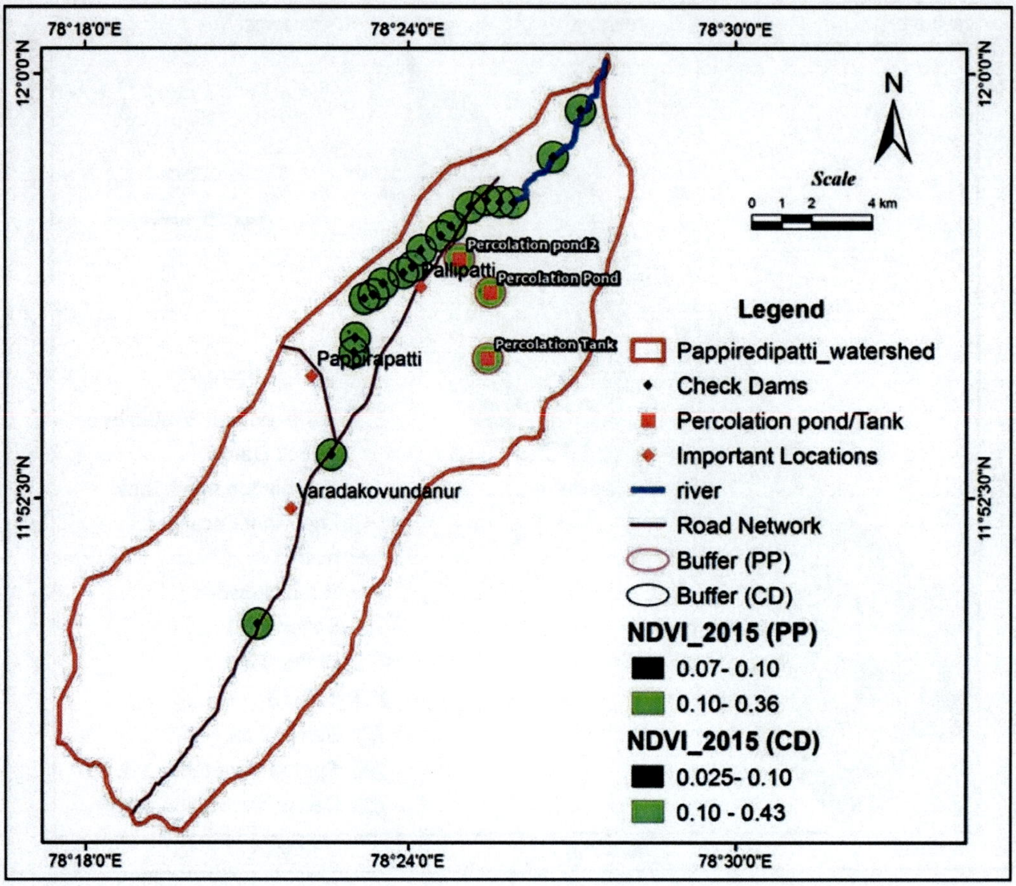

Fig. 12.14 Area of interest buffered (NDVI 2015) in the context of water conservation structures

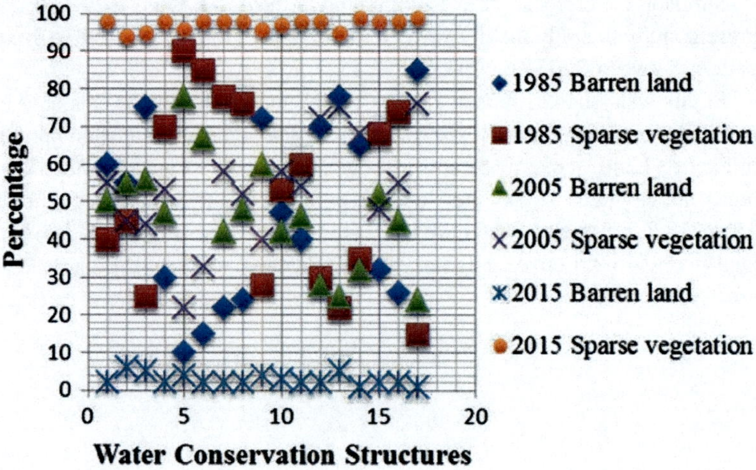

Fig. 12.15 Percentage of NDVI in the context of water conservation structures

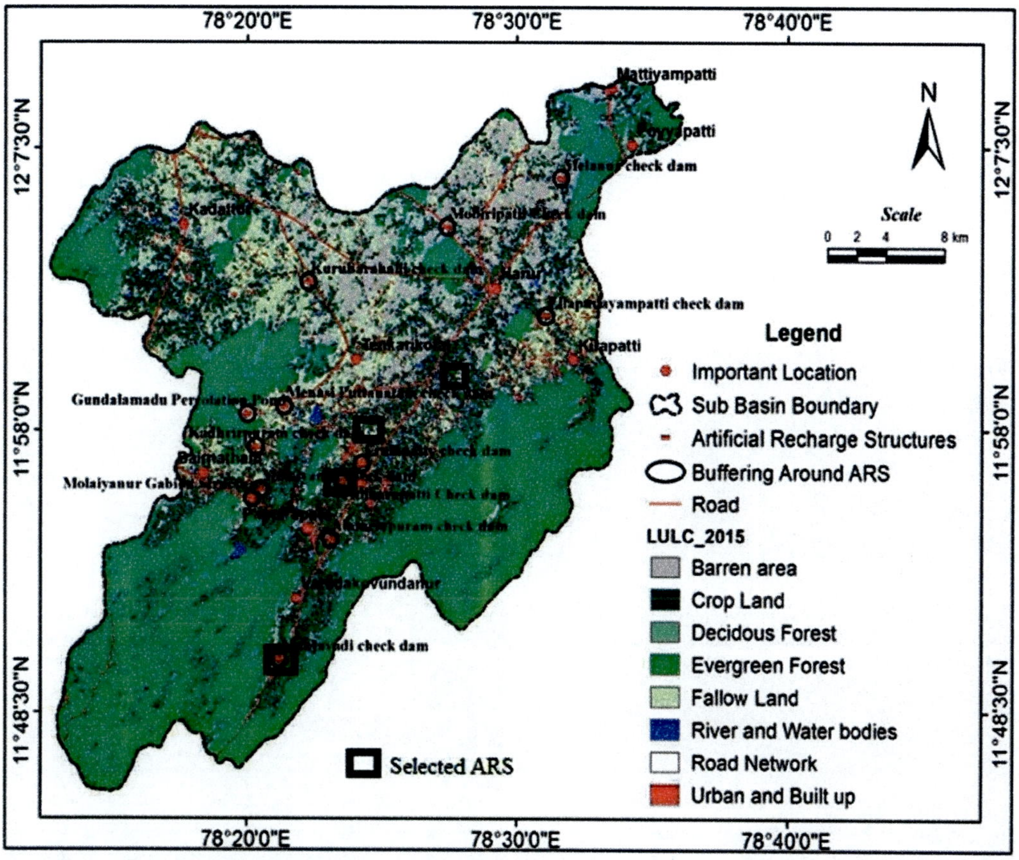

Fig. 12.16 Shown land use and land cover pattern 2015

Table 12.1 Relative abundance of the LU/LC

Check dam ID	2005		2015	
	H	Evenness	H	Evenness
Manjavadi (2-MCD)	1.454	0.734	1.175	0.565
Aalapuram (MCD)	1.399	0.721	1.465	0.704
Nadur (MCD)	1.455	0.735	1.152	0.554
Erumiyampatti (MCD)	1.291	0.620	1.680	0.807

Table 12.2 Showing Simpson's diversity Index around the check dam

Land use/Land cover categories 2005			
Si. no.	Simpson's index (D)	Simpson's index of diversity (1-D)	Simpson's reciprocal index (1/D)
1	1	0	1
2	1	0	1
3	1	0	1
4	1	0	1
Land use/Land cover categories 2015			

(continued)

Table 12.2 (continued)

Land use/Land cover categories 2005			
Si. no.	Simpson's index (D)	Simpson's index of diversity (1-D)	Simpson's reciprocal index (1/D)
1	0.35	0.64	2.78
2	0.31	0.68	3.19
3	0.46	0.53	2.14
4	0.22	0.77	4.43

Fig. 12.17 **a** Showing well 4 at the Manjavadi Open well cutting for the geological formations of conglomerate and silt deposits. **b** Showing well location 3 at the Manjavadi Measuring water level and Open well cutting for the silt deposits with cropland. **c** Erosion control of the small check dam at Kombur. **d** Major check dam at Manjavadi during the heavy rain season

coarse and fine accumulated at the check dam site. Although the upstream gradient is low to moderate, the velocity of water during the monsoon is high which is evident by the large amounts of sediments and boulders behind the check dam.

12.5 Conclusion

This study clearly indicates that the response of land use (LU) and land cover (LC) to increase vegetation is useful for planning and managing sustainable watershed development through natural and artificial conservation structures. It is necessary for land use planner retrieval data about the past, present and future to make appropriate decisions regarding land use for future challenges. Landscape alternation in the urban area is more significant than in the past in this watershed. Deforestation and human impedance are more essential dangers to the exceedingly rate biodiversity assets around there. The simulated land use map is beneficial for water conservation where exact site selection for well development. The above showing the NDVI for 1985, 2015 and 2015 over the Pappiredipatti region clearly indicates that vegetation in 2005 is much less than that in 2015. Cropland analysis indicates the drought conditions during 2005. Generally, it is observed that almost 60% of the total area is affected by a decrease in crop area. A regression equation concerning the yield to maximum NDVI for 2005 was developed for 1985 based on the maximum NDVI values of 2015. In the watershed, the water deficit was estimated, and a good relation with the reduction in NDVI was found. The reduction of the total crop area in 2005 was identified due to deficiency of rainfall in the watershed compared with 2015. The detailed analyses of NDVI clearly indicate drought condition assessments in the watershed for the development of agricultural patterns and predictions of future vegetal cover.

References

Ala-aho P, Rossi PM, Kløve B (2015) Estimation of temporal and spatial variations in groundwater recharge in unconfined sand aquifers using Scots pine inventories. Hydrol Earth Syst Sci 19:1961–1976 (2015). www.hydrol-earth-syst-sci.net/19/1961/2015/. https://doi.org/10.5194/hess-19-1961-2015

Albhaisi M, Brendonck L, Batelaan O (2013) Predicted impacts of land use change on groundwater recharge of the upper Berg catchment, South Africa. Water SA 39(2). https://doi.org/10.4314/wsa.v39i2.4

Bhaduri B, Harbor J, Engel BA, Grove M (2000), Assessing watershed-scale, long-term hydrologic impacts of land use change using a GIS-NPS model. Environ Manag 26(6):643–658

Bosch JM, Hewlett JD (1982) A review of catchment experiments to determine the effect of vegetation changes on water yield and evapotranspiration. J Hydrol 55(1/4):3–23

Brown TC, Foti R, Ramirez JA (2013) Projected freshwater withdrawals in the United States under a changing climate Water Resour. Res 49:1259–1276

Chatterjee R, Purohit RR (2009) Estimation of replenishable groundwater resources of India and their status of utilization. Curr Sci 96(12)

Dams J, Woldeamlak ST, Batelaan O (2008) Predicting land-use change and its impact on the groundwater system of the KleineNete catchment, Belgium. Hydrol Earth Syst Sci 12:1369–1385. www.hydrol-earth-syst-sci.net/12/1369/2008/

Groundwater Resource Estimation Methodology 1997, Report of the Groundwater Resource Estimation Committee (GEC), Central Ground Water Board (CGWB), Ministry of Water Resources, Government of India, New Delhi

Hornbeck JW, Adams MB, Corbett ES, Verry ES, Lynch JA (1993) Long-term impacts of forest treatments on water yield: a summary for northeastern USA. J Hydrol 150(2–4):323–344

Hosseinimarandi H, Mahdavi M, Ahmadi H, Motamed-vaziri B, Adelpur A (2014) Assessment of groundwater quality monitoring network using cluster analysis, Shib-Kuh Plain, Shur Watershed, Iran. J Water Resour Protect 6:618–624. Published Online April 2014. https://doi.org/10.4236/jwarp.2014.66060

Jeyaseelan AT, Suresh Babu AV, Chandrasekar K, Rupeshkumar GV (2008) IRS WIFS data use for 1999 droughts in Rayalaseema districts of Andhra Pradesh, India. In: Conference proceeding ICORG, pp 61–65

Jinno K, Tsutsumi A, Alkaeed Q, Saita S, Berndtsson R (2009) Effects of land-use change on groundwater recharge model parameters. Hydrol Sci J 54:2, 300–315. https://doi.org/10.1623/hysj.54.2.300

Leterme B, Mallants D (2011) Climate and land use change impacts on groundwater recharge. In: Models—repositories of knowledge, proceedings model CARE2011 held at Leipzig, Germany, in September 2011, IAHS Publ. 3XX, 201X

Narjary B, Kumar S, Kamra SK, Bundela DS, Sharma DK (2014) Impact of rainfall variability on groundwater resources and opportunities of artificial recharge structure to reduce its exploitation in fresh groundwater zones of Haryana. Curr Sci 107(8)

Oke MO, Martins O, Idowu O, Aiyelokun O (2013) Comparative analysis of empirical formulae used in groundwater recharge in Ogun—Oshun River Basins. J Sci Res Rep 2(2):692–710; Article no. JSRR.2013.017

Ott B, Uhlenbrook S (2004) Quantifying the impact of land-use changes at the event and seasonal time scale a process-oriented catchment model. Hydrol Earth Syst Sci 8(62–78):2004

Parsa VA, Yavari A, Nejadi A (2016) Spatio-temporal analysis of land use/land cover pattern changes in Arasbaran Biosphere Reserve: Iran, model. Earth Syst Environ 2:178. https://doi.org/10.1007/s40808-016-0227-2

Raposo JR, Dafonte J, Molinero J (2013) Assessing the impact of future climate change on groundwater recharge in Galicia-Costa, Spain. Hydrogeol J 21:459–479. https://doi.org/10.1007/s10040-012-0922-7

Robinson M, Cognard-Plancq AL, Cosandey C, David J, Durand P, Fuhrer HW, Hall R, Hendriques MO, Marc V, McCarthy R, McDonnell M, Martin C, Nisbet T, O'Dea TP, Rodgers M, Zollner A (2003) Studies of the impact of forests on peak flows and base flows: a European perspective. For Ecol 186:85–97

Sashikkumar MC, Selvam S, Lenin Kalyanasundaram V, Colins Johnny J (2017) GIS based groundwater modeling study to assess the effect of artificial recharge: a case study from Kodaganar River Basin, Dindigul District, Tamil Nadu. J Geol Soc India 89:57–64

Sethi RR, Kumar A, Sharma SP (2009) Quantification of groundwater recharge in a hard rock terrain of Orissa: a case study. QIWA Publishing 2009 Water Science & Technology—WST (60.5) 2009

Stiefel JM, Melesse AM, McClain ME et al (2009) Hydrogeol J 17:2061. https://doi.org/10.1007/s10040-009-0491-6

Tang Z, Engel BA, Pijanowski BC, Lim KJ (2005), Forecasting land use change and its environmental impact at a watershed scale. J Environ Manag 76:35–45

Van Ty T, Sunada K, Ichikawa y, Oishi S (2012) Scenario-based Impact assessment of land use/cover and climate changes on water resources and demand: a case study in the Srepok River Basin, Vietnam Cambodia. Water Resour Manage 26:1387–1407. https://doi.org/10.1007/s11269-011-9964-1

Varni M, Comas R, Weinzettel O, Dietrich S (2013) Application of the water table fluctuation method to characterize groundwater recharge in the Pampa plain, Argentina. Hydrol Sci J 58:7, 1445–1455. https://doi.org/10.1080/02626667.2013.833663

13 Morphometric Analysis for Prioritizing Sub-watershed and Management Using Geospatial Technique

Debabrata Ghorai, Gouri Sankar Bhunia, Someswar Devulapalli, and Pravat Kumar Shit

Abstract

Present work analyzes the morphometric characteristics and sub-basin prioritization of coastal regions of Tamil Nadu in India to comprehend the drainage basin's geo-hydrological characteristics using geospatial techniques. Digital Elevation Model (DEM) is obtained from the Shuttle Radar Topographic Mission (SRTM) with 30 m spatial resolution. Overall, seven linear aspects, ten areal aspects, and four relief aspects of the basin's morphometric characteristics were evaluated. Drainage network of the study region shows a dendritic pattern with sixth order stream network. Mean bifurcation ratio and stream length ratio of sub-basins are measured which substantiates the fact that the drainage pattern is not predisposed by geological developments and instabilities in the recent past. The elongation ratio is more than 0.9, inferring the sub-basins to be similar to a circular form. For soil and water conservation practices, the analysis prioritizes 5 sub-basins as high, medium and low. Geospatial technology has thus proven to be highly useful in collecting and haul out the correct information for assessing and investigating the characteristics of the watershed.

Keywords

Morphometric Analysis · Sub-basin Prioritization · Digital Elevation Model · Tamil Nadu Coast

13.1 Introduction

The analysis of watersheds is crucial for any form of growth or long-term maintenance (Sangma and Guru 2020). Since a watershed is composed primarily of natural resources like water, soil, and vegetation, it is important to protect them. People deal with land resources in a watershed, which is the hydrological, biophysical, and socioeconomic unit as a result of their socioeconomic practices. A watershed, for example, intermixes the use of soil, water, and plants if these services are conserved; it maximizes productivity while minimizing environmental impacts (Alemu and Kidane 2014). Watershed morphometry analysis is now the

D. Ghorai
Independent Researcher, Paschim Medinipur, Tajpur, West Bengal, India

G. S. Bhunia (✉)
Department of Geography, Seacom Skill University, Santiniketan, West Bengal, India

S. Devulapalli
Independent Researcher, Amberpet, Hyderabad, Telangana, India

P. K. Shit
Department of Geography, Raja N.L.Khan Women's College (Autonomous), Midnapore, West Bengal, India

P. K. Shit et al. (eds.), *Drainage Basin Dynamics*, Geography of the Physical Environment,
https://doi.org/10.1007/978-3-030-79634-1_13

most fundamental and rational alternative (Imran et al. 2011); it is the most viable tool and comparatively straightforward way to explain quantitative watershed features (Deepak 2015). Morphometry is the calculation and statistical study of Earth's surface structure, shape, and landform measurements (Umamathi and Aruchamy 2014). The morphometric parameters evaluate the structure of the basin quantitatively for an interpretation of stem steepness (initial slope or inequalities), structural controls, recent diastrophisms, geologic, and geomorphic background (Strahler 1964). Although watershed morphometric mapping is solely based on the drainage scheme and the geographical relationship between streams, it has also been distinguished from the use of other data such as soil and land use maps (Kiran and Srivastava 2012). Muluneh and Mamo (2014) investigated the development potential of land and water resources of Didessa River Catchment in Blue Nile Basin based on morphometric analysis. Without incurring high costs or wasting time, subwatersheds can be classified using morphometric criteria (Meshram et al. 2019). Furthermore, morphometric analysis is a comparison of landforms, soil characteristics, hydrological, and erosion features of sub-watersheds, which must be assessed and compared using morphometric parameters (Amulya et al. 2018).

The quantitative dimensions and physical characteristics of a watershed are usually covered by the morphometric element. For efficient natural resource management, it was also essential to prioritize sub-watersheds (Ayele et al. 2017). Welde (2016) tried to classify and prioritize subwatersheds based on annual sediment yield; Abdulkareem et al. 2018 and Singh and Singh (2014) aimed to prioritize sub-watersheds based on peak discharge and lag time. Furthermore, numerous researchers have used morphometry analysis to prioritize and compare subwatersheds (Sharma et al. 2010; Kumar et al. 2015; Thapliyal et al. 2017).

Remote Sensing and GIS techniques were found very useful in the study of morphometric analysis wherein fluvial landforms are determining in a regional scale. The results from the morphometric analysis are considered very helpful for planning and management of the drainage basin. The digital elevation model (DEM) and the Earth's satellite images from the remote sensing systems have proven much more versatile and precise than the use of topographical maps (Lima et al. 2011). Furthermore, spatial analysis approaches focused on applications have become the most sophisticated technologies in the GIS field in order to effectively delineate drainage basins and collect watershed resources and calculate the different indices of morphometry (Romshoo et al. 2012; Muluneh and Mamo 2014). However, there are limited number of studies conducted in Tamil Nadu. The study area, i.e., Tamil Nadu coast intersects five sub-basins. The sub-basins are Vaippar and Others, Pamba and Others, Cauvery Lower, Ponnaiyar and Others and Palar and Others, which have 22, 16, 10, 26 and 34 micro-watersheds, respectively (Central Ground Water Board 2019).

The present study describes the morphometric characteristics and sub-basin prioritization of coastal regions of Tamil Nadu in India to understand the drainage basin's geo-hydrological behavior using geospatial techniques and to illustrate the prevailing climate, structural, geological, and geomorphological conditions.

13.2 Study Area

Five sub-basins along the coast of Tamil Nadu, such as Palar and Others, Ponnaiyar and Others, Cauvery Lower, Pamba and Others, and Vaippar and Others (CGWB 2019) are considered as study area for the present study, is extended between 77° 7′ 44.23″ and 80° 20′ 48.21″ E longitude and 8° 4′ 35.02″ and 14° 25′ 45.08″ N latitude (Fig. 13.1). The Cauvery Lower subbasin has 10 watersheds (CGWB 2019). It is one of the productive sub-basin of Cauvery river basin. The Pennar (Cauvery to Krishna) basin consisting of four sub-basins of which Palar and Others (34 watersheds) situated in Andhra Pradesh and Tamil Nadu coast and Ponnaiyar and Others (26 watersheds) situated in Tamil Nadu coast (CGWB 2019). The other two sub-basins,

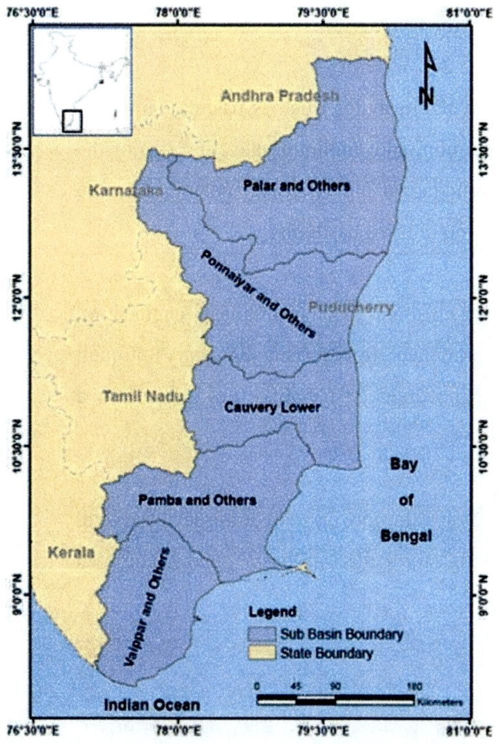

Fig. 13.1 Location map of five sub-basins along the coast of Tamil Nadu

i.e., Pamba and Others and Vaippar and Others are sub-basins of Vaippar (Kanyakumari to Cauvery) basin. Pamba and Others sub-basin has 16 watersheds and Vaippar and Others has 22 watersheds, respectively.

13.3 Materials and Method

13.3.1 Data Used

The primary input for this study is Digital Elevation Model (DEM). The global 1-arc second (30 m) of Shuttle Radar Topography Mission (SRTM) DEM is gathered from the Earth Explorer website (https://earthexplorer.usgs.gov/) of the United States Geological Survey (USGS). Table 13.1 summarizes the specifics of the reference results. For this analysis, the Indian administrative boundary map and watershed map are used as a reference dataset.

13.3.2 Method

13.3.2.1 Delineation of Stream Network

Stream network and watershed boundary are the principle inputs for assessment of the terrain morphometric aspects. In this study, morphometric aspects have been delineated at the sub-basin unit of three major basins namely, Pennar (Cauvery to Krishna), Cauvery, and Vaippar (Kanyakumari to Cauvery). Automatically stream network and watershed boundary were extracted from DEM using ArcGIS software v9.0 (Fig. 13.2). Strahler's stream order number is also assigned automatically to the river network using Python programming language.

Five sub-basins boundary was formed from the catchments boundary by dissolving respective group of catchments. The flowchart of the primary data generation technique and morphometric aspects analysis is illustrated in Fig. 13.2.

13.3.2.2 Quantitative Analysis of Morphometric Characteristics

In the GIS environment, the quantitative morphometric analysis for the delineated sub-basins were analyzed and calculated on the basis of the formula provided by Strahler, Horton, Schumm, etc., for determining their linear, areal, and relief aspects (Table 13.2).

The above morphometric parameter's value was automatically calculated from the primary inputs, namely drainage network, watershed boundary, and elevation of the topography through GIS workflow automation using Python language. The pseudo-code of the Python function that does the automatic calculation for morphometric parameters is presented below.

```
# Pseudo-code of Automatic Morphometric Parameter Extraction
If basin boundary shapefile exist:
    Read geometry of watershed boundary and following attributes, such as basin's unique id,
    basin's area (km2), basin's parameter (km), maximum and minimum elevation (m) of the
    basin (extracted from elevation against basin boundary using GIS s/w and joined to the
    basin boundary), and basin's longest length (km) and return a basin_list
If drainage network shapefile exist:
    Read geometry of drainage network and following attributes, such as drainage length (km),
    Strahler's stream order, and sub-basin id (extracted from basin boundary through
    intersection analysis in GIS s/w and joined to the drainage network file) and return a
    drainage_list
If drainage_list:
    Calculated total number of stream, total length of stream network, stream order wise total
    count, and stream order wise total length for all sub-basin and return a
    drainage_parameter_list
If basin_list:
    If drainage_parameter_list:
        A data frame (df) was created with the columns of basin id, basin area, basin perimeter,
        basin's longest length, basin's total stream network, basin's total stream network lengths,
        basin's stream order wise total count, basin's stream order wise total length, basin's
        maximum elevation, and basin's minimum elevation
If df:
        # Calculate Aerial Aspect
        Calculated aerial aspect values based on the formula presented in Table 13.2.
        # Calculate Linear Aspect
        Calculated linear aspect values based on the formula presented in Table 13.2.
        # Calculate Relief Aspect
        Calculated relief aspect values based on the formula presented in Table 13.2.
```

13.3.2.3 Sub-watershed Prioritization

Prioritization of sub-basin of the study area was carried out based on ranking on the computed morphological parameters wherein lowest ranking is given highest priority in terms of soil erosion and less conservation measure (Choudhari et al. 2018). Following parameters were considered for this analysis, such as drainage density, bifurcation ratio, stream frequency, length of overland flow, texture ratio, elongation

Table 13.1 Data used in this study

Sl. No.	Dataset	Scale/Resolution	Year of published	Source
1	SRTM DEM	30 m	2000	USGS
2	Political Map of India	1:4,000,000	7th Edition, 2017	SOI
3	Watershed Map	1:250,000	2019	CGWB

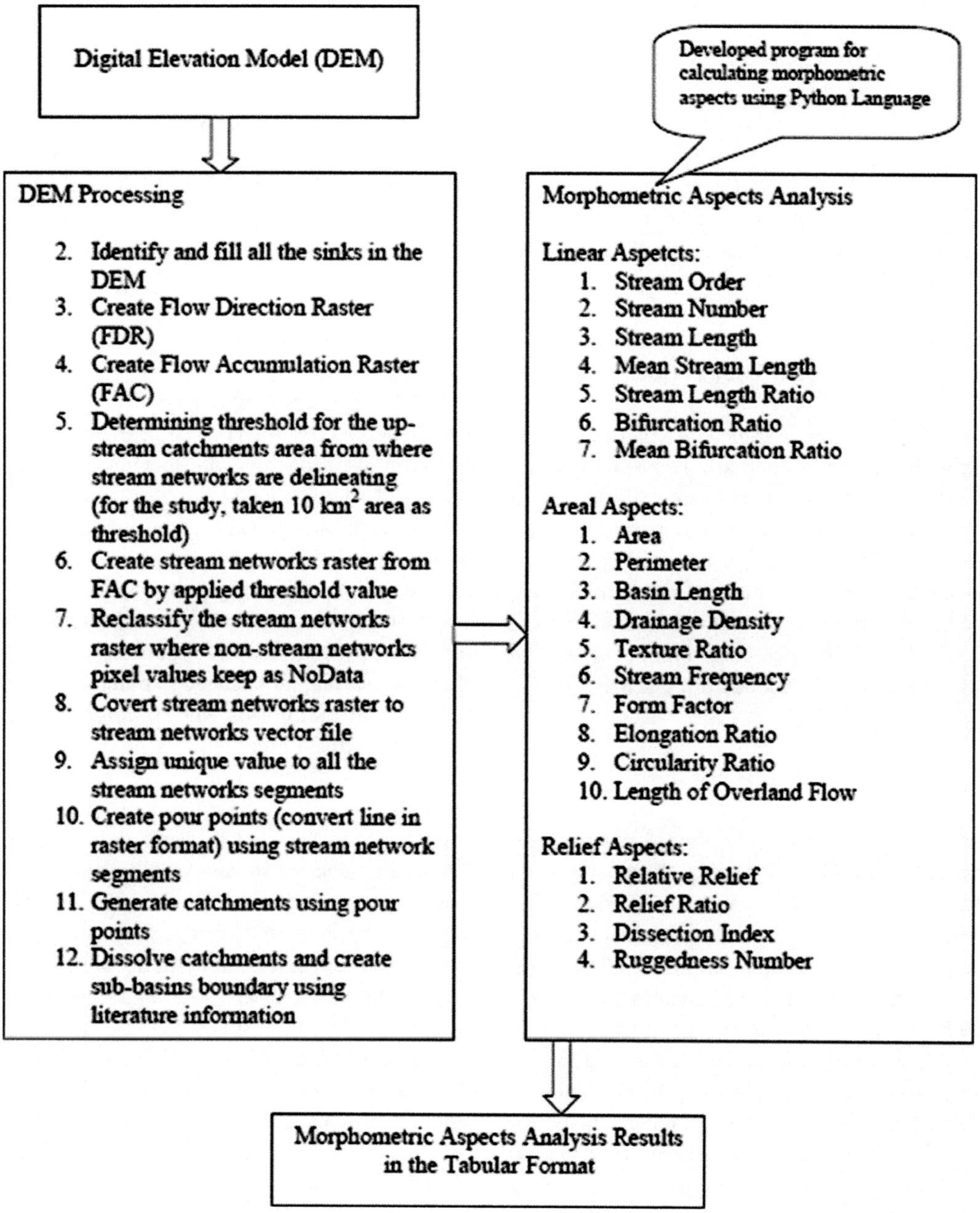

Fig. 13.2 Identification of Stream network and sub-basin boundary extraction and Morphometric Analysis

ratio, circularity ratio, compactness constant, basin shape, and form factor. Ranking of the sub-watersheds has been carried out for giving highest priority based on highest value in case of linear parameters as linear parameter has direct relationship with erodability (Said et al. 2018) and lowest value in case of shape parameters. After the rating has been assigned based on every single parameter, the rating value for all the five sub-watersheds were averaged so as to arrive at a compound value (Cp).

$$\mathrm{Cp} = \sum \mathrm{Rm}/\mathrm{N} \qquad (13.1)$$

where Cp is the compound value, Rm is the rank of morphometric parameter, and N is the total number of morphometric parameters.

13.4 Result and Discussion

The morphometric parameter's value extraction through automatic approach helps in reducing human efforts in calculation through manual or semi-automatic way in GIS environment. It also reduces the chances of mistake and error-prone value estimation. Thereby, it helps in making decision toward watershed prioritization. The detailed results of each morphometric aspect and watershed prioritization are discussed as follows.

13.4.1 Morphometric Aspect

The whole field of study is divided into five sub-basins namely Palar, Ponnaiyar, Lower Cauvery, Pamba, and Vaippar. Figure 13.3 shows the Tamil Nadu (India) coastal stream network and sub-basin boundary extracted from SRTM DEM. The Palar River Basin is situated in the extreme north of the study region, and in the extreme south is the Vippar River Basin. Tables 13.3 and 13.4 outline the morphometric characteristics of the sub-basin region Palar, Ponnaiyar, Lower Cauvery, Pamba, and Vaippar. For the delineated sub-basins, the morphometric parameters such as linear, aerial, and relief aspects are determined using the formulas described in Table 13.2. In this study the linear aspects of the drainage network are considered as stream order, stream length, mean stream length, stream length ratio, bifurcation ratio, mean bifurcation ratio. The basin area, perimeter, and basin length are considered in terms of morphometric characteristics of the regions. Quantitative statistics are measured for each sub-basin about the spatial distribution of aerial aspects such as drainage rate, texture ratio, flow level, shape factor, elongation ratio, circularity ratio, and overland flow volume. Moreover, the relief characteristics, such as relative relief (RR), relief ratio (Rh), dissection index (DI), ruggedness number (Rn) are calculated for each sub-basin of the study area.

13.4.2 Linear Aspects

13.4.2.1 Stream Order (Os)

Stream order (Os) refers to stream numbering based on its properties. In this analysis, the stream ordering method of Strahler is adopted to allocate Os to the drainage channels. According to Strahler (1952) all of the river's smallest tributaries or starting segments are known as streams of first class. A second order stream is formed where two first order streams combine; where two second order streams combine, a third order stream is formed; and so forth. Subsequently the stream order will not be increased if two streams of different order cross. Instead, it is important to retain the highest order stream in the hierarchy until it reaches the same order stream. The total number of streams of all order is estimated to be in Palar sub-basin in 1926, in Ponnaiyar sub-basin in 1525, in Cauvery Lower sub-basin in 958, in Pamba sub-basin in 1189, and in Vaippar sub-basin in 937. The highest order of stream (6th Os) in the study area is observed in the Ponnaiyar and Cauvery Lower sub-basins. The other three highest order sub-basins namely Palar, Pamba and Vaippar are classified as fifth Os. The information about the number of streams for each order is shown in Table 13.3. At Bay-of-Bengal all those highest Os meet.

Table 13.2 Methodology adopted for computations of morphometric parameters

Morphometric parameters	Equation	References
Linear aspects		
Stream Order (O_s)	Hierarchical Rank of Streams	Strahler (1952)
Bifurcation Ratio (R_b)	$R_b = N_u/N_{u+1}$ N_u = Total number of streams segments of order 'u' N_{u+1} = Number of segments of the next higher order	Schumm (1956)
Mean Bifurcation Ratio (R_{bm})	R_{bm} = Average of Bifurcation ratios of all orders	Strahler (1957)
Stream Length (L_u)	Length of the stream (km)	Horton (1945)
Mean Stream Length (L_{sm})	$L_{sm} = L_u/N_u$ L_u = Total stream length of order 'u' N_u = Total number of stream segments or order 'u'	Horton (1945)
Stream Length Ratio (R_I)	$R_I = L_u/L_{u-1}$ L_u = Total mean stream length of order 'u' L_{u-1} = Total mean stream length of its next lower order	Horton (1945)
Areal aspects		
Area (km^2)	A	GIS
Perimeter (km)	P	GIS
Basin Length (km)	L_b	GIS
Drainage Density (D_d)	$D_d = L_u/A$ L_u = Total stream length of all orders (km) A = Area of the Basin (km^2)	Horton (1945)
Texture Ratio (R_t)	$R_t = N_1/P$ N_1 = Total number of streams of first orders P = Perimeter (km)	Waikar and Nilawar (2014)
Stream Frequency (F_s)	$F_s = N_u/A$ N_u = Total number of streams of all orders A = Area of the Basin (km^2)	Horton (1945)
Form Factor (F_f)	$E_f = A/L_b^2$ A = Area of the Basin (km^2) L_b^2 = Square of the basin length (km)	Horton (1932)
Elongation Ratio (R_e)	$R_e = 2\sqrt{(A/Pi)}/L_b$ A = Area of the Basin (km^2) Pi = 3.14 L_b = Basin length (km)	Schumm (1956)
Circularity Ratio (R_c)	$R_c = 4 * Pi * A/P^2$ A = Area of the Basin (km^2) Pi = 3.14 P^2 = Square of the perimeter (km)	Miller (1953)
Length of Overland Flow (L_g)	$L_g = 1/D_d * 2$ D_d = Drainage density	Horton (1945)
Relief aspects		
Relative Relief (R)	$R = H - h$ H = Maximum height (m) h = Minimum height (m)	Strahler (1952)
Relief Ratio (R_h)	$R_h = R/L_b$ R = Relative relief (m)	Schumm (1956)

(continued)

Table 13.2 (continued)

Morphometric parameters	Equation	References
	L_b = Basin length (m)	
Dissection Index (D_{is})	$D_{is} = R/H$ R = Relative relief (m) H = Maximum height (m)	Singh and Dubey (1997)
Ruggedness Number (R_n)	$R_n = D_d * (R/1000)$ D_d = Drainage density R = Relative relief (m)	Fairbridge (1968)

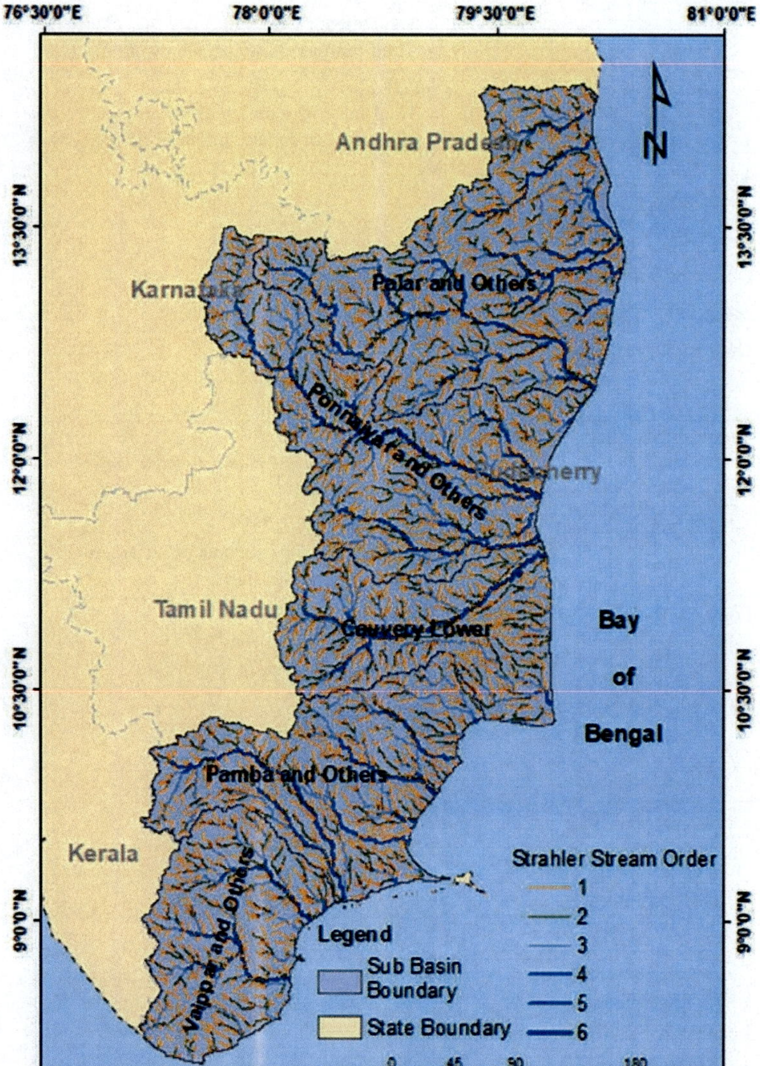

Fig. 13.3 Stream network and sub-basin boundary derived from SRTM DEM

Table 13.3 Linear aspects of the drainage network of the study area

Sub-basin	Stream Order (O_s)	Stream number (S_N)	Stream length (L_u)	Mean stream length	Stream length ratio	Bifurcation ratio	Mean bifurcation ratio
Palar	1	974	4788.73	4.92	–	1.98	2.28
	2	493	2609.10	5.29	1.08	2.16	
	3	228	1137.01	4.99	0.94	1.25	
	4	182	916.22	5.03	1.01	3.71	
	5	49	310.43	6.34	1.26	–	
Ponnaiyar	1	769	4081.76	5.31	–	2.14	1.88
	2	359	1893.49	5.27	0.99	1.79	
	3	201	970.34	4.83	0.92	1.83	
	4	110	482.65	4.39	0.91	2.12	
	5	52	259.67	4.99	1.14	1.53	
	6	34	191.10	5.62	1.13	–	
Cauvery Lower	1	489	2720.59	5.56	–	1.93	2.82
	2	253	1474.21	5.83	1.05	1.60	
	3	158	789.85	5.00	0.86	5.64	
	4	28	173.57	6.20	1.24	4.67	
	5	6	55.45	9.24	1.49	0.25	
	6	24	164.98	6.87	0.74	–	
Pamba	1	613	3366.63	5.49	–	2.20	2.03
	2	279	1666.10	5.97	1.09	1.55	
	3	180	1029.96	5.72	0.96	2.28	
	4	79	456.01	5.77	1.01	2.08	
	5	38	244.23	6.43	1.11	–	
Vaippar	1	485	2377.64	4.90	–	2.06	2.18
	2	236	1230.97	5.22	1.06	1.67	
	3	141	623.32	4.42	0.85	2.71	
	4	52	192.12	3.69	0.84	2.26	
	5	23	125.19	5.44	1.47	–	

13.4.2.2 Stream Length (L_u)

Stream length (L_u) is the measured distance from channel start point to meet point of the next channel start point or outlet of the drainage basin. According to Horton's formula (1945), the L_u is calculated for all the five sub-basins in kilometers. The total stream length of all the sub-basins is 34331.32 km. The highest L_u is recorded as 9761.49 km in Palar sub-basin followed by 7879.01 km in Ponnaiyar sub-basin and 6762.93 km in Pamba sub-basin. The lowest stream length is observed as 4549.23 km in Vaipparsub-basin followed by 5378.65 km in Cauvery Lower sub-basin (Table 13.3). In Ponnaiyar sub-basin, the highest *Os* is sixth and the corresponding stream length is 191.10 km, and the highest *Os* is draining for a longer distance. If L_u is smaller than the corresponding topography, it determines steep slopes and finer texture; while, if the L_u is longer, it indicates flatter topography with low gradients (Said et al. 2018). The drainage of longer distance without a change

Table 13.4 Areal Aspects of drainage basin of Tamil Nadu coast (India)

Sub-basin	Area (sq. km)	Perimeter (km)	Basin length (km)	Drainage density (D_d)	Texture ratio (R_t)	Stream frequency (F_s)	Form factor (F_f)	Elongation ratio (R_e)	Circularity ratio (R_c)	Length of overland flow (L_g)
Palar	34,796.67	1747.61	402.06	0.28053	0.55733	0.05535	0.21526	0.52352	0.14317	7.12938
Ponnaiyar	28,129.98	1970.44	434.23	0.28009	0.39027	0.05421	0.14919	0.43583	0.09104	7.14048
Cauvery Lower	17,511.82	1179.05	287.79	0.30714	0.41474	0.05471	0.21144	0.51885	0.1583	6.5116
Pamba	22,474.04	1458.21	285.66	0.30092	0.42038	0.05291	0.27541	0.59216	0.13282	6.64624
Vaippar	16,907.61	972.78	148.93	0.26906	0.49857	0.05542	0.7623	0.98518	0.22452	7.43317

in *Lu* permit channels to accrue huge volume of water and the dimensions of its contributing area is increased.

13.4.2.3 Mean Stream Length (*Lsm*)

Mean stream length (*Lsm*) shows the relationship between the size of the drainage network components and its contributing basin surfaces (Strahler 1964). The *Lsm* is calculated by dividing the total length of streams of an order by the total number of streams in the order. The highest *Lsm* is calculated as 6.45 km in Cauvery sub-basin, followed by Pamba sub-basin (5.88 km). The lowest *Lsm* is recorded in Vaippar sub-basin (4.73 km), followed by Ponnaiyar sub-basin (5.07 km). Table 13.2 presented the *Lsm* of all order streams for the five sub-basins.

13.4.2.4 Stream Length Ratio (*Rl*)

Stream length ratio (*Rl*) is the ratio of the mean length of the one order to the next lower order of the stream segments (Horton 1945). The value of *Rl* in Palar sub-basin varies from 0.94 to 1.26 followed by 0.91–1.14 in Ponnaiyar sub-basin, 0.74–1.49 in Cauvery Lower sub-basin, 0.96–1.11 in Pamba sub-basin, and 0.84–1.47 in Vaippar sub-basin. The value of *RI* between preceding and following stream orders varies due to differences in terrain slope and topographic condition. This also has a positive relationship with the surface flow discharge and erosional stage of the sub-basin (Magesh et al. 2012).

13.4.2.5 Bifurcation Ratio (R_b)

Bifurcation Ratio (R_b) is defined as the ratio of the number of stream segments of a given order to the number of segments of the next higher order (Schumm 1956). The utility of R_b lies in its ability to concisely express the organization of a drainage basin and level of dissection (Horton 1945). The R_b varies from 0.25 to 5.64 for the entire sub-basins which is generally found in the drainage pattern where geologic structures do not exercise a dominant influence on the drainage pattern. The value of R_b in Palar sub-basin varies from 1.25 to 3.71 followed by 1.53–2.14 in Ponnaiyar sub-basin, 0.25–5.64 in Cauvery lower sub-basin, 1.55–2.28 in Pamba sub-basin, and 1.67–2.71 in Vaippar sub-basin. The higher value of R_b in a sub-basin indicates a strong structural control in the drainage pattern whereas, the lower value indicates that the sub-basin is less affected by structural disturbances (Strahler 1964).

13.4.2.6 Mean Bifurcation Ratio (*Rbm*)

It is the average of bifurcation ratios of all orders for a basin or sub-basin. The *Rbm* characteristically ranges between 3.0 and 5.0 for a basin when the influence of geological structures on the drainage network is negligible (Verstappen 1983; Magesh et al. 2012). *Rbm* of all the sub-basins are found less than 3.0 which indicates the influence of the geological structures on the drainage network is not so dominant.

For the Tamil Nadu sub-basin, analogous geometric relationship is observed between the Os and the SN, and this also indicates the region to have uniform underlying lithology, and geologically, there was no possible upliftment (Fig. 13.4a, b). Results also exemplify that sub-basin hydrological characteristics primarily depend on the characteristics of the drainage. The stream network drainage pattern creates a dendritic form that indicates the sameness in texture and structural control non-existence. The area of study is characterized by tree-like pattern with twigs which interconnect mainly at acute angles.

13.4.3 Areal Aspects

The areal aspect of basin/sub-basin is hydrologically important as it directly affects the size of the storm hydrograph and the magnitude of peak and mean runoff (Rao et al. 2010). All the aerial aspects were calculated and results have been given in Table 13.4. The results show the highest area is recorded for Palar sub-basin (34,796.67 sq km), followed by Ponnaiyar sub-basin (28,129.98 sq km) and the lowest area is recorded for the Vaippar sub-basin (16,907.61 sq km). In terms of basin length, maximum length is estimated for Ponnaiyar sub-basin (434.23 km), followed by Palar sub-basin (402.06). Moreover, the lowest basin length is calculated for Vaippar

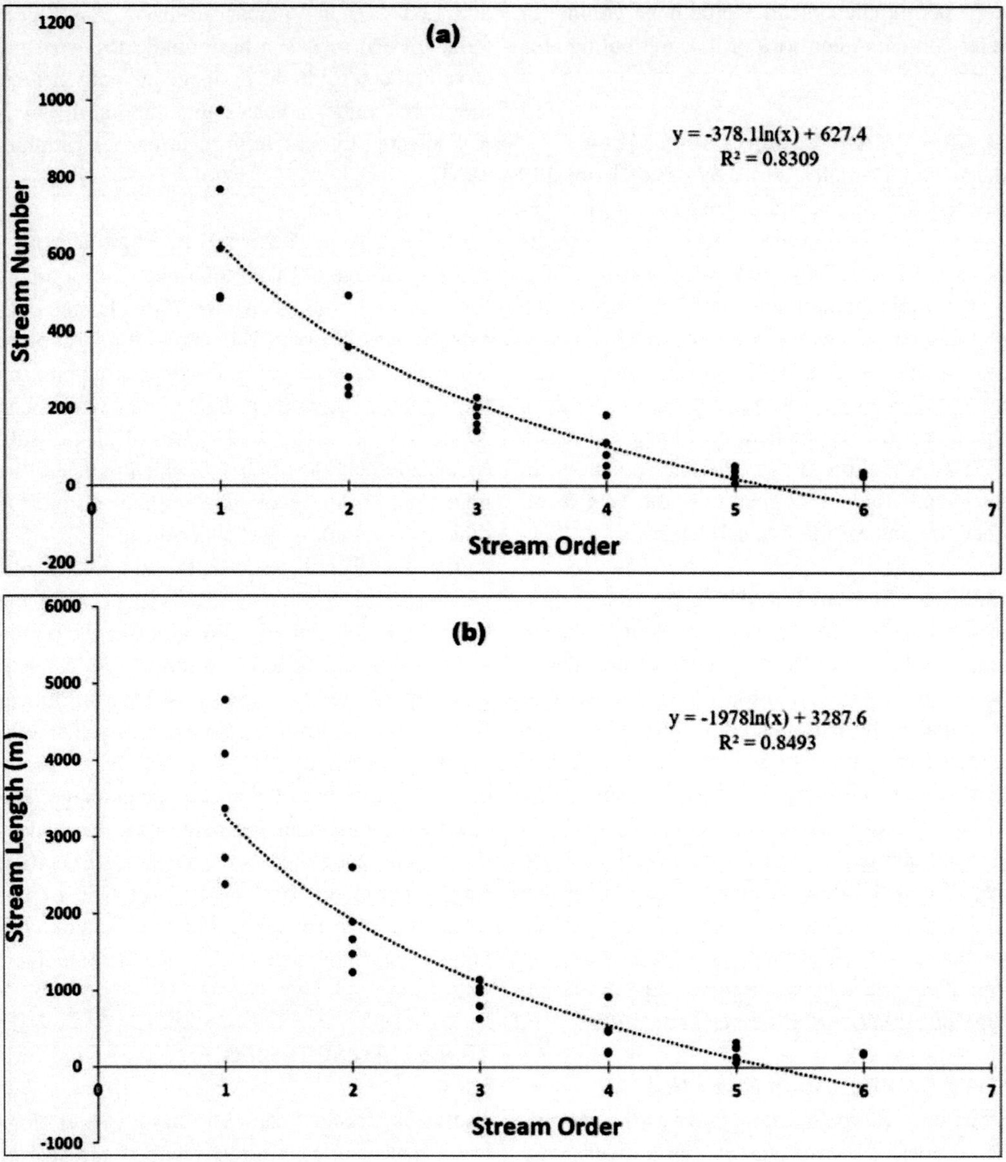

Fig. 13.4 Plot of stream order with: **a** log of stream number, **b** log of stream length

sub-basin (148.93 km), followed by Pamba sub-basin (285.66 km).

13.4.3.1 Basin Length (L_b)

The basin length (L_b) of a watershed refers to the maximum length of the basin measures parallel to the main drainage channel. It determines the shape and extent of the watershed. The study area L_b is presented in Table 13.3 for all the sub-basins. The highest L_b is calculated as 434.23 km (km) in Ponnaiyar sub-basin followed by 402.06 km in Palar sub-basin. The lowest basin length is calculated as 148.93 km in Vaippar sub-basin.

13.4.3.2 Drainage Density (D_d)

It is the ratio of total stream lengths of all order within a basin to the basin area (Horton 1932).

D_d is a measure of fluvial dissection and depends on various factors like climate, slope, rock types, vegetation cover, soil texture, and runoff intensity (Verstappen 1983). All the sub-basins revealed that D_d is very low and ranges between 0.27 and 0.31 which indicates the presence of permeable subsoil materials, low relief, flat terrain, and water infiltration rate is high in the sub-basins. The highest D_d is recorded for the Cauvery Lower sub-basin and the lowest D_d is recorded for Vaippar sub-basin. The higher the infiltration rate, the lower the density of surface water drainage, which is a major control on groundwater recharge (Krishnamurthy et al. 2000). The groundwater potentiality has become low when drainage density is very high and vice-versa.

13.4.3.3 Texture Ratio (R_t)

It is the ratio between total number of first order streams and the basin perimeter (Smith 1950). Smith (1950) has classified $R_t$into five category such as (a) very course (<2), (b) coarse (2 – 4), (c) moderate (4 – 6), (d) fine (6 – 8) and (e) very fine (> 8). It is noted that low drainage density leads to coarse drainage texture, whereas high drainage density results into fine drainage texture. The texture ratio is less than two for all the sub-basins and ranges between 0.39 and 0.56. the maximum value of R_t is recorded for the Palar sub-basin and the lowest value of R_t is recorded for Ponnaiyar sub-basin. The drainage R_t depends upon several natural factors such as climate, rainfall, vegetation, lithology, slope, infiltration capacity, relief and stage of development (Smith 1950; Kalevs 2001). All these sub-basins are fall under very course drainage texture.

13.4.3.4 Stream Frequency (F_s)

Stream Frequency (F_s) refers to the number of stream segments of all orders per unit area (Horton 1945). It depends on lithology of the basin and reflects the texture of the drainage channels. The maximum value of F_s is recorded from Vaippar sub-basin (0.05542), followed by Palar sub-basin (0.05535). The minimum value of $F_s$is documented from Pamba sub-basin (0.05291). The high F_s indicates the high relief and high infiltration capacity of the bedrocks (Magesh et al. 2012). The F_s of all the sub-basins is exhibit positively correlated with the drainage density values (Table 13.4).

13.4.3.5 Form Factor (F_f)

Form Factor (F_f) is the ratio of basin area to square of the basin length (Horton 1932). For the exact circular basin, the value of the form factor is greater than 0.78. The lower value of F_f indicates elongated basin (Magesh et al. 2012). The F_f of all the sub-basin ranges between 0.15 and 0.76 and showing less than 0.78 that reveals that the sub-basins are sub-circular and slightly elongated. The highest value of F_f is recorded from Vaippar sub-basin (0.76230), followed by Palar sub-basin (0.27541). As elongation increases for a given area, F_f value is decreased. The minimum value of F_s is documented from Ponnaiyar sub-basin (0.14919).

13.4.3.6 Elongation Ratio (R_e)

Elongation Ratio (R_e) is the ratio between the diameter of the circle of the same area as the drainage basin and the maximum length of the basin (Schumm 1956). Values of $R_e$generally vary from 0.6 to 1.0 over a wide variety of climate and geologic types (Waikar and Nilawar 2014). The R_e for all the five sub-basins is in the range of 0.43583–0.98518. $R_e$values close to unity correspond typically to regions of low relief, whereas values in the range 0.6–0.8 are usually associated with high relief and steep group slope (Strahler 1964). Results indicated that four sub-basins (Palar, Ponnaiyar, Cauvery Lower, Pamba) care are associated with high relief and Vaippar sub-basin is associated with low relief.

13.4.3.7 Circularity Ratio (R_c)

Circularity Ratio (R_c) is the ratio of the area of the basin to the area of a circle having the same circumference as the perimeter of the basin (Miller 1953). It is influenced by the length and frequency of streams, geological structures, land use/land cover, climate, and slope of the basin (Waikar and Nilawar 2014). The basin of the R_c range 0.4–0.5 which indicates strongly elongated and highly permeable homogenous geologic materials (Miller 1953). All the sub-basins having R_c values of less than 0.4 and ranges between 0.09104 and 0.22452. This indicates that the sub-basins are characterized by below moderate to low relief and drainage systems to be less influenced by structural disturbances. The high value of R_c shows the late maturity stage of topography (Waikar and Nilawar 2014).

13.4.3.8 Length of Overland Flow (L_g)

Length of Overland Flow (L_g) is the length of water over the ground surface before it gets concentrated into definite stream channel (Horton 1945). The L_g value of the sub-basins' ranges between 6.51160 and 7.43317. The L_g is approximately equal to half of the reciprocal of drainage density. This factor is related inversely to the average slope of the channel and is quite synonymous with the length of sheet flow to a large degree (Waikar and Nilawar 2014).

13.4.4 Relief Aspects

The relief of a watershed reflects the difference between the highest and lowest elevation points. It is depended on the elevation values of the watershed. Following relief aspects, such as relative relief (R_R), relief ratio (R_h), dissection index (D_I), and ruggedness number (R_n) are considered (Table 13.5).

13.4.4.1 Relative Relief (R_R)

Relative Relief (R_R) is the elevation difference measured value from the highest elevation to lowest elevation in a watershed (Strahler 1952). In the present study area, it is obtained by visual analysis of the digital elevation model prepared from SRTM DEM data. The elevation varies from 1343 to 2650 m which represents the land with moderate slope. The highest R_R is recorded from Pamba sub-basin and the lowest value of R_R is documented from the Palar sub-basin.

13.4.4.2 Relief Ratio (R_h)

Relief ratio (R_h) is defined as the horizontal distance along the longest dimension of the basin parallel to the principal drainage line (Schumm 1956). It is the unit less measured value that dividing the R_R of a basin by the basin length. R_h is an indicator of the intensity of erosion processes operating on the slopes and overall steepness of a drainage basin (Magesh et al. 2012). The value of the R_h in the study area ranges between 0.00334 and 0.01343 (Table 13.5). The R_h values indicate the sub-basins are low to moderate relief and steep to moderate slope conditions. Low relief ratios also indicate that the drainage capabilities of the watershed are low and chances of groundwater potential are good.

13.4.4.3 Dissection Index (D_I)

Dissection Index (D_I) implies the degree of dissection or vertical erosion and expounds the stages of terrain or landscape development in any given physiographic region or watershed (Singh and Dubey 1997). On an average, the values of D_I vary between 0 and 1. For complete absence of vertical dissection/erosion the D_I is 0 and hence dominance of flat surface. In exceptional cases, vertical cliffs, it may be at vertical escarpment of hill slope or at seashore then the D_I is 1. The D_I for all the sub-basins ranges between 0.99926 and 0.99962 (Table 13.5).

Table 13.5 Relief aspects sub-basin characteristics of Tamil Nadu coast

Sub-basin	Relative relief (R_R)	Relief ratio (R_h)	Dissection index (D_I)	Ruggedness number (R_n)
Palar	1342	0.00334	0.99926	0.37647
Ponnaiyar	1636	0.00377	0.99939	0.45823
Cauvery Lower	1414	0.00491	0.99929	0.4343
Pamba	2650	0.00928	0.99962	0.79744
Vaippar	2000	0.01343	0.9995	0.53813

Therefore, the study area comes under highly dissected index as the D_I value that may be attributed to its adjacent to the seashore location.

13.4.4.4 Ruggedness Number (R_n)

Ruggedness number (R_n) is a combined measure of relief and stream density (Fairbridge 1968). It is calculated as multiplying the drainage density by basin relief. Low R_n value indicates gentler slope and thus less susceptible to soil erosion. The high value of R_n occurs when both D_d and R_R are large and slope is steep (Strahler 1957). The value of R_n for the sub-basin ranges between 0.37647 and 0.79744. As the R_n values are very low, the sub-basins having gentler slope and less susceptible to soil erosion.

13.4.5 Sub-Watershed Prioritization

Five sub-basin prioritizations were made to classify the region of elevated soil erosion activity so that adequate mitigation steps can be taken in order to track soil erosion in that region. Morphometric parameter analysis is critical in the identification and determination of groundwater potential zones and high-risk areas of erosion (Yadav et al. 2018; Said et al. 2018). The compound values (C_p) of all Tamil Nadu coast's twenty-seven sub-watersheds are measured and the final priorities as seen in Table 13.6 are assigned. Figure 13.5 showed the sub-watershed priority zone of Tamil Nadu coast. The 4.6 C_p value sub-watershed SW_9 gains the highest priority led by SW_2. SW_6, SW_{11}, SW_{15}, SW_{18}, SW_{24}, SW_{27} are characterized by high priority of sub-watershed. SW_7 has a high-water holding capacity due to lower runoff density zone, high bifurcation zone, very low stream volume zone, low runoff structure zone, high overland flow zone, and high channel management zone. Sub-watersheds with medium priorities (i.e., SW4, SW5, SW8, SW13, SW14, SW16, SW17, SW19, and SW23) come under mild to extreme farming activities requiring strong irrigation requirements, predominantly flood level, which appears to increase topsoil erosion. Sub-watersheds SW1, SW3, SW21, SW422, and SW26 consisting of mostly barren land, come under low priority areas marked by high drainage density, weak bifurcation zone, high stream volume zone, high runoff texture zone, low overland flow zone, and low river management zone. Highest priority suggests the greater degree of erosion in the individual sub-watershed and is ideal region for soil protection initiatives to be implemented. Soil management steps should then be enforced first at the SW2 and SW9 sub-watersheds; then at other sub-watersheds, based on their priority.

13.5 Conclusions

One of the essential elements of planning for the implementation of its growth and management systems is the prioritization of the basin. This research work establishes the usefulness of geospatial technology for morphometric analysis and prioritizing the Tamil Nadu coast sub-basin in India. The automatic approach for morphometric parameter extraction reduces human efforts enormously and can be performed on several research areas concurrently. The morphometric physiognomies of five sub-basins display their relative characteristics with respect to the hydrological response of the sub-basins. The stream network drainage pattern reveals a dendritic form with a 5th and 6th order stream network that indicates homogeneity in texture and structural control non-existence. The mean bifurcation ratio for Palar, Ponnaiyar, Cauvery Lower, Pamba, and Vaippar sub-basin is determined as 2.28, 1.88, 2.82, 2.03, 2.18, respectively, supporting the fact that the drainage pattern is not affected by a disturbance in recent past geological evolution. Five sub-basin prioritization shows that SW9 and SW2 are included in the high priority group; therefore, hydrologists and decision architects may consider conservation measures in watershed management. The SW1, SW3, SW21, SW422, and SW26 sub-watersheds present steep hills, reduced runoff, and hilly areas with poor drainage and high erosion conditions. The morphometric parameters analyzed by DEM in the GIS platform provided a better understanding of the evolution of

Table. 13.6 Estimated compound parameter with priority ranking

Basin ID	Sub-watershed (SW)	Drainage density (Dd)	Bifurcation ratio	Stream frequency (Fs)	Length of overland flow (Lg)	Texture ratio (Rt)	Elongation ratio (Re)	Circularity ratio (Rc)	Compactness constant	Basin shape	Form factor	C_p value	Final priority
1	1	6.00	4.00	2.00	3.00	3.00	2.00	4.00	6.00	6.00	2.00	3.60	Low
	2	4.00	5.00	2.00	4.00	5.00	6.00	3.00	2.00	2.00	6.00	4.50	Very High
	3	5.00	5.00	2.00	4.00	4.00	4.00	5.00	4.00	4.00	4.00	3.80	Low
	4	4.00	4.00	2.00	5.00	2.00	5.00	5.00	3.00	3.00	5.00	3.90	Medium
	5	4.00	4.00	2.00	4.00	5.00	4.00	3.00	5.00	5.00	4.00	4.10	Medium
	6	4.00	3.00	2.00	4.00	1.00	6.00	7.00	2.00	2.00	6.00	4.20	High
2	7	7.00	1.00	1.00	2.00	3.00	1.00	4.00	6.00	6.00	1.00	2.70	Very Low
	8	5.00	5.00	2.00	4.00	4.00	4.00	4.00	3.00	3.00	5.00	4.00	Medium
	9	6.00	5.00	2.00	3.00	3.00	6.00	5.00	1.00	1.00	7.00	4.60	Very High
	10	3.00	1.00	2.00	5.00	3.00	5.00	5.00	3.00	3.00	5.00	3.80	Low
3	11	3.00	4.00	2.00	5.00	1.00	6.00	6.00	3.00	3.00	6.00	4.20	High
	12	1.00	3.00	2.00	6.00	5.00	1.00	4.00	7.00	7.00	1.00	3.20	Very Low
	13	3.00	3.00	2.00	5.00	4.00	5.00	5.00	2.00	2.00	6.00	3.90	Medium
4	14	5.00	2.00	2.00	4.00	6.00	4.00	3.00	4.00	4.00	4.00	3.90	Medium
	15	3.00	4.00	2.00	5.00	5.00	6.00	3.00	2.00	2.00	7.00	4.30	High
	16	4.00	4.00	2.00	4.00	4.00	5.00	4.00	3.00	3.00	5.00	3.90	Medium
	17	2.00	5.00	2.00	6.00	3.00	5.00	5.00	3.00	3.00	5.00	4.00	Medium
	18	4.00	6.00	2.00	5.00	6.00	3.00	2.00	6.00	6.00	3.00	4.20	High
	19	4.00	1.00	2.00	4.00	3.00	6.00	5.00	2.00	2.00	7.00	4.10	Medium

(continued)

Table. 13.6 (continued)

Basin ID	Sub-watershed (SW)	Drainage density (Dd)	Bifurcation ratio	Stream frequency (Fs)	Length of overland flow (Lg)	Texture ratio (Rt)	Elongation ratio (Re)	Circularity ratio (Rc)	Compactness constant	Basin shape	Form factor	C_p value	Final priority
5	20	6.00	2.00	2.00	3.00	5.00	3.00	2.00	5.00	5.00	3.00	3.40	Very Low
	21	5.00	5.00	2.00	3.00	3.00	4.00	5.00	4.00	4.00	4.00	3.80	Low
	22	7.00	4.00	2.00	2.00	5.00	1.00	3.00	7.00	7.00	1.00	3.60	Low
	23	6.00	5.00	2.00	3.00	2.00	4.00	5.00	5.00	5.00	4.00	3.90	Medium
	24	7.00	7.00	2.00	3.00	6.00	3.00	1.00	5.00	5.00	4.00	4.30	High
	25	7.00	4.00	2.00	3.00	4.00	2.00	3.00	6.00	6.00	2.00	3.40	Very Low
	26	9.00	4.00	2.00	1.00	6.00	1.00	2.00	7.00	7.00	1.00	3.50	Low
	27	8.00	6.00	2.00	2.00	7.00	3.00	1.00	5.00	5.00	4.00	4.20	High

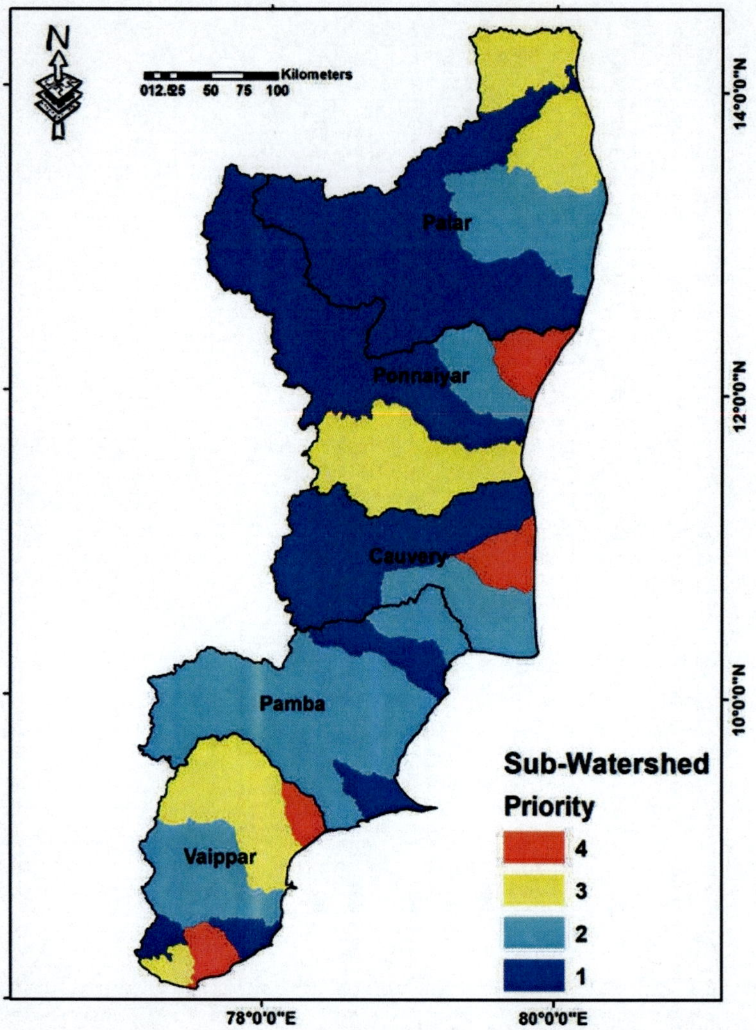

Fig. 13.5 Map showing prioritized sub-watershed with their ranks

the Tamil Nadu basin and its response to hydrological conditions enabling effective management of natural resource strategies even at sub- and micro-watershed levels. In future studies, the integration of morphometric parameters with multi-criteria policy-making will be performed.

References

Abdulkareem JH, Pradhan B, Sulaiman WNA, Jamil NR (2018) Quantification of runoff as influenced by morphometric characteristics in a rural complex catchment. Earth Syst Environ 2(1):145–162. https://doi.org/10.1007/s41748-018-0043-0

Alemu B, Kidane D (2014) The implication of integrated watershed management for rehabilitation of degraded lands: case study of Ethiopian highlands. J Agric Biodivers Res 3(6):78–90

Amulya GV et al (2018) GIS-based morphometric analysis of sub-watershed of Gurupura River, Dakshina Kannada District. Int J Recent Sci Res 9 (6):27545–27549

Ayele AF, Hiroshi Y, Katsuyuki S, Nigussie H, Kifle W (2017) Quantitative analysis and implications of drainage morphometry of the Agula watershed in the semi-arid northern Ethiopia. Appl Water Sci 7:3825–3840

CGWB (2019) Watershed Atlas of India—Basin map. http://cgwb.gov.in/watershed/basinsindia.html. Accessed 9 Aug 2019

Choudhari PP, Nigam GK, Singh SK, Thakur S (2018) Morphometric based prioritization of watershed for

groundwater potential of Mula river basin, Maharashtra, India. Geol Ecol Landsc 2(4):256–267
Deepak KM (2015) The basic concept to study morphometric analysis of river drainage basin: a review. Int J Sci Res 4(7):2277–2280
Fairbridge RW (1968) Quantitative geomorphology. Encyclopedia of Geomorphology, NY, pp 898–912
Horton RE (1932) Drainage basin characteristics. American Geophysical Union, Transactions
Horton RE (1945) Erosional development of streams and their drainage basins; hydrophysical approach to quantitative morphology. GSA Bull 56:275–370
Imran MM, Sultan MB, Kuchay AN (2011) Watershed based drainage morphometric analysis of Lidder catchment in Kashmir valley using geographical information system. Recent Res Sci Technol 3(4):118–126
Kalevs GA (2001) Introduction to geomorphology. Orient Blackswan Private Limited
Kiran VSS, Srivastava YK (2012) Check dam construction by prioritization of micro watershed, using morphometric analysis as a perspective of remote sensing and GIS for Simlapal Block, Bankura, W.B. Bonfring. Int J Ind Engine Manag Sci 2(1):20–31
Krishnamurthy J, Mani A, Jayaraman V, Manivel M (2000) Groundwater resources development in hard rock terrain-an approach using remote sensing and GIS techniques. Int J Appl Earth Obs Geoinf 2(3–4):204–215
Kumar SD, Sharma D, Mundetia N (2015) Morphometric analysis of the Banas River Basin using the geographical information system, Rajasthan India. J Hydrol 3 (5):47–54
Lima CDS, Correa ACDB, Nascimento NRD (2011) Analysis of the morphometric parameters of the rio Preto basin, Serra do Espinhaço (Minas Gerais, Brazil). Geociências, São Paulo 30(1):105–112
Magesh NS, Chandrasekar N, Kaliraj S (2012) A GIS based automated extraction tool for the analysis of basin morphometry. Bonfring Int J Indus Eng Manag Sci 2(1):32–35
Meshram SG, Alvandi E, Singh VP, Meshram C (2019) Comparison of AHP and fuzzy AHP models for prioritization of watersheds. Soft Comput 23:13615–13625. https://doi.org/10.1007/s00500-019-03900-z
Miller VC (1953) A quantitative geomorphic study of drainage basin characteristics in the Clinch Mountain area, Virginia and Tennessee. Office of Naval Research, Geography Branch. Geog Branch Project 389–042:1–30
Muluneh T, Mamo W (2014) Morphometric Analysis of Didessa River Catchment in Blue Nile Basin, Western Ethiopia. Sci Technol Arts Res J 3(3):191–197
Rao N, Latha S, Kumar A, Krishna H (2010) Morphometric analysis of Gostani river basin in Andhra Pradesh State, India. Int J Geomat Geosci 1(2):179–187
Romshoo SA, Bhat SA, Rashid I (2012) Geoinformatics for assessing the morphometric control on hydrological response at watershed scale in the Upper Indus Basin. J Earth Syst Sci 121(3):659–686
Sangma F, Guru B (2020) Watersheds characteristics and prioritization using morphometric parameters and fuzzy analytical hierarchal process (FAHP): a part of lower Subansiri sub-basin. J Indian Soc Remote Sens. https://doi.org/10.1007/s12524-019-01091-6
Said S, Siddique R, Shakeel M (2018) Morphometric analysis and sub-watersheds prioritization of Nagmati River watershed, Kutch District, Gujarat using GIS based approach. J Water Land Develop 39(1):131–139
Schumm SA (1956) Evolution of drainage systems and slopes in badlands at Perth Amboy, New Jersey. Geol Soc Am Bull 67(5):597–646
Sharma SK, Rajput GS, Tignath S, Pandey RP (2010) Morphometric analysis and prioritization of a watershed using GIS. J Ind Water Res Soc 30(2):33–39
Singh Y, Dubey DP (1997) Deep zone karst aquifers as a boon in central India. In: The engineering geology and hydrogeology of karst terranes, pp 245–249
Singh N, Singh KK (2014) Geomorphological analysis and prioritization of sub-watersheds using Snyder's synthetic unit hydrograph method. Appl Water Sci 7:275–283
Smith KG (1950) Standards for grading texture of erosional topography. Am J Sci 248(9):655–668
Strahler AN (1952) Dynamic basis of geomorphology. Geol Soc Am Bull 63(9):923–938
Strahler AN (1957) Quantitative analysis of watershed geomorphology. EOS Trans Am Geophys Union 38 (6):913–920
Strahler AN (1964) Part II. Quantitative geomorphology of drainage basins and channel networks. In: Handbook of applied hydrology. McGraw-Hill, New York, pp 4–39
Thapliyal A, Panwar A, Kimothi S (2017) Prioritization based on morphometric analysis in Alaknanda Basin. Glob J Sci Front Res H Environ Earth Sci 17(3):29–34
Umamathi S, Aruchamy S (2014) Morphometric analysis of Suruliar watershed, Theni district, Tamil Nadu, India: A GIS approach. Indian J Sci Res Technol 2 (1):35–45
Verstappen HT (1983) Applied geomorphology: geomorphological survey for environmental development. Elsevier, Amsterdam
Waikar ML, Nilawar AP (2014) Morphometric analysis of a drainage basin using geographical information system: a case study. Int J Multidiscip Curr Res 2 (2):179–184
Welde K (2016) Identification and prioritization of subwatersheds for land and water management in Tekeze dam watershed, Northern Ethiopia. Int SWC Res 4:30–38
Yadav SK, Dubey A, Szilard S, Singh SK (2018) Prioritisation of sub-watersheds based on earth observation data of agricultural dominated northern river basin of India. Geocarto Int 33(4):339–356

Spatio-temporal Variation of Channel Migration and Vulnerability Assessment: A Case Study of Bhagirathi River Within Barddhaman District, West Bengal, India

14

Debasis Ghosh, Monali Banerjee, Subhadip Pal, and Mrinal Mandal

Abstract

The channel migration across the floodplain areas is a natural dynamic process, mainly for alluvial rivers, and several geo-hydrological factors are involved in this process of shifting. The river Bhagirathi, the westernmost part of the Ganga–Brahmaputra Delta, often experiences more lateral migration along its course from Ketugram-II to Kalna-II Community Development (C.D.) blocks of Barddhaman district. The present study aims to understand the channel migration of the river from 1930 to 2018 based on the collected database, such as the district administrative map of Barddhaman of 1930, satellite images of Landsat 5 Thematic Mapper (TM) of 1987, 1997, 2008, and Landsat 8 Operational Land Imager (OLI) image of 2018. All the data are processed in ArcGIS 10.3.1 software to prepare maps of the river course of different years, and the meandering pattern of the channel is observed. There is no such prominent pattern of bank line shifting observed and the overall average rate of erosion of bank lines from 1930 to 2018 shows a declining trend in general. Moreover, an estimation is also made to know the future pattern of theBhagirathi River course of 2030 adopting the logarithmic regression method. The predicted average bank line erosion rate from 2018 to 2030 is found to be 174 m/year. This possibility of bank line shifting may bring vulnerability to the 17 identified villages adjacent to the river course with a population of around 65,000 of 6 C.D. blocks of Barddhaman district up to 2030.

Keywords

Channel migration · Bhagirathi River · Bank lines · Vulnerability · RS and GIS · Logarithmic regression

14.1 Introduction

Channel migration within its meander belt is one of the common natural processes in the case of an alluvial channel, and the river keeps on shifting its position leaving different imprints (Bhunia et al. 2016; Gregory 1977, 1979, 1983; Hickin and Nanson 1975; Leopold et al. 1995; Lewin and Hughes 1976; Mondal and Mandal 2018; Philip et al. 1989; Thorne 1992, 2002). Several geo-hydrological factors of a river, such as lithology, channel gradient, river discharge, sediment load, velocity and bed materials, etc. control the channel

D. Ghosh (✉) · M. Banerjee · S. Pal
Department of Geography, University of Calcutta, Kolkata, West Bengal 700019, India
e-mail: dggeog@caluniv.ac.in

M. Mandal (✉)
Department of Geography, Sidho-Kanho-Birsha University, Purulia, West Bengal 723104, India
e-mail: mrinal-mandal@skbu.ac.in

P. K. Shit et al. (eds.), *Drainage Basin Dynamics*, Geography of the Physical Environment,
https://doi.org/10.1007/978-3-030-79634-1_14

migration. The cumulative effects resulting from some processes, like higher rate of seepage, direct corrosion, and bank failure may produce lateral shifting of channel or meander migration and channel expansion or channel avulsion (Charlton 2008; Hickin and Nanson 1975; Kotoky et al. 2005; Kumar et al. 2010; Yang et al. 1999). Channel migration is a natural dynamic hydrologic condition that cheers up during high discharge events. The riverbanks, mainly the meander part used to experience rampant erosion in its outer bend, and sometimes the water spills over the floodplain. Channel migration in the Bhagirathi River almost invariably follows the bank failure that mostly takes place on the outer bend of the meanders (Ghosh 2007). Consequently, the morphology of the river channels has been changed with the constant operations of riverbank erosion and accretion processes (Alexeevsky et al. 2013; Chakraborty and Mukhopadhyay 2015; Duran et al. 2011). Many a time, it is observed that the flow patterns of the river get changed due to alterations in channel morphology and fluvial architecture. It is quite obvious that the entire fluvial environment is altered due to channel migration at reach scale (Church 1992; Leopold et al. 1995). Several in-depth studies have been carried out worldwide to evaluate the activities of channel migration and its associated morphological changes (Beechie et al. 2006; Bisson et al. 2011; Das et al. 2007; Dury 1977; Ghosh and Sahu 2018; Gregory 1977; Hickin and Nanson 1975; Knight 1975; Millar 2000; Thorne 1982, 1992; Újvári et al. 2009; Yunus et al. 2019). Channel migration is a dynamic process that changes in response to a variety of inputs. Hydraulic action, abrasion (corrasion) and attrition are well-known geomorphic processes involving running water, which continuously controls the fluid dynamics and sediment entrainment processes. The alluvial river like the Bhagirathi used to keep on shifting its position over the floodplain within the meander belt laterally by the riverbank erosion. Riverbank shift is a subject matter of fluvial morpho-dynamics because fluid dynamics and fluvial architecture are mutually adjusted with each other. Riverbank erosion refers to the separation and entrainment of bank materials in the form of grains, aggregates, or blocks by fluvial, subaerial, and/or geo-tectonic processes. It is a common form of geomorphic hazard associated with meandering or braided river systems and floodplains (Ghosh 2007). The resulting channel migration from combined hydro-geomorphic actions can take a faster speed by human interventions. Therefore, a combination of natural and anthropogenic drivers increases the spatial extension and severity of channel migration (Chakraborty and Mukhopadhyay 2015; Dey and Mandal 2019). On the other hand, certain land-use practices may prepare riverbanks to inhibit erosion. Trampling by grazing animals can reinforce the compactness of the top layer. The right type of vegetation increases resistance to erosion manifold by their elaborate and condense root systems. However, in rivers with high banks, common in the middle and lower reaches of large rivers, the effect of vegetation in arresting bank erosion tends to be limited (Ghosh 2007).

A number of research works have already been done on the geomorphic and environmental effects of the Farakka barrage related to the shifting of the lower course of the river Ganga and its distributaries including Bhagirathi in the downstream areas of India and Bangladesh. Some of the important works carried out on different aspects, such as sedimentation on the river bed of the Bhagirathi (Basu et al. 1996); channel shifting of the Bhagirathi (Bag et al. 2019; Basu 2001; Islam and Guchhait 2017; Laha 2015; Rudra 1996, 2014; Sarkar et al. 2003); channel morphology of the Ganga (Majumder 2004; Sarkar et al. 2003) and the Padma (Ophra et al. 2018; Sarkar 2004); loss and quality deterioration of agricultural field adjacent to the Padma (Gain and Giupponi 2014; Pal et al. 2016). In this context, it becomes more important to generate a digital database based on measuring spatial shifting of riverbanks over a longer period of time to understand lateral migration of river and to estimate the future trend of channel shifting (Das et al. 2012; Philip et al. 1989). It will help to formulate proper management plans for the entire area well in advance. To understand the nature of lateral shifting of the river, the application of Remote Sensing (*RS*) and

Geographical Information System (*GIS*) techniques has been using more popularly worldwide (Aher et al. 2012; Ahmed et al. 2018; Alexeevsky et al. 2013; Beechie et al. 2006; Bisson et al. 2011; Das et al. 2012; Debnath et al. 2017; Dey and Mandal 2019; Duran et al. 2011; Guite and Bora 2016; Hamid et al. 2017; Islam and Guchhait 2017; Kotoky et al. 2005; Laha and Bandyapadhyay 2013; Mandal et al. 2017; Mukherjee et al. 2016; Ophra et al. 2018; Rudra 2014; Saleem et al. 2020; Thakur et al. 2012; Yunus et al. 2019). A number of research works have been carried out by several researchers adopting *RS* and *GIS* techniques to prepare channel migration zones of the river Ganga (Ghosh and Sahu 2018; Philip et al. 1989), the river Torsa (Dey and Mandal 2019), the river Pravara (Aher et al. 2012), the Bhagirathi (Pal et al. 2016) using different techniques.

But there is no such work done on identifying the chronological changes of theBhagirathi River channel for a larger time span along with the future prediction of channel shifting, rate of erosion and accretion of both banks, and assessment of future vulnerable areas with human settlements and agricultural lands. Thus, the amalgamation of *RS* and *GIS* with statistical techniques may be used to assess the channel migration with a scope of estimating future shifting. Here, the present study tries to understand the pattern of channel migration of the river Bhagirathi along the eastern boundary of Barddhaman district for the last 88 years and based on that, a future estimation of shifting of the river course is computed for the year 2030. In addition to this, the vulnerable riparian villages resulting from lateral migration of the river Bhagirathi from 2018 to 2030 of the Barddhaman district are recognized.

14.2 Study Area: General Geographical Settings

West Bengal is susceptible to a varied range of Natural Environmental Hazards (NEH) because of its unique physical setting that stretches from the Himalayas to the Bay of Bengal, which no state of India has to cope with. The Ganga is India's largest river in terms of annual discharge (522×10^9 m^3), length (2,158 km), and drainage area (861.40×10^3 km^2: 78.60% of the total basin area) (Ghosh 2007). The Ganga originates from the Gangotri glacier in the Himalaya at 4,000 m above mean sea level beyond the mountains. It enters into West Bengal at the northern border of Manikchak Community Development (C.D.) block of Malda district (25° 13′ 13″ N, 87° 47′ 40″ E; altitude: 29 m), 19 km north of Rajmahal town. In its deltaic course, the general trend of the Ganga follows a south-southwest ward trend for the first 100 km up to the off-take point of the Bhagirathi and then takes up an east-southeast ward course to its confluence with the Brahmaputra and beyond. The broad physiographic divisions of the area through which the Ganga flows in West Bengal are the lowlands, the Pleistocene terraces/piedmonts, and the hills. In the lowlands, the altitude reduces in the same general direction in which the Ganga flows—from northwest to southeast. It is widely believed that the Bhagirathi-Hugli river represents the earliest course of the Ganga in the delta. Subsequently, the alignment of the river shifted to its present position by successive development and abandonment of several distributary channels that can be traced as off-shoots of the present river. From west to east, they include the Bhagirathi, Kalkali, Bhairab, Silamari, Jalangi (in India), Mathabhanga, Garai-Madhumati, Chandana and Kumar (in Bangladesh). At present, all these rivers are in various stages of degeneration. The only exception is the Bhagirathi-Hugli, which is artificially resuscitated by the Farakka barrage (Ghosh 2007; Ranjan and Ramanathan 2018; Rudra 2012, 2018).

In the present study, the middle stretch of the Bhagirathi River, flowing along the eastern margin of the Barddhaman district of West Bengal is the study of interest (Fig. 14.1). Many geomorphological features of this river (active channel cutoffs, ox-bow lakes, meander scars, etc.) are frequently observed throughout the studied segment. The river Ajay, a right-hand tributary of Bhagirathi flowing over the Rarh

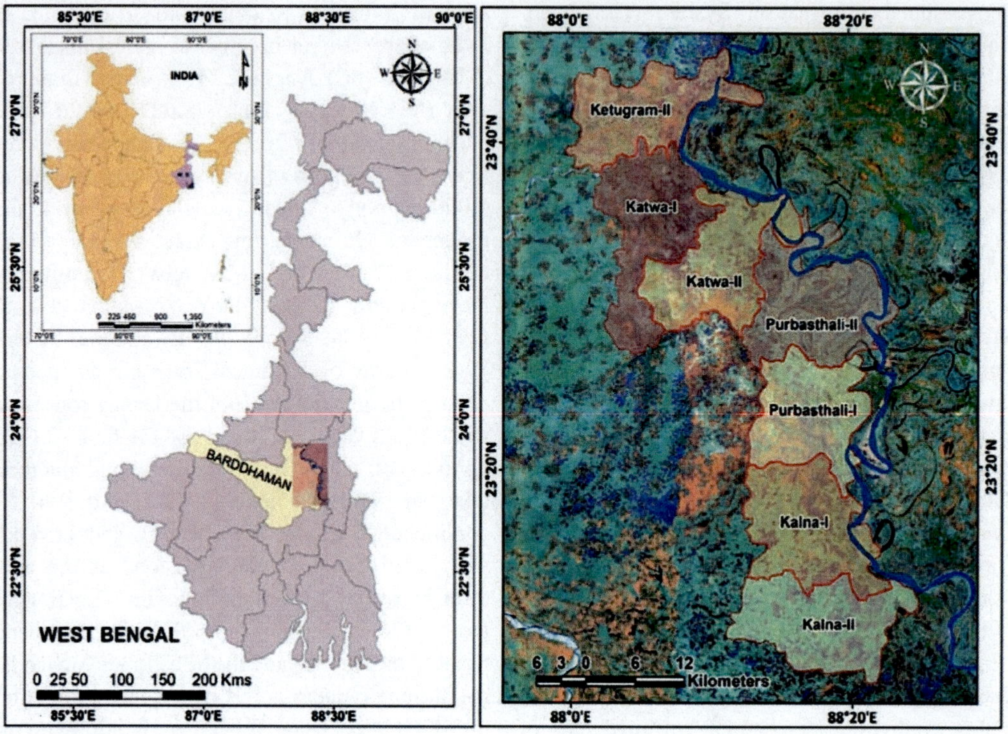

Fig. 14.1 The course of the Bhagirathi River within the study area

Bengal, meets the Bhagirathi near Katwa of Barddhaman district. In the year 2018, the length of the Bhagirathi River within the study area is found to be 114.66 km. The geographical extent of the area of interest lies in between 87° 58′ E to 88° 26′ E and 23° 44′ N to 23° 07′ N. The entire area experiences Monsoon climate, hot and humid summer, and dry mild cold in winter. The high discharge condition is experienced by the river only during the period of south-western Monsoon (July to September) (Ghosh 2007; Rudra 2014). A total number of 7 C.D. blocks, namely: Ketugram-II, Katwa-I, Katwa-II, Purbasthali-I, Purbasthali-II, Kalna-I and Kalna-II come under the studied region. The respective total numbers of villages of the C.D. blocks are 56, 66, 68, 95, 90, 99 and 112. The average population density of all 7 blocks is 1010 person/km^2 in 2011 (Census of India 2011). The eastern portion of the district is a wide alluvial plain enclosed by the river Ajay, the Bhagirathi and the Damodar on the north, south, and east.

14.3 Materials and Methods

14.3.1 Database

The present work has been carried out based on a detailed field survey at several segments of the river Bhagirathi within studied reach. The lateral migration of the river Bhagirathi in its part of the middle reach, mainly in the Barddhaman district, has been portrayed in the present research work using different types of data, such as the district administrative map of Barddhaman, 1930 at a scale of 1.5–10 km collected from Administrative Atlas-West Bengal, 2001 (Census of India 2001); police station map of each C.D. block (total 7 C.D. blocks of the study area) at a scale of 1–10 km taken from Administrative Atlas-West Bengal, 2001 (Census of India 2001); satellite images of Landsat 5 Thematic Mapper (TM) of 1987 (acquired on 24.12.1987), 1997 (acquired on 03.12.1997), 2008 (acquired on

Table 14.1 Detail specifications of the database used in the study

Data type	Map/Satellite/Sensor details	Acquisition date	Path	Row	Map scale/Spatial resolution	Source
District map	Administrative map of Barddhaman	1930	–	–	1.5 cm = 10 km	Administrative Atlas-West Bengal, Census of India 2001 (https://censusindia.gov.in/)
Block map	Police station map along with villages	1930	–	–	1 cm = 1 km	Administrative Atlas-West Bengal, Census of India 2001 (https://censusindia.gov.in/)
Remote sensing images	Landsat 5-TM	24.12.1987	139	44	30 m	USGS (https://earthexplorer.usgs.gov/)
	Landsat 5-TM	03.12.1997	139	44	30 m	
	Landsat 5-TM	16.01.2008	139	44	30 m	
	Landsat 8-OLI	13.12.2018	139	44	30 m	

16.01.2008) and Landsat 8 Operational Land Imager (OLI) image of 2018 (acquired on 13.12.2018) with 30 m spatial resolution downloaded from EarthExplorer Interface (https://earthexplorer.usgs.gov/) developed by United States Geological Survey (USGS) (Table 14.1).

14.3.2 Methodology

14.3.2.1 Data Processing

The district administrative map has been used to extract the blocks of Bardhhaman district and the river channel of 1930, and satellite images have been used to extract river channels of 1987, 1997, 2008 and 2018 to identify the bank line shiftings of the river Bhagirathi. The police station map is used to extract and mark the river adjacent villages, which have more possibility of being vulnerable to erosion during high river discharge. All the maps are transferred into raster format and properly geocoded (Universal Transverse Mercator projection, World Geodetic System 84 datum and zone 45 N) using the georeferencing tool of ArcGIS 10.3.1 software. Now all four images of different years are done Standard False Colour Composition (SFCC) in ArcGIS software to prepare the vector layers of riverbank lines.

14.3.2.2 Data Analysis

Sinuosity index is an important morphometric technique, which indicates the nature and behaviour of a river. It provides an understanding of the irregularity pattern of a river course in terms of channel bending (Charlton 2008; Huggett 2017; Leopold et al. 1995; Schumm 1963; Summerfield 2013) and it can be computed using the following formula (Schumm 1963):

$$\textit{Sinuosity Index} = \frac{OL}{EL} \tag{14.1}$$

where *'OL'* is the length of the observed path and *'EL'* is the length of the expected straight path of a river. A value of less than 1.1 of the sinuosity index refers to the straight path of a river channel, while the value between 1.1 and 1.5 indicates sinuous. In contrast, a value of more than 1.5 is considered as meandering.

Now, the actual area of each C.D. block is measured by subtracting the river area from the total area of the concerned C.D. block. The total area of erosion and deposition within the Barddhaman district of the river Bhagirathi is also calculated. To do this, the vector layer of the river course of 1987 is superimposed on 1930 in ArcGIS software, and the total area of erosion and deposition is computed based on the difference between vector layers of river courses. The same procedure is repeated from 1987 to 1997, 1997 to 2008 and 2008 to 2018. Ultimately, it gives results of erosion and deposition for a particular period. All the vector layers of riverbank lines of 5 different years are superimposed chronologically (the year 1930, 1987, 1997, 2008 and 2018) in ArcGIS software to generate a particular thematic map of change detection of the Bhagirathi River course. Based on the 88 years of digital database for change detection of the river, precise observation is made to identify those hotspots, which experienced more vulnerability in terms of lateral shifting of the channel in different periods in the entire studied river stretch. Therefore, a total number of 23 hotspots (most vulnerable riverbank areas to erosion) along the river are identified for cross-section considering the superimposed map (necessary ground truth verifications are also done) (Fig. 14.8). Based on these drawn cross-sections, the shifting of bank line of either side of the river is measured between two consecutive selected years e.g. 1930 to 1987, 1987 to 1997, 1997 to 2008 and 2008 to 2018 adopting a measuring tool in ArcGIS software. With the help of the measured data, the average erosion rate is calculated for each period.

Here an attempt has been made to estimate the future migration of the river Bhagirathi using statistical techniques. It will help to identify the areas, which might encounter the chance of being vulnerable due to river migration. The dataset of 1997, 2008 and 2018 are taken to predict the bank line shift for 2030 using the logarithmic regression method which has already been applied by many earlier researchers successfully (Das et al. 2012; Rusov et al. 2017). To validate the acceptability of future prediction, linear, logarithmic and polynomial regressions are performed to project the right bank shifting of the river along the cross-section for 2018 based on the right bank shifting database of the river from 1930–1987, 1987–1997 and 1997–2008. The derived results of logarithmic and linear regressions are found to be satisfactory than the polynomial regression when compared with actual bank line shifting data of 2018. Moreover, the logarithmic regression gives more reliable data for 2018 than the linear regression (Table 14.2). Therefore, the future shifting of both the bank lines from 2018 to 2030 is projected by the logarithmic regression model based on the computed database of 23 hotspot cross-sections and their previous shifting of the riverbank lines from 1987–1997, 1997–2008 and 2008–2018. On the basis of projected riverbank shifting, the future vulnerable villages of the selected C.D. blocks of the Barddhaman district have been identified. The formulae of linear regression (formula 14.2) and logarithmic regression (Formula 14.3) are as follows:

$$Y = a + bX \tag{14.2}$$

where *'Y'* is the estimated dependent variable, *'X'* is the independent variable, *'a'* and *'b'* are constant and represent the *Y* intercept and the slope of the line, respectively.

$$Y = b\ln(X) + a \tag{14.3}$$

where *'Y'* refers to the estimated dependent variable, *'X'* means the independent variable and its value must be greater than '0', while *'ln'* is the log_e and is generally called as natural logarithm. The constants of a particular equation are *'a'* and *'b'*.

After the computation of projected riverbank line migration of the Bhagirathi for the year 2030, some of the hotspots of selected cross-sections, which are estimated to be more vulnerable out of total 23, are visited for ground truth verification and to understand the actual trend of bank line shifting at the present day (Fig. 14.15).

14.4 Results and Discussion

14.4.1 Changing Pattern of the Bhagirathi River from 1930 to 1987

The length of the river Bhagirathi was 123.53 km in 1930 and it was flowing along the extreme eastern margin of the Barddhaman district covering all the 7 C.D. blocks (Fig. 14.2). An area of 58.99 km^2 was computed as the total area of the Bhagirathi. The calculated sinuosity

Table 14.2 Actual and predicted right bank line shift of the Bhagirathi River

Cross-section no	Actual right bank line shift in 2018 (m)	Right bank line shift in 2018 by linear regression (x = 31, i.e. no. of years since 1987)	Right bank line shift in 2018 by logarithmic regression (x = 32nd year as 1987 is the 1st year)
1	4.40	23.93	1.66
2	−13.74	−139.98	−71.66
3	25.77	−234.25	−117.54
4	−31.17	131.84	61.18
5	12.07	−42.59	−16.95
6	54.88	−699.76	−207.79
7	−11.40	−385.56	−173.86
8	224.16	−182.90	−58.79
9	−1276.47	459.09	495.57
10	8.58	2375.53	871.75
11	145.47	2006.69	849.38
12	15.06	78.30	0.18
13	−22.50	−1331.09	−710.58
14	−42.57	−562.00	−238.45
15	57.88	692.70	235.82
16	109.82	2374.52	1136.82
17	21.67	−1580.67	−760.94
18	16.93	−1491.38	−557.04
19	−21.26	−1924.39	−815.25
20	8.69	250.52	67.18
21	13.41	534.54	236.99
22	−13.84	−759.82	−312.32
23	76.94	27.67	56.24

Note Negative values indicate erosion and positive values indicate accretion

index of the river is 1.96, which indicates the meandering nature of the Bhagirathi (Fig. 14.6). On the contrary, the river experienced more lateral shifting in 1987 within a time frame of 57 years (from 1930 to1987). The river almost shifted more eastward from Katwa-I, Purbasthali-I and Kalna-II C.D. blocks. The total length of the Bhagirathi was 133.47 km in 1987 with an increase of 34.94 km in length compared to 1930. Despite, an increase in the length of 9.94 km in 1987, the river lost an area of 11.63 km^2. This was due to the decrement in the width of the river from 1930 to 1987 (Fig. 14.10). During this period, a total area of all C.D. blocks increased, but Purbasthali-II experienced more rise in the area compared to others C.D. blocks in 1987 (Fig. 14.7b). A value of 2.13 of the sinuosity index refers to more meander course of the Bhagirathi River in 1987 than in 1930 (Fig. 14.6).

14.4.2 Changing Pattern of the Bhagirathi River from 1987 to 1997

In the case of 1997, the total length of the river was 117.31 km with a sinuosity index of 1.87, though the value depicts the meandering nature of the Bhagirathi River, it was much less than the

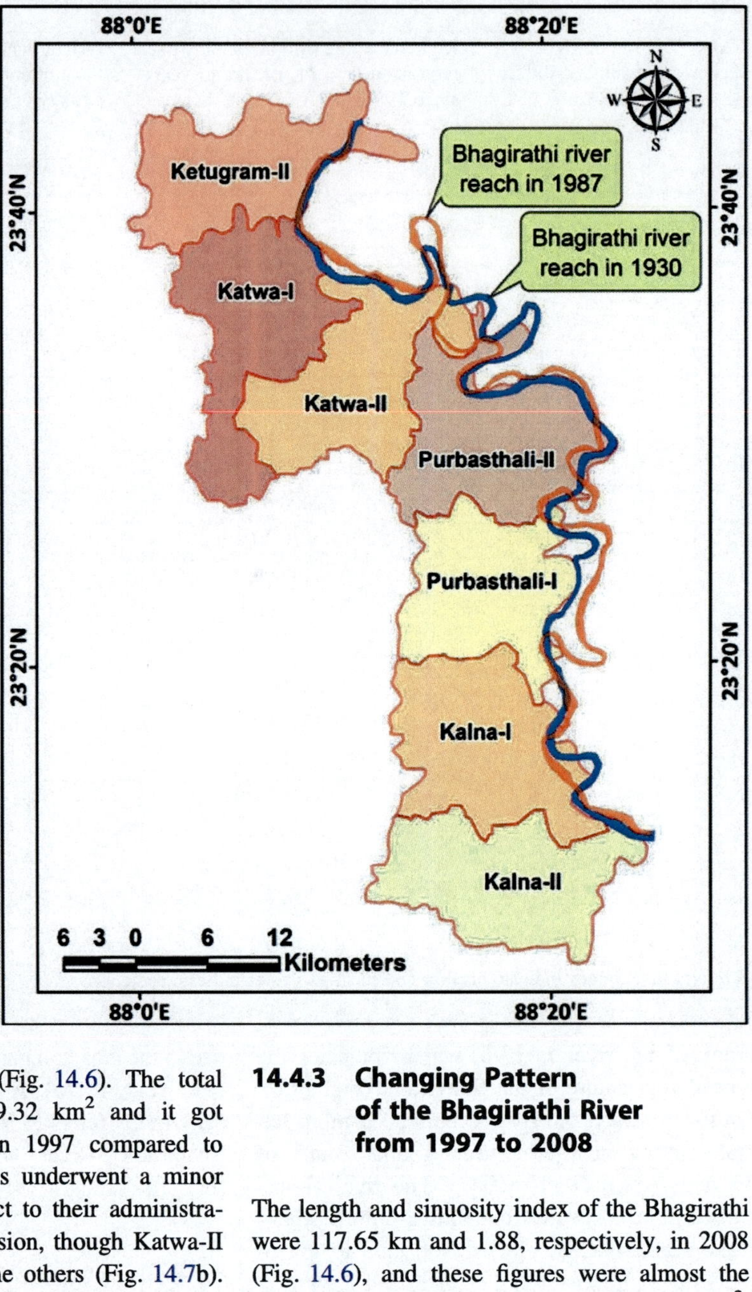

Fig. 14.2 Migration of the Bhagirathi River reach from 1930 to 1987

sinuosity value of 1987 (Fig. 14.6). The total area of the river was 49.32 km^2 and it got increased by 1.96 km^2 in 1997 compared to 1987. All the C.D. blocks underwent a minor transformation with respect to their administrative areas by the river erosion, though Katwa-II was more affected than the others (Fig. 14.7b). The river Bhagirathi took a straight course at Katwa-II and in between Purbasthali-II and Purbasthali-I, leaving behind two ox-bow lakes. The river shifted apart almost from the Katwa-I and Kalna- II C.D. blocks, but overall, the river maintained almost the same course of 1987 (Fig. 14.3).

14.4.3 Changing Pattern of the Bhagirathi River from 1997 to 2008

The length and sinuosity index of the Bhagirathi were 117.65 km and 1.88, respectively, in 2008 (Fig. 14.6), and these figures were almost the same in 1997. The river area was 48.45 km^2, which decreased by 0.87 km^2 from 1997. The course of the river was almost the same as in 1997, but there were some changes observed in the geometry of meanders. The river shifted more towards the eastern boundary of Kalna-II C.D. block (Fig. 14.4). The river Bhagirathi had not

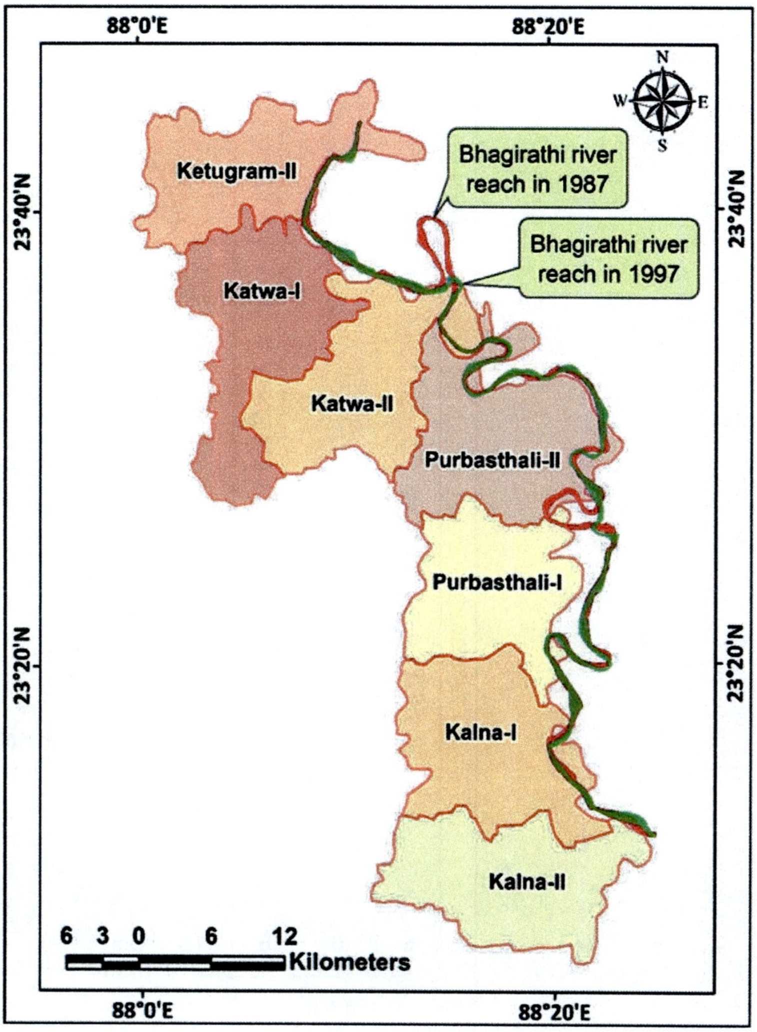

Fig. 14.3 Migration of the Bhagirathi River reach from 1987 to 1997

made any significant changes in areas of C.D. blocks; it made a small loss in terms of area of 6 C.D. blocks (Fig. 14.7b).

14.4.4 Changing Pattern of the Bhagirathi River from 2008 to 2018

The length of the river in 2018 was 114.66 km and it reduced by 2.99 km from 2008. The total area of the river was calculated to be 49.32 km^2, which increased by 0.87 km^2 than the area in 2008. The meandering behaviour of the river was found from the calculated sinuosity index, i.e. 1.84 (Fig. 14.6). The Bhagirathi River changed its course at Katwa-II C.D. block developing an ox-bow lake in 2018. The river maintained almost the samc rivcr course, as it was in 2008. Only some changes took place in river meander geometry (Fig. 14.5). The total areas of all the C.D. blocks remained almost the same in 2018 (Fig. 14.7b).

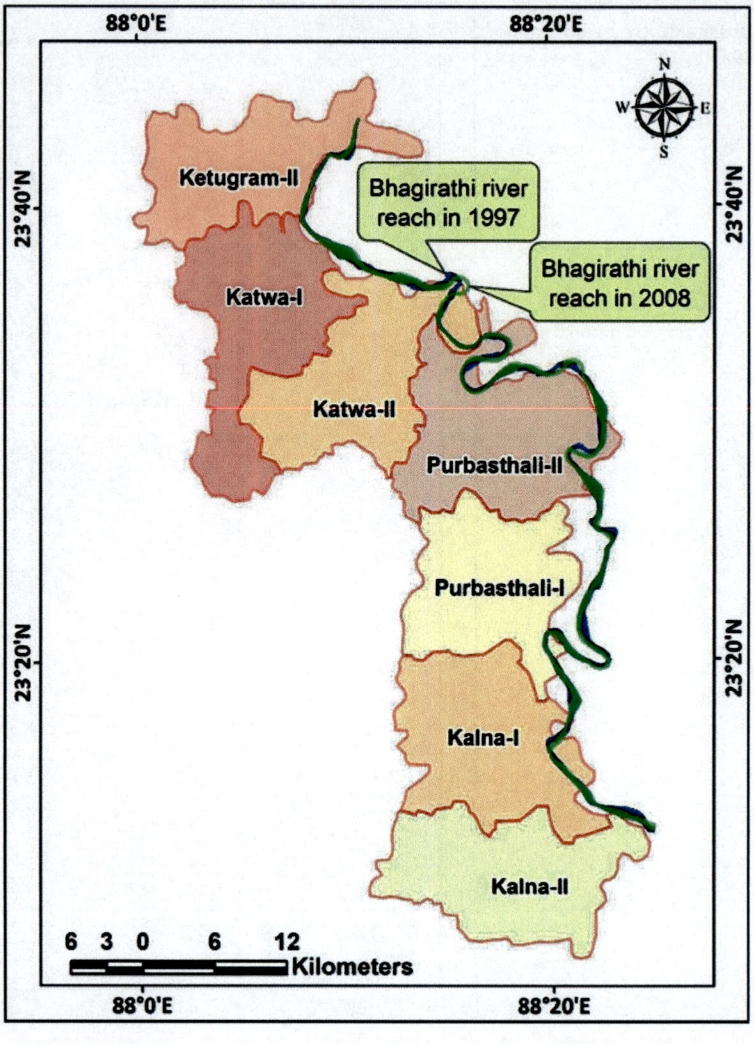

Fig. 14.4 Migration of the Bhagirathi River reach from 1997 to 2008

14.4.5 Bank Line Dynamics of the Bhagirathi River: Socio-economic Impacts

The dynamic nature of the fluvial processes never be stopped and it always tries to keep a balance between its flow conditions and sediment load, and that will give rise to changing the nature of floodplain architecture. This dynamicity of the fluvial processes can better be understood from W. M. Borlaug and E. W. Lane's diagram, where it was explained that when sediment load is powerful than the power of streamflow, the deposition will take place and when the discharge rate is high and powerful than sediment load erosion will take place (Gupta 2011). It is quite relevant in the case of our studied segment. From Tables 14.3 and 14.4, it is evident that in cross-sectional areas, both the processes of accretion and erosion are involved in the channel migration of the Bhagirathi.

The analytical results reveal that the river Bhagirathi has been changing its course from 1930 to 2018 without any specific trend and pattern. The shifting has no particular direction over the years; sometimes both the bank lines of the river have shifted in the same direction and sometimes one bank line has shifted in the

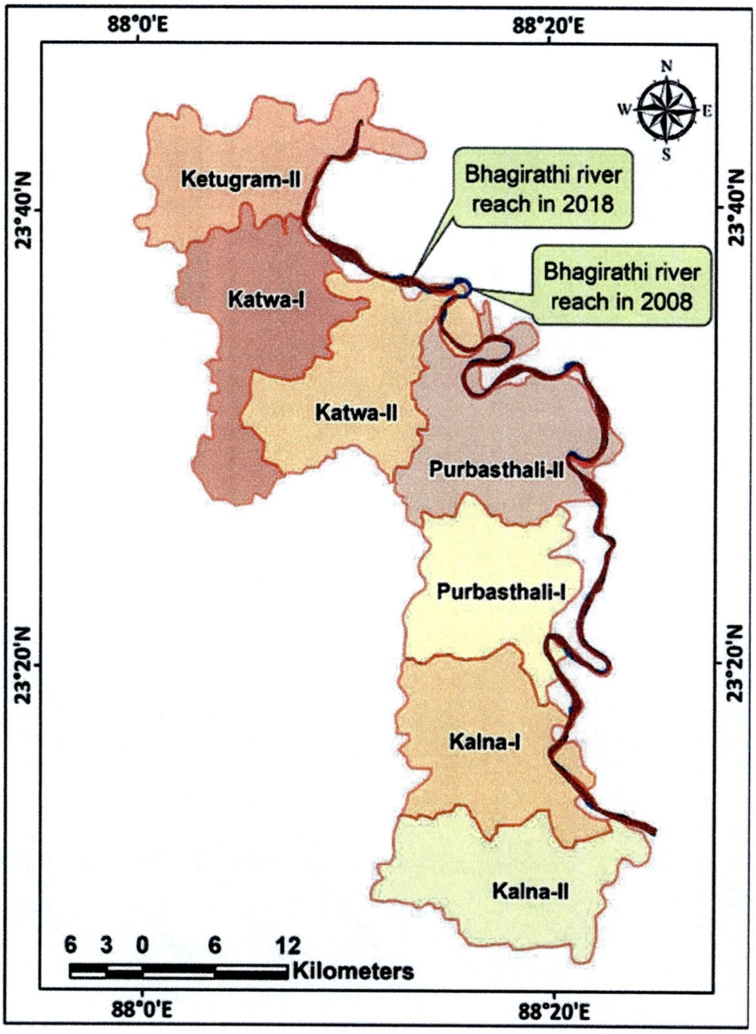

Fig. 14.5 Migration of the Bhagirathi River reach from 2008 to 2018

opposite direction to the other (Fig. 14.9). Thus, the width of the river has also changed (Fig. 14.10). This shifting has resulted in continuous devastation of agricultural lands and settlements due to erosion and also in the generation of new lands due to the aggradation process. The shifting pattern of the channel is more prominent between 1930 and 1987 (Fig. 14.2). The bank line shifting from 1987 to 2018 is not as prominent as 1930 to 1987 (Fig. 14.8). Thus, the river area of each C.D. block changed more from 1930 to 1987 as compared to 1987 to 2018 (Fig. 14.7a). It is also revealed that within the 57-year time span (from 1930 to 1987), the overall total eroded land area was 1054 ha affecting 29 mouzas (i.e. villages as per District Census Handbook, Barddhaman) of 6 C.D. blocks and the depositional area was 2777 ha within 43 mouzas of 7 C.D. blocks of Barddhaman district (Fig. 14.11). During 1987–1997, the total eroded area decreased to 294 ha affecting 18 mouzas of 6 C.D. blocks and deposited 274 ha of land within 19 mouzas of 6 C.D. blocks (Fig. 14.11). From 1997 to 2008 time period, the eroded area slightly increased to 362 ha affecting 19 mouzas and the deposited area increased to 282 ha influencing 16 mouzas of 5 C.D. blocks respectively (Fig. 14.11). In the

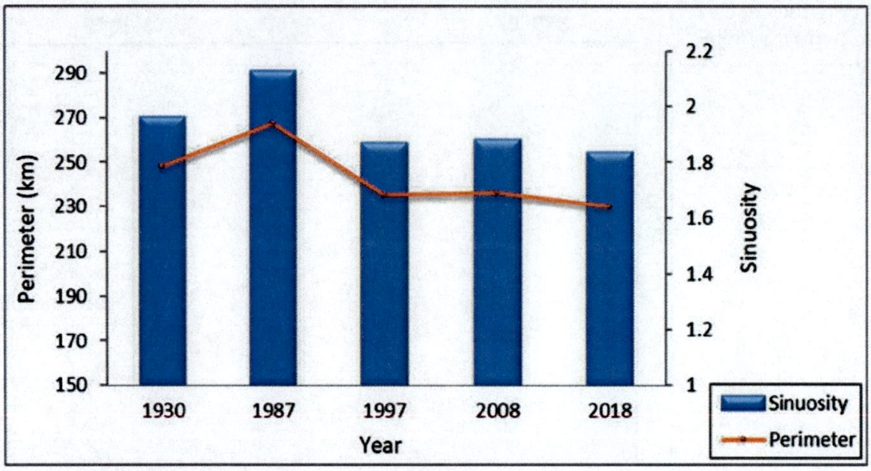

Fig. 14.6 Change in perimeter and sinuosity of the Bhagirathi River from 1930 to 2018

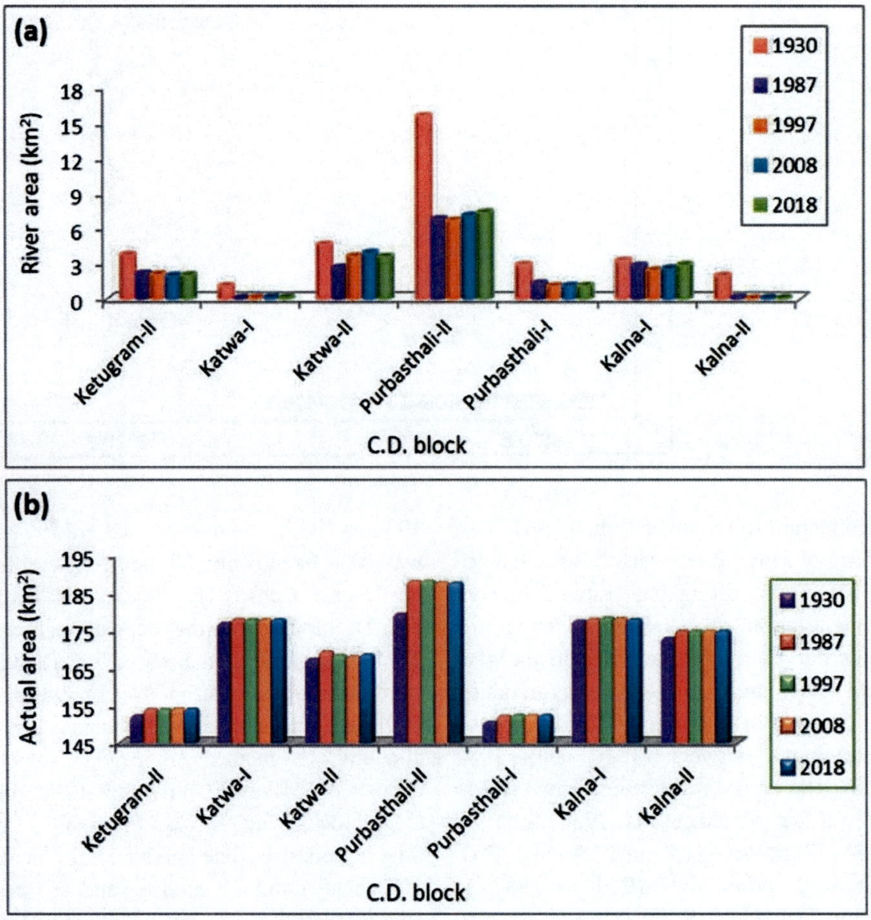

Fig. 14.7 Change in (**a**) river area and (**b**) Actual area within C.D. blocks due to channel shifting from 1930 to 2018

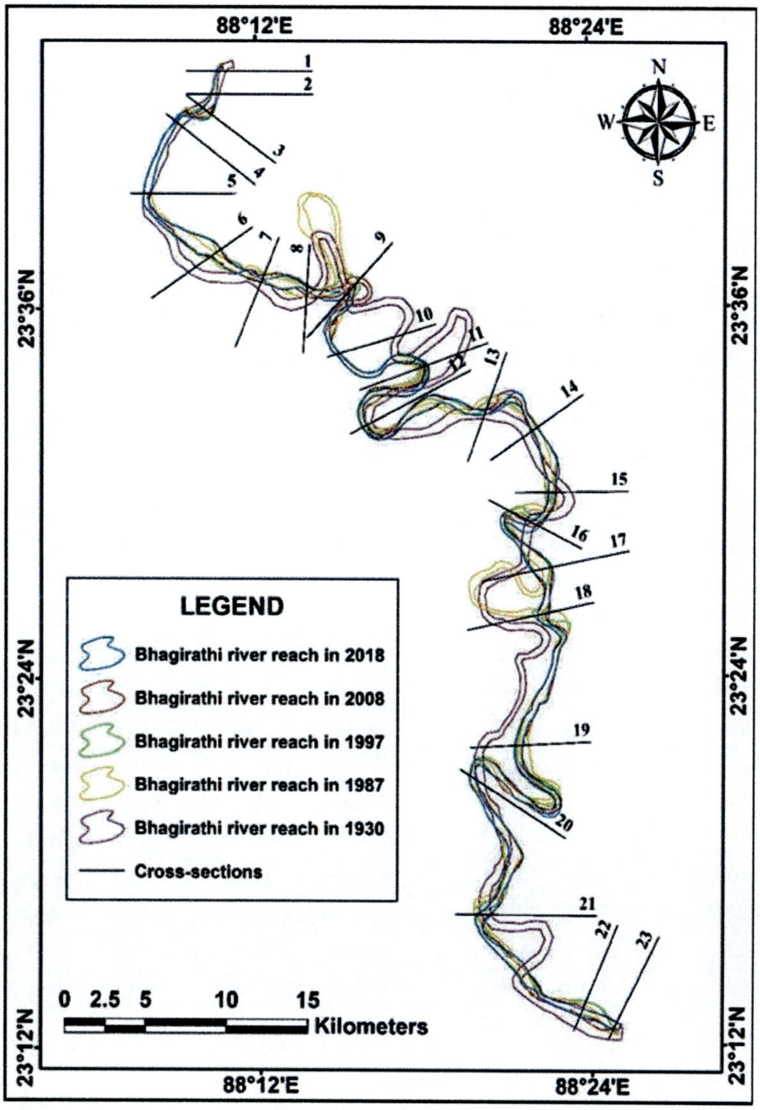

Fig. 14.8 Selected cross-sections across the bank lines of the Bhagirathi River reach

past decade of 2008–2018, the eroded area again reduced to 231 ha of land distributed within 17 mouzas of 5 C.D. blocks and the deposited area also reduced to 201 ha within 22 mouzas of 5 C. D. blocks (Fig. 14.11). Now, observing the overall trend and pattern of total erosion and deposition, the bank failure event may be perceived as insignificant with respect to its gravity and intensity. But the reality is something different from the overall average rate of erosion and accretion within this 114 km long linear stretch of the river Bhagirathi. Thus, the 23 drawn cross-sections (Fig. 14.8) based on identified bank erosion hotspots are assessed and evaluated to know the rate and pattern of bank line shifting for a better understanding of the reality and future trend. The cross-sections data unveils that within 1930–1987, the average rate of erosion due to the bank line shifting was 202 m/year affecting on an average 6 C.D. blocks. From 1987 to 1997, the average rate of erosion due to channel migration reduced to 167 m/year influencing 6 C.D. blocks. During 1997–2008, the average rate of river migration

Table 14.3 Temporal shifting of bank lines of the Bhagirathi River along the cross-sections

Cross-section no	1930–1987				1987–1997			
	Left bank line shift (m)	Direction	Right bank line shift (m)	Direction	Left bank line shift (m)	Direction	Right bank line shift (m)	Direction
1	395.46	Right	−90.77	Right	−4.54	Left	−4.04	Right
2	133.82	Right	137.58	Left	14.40	Right	−30.56	Right
3	−287.51	Left	399.39	Left	−114.04	Left	−97.91	Right
4	280.46	Right	−142.44	Right	19.96	Right	14.58	Left
5	−53.12	Left	10.16	Left	201.40	Right	14.85	Left
6	−1473.65	Left	1451.84	Right	−95.93	Left	39.45	Left
7	−393.87	Left	519.60	Left	−8.12	Left	−60.85	Right
8	−349.94	Left	532.76	Left	−187.16	Left	−51.44	Right
9	250.31	Right	−219.53	Right	−602.38	Left	780.58	Left
10	4152.88	Right	−3790.60	Right	−61.01	Left	−15.27	Right
11	2930.70	Right	−2676.09	Right	−304.75	Left	146.75	Left
12	362.35	Right	−220.89	Right	66.92	Right	−52.63	Right
13	−790.29	Left	995.98	Left	108.47	Right	−275.90	Right
14	−519.59	Left	784.99	Left	−27.63	Left	−53.60	Right
15	1098.72	Right	−859.45	Right	71.84	Right	−135.90	Right
16	1378.04	Right	−1019.48	Right	116.10	Right	−128.10	Right
17	−240.10	Left	2107.33	Left	50.67	Right	−381.10	Right
18	−3113.33	Left	2779.88	Left	935.80	Right	−145.60	Right
19	−2734.25	Left	2674.58	Left	17.05	Right	−177.20	Right
20	676.50	Right	−491.83	Right	23.70	Right	−43.98	Right
21	761.06	Right	−476.09	Right	59.90	Right	−5.18	Right
22	−746.56	Left	875.85	Left	−211.30	Left	15.26	Left
23	−86.76	Left	417.45	Left	−44.36	Left	−13.63	Right

Note Negative values indicate erosion and positive values indicate accretion

through bank erosion increased to 226 m/year affecting 5 C.D. blocks. Interestingly, again the rate of erosion within 2008–2018 reduced to 120 m/year on an average influencing 5 C.D. blocks. So, it could be found that the pattern of change in terms of erosion, where the high rate was replaced by a low rate of erosion and again the same trend was repeated. But the overall general trend shows the gradual decrement in erosion rate over the years (Fig. 14.13).

Based on the analyzed data, the rate, trend and pattern of the channel migration within this reach, it can be said that the river Bhagirathi is going to attain its decaying stage, but now resuscitated by the Feeder canal of the Farakka barrage system. It is a quite normal process of degeneration of the distributary system of the Ganga–Brahmaputra Delta (GBD), where the off-take points of the distributaries were getting defunct. The establishment of the Farakka barrage in Murshidabad in 1975 had affected river hydraulics (Basu and Sen 1997; Pal et al. 2016) and changed the dynamicity of the river. There was a significant change in the Bhagirathi River seen over the long period of 57 years, i.e. in between the courses of 1930 and 1987 (Fig. 14.2). The diverted water from the Ganga into the Bhagirathi increased the

Table 14.4 Temporal shifting of bank lines of the Bhagirathi River along the cross-sections

Cross-section no	1997–2008				2008–2018			
	Left bank line shift (m)	Direction	Right bank line shift (m)	Direction	Left bank line shift (m)	Direction	Right bank line shift (m)	Direction
1	25.78	Right	−24.35	Right	−46.80	Left	4.40	Left
2	−2.38	Left	−32.62	Right	9.78	Right	−13.74	Right
3	190.64	Right	36.57	Left	−125.96	Left	25.77	Left
4	33.74	Right	27.82	Left	19.76	Right	−31.17	Right
5	−63.12	Left	−30.50	Right	−16.04	Left	12.07	Left
6	−280.66	Left	157.20	Left	−48.13	Left	54.88	Left
7	137.65	Right	−28.16	Right	211.27	Right	−11.40	Right
8	−64.21	Left	128.10	Left	−66.61	Left	224.16	Left
9	−202.84	Left	63.04	Left	1321.40	Right	−1276.50	Right
10	12.77	Right	−18.64	Right	−12.17	Left	8.58	Left
11	−103.60	Left	198.55	Left	−62.03	Left	145.47	Left
12	15.86	Right	−34.48	Right	−11.08	Left	15.06	Left
13	683.58	Right	−462.23	Right	−16.98	Left	−22.50	Right
14	79.95	Right	−35.87	Right	2.04	Right	−42.57	Right
15	−232.33	Left	141.47	Left	77.80	Right	57.88	Left
16	−1546.60	Left	1325.80	Left	−154.08	Left	109.82	Left
17	−63.99	Left	−96.50	Right	−60.40	Left	21.67	Left
18	−303.63	Left	237.40	Left	−218.76	Left	16.93	Left
19	344.55	Right	−130.66	Right	53.23	Right	−21.26	Right
20	−4.25	Left	−36.19	Right	−9.66	Left	8.69	Left
21	−708.03	Left	175.68	Left	−107.62	Left	13.41	Left
22	389.50	Right	−156.69	Right	86.91	Right	−13.84	Right
23	−317.40	Left	222.64	Left	−13.69	Left	76.94	Left

Note Negative values indicate erosion and positive values indicate accretion

hydraulic dynamicity and channel oscillation of the river Bhagirathi (mainly for the first two decades after the establishment of Farakka Barrage) to accommodate itself into the altered hydraulic condition (Rudra 2018). The river Bhagirathi with new hydraulic dynamicity developed two ox-bow lakes on the floodplain beheading the meanders. The first one was developed in 1989 upstream of Nabadwip, which was triggered by huge flood water of the same year (Basu et al. 2005; Islam and Guchhait 2017; Rudra 2018), and another one was developed near Shantipur in 1994 (Rudra 2018). The hotspots of the bank erosion remain active in high discharge conditions during the Monsoon period and also aggravates by the local anthropogenic activities in the form of sediment mining from the channel or ferry service operations (Fig. 14.12a, b). On the other hand, the tributary system contributes a lot and plays a vital role in maintaining the Bhagirathi River system. Two major right hand tributaries, namely: Mayurakshi and Ajay meet the river Bhagirathi near Kalyanpur of Murshidabad district and near Katwa of Bardhaman district, more south to Kalyanpur, respectively (Pal et al. 2016; Rudra 2008). Mayurakshi and Ajay discharge a huge volume

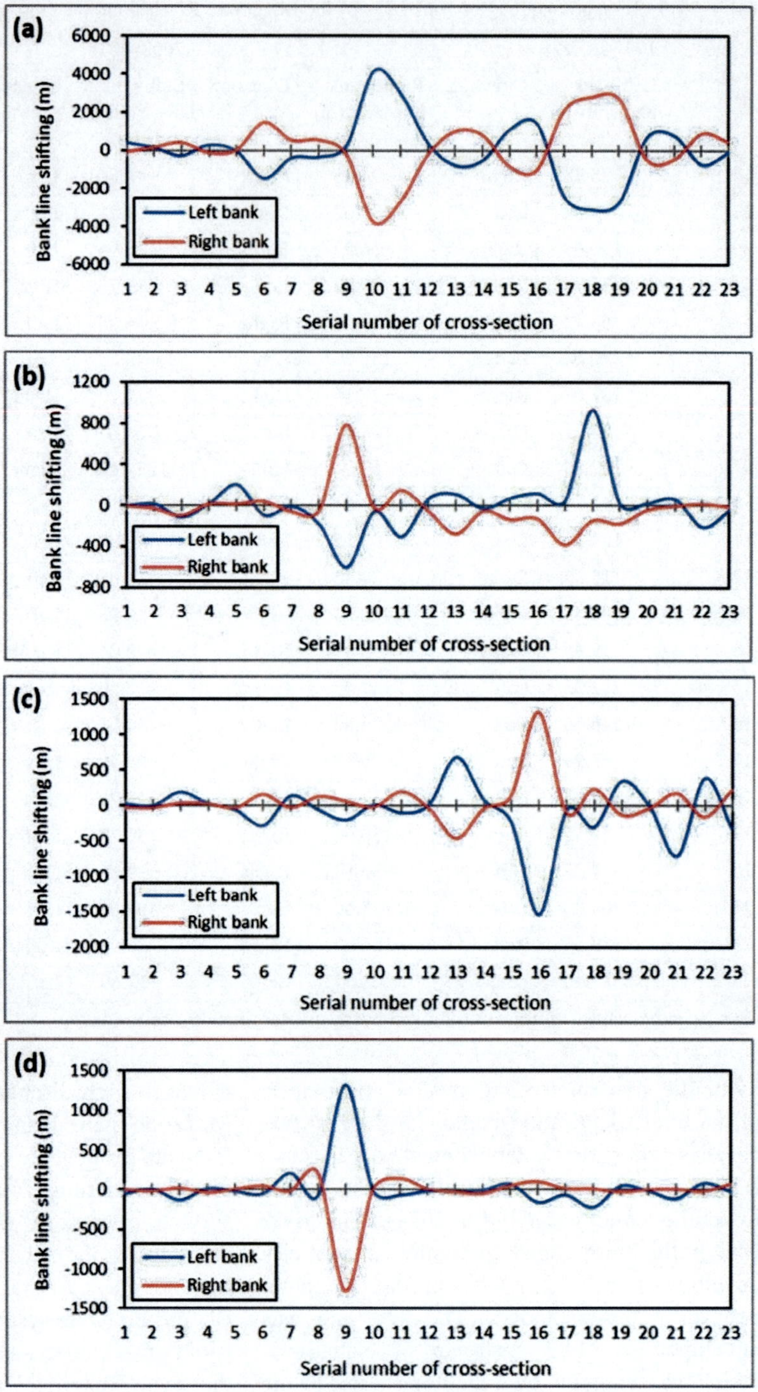

Fig. 14.9 Bank line shifting of the Bhagirathi River due to erosion (negative values) or accretion (positive values) from (**a**) 1930 to 1987, (**b**) 1987 to 1997, (**c**) 1997 to 2008, (**d**) 2008 to 2018

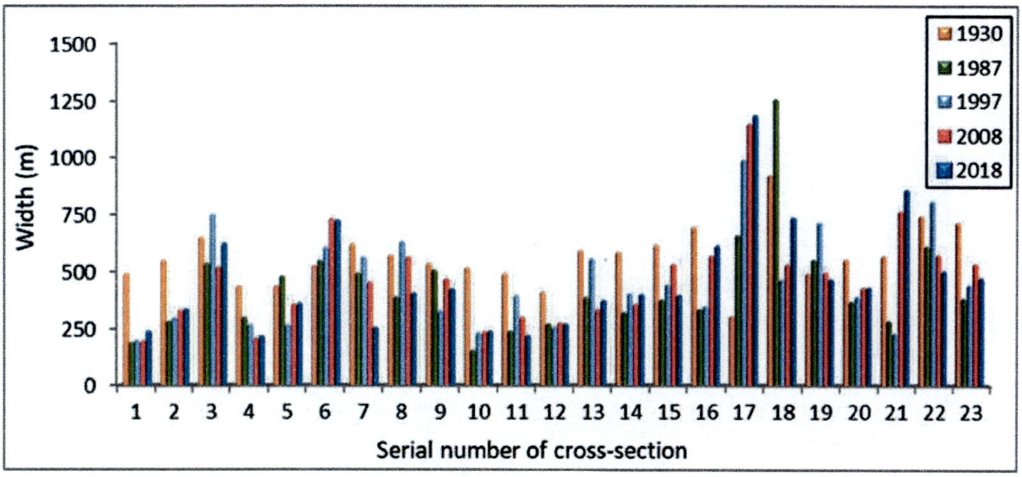

Fig. 14.10 Change in width of the Bhagirathi River along the cross-sections from 1930 to 2018

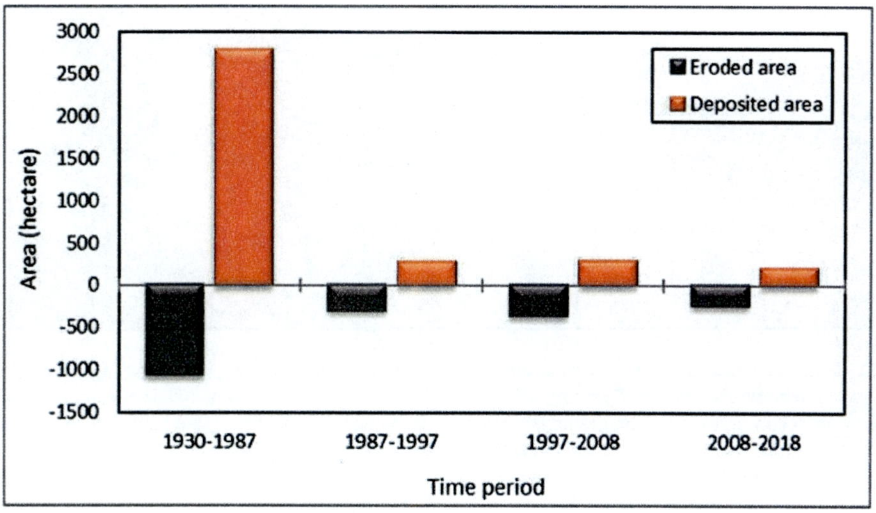

Fig. 14.11 Total eroded and deposited area within Barddhaman district due to lateral migration of the Bhagirathi River in different periods

of water along with sediments to the Bhagirathi during the rainy season, and ultimately this brings significant changes in meander geometry and associated features (Rudra 2018). It is evident from the Google Earth Pro images (images of Landsat/Copernicus satellite) of 1984, 1990, 2000, 2010 and 2018 (all the images are of December month) that a significant variation in alluvial bars in terms of number is observed during this period in the study area. There were 17 alluvial bars of different sizes found in the river channel in 1984 and it got increased over time, such as 20 in 1990, 22 in 2000, 31 in 2010 and 43 in 2018. It simply refers to the dynamicity of the river Bhagirathi.

On the contrary, the process of erosion also gets accelerated due to anthropogenic activities. In the entire stretch of the study area, there are plenty of numbers of brickkiln industries on

Fig. 14.12 Illegal sediment mining sites for **a** brickkiln industry and **b** pottery industry. **c** Dumping of waste materials of brickkiln industries along the Bhagirathi River as an effort to prevent erosion by the local residents. **d** The lady in this picture has lost her in-laws house along with agricultural land and became agricultural labourer, and the man beside the lady has already lost some of his land and the agricultural land on which he standing is also eroding

either side of the river. Many people are engaged in illegal and unscientific sediment mining to supply the same to the brickkiln industry, pottery industry and other works (Fig. 14.12a, b, c). These works suddenly enhance the process of bank erosion modifying the local flow pattern and enhancing the toe erosion. The bank materials along the river are mainly composed of loose soil particles of non-cohesive sand with more pore spaces. During the Monsoon period, the heavy downpour causes abrupt bank erosion, which may incur a bank slump (Bag et al. 2019; Panda and Bandyopadhyay 2011). A sudden high discharge event encourages the chance of bank failure. The fast rates of bank retreat usually seen in the meander apex are also explained by high shear stress generated at the bend by higher flow velocity. Apart from direct entraining materials from the bank face, the flow also scours its base that leads to over steepening and failure. Entrainment of bank materials is also possible by oscillation created by the boats and launches. With the shifting of the Bhagirathi, in most segments, lands continuously get eroded at one bank and accrete on the other. Allotment of the newly emerged land (alluvial bars or *chars*)

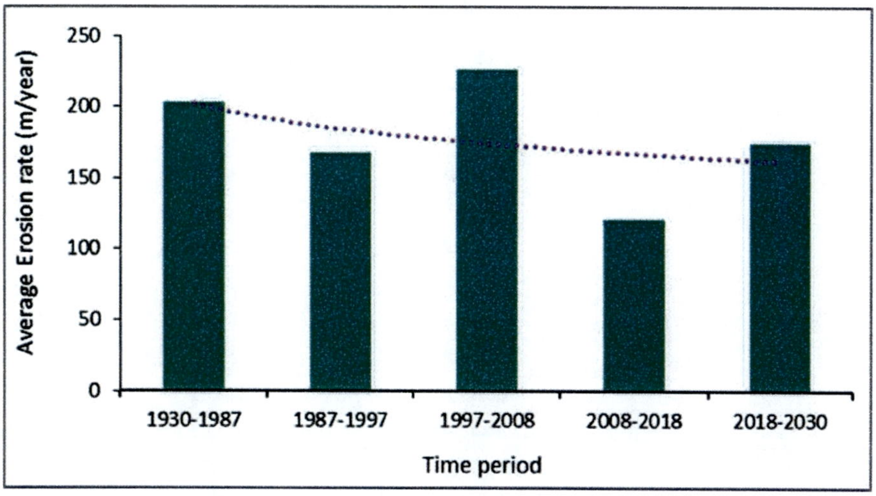

Fig. 14.13 Average rate of erosion of bank lines of the Bhagirathi River in different periods with an overall trend of average erosion rate

among the farmers becomes a major issue in the region when it becomes mature enough for cultivation. The overall economic status is low with low per capita income and below the standard of living because a few numbers of people are landlords. Mental stress and anxiety, arising from the continuous threat of land erosion, are a constant companion for the residents of the riparian settlements (Fig. 14.12d). On the other hand, depositional activity became aggravated because of different engineering structure construction as bank protection measures, like bank paving, spur construction, etc. and communication linkages, like bridge construction. Riverbank protection measures of the river Bhagirathi show encouraging results, and the local administrations are also eager to take necessary steps for bank protection measures. The study of bank erosion helps to bring out relative magnitudes of diverse issues related to bank erosion by the river Bhagirathi from the standpoint of the affected people.

14.4.6 Chances of Vulnerability of the Areas Adjacent to the Bhagirathi River

It has also been computed to know the future trend of lateral migration based on the logarithmic regression model and found that the bank lines will undergo an average rate of erosion of 174 m/year from 2018 to 2030 (Table 14.5 and Fig. 14.13). It reveals that the pattern of erosion of riverbanks constantly changes over each period. Based on the predicted bank line shifting of 2030, the villages belonging adjacent to either side of the river channel within the jurisdiction of Barddhaman district are identified. There may 17 villages be affected by the lateral migration of the river due to the process of erosion in the future, of which 3 villages (namely: Kalyanpur, Bishnupur and Sankhai) are from Ketugram-II C.D. block, 3 villages (namely: Char Brajanathpur, Kalikapur and Gazipur) belong to Katwa-II C.D. block, 6 villages (namely: Sarisha, Jhaudanga, Dampal, Majida, Mertala and Gopipur) come under Purbasthali-II C.D. block, 2 villages (namely: Jaluidanga and Jalahati) are from Purbasthali-I C. D. block, 2 villages (namely: Gramkalna and Krishnadebpur) are located at Kalna-I C.D. block and 1 village (namely: Hanspukur) is from Kalna-II C.D. block (Figs. 14.14 and 14.15). It is estimated that more than 16,000 households with a population of around 65,000 may find themselves vulnerable to those 17 villages. There is a chance of more or less 14,000 households of Kalna-I, Purbasthali-II, Katwa-II and

Table 14.5 Future shifting trend in right and left bank lines of the Bhagirathi River

Cross-section no	Right bank line shift in 2030 (m) by logarithmic regression (x = 34th year as 1997 is the 1st year)	Associated process on river bank	Left bank line shift in 2030 (m) by logarithmic regression (x = 34th year as 1997 is the 1st year)	Associated process on river bank
1	−8.72	Erosion	−18.77	Erosion
2	−19.60	Erosion	2.23	Accretion
3	61.70	Accretion	38.29	Accretion
4	−11.20	Erosion	27.09	Accretion
5	−11.09	Erosion	−94.04	Erosion
6	112.68	Accretion	−159.21	Erosion
7	−8.17	Erosion	226.01	Accretion
8	240.79	Accretion	−35.89	Erosion
9	−1071.82	Erosion	986.71	Accretion
10	0.04	Accretion	12.93	Accretion
11	173.18	Accretion	−24.68	Erosion
12	5.45	Accretion	−15.93	Erosion
13	−192.75	Erosion	322.25	Accretion
14	−36.34	Erosion	50.43	Accretion
15	149.36	Accretion	−84.48	Erosion
16	809.93	Accretion	−955.25	Erosion
17	57.67	Accretion	−89.42	Erosion
18	173.08	Accretion	−545.53	Erosion
19	−40.93	Erosion	215.88	Accretion
20	−2.13	Erosion	−14.94	Erosion
21	103.63	Accretion	−465.54	Erosion
22	−96.36	Erosion	319.56	Accretion
23	176.00	Accretion	−166.53	Erosion

Note Negative values indicate erosion and positive values indicate accretion

Purbasthali-I C.D. blocks to be affected badly by future channel migration (Census of India 2011). Since the river Bhagirathi stays apart from the Katwa-I C.D. block, thus no village of the C.D. block might not face any vulnerability related to the lateral migration of the channel from 2018 to 2030 (Fig. 14.14).

14.5 Data Validation

The common remotely sensed data validation technique includes the formulation of an error matrix using the field data or published data of other reliable sources. However, the digital dataset, such as ASTER (Advanced Spaceborne Thermal Emission and Reflection Radiometer) and SRTM (Shuttle Radar Topography Mission) can also be used as an alternative way for the comparison and validation of satellite image derived data (Hamid et al. 2017). In the present study, the field data cannot be generated as the selected time periods are of the past, and the published data of other sources also cannot be used because of the discrepancy in the absolute location of the drawn cross-sections. Therefore, the only available SRTM data, which was acquired in the year 2000 is used after proper geocoding (Universal Transverse Mercator projection, World Geodetic System 84 datum and

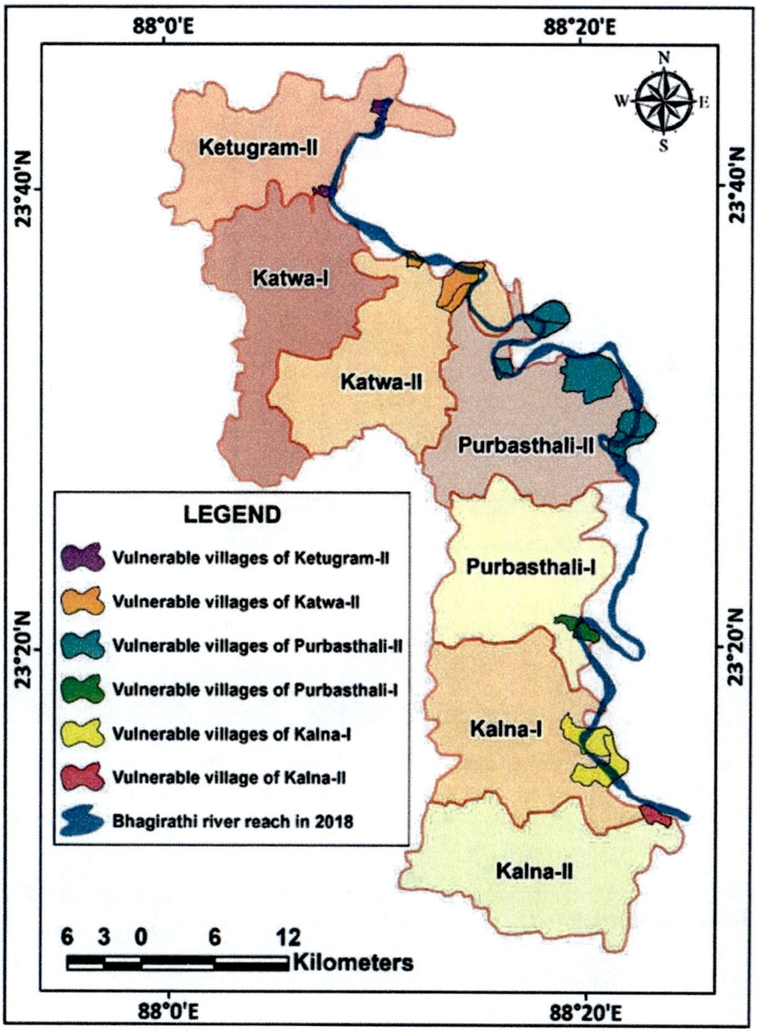

Fig. 14.14 Predicted vulnerable villages of the study area due to channel migration of the Bhagirathi River from 2018 to 2030

zone 45 N) in the ArcGIS environment to validate the satellite image derived Bhagirathi River reach of 1997, as there will be minimal difference between the river reach of 1997 and 2000. Based on the SRTM data, profile graphs are generated one by one along the 23 cross-sections by using the Stack Profile (3D Analyst) tool in ArcGIS. The distance and elevation data of each profile are exported from the attribute table of Stack Profile to MS Excel to generate a profile line graph. The elevation of the intersection points of each cross-section and river bank lines of 1997 are extracted from SRTM, and the distance of these points along each cross-section are measured by adopting a measuring tool in ArcGIS. These points are overlaid on the profile line graph of the respective cross-section with the help of distance and elevation data in MS Excel, which denote the position of the river banks on the profile line graph. Figure 14.16a, b represent the location of the river banks in 1997 that fall almost accurately on the profile line graph generated from SRTM data along the cross-section numbers 11 and 19, respectively.

Fig. 14.15 Erosion prone spots at the bank of the Bhagirathi River creating vulnerability to the local residents in terms of houses, agricultural land and adjacent pack-track of the villages of (**a**) Ketugram-II, (**b**) Katwa-II, (**c**) Purbasthali-II, (**d**) Purbasthali-I, (**e**) Kalna-I and (**f**) Kalna-II C. D. blocks

14.6 Conclusion

The present study unveils that the channel of river Bhagirathi in the studied reach is dynamic. From 1930 to 2018, the average rate of bank line erosion within the studied reach was 192 m/year and it has been observed along with gradual changes in its meander geometry. Due to the process of accretion, many point bars have been evolving and some of them are used in the form of new land for human habitation and agricultural practices. It has come to see that a point bar converted into a village, namely: Char Brajanathpur on the right bank of the river, near the Katwa-II C.D. block, after 1995 (Census of India 2001). The river Bhagirathi flows along the eastern margin of 7 populous C.D. blocks and increasing population density exerts tremendous pressure on land for habitation and food on either side of the river. Therefore, any kind of change in the position of the bank line automatically affects a huge number of riparian inhabitants and making the issue popular to society. The future prediction of channel migration in the form of bank line shifting of 2030 estimates that more or less 65,000 people of 17 riverine villages (Census of India 2011) of Barddhaman district may be vulnerable to erosion in the future. This gradual process of bank line shifting leads to damages and losses of agricultural lands and settlements due to the process of erosion and also creates new hope of lights generating new land by the process of accretion. But the proper planning well in advance with the appropriate management strategy may be a checkmate to combat the future vulnerability of the study area. This study also opens new avenues of researches for researchers to understand the variable factors associated with determining the dynamicity of the present river

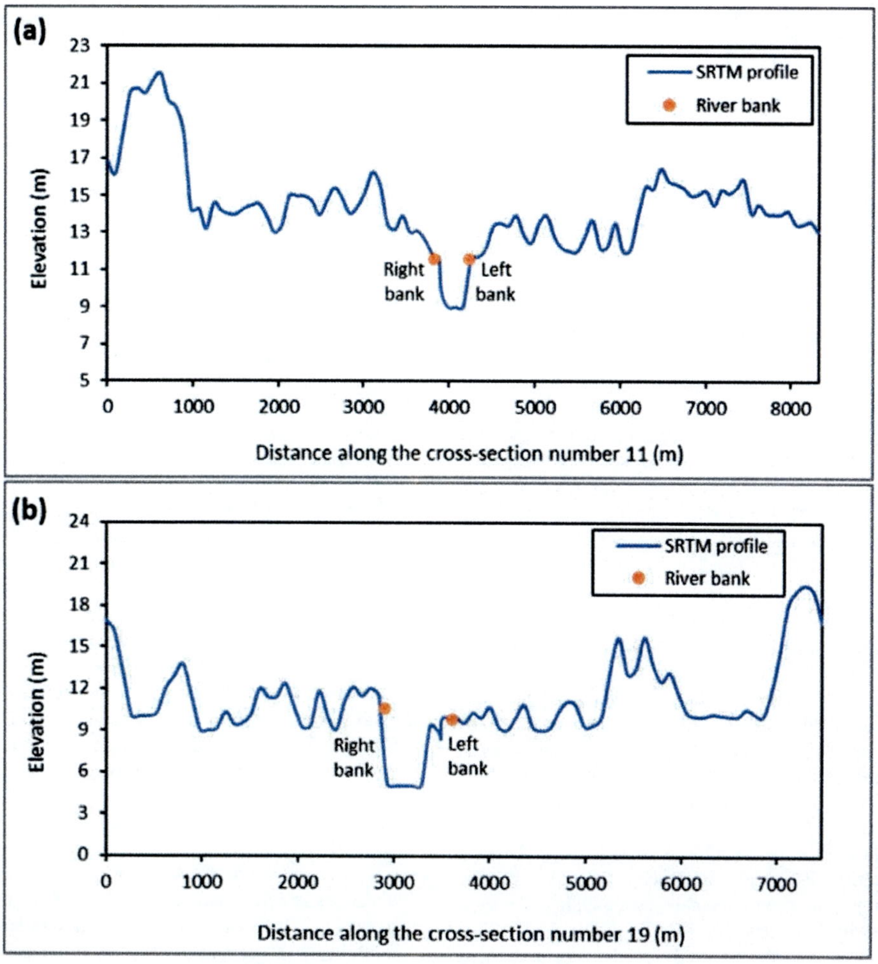

Fig. 14.16 Validation of the river reach generated from satellite image

reach. The long-term assessment of rate, trend and pattern of bank erosion and its digital database generation will help to formulate proper planning and rehabilitation of people up to C.D. block level, mouza level and even up to plot-level monitoring. It will also be able to find out the future vulnerable pockets of bank erosion and to formulate its proper way of mitigation. The plot-level distribution of land to erosion victims and rehabilitation can be made easier if the digital data can be linked with the land record system through e-governance, and ultimately sustainable development plan can be formulated updating the whole system involving digital database and its management plan from time to time.

Acknowledgements The authors express special thanks to Manas Karmakar and Dayamoy Mandal, Research Scholars, Department of Geography, University of Calcutta for their kind help in the preparation of some figures. They are grateful to the villagers of the study area for interacting with us and sharing valuable information. They are also indebted to different national and international organizations for data, maps and images.

Funding No specific grant for this research is received from any funding agencies in the public, commercial or not-for-profit sectors.

Disclosure Statement The authors declare that they have no conflict of interest.

References

Aher SP, Bairagi SI, Deshmukh PP, Gaikwad RD (2012) River change detection and bank erosion identification using topographical and remote sensing data. Int J Appl Inf Syst 2(3):1–7

Ahmed I, Das N, Debnath J, Bhowmik M (2018) Erosion induced channel migration and its impact on dwellers in the lower Gumti river, Tripura India. Spat Inf Res 26(5):537–549. https://doi.org/10.1007/s41324-018-0196-9

Alexeevsky NI, Chalov RS, Berkovich KM, Chalov SR (2013) Channel changes in largest Russian rivers: natural and anthropogenic effects. Int J River Basin Manag 11(2):175–191. https://doi.org/10.1080/15715124.2013.814660

Bag R, Mondal I, Bandyopadhyay J (2019) Assessing the oscillation of channel geometry and meander migration cardinality of Bhagirathi river, West Bengal India. J Geograp Sci 29(4):613–634. https://doi.org/10.1007/s11442-019-1618-z

Basu SR (2001) Impact of the Farakka barrage on the Bhagirathi-Hugli and the port of Calcutta. In: Basu SR (ed) Impact of on man on environment: some cases of concern: professor R. N. Dubey memorial lectures. Professor R.N. Dubey Foundation, Allahabad, pp 39–71

Basu SR, Ghosh A, De SK (2005) Meandering and cut-off of the river Bhagirathi. In: Kalwar SC (ed) Geomorphology and environmental sustainability. Concept Publishing Company, New Delhi, pp 20–37

Basu SR, Sen A (1997) A study on the recent changes (1978–1992) in the morphology of River Hooghly Indian. J Geomorphol 2(1):39–49

Basu SR, Sen A, Ghosh A (1996) Some consideration on the decay of deltaic drainage system of Bengal with special reference to the river Bhagirathi-Hugli. In: Tiwari RC (ed) Proceedings, 6th conference of Indian Institute of Geomorphologists. Allahabad Geographical Society, Allahabad, pp 57–68

Beechie TJ, Liermann M, Pollock MM, Baker S, Davies J (2006) Channel pattern and river-floodplain dynamics in forested mountain river systems. Geomorphology 78(1–2):124–141. https://doi.org/10.1016/j.geomorph.2006.01.030

Bhunia GS, Shit PK, Pal DK (2016) Channel dynamics associated with land use/cover change in Ganges river, India, 1989–2010. Spat Inf Res 24(4):437–449. https://doi.org/10.1007/s41324-016-0045-7

Bisson M, Piccinini S, Zanchetta G (2011) A multidisciplinary GIS-based approach for mapping paleoriver migration: a case study of the Serchio river (Lucca alluvial plain, Tuscany). Gisci Remote Sens 48 (4):566–582. https://doi.org/10.2747/1548-1603.48.4.566

Census of India (2001) Administrative Atlas-West Bengal 2001. Directorate of Census Operations, West Bengal, India

Census of India (2011) District census handbook-Barddhaman 2011. Directorate of Census Operations, West Bengal, India

Chakraborty S, Mukhopadhyay S (2015) An assessment on the nature of channel migration of river Diana of the sub-Himalayan West Bengal using field and GIS techniques. Arab J Geosci 8(8):5649–5661. https://doi.org/10.1007/s12517-014-1594-5

Charlton R (2008) Fundamentals of fluvial geomorphology. Routledge, New York

Church M (1992) Channel morphology and topology. In: Calow P, Petts GE (eds) The rivers handbook: hydrological and ecological principles. Blackwell, Oxford, pp 126–143

Das B, Mondal M, Das A (2012) Monitoring of bankline erosion of river Ganga, Malda district and West Bengal: using RS and GIS compiled with statistical techniques. Int J Geomat Geosci 3(1):239–248

Das JD, Dutta T, Saraf AK (2007) Remote sensing and GIS application in change detection of the Barak river channel, N.E. India. J Indian Soc Remote Sens 35 (4):301–312. https://doi.org/10.1007/BF02990786

Debnath J, Das (Pan) N, Ahmed I, Bhowmik M (2017) Channel migration and its impact on land use/land cover using RS and GIS: a study on Khowai river of Tripura, North-East India. Egypt J Remote Sens Space Sci 20(2):197–210. https://doi.org/10.1016/j.ejrs.2017.01.009

Dey S, Mandal S (2019) Assessing channel migration dynamics and vulnerability (1977–2018) of the Torsa river in the Duars and Tal region of eastern Himalayan foothills, West Bengal India. Spat Inf Res 27(1):75–86. https://doi.org/10.1007/s41324-018-0213-z

Duran R, Beevers L, Crosato A, Wright N (2011) Bank retreat study of a meandering river reach: case study: river Irwell. 7th international symposium on ecohydraulics 2009. Curran Associates Inc., New York, pp 1002–1012

Dury GH (1977) Underfit streams: retrospect and prospect. In: Gregory KJ (ed) River channel changes. Wiley Interscience, Chichester, pp 281–293

Gain AK, Giupponi C (2014) Impact of the Farakka dam on thresholds of the hydrologic flow regime in the lower Ganges river basin (Bangladesh). Water 6 (8):2501–2518. https://doi.org/10.3390/w6082501

Ghosh D (2007) Environmental appraisal of bank erosion of the Ganga in Malda and Murshidabad districts, West Bengal. Unpublished thesis, The University of Burdwan

Ghosh D, Sahu AS (2018) Problem of river bank failure and the condition of the erosion victims: a case study in Dhulian, West Bengal, India. Reg Sci Inq X (2):205–214

Gregory KJ (1977) The context of river channel changes. In: Gregory KJ (ed) River channel changes. Wiley Interscience, Chichester, pp 1–12

Gregory KJ (1979) Fluvial geomorphology. Progress in Physical Geography: Earth and Environment 3 (2):274–282. https://doi.org/10.1177/030913337900300207

Gregory KJ (1983) Fluvial geomorphology. Prog Phys Geograp Earth Environ 7(3):385–396. https://doi.org/10.1177/030913338300700305

Guite LT, Bora A (2016) Impact of river bank erosion on land cover in lower Subansiri river flood plain. Int J Sci Res Publ 6(5):480–486

Gupta A (2011) Tropical geomorphology. Cambridge University Press, Cambridge

Hamid M, Ashraf M, Hamid Q, Sarwar SM, Saqib ZA (2017) Geospatial techniques for assessment of bank erosion and accretion in the Marala Alexandria reach of the river Chenab, Pakistan. Sains Malays 46 (3):413–420. https://doi.org/10.17576/jsm-2017-4603-08

Hickin EJ, Nanson GC (1975) The character of channel migration on the Beatton river, northeast British Columbia. Can Geolog Surv Am Bull 86:487–494. https://doi.org/10.1130/0016-7606(1975)86<487:TCOCMO>2.0.CO;2

Huggett RJ (2017) Fundamentals of geomorphology, 4th edn. Routledge, New York

Islam A, Guchhait SK (2017) Analysing the influence of Farakka barrage project on channel dynamics and meander geometry of Bhagirathi river of West Bengal India. Arab J Geosci 10(11):245. https://doi.org/10.1007/s12517-017-3004-2

Knight MJ (1975) Recent crevassing of the Erap river New Guinea. Aust Geograp Stud 13(1):77–82. https://doi.org/10.1111/j.1467-8470.1975.tb00068.x

Kotoky P, Bezbaruah D, Baruah J, Sarma JN (2005) Nature of bank erosion along the Brahmaputra river channel, Assam India. Curr Sci 88(4):634–639

Kumar A, Jayappa KS, Deepika B (2010) Application of remote sensing and geographic information system in change detection of the Netravati and Gurpur river channels, Karnataka India. Geocarto Int 25(5):397–425. https://doi.org/10.1080/10106049.2010.496004

Laha C (2015) Oscillation of meandering Bhagirathi on the alluvial flood plain of Bengal basin, India; as controlled by the palaeo-geomorphic architecture. Int J Geomat Geosci 5(4):564–572

Laha C, Bandyapadhyay S (2013) Analysis of the changing morphometry of river Ganga, shift monitoring and vulnerability analysis using space-borne techniques: a statistical approach. Int J Sci Res Publ 3(7):1–10

Leopold LB, Wolman MG, Miller JP (1995) Fluvial processes in geomorphology, 2nd edn. Dover publications, INC, New York

Lewin J, Hughes D (1976) Assessing channel change in Welsh rivers. Cambria 3:1–10

Majumder MK (2004) Bangladesh: Ganga under Threat. Dispute over the Ganga: a look in the potential water-related conflicts of the south Asia. Panose Institute South Asia, Kathmandu, pp 22–53

Mandal AC, Patra P, Majumder R, Ghosh DK, Bhunia GS (2017) Evaluating meander shifting dynamics (1977–2017) of the Bhagirathi river course in Murshidabad District, West Bengal India. Spat Inf Res 26(1):33–45. https://doi.org/10.1007/s41324-017-0153-z

Millar R (2000) Influence of bank vegetation on alluvial channel patterns. Water Resour 36(4):1109–1118. https://doi.org/10.1029/1999WR900346

Mondal J, Mandal S (2018) Monitoring changing course of the river Ganga and land-use dynamicity in Manikchak Diara of Malda district, West Bengal, India, using geospatial tools. Spat Inf Res 26(6):691–704. https://doi.org/10.1007/s41324-018-0210-2

Mukherjee R, Bilas R, Biswas SS, Pal R (2016) Bank erosion and accretion dynamics explored by GIS techniques in lower Ramganga river, Western Uttar Pradesh India. Spat Inf Res 25(1):23–38. https://doi.org/10.1007/s41324-016-0074-2

Ophra SJ, Begum S, Islam R, Islam MN (2018) Assessment of bank erosion and channel shifting of Padma River in Bangladesh using RS and GIS techniques. Spat Inf Res 26(6):599–605. 10.1007/s41324-018-0202-2

Pal R, Biswas SS, Pramanik MK, Mondal B (2016) Bank vulnerability and avulsion modeling of the Bhagirathi-Hugli river between Ajay and Jalangi confluences in lower Ganga plain India. Model Earth Syst Environ 2 (2):65. 10.1007/s40808-016-0125-7

Panda S, Bandyopadhyay J (2011) Morphodynamic changes of Bhagirathi river at Murshidabad district using geoinformatics. J GeograpInf Syst3:85–97. 10.436/jgis.2011.31006

Philip G, Gupta RP, Bhattacharya A (1989) Channel migration studies in the middle Ganga basin, India, using remote sensing data. Int J Remote Sens 10 (6):1141–1149. https://doi.org/10.1080/01431168908903953

Ranjan P, Ramanathan A (2018) Hooghly river. In: Singh DS (ed) The Indian rivers: scientific and socio-economic aspects. Springer Nature, Singapore, pp 251–257

Rudra K (1996) The Farakka barrage- an interruption to fluvial regime. Indian J Landsc Syst Ecolog Stud 19 (2):105–110

Rudra K (2008) Banglar nadikatha (in Bengali). Sahitya Samsad, Kolkata

Rudra K (2012) Atlas of the changing river courses in West Bengal. Sea Explorers Institute, Kolkata

Rudra K (2014) Changing river courses in the western part of the Ganga-Brahmaputra delta. Geomorphology 227:87–100. https://doi.org/10.1016/j.geomorph.2014.05.013

Rudra K (2018) Rivers of the Ganga-Brahmaputra-Meghna delta: a fluvial account of Bengal. Springer International Publishing AG, Switzerland

Rusov J, Misita M, Milanovic DD, Milanovic DL (2017) Applying regression models to predict business results. FME Trans 45:198–202

Saleem A, Dewan A, Rahman MM, Nawfee SM, Karim R, Lu XX (2020) Spatial and temporal variations of erosion and accretion: a case of a large tropical river. Earth Syst Environ 4(1):167–181. https://doi.org/10.1007/s41748-019-00143-8

Sarkar MH (2004) Impact of upstream human interventions on the morphology of the Ganges-Garai system. In: Mirza MMQ (ed) The Ganges water diversion: environmental effects and implications. Kluwer Academic Pub, Dordrecht, pp 49–80

Sarkar SK, Bhattacharya A, Bhattacharya B (2003) The river Ganga of northern India: an appraisal of its geomorphic and ecological changes. Water Sci Technol 48(7):121–128

Schumm SA (1963) Sinuosity of alluvial rivers on the great plains. Geol Soc Am Bull 74:1089–1099

Summerfield MA (2013) Global geomorphology: an introduction to study of landforms, 2nd edn. Routledge, New York

Thakur PK, Laha C, Aggarwal SP (2012) River bank erosion hazard study of river Ganga, upstream of Farakka barrage using remote sensing and GIS. Nat Hazards 61:967–987. https://doi.org/10.1007/s11069-011-9944-z

Thorne CR (1982) Processes and mechanisms of river bank erosion. In: Hey RD, Bathurst JC, Thorne CR (eds) Gravel bed rivers. Wiley, Chichester, pp 227–259

Thorne CR (1992) Bend scour and bank erosion on the meandering Red River, Louisiana. In: Carling MA, Petts GE (eds) Lowland floodplain rivers: geomorphological perspectives. Wiley, Chichester, pp 95–116

Thorne CR (2002) Geomorphic analysis of large alluvial rivers. Geomorphology 44(3–4):203–219. https://doi.org/10.1016/S0169-555X(01)00175-1

Újvári G, Mentes G, Bányai L, Kraft J, Gyimóthy A, Kovács J (2009) Evolution of a bank failure along the river Danube at Dunaszekcso Hungary. Geomorphology 109(3–4):197–209. https://doi.org/10.1016/j.geomorph.2009.03.002

Yang X, Damen MCJ, Zuidam RAV (1999) Satellite remote sensing and GIS for the analysis of channel migration changes in the active Yellow river delta, China. Int J Appl Earth Obs Geoinf 1(2):146–157. https://doi.org/10.1016/S0303-2434(99)85007-7

Yunus AP, Jie D, Armugha K, Sravanthi N, Rao LAK, Hao C (2019) Channel migration characteristics of the Yamuna river from 1954 to 2015 in the vicinity of Agra, India: a case study using remote sensing and GIS. Int J River Basin Manag 17(3):367–375. https://doi.org/10.1080/15715124.2019.1566238

15 Alluvial Channel Dynamic Associated with LULC Change in Himalayan Foothill

MD Hasanuzzaman, Aznarul Islam, and Pravat Kumar Shit

Abstract

Dynamicity of the channel is the main characteristic of the Kaljani River in the Himalayan foothill. The present work intends to document the historical changes in the land use and land cover (LULC) pattern driven by channel migration during 1987–2020 at the Kaljani River adjacent village area. In this study, the sinuosity index, the radius of curvature, meander wavelength, amplitude, meander width, channel width, arc angle, direction angle, rate of channel migration, and direction of migration have been calculated for the years of 1987, 2004, and 2020. The historical positions of both bankline and dynamic channel width and meander width indicate that a large portion of the floodplain area depict an erosion-accretion sequence with time. This work also investigated LULC changes in the Kaljani River adjacent village area using supervised image classification with an overall accuracy ranging between 85 and 89%. This research has demonstrated the application and capability of RS and GIS technology and generated a detailed evaluation of temporal and spatial changes in river channel processes and adjustment of LULC types. The LULC results revealed that the water bodies and dense forest are decreased and sandy area and built-up areas are increased. The LULC changes by the direct effect of bankline migration have a bad impact on the dwellers of the floodplain adjacent village area of the Kaljani River. The results of this study can represent an important indicator of the vulnerability of the Kaljani River adjacent village area and also provide information about geomorphological instabilities of the study area.

Keywords

Alluvial channel · Channel migration · Channel width · Sinuosity · Radius of curvature · LULC changes

15.1 Introduction

An alluvial river floodplain area of the Himalayan foothills experiences frequent periodical and seasonal variations as a result of channel migration and conversion in the land use and land cover (LULC) pattern (Bastawesy et al. 2013). The channel migration, high sediment load, frequent floods, and anthropogenic interventions have caused morphological changes in

MD. Hasanuzzaman (✉) · P. K. Shit
Department of Geography, Raja N.L.Khan Women's College (Autonomous), Midnapore, West Bengal, India

A. Islam
Department of Geography, Aliah University, Kolkata, India

P. K. Shit et al. (eds.), *Drainage Basin Dynamics*, Geography of the Physical Environment,
https://doi.org/10.1007/978-3-030-79634-1_15

the river course (Kuehl et al. 2005; Kummu et al. 2008). Meandering rivers signify channel dynamicity which reveals frequent changes in channel patterns through bank erosion, deposition, down cutting, etc. (Uddin et al. 2011). River meander and channel shifting are very complex processes of river floodplains that are triggered by different factors in the fluvial system (Hooke 2013). Channel migration is one of the significant characteristics that can change the morphology of the river buffer area, with major problems for human life (Güneralp et al. 2012). River geomorphological factors are important indicators to investigate environmental changes and are especially related to changes in river buffer areas, such as various LULC types in the river floodplains and drainage basin (Dai et al. 2008). These geomorphological changes are associated with changes in sediment load and river discharge variation (Cserkész-Nagy et al. 2010). Therefore, the Himalayan Foothill Rivers are generally dynamic and riverbank erosion-accretion is strongly related to the formation and shifting of meander bends.

River buffer areas are the most important ecosystems for habitats and food production, but they are considered to be one of the most vulnerable ecosystems in the world as they are facing degradation by channel migration and land-use pressure (James and Lecce 2013). River buffer areas are subjected to periodical and seasonal modifications as a result of channel migration of river driving a change in the LULC pattern (Hazarika et al. 2015). LULC changes are a direct or indirect human activity impact that can be changing river morphology at different spatial and temporal scales (Belletti et al. 2016). Changes in geometric parameters of river meander can be associated with river floodplain management activities (Nelson et al. 2013), and meanders can also naturally derive and change behavior over time (Brice 1960; Hooke 1984). Channel lateral migration processes have been observed through erosion-accretion, and it can change the riparian land use in the river buffer area (Yanan et al. 2011; Riley and Rhoads 2012). Moreover, channel migration and LULC changes have been accentuated due to environmental, social, and economical factors (James and Lecce 2013). The studies concerning the relation between river dynamics and LULC changes are limited; however, they are crucial for the development of floodplain management policies (Hazarika et al. 2015; Yousefi et al. 2016; Bhunia et al. 2016; and Debnath et al. 2017).

For measuring, mapping, representing, and monitoring the geomorphodynamic features, effective tools, and techniques are very important. At present, remote sensing (RS) and geographical information systems (GIS) have enormous importance for change detection and mapping in rivers and their buffer zone dynamics at a different strategic scale (Wang and Mei 2016; Wang and Xu 2018). RS and GIS tools and techniques with field verification can accurately and quickly map and investigate river morphological changes (Rinaldi et al. 2013; Langat et al. 2018). Channel migration and LULC change are integrated. The present study tries to relate both changes using the modern geospatial tools and techniques. The geospatial tools have been adopted for calculating the rate of channel migration, and meander geometry (Lawler 1993; Debnath et al. 2017), and LULC changes (Ahmed 2012; Mondal et al. 2016; Maviza and Ahmed 2020).

The area under study being situated in the seismo-tectonically unstable foothill terrain (Bhutan Himalaya) has produced an array of magnificent landscapes both physical and cultural involving multiple cycles of fluvial migration. The Kaljani River is very important in terms of ecological variation (Dey et al. 2015). However, the previous literature indicates a relative vacuum of work in the geomorphological domain. This research is unique in terms of dynamic river studies in the geological and geomorphologically active or the highly vulnerable areas of the Himalayan foothills of West Bengal. Especially, the effort to measure channel migration, meander geometry parameters, and the LULC change of Kaljani River buffer area is almost absent at the national level or local level. Therefore, this work attempts to measure the channel migration and LULC change in the past 33 years timeframe. This work has been developed for estimating the historical bank-line. Moreover calculating channel migration, channel width, channel sinuosity, the radius of curvature, meander wavelength, amplitude, meander width, arc angle, direction angle, and changing LULC pattern is attempted using geo-spatial techniques coupled with the field verification. Thus, the work intends to address the following objectives:

(a) To study the morphodynamic changes of the Kaljani River, and
(b) To identify the changes in the LULC pattern concerning channel migration.

15.2 Study Area

The Kaljani River is a tributary of Torsha River that has originated from the foothills of the Himalayas in Bhutan and it flows from north to south via Bhutan and India and confluences with the Torsha River that again confluences with the Brahmaputra River which confluences with the Padma River that eventually debouches into the Bay of Bengal. The Kaljani River has negotiated the undulating terrain of the Bhutan Himalaya with the alluvial fans and the Terai plain downstream thus covering both the *'Bhabar'* and the continuous plains of *'Terai'* downhill. The major tributaries of Kaljani River are Dima, Nonai, etc.

The Kaljani River is situated in the Eastern and North-Eastern parts of Alipurduar and Cooch Behar districts in West Bengal. The region is extended from 26° 43′ 08″ N to 26° 16′ 30″ N latitude and 89° 25′ 17″ E to 89° 34′ 56″ E longitude. The length of the river of the study area is 70.5 km. Some selected mouzas (smallest administrative units for revenue collection) are taken into consideration for delineating the floodplain area of the Kaljani River. This adjacent village area is a maximum of 5.88 km and a minimum of 1.64 km area from the Kaljani River (Fig. 15.1).

15.3 Database and Methodology

15.3.1 Database

In the study, TM, ETM+ , and Landsat OLI data sets collected in 1987, 2004, and 2020 have been used to demarcate the channel bank lines and changes in the LULC pattern. These years have been divided into short term (1987–2004 and 2004–2020) and long-term (1987–2020) for a better understanding. For the validation of this work, we have field visits with the GPS of different active migration sites and verified the map migration area to field migration area. Also, we have measured the river embankment length with the help of Google Earth Pro software.

15.3.2 Methodology

The work has been carried out following a systematic methodology (Fig. 15.2). First, all the satellite images were projected in the Universal Transverse Mercator (UTM) projection with zone 45 north and world geodetic survey 1984 (WGS 84) datum and resembled in the ArcGIS environment. Accuracy assessment of this study revealed through the Kappa index, field assessment, Copernicus satellite images, GPS surveys, and other possible ways. The Kaljani River adjacent village area is divided into three zones, namely A (Uttar Latabari to Dakshin Paitkapara), B (Dakshin Paitkapara to Ambari), and C (Ambari to Deocharai) by a length of around 26.8 km (Fig. 15.3).

15.3.2.1 Bankline Extraction

Accurate riverbank line extraction is a complex process. We have used the normalized difference water index (NDWI) (McFeeters 1996) and modified normalized difference water index (MNDWI) (Xu 2006) for bank line extraction (Eqs. 15.1 and 15.2). This method was employed in the NIR band for segregating the land from water. The land and water pixels are assigned as '0' and '1', respectively, to achieve a binary image.

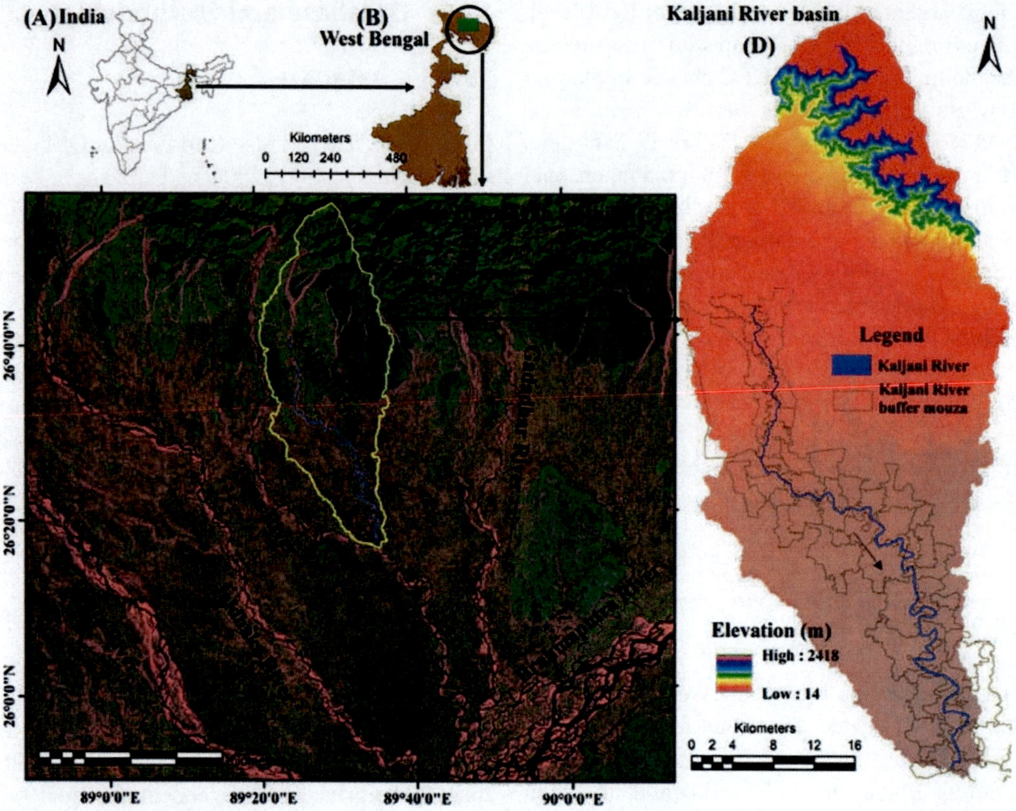

Fig. 15.1 Location of the study area

$$\text{NDWI} = \text{Green} - \text{NIR}/\text{Green} + \text{NIR} \quad (15.1)$$

$$\text{MNDWI} = \text{Green} - \text{MIR}/\text{Green} + \text{MIR} \quad (15.2)$$

To estimate the MNDWI was used MIR band for Landsat 7 and Landsat 5 and 8 was used SWIR band. So Eq. 15.2 is also written as per Eq. 15.3.

$$\text{MNDWI} = \text{Green} - \text{SWIR}/\text{Green} - \text{SWIR} \quad (15.3)$$

To generate the new image NDWI was multiplied by MNDWI. Also, the binary and new images were multiplied and processed to get the final image (de Bethune et al. 1998). The bank-line extraction is the process of transformation of the final image into a vector layer to determine right and left banklines separately on particular imagery (Jana 2019). This process is adopted of all imagery for historical bankline positions.

15.3.2.2 Determination of Channel Migration, and Geometrical Parameters

The entire Kaljani River from Gabaur Bachhra up to Deocharai was divided into 3 zones and 28 reference sites were selected for transects. Following the methods of Chakraborty and Mukhopadhyay (2015), Das et al. (2014), and Debnath et al. (2017) transect-wise bank line migration were measured. The channel width, amount of migration (right and left separately), and the radius of curvature were measured along transects, and the sinuosity index was calculated zone-wise. Therefore, the extracted bank lines were overlapped to calculate the escalation of the channel migration for each transect.

The ArcGIS automate extension 'Channel Migration Toolbox' (CMBT) was used for the measurement of the average length of migration,

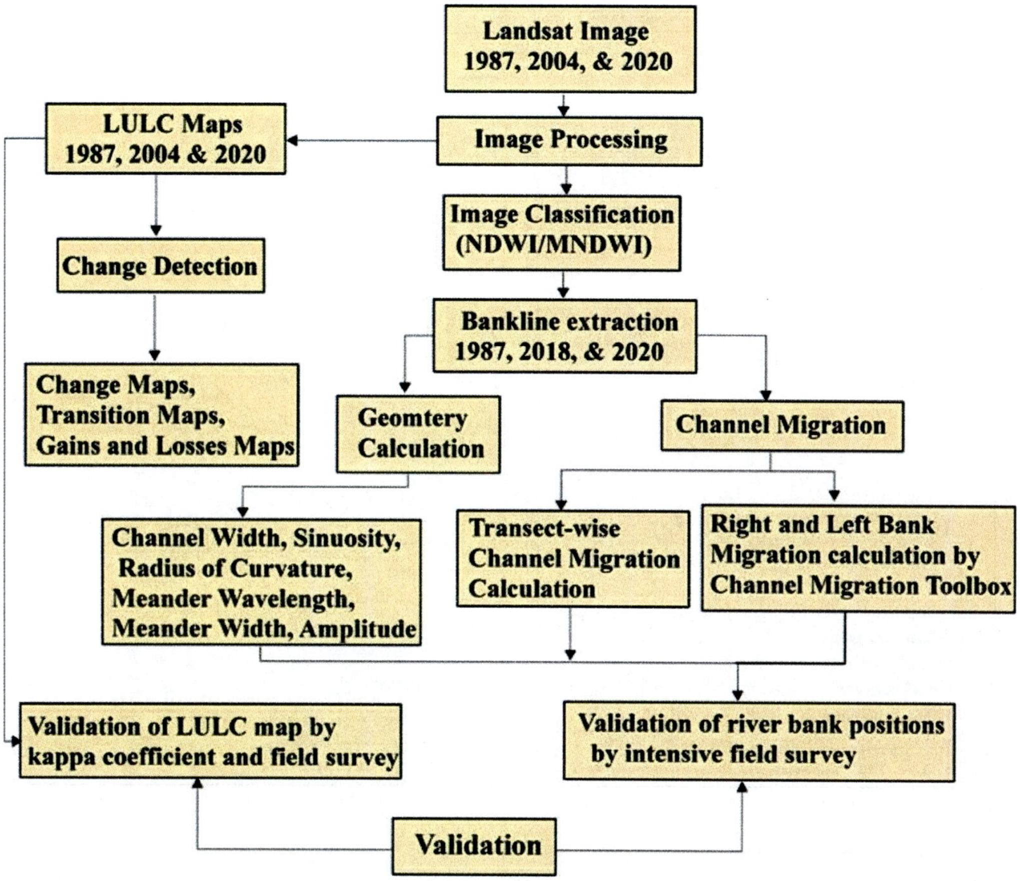

Fig. 15.2 Flowchart of the methodology

and shape area migration of Kaljani River (Roux et al. 2013; Hasanuzzaman and Mandal 2020) by using the right and left bank lines.

Channel sinuosity (S) an important parameter of meander geometry is the ratio between the actual length (AL) and the straight length (SL) of the channel.

$$\text{Channel sinuosity} = \frac{AL}{SL}\ (\text{Schumm 1963}) \tag{15.4}$$

The sinuosity indices of the entire three zones were calculated for the years 1987, 2004, and 2020, respectively. According to the sinuosity index, channels can be categorized into three classes: straight (SI < 1.05), sinuous (SI 1.05–1.5), and meandering (SI > 1.5) (Schumm 1977).

The radius of curvature is another important geometry parameter, which is related to the meander belt and its nature of lateral migration.

$$\text{RC} = \frac{2w\psi}{(lu + ld)}\ (\text{Howard 1992}) \tag{15.5}$$

where RC = Curvature, W = Channel width, ψ = the angular change in direction at the meander, lu and ld = the distance to the adjacent upstream and downstream nodes.

15.3.2.3 LULC Mapping and Their Change Detection

In this study, three Landsat images were used such as 1987 (Thematic Mapper or TM), 2004 (Enhanced Thematic Mapper plus or ETM+), and 2018 (Operational Land Imager or OLI) for

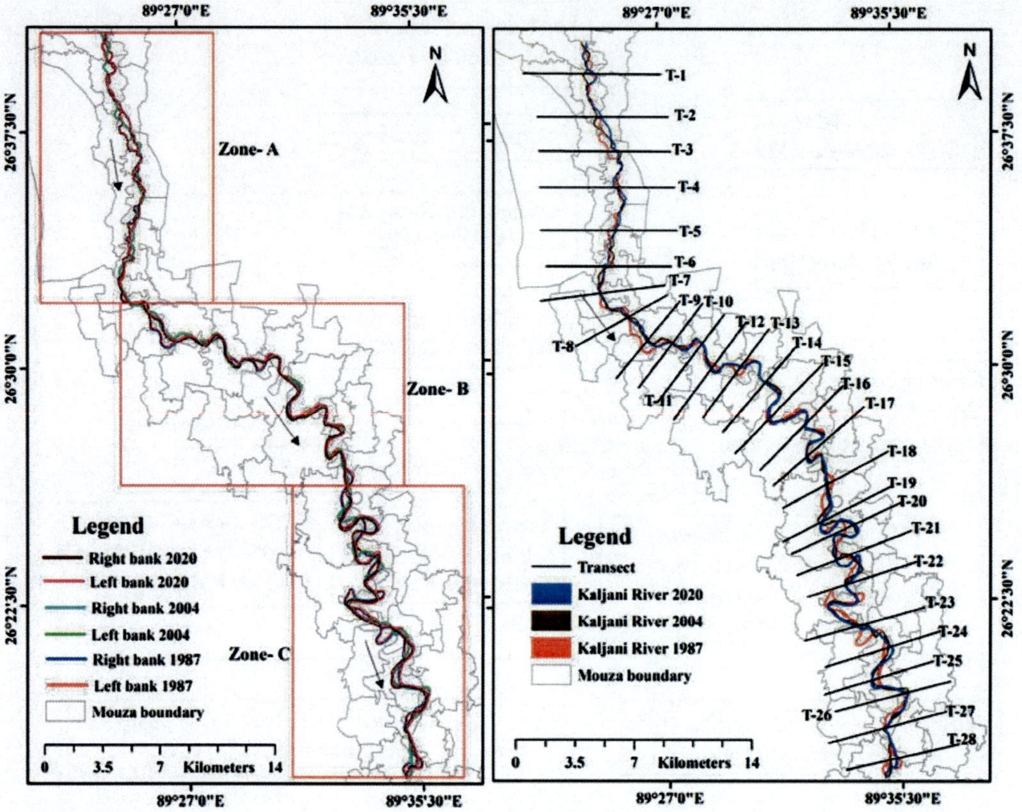

Fig. 15.3 Zone-wise bankline and transects sites of the Kaljani River

LULC classification. All Landsat data were having 30 m resolutions. The buffer village area adjacent to the Kaljani River is classified into six classes: water bodies, sandy area, dense forest, open forest, agricultural land, and built-up area using ArcGIS and Idrisi selva environment (Table 15.1). A supervised classification technique has been employed for getting accurate result of LULC classification. The maximum likelihood classifier technique that classifies the pixels based on the maximum probability of belonging to a definite class (Richards and Jia 2006) has been adopted for classification of the images. For the present study area, 250 spectral signatures were assigned to observe each LULC in the selected study area (Hazarika et al. 2015; Debnath et al. 2017; Wang et al. 2020; Du et al. 2020).

Many LULC change detection techniques are widely used for monitoring land-use change (Kaufmann and Seto 2001; Ahmed 2012; Mondal et al. 2016; Maviza and Ahmed 2020). However, the efficiency of the commonly applied methods to detect change depends on the accuracy of the individual classification (Du et al. 2020), the post-classification based on rectification, and independently classified image. In this method, an array of "from-to" matrixes can be developed by correlating on a pixel by pixel basis, and these matrixes include pixel conversion matrix, area conversion matrix, and percentages conversion matrix. In this study, LULC images of data set during 1987–2004, 2004–2020, and 1987–2020 portray changes in the area conversion matrix. The changing areas of each map were also measured from each LULC map.

15.3.2.4 Validation Techniques

Classification map cannot represent the exact reality of any place. There must be some error.

Table 15.1 Description of the LULC classification in the study

Class No	LULC Types	Description
1	Water bodies	River, permanent open water, lakes, ponds, and reservoirs
2	Sandy area	Point bar and char land along the river
3	Dense forest	Very deep forest area
4	Open forest	Trees, grassland and scrub area
5	Agricultural land	Cropland, and orchards
6	Built-up area	Residential, roads, and constructed area

So, it is essential to check the accuracy to obtain the effective analysis of LULC change (Butt et al. 2015; Mosammam et al. 2016) and it is the most significant part of image processing. Therefore, in this study, Kappa coefficient technique has been employed to check the accuracy assessment test (Congalton and Green 1999). An accuracy assessment was employed through an error or confusion matrix. It is formulated between the classified map and the ground truth (Jensen 2005; Gerard et al. 2010). The producer's accuracy and user's accuracy were measured employing the error matrix table. The Kappa coefficient was calculated using the following formula (Congalton and Green 2009):

$$\text{Kappa coefficient} = \frac{\sum_{i=1}^{k} n_{ii} - \sum_{i=1}^{k} n_{ii}(G_iC_i)}{n - \sum_{i=1}^{k} n_{ii}(G_iC_i)} \quad (15.6)$$

where i is the class number, n is the total number of classified pixels that are being compared to actual data, n_{ii} is the number of pixels belonging to the actual data class i, that were classified with a class i, C_i is the total number of classified pixels belonging to class i and G_i is the total number of actual data pixels belonging to class i. Kappa statistic values range from 0 to 1.

15.4 Results

The present study is represented in two main parts: (1) the detailed analysis of changing channel migration and geometry parameters and (2) the land use and land cover changes in the buffer zone of the study river in relation to channel migration.

15.4.1 Dynamics of the Channel Migration and Meander Geometry

Channel migration is a natural process of the river that changes the meander parameters such as river width, the radius of curvatures, sinuosity, meander wavelength, amplitude, meander width, channel width, arc angle, direction angle, etc. (Ayman and Ahmed 2009).

15.4.2 Channel Migration

This study, has calculated channel migration in two different ways (first, along transects and second, the average length of migration, and shape area migration of entire river by channel migration toolbox) of different periods (1987–2004, 2004–2020, and 1987–2020). The bankline migration of the Kaljani River has been measured on left and right banks at 28 transects with direction (Fig. 15.3).

During 1987–2004, the maximum bankline migration of the right bank was 734.23 m at T-3 (transect) and for the left bank it was 426.56 m at T-1. The minimum bankline migration of the right bank was 5.49 m at T-27 and for the left bank it was 6.94 m at T-14 (Table 15.2). In this period, bankline migration was very high and the maximum direction (transect-wise) of bankline migration was rightward at both the banks. At that time, the Kaljani River was very dynamic due to the high volume of water and impacts of the flood. In the time-frame, 2004–2020, the maximum bankline migration of right and the left bank was 744.81 m (T-3) and 609.39 m (T-1), respectively, while the minimum bankline migration of right and left bank was 13.08 m (T-8), and 5.97 m (T-11), respectively. During that period, transect-wise bankline migaration was maximum at the rightward direction at both the banks. However, in the recent time (1987–2020), the maximum and minimum bankline migration of the right bank are found as 621.28 m on T-6 and 20.61 on T-11, respectively. At this time, the maximum and minimum bankline migration of the left bank are 649.96 on T-20 and 21.07 on T-11, respectively. In this study, the result revealed that the Kaljani River is very dynamic and tends to move to a rightward direction throughout the observation period.

The shifting of the right and left bankline is used to assess the reach average length of migration, and shape area of migration of channel by Channel Migration Toolbox. The average length of migration is the average distance between two intersecting banklines (right and left) of a river reach and the shape area of migration of channel is a dynamic or migrated area between two intersecting (right and left) of a river reach. During 1987–2004, the average length of migration and the shape area of migration of channel was 182.9 m and 12,803,143.4 m^2 at the right bank, and 183.85 m and 12,869,784.4 m^2 at the left bank (Fig. 15.4).

The annual rate of the average length of migration of this river is 10.76 m/year at the right bank and 11.49 m/y at the left bank. The average length of migration and the shape area of migration of channel of the right bank are 119.69 m and 8,378,798.48 m^2, respectively, while the left bank records 117.21 m and 8,204,667.88 m^2 in the study period 2004–2020. The annual rate of average length of migration of this river is 7.04 m/year at the right bank and 7.32 m/y at the left bank. During the past 33 years (1987–2020), the average length of migration and the shape area of migration of the channel of the right bank are 302.59 m and 21,181,941.5 m^2 while the left bank records 301.06 m and 21,074,452.3 m^2 respectively. The annual rate of the average length of migration of this river is 9.17 m/year at the right bank and 9.12 m/y at the left bank during 1987–2020. These changes have followed the specific pattern of migration. The annual rate average length of migration was the maximum during the period 1987–2004. The average length of migration and the shape area of migration of channel have gradually decreased with time from 1987 to 2020 at both the banks. However, the historical dataset and mathematical output have represented transect-wise migration. The average length and shape area of migration of channel are very dynamic in some years while others are exclusively normal. In our analysis, over the study periods Uttar Paitkapara, Jaigir Chilakhana, Chhatoa, Kaljani, Bhelapeta, Dakshin Latabari, Nimtijhora Tea Garden, Kholta, Chalnipak, Bhelakopa Dwitia Khanda, Ambari, Dakshin Paitkapara. Chalnipak, and Amlaguri Dwitia Khanda mouza are very active for channel migration.

15.4.3 Widening of Channel

The most important geomorphic attribute of the alluvial rivers is channel migration which leads to the widening of the river course. Therefore, lateral channel migration of the banks and demarcation of the channel width were carried out from two maps of different years. During the monsoon season, the channel migration rate is very high, which leads the bank erosion ultimately leading to the widening of the channel (Bhowmik and Das (Pan) 2014). In this work, the active widening of the Kaljani River has been calculated at 28 transects across the river from 1987 to 2020. Table 15.3 depicts the transect-wise change of active width during the observation period. The maximum change of width is recorded as 169.91 m at transect 19, and the lowest is 0.2 m at transects 14 during 2004–2020. The overall result indicates that the

Table 15.2 Migration of the Kaljani River during 1987–2020

	Migration (Right bank)						Migration (Left bank)					
Transcets	1987–2004 (m)	SD	2004–2020 (m)	SD	1987–2020 (m)	SD	1987–2004 (m)	SD	2004–2020 (m)	SD	1987–2020 (m)	SD
T-1	474.42	R	622.83	L	163.53	L	426.56	R	609.39	L	182.83	L
T-2	68.31	L	71.97	R	162.86	L	233.18	L	239.19	L	170.17	L
T-3	734.23	R	744.81	L	485.41	L	265.41	R	227.77	R	515.12	L
T-4	137.39	R	124.6	R	388.85	R	345.08	R	243.29	R	386.91	R
T-5	442.09	R	459.85	R	448.92	L	12.7	L	13.43	R	453.68	L
T-6	276.49	R	256.84	R	621.28	R	363.8	R	377.64	R	648.17	R
T-7	62.23	R	36.96	R	32.44	R	31.95	L	10.48	L	29.69	R
T-8	8.56	R	13.01	R	74.83	R	77.82	R	59.82	R	49.53	R
T-9	624.59	R	592.77	R	518.06	L	116.21	L	136.94	L	458.02	L
T-10	97.42	R	90.87	R	79.72	R	19.29	L	39.55	L	51.03	R
T-11	15.23	L	28.1	L	20.61	R	36.46	R	5.97	R	21.07	L
T-12	326.6	R	306.61	R	225.79	R	101.9	L	74.08	L	228.01	R
T-13	493.94	L	68.4	L	459.06	L	71.57	R	48.64	R	460.52	L
T-14	14.12	R	14.97	R	41.29	R	6.94	L	7.54	L	82.72	R
T-15	144.21	L	187.21	L	129.75	L	12.76	R	22.02	L	209.19	L
T-16	7.63	L	15.3	R	54.95	R	56.62	R	12.58	R	33.04	R
T-17	52.36	L	17.7	R	39.48	R	94.27	R	51.18	R	67.49	R
T-18	354.18	R	389.54	R	470.82	R	116.3	L	69.36	L	467.37	R
T-19	252.34	R	330.62	R	108.55	R	139.31	L	299.6	L	14.62	R
T-20	522.8	L	517.93	L	598.42	L	82.35	L	123.47	L	649.96	L
T-21	258.62	L	266.17	L	580.28	R	436.84	R	464.64	R	592.3	R
T-22	124.18	R	126.19	R	148.12	R	23.71	L	6.67	L	121.02	R
T-23	421	R	449.97	R	194.81	R	37.96	R	53.12	R	45.3	R
T-24	6.89	R	11.29	L	445.06	L	273.94	R	222	R	504.1	L
T-25	336.61	R	419.23	R	443.79	R	136.67	R	83.39	R	433.17	R
T-26	342.09	L	370.27	L	153.89	L	178.56	R	143.36	R	248.48	L
T-27	5.49	L	11.15	L	136.81	L	13.33	R	43.95	R	191.62	L
T-28	488.1	R	498.4	R	113.52	R	370.65	L	273.38	L	112.09	R

Note L for left bank and R for right bank, SD = Shifting Direction

widening of the river is very high and gradually width is increasing over the study periods (Table 15.3).

15.4.4 Radius of Curvature

In contrast to sine curves, the loops of the meandering stream are nearly circular. The radius of the loop is considered to be the straight line perpendicular to the down-valley axis intersecting the sinuous axis at the apex (Deb et al. 2012; Debnath et al. 2017). During the study years of 1987, 2004, and 2020, the variations of the radius of curvature have been revealed in Fig. 15.5. This research reflects that the maximum radius of curvature is found to be 548 m at T-9 in 2020 while the minimum was 72.91 m at

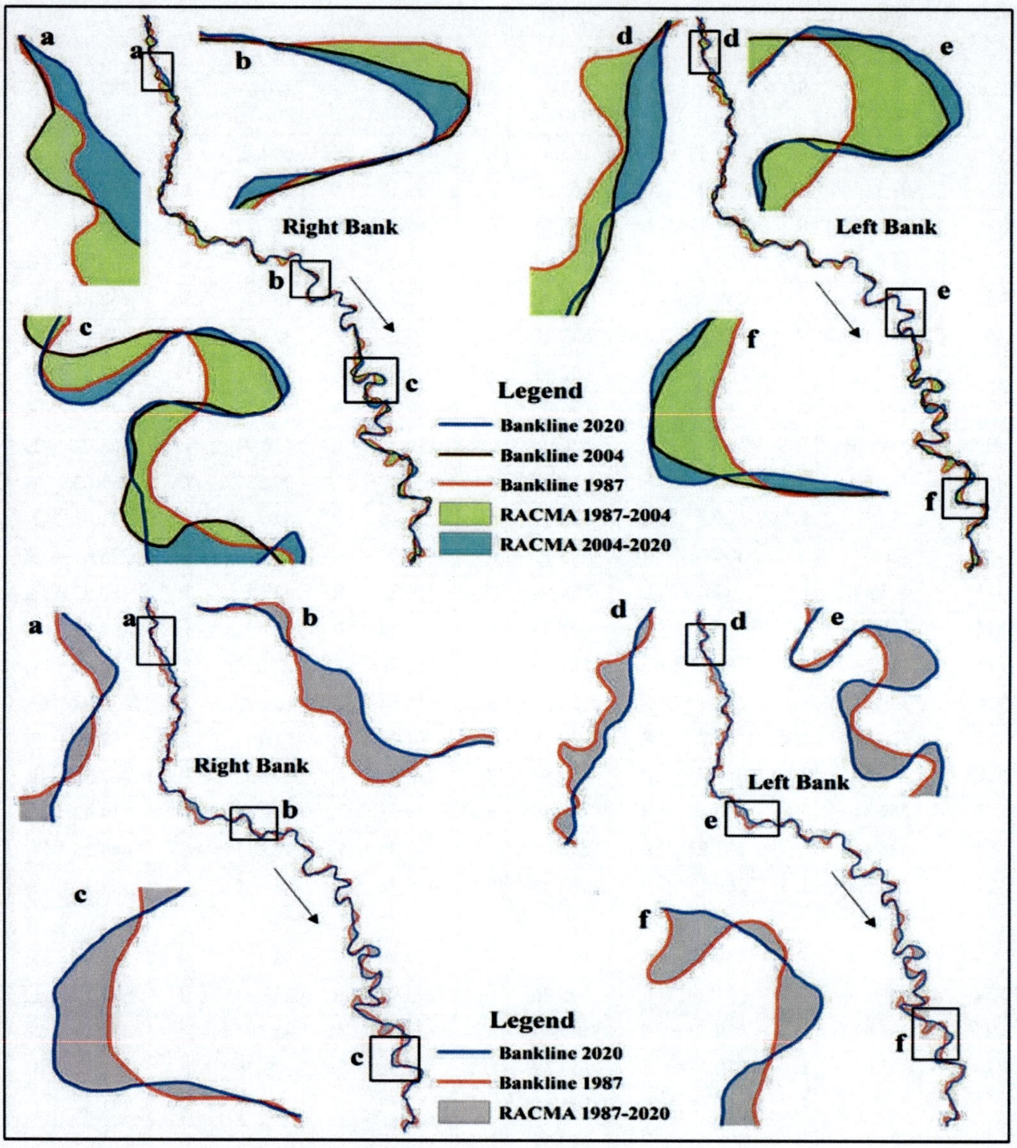

Fig. 15.4 Reach average channel migration of Kaljani River during the study periods

T-6 in 1987. At T-15, T-26, and T-27 the radius of curvature was very high during 1987, however, it decreased in 2020. On the other hand, at T-9, T-10, T-18, T-19, T-20, and T-21 the radius of curvature was very low but increased in 2020 (Fig. 15.5). Thus, this indicates a significant change in the meander geometry of the Kaljani River during 1987–2020.

15.4.5 Sinuosity Index

Meandering is a significant geomorphological process in alluvial rivers, which results in continuous channel migration or especially lateral channel shifting (Ayman and Ahmed 2009). In this study, the meandering of the Kaljani River has been reduced than earlier times. This river

Table 15.3 Variation of active channel width between 1987 and 2020

Transects	Width (m) 1987	2004	2020	Total Width Change (m) 1987–2004	2004–2020	1987–2020
T-1	29.17	73.62	65.55	−44.45	8.07	−36.38
T-2	44.06	36.48	43.86	7.58	−7.38	0.2
T-3	41.23	49.21	89.7	−7.98	−40.49	−48.47
T-4	48.02	62.34	65.8	−14.32	−3.46	−17.78
T-5	43.39	59.99	62.82	−16.6	−2.83	−19.43
T-6	63.5	80.73	64.89	−17.23	15.84	−1.39
T-7	63.46	89.21	68.39	−25.75	20.82	−4.93
T-8	68.08	77.48	97.34	−9.4	−19.86	−29.26
T-9	139.3	108.46	82.25	30.84	26.21	57.05
T-10	72.07	76.55	99.94	−4.48	−23.39	−27.87
T-11	69.8	83.48	113.56	−13.68	−30.08	−43.76
T-12	77.66	101.28	80.42	−23.62	20.86	−2.76
T-13	118.39	104.56	78.58	13.83	25.98	39.81
T-14	87.87	86.91	86.71	0.96	0.2	1.16
T-15	100.91	139.75	176.23	−38.84	−36.48	−75.32
T-16	249.61	231.11	277.94	18.5	−46.83	−28.33
T-17	166.03	100.84	142.23	65.19	−41.39	23.8
T-18	118.97	81.59	130.19	37.38	−48.6	−11.22
T-19	165.35	74.54	244.45	90.81	−169.91	−79.1
T-20	89.41	89.59	132.46	−0.18	−42.87	−43.05
T-21	121.1	133.62	107.7	−12.52	25.92	13.4
T-22	69.47	68.67	99.75	0.8	−31.08	−30.28
T-23	163.17	138	163.27	25.17	−25.27	−0.1
T-24	40.18	55.46	112.65	−15.28	−57.19	−72.47
T-25	143.34	53.71	112.41	89.63	−58.7	30.93
T-26	43.86	70.1	153.41	−26.24	−83.31	−109.55
T-27	68.63	76.91	131.18	−8.28	−54.27	−62.55
T-28	76.1	88.78	132.4	−12.68	−43.62	−56.3

was a highly meandering channel during 1987, but after that period it started straightening the course (Fig. 15.6). Zone-wise sinuosity calculation (Schumm 1977) of the Kaljani River has been documented as 1.32, 1.79, and 1.9 during 1987; 1.2, 1.65, and 1.76 during 2004, and 1.19, 1.63, and 1.85 during 2020. This analysis portrays that the river is primarily losing the meandering character.

15.4.6 Meander Wavelength

Meander wavelength is the distance between two successive meander crests or troughs (Islam and Guchhait 2017). The wavelength of a meander alone does not reveal any quantitative measure of the intensity of a meander. Only there is a direct relation between wavelength and magnitude of volume and width. The longer the wavelength,

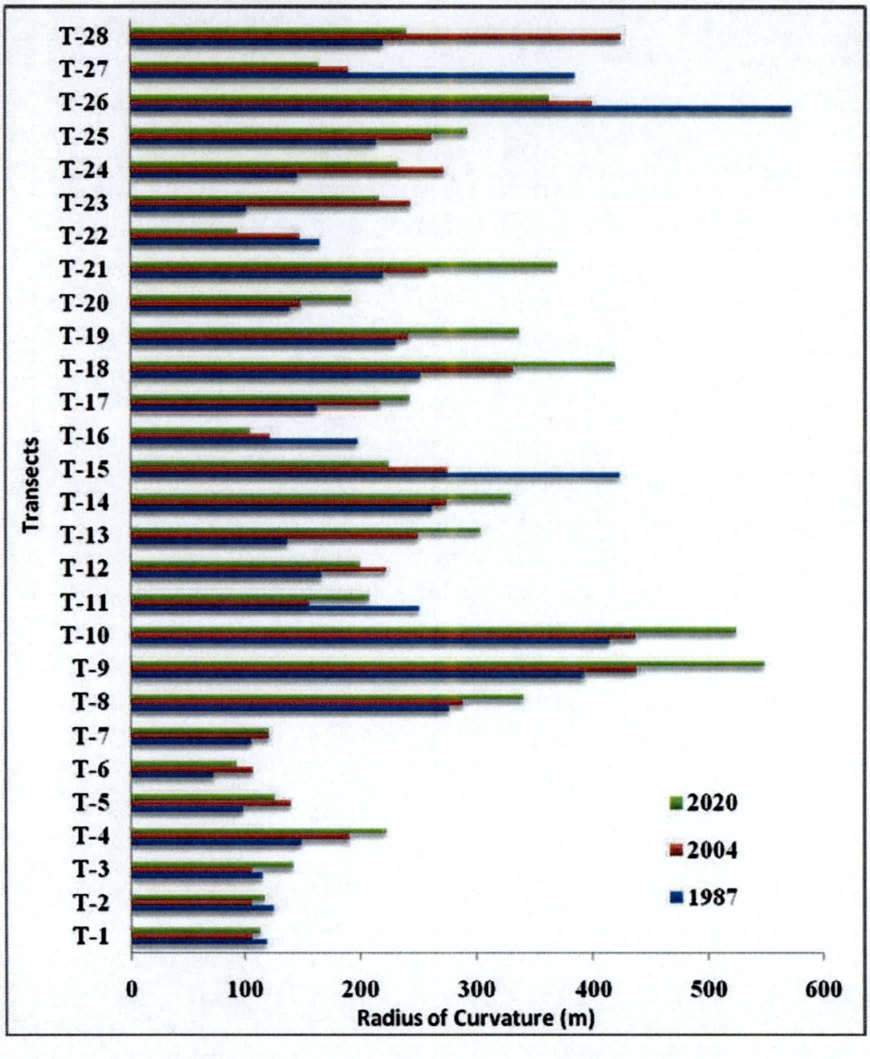

Fig. 15.5 Variation in the radius of curvature of the Kaljani River during 1987 and 2020

the greater is the volume and the larger is the river (Cotton 1952). Various meander wavelengths are shown in Fig. 15.7. This research shows that the maximum wavelength was 3734.81 m at T-27 in 1987 and the minimum was 399.63 m at T-5 in 2020.

15.4.7 Meander Width

Meander width is the lateral distance (perpendicular to the valley) between the outside edges of two meanders that occupy opposite sides of the valley. This is used as an index of the lateral containment or confinement of a stream when compared with the width of the channel (Semwal and Chauniyal 2018). Various meander widths are shown in Fig 15.8. This study depicts that the maximum meander width was 2426.9 m at T-23 in 1987 and the minimum was 100.55 m at T-5 in 2020.

15.4.8 Amplitude

Amplitude is the maximum distance between crest and trough height. A meandering river

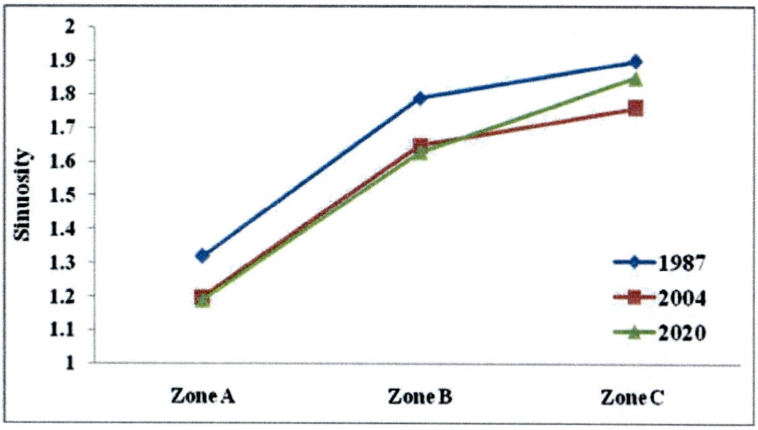

Fig. 15.6 Variation in the sinuosity index of the Kaljani River from 1987 to 2020

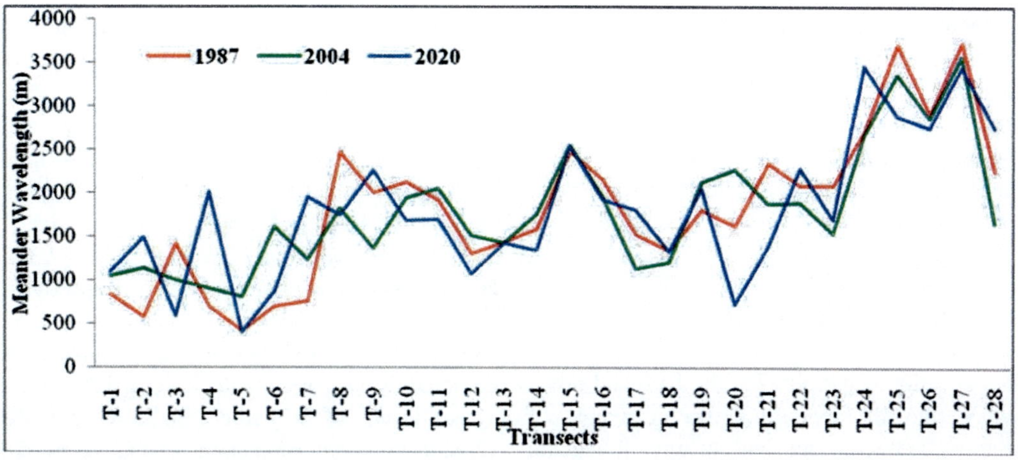

Fig. 15.7 Variation in the meander wavelength of the Kaljani River from 1987 to 2020

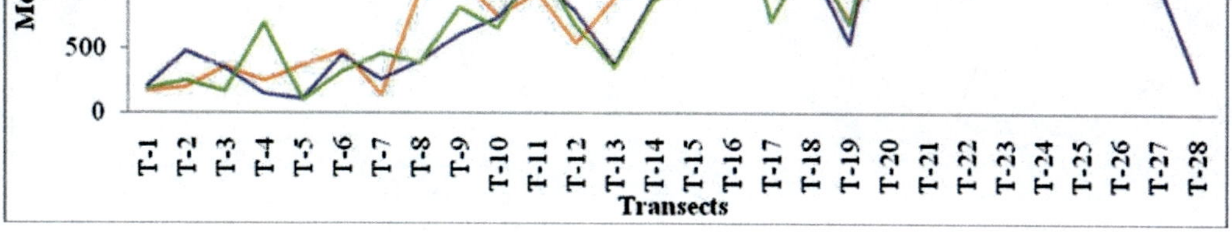

Fig. 15.8 Variation in the meander width of the Kaljani River from 1987 to 2020

consists of a number of consecutive bends. The meander development consists of translation and expansion of their bends due to bank erosion. The amplitudes range from 76.4 to 2343.6 m in the study area (Fig. 15.9). Therefore, this is an important change in the meander geometry of the Kaljani River during 1987–2020.

15.4.9 Arc Angle

The Arc angle is the angle in between lines radiating from the center of the meander arc up to the point of deflection of the meander direction. The higher the arc angle, the more intense is the meander and the increased arc angle tends towards the point of neck-cut off (Langbein and Leopold 1966). Table 15.4 reveals that the arc angle is suddenly decreased during 1987–2004 at T-5 and T-25. At T-26, the arc angle is found to gradually increase which indicates the meander formation.

15.4.10 Direction Angle

The direction angle between the path and down valley direction represents the imbalance of relation between slope and volume (Langbein and Leopold 1966). An increased direction angle also tends towards the point of neck-cut off. This study reveals an increased direction angle as depicted at T-11, T-20, T-21, T-26, and T-27, respectively, during the observation period (Table 15.4).

15.4.11 LULC Change and Channel Migration

15.4.11.1 LULC Change

The dynamic change of LULC categories for the years 1987, 2008, and 2020 are depicted in Fig. 15.10, and Table 15.6 illustrates the LULC change of the Kaljani River adjacent village area. In general, in context of 1987 about 4.11%, 4.89%, 13.76%, 28.71%, 43.91%, and 4.62% area were covered by water bodies, sandy area, dense forest, open forest, agricultural land, and built-up area respectively, whereas in 2004 these areas were changed into 2.17%, 5.58%, 10.36%, 20.87%, 53.45%, and 7.57% area respectively, and in 2020 about 1.71%, 10.88%, 8.93%, 20.98%, 44.87%, and 12.64% area, respectively (Table 15.7). The Kappa coefficient and overall accuracies of the LULC maps and accuracies of the individual classes are depicted in Table 15.5. These levels of agreement indicate that the classification and prediction LULC maps are acceptable. The LULC maps depict overall

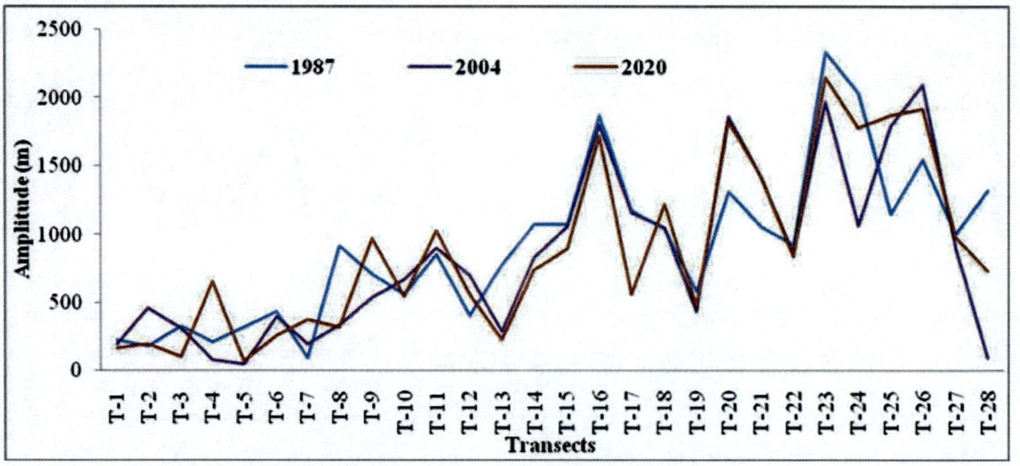

Fig. 15.9 Variation in the amplitude of the Kaljani River from 1987 to 2020

Table 15.4 Variation in the arc angel and direction angel of the Kaljani River from 1987 to 2020

	Arc angel (Degree)			Direction angel (Degree)		
Transects	1987	2004	2020	1987	2004	2020
T-1	129	148	133	50	30	41
T-2	122	108	60	52	33	22
T-3	98	144	141	26	48	25
T-4	225	55	110	86	18	32
T-5	237	59	94	43	15	28
T-6	180	68	206	88	42	89
T-7	107	78	75	17	15	23
T-8	121	65	73	48	25	29
T-9	210	177	79	72	38	21
T-10	175	181	155	65	69	56
T-11	184	191	197	35	49	65
T-12	206	139	152	62	37	35
T-13	148	215	180	84	87	76
T-14	151	144	187	57	46	62
T-15	235	221	202	49	40	51
T-16	170	184	167	67	58	52
T-17	265	215	208	69	62	65
T-18	170	204	212	86	89	75
T-19	217	203	221	50	90	87
T-20	185	195	170	71	91	90
T-21	150	227	159	63	88	92
T-22	207	178	218	70	90	88
T-23	190	156	180	89	90	90
T-24	204	219	236	86	70	63
T-25	231	96	84	93	19	20
T-26	74	78	170	18	35	58
T-27	130	156	180	45	57	79
T-28	110	187	160	50	72	59

accuracies of 87.95%, 85.1%, and 89.36 for the Kaljani River in 1987, 2008, and 2020 with Kappa coefficients of 0.80, 0.83, and 0.88, respectively. The all-over banklines (right and left) positional error is also verified with 100 ground control points (GCPs), collected from the field survey.

Furthermore, the change of all six classes from 1987 to 2004 (Table 15.8a) was very high when compared with the change between 2004 and 2020 (Table 15.8b, c).

On the contrary, sandy area, agricultural land, and built-up areas represent a net increase during the study period of 1987–2020. Sandy area increased by 20.02%, agricultural land by 3.41%, and built-up area by 26, 79%. Sandy area greatly increased during 2004–2020, compared to the 1987–2004 period, whereas the reverse trend was observed for the built-up areas. This analysis has revealed from the area matrixes that the dense forest and water bodies are converted to other land classes; especially water bodies are

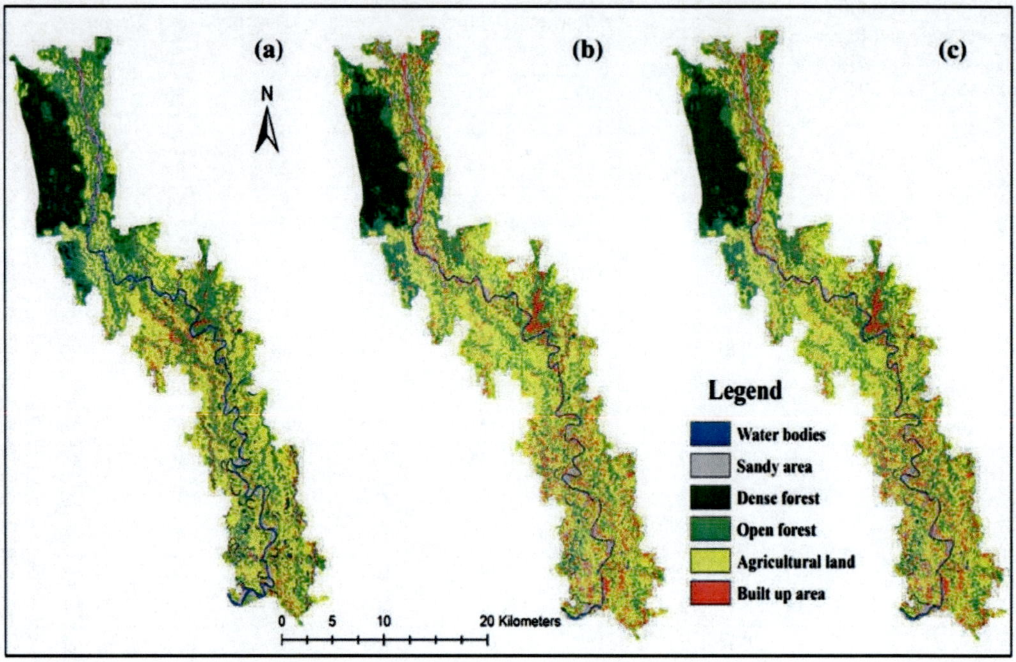

Fig. 15.10 Dynamics of the LULC in different years, **a** 1987, **b** 2004, and **c** 2020

Table 15.5 Classification of the accuracy of the supervised classification and Kappa coefficient for 1987, 2004, and 2020

	1987		2004		2020	
LULC Classes	Producer's	User's	Producer's	User's	Producer's	User's
Water bodies	81.18	90.10	85.91	90.10	90.12	95.23
Sandy area	85.17	80.32	92.13	80.53	92.08	86.76
Dense forest	100	90.21	100	80.32	1000	90.02
Open forest	72.37	80.81	69.53	90.05	69.98	90.42
Agricultural land	86.21	86.76	83.01	83.14	90.17	90.42
Built-up area	86.21	86.76	83.87	86.76	92.01	86.76
Overall Accuracy	87.95		85.1		89.36	
Kappa Coefficient	0.8		0.83		0.88	

gradually converted mainly into the sandy area, agricultural land, and built-up area with time.

15.4.11.2 Effect of Channel Migration on LULC Change

The LULC maps of 1987, 2004, and 2020 revealed the marked characteristics of the Kaljani River adjacent village area. This result of the present study depicts that changes in the pattern of LULC over the short term (1987–2004 and 2004–2020) and long term (1987–2020) are quite similar. The migrating nature of the river has changed the LULC pattern in its floodplain area significantly.

The 'from-to' LULC change map (Fig. 15.11) of all classes during 1987–2020 shows that the maximum change is noted for the classes such as open forest to agricultural land, agricultural land to open forest, agricultural land to sandy area, open forest to built-up area, and agricultural land

Table 15.6 LULC classes in the study area in 1987–2020

	1987		2004		2020	
LULC classes	Area (sq. km)	Area (%)	Area (sq. km)	Area (%)	Area (sq. km)	Area (%)
Water bodies	16.83	4.11	8.89	2.17	7.02	1.71
Sandy area	20.06	4.89	22.91	5.58	44.67	10.88
Dense forest	56.42	13.76	42.55	10.36	36.65	8.93
Open forest	117.72	28.71	85.67	20.87	85.11	20.98
Agricultural land	180	43.91	218.41	53.45	184.19	44.87
Built-up area	18.94	4.62	31.09	7.57	51.87	12.64
Total	409.97	100	409.52	100	409.51	100

Table 15.7 LULC change in the study area

	1987–2004		2004–2020		1987–2020	
LULC classes	Change	%	Change	%	Change	%
Water bodies	−7.94	7.33	−1.87	2.18	−9.81	7.98
Sandy area	2.85	2.63	21.76	25.31	24.61	20.02
Dense forest	−13.87	12.81	−5.9	6.86	−19.77	16.08
Open forest	−32.05	29.6	0.44	0.51	−31.61	25.72
Agricultural land	39.41	36.4	−35.22	40.97	4.19	3.41
Built-up area	12.15	11.22	20.78	24.17	32.93	26.79

to built-up area as it 46.16 km^2, 30.99 km^2, 22.63 km^2, 21.59 km^2, and 19.84 km^2, respectively. The 'from-to' map (Fig. 15.12) of transition all classes to sandy area reveals that the maximum area got converted from agricultural land to the sandy area (22.64 km^2). The 'from-to' map (Fig. 15.12) of transition all classes to built-up area also portrays that the maximum area is converted from open forest to built-up area (21.59 km^2). The 'from-to' map (Fig. 15.12) of transition of dense forest to all class grounds that the maximum area is converted from dense forest to agriculture land (13.24 km^2).

The 'from-to' map (Fig. 15.13) of water bodies explains the maximum transition of water bodies to sandy area, agricultural land, and built-up area as it 4.65 km^2, 4.38 km^2, and 3.06 km^2, respectively, during 1987–2020. In this research, the water bodies have a maximum conversion rate to a sandy area, which indicates the siltation of the river. The 'from-to' gains and losses map (Fig. 15.13) of water bodies analyses that 14.21 km^2 area is lost and 4.41 km^2 is gained during 1987–2020. We found that loss to water bodies' is depositional area while water bodies' gain is erosion area. In this study, the channel dynamicity of the Kaljani River adjacent village area and frequent channel migration is to a great extent related to the LULC pattern.

15.5 Discussion

The alluvial river continuously changed its buffer area LULC through the channel migration with the time. The river-bank channel migration is strongly related to the meander parameter. The continuous floods and channel migration cause diverse natural and socio-economic hazards such as loss to the riparian forest, agricultural land, and built-up area. A general observation from this whole research is that the most dynamic or migrant part of the river is a zone A and zone B compared to zone C which is relatively stable and sequential. In this study, river course in zone C (both banks) is the most dynamic part of this

Table 15.8 **a** Area matrix of LULC in 1987–2004. **b** Area matrix of LULC in 2004–2020. **c** Area matrix of LULC in 1987–2020

a

		2004						
	LULC classes	Agricultural land	Built-up area	Dense forest	Open forest	Sandy area	Water bodies	Grand Total
1987	Agricultural land	151.45	11.41	0.01	9.74	7.49	1.69	181.79
	Built-up area	10.98	2.68	0.00	2.44	1.01	0.66	17.78
	Dense forest	1.35	0.46	40.65	13.68	0.20	0.08	56.41
	Open forest	44.10	10.14	1.87	55.47	4.47	1.23	117.28
	Sandy area	8.36	2.65	0.00	0.93	6.43	1.24	19.61
	Water bodies	6.57	1.88	0.00	1.60	2.76	3.95	16.76
	Grand Total	222.81	29.21	42.53	83.86	22.36	8.86	409.64

b

		2020						
	LULC classes	Agricultural land	Built-up area	Dense forest	Open forest	Sandy area	Water bodies	Grand Total
2004	agricultural land	132.00	25.44	0.27	37.40	25.30	2.35	222.77
	Built-up area	9.66	8.14	0.07	5.86	4.81	0.68	29.22
	Dense forest	5.79	0.21	33.86	1.99	0.67	0.00	42.52
	Open forest	27.96	13.30	2.65	37.63	2.03	0.28	83.86
	Sandy area	9.92	1.58	0.02	1.72	8.13	1.00	22.37
	Water bodies	1.23	1.54	0.01	0.61	2.75	2.72	8.86
	Grand Total	186.56	50.21	36.88	85.22	43.69	7.03	409.59

c

		2020						
	LULC classes	Agricultural land	Built-up area	Dense forest	Open forest	Sandy area	Water bodies	Grand Total
1987	Agricultural land	107.39	19.25	0.26	31.04	22.06	1.73	181.74
	Built-up area	7.00	4.22	0.01	3.28	2.79	0.49	17.79
	Dense forest	12.97	0.38	34.86	7.24	0.95	0.02	56.42
	Open forest	46.87	21.22	1.75	39.98	6.54	0.94	117.28
	Sandy area	7.92	2.19	0.00	1.56	6.69	1.25	19.61
	Water bodies	4.41	2.95	0.01	2.11	4.67	2.60	16.76
	Grand Total	186.56	50.22	36.89	85.21	43.70	7.03	409.60

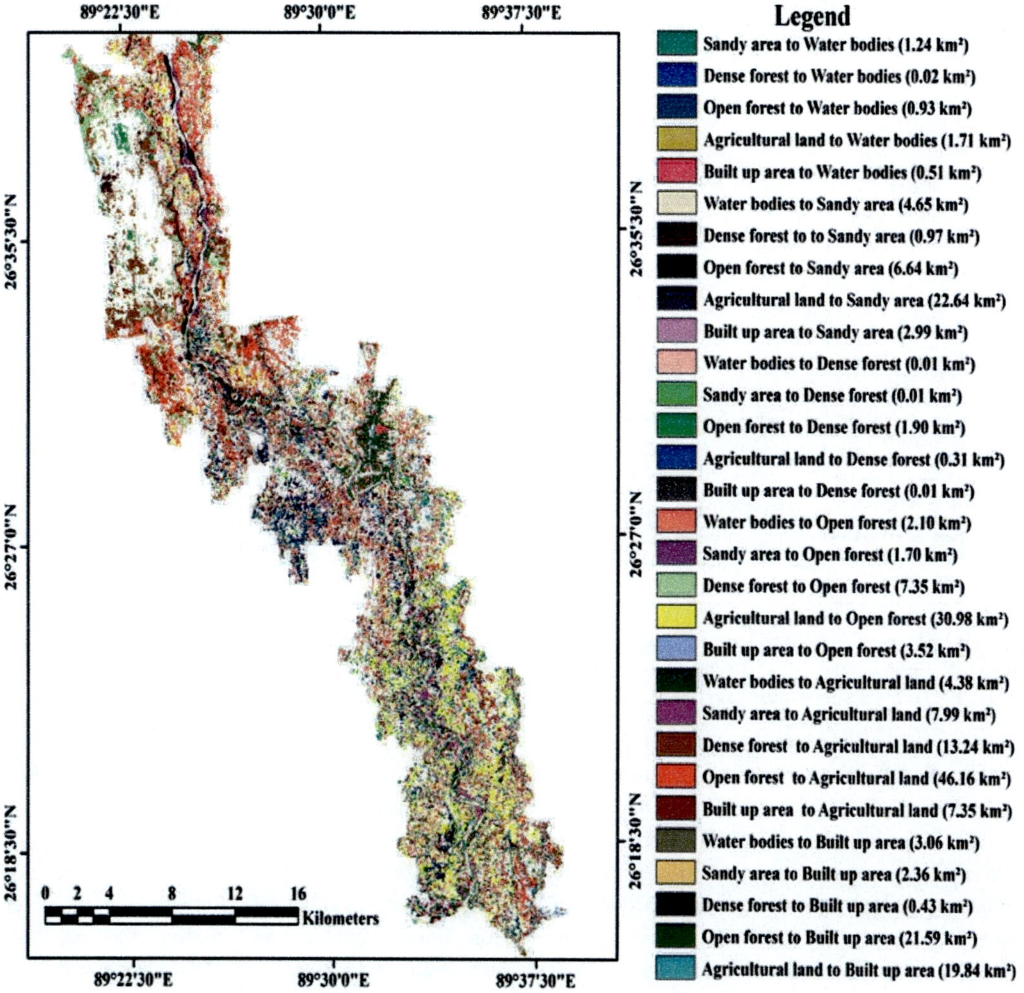

Fig. 15.11 LULC classes change map during 1987–2020

entire river. Jaigir Chilakhana, Chilakhana, Bhelakopa Dwitia Khanda, Amlaguri, Bhelapeta, Panisala mouza of zone C are very active for erosion-accretion and channel migration. In the timeframe of the last 33 years, the Kaljani River has average length migrated at 7.04 m/year in the right bank and 7.32 m/y in the left bank and the shape area migrated at 21,181,941.5 m^2, (right bank) and 21,074,452.3 m^2 (left bank) respectively. However, periodical migration values are regular and very considerable in terms of decreasing trend. This gradual decreasing trend of migration and erosion–deposition assert its formation exclusively to the continuous aggradations and consequence of sand and gravel bar formation (Chakraborty and Mukhopadhyay 2014). This investigation of the historical datasets has depicted the clear picture that the previous day erosion activity and present-day depositional activity are dominated. The Kaljani River course has a high dynamic adjustment for their need and this changed channel behavior is very dramatic.

The adjustment of LULC in the buffer area of the Kaljani River is independently a significant and interesting fact. In this analysis, dense forest and water bodies are reduced over time, which are leading the change of the channel behavioral

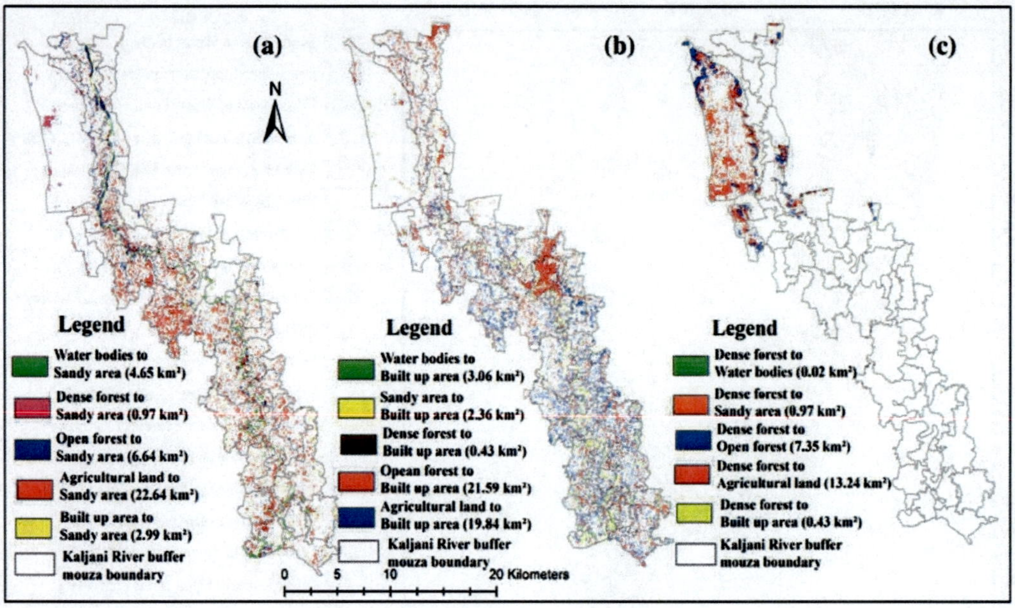

Fig. 15.12 From-to' different classes transition map. **a** Transition map from all classes to sandy area (1987–2020). **b** Transition map from all classes to built-up area (1987–2020). **c** Transition map from dense forest to all classes (1987–2020).

pattern. The built-up area and agricultural land have been eroded away and converted into accretion land which again used for both built-up area and agricultural purposes. But some areas remain as fallow land due to the presence of a high percentage of sand, that cannot be used other purpose. This river has commonly eroded the agricultural land and open forest area, whereas it becomes accretion that took place in the opposite bank. This type of LULC change due to riverbank erosion-accretion is very much significant in the socio-economic aspects. Moreover, in the settled area, people had to migrate to another place due to erosion, and it impacts on changing LULC pattern with change in their livelihood pattern. Afterward, they modify the socio-cultural environment through LULC change that can change the ecological system of the study area.

The investigation has observed that a large number of mouzas with immense population pressure are exerting the impact on the Kaljani River adjacent village area. Most of the people are engaged in the agricultural activities in different zones. Therefore, a large number of embankment installations along the Kaljani River for flood control can modify the erosion-accretion extension and LULC patterns. It could be helped for the planners, environmentalists, administrators' policymakers to understanding and formulating the needed and appropriate channel design schemes of vulnerable areas of Kaljani River. For restraining the hazards related to channel change, sustainable land use planning and constriction of scientific floods protection embankment or river restoration strategy are required. Therefore, this analysis could be guided to understand the channel migration process along the most tectonically active eastern Himalayan foothills. It can be depicted as a guideline for measuring channel migration and LULC change.

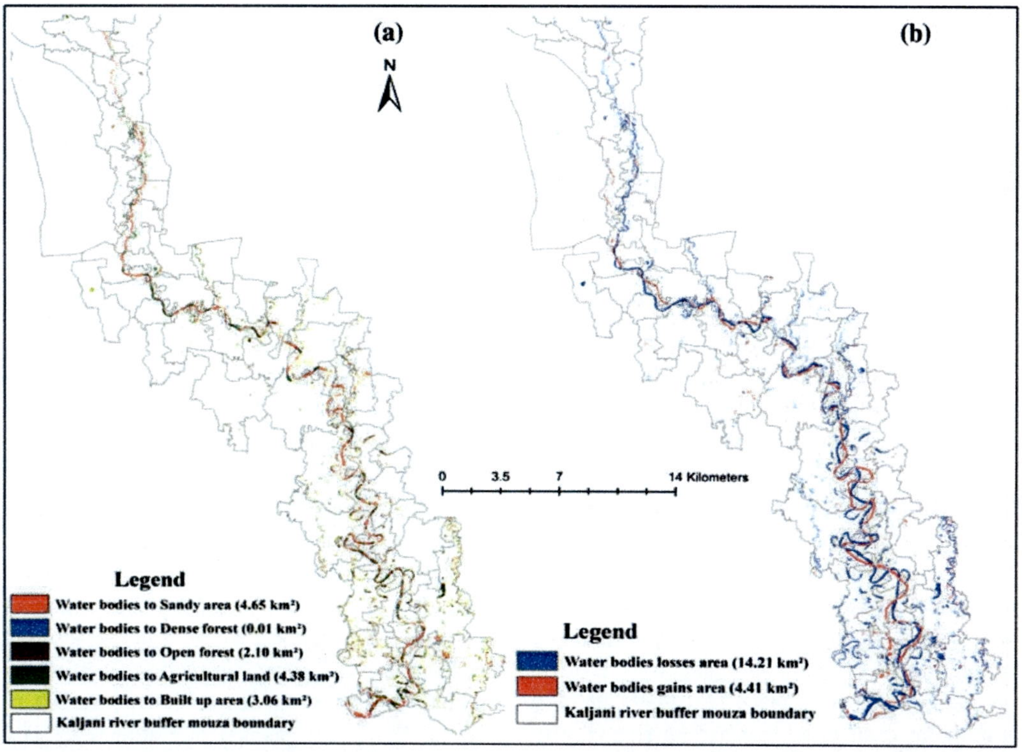

Fig. 15.13 From-to' change map. **a** Transition map from water bodies to all classes (1987–2020). **b** Gains and losses in water bodies (1987–2020)

15.5.1 Impact of Floods on Channel Migration and Changing LULC Pattern

The dynamics of channel migration, meander geometry and LULC reveals that the floods especially episodic floods play an important role in the Kaljani River adjacent village area. The average annual rainfall of the study region is 3444.05 with episodic event of extreme rainfall triggering high magnitude floods in those years (Rudra 2002; Mukhopadhyay 2005). The Kaljani River experiences floods almost every year or every second year (Starkel et al. 2008). Floods play a significant role in the evolution of the fluvial geomorphology of a river (Hooke 2016). The field visit after the flood represented that the number and area of island bars in the Kaljani River decreased due to the flood. During floods, the flow of water and the sediment transport capacity of the river increases to a great extent (Cenderelli and Wohl 2003; Hooke 2013). Moreover, the bank erosion of the river course and scouring of sediment led to severe erosion of the existing island bars. The new river course generally takes the shortest route decreasing the sinuosity index with definite channel widening (Yousefi et al. 2018). The results depicted that the flood deeply affected the channel geometry parameter and channel migration. Therefore, the main effect was the change of channel sediments and severe bank erosion in the meandering reaches. The field visit suggests that the flood frequently occupies the Kaljani River buffer adjacent village area (Fig. 15.14). The results represent that about 14.21 km^2 of the river buffer adjacent village area was affected by erosion processes from 1987 to 2020. The maximum bank retreat was observed through the flood on agricultural land. Therefore, it signifies that one of the most important triggers of change in channel migration and LULC pattern is the recurrent flood.

15.5.2 Effect of Tectonic Activity on Channel Migration and Changing LULC Pattern

The geomorphological adjustment in terms of channel migration and LULC pattern can be undoubtedly depicted through the changes of the river buffer zone that took place over space and time (Wallick et al. 2007; Bolton and Shellberg 2001). Channel migration and geometry may be modified directly or indirectly by an increase or decrease in slope imposed by areal uplift or subsidence. Sufficiently decreased slopes may cause rivers to transform along the path of meandering to the braided channels. The most commonly observed of these adjustments is for a meandering channel to increase its sinuosity in response to the increased slope, or decrease its sinuosity as an adjustment to the decreased slope (Holbrook, and Schumm 1999). For example, the Bhagirathi-Hooghly system of the Bengal Basin also portrays the channel migration and its alignment with respect to the lineaments and subsurface faults (Islam and Guchhait 2020). The Kaljani River is bearing the imprints of active tectonics of the region as they lie in the zone of Himalayan Frontal Fault, the most active thrust belt of the Himalayas (Das 2004; Goswami et al. 2012). Tectonically activity changes are depicted by the responses made in the adjoining morphology of this river behavior. Among the various causes, one of the most important causes of this river dynamicity is tectonic activity. The Bhutan Himalayas boulders, sand, and detritus are continuously deposited in these rivers causing the river beds to gradually rising (Saha and Bhattacharya 2019). This large sedimentation has choked the channel by the formation of countless islands and bars forcing the river to collapse the river banks and thus promoting continuous channel migration. However, the present study finds a decreasing channel migration rate, and increasing sedimentation that stimulated a radical shift in the LULC pattern.

Fig. 15.14 Some field photos during the flood (2020) in the Kaljani River. **a** During flood at Satali Nakadala. **b** During flood at Chapatali. **c** During flood at Uttar Chakoakheti. **d** During flood at Kholta. **e** During flood at Bhelakopa Dwitia Khanda. **f** During flood at Amlaguri

15.5.3 Human Impact on Channel Migration and Changing LULC Pattern

In this analysis, both these processes such as natural (floods, geological activity), and anthropocentric activity (construction embankment and bridges) are playing an important role for change river behavior and LULC patterns. In the Kaljani River adjacent village area, channel management through the construction of bridges and embankments along the river has played an important role in a river channel change. Embankments contain the water flow and sediments in a more confined and uniform channel, and by artificially raising riverbanks, these structures prevent spillover onto the adjacent floodplain. When high flows decrease at the end of the monsoon season, sediments start depositing on the riverbed instead of on the floodplain reducing the dimensions of the channels. Embankments restrict access to floodplain habitat, which is critical for the reproduction and growth of river fish species (Smith et al. 1998). In the study area, the most active banks have been stabilized, especially along the right side. The field survey and Copernicus satellite image show 17.58 km (27 places) of embankments along the right bank and 7.53 km (in 13 places) of embankments along the left bank and 20 bridges (Fig. 15.15). Thus, the erosion rate at the right bank has decreased due to embanking while at the left bank erosion rate is high due to the lesser control through the embankment.

15.6 Conclusion

The present study has demonstrated the application and capability of RS and GIS technology and generated a detailed evaluation of temporal and spatial changes in river channel processes and adjustment of LULC types of the Kaljani River adjacent village area. The Landsat data analysis reveals that the Kaljani River has changed its bankline positions due to the operation of the extensive channel migration processes and significant modification in the buffer area LULC pattern. During 1987–2020, the maximum and minimum bankline migration of the right bank is 621.28 m on T-6 and 20.61 on T-11, respectively, while maximum and minimum bankline migration of the left bank are 649.96 on T-20 and 21.07 on T-11, respectively. The study reveals that the Kaljani River is very dynamic and moving rightward in all periods. The average length of migration and the shape area of migration of channel are gradually decreased during 1987–2020 at both the banks. During 1987–2020, the river sinuosity gradually decreased with an increase in channel width. Therefore, the width is directly dependent on the river bankline migration; the depth of the river has reduced significantly owing to the high amount of deposition of sediment on the river bed. The radius of curvature has significantly changed over the study time. Also, other geometrical parameters such as meander wavelength, amplitude, meander width, channel width, arc angle, and direction angle results show the Kajani River's significant changes.

The results of LULC change detection represent that during 1987–2020, 26.09 km^2 of water bodies and dense forest were converted to a sandy area, agricultural lands, and built-up areas. The study has assessed the land-use change due to the dynamics of river bankline positions in vulnerable areas. The overall classification accuracies using the Kappa coefficient suggest that the LULC classification is reliable with an acceptance level of overall accuracies of 87.95%, 85.1%, and 89.36 in 1987, 2008, and 2020 with Kappa coefficients of 0.80, 0.83, and 0.88, respectively. Moreover, it will be helpful for engineers, planners, and administrators to adopt the river adjacent village area management plans required to ensure minimum human intervention in the river buffer area for its revival of ecological and environmental health. Thus, there is a need for in-depth research for the integration of geological characteristics, fluvial morphology, and geotectonic movement to understand the complex physical processes and to suggest successful management strategies on a sounder basis.

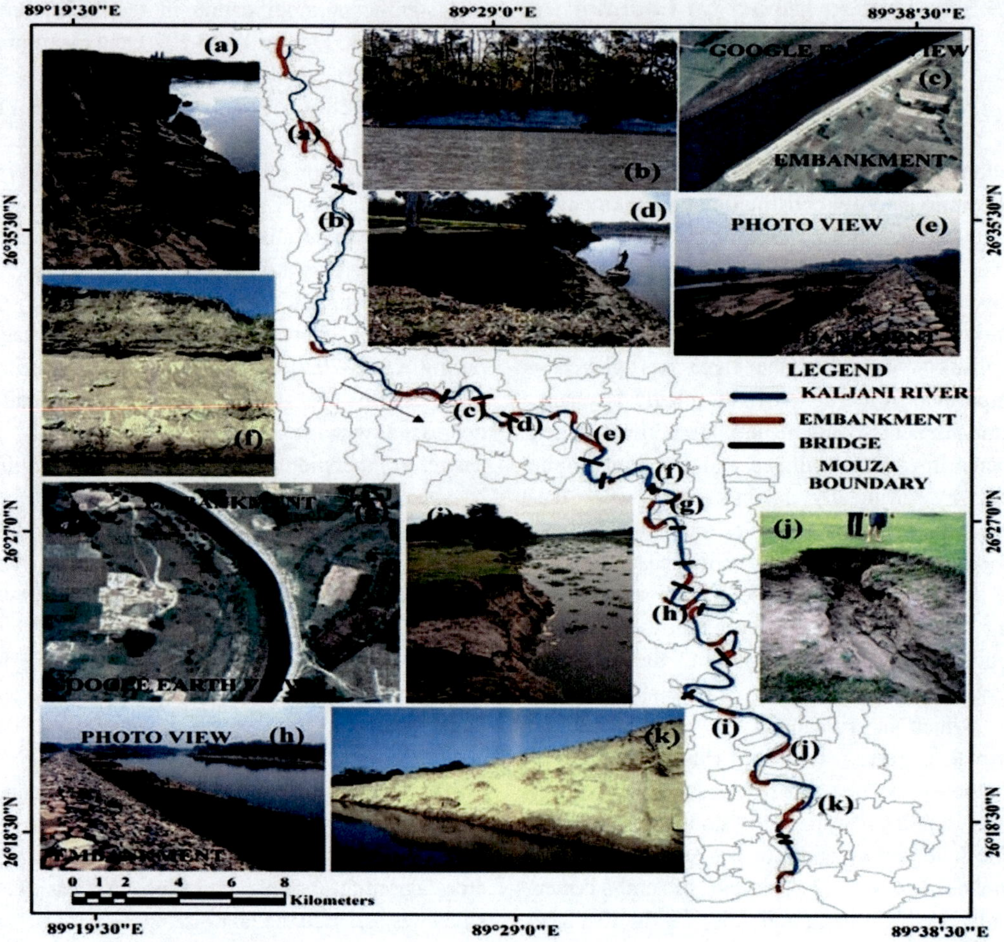

Fig. 15.15 Bridges and embankment along the Kaljani River with field validation photos

References

Ahmed F (2012) Detection of change in vegetation cover using multi-spectral and multi-temporal information for District Sargodha Pakistan. Soc Nat 24:557–572

Ayman AA, Ahmed F (2009) Meandering and bank erosion of the River Nile and its environmental impact on the area between Sohag and El-Minia Egypt. Arab J Geosci 4(1):1–11

Belletti B, Nardi L, Rinaldi M (2016) Diagnosing problems induced by past gravel mining and other disturbances in Southern European rivers: the Magra River Italy. Aquat Sci 78(1):107–119

Brice JC (1960) Index for description of channel braiding. Geol Soc Am Bull 71:1833

Butt A, Shabbir R, Ahmad SS, Aziz N (2015) Land use change mapping and analysis using Remote Sensing and GIS: a case study of Simly watershed, Islamabad, Pakistan Egypt. J Remote Sens Space Sci 18:251–259

Bhowmik M, Das (Pan) N (2014) Qualitative assessment of bank erosion hazard in a part of the Haora River, West Tripura District. In: Singh M et al (eds) Landscape ecology and water management: proceedings of IGU Rohtak conference, 2

Bolton S, Shellberg J (2001) Aquatic habitat guidelines white paper: ecological issues in floodplains and riparian corridors. Prepared for WA State Dept of Fish and Wildlife and others

Bhunia GS, Shit PK, Pal DK (2016) Channel dynamics associated with land use/cover change in Ganges river, India, 1989–2010. Spat Inf Res 24(4):437–449. https://doi.org/10.1007/s41324-016-0045-7

Cenderelli DA, Wohl EE (2003) Flow hydraulics and geomorphic effects of glacial-lake outburst floods in the Mount Everest region Nepal. Earth Surf Process Landf 28:385–407. https://doi.org/10.1002/esp.448

Cotton CA (1952) Geomorphology an introduction to the study of landforms. Wiley, New York, p 100

Cserkész-Nagy Á, Tóth T, Vajk Ö, Sztanó O (2010) Erosional scours and meander development in response to river engineering: Middle Tisza region, Hungary. In: Proceedings of the Geologists' Association 121, pp 238–247

Chakraborty S, Mukhopadhyay S (2014) An assessment on the nature of channel migration of River Diana of the sub-HimalayanWest Bengal using field and GIS techniques. Arab J Geosci. https://doi.org/10.1007/s12517-0141594-5

Chakraborty S, Mukhopadhyay S (2015) An assessment on the nature of channel migration of River Diana of the sub-Himalayan West Bengal using field and GIS techniques. Arab J Geosci 8:5649–5661

Congalton RG, Green K (1999) Assessing the accuracy of remotely sensed data: principles and practices. CRC/Lewis Press, Boca Raton, FL, p 137

Congalton RG, Green K (2009) Assessing the accuracy of remotely sensed data: principles and practices, 2nd edn. Taylor and Francis Group, LLC, Abingdon, UK

Dai SB, Yang SL, Cai AM (2008) Impacts of dams on the sediment flux of the Pearl River, southern China. Catena 76:36–43

Das JD (2004) Active tectonics of the Eastern Himalayan foothills region and adjoining Brahmaputra Basin based on satellite images. Int J Remote Sens 25 (3):549–557. https://doi.org/10.1080/0143116031000148070

Das S, Adak K, Samanta K (2014) Hydrodynamic changes of river course of part of Bhagirathi—Hugli in Nadia district—a geoinformatics appraisal. Int J Geomatics Geosci 5(2):284–299

Debnath J, Das (Pan) N, Ahmed I, Bhowmik M (2017) Channel migration and its impact on land use/land cover using RS and GIS: A study on Khowai River of Tripura, North-East India. Egypt J Remote Sens Space Sci 20(2):197–210. https://doi.org/10.1016/j.ejrs.2017.01.009

Deb M, Das D, Uddin M (2012) Evaluation of meandering characteristics using RS & GIS of Manu River. J Water Resour Prot 4:163–171

de Bethune S, Muller F, Donnay JP (1998) Fusion of multispectral and panchromatic images by local mean and variance matching filtering techniques. Fusion Earth Data: 28–30

Dey A, Nur R, Sarkar D, Barat S (2015) Ichthyofauna diversity of River Kaljani in Cooch Behar District of West Bengal India. Int J Pure Appl Biosci 3(1):247–256

Du X, Zhao X, Liang S, Zhao J, Xu P, Wu D (2020) Quantitatively assessing and attributing land use and land cover changes on China's Loess Plateau. Remote Sens 12(3):353. https://doi.org/10.3390/rs12030353

El Bastawesy M, White KH, Gabr S (2013) Hydrology and geomorphology of the Upper White Nile Lakes and their relevance for water resources management in the Nile basin. Hydrol Process 27:196–205. https://doi.org/10.1002/hyp.9216

Güneralp I, Abad JD, Zolezzi G, Hooke J (2012) Advances and challenges in meandering channels research. Geomorphology: 163–164, 1–9

Gerard F et al (2010) Land cover change in Europe between 1950 and 2000 determined employing aerial photography. Prog Phys Geogr 34(2):183–205

Goswami C, Mukhopadhyay D, Poddar BC (2012) Tectonic control on the drainage system in a piedmont region in tectonically active eastern Himalayas. Front Earth Sci 6(1):29–38. https://doi.org/10.1007/s11707-012-0297-z

Hazarika N, Das AK, Borah SB (2015) Assessing land-use changes driven by river dynamics in chronically flood affected Upper Brahmaputra plains, India, using RS-GIS techniques Egypt. J Remote Sens Space Sci 18:107–118

Hasanuzzaman M, Mandal S (2020) A Morphology-independent methodology to assess erosion, accretion and lateral migration of an alluvial channel using geospatial tools: a study on the Raidak-I River of Himalayan foothills. Sustaine Water Resour Manag 6 (3). https://doi.org/10.1007/s40899-020-00393-9

Howard AD (1992) Modeling Channel Migration and Flood plain Sedimentation in Meandering Streams. In: Carling P, Petts G (eds) Lowland Flooplain rivers: geomorphological perspectives. John Wiley and Sons, Chichester, p 12

Hooke J (1984) Changes in river meanders—a review of techniques and results of analyses. Prog Phys Geograp 8:473–508

Hooke JM (2013) River meandering. Elsevier Ltd., Treatise on Geomorphology. https://doi.org/10.1016/B978-012-374739-6.00241-4

Hooke JM (2016) Geomorphological impacts of an extreme flood in SE Spain. Geomorphology 263:19–38. https://doi.org/10.1016/j.geomorph.2016.03.021

Holbrook J, Schumm SA (1999) Geomorphic and sedimentary response of rivers to tectonic deformation: a brief review and critique of a tool for recognizing subtle epeirogenic deformation in modern and ancient settings. Tectonophysics 305(1–3):287–306. https://doi.org/10.1016/s0040-1951(99)00011-6

Islam A, Guchhait SK (2017) Analysing the influence of Farakka Barrage Project on channel dynamics and meander geometry of Bhagirathi river of West Bengal, India. Arab J Geosci 10(11): 245. https://link.springer.com/article/https://doi.org/10.1007/s12517-017-3004-2

Islam A, Guchhait SK (2020) Characterizing cross-sectional morphology and channel inefficiency of lower Bhagirathi River, India, in post-Farakka barrage condition. Nat Hazards: 1–34. https://link.springer.com/article/https://doi.org/10.1007/s11069-020-04156-9

James L, Lecce S (2013) Impacts of land-use and land-cover change on river systems. In: Shroder Jr J (ed in chief), Wohl E (ed) Treatise on geomorphology 9, pp 768–793

Jana S (2019) An automated approach in estimation and prediction of riverbank shifting for flood-prone middle-lower course of the Subarnarekha River, India. Int J River Basin Manag: 1–49. https://doi.org/10.1080/15715124.2019.1695259

Jensen JR (2005) Introductory digital image processing: a remote sensing perspective. 3rd Prentice Hall, Upper Saddle River, New Jersey

Kuehl SA, Allison MA, Goodbred SL, Kudrass H (2005) The Ganges–Brahmaputra delta. Soc Sedimen Geol 83:413–434

Kummu M, Lu XX, Rasphone A, Sarkkula J, Koponen J (2008) Riverbank changes along the Mekong River: Remote sensing detection in the Vientiane-Nong Khai area. Quatern Int 186(1):100–112

Kaufmann RK, Seto KC (2001) Change detection, accuracy, and bias in a sequential analysis of Landsat imagery in the Pearl River Delta, China: econometric techniques. Agr Ecosyst Environ 85(1–3):95–105. https://doi.org/10.1016/s0167-8809(01)00190-6

Langat PK, Kumar L, Koech R (2018) Understanding water and land use within Tana and Athi River Basins in Kenya: opportunities for improvement. Sustain Water Resour Manag: 1–11

Lawler DW (1993) The measurement of river bank erosion and lateral channel change: a review. Earth Surf Process Landf 18:777–821

Langbein WB, Leopold LB (1966) River meanders—theory of minimum variance. In: United States Government Printing Office, Washington, Hypsographic and Hydraulic studies of rivers geological survey professional Paper 422-h

Maviza A, Ahmed F (2020) Analysis of past and future multi-temporal land use and land cover changes in the semi-arid Upper-Mzingwane sub-catchment in the Matabeleland south province of Zimbabwe. Int J Remote Sens 41(14):5206–5227. https://doi.org/10.1080/01431161.2020.1731001

Mondal MS, Sharma N, Garg PK, Kappas M (2016) Statistical independence test and validation of CA Markov land use land cover (LULC) prediction results. Egypt J Remote Sens Space Sci 19(2):259–272. https://doi.org/10.1016/j.ejrs.2016.08.001

McFeeters SK (1996) The use of the Normalized Difference Water Index (NDWI) in the delineation of open water features. Int J Remote Sens 17(7):1425–1432

Mosammam HM, Nia JT, Khani H, Teymouri A, Kazemi M (2016) Monitoring land use change and measuring urban sprawl based on its spatial forms. Egypt J Remote Sens Space Sci

Mukhopadhyay SC (2005) Combating Disaster Perspective in the New Millenium, 'Flood and its Management in the Brahmaputra Basin', edited by Banerjee A, Mallick B, Sarkar D, Datta H, Chakraborti J, Bhattacharya P, Mandal R, Mandal S. acb Publications, Kolkata, pp 262–270

Nelson NC, Erwin SO, Schmidt JC (2013) Spatial and temporal patterns in channel change on the Snake River downstream from Jackson Lake dam, Wyoming. Geomorphology 200:132–142

Riley JD, Rhoads BL (2012) Flow structure and channel morphology at a natural confluent meander bend. Geomorphology 163–164:84–98

Rudra K (2002) Changing Environmental Scenario of the Indian Subcontinent, 'Floods in West Bengal, 2000 Causes and Consequences', edited by Basu SR. acb Publications, Kolkata, pp 326–347

Richards JA, Jia X (2006) Remote sensing digital image analysis an introduction, 4th edn. Springer, Heidelberg, Germany

Rinaldi M, Surian N, Comiti F, Bussettini M (2013) A method for the assessment and analysis of the hydromorphological condition of Italian streams: the Morphological Quality Index (MQI). Geomorphology 180:96–108

Roux C, Alber A, Piégay H (2013) Centerline guideline for the Fluvial Corridor toolbox, a new ArcGIS toolbox package for exploring multiscale river scape at a network scale. Sedalp (Sediment Management in Alpin Basins) and CNRS (UMR5600)

Semwal S, Chauniyal DD (2018) Meanders geometry of Alaknanda River in Srinagar valley (Garhwal Himalaya). Indian J Geograp Environ: 15–16

Schumm SA (1963) Sinuosity of alluvial rivers on the Great Plains. Bull Geol Soc Am 74:1089–1100

Schumm SA (1977) The fluvial system. Wiley, New York, p 337

Starkel L, Sarkar S, Soja R (2008) Present-day evolution of the Sikkimese-Bhutanese Himalayan piedmont, panigipz Warszawa

Saha UD, Bhattacharya S (2019) Reconstructing the channel shifting pattern of the Torsa River on the Himalayan Foreland Basin over the last 250 years. Bull Geograp Phys Geograp Series No 16:99–114. https://doi.org/10.2478/bgeo-2019-0007

Smith BD, Haque AKMA, Hossain MS, Khan A (1998) PROFILE: river dolphins in Bangladesh: conservation and the effects of water development. Environ Manag 22(3):323–335. https://doi.org/10.1007/s002679900108

Uddin K, Shrestha B, Alam MS (2011) Assessment of morphological changes and vulnerability of river bank erosion alongside the river Jamuna using remote sensing. J Earth Sci Eng 1:29–34

Wang S, Mei Y (2016) Lateral erosion/accretion area and shrinkage rate of the Linhe reach braided channel of the Yellow River between 1977 and 2014. J Geog Sci 26(11):1579–1592

Wang B, Xu YJ (2018) Dynamics of 30 large channel bars in the Lower Mississippi River in response to river engineering from 1985 to 2015. Geomorphology 300:31–44

Wang SW, Gebru BM, Lamchin M, Kayastha RB, Lee W-K (2020) Land use and land cover change detection and prediction in the Kathmandu district of Nepal

using remote sensing and GIS. Sustainability 12 (9):3925. https://doi.org/10.3390/su12093925
Wallick JR, Grant GE, Lancaster ST, Bolte JP, Denlinger RP (2007) Patterns and controls on historical channel change in the Willamette River, Oregon, USA. In: Gupta A (ed) Largerivers: geomorphology and management. Wiley, pp 492–516
Xu H (2006) Modification of normalised difference water index (NDWI) to enhance open water features in remotely sensed imagery. Int J Remote Sens 27 (14):3025–3033
Yanan L, Yuliang Q, Yue Z (2011) Dynamic monitoring and driving force analysis on rivers and lakes in Zhuhai city using remote sensing technologies. Procedia Environ Sci 10:2677–2683
Yousefi S, Pourghasemi HR, Hooke J, Navratil O, Kidová A (2016) Changes in morphometric meander parameters identified on the Karoon River, Iran, using remote sensing data. Geomorphology 271:55–64. https://doi.org/10.1016/j.geomorph.2016.07.034
Yousefi S, Mirzaee S, Keesstra S, Surian N, Pourghasemi HR, Zakizadeh HR, Tabibian S (2018) Effects of an extreme flood on river morphology (case study: Karoon River, Iran). Geomorphology 304:30–39. https://doi.org/10.1016/j.geomorph.2017.12.034

Preliminary Insights on the Dynamics of Flow Regime and Sediment Flux in Drainage Basin Study

16

Suvendu Roy

Abstract

Being in the active phase of quantification in geomorphological research, digitally available data on river flow, sediment flux, and dam statistics has been used to study the characteristic of regional flow regime, sediment flux, and anthropogenic effects on their dynamics, with a special focus on South-East Asian dams and Indian rivers. Mean monthly discharge data of 270 months (January 1998 to June 2020) from Dartmouth Flood Observatory (DFO) from ten gauge stations of major Indian rivers have been used to estimate discharge volume at different return periods (2.3 years (Y), 5Y, 10Y, 20Y, 50Y, 75Y, 100Y, 125Y, and 200Y) using Log Pearson III flood frequency method. Dartmouth Flood Observatory has also been used to monitor the live flooding situation of Brahmaputra River and for flood prone area mapping of West Bengal and surroundings using the maximum extension of floodwater since 1993. In West Bengal, about 17,569 km^2 of land comes under flood prone area with significant variation among the twenty drainage basins. Among the non-classified rivers of India, Godavari carries the maximum sediment load (59 mmt) followed by Narmada (32.86 mmt) and Tapi (32.53 mmt). Dam is the major anthropogenic factor to alter the characteristic of sediment and hydrology of a drainage basin. The countries of South-East Asia are alone holding about 58% of total large dams of the World and contain a huge amount of storage capacity, where China alone holds about 10 million mcm. The geomorphology and hydrology of West Bengal rivers are also affected by dam constructions. The environmental perspectives could help to sustain the neutrality of our rivers by hydro-geomorphological and ecological management.

Keywords

Flow regime · Sediment flux · Return period · Flood prone area · Dam · E-flow

16.1 Introduction

Since the beginning of the systematic study of the Earth's landscape, rivers are the fundamental issue to know their role in landscape development as a prime research interest by the geoscientists in the name of fluvial geomorphology. In such studies, when looking for details of the river, such as location, spatial coverage, quality and quantity of water, sediments, nutrients,

S. Roy (✉)
Department of Geography, Kalipada Ghosh Tarai Mahavidyalaya, Bagdogra, Darjeeling 734014, West Bengal, India

P. K. Shit et al. (eds.), *Drainage Basin Dynamics*, Geography of the Physical Environment,
https://doi.org/10.1007/978-3-030-79634-1_16

ongoing physical processes and landforms, their use and management, watershed and/or drainage basin (both terms are used interchangeably) is the 'fundamental geomorphic unit' (Chorley 1969) to study after dividing the Earth's surface accordingly (Fig. 16.1). A watershed or drainage basin is a part of the earth's surface where water from different sources is collected and/or concentrated to drain at a common outlet through stream networks. Watersheds are separated from each other by a drainage divide, which is a topographical boundary line form by any ridges or summits to separate runoff between adjacent watersheds (Rhoads 2020). Drainage basin mainly works as a precipitation collector and distributor with different hydro-geomorphological processes under the framework of an 'open geomorphic system' (Chorley 1962; Hack 1965; Dury 1966). Apart from energy, the major inputs of this system are precipitation, snow-melt water, and sediments, which are controlling different forms and processes of drainage basins in combination with the topographical characteristics, soil type, bedrock type, climate, and vegetation cover. The different patterns of drainage networks largely depend on those factors.

The numerous dimensions of drainage basin research could be summarized by the term 'fluvial geomorphology', a sub-field of geomorphology. Fluvial geomorphology is an interdisciplinary field of river science and its spectrum of investigation expanding very first with the progress of time. As per Rhoads (2020), the initial recognition of this field comes through the concept of 'uniformitarianism' by Hutton (1795) during the late eighteenth century (although the term was coined by William Whewell in the late eighteenth century), as the first great fluvialist (Chorley and Beckinsdale 1964) and the founder of geomorphology (Orme 2013). However, the awareness about the effective role of fluid on earth's surface present since the fifteenth century, as Leonardo da Vinci (1452–1519) was concerned about the hydraulic geometry of stream and had measured channel reach, cross-section, slope, and velocity (Biswas 1970; Orme 2013). The timeframe on the development of fluvial geomorphology highlights the major milestones in fluvial research (Fig. 16.2). In brief, until 1950, the geomorphological resources were mainly qualitative and descriptive to explain landforms, works of rivers, other changes, etc.

However, the initial imprint of quantitative technique in fluvial research has been given by Gilbert (1914, 1917), and later by Horton (1945), a hydraulic engineer, the main kick to pursue quantitative approach in geomorphological research has been made by Strahler (1952). It was the beginning of a new generation in geomorphology and a time of paradigm shift. Since then, geomorphology becomes a process-based quantitative and analytical field of research and has developed a number of landmarked works. For example, the "at–a–station" hydraulic geometry model by Leopold and Maddock (1953) establishes the relationship between channel morphology (width, depth, and velocity) and discharge. The contemporary fluvial geomorphology was started with the infusion of fundamental fluid dynamics and hydraulics models of Bernoulli Principle (1738), Chezy' Formula (1775), Manning Equation (1891), etc. During the late twentieth century, a number of advanced technologies, e.g., remote sensing, geographical information system (GIS), digital elevation model (DEM), and geographical positioning system (GPS) have been intensively used for drainage basin research and modeling in landform evolution and nonlinear dynamic system (McDowell 2013).

The beginning of the twenty-first century started with the advancement of present technology, e.g., 90 m DEM to sub-meter level interferometric DEM, very high resolution optical remote sensing data, microwave remote sensing, Light Detection and Ranging (LiDAR)-based digital surface model, hyperspectral remote sensing. Historical evolution of fluvial landscape and effect of past climate on geomorphology are now more precise to know with the application of optically stimulated luminescence (OSL) dating technique (Jain and Tandon 2003). During the last two decades, Unmanned Aerial Vehicles (UAVs), informally known as drones, prove their

Fig. 16.1 Major drainage basins of India and its neighboring regions with selected gauging stations of Indian rivers from Dartmouth Flood Observatory (DFO) and their volume (in cumec) of discharge at 50 years return-period

scientific ability to spatial data acquisition in various geoscientific applications (Niedzielski 2018). Piegay et al. (2015) mentioned that in this century our data availability is significantly changed with lower cost and easy accessibility, increasing variability in the source of information by remote sensing (ground, aerial, and space). In such circumstances, the twenty-first century will be the new era of data science and the start of a more active second phase of qualitative geomorphology and internationalization of fields (Piegay et al. 2015).

As mentioned, the present century deals with more quantitative information and data on digital

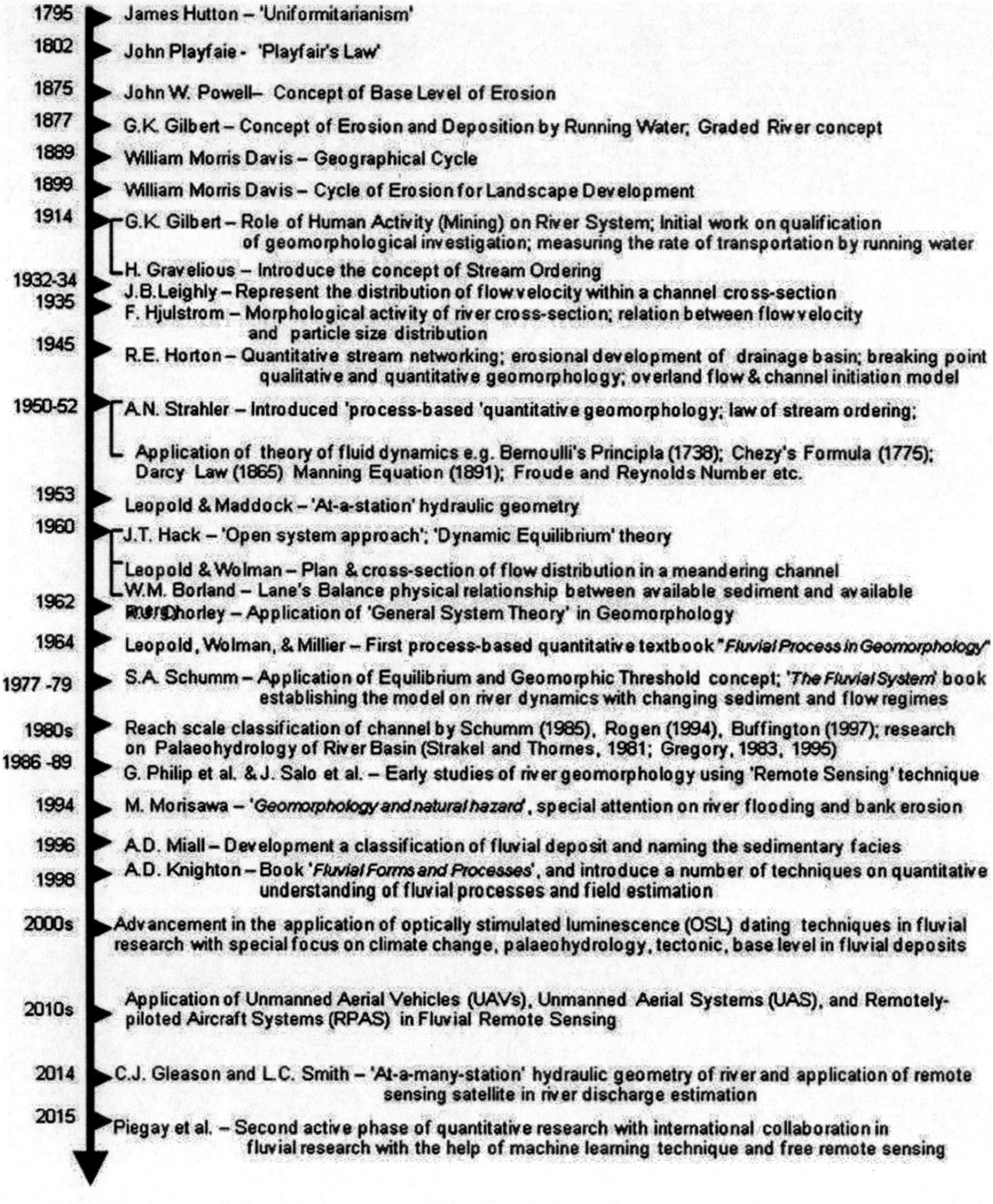

Fig. 16.2 Timeframe of the fluvial research development since late eighteen century by pointing significant events and benchmarks

formats, therefore, the objective of the present chapter is to make a preliminary insight on the fundamental concepts of fluvial geomorphology, especially on the changing flow regime, sediments flux, and anthropogenic influences on their dynamics and management, with digitally available data on river discharge, sediment flux, and dams, in regional as well as global scale.

16.2 Concept of System and Scale in Basin Research

Complexity is more common than simplicity in geomorphology as proposed by Thornbury (1954) due to the assemblage of numerous geomorphological units and their mutual interaction to run geomorphological processes and landform development, which is simply called 'geomorphic system' (Chorley 1962). Since the 1950s, Bertalanffy's (1951) 'general system theory' has been started to introduce in geomorphological research (Chorley 1962), however, the model was famous in geomorphology, as well as in basin research, after the emphasized on the difference between 'open system' and 'close system' by Chorley (1962). Earlier works by Davis (1889, 1899) and his followers support closed system to explain the theory of landscape development over a long period of time without any disturbance, whereas, since the 1960s, open system theory in drainage basin dynamics has been introduced by Hack (1960), Chorley (1962) with the insights of complexity in landscape development by tectonic activity, climatic variability, lithological control, dynamic equilibrium concept, where import and export of matter and energy are going across the system boundary. The open system approach in geomorphological study helps to emphasize the interactions between process and form and the tendency towards steady-state equilibrium (Abrahams 1968). In particular, fluvial geomorphology deals with three types of system as follows (Charlton 2007):

(a) *Morphological (form) System*: In a drainage basin, the form of each geomorphic unit such as hillslopes, streams, floodplain, bars, pools, riffles is related to the form of others. For example, steep barren hillslopes are induced to generate more deep rills and gullies and input for sediments in the fluvial system, which might cause more stable and large channel bars downstream.

(b) *Cascading (process) System*: Ongoing morphological processes within a basin are interconnected to each other like the higher overland flow intensifies the process of soil erosion and eroded soil transfer through flow in the network and increased the rate of sedimentation or aggradations.

(c) *Process-response System*: Morphological and Cascading systems are interacting in the process-response system. Where any morphological process influences the morphological form and any specific form also controls the rate of the process. Steep hillslope as a form could increase the process of soil erosion, whereas a higher rate of soil erosion can also reduce the slope amount. Such circumstances guide the '*Feedback*' mechanism in the fluvial system, where a change in one variable leads to a change in one or more variables (Charlton 2007). Two types of feedback mechanisms are observed: Positive Feedback and Negative Feedback. A positive feedback mechanism acts to reinforce the original change and leads to a new outcome, whereas in negative feedback, the system will recover again after certain changes.

Scale in geomorphology is an essential aspect to understand the process-form interaction in the fluvial system, as well as in all other geomorphological systems, through the dimensions of time and space. In fluvial geomorphology, the space scale ranges from the individual grain of sediment to the entire drainage basin, whereas time scale also varies from few seconds to thousand of millions of years (Fig. 16.3). Geomorphologists choose the frame of time and space as per the objects of research. For example, to know the effect of climate change on the hydrological cycle, the entire basin has to be considered on the spatial scale and thousands of years also need to monitor on the temporal scale. The movement of sediment or grain on river bed could be assessed within an hour, whereas to assess the change in flow regime, thousands of years have to be considered. Nevertheless, to estimate the rate of denudation of the entire landscape, millions of years would be the time frame. The methodology and/or techniques in

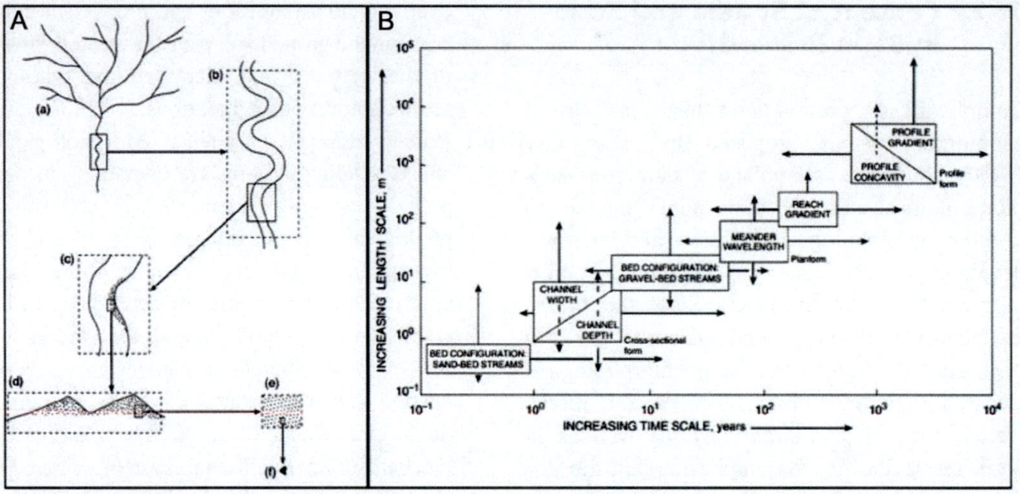

Fig. 16.3 A Range of spatial scale in different components of the fluvial system from entire drainage basin (a), to a reach (b), to meander bend (c), to bed form (d), to sedimentary structure (e), and to individual grain (f). B Changing time frame with the various components of fluvial features (*Source* Adopted from Charlton 2007)

fluvial geomorphological research are directly dependent on these two scale factors.

16.3 Flow Regime, Flood Geomorphology, and Climate Change

Drainage basin is a logical areal unit to study the hydrological characteristic of any region (More 1969). A basin could be used to estimate the regional water balance to assess the water resource and its management. In this procedure, the key requirement is hydrological information about the basin's input, storages, transfer, and output capacity, mainly through estimating the amount of water input by precipitation, surface and sub-surface runoff characteristics, evapo-transpiration, groundwater flow, and other major storages within the basin, and the output of sediment and water at the outlet. Discharge is the volume of water passing through a known cross-section of a channel in a specific time and a key factor to control most of the processes in a drainage basin, basically measure in cubic meters per second (m^{-3} s^{-1}) or 'cumec'. In a natural drainage basin, discharge and/or flow amount defers both spatially and temporally, towards downstream, and with changing basin's climate, as well as land use land cover characteristic, such notion is known as 'flow regime' (Charlton 2007) (Fig. 16.4). As observed in Fig. 16.4, dark blue pixels indicate higher annual discharge, which are concentrated along the major rivers of this region such as Ganga (India), Brahmaputra (India and Bangladesh), Irrawaddy (Myanmar), Mekong (Cambodia, Vietnam, Laos), Xi (China), Yangtze (China), etc.

The accurate measurement of river specific discharge for a long time helps to estimate the probability of extreme events occurrences, such as flood and drought. The discharge amount at different return periods (2.3Y, 5Y, 10Y, 20Y, 50Y, 70Y, 100Y, 125Y, and 200Y) of ten selected gauging stations of Indian Rivers has been calculated based on the monthly mean discharge data from Dartmouth Flood Observatory (DFO 2020) (http://floodobservatory.colorado.edu/index.html) during January 1998 to July 2020 (Table 16.1).

Dartmouth Flood Observatory (University of Colorado) also digitally maintains a "Global Active Archive of Large Flood Events" and maximum extent of floodwater since 1985, which helps to map the flood prone areas and their management (http://floodobservatory.colorado.

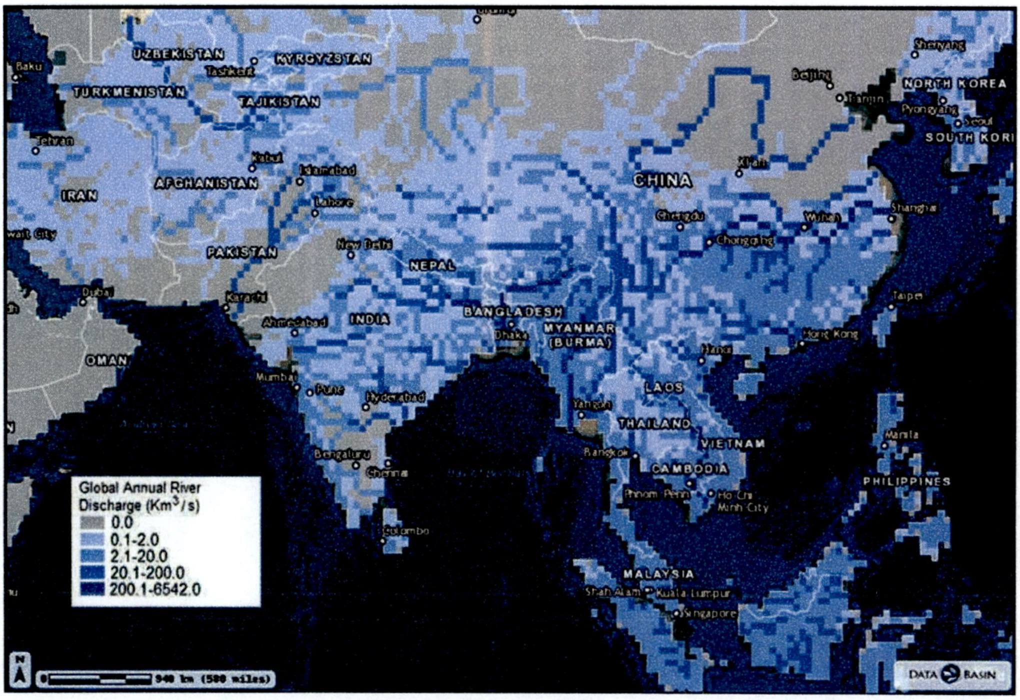

Fig. 16.4 Regional variation of annual discharge (km^{-3} y^{-1}) of South-East Asian rivers (*Source* Global Annual River Discharge, Data Basin, 2020, retrieved from https://databasin.org/maps/new#datasets=3a2afd820d2d4185b5a5903d45c0e741 on 23rd July, 2020)

edu/Archives/index.html). In particular, Fig. 16.5 shows the spatial expansion of flood prone area in West Bengal and its surrounding region, especially in Bihar State and Brahmaputra floodplain, as per the maximum observed flooding area since 1993 by Dartmouth Flood Observatory. The data could be used to estimate the catchment level variation of flood prone area. For example, a significant variation has been observed among the twenty catchments of the West Bengal. Mayurakshi river basin has experienced the maximum area under flood prone, followed by GPBH (Ganga-Padma–Bhagirathi-Hoogly) basin, Sundarban region, Ichhamati-Bidyadhari basin, etc.

A hydrograph is the key tool for research on basin hydrology, which defines as a graphical representation of discharge against a specific timeframe as per data received from a particular gauging station. Hydrograph shows how a river responds to a specific or annual precipitation event within its drainage basin and analysis the time distribution of runoff produced by given precipitation on it. In India, most of the rivers carry their 80–95% of annual discharge and sediments during the three to four months of monsoon, whereas in the rest of the year, most of them are dried up or just carry very low base flow (Kale 2002). Therefore, from the fluvial geomorphological perspective, the major changes, especially the channel morphology, take place in the monsoon period mainly in the Indian River system as well as in World Rivers. Nevertheless, the hazard related to the peak discharge, i.e., flood is also a common phenomenon for Indian rivers corresponding to the same period of high flows. Working Group on Flood Control Programme (WGFCP) of India reported about 45.64 million ha. area in the country is liable to flood, which is about 14% of the geographical area (Das et al. 2020a, b).

For example, the given hydrograph of Brahmaputra River in Assam, India (near Dibrugarh), from 1998 to the present (21st July

Table 16.1 Discharge at different return periods (years) of Indian major rivers based on the monthly mean discharge value for the period of January 1998 to July 2020 using Log Pearson III flood-frequency analysis

River name	Catchment area (km^2)[a]	Nearest city	Lat	Long	Discharge (m^3/s) at different return periods (Log Pearson III)								
					2.3Y	5Y	10Y	20Y	50Y	70Y	100Y	125Y	200Y
Yamuna	59,963	Karnal	29.564	77.085	841	1924	3385	5489	9632	11,658	14,169	15,953	20,327
Ramganga	22,909	Bareilly	28.034	79.515	267	517	812	1193	1865	2170	2534	2784	3375
Ganga-Varanasi	449,883	Varanasi	25.244	82.395	2141	5323	11,638	25,062	68,100	98,015	143,995	183,056	303,066
Ghaghra	87,434	Faizabad	26.864	81.765	1110	2949	6220	12,344	28,841	38,925	53,198	64,516	96,304
Gandak	43,387	Bettiah	26.774	84.375	1240	2706	5363	10,563	25,723	35,625	50,284	62,368	98,112
Kosi	64,711	Saharsa	25.964	86.445	1533	3084	5848	11,178	26,517	36,470	51,165	63,256	98,961
Son	72,642	Dehri	24.884	84.195	696	1390	2274	3503	5861	7000	8403	9397	11,826
Ganga-Farakka	915,908	Malda	24.974	87.975	8352	17,215	33,814	67,405	170,130	239,697	345,181	433,908	703,454
Hugli	18,066	Nabadwip	23.534	88.335	130	354	772	1596	3965	5481	7689	9482	14,670
Brahmaputra	383,449	Guwahati	26.324	92.025	11,566	18,379	25,527	34,053	47,986	54,005	60,991	65,703	76,564
Mahanadi	122,071	Athmallik	20.744	84.465	1202	2926	5637	10,164	20,816	26,739	34,662	40,659	56,552
Godavari	308,946	Rajamahendravaram	17.054	81.675	3238	5735	8313	11,294	15,947	17,879	20,070	21,518	24,772
Krishna	259,597	Vijayawada	16.604	80.415	1415	3909	8679	18,275	46,518	64,888	91,900	114,005	178,624
Cauvery	75,506	Tiruchirappalli	10.844	78.885	905	1456	2011	2647	3639	4053	4524	4836	5542

[a]Values shows the catchment area covered above the data station

Source Values are calculated by author using monthly mean discharge data available from Dartmouth Flood Observatory (DFO) (2020); http://floodobservatory.colorado.edu/index.html

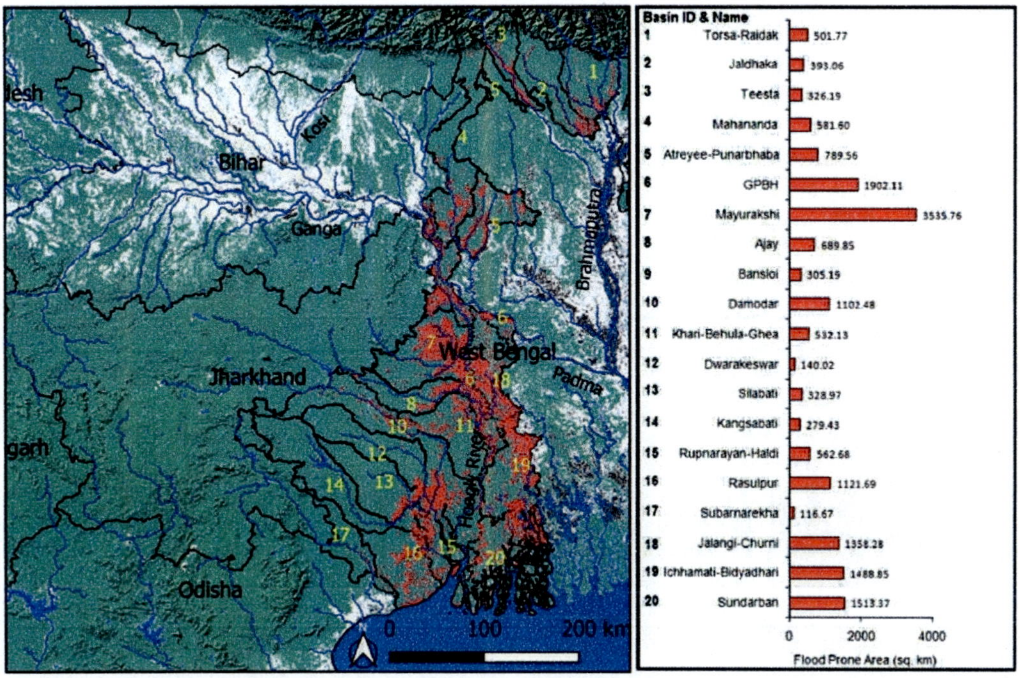

Fig. 16.5 Flood prone area of the West Bengal (in red) and its surrounding (in gray patches) as per maximum observed flooding area since 1993 by Dartmouth Flood Observatory; the inserted diagram also shows the catchment level variation of flood prone area in West Bengal (*Source* of Flood Prone Area: DFO 2020)

2020) represents the flow regime behavior of this flood prone basin with the primary peak in every rainy season (Fig. 16.6). At present, the discharge of this river crosses its 25-years recurrence interval and facing severe floods like other major floods in 1998 and 2000. As per the Daily Flood Report on 21st July (2020) of Assam State Disaster Management Authority (ASDMA), the rivers were staring spilling channel on 20th June 2020, and since then, more than 2.4 million people in 2,323 villages across 24 districts have been affected including 110,323 ha of cropland, whereas the death toll to 84. Central Water Commission (CWC) of India has declared 'Severe Flood Situation' at many stations as river Brahmaputra is flowing at 'High Flood Level (DFL)' and alerting with the chances of water level rising and possible area of inundation at daily basis.

Water is essential for sustainable human development and the fruitful functioning of the Earth's ecosystem (Das et al. 2020a, b). However, the imbalance in its availability creates situations like drought and flood due to deficits and surplus of water, respectively. On an average natural disasters induced death toll to 60,000 people per year, where the water derived disasters e.g. flood and drought are sharing the maximum of it since 1900 (Ritchie 2014; Fig. 16.7).

Worldwide flood is a more frequent phenomenon. Therefore, detailed knowledge about the flow regime of flood prone drainage basins is necessary to understand its typical behavior for sustainable management plans. A sudden increase in the input of water due to heavy rain and/or dam failure within the drainage basin makes the channel inadequate to contain the rising flow and started spill over the bank and makes flooding in the adjacent area. Including climate as the primary one, a range factor leads to floods, such as characteristic of drainage basin, e.g., shape and size of the basin, relief, geology, drainage density, soil type, land use land cover, and channel morphology; anthropogenic

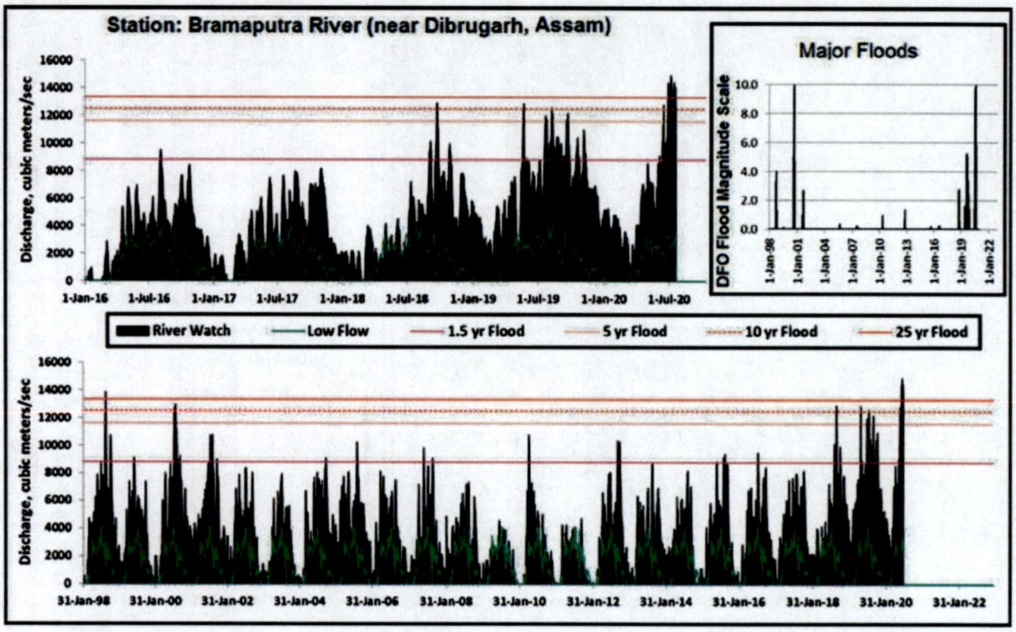

Fig. 16.6 Annual flow regime of Brahmaputra River in Assam, India (near Dibrugarh), for last two decades with flood discharge at different recurrence intervals and major floods as per the Dartmouth Flood Observatory (DFO) flood magnitude scale (*Source* DFO Flood Observatory, 2020; http://floodobservatory.colorado.edu/SiteDisplays/1938.htm, access on 21st July 2020)

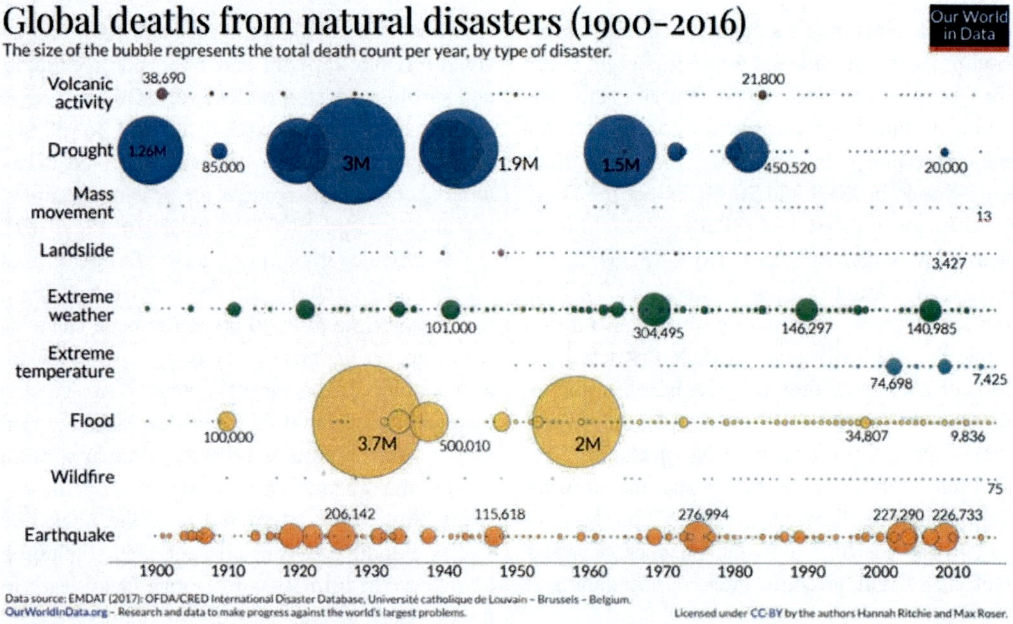

Fig. 16.7 Scenario of global death since 1900 due to natural disasters, where water-related disasters (Drought and Flood) are more dangerous than others (*Source* OurWorldInData.org; retrieved on 23rd July 2020)

disturbances, e.g., urbanization, dam, channelization, embankment, etc. (Raghunath 2013). Benito and Hudson (2010) demonstrated the significant association between cross-section geometry of the channel (including floodplain) and flood process, area of inundation, and flood frequency. Large flood has a significant effect on river morphodynamics, which could be assessed by determining the changes in channel morphology, floodplain features, channel bed materials (Costa and O'Connor 1995). The effectiveness in channel morphodynamics of any flood is referred by 'flood power'—the stream power per unit area (Baker and Costa 1987), which vary with river character, e.g., if the flood power exceeding the 300 W m^{-2} makes a significant change in alluvial and semi-alluvial rivers, whereas no substantial change could make in a channel with resistance boundary (Miller 1990). The common geomorphic responses of a river through floods are channel widening, incision, avulsion, dynamics in channel bars, floodplain sedimentation, sand and gravel splay, the formation of natural levees, and sand ridges, etc. (Rhoads 2020).

The impact and risk assessment by IPCC (Intergovernmental Panel on Climate Change) (2018) on the 1.5 °C rise of global temperature above pre-industrial level shows medium to high risk of severe impact by the coastal and fluvial flooding with less ability to combat it. By influencing the natural characteristics of rainfall patterns, snow melting, temperature, sea level, river channel discharge, sediment production; climate change might be altering the flooding frequency and magnitude (Ministry for the Environment 2010). The significant effect of global climate change on the variation of river discharge globally has been represented by Shi et al. (2019) with 40 major drainage basins as samples across the world. The result shows that the discharge amount significantly varies with the continent and climatic zones. From 1960 to 2010, a downward trend has been observed in the North American and African rivers, whereas an upward trend has been detected in the European rivers. In Asia, the trend varies with drainage basins, positive change has been observed in the Lene, Yenisei, Yana, Olenyok basins, whereas negative trends were detected in the Ganges–Brahmaputra, Yellow, Tigris and Euphrates, Mekong, Pearl, Indus, Amur basins. Nevertheless, as per the climatic zone, the upward trend in river discharge has been shown in the subtropical monsoon humid climate and tropical savanna climate, whereas in the tropical rainforest climate and the tropical monsoon climate zone river, discharge decreased over time. A relatively stable condition has been shown over the temperate monsoon, temperate continental, and temperate marine climates. According to Wasko and Nathan (2019), despite the increasing trend in global rainfall extremes, the magnitude of flooding has reduced in Australia, as well as around the World, due to reducing soil moisture patterns. However, Petrow and Merz (2009) have reported an increasing trend in the frequency and/or magnitude of flooding in Germany from 1951 to 2002 using 145 gauge stations data.

16.4 Sediment Flux in Fluvial System

The discharge of sediment at the outlet of any drainage basin is called sediment load (Q_s) and is generally measured in tons per year unit. To normalize and make a relative comparison of sediment load among the basins of any region, sediment yield (Q_{sy}) data is generally used after dividing the total sediment load by the area of the respective drainage basin (A_d)

$$Q_{sy} = \frac{Q_s}{A_d}$$

Three types of sediment movement are mainly observed in a drainage basin, i.e., sediment in suspension, sediment moving on bed, and dissolved sediments. Quantitatively, by suspension movement process, the river moves its majority amount of sediments from upstream to downstream to the ocean. Global estimations have shown that annually about 25 gigatons (Gt) of sediment transfer to the oceans, within which only 3–6 Gt yr^{-1} is transported by glacial, coastal, and wind actions (Rhoads 2020).

Spatially, about 30% of the total sediment flux comes from only the fourteen major river basins of the world, where the Amazon River is delivering the largest amount (1200 Mt yr^{-1}) followed by Huanghe (1100 Mt yr^{-1}), Brahmaputra, Ganga, Changjian, and others (Milliman and Meade 1983; Rhoads 2020). However, Ganga–Brahmaputra collectively delivers the highest amount of sediment (1670 Mt yr^{-1}) to the Bay of Bengal based on the area of the outlet (Milliman and Meade 1983). The role of climate on sediment yield is crucial as the rate of erosion varies with climatic region. Jansson (1988) shown that sediment yield is very high in a tropical humid climate without a dry period (Af) followed by a tropical climate with a dry season, temperate climates with a dry season (Cs and Cwa). On the continental scale, rivers in Asia and Oceania generate and deposit almost two-third (12,500 Mt yr^{-1}) of the total sediment flux (Milliman and Meade 1983; Rhoads 2020).

Like the discharge of river water, sediment is also an essential element in the fluvial system, and geomorphologists are immensely interested in its redistribution across the basin and in the ocean through different fluvial processes (Rhoads 2020). In association with the hydrological actions, sediment dynamics is a primary cause to develop different erosional and depositional landforms in a drainage basin. Erosional landforms are the by-product of erosion over the catchment area, especially on the hillslope, river bed, underlying rock, riverbank, etc., and produced sediments, whereas depositional landforms are produced by the deposition of those eroded sediments, subject to the condition of channel slope. Based on the sediment dynamics within a drainage basin, the basin area is classified into three primary zones, e.g., sediment production zone (SPZ), sediment transport zone (STZ), and sediment storage zone (SSZ) (Schumm 1977). The headwater regions of the trunk river and tributaries are mainly considered as SPZ and supply maximum sediments in the river system. Schrott et al. (2003) reported that sediment in a drainage basin spends more time as storage in different short-live landforms (bars, beaches, islands) and long-live landforms (terraces, alluvial fan, and floodplain) than in the mode of transport. Sediment flux from source to sink depends on multiple factors (natural and anthropogenic), especially on the rainfall variability, magnitude of slope, relief ratio, soil characteristics, lateral and longitudinal connectivity, coupling between hillslope and channel, land use and land cover characteristics, etc. (Luca et al. 2020). The sediment load also significantly varies with season, maximum and minimum loads have been observed in the monsoon and non-monsoon season, respectively (Table 16.2).

Table 16.2 Sediment load of major river basins (non-classified) of India and their seasonal variation

Name of the basin	Sediment load (million metric tonnes), 2012–2013		
	Monsoon	Non-monsoon	Annual
Mahanadi	0.018–9.594	0.000–0.016	0.018–9.610
Brahmani	1.039–7.530	0.000–0.169	0.000–7.546
Godavari	0.000–59.332	0.000–0.126	0.000–59.391
Krishna	0.000–0.865	0.000–0.019	0.000–0.865
Cauvery	0.000–0.069	0.000–0.017	0.000–0.081
West flowing rivers	0.001–0.477	0.000–0.002	0.001–0.477
Tapi	1.084–32.525	0.000	1.084–32.525
Narmada	0.011–32.740	0.000–0.257	0.000–32.863
Mahi, Sabarmati, and others	0.000–1.088	0.000	0.000–1.088

Source CWC, Integrated Hydrological Data Book, 2016

16.5 Human Interference on River System

The legendary books '*Man and Nature*' of Marsh (1864) and '*Man as Geological Agent*' of Sherlock (1922) are initially recognized the significance of anthropogenic activities on changing earth's surface. From a quantitative perspective, Gilbert (1917) estimated first the effect of gold mining in the rivers of Sierra Nevada and an abrupt increase in sediment load. Since 1960, the research on human-induced alternation in the fluvial system has started to focus on the dynamics of channel morphology, sediment, landform, hydraulics, while the researches of pre-1960s emphasized the effects on hydrological behaviour upon channels (Gregroy 2006). The term 'anthropogeomorphology' was coined by Golomb and Eder (1964) in their research paper 'Landform made by Man' [Landscape 14(1): 4–7]. '*The Human Impact*' is also a remarkable book by Andrew Goudie (1981, 2006), which enriches this field with detailed documentation on the effect of direct and indirect human activities on vegetation, animals, soils, water, landforms, and climate and their role on changing geomorphological processes. The anthropogenic changes in the river system have been divided into two major categories by Knighton (1984): (a) Direct or channel-phase changes and (b) Indirect or land-phase changes (Table 16.3). Geoscientists have started to give more emphasis on anthropogeomorphological research from twenty-first Century, when the world's distinguished scientists, institutions, and academic societies are under ongoing discussion for '*Anthropocene*' as a new epoch in the Earth's Geological Time Scale. In 2006, the Binghamton Geomorphology Symposium has specially emphasized on the human impact on fluvial system and named the symposium by '*Human Role in Changing Fluvial System*' (James and Marcus 2006), which was celebrating the 50th anniversary of 1955 *Man's Role in Changing the Face of the Earth Symposium.* In a recent book '*Anthropogenic Geomorphology: A Guide to Man-Made Landforms*' by Szabo et al. (2010) stated that the anthropogeomorphological deals with the landforms that produced by human activities (directly and/or indirectly) and also about the modification of geomorphological processes such as weathering, erosion, transport, and deposition (Szabo 2010; Szabo et al. 2010). Goudie and Viles (2016) excellently have synthesized the variability of human impact on geomorphological processes across the world in their book "*Geomorphology in the Anthropocene*". The book '*Anthropogeomorphology of Bhagirathi-Hooghly River System in India*' by Das et al. (2020a, b) also shows the regional level implementation of this approach.

16.5.1 Dam: As Major Anthropogenic Factor in the Dynamics of Fluvial System

Dam is the most prominent and oldest anthropogenic interference in rivers by restricting the natural flow of channels. The primary purpose of

Table 16.3 Different man-induced changes in channel morphology (*Source* after Knighton 1984)

Direct or channel-phase changes	Indirect or land-phase changes
River regulation-	Land use changes-
Water storage by reservoirs	Removal of vegetation, especially deforestation
Diversion of water (Dam)	Afforestation
Channel changes-	Changes in agricultural practices
Bank stabilisation	Urbanisation
Channel straightening	Mining activity
Stream gravel extraction	Land drainage-
	Agricultural drains
	Storm-water sewerage systems

this across-channel structure is controlling floods and increasing the utility of water resources by irrigation, hydropower, industrial use, human consumption, aquaculture, and navigability. Almost 5000 years ago, man had first attempted to construct a dam in Egypt, to regulate the water of the Nile River (Smith 1971). But the effect of man-induced changes on river systems in general and particularly dams on river geomorphology has been addressed since the beginning of the rapid worldwide development of large reservoirs around the 1950s (Ronco et al. 2010). In the World, there are countless dams of different sizes and storage capacities, while the International Commission on Large Dam (ICOLD), as of September 2019, has listed around 57,985 numbers of 'large dams' only across the world. A large dam is defined as a dam with "a height of 15 metres or greater from lowest foundation to crest or a dam between 5 metres and 15 metres impounding more than 3 million cubic metres" (https://www.icold-cigb.org/GB/world_register/general_synthesis.asp). Among the total large dams of the World, only South-East Asian countries are accounting about 58% of it, leading by China (23841) and followed by India (4408), Japan (3180), Korea (Rep. of) (1338), Thailand (220), etc. (Fig. 16.8). The United States of America also holds about 9263 large dams. However, National Register of Large Dam (NRLD), Govt. of India shows a total 5745 number of large dams (existing and ongoing) in India with a major concentration in the State of Maharashtra (2394) followed by Madhya Pradesh (906), Gujarat (632), Chhattisgarh (258), etc. The storage capacity of these dams is nearly 10 million mcm and China leads with almost 50% of this water (Fig. 16.8).

Humans have started to construct dams very first to utilized more water and related facility without focusing its long-term multidimensional effects on landscape dynamics, river geomorphology and ecosystem, human health, socio–economic structure, geopolitics, etc. A detailed review by Graf (2005, 2006) on the effect of large dam construction on the downstream geomorphological and hydrological alternation on the American rivers shows that "the hydrology and geomorphology of large rivers in America reflect the pervasive influence of an extensive water control infrastructure including more than 75,000 dams". Kirchherr and Charles (2016) represent a comprehensive review of the social impact of dams based on 217 scholarly articles across the globe, with special focus on the key dimensions like infrastructure, community, and livelihood. Kondolf (1997) precisely shows the 'hungry water' effects on the river due to the construction of dam and gravel mining. Primarily, the effect of dam on any river system could be classified into two categories: (a) Upstream Impacts and (b) Downstream Impacts, which includes multiple factors, of different major aspects, like hydrological processes, fluvial geomorphology, water quality, riverine ecosystem, local climate, socio-economic and cultural characteristics, human health, and overall degradation of the environment (Schreier and Pang 2015) (Fig. 16.9).

The regional level study on the river system of West Bengal shows the significant effect of dam construction on the hydrology and fluvial geomorphology of studied rivers (Table 16.4). As per National Register for Large Dams (2019), a total of 30 major dams have been constructed across the different rivers of West Bengal for the purpose of flood control, irrigation, hydroelectricity, and water supply.

16.6 Environmental Perspectives to Manage Rivers

The in-depth studies on fluvial geomorphology to know fluvial forms and processes well are primarily intended to manage our rivers, in recent decades the integration with engineering and ecological understanding helps a lot to improve the management strategies (Rhoads 2020). Despite the contemporary concept like rivers is a fundamental exogenetic agent to shape the Earth's landscape, the study of river from an ecological perspective is a widely accepted theoretical concept to link the river with different biotic and abiotic components in its surroundings and within it also. Based on the 'dynamics

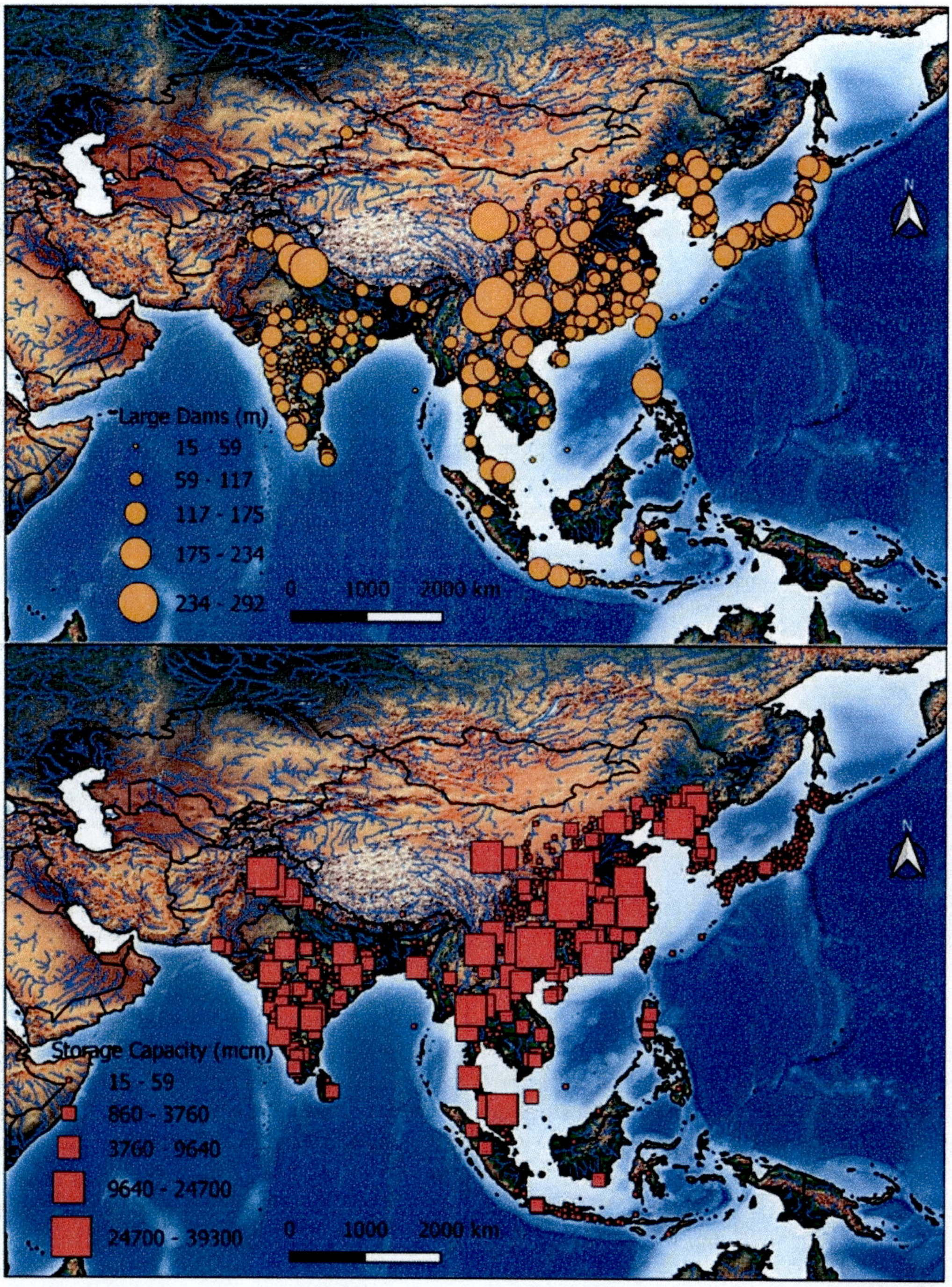

Fig. 16.8 Major large dams (>15 m) of South-East Asia with their proportional height (metre) and storage capacity (million cubic metres) (*Data Source* Food and Agriculture Organization, 2014)

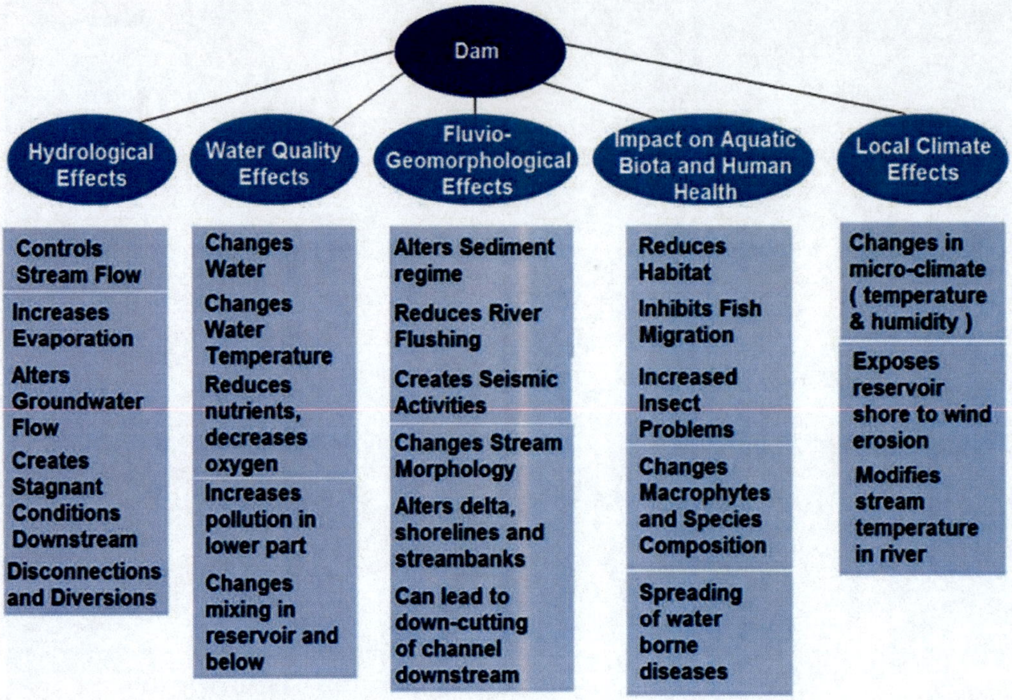

Fig. 16.9 Multidimensional effect of dam in river system (*Source* Schreier and Pang 2015)

equilibrium' model of Leopold and Langbein (1962) of pure geomorphology, Vannote et al. (1980) initially demonstrate the 'River Continuum Concept' to acknowledge the structural and functional relationships of riverine communities along the river from its source to sink. As the physical properties of a river system change along its flow path, the biological characteristics, especially the energy input, transportation of organic matter, function of macro, and micro invertebrate also become dynamics along the gradients of the river system. However, Ward and Stanford (1983, 1995) introduced the 'Serial Discontinuity Concept' for the regulated rivers, like damming across the channel severely disconnected the longitudinal connectivity of free flow of water and all important physical, biological ingredients of the river ecosystem. Stanford and Ward (1988) and Ward (1989) also noticed over a time frame that there is three-dimensional viz. lateral, longitudinal, and vertical connectivity within a river system (Fig. 16.10), which plays a significant role in hydrology, geomorphology, and ecology (Brierley et al. 2006; Kondolf et al. 2006; Wohl 2017). The 'Flood Pulse Concept' of Junk et al. (1989) strongly supports the connectivity concept and intensifies the relationship of river-floodplain systems. Haslam (2008) and Stanford et al. (2017) believes in the concept of 'Riverscape' and defines it as the 'part of landscape which has (or had) a watercourse as its focus'. The concept tries to highlight the river and its surroundings as an integrated part. Another important and prosperous concept in river management is Integrated River Basin Management (IRBM), which is "the process of coordinating conservation, management and development of water, land and related resources across sectors within a given river basin, to maximize the economic and social benefits derived from water resources in an equitable manner while preserving and, where necessary, restoring freshwater ecosystems" (Global Water Partnership 2000). World Wide Fund (WWF) for nature sets seven key elements for a successful IRBM: Vision, Integration,

Table 16.4 Dam induced hydrological and geomorphological changes in the different river systems of West Bengal

Name of the dam	Location	Post-dam/reservoir/barrage changes		Source(s)
		Hydrological changes	Geomorphological changes	
Komardanga dam (Bangladesh)	**River**: Dhepa River (in Bangladesh) of Punarbhaba River Basin **Year**: 1992 **Lat/Long**: 25° 51′ 46″ N/88° 39′ 52″ E	Reducing in the average water level of pre-monsoon and post-monsoon by 52.24% and 32.34%, respectively	Reducing of floodplain area by squeezing the river corridor and about 40% reduction in flood water extension over the basin; Disconnection between active channel, floodplain, and wetlands Water crisis for the wetland habitats	Talukdar and Pal (2017)
Massanjore dam (Jharkhand, India) and Tilpara reservoir (WB)	**River**: Mayurakshi River (Jharkhand) and Kushkarni River (WB) **Year**: 1955 and 1976 **Lat/long**: 24° 06′ 25″ N/87° 18′ 31″ E and 23° 56′ 46″ N/87° 31′ 30″ E	Decreasing monsoon and pre-monsoon water level and about 34% (7.73–4.96 mg/l) reduction in suspended sediment load below the dam and reservoir	Reducing the carrying capacity of upstream channels e.g. 26% for Kushkarni River and declining the longitudinal bed slope and velocity Experience of river bank erosion	Pal (2016a, b)
Mohanpur dam and reservoir (Bangladesh)	**River**: Atreyee **Year**: 2012 **Lat/long**: 25° 32′ 23.28″ N/88° 45′ 35.39″ E	Reducing of seasonal discharge by 30.97, 66.86 and 64.01% during pre-monsoon, monsoon and post monsoon periods and bbout 18.26% negative change in base flow, immediate after the dam construction		Pal (2016c)
Farakka barrage (WB)	**River**: Ganga **Year**: 1975 **Lat/long**: 24° 48′ 16.76″ N/87° 55′ 50.73″ E	**On Padma**: At Hardinge bridge station (Bangladesh), the average dry-season (Jan-may) discharge 2340 $m^3\ s^{-1}$ of pre-Farakka (1934–1975) reduced to 1236 $m^3\ s^{-1}$ during post-Farakka (1975–1995). In particular, the maximum, average and minimum discharges have been reduced around 22, 48 and 72%, respectively, in dry-season; About 13% increase in peak-	Huge sediment load has been trapped from the upstream of the Ganga Basin and about 87 million cubic metres of water was impounded and the effect exhibit through changing course and severe bank erosion in Malda district (WB) **On Bhagirathi-Hugli**: Increasing the formation rate of cutoffs and oxbox lakes	Rudra (2014, 2016, 2018), Rahaman and Rahaman (2018), Rahaman (2009), Rahaman and Asaduzzaman (2010), Mirza (1997)

(continued)

Table 16.4 (continued)

Name of the dam	Location	Post-dam/reservoir/barrage changes		Source(s)
		Hydrological changes	Geomorphological changes	
		discharge during post-Farakka Period The salinity of the Padma in Bangladesh also increased from 380 μΩ/cm during the pre diversion period in 1974 to about 29,500 μΩ/cm in 1992		
Maithon and Panchet dams (Jharkhand) and Durgapur Barrage (WB)	**River**: Damodar **Year**: 1957, 1959, 1955 **Lat/long**: 23° 47′ 7″ E/86° 48′ 43″ N; 23° 40′ 51″ E/86° 44′ 50″ N; 23° 28′ 35.95″ N/87° 18′ 5.16″ E	The monsoon discharge of 6081–10,676 $m^3\ s^{-1}$ is reduced up to 2574–4470 $m^3\ s^{-1}$ due to reservoir storage and diversion of flow through canals; forwarding the period of peak flood	The dominance of aggradational landforms, braiding, avulsion, high width–depth ratio, breaching of right bank, and valley widening up to 82 km from Durgapur Barrage then phenomena of bank erosion, confined sinuosity, low width–depth ratio, and narrowness are more pronounced up to the confluence	Ghosh and Guchhait (2014, 2016)
Kangsabati dam (WB)	**River**: Kangsabati Kumari **Year**: 1965 **Lat/long**: 86° 47′ 20.14″ E/22° 57′ 49.90″ N	Significant reduction in peak flow and total annual discharge; non-monsoonal low flow increasing due to irrigational water supply; The frequency of low flood (2–10 years return period) reduce and large flood (>10 years) has been eliminated	Changes in river bed elevation by huge sedimentation and loss of habitat	Mittal et al. (2014)
Teesta Low dam III and IV (WB) and all other upstream dams in Sikkim	**River**: Teesta **Year**: 2013 and 2016 **Lat/Long**: 88° 27′ 30″ E/27° 00′ 00″ N; 88° 27′ 21″ E/26° 55′ 04″ N	About 52–88% of the free-flowing river stretches will be diverted Post-dam reduction of peak discharge from 2456 to 2033 $m^3\ s^{-1}$	Raising the bed elevation with smoother or decreasing cross-section area with a gentle gradient in the downstream of the dam in comparison with pre-dam condition Ecology: 7.6–24% of the riverine ecosystems will be converted into semi lacustrine ecosystems in different rivers and likely to affect more than 100 fish species	Bhatt et al. (2017), Sanyal (2017)

Scale, Timing, Participation, Capacity, and Knowledge.

In the twenty-first Century, the importance of flowing water on ecological sustainability and hydro-geomorphological stability has been realized seriously by introducing concepts like 'ecological flow', 'e-flow', or 'environmental flow'. In an international conference at Brisbane, Australia, in 2007, over 800 scientists, economists, policymakers, researchers, and resource managers concluded the definition of e-flows as 'the quantity, timing, and quality of water flows required to sustain freshwater and estuarine ecosystems and the human livelihoods and well-being that depend on these ecosystems' (International Water Centre 2007). In Indian, the Ministry of Environment Forest and Climate Change (MoEFCC) is also paying attention to the 'minimum flows' criteria to give clearance to any river valley project since 2007–08, initially it was 10% of the observed minimum flow of river, later it has revised with '20% average flow observed in the four lean season months at 90% dependability' by 2010 (Water Conflicts in India 2017). However, there are lots of controversies regarding the required minimum flow based on seasonal variability of discharge, huge diversity in river basin character of the country, lack of hydrological data, problem in generalization, and overall it is ultimately a social choice with a priority on how human use the water with lack of importance on river ecology and geomorphology (Water Conflicts in India 2017).

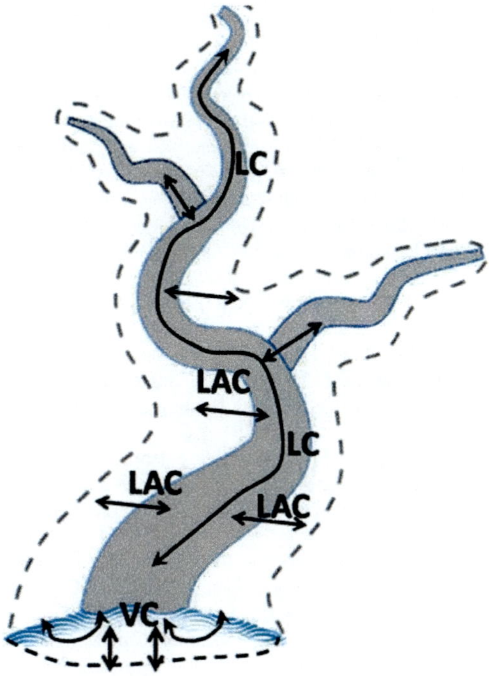

Fig. 16.10 Schematic diagram to show the lateral (LAC), longitudinal (LC), and vertical (VC) connectivity within a river system (*Source* Conceptually modified after Stanford and Ward 1988)

16.7 Concluding Remarks

The theoretical development of fluvial geomorphology has a long history since the fifteenth century and has switched its mode of analysis from descriptive to quantitative since the 1950s. The spatial scale of research also focuses more on the micro-geomorphological changes and processes behind it with the help of advanced surveying instruments and geospatial technology. Therefore, the twenty-first century has started to entitle as 'second phase of active quantitative geomorphology' (Piegay et al. 2015) with easy availability of very high-resolution geospatial data and modeling algorithms. Researches on the flow regime show globally it varies in different river basins and significantly influenced by climate change and changing the magnitude and frequency of floods over the time frame. Flood prone area mapping shows that about 18,000 km^2 land of WB has experienced flood since 1993 with significant variation among the catchments of the state. A current flood event in Assam by Brahmaputra River shows the severity and effectiveness of a flood with a 25-years return period only. Remotely sensed flood data by DFO could be used for live-flood condition monitoring of major rivers, and daily basis discharge data are also useful in flood frequency analysis to estimate the discharge amount at different return periods. More than 25 Gt per year of sediment load could be a proxy indicator of land degradation in the upstream catchment of World Rivers. In such circumstances, the drainage basins of south-east Asia are more vulnerable than others, as they hold the maximum share of sediment load from major rivers of India (Ganga, Brahmaputra, Indus) and China (Yellow, Yangtze). Among the non-classified rivers of India, Godavari generates maximum sediment ($\sim$60 mcm) from its catchment followed by Narmada ($\sim$33 mcm), Tapi ($\sim$32 mcm), and Mahanadi ($\sim$10 mcm). Dam is the most effective driver to change river system effectively among the anthropogenic activities followed by land use land cover change, mining, urbanization, construction of in-stream or out-stream infrastructure, etc. River systems of WB are also significantly affected by dams and reservoirs. To manage our rivers, scientists have developed a number of effective models and concepts to restore the neutrality of the river system. The ultimate positive sign is that since the beginning of the current century, an active environmental perspective has been started to follow by the policymakers, planners, and governments.

References

Abrahams AD (1968) Distinguishing between the concepts of steady state and dynamic equilibrium in geomorphology. Earth Sci J 2(2):160–166

Baker VR, Costa J (1987) Flood power. In: Mayer L, Nash DB (eds) Catastrophic flooding. Allen and Unwin, London, pp 1–22

Benito G, Hudson PF (2010) Flood hazards: the context of fluvial geomorphology. In: Alcántara-Ayala I, Goudie AS (eds) Geomorphological hazards and disaster prevention. Cambridge University Press

Bertalanffy V (1951) An outline of general system theory. J Br Philos Sci 1:134–165

Bhatt JP, Tiwari S, Pandit MK (2017) Environmental impact assessment of river valley projects in upper Teesta basin of Eastern Himalaya with special reference to fish conservation: a review. Impact Assess Proj Apprais 35(4):340–350. https://doi.org/10.1080/14615517.2017.1354642

Biswas AK (1970) History of hydrology. North Holland, Amsterdam

Brierley G, Fryirs K, Jain V (2006) Landscape connectivity: the geographic basis of geomorphic applications. Area 38(2):165–174. https://doi.org/10.1111/j.14754762.2006.00671.x

Charlton R (2007) Fundamentals of fluvial geomorphology. Routledge, Oxon, USA

Chorley RJ (1962) Geomorphology and general systems theory. U.S. Geological Survey Professional Paper 500-8

Chorley RJ (1969) The drainage basin as the fundamental geomorphic unit. In: Choeley RJ (ed) Water, earth, and man: a synthesis of hydrology, geomorphology, and socio-economic geography. Methuen & Co., London, pp 77–99

Chorley RJ, Beckinsdale RP (1964) The history of the study of landforms or the development of geomorphology. Volume One: geomorphology before Davis. Methuen, London

Costa JE, O'Connor JE (1995) Geomorphically effective floods. In: Costa JE, Miller AJ, Potter KP, Wilcock PR (eds) Natural and anthropogenic influences in fluvial geomorphology (the Wolman volume). American Geophysical Union, Washington, DC, pp 45–56

Das BC, Ghosh S, Islam A, Roy S (2020a) Anthropogeomorphology of Bhagirathi-Hooghly River System in India. CRC Press, Boca Raton and London

Das SK, Gupta RK, Varma HK (2020b) Flood and drought management through water resource development in India. World Meteorol Organ Bull 56(3) (Online Resource)

Davis WM (1889) The rivers and valleys of Pennsylvania. Natl Geogr Mag 1:11–26

Davis WM (1899) The geographical cycle. Geogr J 14:481–504

Dury GH (1966) Open systems and the erosion cycle, Chapter 1. In: Aspects of the content of geography in fifth and sixth forms. University of Sydney

Ghosh S, Guchhait SK (2014) Hydrogeomorphic variability due to dam constructions and emerging problems: a case study of Damodar River, West Bengal, India. Environ Dev Sustain 16(3):769–796

Ghosh S, Guchhait SK (2016) Dam-induced changes in flood hydrology and flood frequency of tropical river: a study in Damodar River of West Bengal, India. Arab J Geosci. https://doi.org/10.1007/s12517-015-2046-6

Gilbert GK (1914) The transportation of debris by running water. U.S. Geological Survey Professional Paper 86. U.S. Government Printing Office, Washington, DC

Gilbert GK (1917) Hydraulic-mining debris in the Sierra Nevada. U.S. Geological Survey Professional Paper 105, U.S. Government Printing Office, Washington, DC

Global Water Partnership (2000) Integrated water resources management. Technical Advisory Committee (TAC) Background Papers Series, No. 4. Sweden

Golomb B, Eder HM (1964) Landform made by man. Landscape 14(1):4–7

Goudie AS (1981) The human impact. Basil Blackwell, Oxford

Goudie AS (2006) The human impact on the natural environment: past, present and future. Blackwell, Oxford
Goudie AS, Viles HA (2016) Geomorphology in the anthropocene. Cambridge University Press, UK
Graf WL (2005) Geomorphology and American dams: the scientific, social, and economic context. Geomorphology 71:3–26
Graf WL (2006) Downstream hydrologic and geomorphic effects of large dams on American rivers. Geomorphology 79:336–360
Gregroy KJ (2006) The human role in changing river channels. Geomorphology 79:172–191. https://doi.org/10.1016/j.geomorph.2006.06.018
Hack JT (1960) Interpretation of erosional topography in humid temperate regions. Am J Sci 258-A:80–97
Hack JT (1965) Geomorphology of the Shenandoah Valley, Virginia and West Virginia and origin of the residual ore deposits. U.S. Geological Survey Professional Paper 484
Haslam SM (2008) The riverscape and the river. Cambridge University Press, UK
Horton RE (1945) Erosional development of streams and their drainage basins: a hydrophysical approach to quantitative morphology. Geol Soc Am Bull 56:275–370
Hutton J (1795) Theory of the earth, with proofs and illustrations, vols 1 and 2. Printed for Cadell and Davies; William Creech. Edinburgh; London
International Water Centre (2007). The Brisbane declaration. Retrieved from http://www.watercentre.org/news/declaration
IPCC (Intergovernmental Panel on Climate Change) (2018) Summary for policymakers. In: Global warming of 1.5 °C: an IPCC special report on the impacts of global warming of 1.5 °C above pre-industrial levels and related global greenhouse gas emission pathways, in the context of strengthening the global response to the threat of climate change, sustainable development, and efforts to eradicate poverty. World Meteorological Organization, Geneva, Switzerland, 32 pp
Jain M, Tandon SK (2003) Fluvial response to late Quaternary climate changes, western India. Quat Sci Rev 22:2223–2235
James LA, Marcus WA (2006) The human role in changing fluvial system: retrospect, inventory and prospect. Geomorphology 79:152–171
Jansson MB (1988) A global survey of sediment yield. Geogr Ann 70A(1–2):81–98
Junk WJ, Bayley PB, Sparks RE (1989) The flood pulse concept in river-floodplain systems. Can Spec Publ Fish Aquat Sci 106:110–127
Kale VS (2002) Fluvial geomorphology of Indian rivers: an overview. Prog Phys Geogr 26(3):400–433. https://doi.org/10.1191/0309133302pp343ra
Kirchherr J, Charles KJ (2016) The social impacts of dams: a new framework for scholarly analysis. Environ Impact Asses Rev 60:99–114. https://doi.org/10.1016/j.eiar.2016.02.005
Knighton AD (1984) Fluvial forms and processes. Edward Arnold, Sheffield
Kondolf GM, Boulton AJ, O'Daniel S, Poole GC, Rahel FJ, Stanley EH, Wohl E, Bång A, Carlstrom J, Cristoni C, Huber H, Koljonen S, Louhi P, Nakamura K (2006) Process-based ecological river restoration: visualizing three-dimensional connectivity and dynamic vectors to recover lost linkages. Ecol Soc 11 (2):5. http://www.ecologyandsociety.org/vol11/iss2/art5/
Kondolf GM (1997) Hungry water: effects of dams and gravel mining on river channels. Environ Manag 21 (4):533–551
Leopold LB, Maddock T (1953) The hydraulic geometry of stream channels and some physiographic implications. U.S. Geological Survey Professional Paper 252. U.S. Government Printing Office, Washington, DC
Leopold LB, Langbein WB (1962) The concept of entropy in landscape evolution. U.S. Geological Survey Professional Paper 500-A:20 p
Luca C, David C, Sergio A (2020) Sediment generation and sediment routing systems. Earth Sci Rev. https://doi.org/10.1016/j.earscirev.2020.103221
Marsh GP (1864) Man and nature or physical geography as modified by human action. Harvard University Press, Cambridge
McDowell PF (2013) Geomorphology in the late twentieth century. In: Shroder J (Editor in Chief), Orme AR, Sack D (eds) Treatise on geomorphology, vol 1. The foundations of geomorphology. Academic Press, San Diego, CA, pp 108–123
Miller AJ (1990) Flood hydrology and geomorphic effectiveness in the central Appalachians. Earth Surf Process Landf 15(2):119–134
Milliman JD, Meade RH (1983) World-wide delivery of river sediment to the oceans. J Geol 91(1):1–21
Ministry for the Environment (2010) In: Woods R, Mullan AB, Smart G, Rouse H, Hollis M, McKerchar A, Ibbitt R, Dean S, Collins D (NIWA) (eds) Tools for estimating the effects of climate change on flood flow: a guidance manual for local government in New Zealand. Prepared for Ministry for the Environment, Wellington, New Zealand
Mirza MMQ (1997) Hydrological changes in the Ganges system in Bangladesh in post-Farakka period. Hydrol Sci J Des Sci Hydrol 42(5):613–631
Mittal N, Mishra A, Singh R, Bhave AG, van der Valk M (2014) Flow regime alteration due to anthropogenic and climatic changes in the Kangsabati River, India. Ecohydrol Hydrobiol 14:182–191. https://doi.org/10.1016/j.ecohyd.2014.06.002

More RJ (1969) The basin hydrological cycle. In: Choeley RJ (ed) Water, earth, and man: a synthesis of hydrology, geomorphology, and socio-economic geography. Methuen & Co., London, pp 77–99

Niedzielski T (2018) Applications of unmanned aerial vehicles in geosciences: introduction. Pure Appl Geophys 175:3141–3144. https://doi.org/10.1007/s00024-018-1992-9

Orme AR (2013) The scientific roots of geomorphology before 1830. In: Shroder JW (ed) Treatise on geomorphology, vol 1. The foundations of geomorphology, AR Orme, Sack D (vol eds). Elsevier, New York, pp 11–36

Pal S (2016a) Impact of Massanjore Dam on hydro-geomorphological modification of Mayurakshi River, Eastern India. Environ Dev Sustain 18:921–944. https://doi.org/10.1007/s10668-015-9679-1

Pal S (2016b) Impact of Tilpara barrage on backwater reach of Kushkarni River: a tributary of Mayurakshi River. Environ Dev Sustain 19(5):2115–2142. https://doi.org/10.1007/s10668-016-9833-4

Pal S (2016c) Impact of water diversion on hydrological regime of Atreyee River of Indo-Bangladesh. Int J River Basin Manag 14(4):459–475. https://doi.org/10.1080/15715124.2016.1194282

Petrow T, Merz B (2009) Trends in flood magnitude, frequency and seasonality in Germany in the period 1951–2002. J Hydrol 371(1–4):129–141. https://doi.org/10.1016/j.jhydrol.2009.03.02

Piegay H, Kondolf GM, Minear JT, Vaudor L (2015) Trends in publications in fluvial geomorphology over two decades: a truly new era in the discipline owing to recent technological revolution? Geomorphology. https://doi.org/10.1016/j.geomorph.2015.07.039

Raghunath HM (2013) Hydrology: principles, analysis, design. New Age Publication, New Delhi

Rahaman MM (2009) Integrated Ganges basin management: conflict and hope for regional development. Water Policy 11:168–190

Rahaman MM, Rahaman MM (2018) Impacts of Farakka barrage on hydrological flow of Ganges river and environment in Bangladesh. Sustain Water Resour Manag 4:767–780. https://doi.org/10.1007/s40899-017-0163-y

Rahman MR, Asaduzzaman M (2010) Ecology of Sundarban, Bangladesh. J Sci Found 8(1 & 2):35–47

Rhoads BL (2020) River dynamics: geomorphology to support management. Cambridge University Press, UK

Ritchie H (2014) Natural disasters. Published online at OurWorldInData.org. https://ourworldindata.org/natural-disasters (Online Resource)

Ronco P, Fasolato G, Nones M, Silvio GD (2010) Morphological effect of damming on lower Zamebezi River Basin. Geomorphology 115:43–45

Rudra K (2014) Changing river courses in the western part of the Ganga-Brahmaputra delta. Geomorphology 227:87–100

Rudra K (2016) State of India's rivers: West Bengal. India River Week 2016. https://sandrp.files.wordpress.com/2017/03/west-bengal.pdf. Accessed 11 May 2020

Rudra K (2018) Rivers of the Ganga–Brahmaputra–Meghna delta: a fluvial account of Bengal. Geography of the physical environment. Springer International Publishing, Switzerland. https://doi.org/10.1007/978-3-319-76544-0

Sanyal J (2017) Predicting possible effects of dams on downstream river bed changes of a Himalayan river with morphodynamic modeling. Quat Int 453:48–62. https://doi.org/10.1016/j.quaint.2017.03.063

Schreier H, Pang G (2015) Hydro-power & reservoirs: impact of dams, water in international development. Faculty of Land and Food Systems, The University of British Columbia. Webpage: http://ubclfs-wmc.landfood.ubc.ca/webapp/WID/course/hydro-power-reservoirs-9/impacts-33/. Accessed 1 Aug 2020

Schrott L, Hufschmidt G, Hankammer M, Hoffmann T, Dikau R (2003) Spatial distribution of sediment storage types and quantification of valley fill deposits in an Alpine basin, Reintal, Bavarian Alps, Germany. Geomorphology 55:45–63. https://doi.org/10.1016/S0169-555X(03)00131-4

Schumm SA (1977) The fluvial system. Blackburn Press, Caldwell, NJ, p 338

Sherlock RL (1922) Man as a geological agent. Witherby, London

Shi X, Qin T, Nie H, Weng B, He S (2019) Changes in major global river discharges directed into the ocean. Int J Environ Res Public Health 16:1469. https://doi.org/10.3390/ijerph16081469

Smith N (1971) A history of dams. Peter Davies, London

Stanford JA, Ward JV (1988) The hyporheic habitat of river ecosystems. Nature 335:64–66

Stanford JA, Alexander LC, Whited DC (2017) Riverscapes. In: Hauer FR, Lamberti GA (eds), Methods in stream ecology: vol I: ecosystem structure, pp 3–19. https://doi.org/10.1016/B978-0-12-416558-8.00001-9

Strahler AN (1952) Dynamic basis of geomorphology. Geol Soc Am Bull 63:923–938

Szabo J (2010) Anthropogenic geomorphology: subject and system. In: Szabo J, David L, Loczy D (eds) Anthropogenic geomorphology: a guide to man-made landforms. Springer, Netherlands, pp 3–10

Szabo J, David L, Loczy D (eds) (2010) Anthropogenic geomorphology: a guide to man-made landforms. Springer, Netherlands

Talukdar S, Pal S (2017) Impact of dam on inundation regime of flood plain wetland of punarbhaba river basin of barind tract of Indo-Bangladesh. Int J Soil Water Conserv Res 5(2):109–121. https://doi.org/10.1016/j.iswcr.2017.05.003

Thornbury WD (1954) Principals of geomorphology. Wiley, New York, p 618

Vannote RL et al (1980) The river continuum concept. Can J Fish Aquat Sci 37:130–137

Ward JV (1989) The four-dimensional nature of the lotic ecosystem. J N Am Benthol Soc 8:2–8

Ward JV, Stanford JA (1983) The serial discontinuity concept of lotic ecosystem. In: Fontaine TD, Bartell SM (eds) Dynamics of lotic ecosystems. Ann Arbor Science, Ann Arbor, pp 29–42

Ward JV, Stanford JA (1995) The serial discontinuity concept: extending the model to floodplain rivers. Regul Rivers: Res Manag 10:159–168. https://doi.org/10.1002/rrr.3450100211

Wasko C, Nathan R (2019) Influence of changes in rainfall and soil moisture on trends in flooding. J Hydrol 575:432–441

Water Conflicts in India (2017) E-flows in Indian rivers—methodologies, issues, indicators and conditions: learnings from Hasdeo Basin. Forum for Policy Dialogue on Water Conflicts in India. https://waterconflictforum.org/lib_docs/E-flows-in-Indian-Rivers.pdf. Accessed 1 Aug 2020

Wohl E (2017) Connectivity in rivers. Prog Phys Geogr 41(3):345–362. https://doi.org/10.1177/0309133317714972

17 An Integrated Approach of River Health Assessment Based on Physico-chemical Parameters of the River Subarnarekha, India

Ujjwal Bhandari and Uttam Mukhopadhyay

Abstract

In fluvial geomorphology, dynamics of process and structure are considered as prime controlling attributes for all river channels. However, if there is any kind of change in river flow and sediment regime condition, then obviously it modifies the entire catchment system. As a result, we generally see the alteration of habitats mainly because of changes in the physical and behavioral pattern of channels. This study has been carried out in an integrated way to assess river health for the key issues of effective river management and has also been used as a tool for identifying different factors in moribund ecosystems. Rapid alterations in land use and land cover like urbanization, industrialization, intense agriculture, etc., have degenerated the condition of river health. As the health of the river Subarnarekha has also deteriorated, thus we consider analyzing all of the above components as important parameters to diagnose the river system. Traditionally, the majority of these kinds of studies have focused only on chemical parameters, but in the present context, complex outcomes on habitat modifications by urbanization, industrialization, and a barrier (like dams, embankments, bridges, transport networks, etc.) also alter flow regimes. It demonstrates that an integrated approach is necessary to identify river health through chemical quality, channel planform adjustment, and anthropogenic influences, respectively. To understand the condition of physical health and the intensity of anthropogenic effects, multi-temporal satellite imageries for two years 1990 and 2014 have been used. Besides, in this study, 21 sampling sites were selected in the river for three times, namely, pre-monsoon, monsoon, and post-monsoon periods for the year 2014 to delineate chemical health. The overall analysis showed a seasonal concentration pattern of metals, and which is higher in the pre-monsoon season but lower in the monsoon period due to the dilution effect.

Keywords

River health · Physical health · Chemical health · Anthropogenic influences · Seasonal fluctuation · Subarnarekha River

U. Bhandari (✉)
Department of Geography, University of Calcutta, 35 B. C. Road, Kolkata 700019, India

U. Mukhopadhyay
Department of Geography, Vidyasagar College, 39 Sankar Ghose Lane, Kolkata 700006, India

17.1 Introduction

Perceptive and insightful measures of riverine ecosystem quality are required to assess, sustain, or restore ecological conditions of water bodies

P. K. Shit et al. (eds.), *Drainage Basin Dynamics*, Geography of the Physical Environment,
https://doi.org/10.1007/978-3-030-79634-1_17

(Lettenmaier 1976; Karr 1991; Norris and Norris 1995; Wright 1995; Resh et al. 1995, 1996). Bioindicators give a simulated view of the intricacy of the health of the river ecosystem and are considered highly imperative towards the wise and scientific management of a river (Gore 1985; Karr 1991; Rapport 1989, 1991). Alongside biological parameters, basin structure and bed configuration along with channel pattern and its behavior has remarkable control towards maintaining a healthy river channel (Schumm 1988; Chapman 1992; Gurnell and Petts 1995; Hart and Fonseca 1996). In the case of chemical indicators, different types of water quality indices are used to identify river health (Montgomery and Reckhow 1984; Environment Canada 1987; Meybeck et al. 1989; Osborne and Kovacic 1993; Hart et al. 1999; Maher et al. 1999; Chang and Carlson 2005). In health assessment, physical and chemical parameters are most widely used all over the world although biological aspects are coming into play in recent times. (ANZECC 1992; Hart et al. 1999; Maher et al. 1999).

In the era of Anthropocene, the river health assessment is a growing concern to the environmentalists and ecologists of the world, because biodiversity and human society are severely affected by toxic metals and polluted water. Many important reports and studies are analyzed here to judge the environmental health and toxicity of river water in India and abroad (Global Water Partnership 2000; Ahearn et al. 2005; Jain and Sharma 2006; Hamner et al. 2007; Ahmed et al. 2009; Trivedi 2010; Asian Development Bank 2013; Giri and Singh 2014; Satya and Narayan 2018; Asim and Rao 2021; Mishra and Kumar 2021). Using geo-accumulation indices, Singh et al. (2005) proved that the heavy metals, viz., Cd, Cu, Fe, Pb, Mn, Ni, and Zn, were concentrated in the water and bed sediments of the Gomti River (a tributary of Ganga) in between 2002 and 2003. Jain and Sharma (2006) found that the contribution of monsoon months to the total transported load was calculated as very high in the Hindon River (a tributary of Yamuna) and in the rainy season, contributed more than 40% of total heavy metal loading annually. River water monitoring of Ganga over the past years demonstrated Fecal Coliform counted up to 10^8 MPN (most probable number) per 100 ml and BOD levels averaging over 40 mg l^{-1} in the polluted stretch of Varanasi (Hamner et al. 2007). The overall rate of waterborne or enteric diseases (cholera, dysentery, hepatitis-A, typhoid, etc.) was estimated to be about 66 percent increase due to water pollution in the Ganga River of Varanasi (Hamner et al. 2007). The environmental study, done by Trivedi (2010), revealed that in the Ganga River stretch, in between Kannauj and Allahabad (350 km long), about 12,222 mld (million litres per day) of domestic and 2500 mld of industrial waste water was mixed. The quality assessment of Damodar River revealed the contamination of Coliform counts, metals, and organic pollutants at a level above the permissible limits for domestic usage (Chatterjee et al. 2010). The oil and grease levels along with TDS and TSS indicate the possibility of high contaminants in the river water of Damodar because the river is surrounded by large-scale heavy industries, municipalities, and coal washeries (Chatterjee et al. 2010).

Due to multifarious anthropogenic activities, the water of Chambal River (covering the sanctuary part of Chambal River) was very much polluted by industrial effluents and it created havoc pressure on biological diversity (Yadav et al. 2014). It is reasonable to assess that the increased concentrations of metals in water of the Subarnarekha River is considerably high due to direct discharge of industrial, urban, and mining wastes into the mighty river, having high concentration of As, Co, and V (Giri and Singh 2014). Chaudhary et al. (2016) indicated that the water quality of River Ganga is unsuitable (between Haridwar and Garhmukeshwar) for drinking purpose, because the average NSFWQI (National Sanitation Foundation Index) was found to be greater than 40 in both pre- and post-monsoon months. The river bed mining of the Ganga River had a negative impact on the surface water quality based on physicochemical parameters (Kamboj and Kamboj 2019). The riparian land use pattern was an important factor that influenced changes in the aquatic ecosystem health grade, and total phosphorus and potassium bichromate affected the benthic macroinvertebrates and the aquatic ecosystem health to a great extent (Kerans and Karr 1994; Scrimgeour and Wicklum 1996; Chen et al. 2019). It is a very alarming fact that the direct inflow of waste water into the Narmada River has deteriorated the ecological health and drinking water quality. Mishra and Kumar (2021) have assessed the water quality status in the Narmada River using key indices, viz., comprehensive pollution index (CPI), heavy metal pollution index (HPI), risk assessment index (RAI), and cancer risk index (CRI). The River Yamuna, in between Baghpat and Chhainssa (125 km), was assessed by a heavy metal pollution index using atomic absorption spectrophotometer (Asim and Rao 2021). The average concentration of heavy metals was recorded as very high and 85 percent of the river was classified as highly polluted, due to the heavy load of metals (Asim and Rao 2021).

The popular method of river health assessment is characterized by spatiotemporal variability in fluvial regimes and ecological flows by adding biological attributes along with physical and chemical parameters (Hynes 1975; Richter et al. 1996; Simpson et al. 1997; Thoms and Swirepik 1998). However, Rosgen (1994, 1996a, b) does not link biological health assessment indicators in fluvial geomorphology. Rather, he included channel pattern, slope, particle size, cross-section area, entrenchment, bank full stage, flood prone area, etc., in appropriate spatial scale (Falkenmark 1989; Gleick 1990; Miller and Ritter 1996; Knight and Morris 1996; Allan and Johnson 1997; Heap et al. 1998; GWP 2000; Sullivan et al. 2006; Donohue et al. 2006; Zeitoun 2011; ADB 2013; Srinivasan et al. 2017; Varis et al. 2017; Zende et al. 2018; Jensen and Wu 2018). It is also argued that if any river catchment or its buffer area habitat is in insignificant condition, the river is definitely in poor health condition (Brookes and Shields 1996; Plafkin et al. 1989). Land use and land cover changes by anthropogenic activities can deflect the chemical, physical, and biological health of a river system (Karr 1991). Also, river health changed due to any kind of human alteration in the riparian system (Zhou et al. 2015; Sun et al. 2016; Spanò et al. 2017; Bandyopadhyay and De 2017).

Thus, the present study attempts to examine the physicochemical quality of water and river health in the Subarnarekha River. Both geospatial technologies and field-based investigations have been taken into account to assess this sensitive river. This type of scientific research helps our understanding towards river water quality triggered by its adjoining areas' physical condition, as well as predominant sociocultural activities. The river health is being controlled by the relationship between the river and human values, thereby justifying the suitability of choosing this river (Nandi et al. 2016).

17.2 Study Area

The Subarnarekha basin, a part of the Chhotanagpur Plateau of India, is deemed to be a small museum of Indian geology and geomorphology. The different magnitude of rock resistance and spatial variability of tectonic, lithological, erosional, and climatic processes play a critical role in the evolution of this polycyclic landform. Representing a varying degree of the landscape from Archaean to Quaternary landforms, it is mainly formed by Archaean granite and gneiss rocks but sometimes by Dharwar rocks, patches with mica-schist and phyllite. The Subarnarekha basin (24,196 km^2: 85° 08′ to 87° 32′ E, 21° 15′ to 23° 34′ N) extends over the states of Jharkhand, Orissa, and West Bengal with the maximum length and width of 297 km and 119 km, respectively (Fig. 17.1). The basin is bounded by the Chhotanagpur Plateau in the north and west, ridges spreading from the Baitarani basin in the south, by the Kangsabati water divide in the east, and the Bay of Bengal in the southeast. It rises near Nagri village in the Ranchi district of Jharkhand at an altitude of 600 m and flows for 395 km and after that it confluences with the sea. In this context, an attempt has been made to assess the process-based spatiotemporal dynamics of river health of the entire channel and with special emphasis on the 10 km buffer region which happens to be most affected.

17.3 Methodology

17.3.1 River Health Indicators

Multiple indices have been used for the assessment of river health by several researchers. However, in this case, the selection of the indices has been considered taking into account the prevailing condition of River Subarnarekha. From the detailed survey on the River Subarnarekha, three prominent problems, namely, physical, chemical, and biological health parameters have been considered. Taking into account the available data and study of other researchers in a different environment, the indicators are selected which reflects the overall health status of the River Subarnarekha (Boulton 1999; Karr 1999; Feng et al. 2012; Huaibin and Jianping 2014; Deng et al. 2015; Zhang et al. 2014; Nandi et al. 2016). The sub-factors of health assessment have been summarized in Table 17.1.

The roadmap for this assessment has been given (Fig. 17.2) to present the flowchart of the river health assessment procedure.

17.3.2 Chemical Health Analysis

Several attributes, namely, Aluminum (Al), Barium (Ba), Arsenic (As), Chromium (Cr), Cobalt (Co), Copper (Cu), Manganese (Mn), Nickel (Ni), Iron (Fe), Strontium (Sr), Vanadium (V), and Selenium (Se), have been tested by the Public Health Engineering Department, West Bengal to assess chemical health. It is such that changes in the proportions of these elements trigger the fluvial health condition by augmenting bioaccumulation and biomagnifications.

17.3.3 Sampling Procedure

Recommended standard procedures should be carefully followed for better sampling. Each method of sampling approaches should be followed accurately every time. One must stand downstream and collect samples from upstream direction (Bartram and Balance 1996). Since the representativeness of the sample is of utmost importance, a standard type of sampler, depth, position, and time of sampling is extremely required. To assure it, the samples must be replicated spatially from different points in the river and temporally from one point of different time intervals (preferably seasonally) and it must be without contamination. For analysis, specific post-collection operations like filtration, preservation, transportation to the laboratory in permissible time, etc., must be taken into due consideration. During analysis, we must take high care while washing equipment and the purity of chemicals must be maintained.

The samples have been collected three times a year covering pre-monsoon, monsoon and post-monsoon months for the respective year. These seasons were selected based on the hydrological regime of the basin (Giri and Singh 2014). Thus, the model will give reliable and dynamic results within these seasons. The pre-monsoon sampling runs from February to May; the monsoon sampling runs from June to September and the post-monsoon sampling runs from October to January. Sampling sites were chosen to fulfill the monitoring of river health within a basin as the selected sites were representing the larger segment of the studied river. For such analysis, it is best to aim the sample sites by observing the nature of river morphology with special attention to its habitat, urban and industrial agglomeration in the buffer region. To highlight the above influences on the riparian system, 21 sample stations have been randomly selected to collect water samples throughout the whole channel in the pre-monsoon, monsoon and post-monsoon

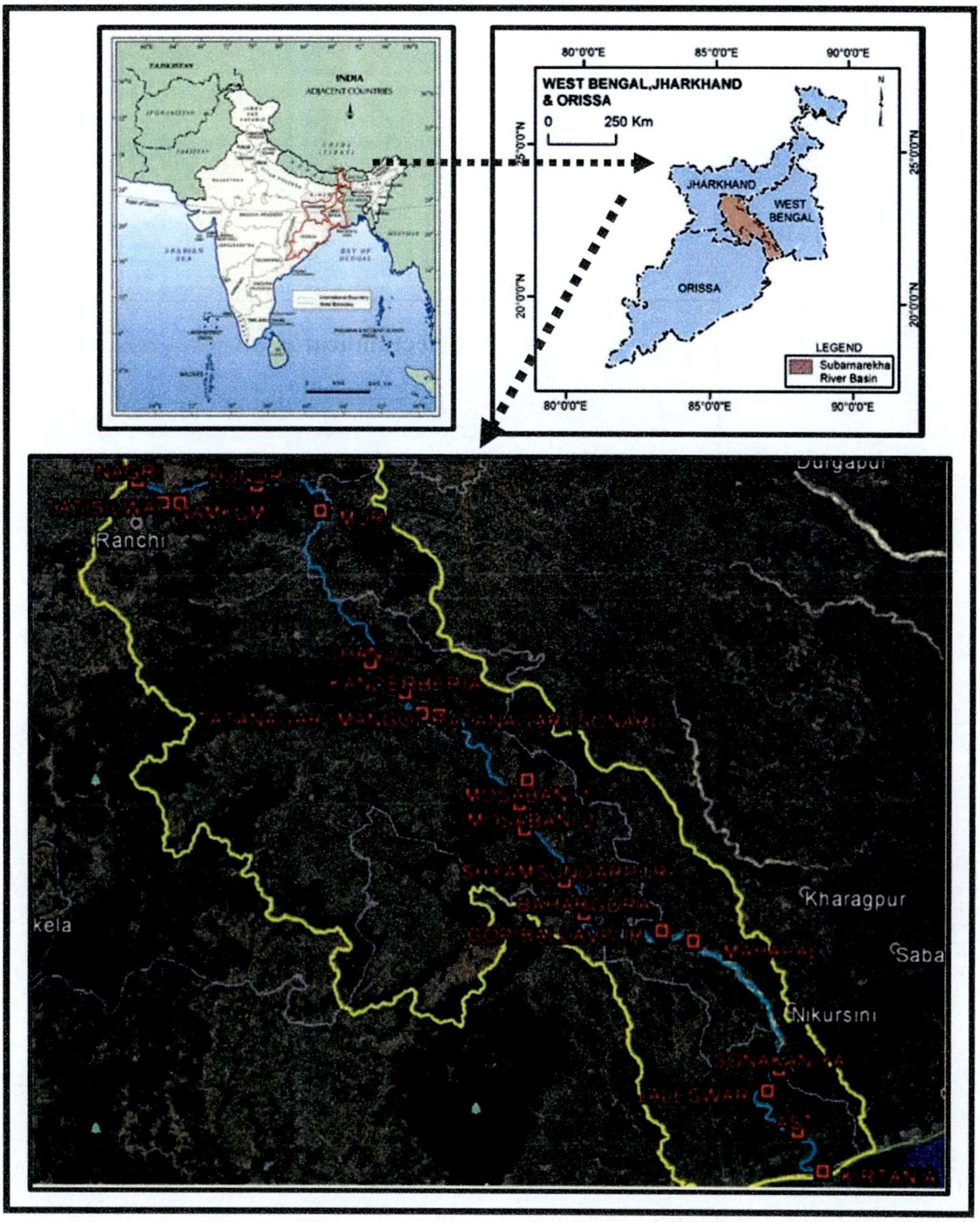

Fig. 17.1 Location map showing the study area

months of 2014. We collect the water samples by HDPE (High-Density Polyethylene) containers, which are actually preconditioned with acid wash. After that, they were again filtered by 0.45 μm Millipore nitrocellulose filters. To determine the metals, Inductively Coupled Plasma-Mass Spectrometry (ICP-MS, Perkin Elmer Elan DRC- e) (Beauchemin et al. 1989; Longerich et al. 1987; Buiteveld et al. 1994; Jackson and Marmulla 2011; Yamanaka

Table 17.1 Parameters and sub-parameters of river health analysis

Parameters	Sub-parameters	Parameters	Sub-parameters		
Physical	Temperature	*Chemical*	Aluminum (Al)	Chromium (Cr)	Nickel (Ni)
	Salinity		Arsenic (As)	Copper (Cu)	Selenium (Se)
	Suspended solids		Barium (Ba)	Manganese (Mn)	Strontium (Sr)
Biological	Chlorophyll-a		Cobalt (Co)	Iron (Fe)	Vanadium (V)

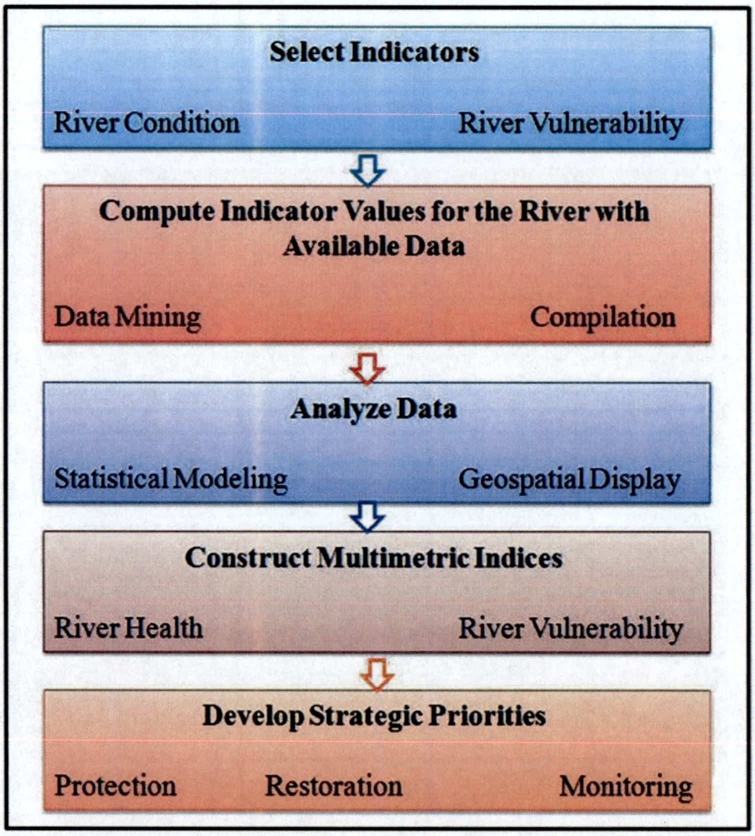

Fig. 17.2 Flowchart of river health assessment

et al. 2007; Philp et al. 2005; Djedjibegovic et al. 2012; Giri and Singh 2014) was used and it was tested by Public Health Engineering Department, Govt. of West Bengal. The overall accuracy level of the test was 94.5–105.8 as calculated by Student's t-test (Lettenmaier 1976; Alther 1979; Hirsch et al. 1993; Thornton 1980; Harris and Silveira 1999; Harris 2001; Reckhow 1994). The method was further validated by NIST 1640 (a) certified (Bashir and Paull 2001; Stosnach 2005; Martendal et al. 2009; Giri and Singh 2014) reference material supplied by the National Institute of Standards and Technology (NIST, USA) (Table 17.2).

17.3.4 Physical and Biological Health Analysis

In the last fifty years, numerous scientists have published different models to demarcate chemical, physical, and biological parameters of water

Table 17.2 Sample stations for assessing chemical health

ID	Station name	ID	Station name	ID	Station name
1	Nagri	8	Kanderbera	15	Bahargora
2	Namkom	9	Tatanagar (Sonari)	16	Gopiballavpur
3	Tatisilwai	10	Tatanagar (Mango)	17	Mahapal
4	Hundru	11	Galudih	18	Sonakaniya
5	Muri	12	Mosabani (U/S Sankh)	19	Jaleswar
6	Chandil	13	Mosabani (D/S Sankh)	20	Pontai
7	Barbinda	14	Shyamsundarpur	21	Kirtania

bodies using remote sensing (Facchinelli et al. 2001; Liu and Dogg 2003; Matthews 2011; Odermatt et al. 2010; Ahmad et al. 2018; Simon et al. 2020; Topp et al. 2020) owing to the huge limitations of field investigation (Tyler et al. 2006; IOCCG 2018; Giardino et al. 2019). Thus, for biophysical health analysis, satellite images are required (Weng 2009; Nouri et al. 2016; Rosli et al. 2015). Landsat series Thematic Mapper (TM) (Dekker and Peters 1993; Yacobi et al 1995; Wang et al. 2004) and Operational Land Imager (OLI) (Watanabe et al. 2015; Lee et al. 2016) of spatial resolution (30 m/30 m) was selected to detect physical and biological health conditions. The whole analysis was done using Arc GIS software version 10.1 for pre-processing like georeferencing, mosaicking, sub setting, etc. The determination of the degree of biophysical health parameters has been done by TNT MIPS of micro images version 2014. A list of satellite data used for the above analysis has been given in Table 17.3.

As near and shortwave infrared spectrum is highly absorptive in the case of water, thus its radiance in the visible spectrum has limited distortion to assess temperature and salinity (Buiteveld et al. 1994; IOCCG 2000; Doxaran et al. 2002; Rottgers et al. 2014). Though in recent years, hyper spectral sensors are used to identify such kinds of parameters with high accuracy levels (Govender et al. 2007; Bioucas-dias et al. 2013). Total Suspended Solids (TSS) are fine particles that constitute organic and inorganic components in the water column. The monitoring of these suspended sediment concentrations for biogeochemical assessment is necessary for knowing physical-biochemical health (Sweeney 1992; Julian et al. 2008; Rügner et al. 2013). TSS reduces the concentration of sunlight which results in more turbidity into the water which

Table 17.3 List of satellite data used for analysis

Data type				Date of acquisition		
Satellite	Sensors	Path/row	Spatial resolution (mts)			
Landsat 5	TM	139/44	30/30	11_04_1990	15_07_1990	20_10_1990
Landsat 5	TM	139/45	30/30	11_04_1990	15_07_1990	20_10_1990
Landsat 5	TM	140/44	30/30	18_04_1990	22_08_1990	30_12_1990
Landsat 5	TM	140/45	30/30	18_04_1990	22_08 1990	30 12 1990
Landsat 8	OLI/TIRS	139/44	30/30	29_04_1990	23_08_2014	11_10_2014
Landsat 8	OLI/TIRS	139/45	30/30	29_04_1990	23_08_2014	11_10_2014
Landsat 8	OLI/TIRS	140/44	30/30	04_04_1990	18_09_2014	15_10_2014
Landsat 8	OLI/TIRS	140/45	30/30	06_05_1990	18_09_2014	15_10_2014

leads to hampering of primary producers into the water, and it also controls the erosional regime of the river (Bilotta and Brazier 2008; Kefford et al. 2010; Overeem et al. 2017). To identify TSS concentration from remotely sensed data of spectral signature, 550 nm red or near-infrared wavelengths are considered as inorganic TSS. But organic TSS are mostly shown as backscatter peaks in chlorophyll-II (Walker 1996; Doxaran et al. 2002; Telmer et al. 2006; Volpe et al. 2012; Shi et al. 2013; Brando et al. 2015; Pereira-Sandoval et al. 2019; Overeem et al. 2017).

Chlorophylls convert energy with the help of photosynthesis and their components transform light into energy. In riparian system, it is considered that this chlorophyll-a concentration is a proxy data which informs us about the measurement of total algal biomass in the region (McCormick and Cairns 1994; Gitelson et al. 1994; Paerl and Huisman 2009; Kutser 2009; Svircev et al. 2007; Carvalho et al. 2013; Zhou et al. 2019). In recent days, climate change and anthropogenic influences increase the percentage of these harmful algal blooms in fluvial systems that degenerate the biological health condition (Paerl and Huisman 2009). By the optical sensors, we can detect the Chlorophyll-through spectral signatures (Gitelson et al. 1994; Dierssen 2010; Zhou et al. 2019). In the case of low concentrations, sun-induced fluorescence peak is near about 680 nm but in high concentrations, it varies between 665 and 710 nm (Gower et al. 1999, 2006; Matthews et al. 2012; Oyama et al. 2015). It is by this ratio of two wavelengths of remotely sensed data, Chlorophyll-a or biological health condition of the study area has been retrieved.

17.3.5 Multi-Criteria Decision Analysis (MCDA)

MCDA is a structural pathway which is applied if there is multiple set of attributes in measurement and its support to select a choice from multiple alternatives (Reynoldson et al. 1997; Marttunen and Suomalainen 2005; Hajkowicz and Collins 2007; Lai et al. 2008). In a complex decision-making process, it has a potentiality because here one can introduce a rank system by using the weight method by selecting some criteria (Castillo 2006). So, naturally, in resource planning and water management purposes, it has been widely applied all over the world for better decision-making strategies (Brown 1984; Allett 1986; Roy 1996; Bella et al. 1996; Clemen 1996; Hobbs and Meier 2000; Raju et al. 2000; Belton and Stewart 2002; Ning and Chang 2002; Ganoulis 2003; Herath 2004; Marttunen and Suomalainen 2005; Hazkowicz and Higgins 2008; Jato-Espino et al. 2014; Kabir et al. 2014).

In this method, using several criteria that include water conditions for human health and others, the results of alternative decisions have been scored. Then by applying the weighted sum method, ranks have been assigned for all criteria results through expert opinion. Here, the best suitable water quality has been given a standardized score of 1 and worse vulnerable water quality has been given a standardized score of 6. Thus, these score values depict the sensitivity of the alternative criterion.

17.4 Results and Discussion

17.4.1 Chemical Health Status

Seasonal variations of metal data analysis have been given in Table 17.4. The average concentration of diluted and non-diluted metals for premonsoon Al, Ba, As, Cr, Co, Cu, Mn, Ni, Fe, Sr, V, and Se was found to be 60.8, 35.1, 7.3, 1.3, 1.2, 22.8, 10.1, 12.7, 127.3, 4.8, 18, and 5.3 μg/L, but in monsoon, the values were 59.4, 26.9, 5.8, 0.7, 0.8, 32.4, 3.7, 3.5, 101.8, 2.6, 4.3, and 1.7 μg/L. However, in post-monsoon, mean concentrations were 47.2, 54.6, 2.9, 0.9, 0.6, 24.7, 6.6, 6.5, 78.1, 3.7, 5.4, and 0.9 μg/L. In a nutshell, the concentration of metals in the monsoon season has been showing lower than all other seasons. If we consider the permissible limit of drinking water, most of the metals showed higher value than the standard value as referred by WHO (1991, 1992, 2006) and USEPA (2009, 2012). These toxic metals were also affecting aquatic life which triggered the hydrological health regime of the fluvial system.

From the overall analysis of the chemical condition, metal concentration in the pre-monsoon season showed highest because of the terrain and climate regime of the study area. It is characterized by high evaporation with numerous human activities like agriculture, industry, mining, brick kiln, etc. But due to the heavy rainfall in the monsoon season, the dilution effect has been predominated in the river thereby showing low metal concentration except for pointed sources like mining, industrial, or urbanized areas (Table 17.5 and Fig. 17.3).

From the MCDA analysis, the chemical health zones have been classified into five classes, namely, most suitable, vulnerable followed by suitable, moderately suitable, suitable followed by vulnerable, and vulnerable. The upstream areas showed suitable health conditions, whereas downstream or estuary shows the more vulnerable situation in terms of the chemical health status of three seasons. Vulnerable followed by suitability areas were distributed along with mining and industrial belts and non-pointed areas showed suitable followed by a vulnerable condition. The Subarnarekha River has carried high discharge mainly in monsoon season, and during that time maximum metals are concentrated in the terminal estuary section due to rapid transportation (Table 17.6 and Fig. 17.4).

17.4.2 Physico-biological Health Status

Physical and biological health conditions were assessed from remotely sensed data for the years 1990 and 2014. Physical parameters like temperature, salinity, and total suspended solids were measured and algal bloom or chlorophyll-a concentration was measured as biological health status. The average range of dynamicity of these parameters has been shown in Table 17.7 (Fig. 17.5).

Here, all the values have been classified and it ranges from one to six—One indicates the lower value and four indicates a higher amount of parameters. From 1990 to 2014, temperature value increases in the buffer zone of river Subarnarekha mainly due to the rapid change of land use and land cover. The salinity amount is higher in 2014 and it is shifted towards the upstream side in recent years due to the low amount of discharge and sea-level change. In both the years,

Table 17.4 Seasonal variation of the various metal conditions in the river Subarnarekha

Metals	The seasonal average concentration of metals (μg/L)			References	
	Pre-monsoon	Monsoon	Post-monsoon	WHO (2006)	USEPA (2009)
Al	60.8	59.4	47.2	200	50–200
As	7.3	5.8	2.9	10	10
Ba	35.1	26.9	54.6	700	2000
Co	1.2	0.8	0.6	0.32	2.79
Cr	1.3	0.7	0.9	50	100
Cu	22.8	32.4	24.7	2000	1300
Mn	10.1	3.9	6.6	400	50
Fe	127.3	101.8	78.1	2000	300
Ni	12.7	3.5	6.5	70	20
Se	5.3	1.7	0.9	40	50
Sr	4.8	2.6	3.7	4	90
V	18.0	4.3	5.4	4	140

Table 17.5 Overall ranges of different seasonal metals (µg/L) from the selected stations for the Subarnarekha River

Chemicals	Pre-monsoon	Monsoon	Post-monsoon
Al	(−1.9367)–(−0.9218)	(−1.48)–(−0.52)	(−0.62)–0.27
	(−0.9218)–0.0932	(−0.52)–0.45	0.27–1.16
	0.0932–1.1081	0.45–1.42	1.16–2.04
	1.1081–2.1230	1.42–2.38	2.04–2.93
	2.1230–3.1380	2.38–3.35	2.93–3.82
	3.1380–4.1529	3.35–4.32	3.82–4.71
As	(−0.57)–0.07	(−0.58)–0.02	(−0.45)–0.24
	0.07–0.71	0.02–0.61	0.24–0.93
	0.71–1.35	0.61–1.20	0.93–1.63
	1.35–1.99	1.20–1.80	1.63–2.32
	1.99–2.63	1.80–2.39	2.32–3.02
	2.63–3.27	2.39–2.99	3.02–3.71
Ba	(−1.10)–(−0.44)	(−0.82)–(−0.09)	(−2.90)–(−2.13)
	(−0.44)–0.22	(−0.09)–0.64	(−2.13)–(−1.36)
	0.22–0.88	0.64–1.38	(−1.36)–(−0.59)
	0.88–1.54	1.38–2.11	(−0.59)–0.18
	1.54–2.20	2.11–2.84	0.18–0.95
	2.20–2.86	2.84–3.58	0.95–1.72
Co	(−0.82)–(−0.02)	(−0.81)–(−0.07)	(−0.6283)–0.5033
	(−0.02)–0.77	(−0.07)–0.67	0.5033–1.6349
	0.77–1.57	0.67–1.41	1.6349–2.7665
	1.57–2.37	1.41–2.15	2.7665–3.8981
	2.37–3.16	2.15–2.89	3.8981–5.0297
	3.16–3.96	2.89–3.63	5.0297–6.1613
Cr	(−1.3423)–(−0.0564)	(−1.3393)–(−0.1179)	(−2.21)–(−1.62)
	(−0.0564)–1.2295	(−0.1179)–1.1034	(−1.62)–(−1.02)
	1.2295–2.5154	1.1034–2.3248	(−1.02)–(−0.43)
	2.5154–3.8013	2.3248–3.5462	(−0.43)–0.16
	3.8013–5.0872	3.5462–4.7675	0.16–0.75
	5.0872–6.3731	4.7675–5.9889	0.75–1.35
Cu	(−0.72)–(−0.00)	(−0.70)–(−0.00)	(−0.92)–(−0.38)
	(−0.00)–0.72	(−0.00)–0.69	(−0.38)–0.15
	0.72–1.43	0.69–1.39	0.15–0.68
	1.43–2.15	1.39–2.08	0.68–1.21
	2.15–2.86	2.08–2.78	1.21–1.75
	2.86–3.58	2.78–3.48	1.75–2.28
Mn	(−1.25)–(−0.43)	(−1.10)–(−0.15)	(−0.73)–0.03
	(−0.43)–0.39	(−0.15)–0.80	0.03–0.78
	0.39–1.21	0.80–1.75	0.78–1.53
	1.21–2.03	1.75–2.70	1.53–2.28

(continued)

Table 17.5 (continued)

Chemicals	Pre-monsoon	Monsoon	Post-monsoon
	2.03–2.85	2.70–3.65	2.28–3.03
	2.85–3.67	3.65–4.59	3.03–3.79
Fe	(−0.84)–(−0.24)	(−1.25)–(−0.47)	(−1.06)–(−0.08)
	(−0.24)–0.35	(−0.47)–0.31	(−0.08)–0.91
	0.35–0.95	0.31–1.09	0.91–1.89
	0.95–1.55	1.09–1.87	1.89–2.87
	1.55–2.14	1.87–2.65	2.87–3.86
	2.14–2.74	2.65–3.43	3.86–4.84
Ni	(−0.87)–(−0.08)	(−1.0312)–0.0037	(−2.03)–(−1.21)
	(−0.08)–0.72	0.0037–1.0385	(−1.21)–(−0.39)
	0.72–1.52	1.0385–2.0734	(−0.39)–0.43
	1.52–2.31	2.0734–3.1082	0.43–1.25
	2.31–3.11	3.1082–4.1431	1.25–2.06
	3.11–3.91	4.1431–5.1779	2.06–2.88
Se	(−0.42)–0.27	(−0.44)–0.25	(−0.38)–0.30
	0.27–0.96	0.25–0.94	0.30–0.99
	0.96–1.65	0.94–1.64	0.99–1.67
	1.65–2.34	1.64–2.33	1.67–2.35
	2.34–3.03	2.33–3.02	2.35–3.03
	3.03–3.72	3.02–3.72	3.03–3.71
Sr	(−0.52)–0.07	(−0.55)–0.08	(−0.52)–0.19
	0.07–0.66	0.08–0.71	0.19–0.89
	0.66–1.25	0.71–1.34	0.89–1.60
	1.25–1.84	1.34–1.96	1.60–2.30
	1.84–2.43	1.96–2.59	2.30–3.01
	2.43–3.02	2.59–3.22	3.01–3.71
V	(−0.64)–0.15	(−0.78)–0.05	(−0.81)–(−0.07)
	0.15–0.95	0.05–0.87	(−0.07)–0.66
	0.95–1.75	0.87–1.70	0.66–1.39
	1.75–2.54	1.70–2.52	1.39–2.13
	2.54–3.34	2.52–3.35	2.13–2.86
	3.34–4.13	3.35–4.17	2.86–3.60

estuaries show higher salinity because of the mixing zone with the Bay of Bengal, and the upper part shows a very poor amount as the area is rain-fed. The concentration of suspended solids has decreased in 2014 although a moderate concentration zone has increased all over the study area. Algal bloom concentration showed a scattered distribution in pockets. In the places where the sewage and other nutrients were added into the river course, the concentration was normally high which was typically observed in the upper and middle course of river Subarnarekha.

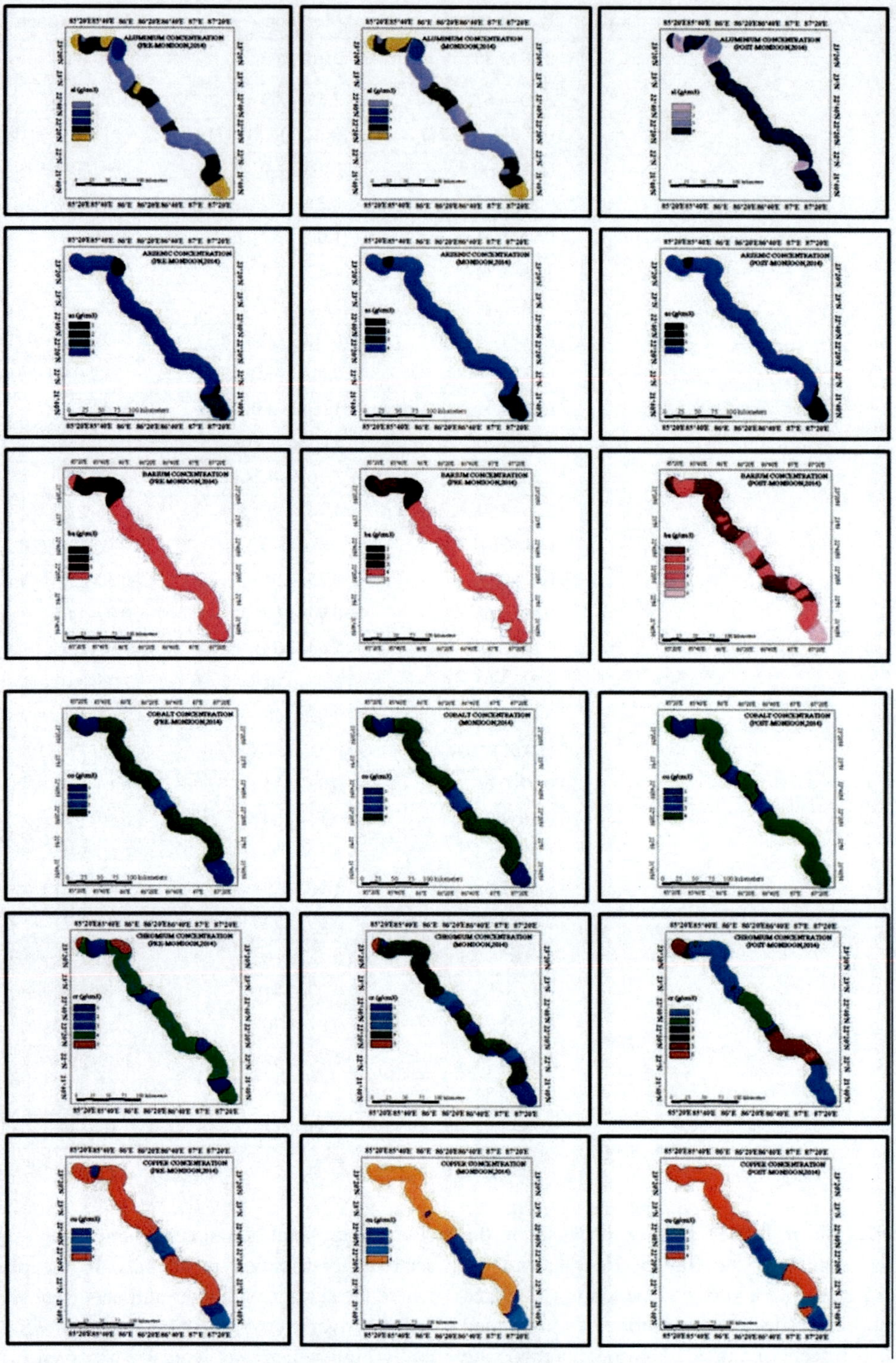

Fig. 17.3 Zonation of seasonal metal concentration of Subarnarekha River

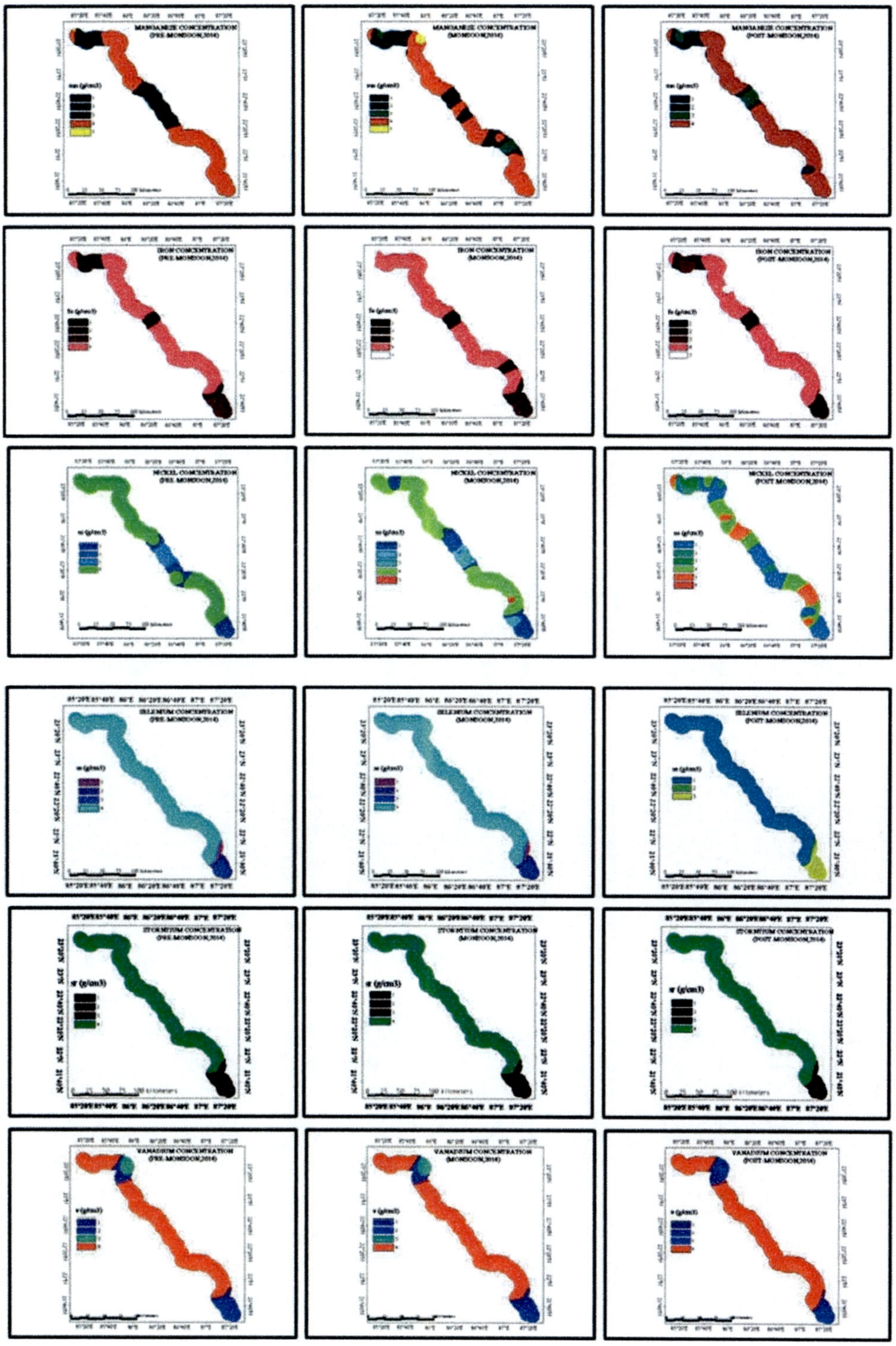

Fig. 17.3 (continued)

Table 17.6 Multi-Criteria Decision-Making Analysis of different chemical components for the river Subarnarekha

Value	Count	MON_RE	PRE_RE	POST_RE	Combination_Class	MCDA
1	222,339	1	1	1	Most suitable	1
2	423	4	2	4	Vulnerable followed by suitability	2
3	69	4	2	3	Vulnerable followed by suitability	2
3	32	4	2	3	Moderately suitable	3
4	69	3	3	3	Moderately suitable	3
5	842	4	3	3	Moderately suitable	3
6	288	3	3	2	Moderately suitable	3
8	518	4	3	2	Moderately suitable	3
9	1363	4	3	2	Moderately suitable	3
10	3340	3	3	3	Moderately suitable	3
11	2131	3	3	3	Moderately suitable	3
12	251	3	4	3	Moderately suitable	3
13	422	4	3	3	Moderately suitable	3
13	4133	4	3	3	Vulnerable followed by suitability	2
14	1462	2	4	2	Suitable followed by vulnerable	4
15	333	2	4	3	Suitable followed by vulnerable	4
16	533	4	4	3	Moderately suitable	3
18	159	3	3	3	Vulnerable followed by suitability	2
19	386	3	3	2	Moderately suitable	3
20	1529	3	3	2	Moderately suitable	3
21	15	2	4	1	Moderately suitable	3
22	1599	3	4	3	Moderately suitable	3
23	342	4	4	2	Moderately suitable	3
23	30	3	4	2	Moderately suitable	3
24	680	4	3	3	vulnerable	5
25	2322	4	3	4	Vulnerable followed by suitability	2
26	1115	3	3	4	Vulnerable followed by suitability	2
28	1002	4	3	4	Vulnerable followed by suitability	2
29	1891	3	3	3	vulnerable	5
30	320	3	3	4	Increasing vulnerability	6
31	3464	3	3	4	Increasing vulnerability	6
32	305	3	3	3	Increasing vulnerability	6
33	623	3	3	4	Increasing vulnerability	6
33	32	3	3	3	Increasing vulnerability	6
34	85	2	3	4	Increasing vulnerability	6
35	125	2	3	2	Moderately suitable	3
36	260	2	3	3	Suitable followed by vulnerable	4
38	15	2	3	3	Increasing vulnerability	6
39	19	1	4	2	Suitable followed by vulnerable	4

(continued)

Table 17.6 (continued)

Value	Count	MON_RE	PRE_RE	POST_RE	Combination_Class	MCDA
30	436	2	4	3	vulnerable	5
31	131	2	4	4	Increasing vulnerability	6
32	516	3	4	4	Increasing vulnerability	6
33	2309	3	4	3	Suitable followed by vulnerable	4

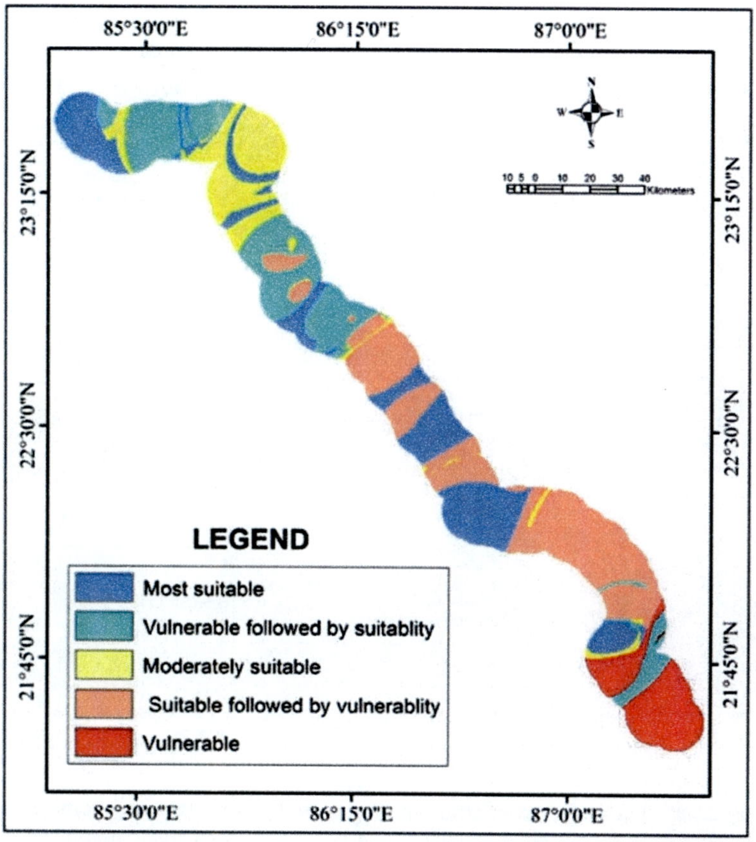

Fig. 17.4 Chemical health zonation map of Subarnarekha River from MCDA analysis

Surprisingly, it was also high in the estuarine environment because transported nutrients were accumulated here (Table 17.8 and Fig. 17.6).

After combining all sub-parameters of physical and biological health indicators by the MCDA technique, they were further reclassified into the same five classes as for chemical health. The most suitable class was found in upper-middle and lower reach and in some pockets where there was low anthropogenic pressure and the river controlled its regime. But the maximum upper reaches showed decreasing suitability because of huge alterations due to anthropogenic alterations and interventions. Although some reaches of the upper area indicated increasing suitability because nowadays, the amount of afforestation and purification of mining and industrial sewage has increased due to consciousness and strict government policies. But the major part of the plain land showed moderate suitable conditions, whereas the extreme lower part was vulnerable because of anthropogenic

Table 17.7 Overall ranges of the seasonal physical and biological health status of the river Subarnarekha

Parameters	1990	2014
Temperature	(−18.7380)–(−14.1042)	(−10.6194)–(−7.6299)
	(−14.1042)–(−9.4704)	(−7.6299)–(−4.6404)
	(−9.4704)–(−4.8366)	(−4.6404)–(−1.6509)
	(−4.8366)–(−0.2028)	(−1.6509)–1.3386
	(−0.2028)–4.4310	1.3386–4.3281
	4.4310–9.0648	4.3281–7.3176
Algal bloom	(−8.7069)–(−4.2621)	(−21.2756)–(−17.5127)
	(−4.2621)–0.1828	(−17.5127)–(−13.7497)
	0.1828–4.6277	(−13.7497)–(−9.9867)
	4.6277–9.0725	(−9.9867)–(−6.2237)
	9.0725–13.5174	(−6.2237)–(−2.4607)
	13.5174–17.9622	(−2.4607)–1.3022
Salinity	(−3.7014)–(−1.5606)	(−1.1618)–2.1477
	(−1.5606)–0.5803	2.1477–5.4572
	0.5803–2.7212	5.4572–8.7667
	2.7212–4.8621	8.7667–12.0762
	4.8621–7.0030	12.0762–15.3857
	7.0030–9.1439	15.3857–18.6952
Suspended solids	(−5.8497)–(−3.2856)	(−1.1487)–3.6217
	(−3.2856)–(−0.7215)	3.6217–8.3921
	(−0.7215)–1.8427	8.3921–13.1625
	1.8427–4.4068	13.1625–17.9329
	4.4068–6.9709	17.9329–22.7034
	6.9709–9.5351	22.7034–27.4738

pressure in coastal belts and narrowing of estuary in the last years (Table 17.9).

From the overall analysis, it has been observed that observed stations of Nagri, Namkom, Mosabani show the most suitable health conditions but Tatanagar, Tatisilwai, Hundru, Chandil, and Barbinda stations show a decreasing trend of suitable health conditions. The stations of Muri, Kanderbera, and Sonakanai have good health suitability, and Galudih, Shyamsundarpur, Bahargora, and Gopiballavpur represent moderately suitable heath status. The worse scenario was observed in Jaleswar, Pontai, and Kirtania stations of lower reach which depict the most vulnerable health scenario.

17.5 Conclusions

From the chemical test, it was observed that dissolved metal concentration (Al, Ba, As, Cr, Co, Cu, Mn, Ni, Fe, Sr, V, and Se) in the River Subarnarekha shows spatial variability including seasonality. It is lower in the period of monsoon than the other two seasons because of the immense dilution effect. Major metal concentrations are found beside the mining and heavy industrial belts, and it indicates the strong influences of anthropogenic pressure along the river. To know the combined effect of natural processes, such as a change in temperature, salinity,

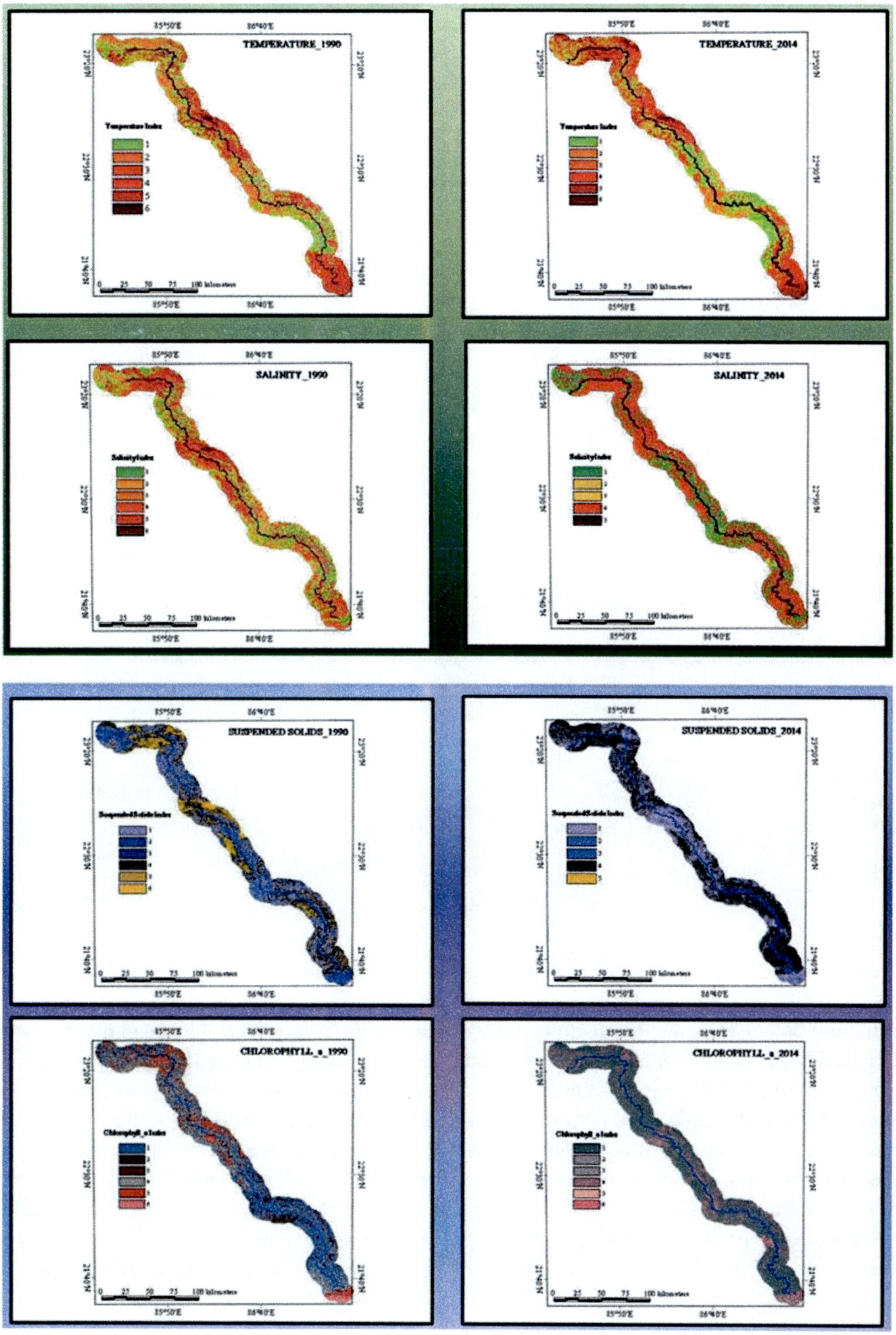

Fig. 17.5 Physical and biological health zonation map of Subarnarekha River

Table 17.8 Multi-Criteria Decision-Making Analysis of different physical and biological components for the river Subarnarekha

Value	Count	Combine_14	Combine_90	Combination_Class	MCDA
1	5,731,389	1	1	Most suitable	1
2	482,301	2	1	Most suitable	1
3	1,062,224	1	2	Most suitable	1
4	41,404,107	2	2	Most suitable	1
5	18,973	2	3	Decreasing suitability	2
6	42,934	3	1	Increasing suitability	3
7	17,352	3	2	Increasing suitability	3
8	25,511	1	3	Decreasing suitability	2
9	32,189	3	3	Moderately suitable	4
10	4315	1	4	Vulnerable	5
11	19,488	3	4	Vulnerable	5
12	18,042	2	4	Vulnerable	5
13	62	4	1	Increasing suitability	3
14	1456	4	2	Increasing suitability	3
15	23,200	4	4	Vulnerable	5
16	14,291	4	3	Moderately suitable	4
17	17,120	3	5	Vulnerable	5
18	13,404	2	5	Vulnerable	5
19	1225	1	5	Vulnerable	5
20	30,588	4	5	Decreasing suitability	2
21	11,621	5	4	Vulnerable	5
22	11,370	5	5	Vulnerable	5
23	3	5	1	Increasing suitability	3
24	5052	5	3	Moderately suitable	4
25	91	5	2	Increasing suitability	3

suspended solids, chlorophyll-a and anthropogenic activities like industrial, urban, and agricultural processes which results lowering of water level, decreasing volume of sand and gravels and which left major distinct imprints upon the10 km buffer region of the Subarnarekha River Basin. Now it can strongly prove that the increasing trend of heavy metals in Subarnarekha River has a direct link with its high number of affluent discharge points from mining, urban, and industrial complexes. Thus, the risk factor for human health increases day by day as the majority of the people are exposed to heavy metals, especially for children and older persons, and here the oral intake happens to be the primary exposure pathway. It can be concluded that though the metal concentration fluctuates in different seasons, incurable risks are spread in all seasons mainly due to rapid increases of As, Co, and V. Therefore, proper strategy and scientific planning are required to reduce and maintain the health condition of the Subarnarekha River.

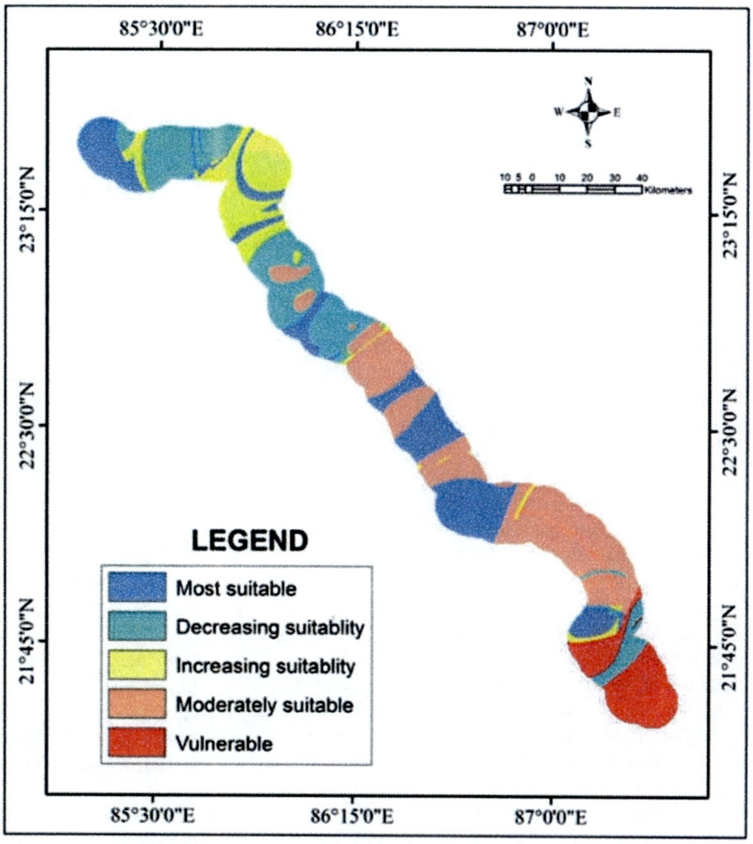

Fig. 17.6 Physical and biological health zonation map of Subarnarekha River from MCDA analysis

Table 17.9 Combined health status of the River Subarnarekha

Health class	Stations
Most suitable	Nagri, Namkom, Mosabani (U/S Sankh), Mosabani (D/S Sankh)
Decreasing suitability	Tatanagar (Sonari), Tatanagar (Mango), Tatisilwai, Hundru, Chandil, Barabinda
Increasing suitability	Muri, Kanderbera, Sonakania
Moderately suitable	Galudih, Shyamsundarpur, Bahargora, Gopiballavpur
Vulnerable	Jaleswar, Pontai, Kirtania

Acknowledgements The authors would like to thank the Department of Public Health Engineering, Govt. of West Bengal for conducting chemical analysis, and the United States Geological Survey (USGS) Earth Explorer for providing satellite imageries. We appreciate and thank the reviewers for their valuable comments and suggestions that helped us to improve our manuscript.

References

Ahearn DS, Sheibley RW, Dahlgren RA, Anderson M, Johnson J, Tate KW (2005) Land use and land cover influence on water quality in the last free flowing river draining the western Sierra Nevada, California. J Hydrol 234–247

Ahmad M, Protasov S, Khan AM, Hussain R, Khattak AM, Khan WA (2018) Fuzziness-based active learning framework to enhance hyperspectral image classification performance for discriminative and generative classifiers. PLoS One 13

Ahmed MK, Ahmed S, Rahman S, Haque MR, Islam MM (2009) Heavy metals concentration in water, sediments and their bioaccumulations in some freshwater fishes and mussel in Dhaleshwari River, Bangladesh. Terr Aquat Environ Toxicol 3(1):33–41

Allan JD, Johnson LB (1997) Catchment-scale analysis of aquatic ecosystems. Freshw Biol 37:107–111

Allett EJ (1986) Environmental impact assessment and decision analysis. J Oper Res Soc 37(9):901–910

ANZECC (1992) Australian water quality guidelines for fresh and marine waters, National Water Quality Management Strategy, Australian and New Zealand Environment and Conservation Council

Asian Development Bank (2013) Asian water development outlook: measuring water security in Asia and the Pacific, Manila, Philippines

Asim M, Rao KN (2021) Assessment of heavy metal pollution in Yamuna River, Delhi-NCR, using heavy metal pollution index and GIS. Environ Monit Assess 193(2). https://doi.org/10.1007/s10661-021-08886-6

Bandyopadhyay S, De S (2017) Human interference on river health: a study on the Haora River, Tripura, India. Advances in Asian human-environmental research. Springer, pp 1–193

Bartram J, Ballance R (eds) (1996) Water quality monitoring: a practical guide to the design and implementation of freshwater quality studies and monitoring programmes. Chapman & Hall, London

Bashir W, Paull B (2001) Determination of trace alkaline earth metals in brines using chelation ion chromatography with an iminodiacetic acid bonded silica column. J Chromatogr A 907(1–2):191–200

Beauchemin KA, Zelin S, Genner D, Buchanan-Smith JG (1989) An automatic system for quantification of eating and ruminating activities of dairy cattle housed in stalls. J Dairy Sci 72:2746–2759

Bella A, Duckstein L, Szidarovszky F (1996) A multicriterion analysis of the water allocation conflict in the upper rio Grande river basin. Appl Math Com 77(2–3):245–265

Belton V, Stewart TJ (2002) Multiple criteria decision analysis—an integrated approach. Kluwer Academic Publishers, Boston, Dordrecht, London

Bilotta GS, Brazier RE (2008) Understanding the influence of suspended solids on water quality and aquatic biota. Water Res 41:2849–2861

Bioucas-Dias JM, Plaza A, Camps-Valls G, Scheunders P, Nasrabadi NM, Chanussot J (2013) Hyperspectral remote sensing data analysis and future challenges. IEEE Geosci Remote Sens Mag 1(2):6–36

Boulton AJ (1999) An overview of river health assessment: philosophies, practice, problems and prognosis. Freshw Biol 41:469–479

Brando VE, Braga F, Zaggia L, Giardino C, Bresciani M, Matta E, Bellafiore D, Ferrarin C, Maicu F, Benetazzo A, Bonaldo D (2015) High-resolution satellite turbidity and sea surface temperature observations of river plume interactions during a significant flood event. Ocean Sci 11(6):909

Brookes A, Shields FD (eds) (1996) River channel restoration: guiding principles for sustainable projects. Wiley, Chichester, p 433

Brown CA (1984) The central Arizona water control study: a case for multiobjective planning and public involvement. Water Resour Bull 20(3):331–337

Buiteveld H, Hakvoort JHM, Donze M (1994) The optical properties of pure water. In: SPIE ocean optics XII, vol 2258, pp 174–183

Carvalho L, McDonald C, de Hoyos C, Mischke U, Phillips G, Borics G (2013) Sustaining recreational quality of European lakes: minimising the health risks from algal blooms through phosphorus control. J Appl Ecol 50:315–323

Chang H, Carlson TN (2005) Water quality during winter storm events in Spring Creek, Pennsylvania, USA. Hydrobiologia 321–332

Chapman PM (1992) Ecosystem health synthesis: can we get there from here? J Aquat Ecosyst Health 1:69–79

Chatterjee SK, Bhattacharjee I, Chandra G (2010) Water quality assessment near an industrial site of Damodar River, India. Environ Monit Assess 161:177–189

Chaudhary M, Mishra S, Kumar A (2016) Estimation of water pollution and probability of health risk due to imbalanced nutrients in River Ganga, India. Int J River Basin Manag 15(1):53–60

Chen J, Wang Y, Li F, Liu Z (2019) Aquatic ecosystem health assessment of a typical sub-basin of the Liau River based on entropy weights and a fuzzy comprehensive evaluation method. Sci Rep 9:14045. https://doi.org/10.1038/s41598-019-50499-0

Clemen RT (1996) Making hard decision: an introduction to decision analysis. Duxbury Press, Wadsworth Publishing Company, New York

Dekker AK, Peters S (1993) The use of the Thematic Mapper for the analysis of eutrophic lakes: a case study in the Netherlands. Int J Remote Sens 14 (5):799–822

Deng X, Xu Y, Han L (2015) Assessment of river health based on an improved entropy-based fuzzy matter-element model in the Taihu Plain, China. Ecol Indic 57:85–95

Dierssen HM (2010) Perspectives on empirical approaches for ocean color remote sensing of chlorophyll in a changing climate. Proc Natl Acad Sci U S A 107:17–78

Djedjibegovic J, Larssen T, Skrbo A, Marjanovic A, Sober M (2012) Contents of cadmium, copper, mercury and lead in fish from the Neretva river (Bosnia and Herzegovina) determined by inductively coupled plasma mass spectrometry (ICP-MS). Food Chem 131:469–476

Donohue I, McGarrigle ML, Mills P (2006) Linking catchment characteristics and her chemistry with the ecological status of Irish rivers. Water Res 40(1):91–98

Doxaran D, Froidefond JM, Lavender S, Castaing P (2002) Spectral signature of highly turbid water application with SPOT data to quantify suspended particulate matter concentration. Remote Sens Environ 81(2):149–161

Environment Canada (1987) Canadian water quality guidelines. Prepared by the Task Force on Water Quality Guidelines of the Canadian Council of Resource Ministers, Environment Canada, Ottawa

Facchinelli A, Sacchi E, Mallen L (2001) Multivariate statistical and GIS-based approach to identify heavy metal sources in soils. Environ Pollut 114(3):24–313

Falkenmark M (1989) The massive water scarcity now threatening Africa: why isn't it being addressed? Ambio 18:112–118

Feng Y, Da-Ming HE, Yang LP (2012) Selection of major evaluation indicators on river health evaluation. Geogr Res 31(3):389–398

Ganoulis J (2003) Evaluating alternative strategies for wastewater recycling and reuse in the mediterranean area. Water Sci Tech: Water Supply 3(4):11–19

Giardino C, Brando VE, Gege P, Pinnel N, Hochberg E, Knaeps E, Reusen I, Doerffer R, Bresciani M (2019) Imaging spectrometry of inland and coastal waters: state of the art, achievements and perspectives. Surv Geophys 40:401–429

Giri S, Singh AK (2014) Risk assessment, statistical source identification and seasonal fluctuation of dissolved metals in the Subarnarekha River, India. J Hazard Mater 265:305–314

Gitelson A, Mayo M, Yacobi YZ et al (1994) The use of high spectral radiometer data for detection of low chlorophyll concentrations in Lake Kinneret. J Plankton Res 16:993–1002

Gleick PH (1990) Global climatic changes: a summary of regional hydrologic impacts. Civ Eng Pract 5(1):53–68

Global Water Partnership (2000) Towards water security: a framework for action. Stockholm, Sweden

Gore JA (1985) The restoration of rivers and streams. Butterworth, Stoneham, p 280

Govender M, Chetty K, Bulcock H (2007) A review of hyperspectral remote sensing and its application in vegetation and water resource studies. Water SA 33:145–151

Gower JFR, Doerffer R, Borstad GA (1999) Interpretation of the 685 nm peak in water-leaving radiance spectra in terms of fluorescence, absorption and scattering and its observation by MERIS. Int J Remote Sens 9:1771–1786

Gower JFR, Hu C, Borstad GA, King S (2006) Ocean color satellites show extensive lines of floating Sargassum in the Gulf of Mexico. IEEE Trans Geosci Remote Sens 44:3619–3625

Gurnell AM, Petts GE (eds) (1995) Changing river channels. Wiley, Chichester, p 442

Hajkowicz S, Collins K (2007) A review of multiple criteria analysis for water resource planning and management. Water Resour Manag 21:1553–1566

Hamner S, Tripathi A, Mishra RK, Bouskill N, Broadway SC, Pyle BH, Ford TE (2007) The role of water use patterns and sewage pollution in incidence of water-borne/enteric diseases along the Ganges River in Varanasi, India. Int J Environ Health Res 16 (2):113–132

Harris JH, Silveira R (1999) Large-scale assessments of river health using an Index of Biotic Integrity with low-diversity fish communities. Freshw Biol 41:235–252

Harris RJ (2001) A primer of multivariate statistics. Lawrence Erlbaum Associates, London, p 3

Hart BT, Maher W, Lawrence I (1999) New generation water quality guidelines for ecosystem protection. Freshw Biol 41:347–359

Hart DD, Fonseca DM (1996) An important confluence for stream ecology. Trends Ecol Evol 11:272–273

Hazkowicz S, Higgins A (2008) A comparison of multiple criteria analysis techniques for water resource management. Eur J Oper Res 184:255–265

Heap C, Kemp-Benedict E, Raskin P (1998) Conventional worlds: technical description of bending the curve scenarios. Polestar series report, Stockholm Environmental Institute, Boston

Herath G (2004) Incorporating community objectives in improved wetland management: the use of the analytic hierarchy process. J Environ Manag 70(3):263–273

Hirsch RM, Helsel DR, Cohn TA, Gilroy EJ (1993) Statistical analysis of hydrologic data. In: Maidment DR (ed) Handbook of hydrology. McGraw-Hill, New York, USA, Ch. 17, pp 17.11–17.37

Hobbs BF, Meier P (2000) Energy decisions and the environment: a guide to the use of multicriteria methods. Kluwer Academic Publishers, Boston, Dordrecht, London, p 257

Huaibin W, Jianping Y (2014) Research on theory and method of river health assessment. Open Cybern Syst J 8(1):1166–1174

Hynes HBN (1975) Edgardo Baldi Memorial Lecture, The stream and its valley. Verhandlungen der Internationalen Vereinigungfür Theoretische und Angewandte Limnologie 19:1–15

IOCCG (2000) In: Sathyendranath S (ed) Remote sensing of ocean colour in coastal, and other optically-complex, waters. Reports of the International Ocean-Colour Coordinating Group, No. 3

IOCCG (2018) In: Greb S, Dekker A, Binding C (eds) Earth observations in support of global water quality monitoring. IOCCG Report Series, No. 17, International Ocean Colour Coordinating Group, Dartmouth, Canada

Jackson DC, Marmulla G (2011) The influence of dams on river fisheries. In: Dams, fish and fisheries: opportunities, challenges and conflict resolution. FAO fisheries technical paper, Rome

Jain CK, Sharma MK (2006) Heavy metal transport in the Hindon River Basin, India. Environ Monit Assess 112:255–270

Jato-Espino D, Castillo-Lopez E, Rodriguez-Hernandez J, Canteras-Jordana JC (2014) A review of application of multi-criteria decision making methods in construction. Autom Constr 45:151–162

Jensen O, Wu H (2018) Urban water security indicators: development and pilot. Environ Sci Policy 83:33–45

Julian JP, Doyle MW, Powers SM, Stanley EH, Riggsbee JA (2008) Optical water quality in rivers. Water Resour Res 44

Kabir G, Sadiq R, Tesfamariam S (2014) A review of multi-criteria decision-making methods for infrastructure management. Struct Infrastruct Eng 10(9):1176–1210

Kamboj N, Kamboj V (2019) Water quality assessment using overall index of pollution in river bed-mining area of Ganga River, Haridwar, India. Water Sci 33 (1):65–74

Karr JR, Chu EW (1998) Restoring life in running waters: better biological monitoring. Island Press, Washington, DC

Karr JR (1981) Assessment of biotic integrity using fish communities. Fisheries 6:21–27

Karr JR (1991) Biological integrity: a long-neglected aspect of water resource management. Ecol Appl 1:66–84

Karr JR (1999) Defining and measuring river health. Freshw Biol 41:221–234

Kefford B, Zalizniak L, Dunlop J et al (2010) How are macro-invertebrates of slow flowing lotic systems directly affected by suspended and deposited sediments? Environ Pollut 158:543–550

Kerans BL, Karr JR (1994) A benthic index of biotic integrity (B-IBI) for rivers for the Tennessee Valley. Ecol Appl 4:768–785

Knight TW, Morris DW (1996) How many habitats do landscapes contain? Ecology 77:1756–1764

Lai E, Lundie S, Ashbolt NJ (2008) Review of multi-criteria decision-aid for integrated sustainability assessment of urban water systems. Urban Water J 5 (4):315–327

Lee ZP, Shang S, Qi L, Yan J, Lin G (2016) A semi-analytical scheme to estimate Secchi disk depth from Landsat 8 measurements. Remote Sens Environ 177:101–106

Lettenmaier DP (1976) Detection of trends in water quality data from records with dependent observations. Water Resour Res 12:1037–1046

Liu H, Dogg M (2003) Interactions between nutrients, phytoplankton growth, and micro- and mesozooplankton grazing in the plume of the Mississippi River. Mar Ecol Prog Ser 258:31–42

Longerich HP, Fryer BJ, Strong DF (1987) Determination of lead isotopes ratios by inductively coupled plasma-mass spectrometry (ICP-MS). Spectrochim Acta 42B:39–48

Maher W, Batley GE, Lawrence I (1999) Assessing the health of sediment ecosystems: use of chemical measurements. Freshw Biol 41:361–372

Martendal E, Maltez HF, Carasek E (2009) Speciation of Cr (III) and Cr (VI) in environmental samples determined by selective separation and pre concentration on silica gel chemically modified with niobium (V) oxide. J Hazard Mater 161:450–456

Marttunen M, Suomalainen M (2005) Participatory and multiobjective development of watercourse regulation —creation of regulation alternatives from stakeholder's preferences. J Multi-Criteria Decis Anal 13:29–49

Matthews MW (2011) A current review of empirical procedures of remote sensing in inland and near-coastal transitional waters. Int J Remote Sens 32:6855–6899

Matthews W, Bernard S, Robertson L (2012) An algorithm for detecting trophic status (chlorophyll-a), cyanobacterial-dominance, surface scrums and floating vegetation in inland and coastal waters. Remote Sens Environ 124:637–652

McCormick PV, Cairns J Jr (1994) Algae as indicators of environmental change. J Appl Phycol 6:509–526

Meybeck M, Chapman D, Helmer R (eds) (1989) Global freshwater quality: a first assessment. Blackwell Reference, Oxford, p 306

Miller JR, Ritter JB (1996) An examination of the Rosgen classification of natural rivers. CATENA 27:295–299

Mishra S, Kumar A (2021) Estimation of physicochemical characteristics and associated metal contamination risk in river Narmada, India. Environ Eng Res 26 (1):190521. https://doi.org/10.4491/eer/2019/521

Montgomery RH, Reckhow KH (1984) Techniques for detecting trends in lake water quality. Water Resour Bull 20(1):43–52 (Paper No. 82105)

Nandi I, Tewari A, Shah K (2016) Evolving human dimensions and the need for continuous health assessment of Indian rivers. Curr Sci 263–271

Ning SK, Chang N (2002) Multi-objective, decision-based assessment of a water quality monitoring network in a river system. J Environ Monit 4 (1):121–126

Norris RH, Norris KH (1995) The need for biological assessment of water quality: Australian perspective. Aust J Ecol 20:1–6

Nouri H, Glenn E, Beecham S, ChavoshiBoroujeni S, Sutton P, Alaghmand S, Noori B, Nagler P (2016) Comparing three approaches of evapotranspiration estimation in mixed urban vegetation, field-based, remote sensing-based and observational-based methods. Remote Sens 8(6):492

Odermatt D, Giardino C, Heege T (2010) Chlorophyll retrieval with MERIS Case-2-Regional in perialpine lakes. Remote Sens Environ 114:607–617

Osborne LL, Kovacic DA (1993) Riparian vegetated buffer strips in water-quality restoration and stream management. Freshw Biol 29:243–258

Overeem I, Hudson BD, Syvitski JPM, Mikkelsen AB, Hasholt B, Brooke MRV, Noel BPY, Morlighem M (2017) Substantial export of suspended sediment to the global oceans from glacial erosion in Greenland. Nat Geosci 10:859–863 (Macmillan Publishers Limited, part of Springer Nature)

Oyama Y, Mitsushita B, Fukushima T (2015) Distinguishing surface cyanobacterial blooms and aquatic macrophytes using Landsat/TM and ETM + shortwave infrared bands. Remote Sens Environ 157:35–47

Paerl HW, Huisman J (2009) Climate change: a catalyst for global expansion of harmful cyanobacterial blooms. Environ Microbiol Rep 1:27–37

Pereira-Sandoval M, Urrego EP, Ruiz-Verdú A, Tenjo C, Delegido J, Sòria-Perpinyà X, Vicente E, Soria J, Moreno J (2019) Calibration and validation of algorithms for the estimation of chlorophyll-a concentration and Secchi depth in inland waters with Sentinel-2. Limnetica 38(1):471–487

Philp JC, Whitely AS, Ciric L, Bailey MJ (2005) Monitoring bioremediation. In: Atlas RM, Philp J (eds) Bioremediation: applied microbial solutions for real-world environmental cleanup. American Society for Microbiology (ASM) Press, Washington, DC, pp 237–292

Plafkin JL, Barbour MT, Porter KD, Gross SK, Hughes RM (1989) Rapid bioassessment protocols for use in streams and rivers: benthic macroinvertebrates and fish. United States Environmental Protection Agency, Washington, DC, EPA/440, pp 4–89

Raju KS, Duckstein L, Arondel C (2000) Multicriterion analysis for sustainable water resources planning: a case study in Spain. Water Resour Manag 14(6):435–456

Rapport DJ (1989) What constitutes ecosystem health? Perspect Biol Med 33:120–132

Rapport DJ (1991) Myths in the foundations of economics and ecology. Biol J Linn Soc Lond 44:185–202

Resh VH, Myers MJ, Hannaford MJ (1996) Macro invertebrates as biotic indicators of environmental quality. In: Methods in stream ecology. Academic Press, San Diego, CA, pp 647–667

Resh VH, Norris RH, Barbour MT (1995) Design and implementation of rapid assessment approaches for water resource monitoring using benthic macroinvertebrates. Aust J Ecol 20:108–121

Reynoldson TB, Norris RH, Resh VH, Day KE, Rosenberg D (1997) The reference condition: a comparison of multimetric and multivariate approaches to assess water-quality impairment using benthic macroinvertebrates. J N Am Benthol Soc 16:833–852

Richter BD, Baumgartner JV, Powell J, Braun DP (1996) A method for assessing hydrologic alteration within ecosystems. Biol Conserv 10:1163–1174

Rosgen DL (1994) A classification of natural rivers. CATENA 22:169–199

Rosgen DL (1996a) A classification of natural rivers: reply to the comments by J.R. Miller & J.B. Ritter. CATENA 27:301–307

Rosgen DL (1996b) Applied river morphology—wildland hydrology. Pagosa Springs, CO

Rosli SN, Aris AZ, Majid NM (2015) Spatial variation assessment of Malacca River water quality using multivariate statistical analysis. Malays Appl Biol 44 (1):13–18

Rottgers R, Dupouy C, Taylor BB, Bracher A, Wozniak SB (2014) Mass-specific light absorption coefficients of natural aquatic particles in the near-infrared spectral region. Limnol Oceanogr 59:1449–1460

Roy B (1996) Multicriteria methodology for decision aiding. Kluwer Academic Publishers, Boston, Dordrecht, London

Rügner H, Schwientek M, Beckingham B, Kuch B, Grathwohl P (2013) Turbidity as a proxy for total suspended solids (TSS) and particle facilitated transport in catchments. Environ Earth Sci 69(2):373–380

Satya K, Narayan C (2018) Study of physic-chemical and biological characteristics of the water of River Ganga at Patna, India. Curr World Environ 13(3). https://doi.org/10.12944/CWE.13.3.10

Schumm SA (1988) Variability of the fluvial system in space and time. In: Rosswall T, Woodmansee RG (eds) Scales and global change: spatial and temporal variability in biospheric and geospheric processes

Scrimgeour GJ, Wicklum D (1996) Aquatic ecosystem health and integrity: problems and potential solutions. J N Am Benthol Soc 15:2154–2261

Shi K, Li Y, Li L, Lu H, Song K, Liu Z, Xu Y, Li Z (2013) Remote chlorophyll-a estimates for inland waters based on a cluster-based classification. Sci Total Environ 444:1–15

Simpson J, Norris RH, Barmuta L, Blackman P (1997) Australian River Assessment System: National River Health Program predictive model manual

Singh VK, Singh KP, Mohan D (2005) Status of heavy metals in water and bed sediments of River Gomti—a tributary of the Ganga River, India. Environ Monit Assess 105:43–67

Spanò M, Gentile F, Davies C, Lafortezza R (2017) The DPSIR framework in support of green infrastructure planning: a case study in Southern Italy. Land Use Policy 61:242–250

Srinivasan V, Konar M, Sivapalan M (2017) A dynamic framework for water security. Water Secur 1:12–20

Stosnach H (2005) Trace element analysis using a bench top TXRF-spectrometer 19(2)

Sullivan C, Meigh J, Lawrence P (2006) Application of the water poverty index at different scales: a cautionary tale. Water Int 31:412–426

Sun S, Wang Y, Liu J, Cai H, Xu L (2016) Sustainability assessment of regional water resources under the DPSIR framework. J Hydrol 532:140–214

Svirčev Z, Simeunović J, Subakov-Simić G, Krstić S, Vidović M (2007) Freshwater cyanobacterial blooms and cyanotoxin production in Serbia in the past 25 years. Geogr Pannonica 11:12–21

Sweeney BW (1992) Streamside forest and the physical, chemical, and trophic characteristics of Piedmont streams in eastern North America. Water Sci Technol 26:2653–2673

Telmer K, Costa M, SimõesAngélica R, Araujo ES, Maurice Y (2006) The source and fate of sediment and mercury in the Tapajós River, Pará, Brazilian Amazon, Ground- and space-based evidence. J Environ Manag 81(2):101–113

Thoms MC, Swirepik J (1998) Environmental flow management in New South Wales, Australia. In: Harding JK, Gilvear DG (eds) Hydrology in a changing environment: ecological and hydrological interactions. Wiley, Chichester, pp 281–288

Thornton KW (1980) Reservoir sedimentation and water quality—an heuristic model. In: Proceedings of the symposium on surface water impoundments, American Society of Civil Engineers, Minneapolis, pp 61–645

Topp S, Pavelsky T, Jensen D, Simard M, Ross MRV (2020) Research trends in the use of remote sensing for inland water quality science: moving towards multidisciplinary applications. Earth ArXiv Reprints, pp 1–40

Trivedi RC (2010) Water quality of the Ganga River—an overview. Aquat Ecosyst Health Manag 13(4):347–351

Tyler A, Svab E, Preston T, Presing M (2006) Remote sensing of the water quality of shallow lakes: a mixture modelling approach to quantifying phytoplankton in water characterized by high suspended sediment. Int J Remote Sens 27(8):1521–1537

U.S. Environmental Protection Agency (2009) Factoids: drinking water and ground water statistics for 2009. U. S. EPA, Office of Water, Washington, D.C.

U.S. Environmental Protection Agency (2012) Guidelines for water reuse. U.S. Agency for International Development, Washington, D.C., pp 2–38

Varis O, Keskinen M, Kummu M (2017) Four dimensions of water security with a case of the indirect role of water in global food security. Water Secur 1:36–45

Volpe G, Colella S, Forneris V, Tronconi C, Santoleri R (2012) The mediterranean ocean colour observing system—system development and product validation. Ocean Sci 8:869–883

Walker ND (1996) Satellite assessment of Mississippi River plume variability: causes and predictability. Remote Sens Environ 58:1–35

Wan W, Wang S (2013) Areal (2D) simulation of water flood process in unit well pattern. Int J Chem Pet Sci 2 (2):1–10

Wang Y, Xia H, Fu J, Sheng J (2004) Water quality change in reservoirs of Shenzhen, China: detection using LANDSAT/TM data. Sci Total Environ 328(1–3):195–206

Watanabe F, Alcântara EH, Rodrigues T, Imai NN, Barbosa C, Rotta L (2015) Estimation of chlorophyll-a concentration and the trophic state of the Barra Bonita hydroelectric reservoir using OLI/Landsat-8 images. Int J Environ Res Public Health 12:10391–10417

Weng Q (2009) Thermal infrared remote sensing for urban climate and environmental studies: methods, applications, and trends. ISPRS J Photogramm Remote Sens 64:335–344

WHO (1991) GEMS/WATER 1991–2000, the challenge ahead, WHO/PEP/91.2. World Health Organization, Geneva

WHO (1992) GEMS/water operational guide. World Health Organization, Geneva, p 3

WHO (World Health Organization) (2006) Guidelines for drinking water quality, Geneva, vol 3, pp 1–459

Wright JF (1995) Development and use of a system for predicting macroinvertebrates in flowing waters. Aust J Ecol 20:181–197

Yacobi YZ, Gitelson A, Mayo M (1995) Remote sensing of chlorophyll in Lake Kinneret using high spectral resolution radiometer and Landsat TM: spectral features of reflectance and algorithm development. J Plankton Res 17:2155–2173

Yadav NS, Kumar A, Sharma MP (2014) Ecological health assessment of Chambal River using water quality parameters. J Integr Sci Technol 2(2):52–56

Yamanaka M, Okumura M, Nakano T (2007) Isotopic altitude effect and discharge characteristics of river water in Yakushima Island, southwestern Japan. J Jpn Assoc Hydrol Sci 37:41–54

Zeitoun M (2011) The global web of national water security. Glob Policy 2:286–296

Zende AM, Patil RA, Patil VM (2018) Water security assessment in semi-arid region using geospatial techniques. Mater Today: Proc 5(1, Part 1):620–627

Zhang Y, Ma R, Li Z (2014) Human health risk assessment of groundwater in Hetao Plain (Inner Mongolia Autonomous Region, China). Environ Monit Assess 186:4669–4684

Zhou G, Singh J, Wu J, Sinha R, Frostell B (2015) Evaluating low-carbon city initiatives from the DPSIR framework perspective. Habitat Int 50:289–299

Zhou J, Li E, Wei H, Li C, Qiao Q, Armaghani DJ (2019) Random forests and cubist algorithms for predicting shear strengths of rock fill materials. Appl Sci 9

18 Automatic Strahler's Stream Order Computing on Digital Stream Network Dataset

Debabrata Ghorai, Gouri Sankar Bhunia, and Pravat Kumar Shit

Abstract

Stream order is an important parameter in the drainage network system which supports a wide range of applications such as flood risk assessment, water resource management, flood inundation mapping, watershed management and many more. The methods that determine stream order proposed by Strahler have been selected for stream order classification. This paper introduces an automatic approach to classify the river network based on Strahler's order using Python language which is an object-oriented, general-purpose and high-level programming language. The approach is an end to end execution wherein the Python program needs only digital river network data as input and outputs the same digital river network data with added a field of Strahler's order. The tool was developed with sample river network data which was further tested on several geographical areas. The river network datasets for these areas were collected from HydroSHEDS and performed the analysis. The tool provided results for these river networks were accurately matching with the HydroSHEDS provided Srrahler's order for the river segments; hence the tool provided result is satisfactory.

Keywords

Stream order · Strahler method · Python programming language · River network · Drainage network

18.1 Introduction

Digital stream networks have traditionally been formed from digital mapping in combination with several topographical series of national maps. This map series and related data in finer and finer resolutions were generated over time. The stream network has defined various forms of dendritic or network, radial or clustered, longitudinal, linear, deranged, centripetal and centrifugal stream network (Pradhan et al. 2012). The formation of these drainage networks depends upon the morphological aspect of the terrain, such as slope, varied resistance of rocks and its geological and geomorphological history (Garde 2011). Research of the drainage network entails evaluating the topological structure of the streams within a drainage basin (Lindsay et al. 2019). In general, the digital stream data used in Geographic Information

D. Ghorai
Tajpur, Paschim Medinipur, West Bengal, India

G. S. Bhunia (✉)
Department of Geography, Seacom Skill University, Santiniketan, West Bengal, India

P. K. Shit
Department of Geography, Raja N L Khan Women's College (Autonomous), Midnapore, West Bengal, India

P. K. Shit et al. (eds.), *Drainage Basin Dynamics*, Geography of the Physical Environment,
https://doi.org/10.1007/978-3-030-79634-1_18

System (GIS) are extracted from existing maps, aerial photos and satellite imagery by digitization of water lines or can be inferred instantaneously from digital elevation models (DEMs). Analysis of vector-based drainage networks is a practical choice when finer-resolution DEM is not available or when operating on very large geographical scales. GIS is widely used for evaluating drainage network tasks. There are multiple methodologies of drainage ordering designed by various scholars, namely, Horton (1945), Strahler (1952), Scheidegger (1965) and Shreve (1967) for associating orders with stream segments (Pradhan et al. 2012).

Lanfear (1990) was one of the first scholars to define the usage of GIS vector analytical techniques for topological drainage network operations. Dawson et al. (2002) defined a comprehensive GIS framework that can extract various topological features of large-river systems. Pradhan and Ghose (2012) have developed a spiral crossing technique for the conduct of vector hydrography research. Traditional fluvial ordering techniques or methods require large amounts of time, expenses and effort. That is because as contrasted with manual processes, the computerized approaches of GISbased investigation are effective and consistent (Dawson et al. 2002). Developing automated processes would significantly reduce investments and also improve the quality and reliability of the results (Pradhan et al. 2012). Many Algorithms likewise presume that the vertical start and end of each function of a vector stream file is balanced by stream network nodes (e.g., confluences, diffluences, and channel heads) and that the relations bring into line end to end precisely. Therefore, many of the current drainage network vector analysis methods are best defined as semi-automated, and substantial handheld preprocessing of the input data sets can be needed when they were not prepared to meet the requirements of the techniques. This paper intends to address the Strahler method for stream order classification automatically using the advantage of Python programming language. The projected work mechanically identifies the river channel from the drainage network and assigns Strahler's stream order number.

River network is required for geomorphological and hydrological application, hence understanding topological structure (river ordering) is a growing concern of the river networks (Ranalli and Scheidegger 1968; Carraro et al. 2020). River ordering is generally describing the component of a watershed and establishing a relationship of its physical and biological attributes (CAE 2020). This network is very important for hydrologists, geographers, geologists and GIS specialists in order to measure the size of the waterways and find the probable flood extent for disaster management. Apart from this, it also helps biogeographers/biologists to link up their findings (types of life present) from the river to the stream segments for further analysis. This network is also using in many other applications like water resource management, flood hazard loss estimation and topographical analysis. Stream ordering methods are established by various researchers in the recent past. The details of these methods are presented in the following articles such as Hodgkinson (2009) and Jasiewicz (n.d.). Strahler's method is a simple hierarchical-based mathematical ordering system within a watershed and very popular in the recent times (Hughes et al. 2011; Pradhan et al. 2012; Gleyzer et al. 2004; Prem et al. n.d.; Chougale and Sapkale 2017).

18.2 Strahler's Stream Ordering

The stream order hierarchy was projected by Arthur Newell Strahler in 1952 in his article "Hypsometric (Area Altitude) Analysis of Erosional Topology." This method is the modification of Horton's order wherein it fixes the ambiguity of Horton's ordering method (Jasiewicz n.d.). In Strahler's ordering the main channel is not determined; instead the ordering is based on the hierarchy of tributaries. Strahler says that the first-order stream is the smallest affluent or beginning segment in the river. After that, a second-order stream is formed when the two initial streams meet in another downstream segment. Similarly, when two second-order streams combine to a downstream segment, they form a third-order stream and so on. However, if a first-order stream joins a second-order stream, the latter remains a second-order stream. It is not until one stream combines with another stream of the same order that the resulting stream increases by an order of magnitude. Instead, the maximum order stream should be maintained in the hierarchy until it meets the same order stream. For instance, if a second-order drainage channel integrates with a third- order drainage channel, the second-order stream just ends with streams of the third order which keeps the hierarchy (Briney 2020) (Fig. 18.1).

18.3 Materials and Method

The primary input for this study is digital river network data. This data was generated automatically from the SRTM 90 m Digital Elevation Model (DEM) in ArcGIS (9.0) environment for Vaippar and other sub-basin regions in Tamil Nadu. A well connected, proper flow direction and topological error-free river segments (Fig. 18.2) were selected from the generated river network dataset for the development of the automatic Strahler's order classification tool in Python programming language.

In order to resolve the stream code automatically, the sample network is first represented by segments and nodes. The interaction between segment and node forms the basis for the determination of other segments. Our techniques dictate that, in any iteration, we must consider the outer branches of the stream network. The method consisting of the development of the following function in Python language such as (i) From-Node and To-Node ID generation (Func_FromToNode), (ii) upstream to downstream segment's connectivity point identification (Func_Junction), (iii) group downstream segments with same Strahler's order (Func_GroupSegments) and (iv) assign Strahler's stream order number to the group (Func_AssignNumbers).

18.4 Results and Discussion

River networks are modeled as streams and nodes in the form of an acyclic graph. Linkages are streams flowing between upstream and downstream network nodes. The channel heads (i.e., the upstream largest of a channel), confluences (where two streams combine into one), diffluences (with one stream diverging into another) and an outlet are part of the stream network's nodes. The vector line information used for river networks in GIS is similar to its natural analogs. The vector line data consists of arcs connected by straight lines, which consist of a set of x–y coordinates (i.e., vertices). The pseudocode of Python programming language and derivative output of stream network is described below.

Fig. 18.1 Strahler's ordering method

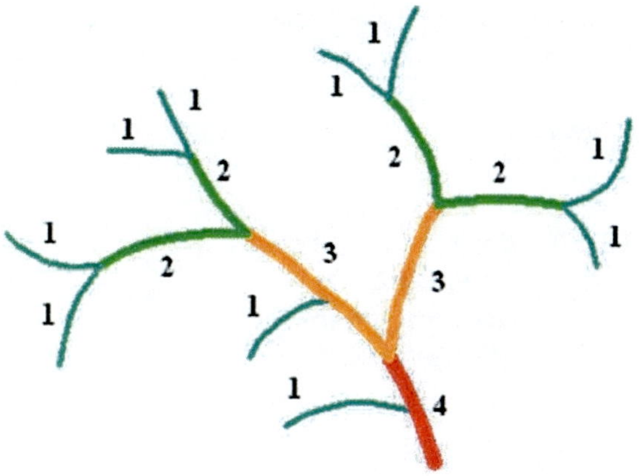

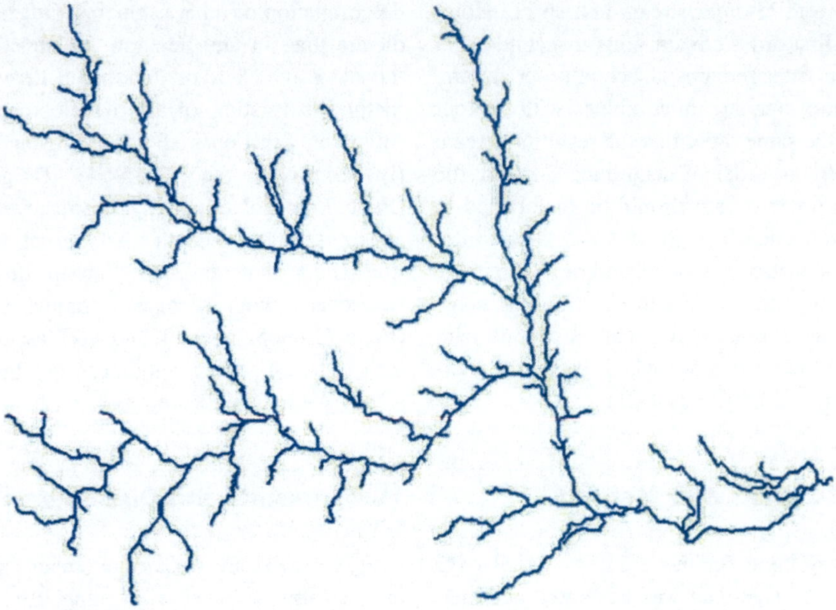

Fig. 18.2 River network vector dataset

18.4.1 Define the Function 'Junction1' with an Argument 'FTNodeList'

From-Node and To-Node of a river segment represent the start-point and end-point of the segment. Here, From-Node is identified by a given segment ID (or OBJECTID) and To-Node is identified by the next segment ID which is connected to the given segment's end-point. This approach was used to generate From-Node and To-Node ID for the entire river network (Fig. 18.3).

 # Pseudocode

18.4.2 Define the Function 'Junction2' with an Argument 'HigherOrder1'

 # Pseudocode

18.4.3 Define the Function 'GroupSegment' with Two Arguments 'HigherOrder1' and 'JunList3'

From-Node and To-Node ID of the river network is the building block for the next function. These IDs are used for the river network's Strahler's order classification. The following function, such as Func_Junction and Func_GroupSegments is built on top of Func_FromToNode and these processes are repeated until unless all the river segments are classified properly (Fig. 18.4). Func_Junction can identify all downstream junction segments which are connected by two or more upstream To-Nodes. This function can identify all the downstream segments which are connected by upstream To-Nodes at least once (Fig. 18.5).

 # Pseudo code—First Order
 # Pseudo code—Second Order

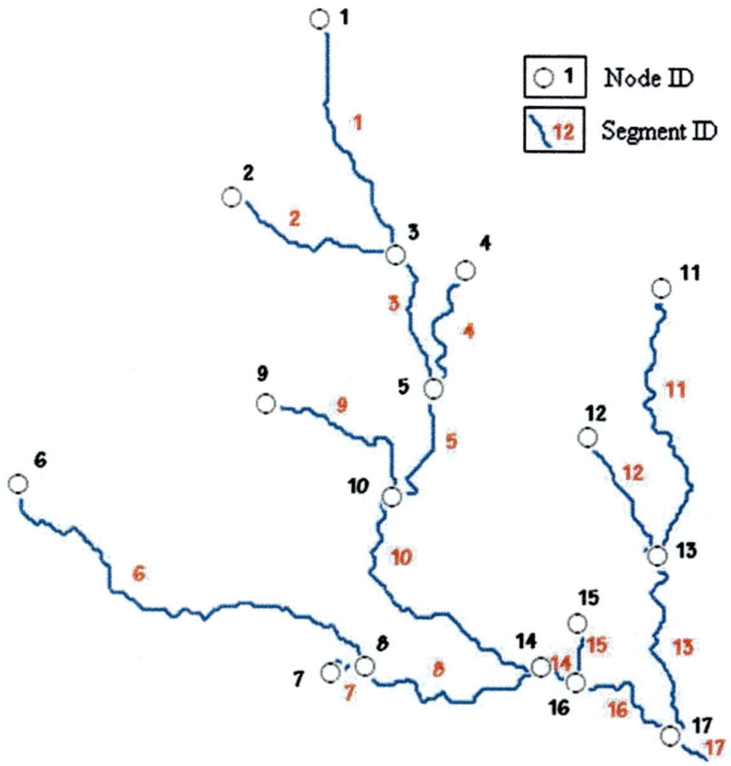

Fig. 18.3 Representing From-Node, To-Node and Segment ID through Python programming

18.4.4 Identify Stream Order Third and Above

The beauty of Func_GroupSegments is that it efficiently identifies river segment ID which is followed up the same type of Strahler's river order and keeps them in a separate list as a group. While grouping these segments above function are iterating "n" number of times until unless all the same Strahler's order stream ID has been found (Fig. 18.6).

 # Pseudo Code

18.4.5 Update Attribute of Stream Order

According to Strahler's stream order, a first-order stream defines those streams whose From-Node has no confluences or else, from where the river starts flowing downwards. All the connected downstream segments were identified and append

Fig. 18.4 Derivation of first and second-order stream network through Python programming

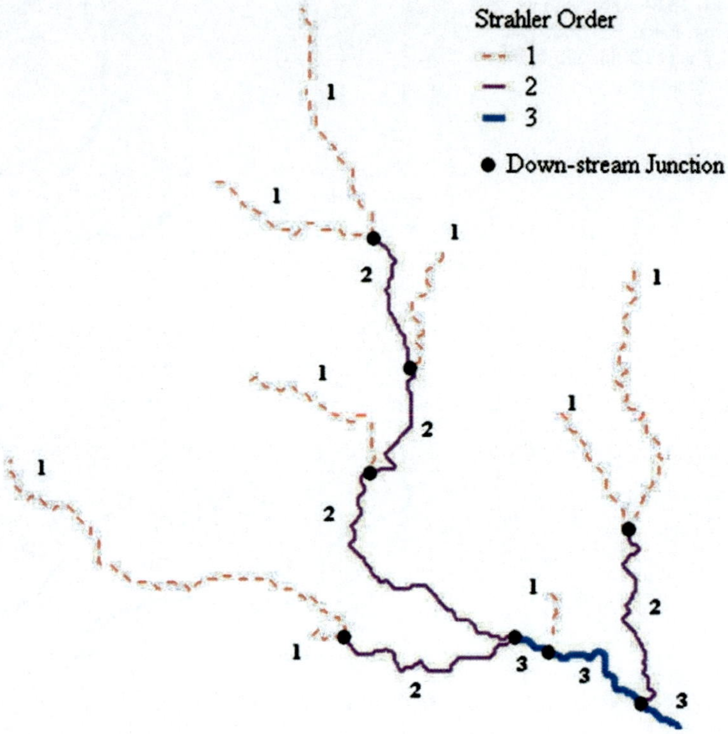

Fig. 18.5 Stream segment's junction through Python programming

to a temporary memory space (say, List-A) using Func_Junction. List-A holds stream ID was temporarily removed from stream networks from/To Node ID list and the remaining stream ID was appending to another temporary space (say, List-B). It has been found that if List-B stream ID has no connectivity at From-Node, then it is a first-order stream segment of the river network which is further coded as "1". Once the first-order stream is identified, Func_Junction was a sprint on List-A to identify all downstream junctions ID and then run Func_GroupSegments identify all upstream ID. In this process two lists of segment ID returned, one is for second-order stream segment which is further coded as "2" and another segment which is considered for higher-order is store in temporary memory space (say, List-C). All these functions are re-run in the same sequence within an infinite loop to identify all higher-order streams from List-C which is starting from 3rd order and finishing at nth higher-order (Fig. 18.7). The infinite loop is automatically stopped when it gets the final river order segment.

 # Pseudo Code

The highest Strahler's stream order was 4th for the selected river network (Fig. 18.8) which was visually verified on the GIS platform and it was found that stream order classification by the above program is correct. The program was also tested with large drainage network datasets of a wider geographical region that had more than ten thousand stream segments. The classification result was accurate. The authors also measure the time of completion for the huge data classification and found a satisfactory result. The program took a couple of minutes to classify the said drainage network.

18.4.6 Validation of Result

HydroRIVERS Version 1.0 (Lehner and Grill 2013) was downloaded from HydroSHEDS (https://www.hydrosheds.org/) for the Asia region. The full form of HydroSHEDS is Hydrological data and maps based on SHuttle

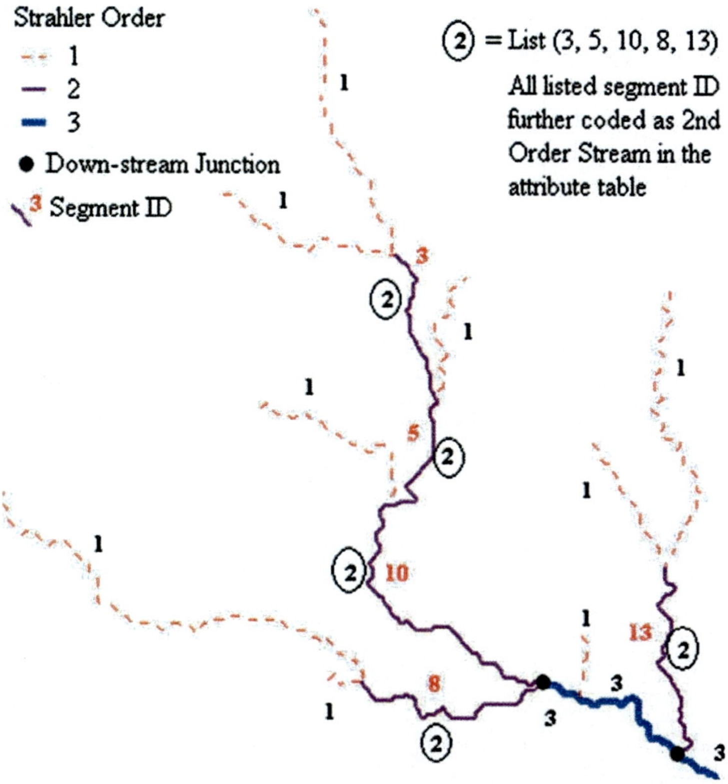

Fig. 18.6 Same type of Strahler's order segment ID grouping into a list

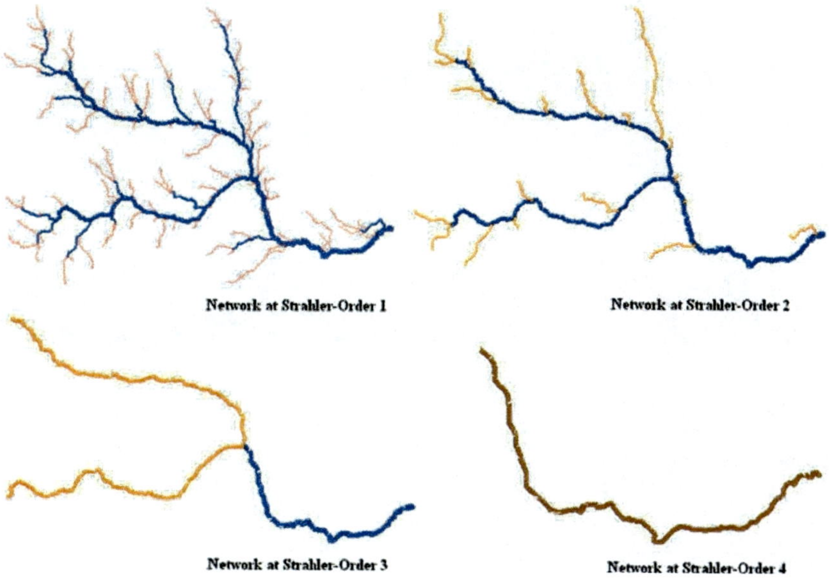

Fig. 18.7 Strahler-order numbering

Elevation Derivatives at multiple Scales (Lehner et al. 2008). The River network dataset for these areas were extracted from the river network dataset of Asia. Strahler's stream order classification was performed for these networks with the proposed tool to test the accuracy based on true classification of the river order (source data attributes) and performance with respect to time and system configuration. This tool was executed on the following system configuration such as 4 GB RAM, Intel® Core™ i3-3217U Processor and 64-bit OS (Operating System). The details of the results are presented in Table 18.1 for all the study areas.

It has been observed from the above table that the time to accomplish the result is varying from region to region which is depended on the count of the river segments. When river segment count is less for a region, the time taken is also less and vice versa.

18.5 Conclusions

The proposed algorithm correctly assigns Strahler's stream order number when there is a complex topologically surface such as braided streams or multiple drainage outlets. It also tremendously reduces human effort and important time for manual Strahler's order classification of the drainage network. The method is easy to implement and programming and does not require accuracy checking if the input river network is error-free and node to node connected. It is anticipated that this technique will be especially useful in the topological analysis of dense

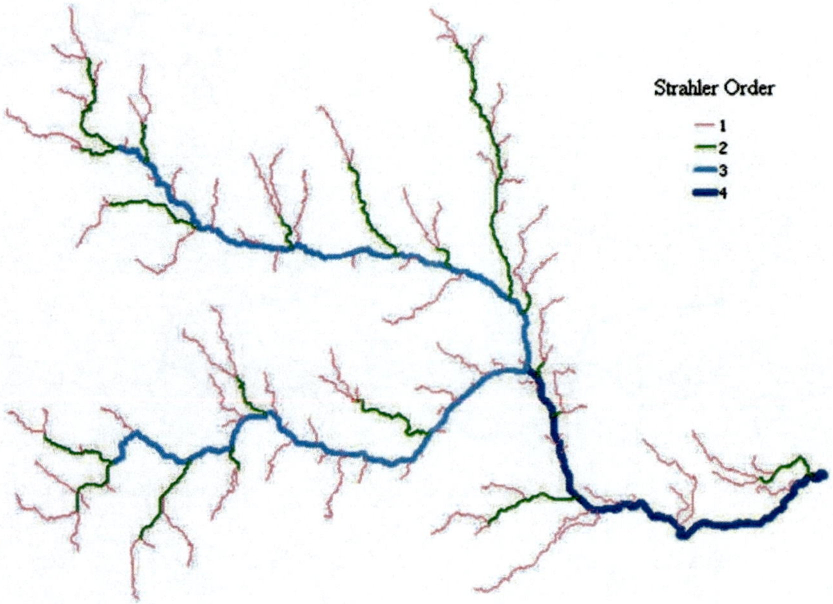

Fig. 18.8 Strahler-order classification

19 Morphometry-Based Subwatershed Prioritization for Flood Potentiality Analysis of the Gumani River Basin (India) Using TOPSIS

Sadik Mahammad, Md. Mofizul Hoque, and Aznarul Islam

Abstract

Flood is an extreme environmental threat amongst the natural hazards that affects the livelihood of millions of people across the world, especially in the flood plains and deltas. The conservation measures should be implemented to prevent the unlikely loss based on the physical, socio-economic, and cultural characteristics of a region. The present paper aims to analyze the morphometric parameters, and therefore, to find out the prioritization of the sub-watersheds of the Gumani River Basin (GRB) to induce flood hazards in the study area. The study area having an area of ~ 1307 km^2 is divided into 34 sub-watersheds. The Shuttle Radar Topographic Mission (SRTM) Digital Elevation Model (DEM) (30 m) and Survey of India (SoI) topographical maps (1:50,000) have been used to delineate the sub-watersheds and assess the morphometric parameters. The two linear (stream order and mean bifurcation ratio), seven areal (form factor, elongation ratio, circularity index, drainage texture, stream frequency, drainage density, and compactness coefficient), and five relief (basin relief, relief ratio, ruggedness number, slope, and hypsometric integral) aspects have been measured in this study. The relative weights of the sub-watersheds have been calculated using the Technique for Order of Preference by Similarity to Ideal Solution (TOPSIS) which ranks the alternative based on the closest distance to the ideal solution and the farthest distance to the negative ideal solution. The produced performance score of the sub-watersheds has been categorized into five classes such as very low (0.36–0.39), low (0.39–0.44), moderate (0.44–0.49), high (0.49–0.55), and very high (0.55–0.73). The very low class includes the five sub-basins, while the very high class consists of the three sub-basins. The low priority class is confined to the twelve sub-basins, while the high class encompasses the five sub-basins. The medium class includes the ten sub-basins of GRB. The present study would address the micro-level planning and development activity of the study area.

Keywords

Micro-level planning · Sub-watersheds · Flood hazards · Relative weights · Basin morphometry

S. Mahammad · Md.Mofizul Hoque · A. Islam (✉)
Department of Geography, Aliah University, 17
Gora Chand Road, Kolkata 700014, India

P. K. Shit et al. (eds.), *Drainage Basin Dynamics*, Geography of the Physical Environment,
https://doi.org/10.1007/978-3-030-79634-1_19

19.1 Introduction

Flood, a hydro-meteorological hazard is probably the more devastating, widespread, and frequent among all the natural hazards (Sanyal and Lu 2004) causing the highest deaths and property damages (CEOS 2003). The conservation measures and mitigation should be implemented to prevent the devastating loss based on the physical, socio-economic, and cultural characteristics of a region. Therefore, the proper strategy for the optimum uses of natural resources is a crucial need for the sustainable development of water resources. Watershed, a catchment area from which all water drains into a common point, is an ideal unit for the management of land and water resources and mitigation of the impact of natural disasters to achieve sustainable developments (Ratnam et al. 2005; Rahaman et al. 2015). Subwatershed prioritization is the ranking of subwatershed of a principal watershed according to the order in which they are treated for soil or water conservation measures or to improve crop productivity (Wani et al. 2011). It plays a crucial role to identify the flood vulnerable zone within the drainage basin and helps to adopt the valuable strategy for the mitigation of flood hazard. Moreover, subwatershed prioritization is also useful in the ungagged river basin where no or limited data is available (Akay and Koçyiğit 2020).

Morphometric analysis in the form of linear, areal, and relief aspects provides the hydrological behaviour of the subwatershed (Balasubramanian et al. 2017). Morphometric analysis can be utilized to derive the characteristics of channel networks of the different drainage basins and their comparison and find out the effects of different sub-surface and climatic variables (Kale and Gupta 2010). Morphological study plays an important role in the prioritization of subwatershed without considering soil maps (Biswas et al. 1999; Thakkar and Dhiman 2007). Morphometric analysis of the drainage basin can be assessed from the traditional dataset such as topographical map and field observations which is time-consuming and cumbersome. Remote sensing (RS) and geographic information system (GIS) provide efficient tools to determine the morphometric parameters of the drainage basin and the prioritization of the sub-basins (Aher et al. 2014; Mahammad and Islam 2021a). Horton (1932, 1945) carried out pioneering works on drainage basin morphometric analysis. However, Horton's law has been updated and applied by many researchers throughout the world (Strahler 1952, 1957, 1968; Smith 1950; Miller 1953; Chorley et al. 1957; Clarke 1966; Dury 1952; Gregory and Walling 1968). Prioritization of the drainage basin has been carried out by numerous researchers in different approaches across the world based on the morphometric parameters of the drainage basin (Biswas et al. 1999; Ratnam et al. 2005; Javed et al. 2009; Yogesh et al. 2016; Farhan and Anaba 2016; Sahu et al. 2018). Many enthusiastic researchers have attempted to analyze the effects of disastrous flood and flash flood hazards using morphometric parameters of a drainage basin (Omran et al. 2011; Aher et al. 2014; Elewa et al. 2016; Islam and Deb Barman 2020).

The Multi-criteria decision-making (MCDM) method has been widely used throughout the world in different fields concerning the problem of having multiple conflicting criteria in decision-making. The researchers have used various MCDM methods for watershed management including soil as well as land conservation, plant growth supervision, and water resource management (Hembram and Saha 2020; Mahammad and Islam 2021b). Several MCDM techniques such as Analytical Hierarchic Process (AHP) (Chowdary et al. 2013; Rahmati et al. 2016; Jaiswal et al. 2014), Frequency ratio (FR) (Samanta et al. 2018), Weighted linear combination (WLC) method (Sujatha and Sridhar 2019), Technique for Order of Performance by Similarity to Ideal Solution (TOPSIS) (Aouragh and Essahlaoui 2018; Nitheshnirmal et al. 2019; Amiri et al. 2019), Fuzzy Analytical Hierarchy

Process (FAHP) (Jaiswal et al. 2015; Meshram et al. 2019), have been widely used for the prioritization of the subwatershed. Many of the researchers have also used comparative analysis to compare the outcomes from the different models and define the efficient results from them such as TOPSIS and Weighted Aggregated Sum Product Assessment (WASPAS) (Bid and Siddique 2019), Simple Additive Weighting (SAW) and TOPSIS (Meshram et al. 2020), Compound Factor (CF), AHP and TOPSIS-AHP (Sadhasivam et al. 2020). In the present study, TOPSIS techniques have been used for the prioritization of the subwatershed.

The state of Jharkhand experienced climatic variability and climatic extremes in terms of maximum temperature, minimum temperature, rainfall, and solar radiation from 1984 to 2014 (Tirkey et al. 2018). In the north-eastern part of the state, the Sahibganj district that mostly covers the present study area was affected by the devastating flood since 1995 (District Disaster Management Plan of Sahebganj 2017). In the recent past, from 2007–08 to 2015–16, the devastating flood hit the district four times (2007–08, 2011–12, 2012–13, and 2015–16) causing the crop damage of Indian National Rupees (INR) 63,534,500 and affecting 718,026 people (District Disaster Management Plan of Sahebganj 2017). In the present study area, various works related to geotectonic geomorphology, historical geography, groundwater potentiality and deforestation probability (e.g. Ball 1877; Rao et al. 2003; Khan 1987; Singh and Singh 1996; Sanyal and Sengupta 2012; Bhattacharji 2012; Mahammad and Islam 2021c; Saha et al. 2020) have been done. However, a review of the previous literature depicts that there is a research gap related to flood potentiality in the Gumani River basin using TOPSIS. Therefore, the main objectives of the present study are as follows:

1. To analyze the basin morphometry of the Gumani River basin in terms of linear, areal, and relief aspects, and
2. To attempt the prioritization of the sub-basin based on the basin morphometry concerning the flood potentiality using TOPSIS.

19.2 Study Area

The study area, the Gumani river basin (GRB) is located on the north-eastern part of Jharkhand state, India, comprising some parts of the three districts (Sahibganj, Godda, and Pakur) (Fig. 19.1). It extends from 24° 37′ 20″ N to 25° 13′ 20″ N latitude and 87° 21′ 14″ E 87° 54′ 40″ E longitude covering the area ~1307 km^2. The topographic elevation of the study area ranges from 20 to 590 m above mean sea level (MSL). The Gumani is the main river of the study area originating from the Chotonagpur plateau region flows towards the north-east direction. Morang River, the main tributary of the Gumani River, flows from North to south and joins the Gumani River at Barhait. The combined flow run towards the east and finally falls into the river Ganga. The drainage pattern of the study area is basically dendritic.

The general geology of the basin is composed of different types of igneous, sedimentary, and metamorphic rock with various geological formations in different geological periods (Fig. 19.2). Broadly the geology of the lower Gumani basin is quite distinct from the middle and upper basin. The upper and middle basin is more complex than the lower basin from a geological perspective. The lower basin is more or less wholly covered with a thick veneer of older alluvium over the solid rocks of Tertiary age. The older flood plain depositions with alternative layers of sands, silts, clays, and unoxidized older alluvium soils are found in the lower edge of the basin, deposited in the Holocene age. And Khaki green clay with brown silts and sands, highly oxidized, impregnated with caliche nodules older alluvium soils are found in the lower and middle of the basin deposited in the late Pleistocene to early Holocene period. Cainozoic (Undiffenciated) laterite and lateritic soils are scattered in the middle and upper basin. More than half of the basin area cover by the Rajmahal trap with intertrappean beds of basalt, claystone, siltstone tuff, and trap dykes that formed in the Upper Jurassic to cretaceous periods. The Dubrajpur formation of the Gondwana supergroup over the Archean

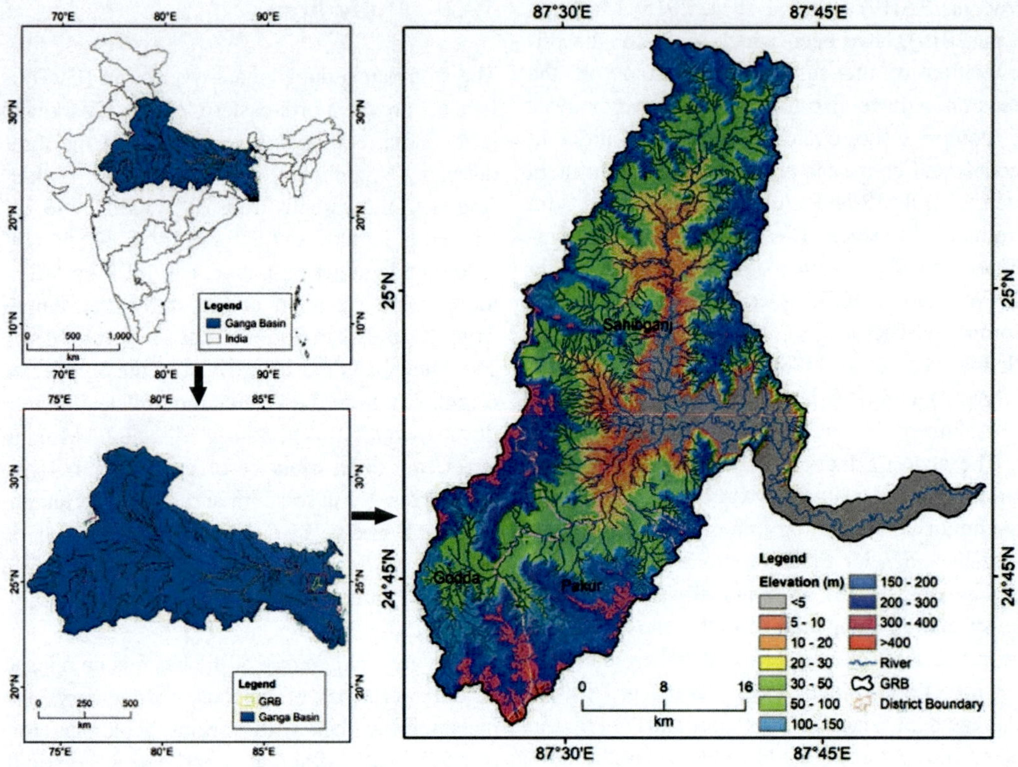

Fig. 19.1 Location of the study area

basement is the sandstone and shale in the upper basin that formed in Triassic to Jurassic periods. And another Gondwana supergroup formation is the Barakar formation. This formation is consisted of the siltstone, sandstone, and carbonaceous with coal seams in the upper part of the basin that formed in the Permian period. The Chhotanagpur gneissic complex composite with unclassified granite gneiss with enclaves of metamorphic or granite gneiss and migmatites is found in the upper basin formed in the Archaean to Proterozoic periods. And metamorphic of Chhotanagpur mica belt is found over the Archean basement in the south-west edge of the basin. Besides that Eastern Ghat supergroup formation is also found in the upper basin.

The soil types of the study area characterized by mainly fine, fine loamy, and loamy skeletal in texture. The GRB is characterized by humid to sub-humid climate with a normal average rainfall of about 1345 mm and maximum temperature recorded up to 45 °C during March to May and minimum temperatures up to 7 °C in December. The area receives rainfall by South-West monsoon.

19.3 Datasets and Methodology

19.3.1 Datasets

In the present study, quantitative information of basin parameters, as well as thematic layers, have been applied from both RS data conventional data in the GIS domain. The study utilized SRTM DEM with 30 m resolution downloaded from USGS earth explorer (https://earthexplorer.usgs.gov/) to delineate the basin relief and slope

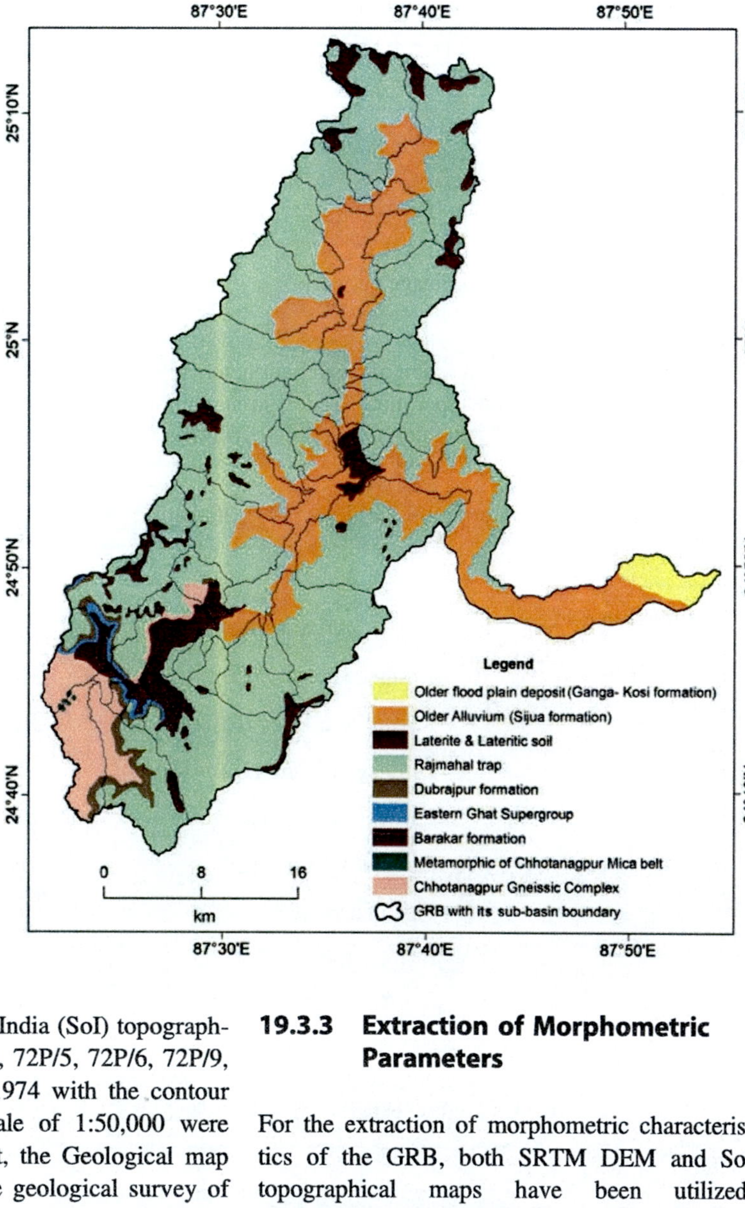

Fig. 19.2 Geological configuration of GRB

properties. The Survey of India (SoI) topographical maps (72O/8, 72O/12, 72P/5, 72P/6, 72P/9, 72P/10, and 72P/13) of 1974 with the contour interval of 20 m and scale of 1:50,000 were collected. Apart from that, the Geological map was downloaded from the geological survey of India.

19.3.2 Methodology

In the present study, to fulfil the objectives a robust and integrated methodology has been adopted from data collection to sub-basin prioritization (Fig. 19.3).

19.3.3 Extraction of Morphometric Parameters

For the extraction of morphometric characteristics of the GRB, both SRTM DEM and SoI topographical maps have been utilized. The SRTM DEM has been corrected using the fills tool in ArcGIS 10.4.1 software to avoid the local effect on elevation. In the GIS environment, the corrected DEM has been employed to delineate the basin boundary of the Gumani River and its sub-basins using the hydrology toolbox of ArcGIS 10.4.1 software. In the delineations of the sub-basins boundary of GRB

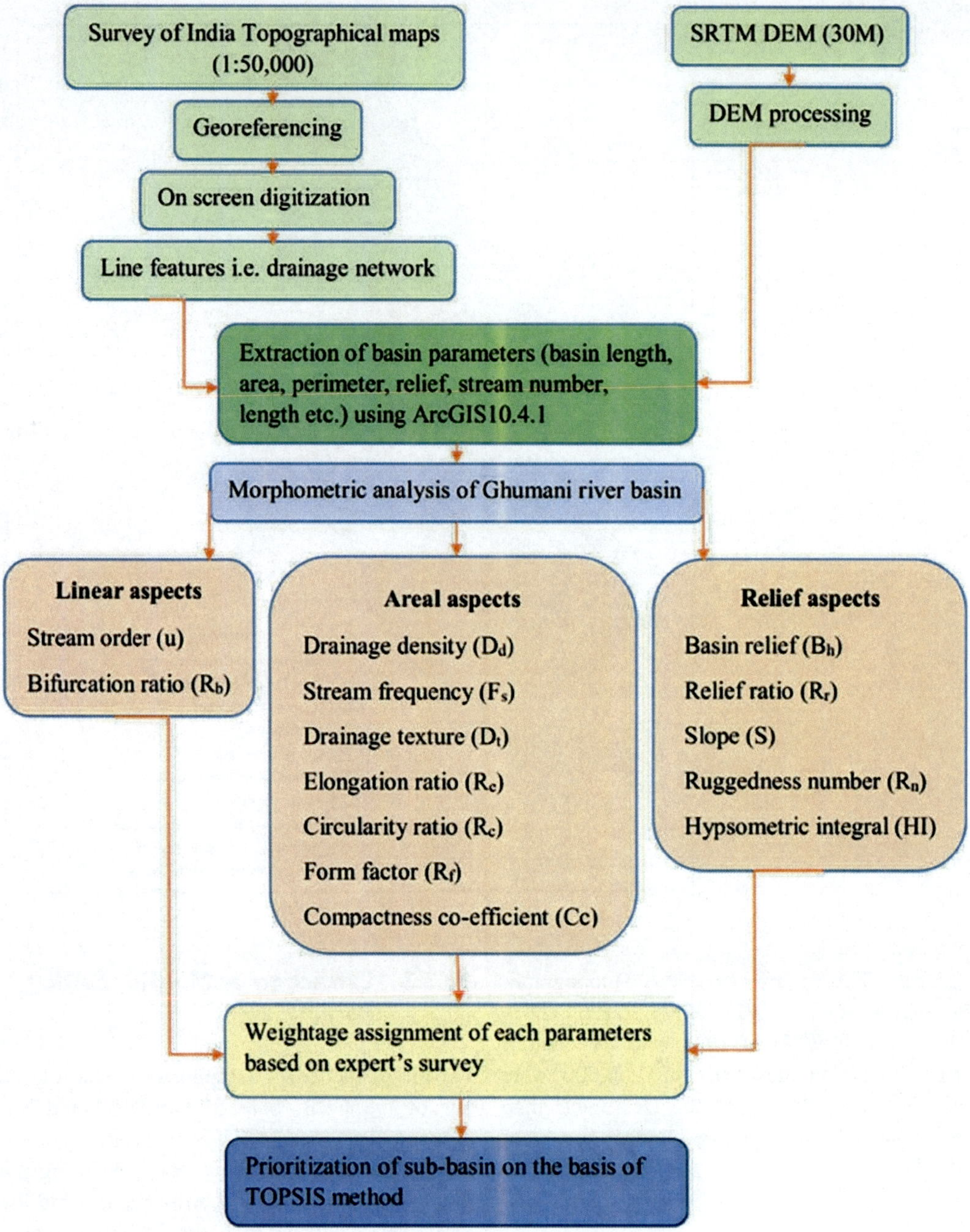

Fig. 19.3 The methodological procedure of the study

minimum, the third-order stream has been taken into consideration. All the topographical maps, geological maps, and soil maps have been georeferenced by adding the longitudinal and latitudinal values of ground control points (GCPs) with a total root mean square (RMS) error of <0.02 for better accuracy. The projection systems of the topographical maps and conventional maps have been defined as Projected Coordinate system (PCS) with reference to WGS

1984, Zone 45 N, and finally rectified in IMAGINE Image format. All the rectified topographical maps are used to make a combined map. The drainage network of the basin, as well as all sub-basins, has been digitized from the topographical maps. Apart from that, the relief and slope of the sub-basins have been extracted from the DEM and are applied to formulate the relief aspects of the sub-basins. For the present study area, 14 morphometric parameters in which two linear aspects (Stream order and mean bifurcation ratio), seven area aspects (form factor, elongation Ratio, circularity ratio, drainage texture, drainage density, Stream frequency, and compactness coefficient), and five relief aspects (basin relief, relief ratio, doggedness number, slope, and hypsometric integral) have been taken to define the geometric characteristics of the drainage basin. The calculations of the parameters have been done using the formula expressed in Table 19.1.

19.3.4 Morphometric Indices

The present study utilized the morphometric parameters in terms of linear, areal, and relief aspects for the prioritization of the drainage basin using the following indices illustrated in Table 19.1.

Table 19.1 Morphometric parameters

Morphometric parameters	Formulas	References
Stream order (u)	Strahler method	Strahler (1964)
Bifurcation ratio (R_b)	$R_b = N_u/N_{u+1}$, where, N_u = No. of Streams of a given order, N_{u+1} = No. of Streams of next higher order	Schumm (1956)
Mean Bifurcation ratio (RB)	RB = Average of bifurcation ratio of all order	Strahler (1957)
Form factor (Rf)	$R_f = A/L^2$, where, A = Area of the basin (km^2), L = Basin length (km)	Horton (1932, 1945)
Elongation ratio (Re)	$R_e = 1.128\sqrt{(A/L^2)}$, where, A = Area of the basin (km^2), L = Basin length (km)	Schumm (1956)
Circularity index (Rc)	$Rc = 4\pi A/P^2$ A, where, A = Area of the basin (km^2), P = Perimeter (km), $\pi = 22/7$	Miller (1953) Strahler (1964)
Drainage texture (D_t)	$D_t = D_d \times F_s$, where, D_d = Drainage density, $F_{s=}$ Stream frequency	Horton (1945)
Stream frequency (F_s)	$F_s = \Sigma N_u/A$, where, ΣL_u = Total no. of streams of all order (km), A = Area of the basin (km^2	Horton (1932, 1945)
Drainage density (D_d)	$D_d = \Sigma L_u/A$, where, ΣL_u = Total length of streams of all order (km), A = Area of the basin (km^2)	Horton (1932, 1945)
Compactness Co-efficient (Cc)	$C_c = 0.2821\ P^{0.5}$, where P = Perimeter (km), A = Area of the basin (km^2	Gravelius (1914)
Basin relief (B_h)	$B_h = H - h$, where, H = Maximum elevation (m), h = Minimum elevation (m) of the basin	Hadley and Schumm (1961)
Relief ratio (R_r)	$R_r = B_h/L$, where, B_h = Basin relief (m), L = Basin length (km)	Schumm (1956)
Slope (S)	From SRTM DEM	Arc GIS 10.3.1
Ruggedness number (R_n)	$R_n = B_h \times D_d$, where, B_h = Basin relief (m), D_d = Drainage density	Schumm (1956)
Hypsometric integral (HI)	$HI = \Sigma X_i \cdot Y_{i+1} - \Sigma X_{i+1} \cdot Y_i/2$	Strahler (1952)

19.3.5 TOPSIS

The Technique for Order of Performance by Similarity to Ideal Solution (TOPSIS), a multi-criteria decision-making (MCDM) technique was first presented by Hwang and Yoon (1981) and further developed by Yoon (1987) and Hwang et al. (1993). It chooses the alternatives from the data sets based on the concept that the alternatives should have a minimum distance from the positive ideal solution and maximum distance from the negative ideal solution. The steps of the TOPSIS method are illustrated using the following equations (Roszkowska 2011).

Step 1: Formation of decision matrix

The multi-criteria problems can be expressed in the following matrix format:

$$\begin{bmatrix} & C_1 & C_2 & \cdots & C_n \\ A_1 & x_{11} & x_{12} & \cdots & c_{1n} \\ A_2 & x_{21} & x_{22} & \cdots & c_{2n} \\ \vdots & \vdots & \vdots & \ddots & \vdots \\ A_m & x_{m1} & x_{m2} & \cdots & x_{mn} \end{bmatrix} \quad (19.1)$$

where

$A_1, A_2, \ldots, A_m$ are the positive alternatives that decision maker have to choose from,

C_1, C_2, …,C_n are the criteria for which the alternative performance is measured,

x_{ij} is the rating of alternatives A_i with respect to the criterion C_j.

Step 2: Calculation of normalized decision matrix

The vector normalization has been done using the following formula (Eq. 19.2):

$$\overline{X}_{ij} = \frac{X_{ij}}{\sqrt{\sum_{i=1}^{n} X_{ij}^2}} \quad (19.2)$$

Step 3: Calculation of weighted normalized matrix

To get the weighted normalized value, the weight of each criterion is multiplied by the normalized value of each cell (Eq. 19.3).

$$V_{ij} = \overline{X}_{ij} \times W_j \quad (19.3)$$

where V_{ij} is the weighted normalized value and W_j is the weight of ith criterion.

Step 4: Determining the positive and negative ideal solution

The positive ideal solution can maximize the benefit criteria and minimize the cost criteria, whereas the negative ideal solution can maximize the cost criteria and minimizes the benefit criteria. The positive ideal solution (A^+) and negative ideal solution (A^-) are calculated in the following way (Eq. 19.4 and Eq. 19.5, respectively).

$$A^+ = \left\{ \left(\left(\max_i v_{ij} | j \in J' \right), \left(\min_i v_{ij} | j \in I \right) \right) \middle| i = 1, 2, \ldots, m \right\} = \{ v_1^+, v_2^+, \ldots, v_m^+ \} \quad (19.4)$$

$$A^- = \left\{ \left(\left(\min_i v_{ij} | j \in J' \right), \left(\max_i v_{ij} | j \in J' \right) \right) \middle| i = 1, 2, \ldots, m \right\} = \{ v_1^-, v_2^-, \ldots, v_m^- \} \quad (19.5)$$

where I is associated with the positive criteria and J' associated with the negative criteria.

Step 5: Calculating the separation measure from the positive ideal solution and negative ideal solution

The calculation of Euclidean distance from the positive ideal solution is given as (Eq. 19.6)

$$S_i^+ = \sqrt{\left[(V_{ij} - V_i^+)^2 \right]} \quad i = 1, 2, \ldots, m \quad (19.6)$$

The calculation of Euclidean distance from the negative ideal solution is given as (Eq. 19.7)

$$S_i^- = \sqrt{\left[(V_{ij} - V_i^-)^2 \right]} \quad i = 1, 2, \ldots, m \quad (19.7)$$

Step 6: Calculating the relative closeness to the positive ideal solution

The relative closeness of ith alternative A_j with respect to A^+ is calculated as follows (Eq. 19.8):

$$P_i = \frac{S_i^-}{S_i^+ + S_i^-} \quad i = 1, 2, \ldots, m \quad (19.8)$$

Step7: Ranking the performance order

The relative closeness value can be ranked in descending order. The larger the P_i value, the better the performance of the alternatives. Therefore, the highest value of relative closeness is assigned as the high priority and the least value assigned as the low priority of the alternatives.

19.4 Results and Discussion

19.4.1 Morphometric Parameters

19.4.1.1 Linear Aspect

Stream order (u) is the relative position of the stream segments in the drainage basin according to the hierarchical order (1st, 2nd, 3rd, and so on). And it's directly proportional to stream discharge, channel dimensions, and the watershed contributing size (Strahler 1964). The stream ordering of the sub-basins of the GRB has been done according to Strahler's method (1952). According to this method, a stream with no tributaries is considered as a first-order stream which is the source of water of the basin, and a stream with only 1st order tributaries is a 2nd order stream, order increase only after the junction of the two same order streams and so on throughout the basin. The relationship between the stream order and flood is inversely related because when flooding occurs, higher-order streams take a longer time to develop flood stage than lower-order streams, and higher-order streams flooding longer for the flood to subside. On the other hand, lower-order streams flood occur rapidly during or immediately after rainfall resulting in flash floods. In the present study, the stream ordering of all the 34 sub-basins extends from 3rd to 5th order (Fig. 19.4a). As per stream ordering of the sub-basin, it reveals that the highest (5th order) stream order basin is noted for the five sub-basins (Fig. 19.4a) which means they take a longer time to flood and they also reduce flood magnitude in the basin area, while the eleven sub-basins having the lowest (3rd order) stream order take a short period to flood and increase flood intensity in the basin area. And the eighteen sub-basin (Fig. 19.4a) having the moderate (4th order) nature are held moderately responsible for flooding in the basin area.

Bifurcation ratio (R_b) is the ratio of stream length between the given order (N_u) and to the

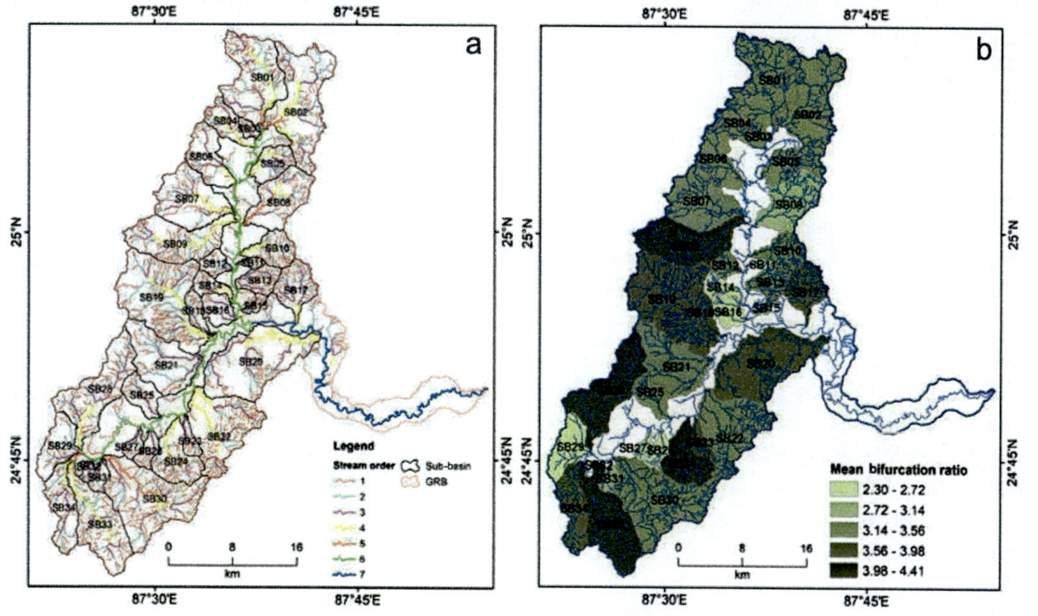

Fig. 19.4 Linear aspects. **a** Stream ordering after Strahler, **b** mean bifurcation ratio

next higher-order ($N_u + 1$) of a drainage basin. The Bifurcation ratio (R_b) was introduced by Horton (1945) and after that modified by Strahler (1952). Bifurcation ratio plays a significant role to control the peakedness of the runoff hydrograph and influences the landscape morphometry in the homogeneous bedrock (Chorley 1969). According to Waugh (2000), there is a negative relationship between flood and bifurcation ratio as the bifurcation ratio reduces the flooding risk within the basin and vice versa. It also directs the flood probability for the part, not for the whole of the basin (Chorley 1969). Flood probability also depends on basin shape. For example, long narrow basins with high bifurcations would portray reduced flood-discharge period, while round basins with low bifurcations ratio would show peaked flood discharge. In the present work, the mean R_b ranges from 2.30 to 4.40 of all the 34 sub-basins of GRB (Fig. 19.4b). Based on the bifurcation ratio value, five flood potentiality zones have been shown such as very low (3.98–4.41), low (3.56–3.98), moderate (3.14–3.56), high (2.72–3.14), and very high (2.30–2.72).

19.4.1.2 Areal Aspect

Drainage density (D_d) is the average length of a stream within a drainage basin (Horton 1932). It is the amount of the dissection of the watershed and is defined as the total length of all streams in the basin divided by its basin area. There is a positive relationship between drainage density and flooding. The higher the drainage density, the greater probability of flash floods. High drainage density decreases the discharge of a single stream, while low drainage density makes it highly liable to flooding. In the present study, the drainage density ranges from 0.13 to 4.09 concerning the 34 sub-basins of GRB (Fig. 19.5a). Based on the drainage density value, five zones have been created such as very low (0.13–0.92) recorded by eleven sub-basins, low (0.92–1.71) by eight sub-basins, moderate (1.71–2.50) by nine sub-basins, high (2.50–3.30) by two sub-basins, and very high (3.30–4.09) by four sub-basins (Fig. 19.5a). The sub-basins having a very high drainage density also have higher flood potentiality in the basin.

Stream frequency (S_f) is the number of streams per unit area (Horton 1945). It is related to the relief, permeability, and infiltration capacity of the basin area. It is necessary to say that according to Horton's Law, the number of stream segments decreases exponentially with increasing order. The higher the stream frequency, the higher the floods capability in the basin. It is an important variable of flood prediction of any basin. And low stream frequency makes it highly liable to flooding. So, there is a positive relation between stream frequency and flooding. In the study area, the stream frequency value of 34 sub-basins has been calculated and produced in the thematic map (Fig. 19.5b). It ranges from 0.85 to 4.48. Based on the stream frequency value, five zones have been created such as very low (0.85–1.57) portrayed by nine sub-basins, low (1.57–2.30) by 14 sub-basins, moderate (2.30–3.03) by five sub-basins, high (3.03–3.75) by four sub-basins, and very high (3.75–4.48) by two sub-basins with their corresponding effects on flood potentiality.

Drainage texture (Dt) includes drainage density and drainage frequency (Horton 1945). The drainage texture is proportionally related to flooding. The high drainage texture value implies a higher probability of flooding or high risk of flooding within the basin. The drainage texture value of 34 sub-basins of GRB is presented in the thematic map and it ranges from 0.12 to 9.26 (Fig. 19.6c). Based on the drainage texture value, five zones have been created such as very low (0.12–1.94) for fifteen sub-basins, low (1.94–3.77) for seven sub-basins, moderate (3.77–5.60) for two sub-basins, high (5.60–7.43) for four sub-basins, and very high (7.43–9.26) for six sub-basins (Fig. 19.6c).

Elongation ratio (R_e) of the diameter of the circle equal to the basin area to the maximum length of the basin (Schumm 1956). The lower the value of the ratio, the greater is the elongated character of the basin whereas the value approaching unity indicates the narrow shape of the basin. According to Mustafa and Yusuf (1999), the elongation ratio values range from 0.4 to 1.0. According to VenTe (1964), the circularity ratios of the basins range from 0.40 to

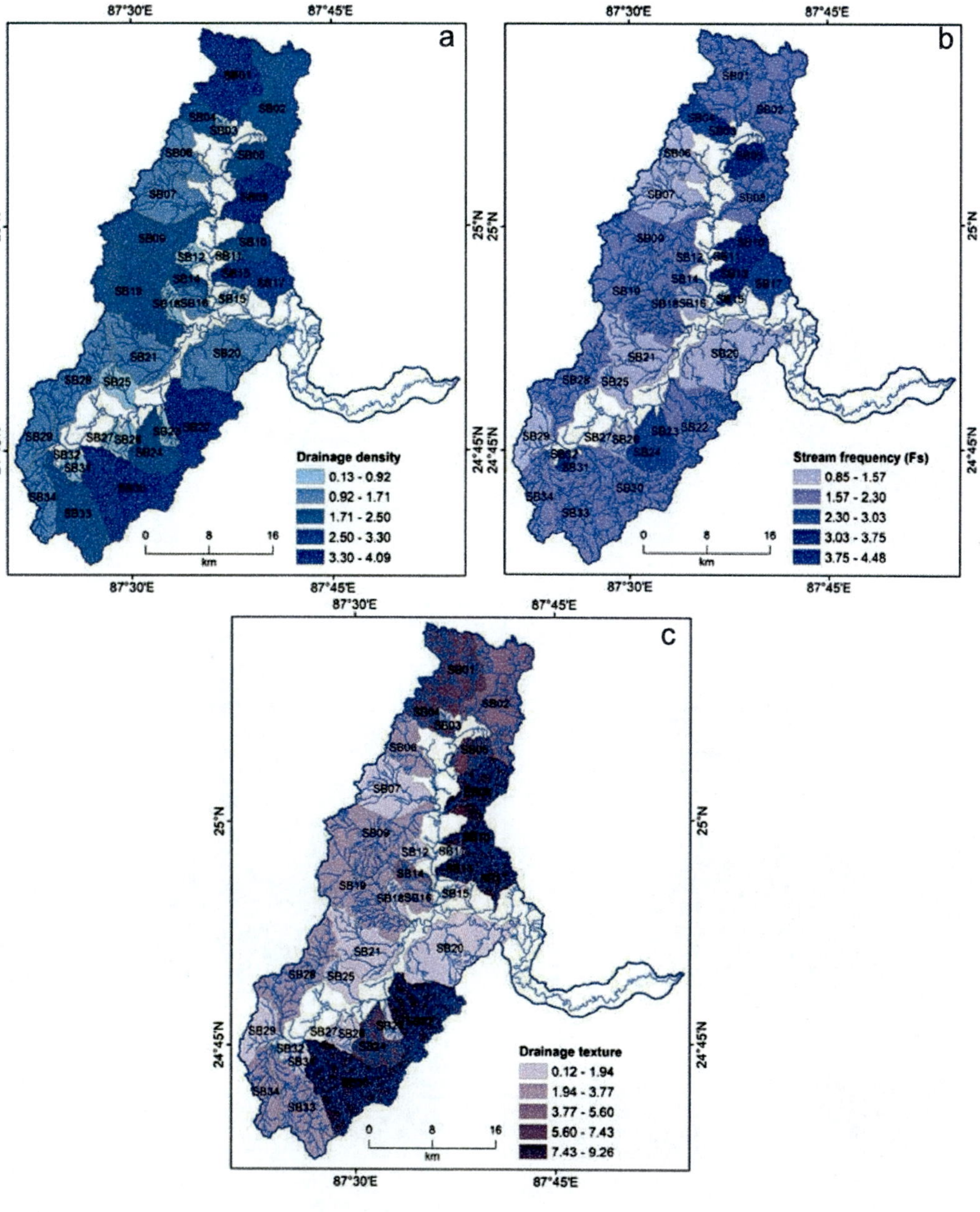

Fig. 19.5 Areal aspects. **a** Drainage density, **b** stream frequency, and **c** drainage texture

0.50 that indicate strongly elongated basins. The elongation ratio is proportionally related to the flooding in the basin area. The distribution of water flow in elongated basins takes a longer time than in a circular basin (Mustafa and Yusuf 1999). In the present study, the elongation ratio value of 34 sub-basins of GRB ranges from 0.51 to 0.78 (Fig. 19.6a). The value of the elongation ratio of 34 sub-basins has been measured and presented in the thematic map. Based on the elongation ratio value, five zones have been created such as very low (0.51–0.57) for eight sub-basins, low (0.57–0.62) for five sub-basins, moderate (0.62–0.67) for five sub-basins, high

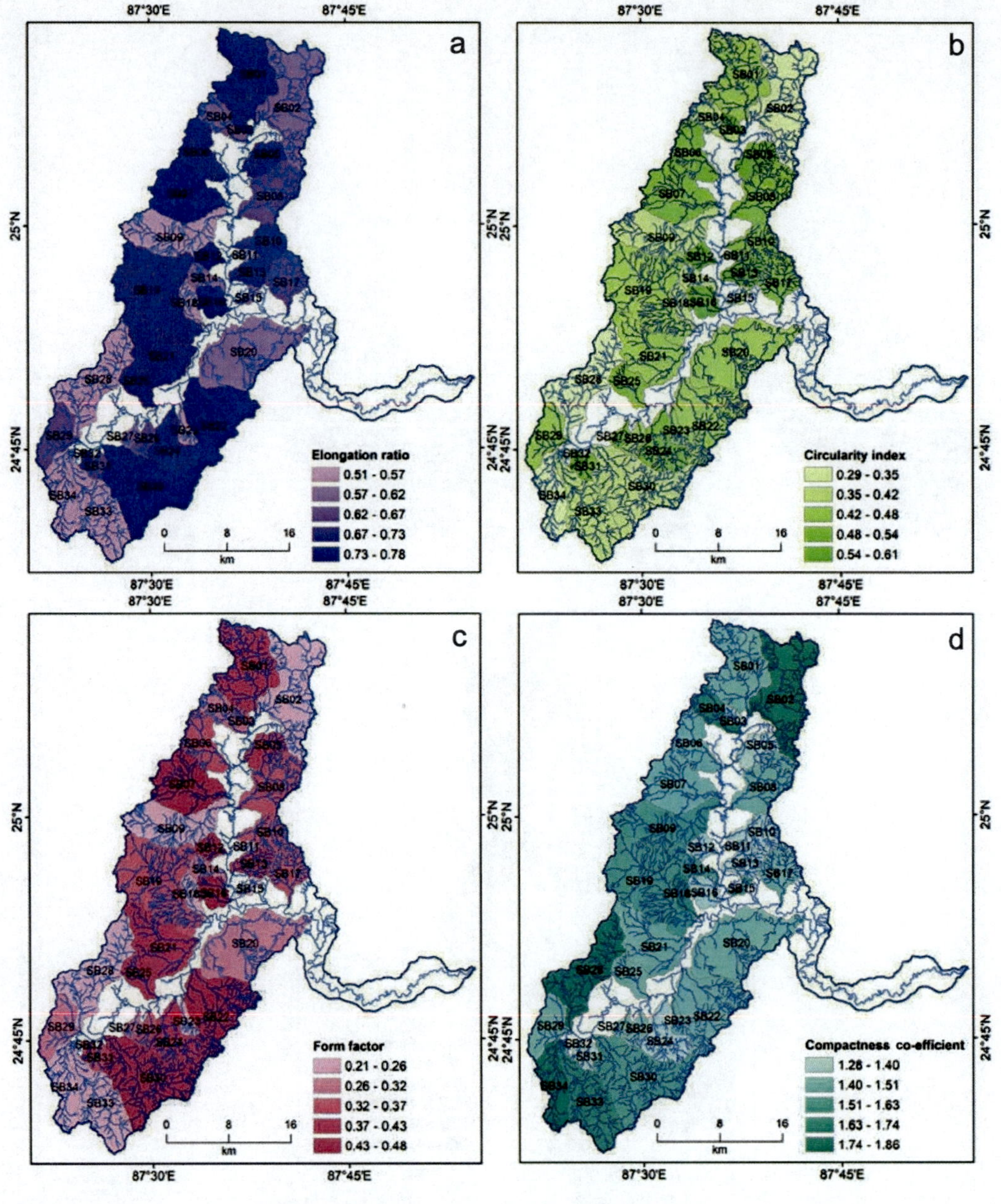

Fig. 19.6 Areal aspects. **a** Elongation ratio, **b** circularity ratio, **c** form factor, and **d** compactness coefficient

(0.67–0.73) for eleven sub-basins, and very high (0.73–0.78) for five sub-basins (Fig. 19.6a).

Circularity ratio (R_c) is the ratio of basin area to the area of a circle of the basin parameter (Miller 1953). The value of the circularity index ranges from 0 to 1. The value 1 denotes the perfect circularity of the basin and the value towards 0 indicates trending to the elongated or oval shape of the basin. The circularity ratio and the flooding are proportionally related to each other. A high circularity ratio has the maximum potentiality to create flooding in the basin area and vice versa. The circularity ratio of the basin lies between 0.40 and 0.50 and signifies the strongly elongated basin (VenTe 1964). In the present study, the circularity ratio value of 34 sub-basins of GRB ranges from 0.29 to 0.61 (Fig. 19.7b). Based on the circularity ratio, five zones have been created such as very low (0.29–0.35) registered by three sub-basins, low (0.35–0.42) by eight sub-basins, moderate (0.42–0.48) by seven sub-basins, high (0.48–0.54) by ten sub-basins, and very high (0.54–0.61) by six sub-basins with their corresponding effects on flood potentiality of the concerned area (Fig. 19.7b).

Form factor (R_f) is calculated as the ratio of the area of the basin to the square length of the basin (Horton 1932). The form factor value of more than 1 indicates the bulging or oval-shaped nature of the basin and the value of less than 1 indicates a narrow elongated nature of the basin. The form factor is proportionally related to the flooding in the basin area. In the present study, the form factor value of 34 sub-basins of GRB ranges from 0.21 to 0.48 which indicates that all the sub-basins are narrow and elongated. The form factor value of 34 sub-basin is produced in the thematic map (Fig. 19.7c). Based on the form factor value, five zones have been created such as very low (0.21–0.26) recorded by nine sub-basins, low (0.26–0.32) by five sub-basins, moderate (0.32–0.37) by eight sub-basins, high (0.37–0.43) by seven sub-basins, and very high (0.43–0.48) by five sub-basins (Fig. 19.7c).

Compactness coefficient (Cc) is defined as the perimeter of the basin divided by the circumference of an equivalent circular area (Horton 1932). The compactness coefficient and flooding are inversely related to each other. A higher compactness coefficient value indicates the lower risk of flooding and vice versa. In the study area, the value of the compactness coefficient ranges from 1.28 to 1.86 (Fig. 19.7d). Based on the value of compactness coefficient, five zones have been generated such as very low (1.74–1.86) for three sub-basins, low (1.63–1.74) for three sub-basins, moderate (1.51–1.63) for five sub-basins, high (1.40–1.51) for twelve sub-basins, and very high (1.28–1.40) for eleven sub-basins (Fig. 19.7d).

19.4.1.3 Relief Aspect

Basin relief (B_h) is the difference in elevation from the lowest to the highest point of a basin. The basin relief and the floodwater discharge are proportionally related to each other which means the maximum value of basin relief indicates the maximum potentiality of floodwater discharge in the watershed and vice versa. In the present study, the basin relief value ranges from 148 to 469 m in all the 34 sub-basins of GRB (Fig. 19.8a). Based on the value of basin relief, five zones have been created such as very low (148–212) depicted by four sub-basins, low (213–276) by five sub-basins, moderate (277–341) by thirteen sub-basins, high (342–405) by eight sub-basins, and very high (406–469) by four sub-basins.

Relief ratio (R_r) is defined as the ratio between total relief and the longest dimension of the basin parallel to the principle drainage line (Schumm 1956). The relief ratio is proportionally related to floodwater discharge. The relief ratio value has been calculated and presented in a thematic map (Fig. 19.8b). The floodwater discharge of the sub-basins has been categorized into very high (59.16–69.81) as in the case of the five sub-basins, high (48.51–59.16) for six sub-basins, moderate (37.85–48.51) for seven sub-basins, low (27.19–37.85) for eleven sub-basins, and very low (16.54–27.19) for five sub-basins (Fig. 19.8b).

The mean slope (S) of all the sub-basin is measured from the DEM and prepared a thematic map (Fig. 19.8c). The slope is a dominant factor that affects mostly the flood water discharge. The area comprising a steep slope increases the surface run-off but the gentle slope region decreases the surface run-off. So there is a positive relationship between slope and floodwater discharge. Based on the degree of slope, five zones are created such as very low (2.38°–4.10°) sub-basins ($n = 3$), low (4.10°–5.81°) sub-basins (8), moderate (5.81°–7.53°) sub-basins (9), high (7.53°–9.25°) sub-basins (10), and very high (9.25°–10.97°) sub-basins (4) (Fig. 19.8c).

Ruggedness number (R_n) is the multiplication of basin relief and drainage density. It is proportionally related to floodwater discharge. The

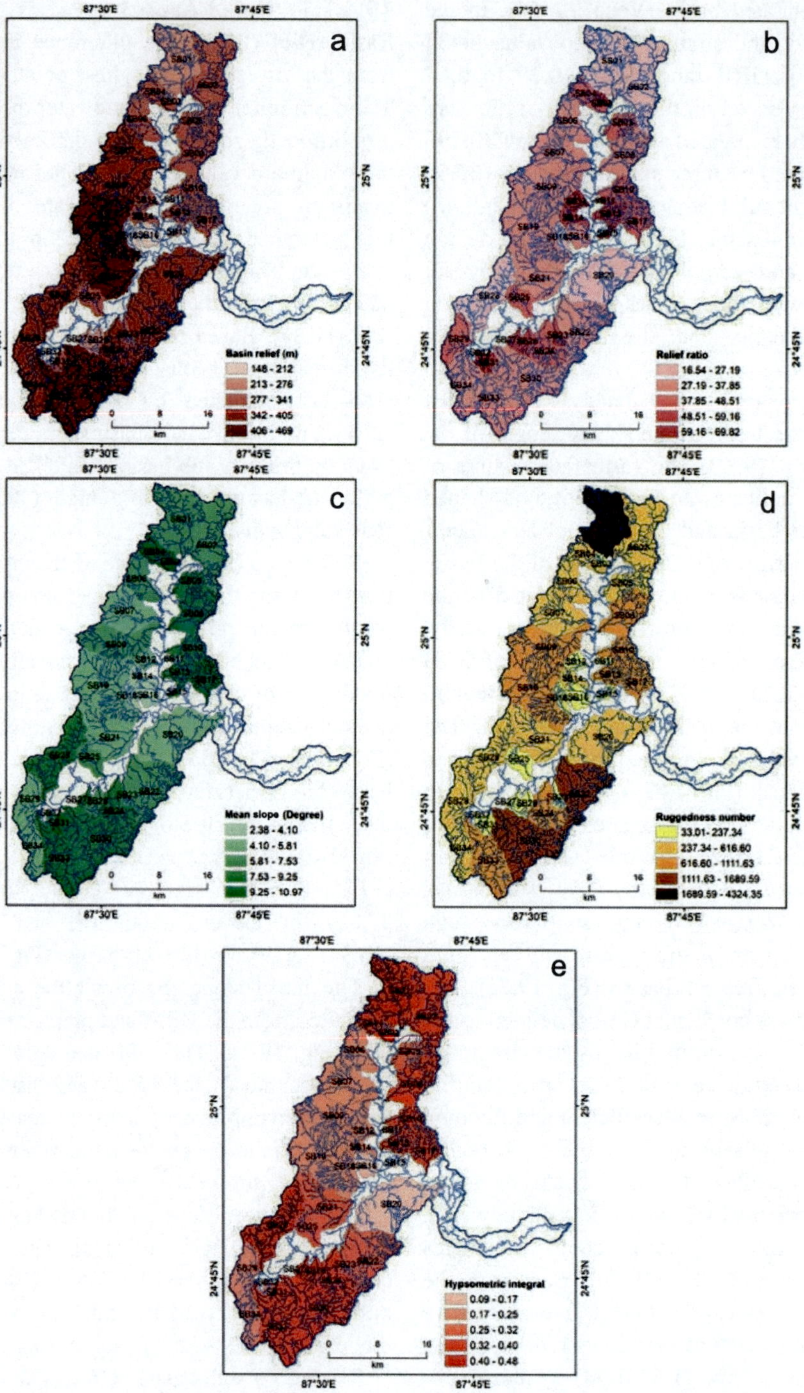

Fig. 19.7 Relief aspects. **a** Basin relief, **b** relief ratio, **c** mean slope, **d** ruggedness number, and **e** hypsometric integral

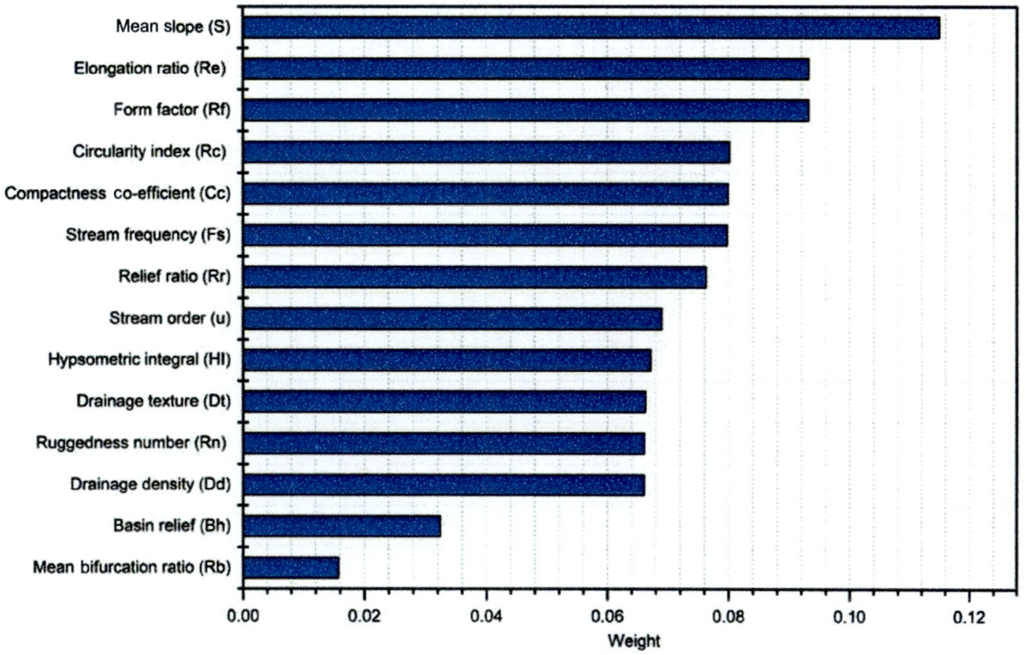

Fig. 19.8 Weights of the morphometric parameters related to flood potentiality

ruggedness number of the sub-basins ranges from 33.01 to 4324.35 (Fig. 19.8d). Based on the ruggedness number, the sub-basins are grouped into five zones such as very low (33.01–237.34) sub-basins (n = 12), low (237.34–616.60) sub-basins (n = 11), medium (616.60–1111.6) sub-basins (n = 8), high (1111.63–1689.59) sub-basins (n = 2), and very high (1689.59–4324.35) sub-basins (n = 1) with their corresponding effects on floodwater discharge (Fig. 19.8d).

Hypsometric integral (HI) portrays the ratio between the area of the land which was eroded and the area of the land to be eroded. Therefore, it gives an idea about the stage of development in which the landmass is situated under such degradation process. Based on HI value, the sub-basins are categorized into five zones in which the relationship between HI value and floodwater discharge is proportionally related to each other. The very high floodwater discharge potential zone is associated with the high HI value, on the other hand, the very low floodwater discharge zone is characterized by the low value of HI. Based on the HI value, five zones are created such as very low (0.09–0.17) as found for three sub-basin, low (0.17–0.25) for ten sub-basin, moderate (0.25–0.32) for the seven sub-basins, high (0.32–0.40) for the ten sub-basins, and very high (0.40–0.48) for the four sub-basins (Fig. 19.8e). According to the study or hypsometric integral, the highest floodwater discharge basin is the SB28 (0.48 HI value) and the lowest one is the SB115 (0.09 HI value).

19.4.2 Prioritization of Drainage Basins

Prioritization of the drainage basin is crucial for mitigation and management purposes concerning natural disastrous events. In the present study morphometric techniques including linear, areal, and relief aspects have been analyzed using TOPOSIS MCDM method to define the priority of the sub-basins related to flood events. In the present study, the 34 Sub-basins of the GRB are presented as the alternatives and 14

Table 19.2 Morphometric parameters

Sub-basin	u	R_b	Rf	R_e	R_c	D_t	F_s	D_d	Cc	B_h	R_r	S	R_n	HI
SB01	5	3.32	0.38	0.70	0.44	7.27	2.10	3.46	1.51	259	20.63	6.96	4324.35	0.41
SB02	5	3.30	0.26	0.57	0.29	3.89	2.15	1.81	1.86	278	16.54	6.77	502.29	0.33
SB03	3	3.50	0.36	0.68	0.56	0.71	3.17	0.22	1.34	226	66.89	9.33	50.41	0.27
SB04	4	3.34	0.28	0.60	0.37	6.10	2.95	2.07	1.64	298	38.18	9.53	616.60	0.32
SB05	4	3.36	0.42	0.73	0.59	6.81	3.17	2.15	1.30	286	44.66	7.96	615.12	0.25
SB06	4	3.25	0.36	0.68	0.49	2.20	1.49	1.48	1.43	309	35.44	7.40	457.80	0.24
SB07	4	3.37	0.43	0.74	0.50	1.51	1.09	1.38	1.42	342	32.15	5.60	472.36	0.18
SB08	5	2.82	0.34	0.66	0.51	8.30	2.28	3.64	1.40	305	28.64	7.88	1111.63	0.34
SB09	4	4.41	0.24	0.55	0.39	3.60	1.88	1.91	1.60	400	27.61	6.23	764.49	0.24
SB10	4	3.53	0.37	0.68	0.53	7.49	3.04	2.47	1.37	310	43.02	8.64	764.57	0.36
SB11	3	2.50	0.23	0.55	0.52	0.67	2.68	0.25	1.38	200	52.68	5.46	50.13	0.26
SB12	3	3.50	0.44	0.75	0.61	0.35	1.85	0.19	1.28	280	63.53	4.62	53.15	0.19
SB13	4	3.46	0.37	0.69	0.59	8.96	3.35	2.68	1.30	310	50.63	8.72	829.67	0.34
SB14	4	2.58	0.26	0.58	0.39	4.52	2.27	1.99	1.60	310	50.90	3.78	615.62	0.19
SB15	3	3.17	0.25	0.57	0.42	1.09	3.83	0.28	1.55	237	62.22	8.40	67.40	0.09
SB16	4	2.30	0.48	0.78	0.57	2.38	1.52	1.56	1.32	152	29.14	2.38	237.34	0.12
SB17	4	3.82	0.35	0.67	0.49	7.57	3.01	2.51	1.43	328	40.08	9.59	824.81	0.35
SB18	3	3.63	0.22	0.53	0.36	0.43	1.91	0.22	1.67	148	22.63	2.81	33.01	0.18
SB19	4	3.86	0.36	0.68	0.42	3.56	1.83	1.94	1.54	422	27.67	5.56	819.56	0.20
SB20	4	3.97	0.29	0.61	0.46	1.46	1.09	1.33	1.48	357	21.95	5.12	475.11	0.15
SB21	4	3.25	0.37	0.69	0.46	1.09	0.85	1.29	1.48	411	36.16	5.60	528.41	0.26
SB22	5	3.42	0.47	0.77	0.46	9.26	2.26	4.09	1.47	403	35.44	7.09	1648.25	0.26
SB23	3	3.60	0.21	0.51	0.45	0.59	2.67	0.22	1.50	316	56.78	7.02	69.59	0.29
SB24	4	4.07	0.39	0.71	0.52	5.79	2.90	1.99	1.39	380	43.65	8.55	758.10	0.36
SB25	3	3.33	0.37	0.69	0.51	0.12	0.91	0.13	1.41	315	47.52	6.58	39.66	0.27
SB26	3	2.50	0.34	0.65	0.46	0.16	1.13	0.14	1.48	247	50.76	6.89	34.56	0.41
SB27	3	2.50	0.30	0.61	0.49	0.16	1.13	0.14	1.48	247	50.76	8.57	34.56	0.41
SB28	4	4.39	0.21	0.52	0.30	2.90	1.73	1.67	1.81	367	22.65	8.47	614.71	0.48
SB29	4	2.63	0.30	0.62	0.45	1.17	1.04	1.12	1.49	346	38.86	5.25	386.70	0.19
SB30	5	3.44	0.43	0.74	0.41	8.72	2.19	3.98	1.57	424	31.20	8.25	1689.59	0.39
SB31	3	3.00	0.32	0.64	0.54	0.30	1.84	0.16	1.37	301	69.82	10.97	49.25	0.38
SB32	3	4.25	0.41	0.72	0.61	0.98	4.48	0.22	1.28	198	67.15	5.96	43.18	0.20
SB33	4	4.18	0.23	0.54	0.37	3.42	1.88	1.82	1.64	469	31.51	8.92	852.14	0.40
SB34	4	3.87	0.22	0.53	0.31	3.45	2.04	1.69	1.79	361	31.27	5.46	609.60	0.24

Computed by the authors 2020

parameters related to the linear, areal, and relief aspects of the drainage morphometry are expressed as the criteria in the matrix table (Table 19.2). The normalization value of the present study has been done using Eq. 19.2 shown in Table 19.3.

Table 19.3 Normalized decision matrix

Sub-basin	(u)	(R_b)	(Rf)	(R_e)	(R_c)	(D_t)	(F_s)	(D_d)	(Cc)	(B_h)	(R_r)	(S)	(R_n)	(HI)
SB01	0.22	0.17	0.19	0.18	0.16	0.27	0.15	0.31	0.17	0.14	0.08	0.17	0.75	0.24
SB02	0.22	0.16	0.13	0.15	0.11	0.15	0.16	0.16	0.21	0.15	0.07	0.16	0.09	0.19
SB03	0.13	0.17	0.18	0.18	0.20	0.03	0.23	0.02	0.15	0.12	0.26	0.22	0.01	0.16
SB04	0.18	0.17	0.14	0.16	0.14	0.23	0.22	0.18	0.19	0.16	0.15	0.23	0.11	0.19
SB05	0.18	0.17	0.21	0.19	0.21	0.26	0.23	0.19	0.15	0.15	0.18	0.19	0.11	0.14
SB06	0.18	0.16	0.18	0.18	0.18	0.08	0.11	0.13	0.16	0.17	0.14	0.18	0.08	0.14
SB07	0.18	0.17	0.22	0.19	0.18	0.06	0.08	0.12	0.16	0.18	0.13	0.13	0.08	0.11
SB08	0.22	0.14	0.17	0.17	0.19	0.31	0.17	0.33	0.16	0.16	0.11	0.19	0.19	0.19
SB09	0.18	0.22	0.12	0.15	0.14	0.14	0.14	0.17	0.18	0.21	0.11	0.15	0.13	0.14
SB10	0.18	0.18	0.18	0.18	0.19	0.28	0.22	0.22	0.16	0.17	0.17	0.21	0.13	0.21
SB11	0.13	0.12	0.12	0.14	0.19	0.03	0.20	0.02	0.16	0.11	0.21	0.13	0.01	0.15
SB12	0.13	0.17	0.22	0.20	0.22	0.01	0.14	0.02	0.15	0.15	0.25	0.11	0.01	0.11
SB13	0.18	0.17	0.19	0.18	0.21	0.34	0.25	0.24	0.15	0.17	0.20	0.21	0.14	0.20
SB14	0.18	0.13	0.13	0.15	0.14	0.17	0.17	0.18	0.18	0.17	0.20	0.09	0.11	0.11
SB15	0.13	0.16	0.13	0.15	0.15	0.04	0.28	0.03	0.18	0.13	0.25	0.20	0.01	0.05
SB16	0.18	0.11	0.24	0.21	0.21	0.09	0.11	0.14	0.15	0.08	0.12	0.06	0.04	0.07
SB17	0.18	0.19	0.18	0.18	0.18	0.28	0.22	0.22	0.16	0.18	0.16	0.23	0.14	0.20
SB18	0.13	0.18	0.11	0.14	0.13	0.02	0.14	0.02	0.19	0.08	0.09	0.07	0.01	0.10
SB19	0.18	0.19	0.18	0.18	0.15	0.13	0.13	0.17	0.18	0.23	0.11	0.13	0.14	0.11
SB20	0.18	0.20	0.15	0.16	0.17	0.05	0.08	0.12	0.17	0.19	0.09	0.12	0.08	0.09
SB21	0.18	0.16	0.19	0.18	0.17	0.04	0.06	0.11	0.17	0.22	0.14	0.13	0.09	0.15
SB22	0.22	0.17	0.24	0.20	0.17	0.35	0.17	0.36	0.17	0.22	0.14	0.17	0.29	0.15
SB23	0.13	0.18	0.10	0.13	0.16	0.02	0.20	0.02	0.17	0.17	0.22	0.17	0.01	0.17
SB24	0.18	0.20	0.20	0.19	0.19	0.22	0.21	0.18	0.16	0.20	0.17	0.20	0.13	0.21
SB25	0.13	0.17	0.19	0.18	0.18	0.00	0.07	0.01	0.16	0.17	0.19	0.16	0.01	0.16
SB26	0.13	0.12	0.17	0.17	0.17	0.01	0.08	0.01	0.17	0.13	0.20	0.16	0.01	0.24
SB27	0.13	0.12	0.15	0.16	0.18	0.01	0.08	0.01	0.17	0.13	0.20	0.20	0.01	0.24
SB28	0.18	0.22	0.11	0.14	0.11	0.11	0.13	0.15	0.21	0.20	0.09	0.20	0.11	0.28
SB29	0.18	0.13	0.15	0.16	0.16	0.04	0.08	0.10	0.17	0.19	0.15	0.12	0.07	0.11
SB30	0.22	0.17	0.22	0.20	0.15	0.33	0.16	0.36	0.18	0.23	0.12	0.20	0.29	0.23
SB31	0.13	0.15	0.16	0.17	0.19	0.01	0.14	0.01	0.16	0.16	0.28	0.26	0.01	0.22
SB32	0.13	0.21	0.21	0.19	0.22	0.04	0.33	0.02	0.15	0.11	0.27	0.14	0.01	0.12
SB33	0.18	0.21	0.12	0.14	0.14	0.13	0.14	0.16	0.19	0.25	0.12	0.21	0.15	0.23
SB34	0.18	0.19	0.11	0.14	0.11	0.13	0.15	0.15	0.21	0.19	0.12	0.13	0.11	0.14

Computed by the authors 2020

In the present study, the weights of the criteria have been assigned based on the expert's observations (Fig. 19.4). The parameters such as slope, elongation ratio, and form factor had the highest weights (0.11, 0.93, and 0.93, respectively) in the present study. The weighted normalized matrix of the current study is presented in Table 19.4. In the present study,

Table 19.4 Weighted normalized decision matrix

Sub-basin	(u)	(R_b)	(Rf)	(R_e)	(R_c)	(D_t)	(F_s)	(D_d)	(Cc)	(B_h)	(R_r)	(S)	(R_n)	(HI)
SB01	0.015	0.003	0.018	0.017	0.013	0.018	0.012	0.020	0.014	0.005	0.006	0.019	0.049	0.015
SB02	0.015	0.003	0.012	0.014	0.008	0.010	0.013	0.011	0.017	0.005	0.005	0.018	0.006	0.012
SB03	0.009	0.003	0.017	0.017	0.016	0.002	0.019	0.001	0.012	0.004	0.020	0.025	0.001	0.010
SB04	0.012	0.003	0.013	0.015	0.011	0.015	0.017	0.012	0.015	0.005	0.012	0.026	0.007	0.012
SB05	0.012	0.003	0.019	0.018	0.017	0.017	0.019	0.013	0.012	0.005	0.013	0.022	0.007	0.009
SB06	0.012	0.003	0.017	0.017	0.014	0.005	0.009	0.009	0.013	0.005	0.011	0.020	0.005	0.009
SB07	0.012	0.003	0.020	0.018	0.014	0.004	0.006	0.008	0.013	0.006	0.010	0.015	0.005	0.007
SB08	0.015	0.002	0.016	0.016	0.015	0.021	0.013	0.021	0.013	0.005	0.009	0.021	0.013	0.013
SB09	0.012	0.003	0.011	0.014	0.011	0.009	0.011	0.011	0.015	0.007	0.008	0.017	0.009	0.009
SB10	0.012	0.003	0.017	0.017	0.015	0.019	0.018	0.015	0.013	0.005	0.013	0.024	0.009	0.014
SB11	0.009	0.002	0.011	0.013	0.015	0.002	0.016	0.001	0.013	0.003	0.016	0.015	0.001	0.010
SB12	0.009	0.003	0.021	0.018	0.018	0.001	0.011	0.001	0.012	0.005	0.019	0.013	0.001	0.007
SB13	0.012	0.003	0.018	0.017	0.017	0.022	0.020	0.016	0.012	0.005	0.015	0.024	0.009	0.013
SB14	0.012	0.002	0.012	0.014	0.011	0.011	0.013	0.012	0.015	0.005	0.015	0.010	0.007	0.007
SB15	0.009	0.002	0.012	0.014	0.012	0.003	0.022	0.002	0.014	0.004	0.019	0.023	0.001	0.003
SB16	0.012	0.002	0.023	0.019	0.017	0.006	0.009	0.009	0.012	0.003	0.009	0.006	0.003	0.005
SB17	0.012	0.003	0.017	0.016	0.014	0.019	0.018	0.015	0.013	0.006	0.012	0.026	0.009	0.013
SB18	0.009	0.003	0.010	0.013	0.010	0.001	0.011	0.001	0.015	0.003	0.007	0.008	0.000	0.007
SB19	0.012	0.003	0.017	0.017	0.012	0.009	0.011	0.011	0.014	0.007	0.008	0.015	0.009	0.007
SB20	0.012	0.003	0.014	0.015	0.013	0.004	0.006	0.008	0.014	0.006	0.007	0.014	0.005	0.006
SB21	0.012	0.003	0.018	0.017	0.013	0.003	0.005	0.008	0.014	0.007	0.011	0.015	0.006	0.010
SB22	0.015	0.003	0.022	0.019	0.013	0.023	0.013	0.024	0.013	0.007	0.011	0.019	0.019	0.010
SB23	0.009	0.003	0.010	0.013	0.013	0.001	0.016	0.001	0.014	0.006	0.017	0.019	0.001	0.011
SB24	0.012	0.003	0.018	0.017	0.015	0.014	0.017	0.012	0.013	0.007	0.013	0.023	0.009	0.014
SB25	0.009	0.003	0.017	0.017	0.015	0.000	0.005	0.001	0.013	0.005	0.014	0.018	0.000	0.010
SB26	0.009	0.002	0.016	0.016	0.013	0.000	0.007	0.001	0.014	0.004	0.015	0.019	0.000	0.015
SB27	0.009	0.002	0.014	0.015	0.014	0.000	0.007	0.001	0.014	0.004	0.015	0.023	0.000	0.015
SB28	0.012	0.003	0.010	0.013	0.009	0.007	0.010	0.010	0.017	0.006	0.007	0.023	0.007	0.018
SB29	0.012	0.002	0.014	0.015	0.013	0.003	0.006	0.007	0.014	0.006	0.012	0.014	0.004	0.007
SB30	0.015	0.003	0.020	0.018	0.012	0.022	0.013	0.023	0.014	0.007	0.009	0.023	0.019	0.015
SB31	0.009	0.002	0.015	0.016	0.016	0.001	0.011	0.001	0.013	0.005	0.021	0.030	0.001	0.014
SB32	0.009	0.003	0.019	0.018	0.018	0.002	0.026	0.001	0.012	0.003	0.020	0.016	0.000	0.008
SB33	0.012	0.003	0.011	0.013	0.011	0.009	0.011	0.011	0.015	0.008	0.010	0.024	0.010	0.015
SB34	0.012	0.003	0.010	0.013	0.009	0.009	0.012	0.010	0.016	0.006	0.009	0.015	0.007	0.009

Computed by the authors 2020

14 criteria have been adopted out of which the criteria having a positive relationship with flood have been chosen as the benefit criteria and the criteria that are inversely related to flood have been chosen as the cost criteria (Table 19.5).

Finally, the performance rank (Pi) has been computed which depicts the higher the value of

Table 19.5 Distribution of the positive and negative ideal solution

	(u)	(R_b)	(Rf)	(R_e)	(R_c)	(D_t)	(F_s)	(D_d)	(Cc)	(B_h)	(R_r)	(S)	(R_n)	(HI)
A+	0.0152	0.0018	0.0226	0.0193	0.0177	0.0231	0.0262	0.0241	0.0118	0.0082	0.0211	0.0299	0.0495	0.0180
A−	0.0091	0.0034	0.0096	0.0126	0.0084	0.0003	0.0050	0.0007	0.0170	0.0026	0.0050	0.0065	0.0004	0.0034

the Pi, the higher priority, and the least value assigned as the low priority of the alternatives. For the present case, Pi of the sub-basins is shown in Table 19.6. Thus, the final score of the relative closeness of the ideal solution, ranging from 0.36 to 0.73 has been defined (Fig. 19.9).

The lowest value of the final score (0.36) stands for sub-basin SB15, whereas the maximum value (0.73) of the final score has been confined in the sub-basin SB01. The final score also be grouped into five classes such as very (0.36–0.39) low (0.39–0.44), medium (0.44–0.49), high (0.49–0.55), and very high (0.55–0.73) potentiality using the natural breaks intervals. The very low class includes the five sub-basins SB11, SB15, SB18, SB23, and SB32, whereas the very high class consists of the three sub-basins SB01, SB22, and SB30. The low priority class is confined into the twelve sub-basins SB02, SB03, SB12, SB14, SB16, SB20, SB25, SB26, SB27, SB29, and SB34 whereas, high class encompasses the five sub-basins SB08, SB10, SB13, SB17, and SB24. The medium class includes the ten sub-basins SB04, SB05, SB06, SB07, SB09, SB19, SB21, SB28, SB31, and SB33.

In addition, the spatial extension of the flood potentiality classes has also been extracted (Table 19.7). From the following table, it is observed that the very low class accounts for 26.36 km^2 (2.57%) area which is the minimum of the total sub-basin area, whereas the medium class contains the areal extensions of 406 km^2 (39.63%) which is the maximum among the flood potentiality classes. The low class, very high class, and high class stand for moderate spatial extensions having an area of 267.91 km^2 (26.10%), 200.97 km^2 (19.58%), and 124.55 km^2 (12.13%), respectively.

19.5 Conclusion

With the specific contribution of the morphometric parameters regarding linear, areal, and relief aspects, the potentiality of the flood of the GRB has been analyzed. The TOPSIS MCDM techniques have been efficiently used to demarcate the rank of the sub-basins based on the

Table 19.6 Euclidean distance from positive and negative ideal solution and relative closeness of the positive ideal solution

Sub-basin	Si+	Si-	Pi	Sub-basin	Si+	Si-	Pi
SB01	0.025	0.071	0.735	SB18	0.069	0.039	0.3593
SB02	0.056	0.043	0.435	SB19	0.053	0.045	0.4597
SB03	0.06	0.042	0.412	SB20	0.061	0.046	0.4293
SB04	0.049	0.045	0.477	SB21	0.059	0.049	0.4533
SB05	0.048	0.045	0.486	SB22	0.038	0.057	0.6025
SB06	0.056	0.047	0.454	SB23	0.063	0.04	0.3864
SB07	0.059	0.048	0.449	SB24	0.046	0.046	0.4973
SB08	0.043	0.053	0.551	SB25	0.065	0.049	0.4281
SB09	0.054	0.044	0.451	SB26	0.064	0.049	0.4296
SB10	0.045	0.047	0.512	SB27	0.064	0.05	0.437
SB11	0.064	0.038	0.376	SB28	0.056	0.048	0.4612
SB12	0.064	0.045	0.413	SB29	0.061	0.046	0.4322
SB13	0.043	0.049	0.531	SB30	0.037	0.058	0.6117
SB14	0.055	0.042	0.436	SB31	0.062	0.051	0.4511
SB15	0.062	0.035	0.363	SB32	0.061	0.034	0.356
SB16	0.063	0.046	0.423	SB33	0.051	0.048	0.4827
SB17	0.044	0.048	0.52	SB34	0.056	0.042	0.4273

Computed by the authors 2020

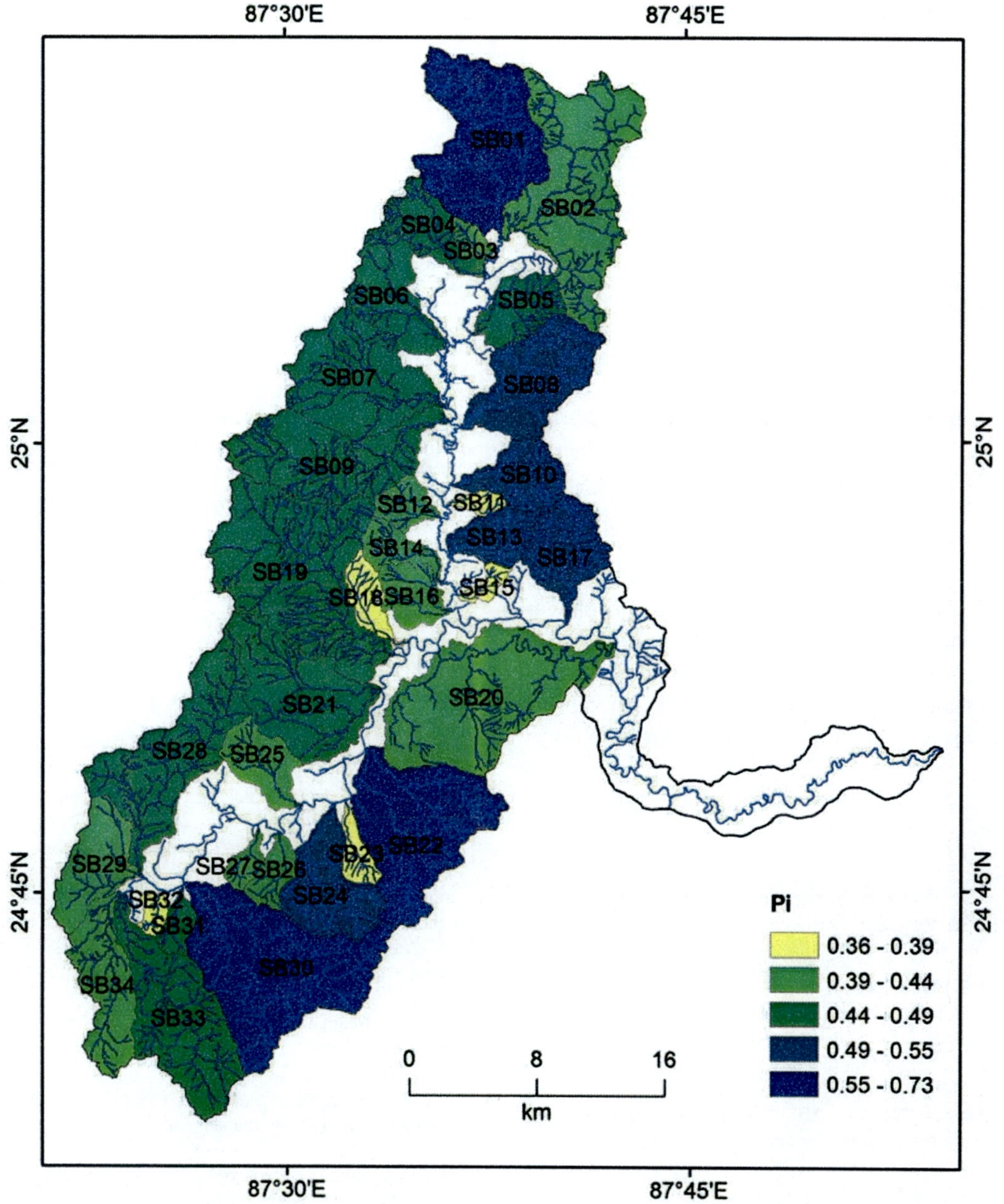

Fig. 19.9 Sub-basin prioritization map

characteristics of the parameters and their role in triggering a flood event. The weightage of the morphometric parameters has also been calculated that have been used in the TOPOSIS method. The final score of the relative closeness from the positive ideal location has been categorized into five classes such as very low (0.36–0.39), low (0.9–0.44), medium (0.44–0.49), high (0.49–0.55), and very high (0.55–0.73). The very low class covers 2.57% area of having an area of 26.38 km^2. The low class accounts for 267.91 km^2 which is 26.10%. The

Table 19.7 Distribution of flood potentiality of the sub-basins

Class	Sub-basin	Area (km^2)	Area (%)
Very low	SB11, SB15, SB18, SB23, and SB32	26.38	2.57
Low	SB02, SB03, SB12, SB14, SB16, SB20, SB25, SB26, SB27, SB29 and SB34	267.91	26.10
Medium	SB04, SB05, SB06, SB07, SB09, SB19, SB21, SB28, SB31 and SB33	406.80	39.63
High	SB08, SB10, SB13, SB17, and SB24	124.55	12.13
Very high	SB01, SB22 and SB30	200.97	19.58

Medium class includes an area of 406.80 km^2 which is 39.63%. The high class consists of 124.55 km^2 which is 12.13%. The very high class covers the area of 200.97 km which is 19.58% of the total sub-basin area. From the final map of the flood potentiality, it is observed that relief aspects are the dominant factors followed by the areal and liner aspects. It is also observed that the RS and GIS along with the statistical technique are curial for the monitoring and analysis of the area that has the possibility of natural calamities. In addition, the present study may be helpful for assessment and monitoring of flood hazard.

References

Aher PD, Adinarayana J, Gorantiwar SD (2014) Quantification of morphometric characterization and prioritization for management planning in semi-arid tropics of India: a remote sensing and GIS approach. J Hydrol 511:850–860

Akay H, Koçyiğit MB (2020) Flash flood potential prioritization of sub-basins in an ungauged basin in Turkey using traditional multi-criteria decision-making methods. Soft Comput 1–13

Amiri M, Pourghasemi HR, Arabameri A, Vazirzadeh A, Yousefi H, Kafaei S (2019) Prioritization of flood inundation of Maharloo Watershed in Iran using morphometric parameters analysis and TOPSIS MCDM model. In: Spatial modeling in GIS and R for earth and environmental sciences. Elsevier, pp 371–390

Aouragh MH, Essahlaoui A (2018) A TOPSIS approach-based morphometric analysis for sub-watersheds prioritization of high Oum Er-Rbia basin, Morocco. Spat Inf Res 26(2):187–202

Bhattacharji M (2012) Topographic features of the Gumani River Basin, Rajmahal volcanic province, Jharkhand: an example of exhumed of topography or palaeotopographic control? J Indian Geomorphol 1:45–54

Balasubramanian A, Duraisamy K, Thirumalaisamy S, Krishnaraj S, Yatheendradasan RK (2017) Prioritization of subwatersheds based on quantitative morphometric analysis in lower Bhavani basin, Tamil Nadu, India using DEM and GIS techniques. Arab J Geosci 10(24):552

Ball V (1877) Geology of Rajmahal Hills. Mem Geol Surv India 13(1)

Bid S, Siddique G (2019) Human risk assessment of Panchet dam in India using TOPSIS and WASPAS multi-criteria decision-making (MCDM) methods. Heliyon 5(6):e01956

Biswas S, Sudhakar S, Desai VR (1999) Prioritisation of subwatersheds based on morphometric analysis of drainage basin: a remote sensing and GIS approach. J Indian Soc Remote Sens 27(3):155

CEOS (2003) The use of earth observing satellites for hazard support: assessments and scenarios, final report of the CEOS Disaster Management Support Group (DMSG). Helen M. Wood, Chair. National Oceanic and Atmospheric Administration (NOAA) United States Department of Commerce

Chorley RJ, Malm DE, Pogorzelski HA (1957) A new standard for estimating drainage basin shape. Am J Sci 255(2):138–141

Chorley RJ (1969) The drainage basin as the fundamental geomorphic unit. Water Earth Man 77–98

Chowdary VM, Chakraborthy D, Jeyaram A, Murthy YK, Sharma JR, Dadhwal VK (2013) Multi-criteria decision making approach for watershed prioritization using analytic hierarchy process technique and GIS. Water Resour Manag 27(10):3555–3571

Clarke JI (1966) Morphometry from maps. Essays in geomorphology. Heinmann, London, pp 235–274

District Disaster Management Plan of Sadebganj (2017) District Disaster Management Authority, Sahibganj

Dury GH (1952) Methods of cartographical analysis in geomorphological research. Silver Jubilee Volume, Indian Geographical Society, Madras, pp 136–139

Elewa HH, Ramadan ESM, Nosair AM (2016) Spatial-based hydro-morphometric watershed modeling for the assessment of flooding potentialities. Environ Earth Sci 75(10):927

Farhan Y, Anaba O (2016) A remote sensing and GIS approach for prioritization of WadiShueib Mini-Watersheds (Central Jordan) based on morphometric and Soil erosion susceptibility analysis. J Geogr Inf Syst 8(1):1–19

Gregory KJ, Walling DE (1968) The variation of drainage density within a catchment. Hydrol Sci J 13(2):61–68

Gravelius H (1914) Grundrifi der gesamtenGewcis-serkunde. Band I: Flufikunde (Compendium of hydrology, vol I. Rivers, in German). Goschen, Berlin, Germany

Hadley RF, Schumm SA (1961) Sediment sources and drainage basin characteristics in upper Cheyenne River basin. US Geol Surv Water Supply Pap 1531:198

Hembram TK, Saha S (2020) Prioritization of sub-watersheds for soil erosion based on morphometric attributes using fuzzy AHP and compound factor in Jainti River basin, Jharkhand, Eastern India. Environ Dev Sustain 22(2):1241–1268

Horton RE (1932) Drainage-basin characteristics. Eos Trans Am Geophys Union 13(1):350–361

Horton RE (1945) Erosional development of streams and their drainage basins; hydrophysical approach to quantitative morphology. Geol Soc Am Bull 56 (3):275–370

Hwang CL, Yoon K (1981) Methods for multiple attribute decision making. In: Multiple attribute decision making. Springer, Berlin, Heidelberg, pp 58–191

Hwang CL, Lai YJ, Liu TY (1993) A new approach for multiple objective decision making. Comput Oper Res 20(8):889–899

Islam A, Barman SD (2020) Drainage basin morphometry and evaluating its role on flood-inducing capacity of tributary basins of Mayurakshi River, India. https://link.springer.com/content/pdf/10.1007/s42452-020-2839-4.pdf

Jaiswal RK, Thomas T, Galkate RV, Ghosh NC, Singh S (2014) Watershed prioritization using Saaty's AHP based decision support for soil conservation measures. Water Resour Manag 28(2):475–494

Jaiswal RK, Ghosh NC, Lohani AK, Thomas T (2015) Fuzzy AHP based multi criteria decision support for watershed prioritization. Water Resour Manag 29 (12):4205–4227

Javed A, Khanday MY, Ahmed R (2009) Prioritization of sub-watersheds based on morphometric and land use analysis using remote sensing and GIS techniques. J Indian Soc Remote Sens 37(2):261

Kale VS, Gupta A (2010) Introduction to geomorphology. Universities Press (India) Private Limited, Hyderabad

Khan ZA (1987) Paleodrainage and paleochannel morphology of a Barakar River (Early Permian) in the Rajmahal Gondwana Basin, Bihar, India. Palaeogeogr Palaeoclimatol Palaeoecol 58(235):247

Mahammad S, Islam A (2021a) Identification of palaeochannels using optical images and radar data: a study of the Damodar Fan Delta, India. Arab J Geosci 14(17):1–22. https://doi.org/10.1007/s12517-021-07818-5

Mahammad S, Islam A (2021b) Evaluating the groundwater quality of Damodar Fan Delta (India) using fuzzy-AHP MCDM technique. Appl Water Sci 11 (7):1–17. https://doi.org/10.1007/s13201-021-01408-2

Mahammad S, Islam A (2021c) Assessing the groundwater potentiality of the Gumani river basin, India, using geoinformatics and analytical hierarchy process. Groundw Soc Appl Geospatial Technol, 161–187. https://doi.org/10.1007/978-3-030-64136-8_8

Meshram SG, Alvandi E, Singh VP, Meshram C (2019) Comparison of AHP and fuzzy AHP models for prioritization of watersheds. Soft Comput 23 (24):13615–13625

Meshram SG, Alvandi E, Meshram C, Kahya E, Al-Quraishi AMF (2020) Application of SAW and TOPSIS in prioritizing watersheds. Water Resour Manag 34(2):715–732

Miller VC (1953) Quantitative geomorphic study of drainage basin characteristics in the Clinch Mountain area, Virginia and Tennessee. Technical report, Columbia University. Department of Geology; no. 3

Mustafa S, Yusuf MIY (1999) A textbook of hydrology and water resources. Craft Technics Publishers

Nitheshnirmal S, Bhardwaj A, Dineshkumar C, Rahaman SA (2019) Prioritization of erosion prone micro-watersheds using morphometric analysis coupled with multi-criteria decision making. In: Multidisciplinary Digital Publishing Institute proceedings, vol 24, no 1, p 11

Omran A, Schroder D, El Rayes A, Geriesh M (2011) Flood hazard assessment in Wadi Dahab, Egypt based on basin morphometry using GIS techniques. GI_Forum Program Committee

Rahaman SA, Ajeez SA, Aruchamy S, Jegankumar R (2015) Prioritization of sub watershed based on morphometric characteristics using fuzzy analytical hierarchy process and geographical information system – a study of Kallar watershed, Tamil Nadu. Aquatic Procedia 4:1322–1330. https://doi.org/10.1016/j.aqpro.2015.02.172

Rahmati O, Haghizadeh A, Stefanidis S (2016) Assessing the accuracy of GIS-based analytical hierarchy process for watershed prioritization; Gorganrood River Basin, Iran. Water Resour Manag 30(3):1131–1150

Rao PJ, Rao BS, Rao MJ, Harikrishna P (2003) Geo-Electrical data analysis to demarcate groundwater pockets and recharge zones in Champavathi River Basin, Vizianagaram District, Andhra Pradesh. J Indian Geophys Union 7(2):105–113

Ratnam KN, Srivastava YK, Rao VV, Amminedu E, Murthy KSR (2005) Check dam positioning by prioritization of micro-watersheds using SYI model and morphometric analysis—remote sensing and GIS perspective. J Indian Soc Remote Sens 33(1):25

Roszkowska E (2011) Multi-criteria decision making models by applying the TOPSIS method to crisp and interval data. Mult Criteria Decis Mak/univ Econ Katow 6:200–230

Sadhasivam N, Bhardwaj A, Pourghasemi HR, Kamaraj NP (2020) Morphometric attributes-based soil erosion susceptibility mapping in Dnyanganga watershed of India using individual and ensemble models. Environ Earth Sci 79(14):1–28

Saha S, Saha M, Mukherjee K, Arabameri A, Ngo PTT, Paul GC (2020) Predicting the deforestation probability using the binary logistic regression, random forest, ensemble rotational forest and REPTree: a case study at the Gumani River Basin, India. Sci Total Environ 139197

Sahu U, Panaskar D, Wagh V, Mukate S (2018) An extraction, analysis, and prioritization of Asna river sub-basins, based on geomorphometric parameters using geospatial tools. Arab J Geosci 11(17):517

Samanta RK, Bhunia GS, Shit PK, Pourghasemi HR (2018) Flood susceptibility mapping using geospatial frequency ratio technique: a case study of Subarnarekha River Basin, India. Model Earth Syst Environ 4 (1):395–408

Sanyal J, Lu XX (2004) Application of remote sensing in flood management with special reference to monsoon Asia: a review. Nat Hazards 33(2):283–301

Sanyal S, Sengupta P (2012) Metamorphic evolution of the Chotanagpur Granite Gneiss Complex of the East Indian Shield: current status. Geol Soc Lond Spec Publ 365(1):117–145

Schumm SA (1956) Evolution of drainage systems and slopes in badlands at Perth Amboy, New Jersey. Geol Soc Am Bull 67(5):597–646

Singh MP, Singh PK (1996) Petrographic characterization and evolution of the Permian coal deposits of the Rajmahal basin, Bihar, India. Int J Coal Geol 29(1–3):93–118

Smith KG (1950) Standards for grading texture of erosional topography. Am J Sci 248(9):655–668

Strahler AN (1952) Dynamic basis of geomorphology. Geol Soc Am Bull 63(9):923–938

Strahler AN (1957) Quantitative analysis of watershed geomorphology. Eos Trans Am Geophys Union 38 (6):913–920

Strahler AN (1964) Part II. Quantitative geomorphology of drainage basins and channel networks. In: Handbook of applied hydrology. McGraw-Hill, New York, pp 4–39

Strahler AN (1968) Quantitative geomorphology. In: The encyclopedia of geomorphology, pp 898–912

Sujatha ER, Sridhar V (2019) Mapping of erosion susceptibility using a weighted linear combination model: a case study of a hill sub-watershed in Kodaikkanal, Western Ghats, South India. Remote Sens Appl: Soc Environ 14:34–45

Thakkar AK, Dhiman SD (2007) Morphometric analysis and prioritization of miniwatersheds in Mohr watershed, Gujarat using remote sensing and GIS techniques. J Indian Soc Remote Sens 35(4):313–321

Tirkey AS, Ghosh M, Pandey AC, Shekhar S (2018) Assessment of climate extremes and its long term spatial variability over the Jharkhand state of India. Egypt J Remote Sens Space Sci 21(1):49–63

VenTe C (1964) Handbook of applied hydrology: a compendium of water-resources technology

Wani SP, Anantha KH, Sreedevi TK, Sudi R, Singh SN, D'Souza M (2011) Assessing the environmental benefits of watershed development: evidence from the Indian semi-arid tropics. J Sustain Watershed Sci Manag 1(1):10–20

Waugh D (2000) Geography: an integrated approach. Nelson Thornes

Yogesh D, Mahesh S, Ravindra J, Sanjay P (2016) Application of watershed Erosion response model in planning resource conservation of Dehrang catchment, district Raigad. Univers J Environ Res Technol 6(1)

Yoon K (1987) A reconciliation among discrete compromise solutions. J Oper Res Soc 38(3):277–286

20 Runoff Estimation of the Kolong River Basin in Assam, India Using NRCS-Curve Number Method and Geospatial Techniques

Manash Jyoti Bhuyan, Debashree Borah, Binod Kumar Nath, Nityananda Deka, and Ashok Kumar Bora

Abstract

Rainfall and runoff contribute significantly to the functioning of the hydrological cycle and thus constitute the most integral components of the hydrological environment of a region. The surface runoff generated basically by rainfall is highly responsible for floods in the floodplains of a river basin. Therefore, it is very important to find out the complex and intricate rainfall-runoff relationship of a river basin in order to understand its hydrological environment, on the one hand, and to manage the associated fluvio-geomorphic problems on the other. The estimation of runoff also helps in watershed management practices. The present paper is, therefore, an attempt to investigate and estimate the surface runoff of the Kolong river basin in Assam considering the rainfall data series of 2004–2018. An analysis of runoff frequency assessment has been carried out in the study to examine the probabilities of occurrence and their corresponding recurrence intervals attached to the estimated runoff magnitudes of the basin. The Natural Resource Conservation Service Curve Number (NRCS-CN) model has been applied integrating with the Remote Sensing and GIS techniques to estimate and predict the runoff volume based on the rainfall pattern of the given years. The curve number (CN) method, also known as the hydrological soil cover complex, takes into consideration several properties of a basin, like soil permeability, land use, and antecedent moisture conditions (AMCs). In this regard, streamflow, hydrologic soil groups (HSGs), slope, and land use land cover maps have been generated using satellite images in a GIS environment. However, the CN parameter values corresponding to various HSGs and land use and land cover conditions of the basin have been selected from the NRCS standard table.

Keywords

Water resource management · Runoff estimation · Kolong river basin · NRCS-CN method

M. J. Bhuyan (✉)
Department of Geography, Nowgong Girls' College, Nagaon, India

D. Borah
Department of Geography, Arya Vidyapeeth College, Guwahati, India

N. Deka · A. K. Bora
Department of Geography, Gauhati University, Guwahati, India

B. K. Nath
Assam Survey & Settlement Training Centre, Guwahati, India

P. K. Shit et al. (eds.), *Drainage Basin Dynamics*, Geography of the Physical Environment,
https://doi.org/10.1007/978-3-030-79634-1_20

20.1 Introduction

Runoff is one of the important hydrologic variables adopted in most of the water resource studies (Nikam et al. 2010; Kumar and Viswanadh 2017) due to its significant contribution in balancing the hydrological cycle (Sitterson et al. 2017). The surface runoff is basically generated by rainfall and as such, its variation happens according to the spatio-temporal variations of rainfall across the basin. It is noteworthy that about one-third of the precipitation that falls on the earth's surface turns into runoff while the other two-thirds is lost to the atmosphere through evaporation, transpiration, and infiltration into the soil (Jain et al. 2007; Perlman 2016; Sitterson et al. 2017). Runoff plays an important role in balancing the hydrological cycle by returning the excess precipitation to the oceans and also controlling the amount of water that flows into stream systems (Perlman 2016). Thus, surface runoff forms an important area of interest for monitoring water resources and their management, flood assessments and mitigation, and ecohydrological relationships in a watershed.

On the contrary, it is necessary to understand the intricate and complex relationship of rainfall-runoff because surface runoff is directly responsible for the occurrence of floods in a watershed (Aghil and Rajashekhar 2018). Hence, the status of water availability and its variations in the spatio-temporal context within a river basin is essential for planning and decision-making on water resource management (Zade et al. 2005). Since runoff occurs when the infiltration capacity of the land exceeds, it is an obvious indication of water availability (Rama et al. 2014). Thus, in situ runoff measurement is vital for such planning and management purposes. However, most of the watersheds in India are ungauged due to several socio-economic constraints (Sarangi et al. 2005; Zade et al. 2005; Ningaraju et al. 2016). Again, runoff measurements using conventional techniques are expensive, time-consuming, and difficult (Zade et al. 2005), for which rainfall-runoff models using geo-spatial tools and techniques are, at present, commonly used for computing runoff (Chattopadhyay and Choudhury 2006). It is noteworthy that besides water resource management and flood forecasting (Kokkonen et al. 2001), the modeling surface runoff can also help in understanding, controlling, and monitoring the quality and quantity of water resources of a catchment (Sitterson et al. 2017).

It should be put to notice that although there are several models used for estimating direct runoff in a river basin, most of them are input data-intensive and need large calibration (Nikam et al. 2010). Therefore, such models used for planning and decision-making should be simple ones (Grayson et al. 1992) demanding few data requirements and clearly stated assumptions (Nikam et al. 2010). The Soil Conservation Service Curve Number (SCS-CN), presently known as Natural Resource Conservation Service Curve Number (NRCS-CN) model, developed by the United States Department of Agriculture (USDA)-Soil Conservation Service (USDA-SCS 1972), has been widely used and largely accepted by fluvio-geomorphologists, hydrologists, water resource planners, and engineers for the estimation of surface run-off (Pandey and Stuti 2017). Therefore, this model has been executed in the Kolong river basin of Assam to estimate its runoff and to analyze its probability.

20.2 Materials and Methods

20.2.1 Study Area

The Kolong river basin taken is a sub-basin of the river Brahmaputra. It covers parts of Nagaon, Morigaon, and Kamrup Metro districts of Assam. The basin is confined within 26° 4′ 12″ N to 26° 6′ 12″ N latitudes and from 91° 55′ 55″ E to 93° 5′ 53″ E longitudes (Fig. 20.1). It is bounded by Sonitpur district in the north, Darrang in the northwest, Kamrup (M) toward west, West Karbi Anglong in South, Hojai district toward south-east, and Karbi Anglong in the eastern side. It has a total area of 3478.5 km^2 with elevation varying

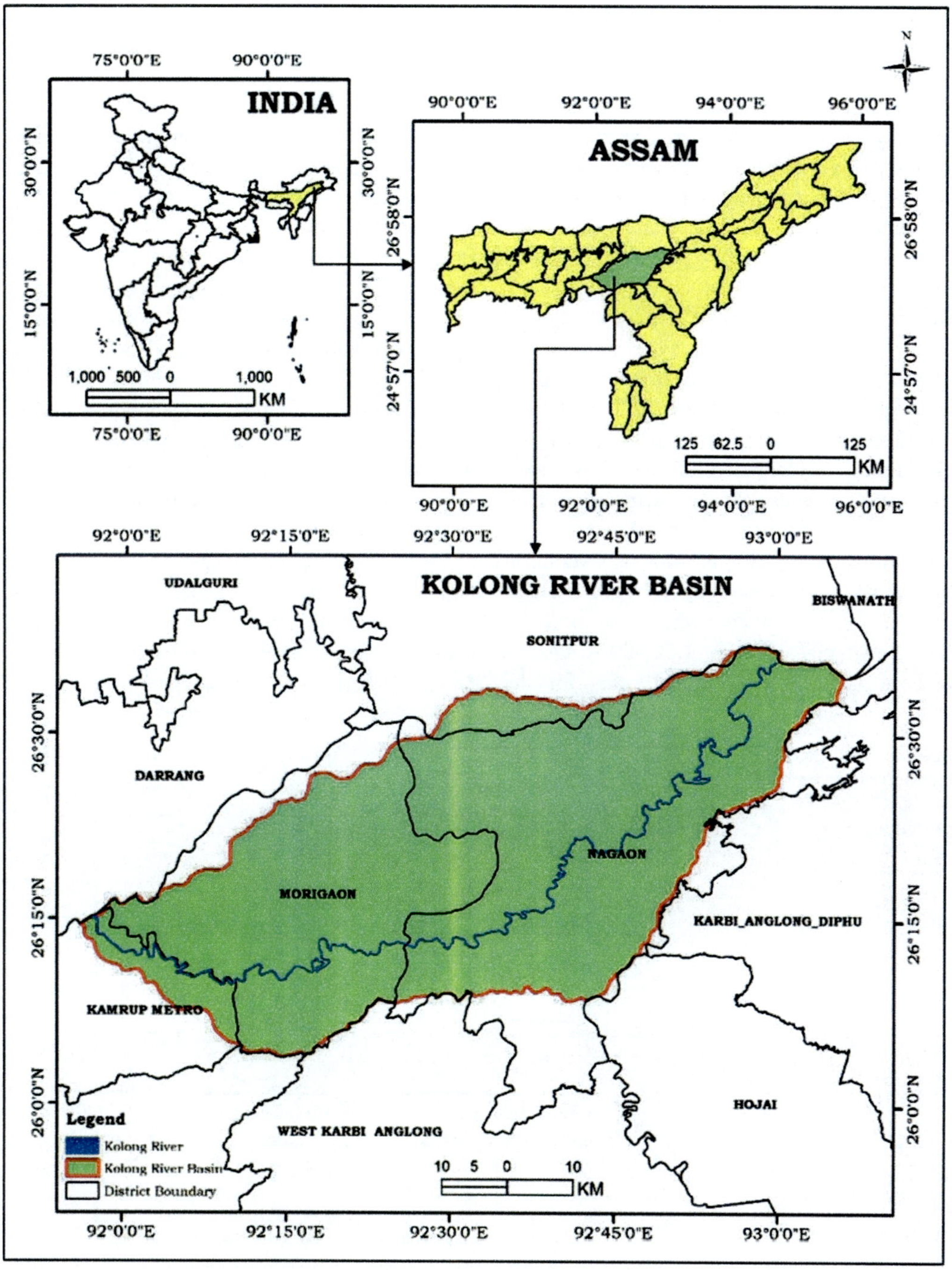

Fig. 20.1 Location of the Kolong river basin

from 0 to 607 m. The climate of the region is characterized by sub-tropical humid climate wherein rainfall reaches its peak during the monsoon with an average annual rainfall of 1541 mm. During the summers, the mean daily temperature ranges from 34 to 24 °C while the winter season experiences a maximum of 25 °C to a minimum of 11 °C.

20.2.2 Database and Methodology

A detailed methodology has been undertaken for the present study wherein a series of data sets are adopted in order to develop a runoff model. For this purpose, demarcation of the river basin, collection of rainfall data, and mapping of land use and land cover, slope, and soil texture have been considered. Initially, the Kolong river basin has been delineated from ALOS PALSAR DEM (30 m) and then verified with Survey of India (SOI) topographical map of 1:50,000 scale. The foremost parameter, i.e. rainfall data for the period 2004–2018 of the study area, has been collected from the Indian Meteorological Department (IMD). Secondly, based on LANDSAT-8 satellite imagery of 2019, the LULC map has been generated wherein supervised classification applying maximum likelihood classifier is adopted in ArcGIS. Another parameter, i.e. soil texture map, has been constructed based on the National Bureau of Soil Survey and Land Use Planning (NBSS&LUP), and then the derived map is converted into hydrologic soil groups as A, B, C, and D according to their soil infiltration capacity. The LULC map has been superimposed over a hydrologic soil map and on this basis, a Curve Number (CN) is assigned for the construction of a soil cover complex map. Further, this process is followed by integrating the derived soil cover complex map and Antecedent Moisture Condition (AMC), allotting actual curve number (CN) values thereby calculating maximum retention (S) and initial abstraction (*Ia*) that finally leads to runoff estimation (Fig. 20.2).

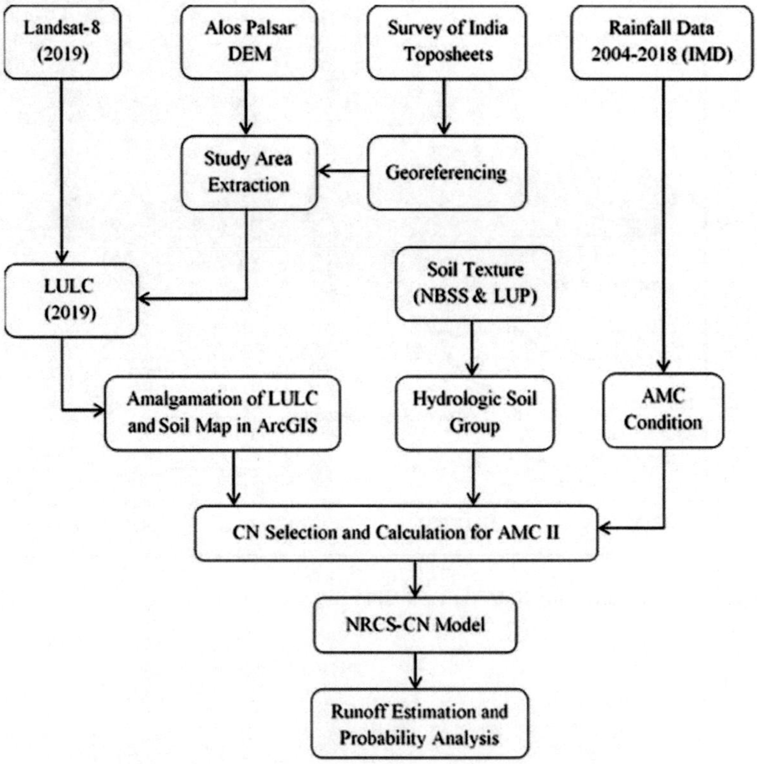

Fig. 20.2 Methodology adopted to estimate runoff using NRCS-CN model and probability analysis

20.2.3 NRCS-CN Model

The NRCS-CN model previously known as the SCS-CN model has been established in 1954 by the United States Department of Agriculture (USDA)-Soil Conservation Society (SCS) (Rallison 1980) and has been widely used for predicting direct runoff or infiltration from rainfall after initial losses (USDA 1986; Zhang and Peralta 2019). This model takes into account the major factors affecting runoff in a basin like soil types, land use, slope, and antecedent moisture conditions (AMCs), and this becomes the primary reason for its wide applicability and acceptability (Mishra and Singh 2003).

The CN method is based on the assumption of proportionality between retention and runoff (Bansode and Patil 2014). The mathematical relation for runoff is

$$Q = \frac{(P - Ia)^2}{P - Ia + S} \quad (20.1)$$

where Q = runoff (mm),

P = rainfall depth (mm),

S = potential maximum retention after runoff begins (mm), and

Ia = initial abstraction (mm).

Initial abstraction (*Ia*) is all the losses before runoff starts. It thus includes surface storage, interception, evaporation, and infiltration prior to the runoff in the watershed (Satheeshkumar et al. 2017). *Ia* has been taken as $0.3S$ for Indian condition (Ahmad et al. 2015) and thus, the empirical relationship can be expressed as

$$Ia = 0.3S \quad (20.2)$$

Substituting Eq. (20.2) in Eq. (20.1), we get

$$Q = \frac{(P - 0.3S)^2}{P + 0.7S} \quad (20.3)$$

S = the potential infiltration after the runoff begins given by the following equation:

$$S = \left(\frac{25400}{\text{CN}}\right) - 254 \quad (20.4)$$

where CN is Curve Number.

The CN values (a dimensionless number ranging from 0 to 100) have been taken from the SCS Handbook of Hydrology (NEH-4), (USDA-SCS 1972) based on land use/land cover (Fig. 20.3), HSG, and AMC. Although the NRCS-CN method is originally designed for use in small river basins having an area of less than 40 square kilometers or 4000 hectares (Roy et al. 2010), it has been modified for runoff estimation in larger watersheds by weighing curve numbers with respect to basin land cover area. The equation for computing the weighted curve number is given below

$$\text{CN}_w = \sum \text{CN}_i \times \frac{A_i}{A} \quad (20.5)$$

where CN_w is the weighted curve number;

CN_i is the curve number from 1 to any number N;

A_i is the area with curve number CN_i; and

A is the total area of the watershed.

20.2.4 Antecedent Moisture Condition

Antecedent Moisture Condition (AMC) is an indicator of the availability of water content in the soil prior to a storm event. It is determined by total rainfall in a 5-day period preceding a storm (Vinithra and Yeshodha 2013). The physical characteristics of the basin, namely LU/LC and soil types of the basin AMC and recharge capacity of the basin, are the basic requirement for the determination of the curve number method (Jasrotia et al. 2002). The Natural Resource Conservation Service (NRCS), previously known as Soil Conservation Service (SCS), had developed three antecedent soil moisture conditions such as AMC-I, AMC-II,

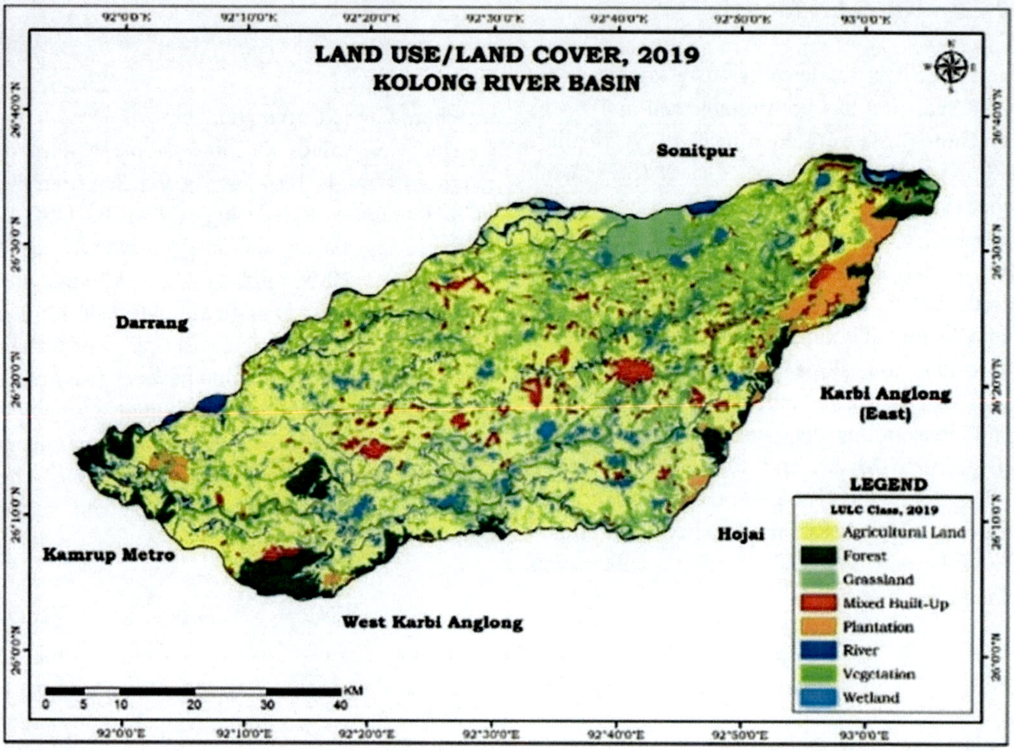

Fig. 20.3 LULC of the basin

and AMC-III which represent dry, normal, and wet conditions, respectively (Amutha and Porchelvan 2009; Pandey and Stuti 2017) for dormant and growing seasons as shown in Table 20.1. However, in the present study, the average condition, i.e. AMC-II is selected to derive the CN value for this purpose.

20.2.5 Hydrologic Soil Group (HSG)

Soil types play an important role in the generation of runoff as different soil types have different rates of infiltration. Infiltration rates of soils vary accordingly with the nature of subsurface permeability. For the estimation of runoff, curve number (CN) values for each soil type are required to be assigned.

Table 20.1 AMC classes for determination of CN values

AMC group	Soil characteristics	Five-day antecedent rainfall in mm	
		Dormant season	Growing season
I	Soils are dry but not to the wilting point; satisfactory cultivation has taken place	Less than 13	Less than 36
II	Average condition	13–28	36–53
III	Heavy rainfalls or light rainfall and low temperatures have occurred within the last 5 days; stared soil	Over 28	Over 53

Source Amutha and Porchelvan (2009)

As per National Engineering Handbook (NEH) developed by the USDA, HSGs are classified into four groups, namely A, B, C, and D according to the soil's infiltration rate, texture, and water transmission (Table 20.2). In this study, the dominant HSG is B representing loam and silt loam soils with moderate infiltration capacity as shown in Fig. 20.4.

20.2.6 Area Weighted Curve Number

The different layers of HSGs, LULC, and AMC are overlaid one by one, and a new polygon attribute table (PAT) was obtained using ArcGIS. The result obtained from the new PAT is used to calculate the total area-weighted curve number (WCN) of the basin. The computed weighted curve number (WCN) is 74 as presented in Table 20.3.

Moreover, slope also influences substantially in surface runoff generation of a particular region. The slope map of the Kolong basin (Fig. 20.5) depicts that it consists of four slope classes ranging from nearly plain to steep slope. Among them, the highest area of 3281.54 km^2 (Table 20.4) falls under nearly level to a gentle slope category indicating a short-time water holding capacity. On the other hand, the moderate and steep category constitutes 198.145 km^2 representing a longer duration of water holding capacity. The areas covered under moderate to steep slopes will help in enhancing future prospects for artificial recharge sites.

20.2.7 Weibull's Plotting Position

The computed yearly runoff data can further be used for analyzing their frequencies in the basin as it will help in the prediction of flood flows. Though there are several statistical methods of flow frequency analysis, Weibull's Plotting Position method (1939) has been adopted for the study. For the analysis, the data series has been arranged in the decreasing order of magnitude and then ranked. The plotting position formula is

$$T = \frac{n+1}{m}$$

where n = number of events or the years of record,

m = order or rank of the event (flood), and

T = recurrence interval.

The probability of occurrence of a flood (having a recurrence interval T-year) in any year, i.e. the probability of exceedance, is

$$P = \frac{1}{T}$$

or, the percentage chance of its occurrence in any one year, i.e. frequency (F), is

Table 20.2 Soil conservation service classification (USDA-SCS 1974)

Hydrologic soil groups (HSGs)	Soil textures	Water transmission	Runoff potential	Final infiltration
A	Deep, well-drained to excessively drained sands or gravel	Rapid rate	Low	>7.5
B	Moderately deep to deep, moderately well-drained to well-drained soils of moderately fine to moderately coarse texture	Moderate rate	Moderate	3.8–7.5
C	Clay loams, shallow sandy loam, soils with moderately fine to fine textures	Slow rate	Moderate	1.3–3.8
D	Clay soils that swell significantly when wet, heavy plastic and soils with a permanent high water table	Very slow rate	Rapid/high	<1.3

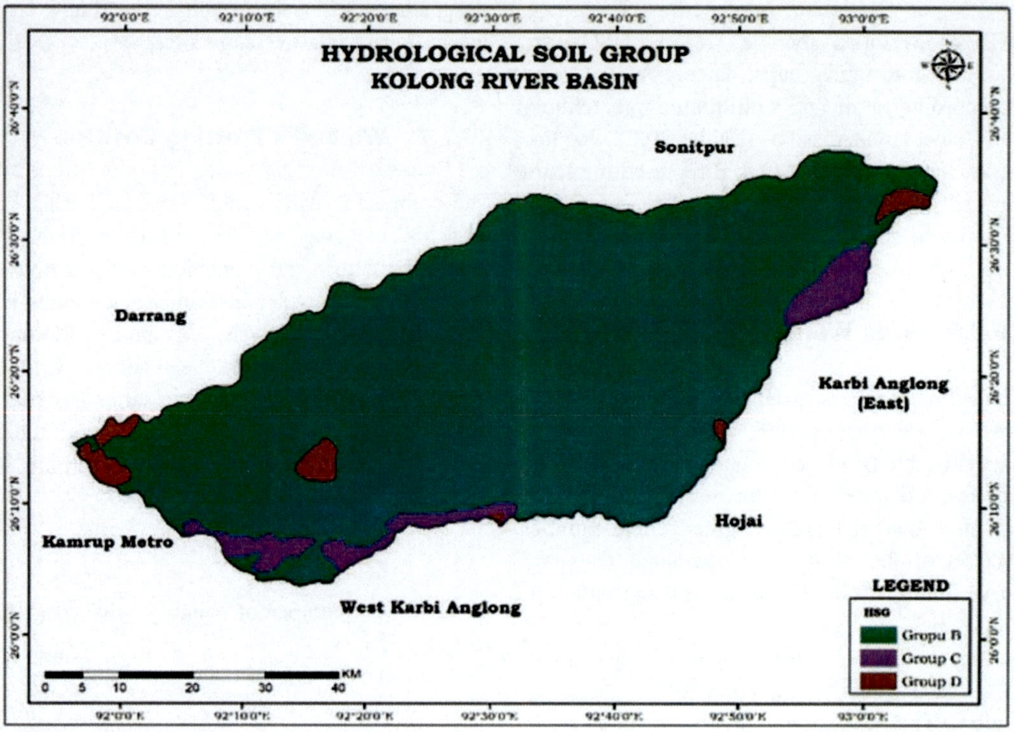

Fig. 20.4 HSG of Kolong river basin

$$F = \frac{1}{T} \times 100$$

20.3 Results and Discussions

The land use and land cover (LULC) map of the Kolong river basin is prepared and interpreted from the satellite imagery (Fig. 20.3). Based on LULC classification, the entire area has been reclassified into eight categories, viz., agricultural lands, forests, grasslands, mixed build-up, vegetation, plantations, rivers, and wetlands (Table 20.3). Among all, the dominant class is found as agricultural lands (50.23%) followed by vegetation-covered areas (24.87%).

The application of the NRCS-CN model in the Kolong river basin taking into consideration several factors like soil types, land use, slope, and antecedent moisture conditions (AMCs) has resulted in runoff estimation from 2004 to 2018. It has been found that the rainfall in the basin varies from 1009.35 to 1723.1 mm, while the runoff varies from 900.76 to 1611.54 mm (2004–2018) (Fig. 20.7). The mean annual runoff and mean runoff volume of the basin for a duration of 15 years are calculated as 1302.546 mm and 4,530,906,261 m^3, respectively. The rainfall-runoff relationship of the Kolong river basinis perfectly positively correlated with a correlation coefficient (r) value being 1 as shown in Fig. 20.6 (Table 20.5).

20.3.1 Runoff Frequency Analysis

It is known that the higher the value of runoff, the less is the probability of the occurrence with the same intensity. In the Kolong river basin, the highest runoff is found in the year 2004 measuring about 1611.54 mm showing a recurrence interval of 16 years, and the frequency of occurrence is 6.25%. This means that such runoff depth may occur in the basin after a recurrence

Table 20.3 Weighted curve number for the Kolong river basin (for AMC II)

Land use/land cover (LULC)	HSG	Area (km^2)		CN	% area	% area × CN	Weighted curve number (WCN)
Agricultural lands	B	1628.91	1747.2 (50.23%)	81	46.82794	3793.063	74
	C	33.9		88	0.974558	85.7611	
	D	84.39		91	2.426046	220.7702	
Forests	B	70.67	160.89 (4.63%)	40	2.031623	81.26491	
	C	90.22		58	2.593647	150.4315	
Grasslands	B	75.5	76.31 (2.19%)	69	2.170476	149.7628	
	D	0.81		84	0.023286	1.956016	
Mixed build-up	B	137.56	148.79 (4.28%)	85	3.954578	336.1391	
	C	10.69		90	0.307316	27.65847	
	D	0.54		95	0.015524	1.474774	
Plantation	B	33.24	100.65 (2.89%)	53	0.955584	50.64597	
	C	64.82		67	1.863447	124.8509	
	D	2.59		71	0.074457	5.286474	
Rivers	B	170.64	253.72 (7.29%)	100	4.905563	490.5563	
	C	4.03		100	0.115855	11.58545	
	D	79.05		100	2.272531	227.2531	
Vegetation	B	847.39	865.22 (24.87%)	50	24.36079	1218.039	
	C	15.06		62	0.432945	26.8426	
	D	2.77		68	0.079632	5.414978	
Wetlands	B	123.8	126.38 (3.63%)	100	3.559005	355.9005	
	C	1.12		100	0.032198	3.219779	
	D	1.46		100	0.041972	4.197211	

Note Figures in the parentheses indicate the percentage to the total basin area

interval of 16 years with the probability of occurrence of 6.25%. Similarly, the lowest runoff has been found as 900.76 mm in 2009 with a recurrence interval of 1 year and the probability of such occurrence is 94.33% (Table 20.6).

It has been found in the study that a 50% probability of occurrence is attached with the runoff magnitude of 1251.37 mm. The Kolong river basin having a mean annual runoff of 1302.546 mm has a probability percentage of 46.92 with a recurring interval of 2 years. The runoff magnitudes having less than 50% probability of occurrence are found in 2005, 2006, 2009, 2012, 2014, 2015, and 2018, while the runoff magnitudes having more than 50% probability are found to occur in 2004, 2007, 2008, 2010, 2011, 2016, and 2017. Thus, the runoff magnitudes having more than 50% probability of occurrence are found in 50% of the total period, i.e. 15 years. On the other hand, the runoff magnitudes having less than 50% probability of occurrence are found in the other 50% of the total period of the study.

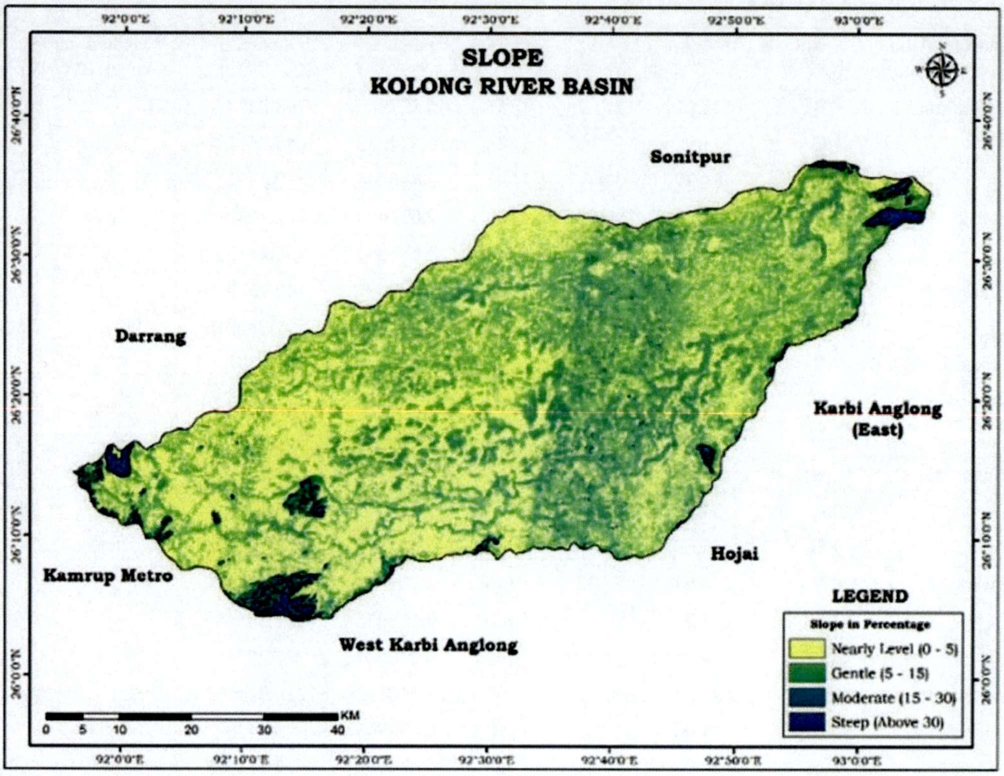

Fig. 20.5 Slope classes of the basin

Table 20.4 Slope classes of Kolong river basin

Slope character	Area covered (km^2)	Status of surface runoff potential
Nearly level	2445.044	Low
Gentle	836.5024	Low
Moderate	143.8922	Moderate
Steep	54.25337	High

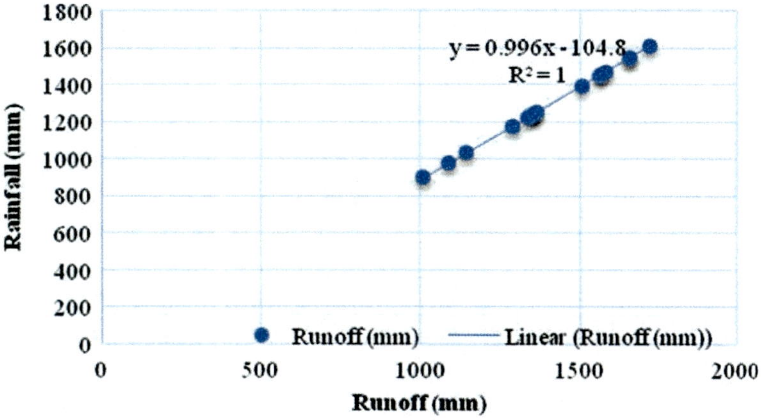

Fig. 20.6 Scatter plot between the rainfall and calculated runoff

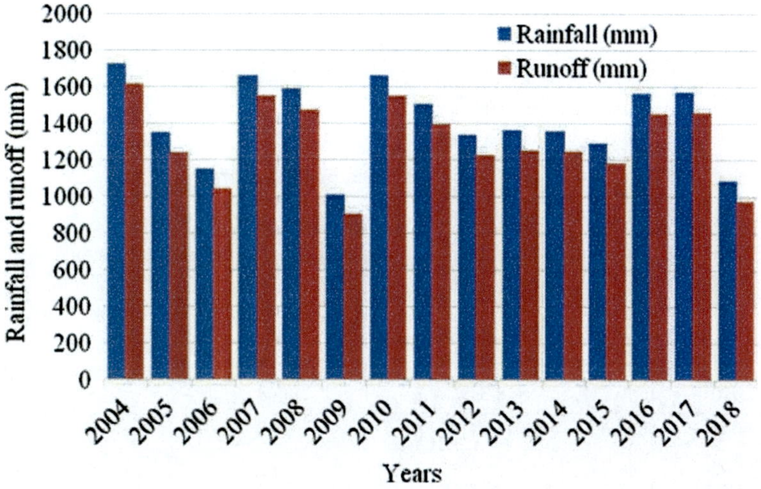

Fig. 20.7 Yearly rainfall and computed runoff using NRCS-CN model

20.4 Conclusion

The present research explains the application of the GIS-based NRCS-Curve Number method to evaluate surface runoff of the Kolong river basin in Assam. In executing the aforesaid model, the considered influencing parameters, viz., LULC, soil, and rainfall have been applied. The rainfall and runoff data observed for 15 years reveals that during 2004, the maximum rainfall and runoff were recorded as 1723.1 mm and 1611.54 mm, respectively. On the contrary, minimum rainfall and runoff are observed in 2009 as 1009.35 mm and 900.76 mm, respectively, and the annual average runoff depth calculated is 1302.543 mm. The rainfall-runoff values represent a perfectly positive correlation between the two and with the rise in slope values, the runoff also increases. During 50% of the time, i.e. the period of study, the runoff in the basin remains above 1251.37 mm. The outcome of the study also provides evidence regarding the flood prediction in the river basin. Thus, this analysis might be input for those who are interested in watershed management practices.

Table 20.5 Calculation of runoff and volume (2004–2018)

Years	Rainfall (mm)	Runoff (mm)	Volume (m^3) = Runoff × Area
2004	1723.1	1611.54	5,605,741,890
2005	1350.15	1239.77	4,312,539,945
2006	1145.7	1036.27	3,604,665,195
2007	1658.6	1547.21	5,381,969,985
2008	1584.35	1473.16	5,124,387,060
2009	1009.35	900.76	3,133,293,660
2010	1659.71	1548.31	5,385,796,335
2011	1506.2	1395.26	4,853,411,910
2012	1337.05	1226.72	4,267,145,520
2013	1361.8	1251.37	4,352,890,545
2014	1353.8	1243.4	4,325,166,900
2015	1288.4	1178.27	4,098,612,195
2016	1563.5	1452.38	5,052,103,830
2017	1567.6	1456.46	5,066,296,110
2018	1086.4	977.31	3,399,572,835
Average	1413.047	1302.546	4,530,906,261

Source Calculated by the authors

Table 20.6 Runoff frequency analysis using Weibull's plotting position method

Years	Rainfall (mm)	Runoff (mm)	Rank of runoff (m)	Recurrence interval (Years) (T)	Percent probability (P)	Frequency (F)
2004	1723.1	1611.54	1	16	0.0625	6.25
2010	1659.71	1548.31	2	8	0.125	12.5
2007	1658.6	1547.21	3	5.3	0.1886	18.86
2008	1584.35	1473.16	4	4	0.25	25
2017	1567.6	1456.46	5	3.2	0.3125	31.25
2016	1563.5	1452.38	6	2.6	0.3846	38.46
2011	1506.2	1395.26	7	2.2	0.4385	43.85
2013	1361.8	1251.37	8	2	0.5	50
2014	1353.8	1243.4	9	1.7	0.5649	56.49
2005	1350.15	1239.77	10	1.6	0.625	62.5
2012	1337.05	1226.72	11	1.4	0.6896	68.96
2015	1288.4	1178.27	12	1.3	0.7692	76.92
2006	1145.7	136.27	13	1.2	0.8130	81.30
2018	1086.4	977.31	14	1.1	0.8771	87.71
2009	1009.35	900.76	15	1	0.9433	94.33

Source Calculated by the authors

References

Aghil, Rajashekhar SL (2018) Estimation of runoff using SCS-CN method for Yelahanka region. Int J Appl Eng Res 13(7):229–233

Ahmad I, Verma V, Verma MK (2015) Application of curve number method for estimation of runoff potential in GIS environment. In: 2nd international conference on geological and civil engineering, vol 80(4), pp 16–20

Amutha R, Porchelvan P (2009) Estimation of surface runoff in Malattar sub-watershed using SCS-CN method. J Indian Soc Remote Sens 37(2):291–304

Bansode A, Patil KA (2014) Estimation of runoff by using SCS curve number method and arc GIS. Int J Sci Eng Res 5(7):1283–1287

Chattopadhyay GS, Choudhury S (2006) Application of GIS and remote sensing for watershed development project—a case study. In: Map India 2006

Grayson RB, Moore ID, McMahon TA (1992) Physically based hydrologic modeling: 2. Is the concept realistic? Water Resour Res 28(10):2659–2666

Jain SK, Agarwal PK, Singh VP (2007) Hydrology and water resources of India, vol 57. Springer Science & Business Media

Jasrotia AS, Dhiman SD, Aggarwal SP (2002) Rainfall-runoff and soil erosion modeling using remote sensing and GIS technique—a case study of tons watershed. J Indian Soc Remote Sens 30(3):167–180

Kokkonen T, Koivusalo H, Karvonen T (2001) A semi-distributed approach to rainfall-runoff modelling—a case study in a snow affected catchment. Environ Model Softw 16(5):481–493. https://doi.org/10.1016/S1364-8152(01)00028-7

Kumar PA, Viswanadh G (2017) Estimation of runoff by using SCS curve number method integrated with GIS. Int Adv Res J Sci Eng Technol 4(7):34–38

Mishra SK, Singh VP (2003) SCS-CN method. Soil conservation service curve number (SCS-CN) methodology. Springer, Dordrecht, pp 84–146

Nikam BR, Garg V, Thakur PK, Aggarwal SP (2010) Rainfall-runoff modelling of Solani watershed using modified NRCS-CN approach in geospatial environment. In: Case studies on real time hydrological modelling for Ganga-Brahmaputra Basins, National Institute of Technology (NIT) Rourkela, India, pp 311–324

Ningaraju HJ, Ganesh Kumar SB, Surendra HJ (2016) Estimation of runoff Using SCS-CN and GIS method in ungauged watershed: a case study of Kharadya mill watershed, India. Int J Adv Eng Res Sci (IJAERS) 3 (5):36–42

Pandey AC, Stuti (2017) Geospatial technique for runoff estimation based on SCS-CN method in upper South Koel River Basin of Jharkhand (India). Int J Hydrol 1 (7):213–220. https://doi.org/10.15406/ijh.2017.01.00037

Perlman H (2016) The water cycle: USGS water science school. Accessed 5 Jan 2017

Rallison RE (1980) Origin and evolution of the SCS runoff equation. In: Symposium on watershed management. American Society of Civil Engineering, pp 912–924, July 1980

Rama SS, Debahsishbhakta, Singh A, Salunkhe SS, Jeyaseelan AT, Sharma JR (2014) Spatial variability of runoff in Cauvery Basin using geo-spatial techniques. In: International symposium on integrated water resources management, CWRDM, Kozhikode, Kerala, India

Roy PS, Dwivedi RS, Vijayan D (2010) Remote sensing applications. National Remote Sensing Centre, Indian Space Research Organisation, Hyderabad, India

Satheeshkumar S, Venkateswaran S, Kannan R (2017) Rainfall–runoff estimation using SCS–CN and GIS approach in the Pappiredipatti watershed of the Vaniyar sub-basin, South India. Model Earth Syst Environ 3(1):24

Sarangi A, Bhattacharya AK (2005) Comparison of artificial neural network and regression models forsediment loss prediction from Banha watershed in India. Agric Water Manag 78(3):195–208

Sitterson J, Knightes C, Parmar R, Wolfe K, Muche M, Avant B (2017) An overview of rainfall-runoff model types. U.S. Environmental Protection Agency

USDA-SCS (1972) National engineering handbook. Hydrology Section 4, Chapters 4–10. United States Department for Agriculture, Washington, D.C.

USDA-SCS (1974) Soil survey of Travis County, Texas. Texas Agricultural Experiment Station, College Station, TX, United States Department for Agriculture, Washington, D.C.

USDA (1986) Urban hydrology for small watersheds TR-55. Technical release. United States Department for Agriculture Soil Conservation Service, Washington, D.C.

Vinithra R, Yeshodha L (2013) Rainfall–runoff modelling using SCS–CN method: a case study of Krishnagiri District, Tamilnadu. Int J Sci Res 6:35–39

Zade M, Ray SS, Dutta S, Panigrahy S (2005) Analysis of runoff pattern for all major basins of India derived using remote sensing data. Curr Sci 1301–1305

Zhang J, Peralta RC (2019) Estimating infiltration increase and runoff reduction due to green infrastructure. J Water Clim Change 10(2):237–242

21 Geomorphological Analyses of Third-Order Basins in Southwestern Nigeria

A. O. Olusola, O. D. Onafeso, O. A. Fashae, and S. Adelabu

Abstract

Drainage basins play a major role in the understanding of process and form dynamics within a landscape and they are referred to as fundamental physical units recognized in hydro-geomorphological studies as having a network of inter-related attributes. Until recently, information about streams in humid tropics was limited to their description in relation to geological structure. There are gaps to fill as regards but not limited to the mechanics of how river channels and networks evolve, lithological units and channel network patterns. This study aims to provide an answer to the question 'what role do lithology play in channel network orientation and pattern?' Answers will be provided using slope–area curves, rose diagrams and background information on long profiles extracted from Digital Elevation Models, DEMs (30 m) within sub-basins in Upper Ogun River Basin. Six third-order basins were selected systematically and randomly for the study. Topographic properties were extracted using System for Automated Geoscientific Analysis (SAGA) version 4.0.0, a GIS software. RockWare software was used in creating rose diagrams for lineament patterns and channel networks. S–A plots show that within the bedrock channels, sediment is limited and processes operating within these channels are sometimes non-fluvial and are most often than not well defined. Therefore, transiting into bedrock channels could produce inflection points that are not quite distinct as slope–area solutions might not have enough to account for the inherent hydro-dynamic variations operating therein.

Keywords

Drainage basins · Channel network pattern · Digital Elevation Models · Upper Ogun Basin

21.1 Introduction

Central to any fluvial geomorphological investigations are forms on the earth's surface and the processes acting on them to produce these forms as well as destroy or (de)shape them. Essentially, the constituents of form or morphology can be listed as (1) constitution (2) configuration (size, shape and other geomorphic properties) and

A. O. Olusola (✉) · O. A. Fashae
Department of Geography, University of Ibadan, Ibadan, Nigeria

O. D. Onafeso
Department of Geography, Olabisi Onabanjo University, Ago-Iwoye, Ogun-State, Nigeria

A. O. Olusola · S. Adelabu
Deparment of Geography, University of the Free State, Bloemfontein, South Africa

P. K. Shit et al. (eds.), *Drainage Basin Dynamics*, Geography of the Physical Environment,
https://doi.org/10.1007/978-3-030-79634-1_21

(3) mass-flow characteristics (discharge, velocity, etc.) (Huggett 2007). These form variables contrast with dynamic variables (representing an expenditure of energy) associated with geomorphic processes, such as power, energy flux, force, stress and momentum (Huggett 2007). While geomorphic processes are complex interrelationships between the chemical and physical means by which the earth's surface undergoes modification (Thornbury 1954), they are driven by controls emanating from inside the earth (endogene processes); originating at or near the earth's surface (exogene processes); and coming from outside the earth (extra-terrestrial processes) (Huggett 2007).

Drainage basins play a major role in the understanding of the process and form dynamics within a landscape and they are referred to as fundamental physical units recognized in hydro-geomorphological studies as having a network of inter-related attributes (Gregory and Walling 1973). The terrain of most humid tropical environments are mostly low-relief except for tectonically induced features such as fold mountain chains, fault scarps, fault valleys, horsts and rift valleys found in tectonically unstable parts. The terrain consists of plains on which varying proportions of the basal surface of weathering are exposed so that the plains are diversified by different elevations. Due to the location of humid tropic environments, most streams flowing within this region are perennial with cases of ephemeral streams. The main provider of water to rivers in humid tropics is rainfall: rivers are, therefore sometimes ephemeral—that is, they carry water only during and immediately after a rain; intermittent—flow during part of the year (the wet season streams), or perennial—flow all year round, as in the rainforest and other continuously wet areas (Faniran and Jeje 2000). Sources of water in channels in the area include karst, basaltic lava, coalescing lines of subsurface water, mass rock debris of forested granitic terrains, etc. (Douglass 1977; Faniran 1982).

Until recently, information about streams in humid tropics was limited to their description about geological structure; leaving out gaps to fill as regards but not limited to the mechanics of how river channels and networks evolve, lithological units and channel network patterns. Most studies especially in temperate regions or river channels with(out) glacial history have examined underlying interactions between process-form dynamics as presented in several works of literature (Olusola 2019). However, the inherent characteristics observed in humid tropical regions as a result of differential weathering suggest that most valley settings and fluvial processes are unique in this area (Olusola et al. 2020; Adeyemi et al. 2020). Therefore, narratives concerning humid tropical basins without glacial history are still growing. The problem then is that available information on basins within a humid tropical region, especially ungauged small headwater basins is too limited for any reasonable and definitive inferences and deductions to be made (Twidale 1976; Faniran et al. 2006). From the foregoing, therefore, there is the need to take advantage of big data (available in Digital Elevation Models (DEMs)) coupled with field-based studies and emerging geo-computational tools to provide an understanding of small headwater basins that are largely ungauged in humid tropical environments. This study seeks to contribute to geomorphic understanding by using Digital Elevation Models (DEMs) to provide interactions between process and form within a drainage basin in Southwestern Nigeria. This will help in pushing the frontier of knowledge as regards locations lagging in geomorphological understandings. Based on the assumption that the hydro-geomorphological characteristics of humid tropical basins are distinct in some ways, the effect, all other things being equal, should be reflected in the forms present in such basins (Wigwe 1966; Jeje 1970; Thorp 1970; Thomas 1974). This study aims to provide an answer to the question 'what role do lithology play in channel network orientation and pattern?' Answers will be provided using slope–area curves, rose diagrams and background information on long profiles.

In interpreting landscape diversity, the concavity index is largely employed (Hack 1973), in relation to tectono-geomorphological, and the influence of base-level on profile morphogenesis. Hence, deviations from this assumption could be said to represent disturbances to an idealised profile condition. Environmental factors such as climate, lithology, vegetation, etc., exercise control through their direct and indirect impact on fluvial process and dynamics. Within a homogenous climatic zone, lithology stands to be one of the most important environmental factors to influence river channel morphology. Rose diagrams present the influence of underlying lithological units on river network orientation and initiation.

21.2 Study Area

21.2.1 Location

The entire Upper and Lower Ogun River Basins, hereafter referred to as the Ogun River Basin (ORB) is within the Southwestern part of Nigeria (latitudes 6° 26′ N and 9° 10′ N; longitudes 2° 28′ E and 4° 8′ E) (Fig. 21.1). The land area is about 23,000 km^2. Generally, the area is underlain by Undifferentiated Basement Complex, intermittent occurrence of Older Granite, and some Sedimentary formations including widespread Alluvium adjoining the Atlantic coast (Faniran 1982). However, this study was conducted within the Upper Ogun River Basin (UORB) made up essentially of the Oyan and Ofiki river systems. It is approximately 200 km long and 140 km wide at its extreme points and basically in a low-relief area (Fig. 21.1). UORB, a low-relief basin in Southwestern Nigeria, presents itself as a fitting setting to bridge the gap and push further the frontier of fluvial geomorphological research. The low-relief landscape of the UORB is not different from other low-relief landscapes in Africa. These landscapes have been carved by the process of deep-weathering leaving out stripped geomorphic features such as hills and mountains associated with the formation of thick

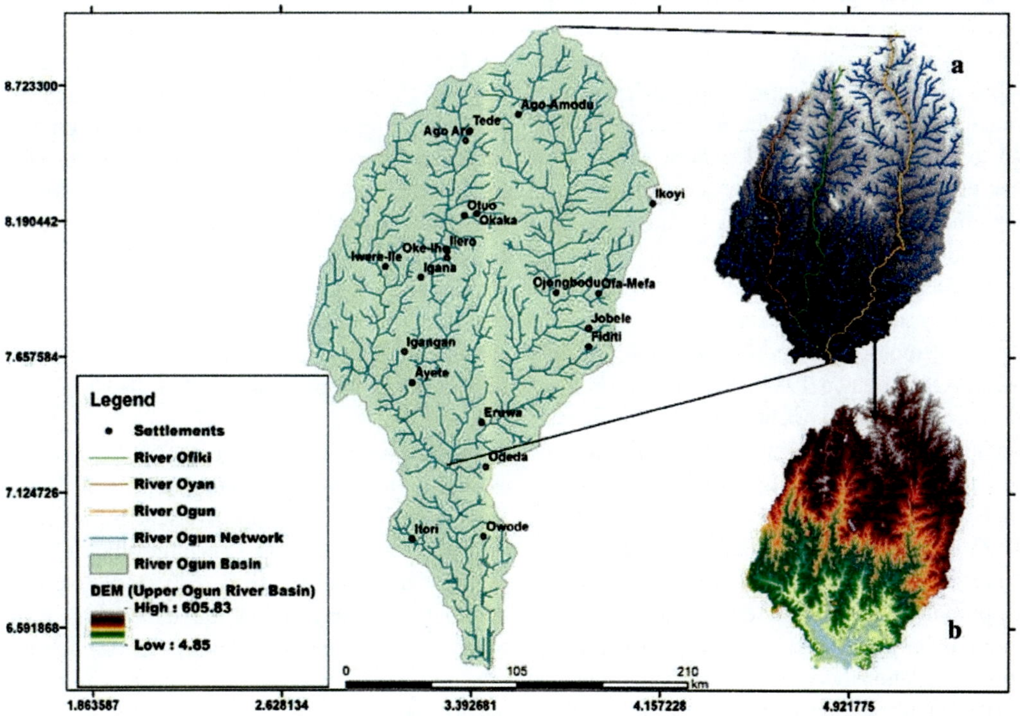

Fig. 21.1 Extent of the Ogun River basin showing a section of the UORB. **a** River network of the UORB with the main river channels. **b** Digital Elevation Model of the UORB. *Source* Olusola (2020)

regolith possessing a highly variable lithology. Across, the entire basin, no deep and complex gullies were observed and more often than not the plains in some instances are truncated by isolated hills (Inselbergs).

21.2.2 Drainage and Relief

The relief is low and rises gently northwards (Fig. 21.1). The relief across the basin from east–west shows a rolling plain consistently dissected by the tributaries of Ogun River (Fig. 21.1), which takes its source at an elevation of about 530 m above mean sea level around Iganra and flows southwards covering a distance of about 480 km, before emptying into the Lagos lagoon. The major tributaries of the Ogun River are the Oyan, Ofiki and Opeki Rivers (Fig. 21.1).

21.2.3 Climate and Vegetation

The climate of the Western Littoral region of Nigeria falls under a broad homogenous condition of tropical climate (Aw), according to Koppen climate classification. Although, there is a marked distinction into the wet (rainy) season and dry season, the rainfall pattern shows two peak periods with the little dry season in August when the rains break. The long-term average air temperature for the entire region ranges within 23 °C mean minimum in June–August, the peak of the rainy season and a mean maximum of 28 °C in December–February the peak of the dry season (Oguntoyinbo 1982). Rainfall varies in a north-west/south-east orientation with the highest rainfall of above 1800 mm/year experienced in the south-east corner of the region towards the coast and the least rainfall of less than 1000 mm/year experienced in the north-west axis of the region. Temporally, however, the rainfall distribution follows a pattern similar to other atmospheric parameters with clear month-by-month variations. As the rainfall amounts become significant reaching up to 40 mm around March and exceeding 100–150 mm by July with an August break around August and then increasing again to the 100–150 mm and above peak by September–October, thereby causing what is known as the "double peak". In the same vein, the potential evapotranspiration (PET) being the true index of wetness and dryness of an area shows a similar trend to that of the precipitation. The PET ranges between a little above 1500 mm/year and a little less than 1000 mm/year (Oguntoyinbo 1982).

The two major vegetation zones that can be identified within UORB are wide expanses of savanna fringe land in the Oke-Ogun area (from Iseyin moving towards Saki) in the north and central parts of the basin and the rainforest that covers the remaining portion of the basin towards the south. Economic trees found across the UORB includes but are not limited to Iroko, Teak, Mahogany, Palm, Gmelina, etc.

21.2.4 Geology

The geology of the study area is an extension of the Precambrian Basement within southwestern Nigeria (Jones and Hockey 1964; Oyawoye 1972; Burke et al. 1972). The UORB sits on the basement complex (Precambrian) which are largely metamorphic rocks in origin (Faniran 1982). Generally, the basement complex of southwestern Nigeria reveals spatial differentiation as regards grain size distribution and mineral composition. The rocks are quartzite, schist consisting essentially of quartz with small amounts of white micaceous minerals, folded gneiss, schist and quartzite complexes (Fig. 21.2), which belong to the older intrusive series (De Swardt et al. 1965). Although in many places outcrops are visible, large areas are overlain by regolith formed by weathering and decomposition of the parent rock material. It is expected that the lithology and structure of the underlying rock control the entire morphology of the river; causing them to respond in varying degrees in relation to process and form (Fig. 21.2).

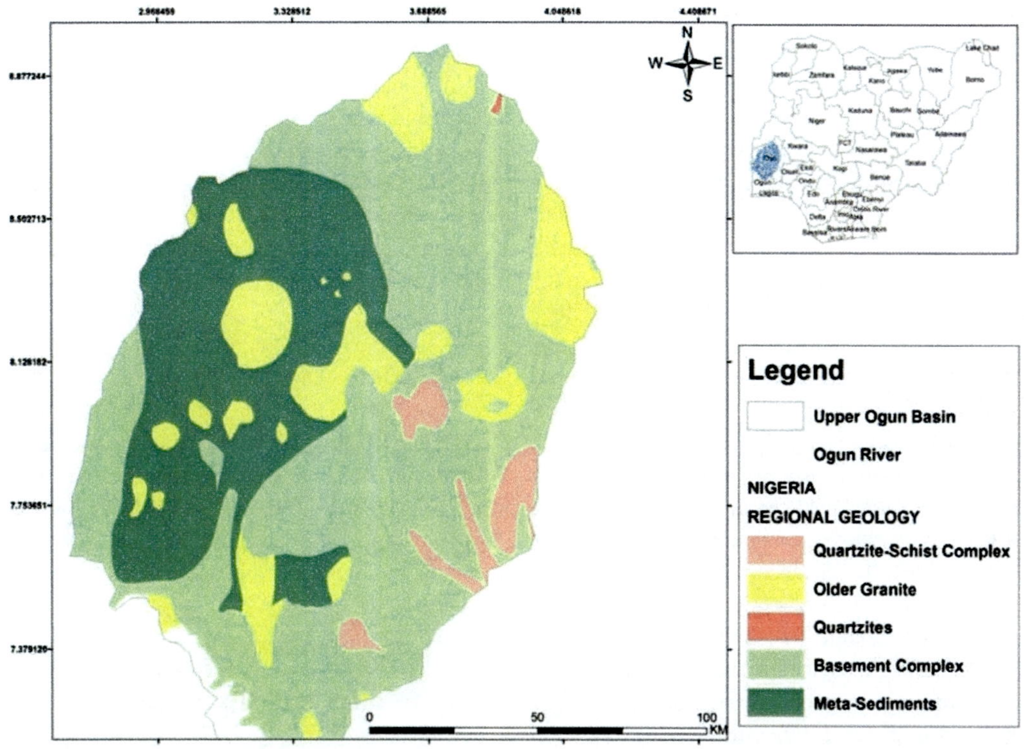

Fig. 21.2 Regional Geology of the UORB

21.3 Methodology

21.3.1 Types and Sources of Data

The study employed archival (secondary) and field (primary) data. The archival data used were obtained from the following sources: (i) topographical maps at 1:50,000, (ii) geological maps of Saki and Ibadan at 1:250,000 and Digital Elevation Models, DEMs (30 m resolution). Primary data consisted of extracted variables from the DEMs.

21.3.2 Extraction of Channel Network Using Digital Elevation Models (DEMs)

The base data layer for analysis was a 30 m Digital Elevation Model (Shuttle Radar Topography Mission, DEM) projected in Universal Traverse Mercator Zone 31 N based on Nigerian Grid Datum. The directions of surface water flow, flow accumulation area and stream network coverage were obtained from a filled DEM based on the methods of Jenson and Domingue (1988), while a variety of methods were used to determine an appropriate flow accumulation area for a stream (Montgomery and Dietrich 1988; Reinfelds et al. 2004; Hayakawa and Oguchi 2006). The study extracted channel profiles from the DEM (Fig. 21.1). The extracted profiles from each basins were validated by comparing them with the 1:50,000 maps information. The steps used in extracting the channel networks from UORB were carried out using ArcGIS 10.2 by importing the DEM ArcGIS, clipping the extent of the basin, and extracting stream network involves filling of depression in DEMs, assignment of flow directions on filled DEM and calculation of flow accumulation on flow direction DEM. Using hydrological tools of the ArcToolbox. Once the stream network was extracted, the Strahler method for ordering was employed to order the stream networks. After ordering, the raster was converted into a vector, and 17 third-order basins were identified across the UORB.

Drainage area (DA), for each site was calculated using DEMs downloaded from SRTM. DA was calculated using the Watershed function (Hydrology tool) in ArcGIS 10.1. The tool uses a flow direction raster in conjunction with pour points to delineate drainage basins and their associated drainage areas. The distance from the headwaters (D) was calculated using the Flow Length Up function in the Hydrology tool. Slope (per cent) was extracted from the DEM in ArcGIS 10.1 environment. A slope surface was created from the DEM using the spatial analyst tool. To extract slope and drainage area along the main channel, the distance tool was used in the spatial analyst toolbox using the cost path and cost backlink module. These modules will extract area and flow distance downstream, while the extraction tool using the sample module will help generate an Excel (*.dbf*) table having drainage area, slope, elevation, longitude and latitude.

21.3.3 Basin Determination and Procedure

To determine the basins to sample out of the 17 third-order basins, a systematic sampling technique was employed. The first step was to identify basins that were heterogeneous based on the lithological variation. For a basin to be heterogeneous, it must have more than two lithological units within the basin, none of which must be more than 65% in terms of areal coverage across the basin (Olusola et al. 2020). Applying this yardstick, six basins (Table 21.1) were selected from the outcome (Fig. 21.3).

21.3.4 Data Analyses

Topographic properties were extracted using System for Automated Geoscientific Analysis (SAGA) version 4.0.0, a GIS software. RockWare software was used in creating rose diagrams for lineament patterns and channel networks.

21.4 Results and Discussion

21.4.1 Geomorphic Description of the Selected Six Third-Order Basins

The entire Upper Ogun Basin in terms of elevation is within 3m and 607m. The physiography is a rolling terrain including Inselbergs and hills scattered across the entire landscape. To a large

Table 21.1 Selected third-order basins

Basin ID	Basin name	Geology
1	SAKI-OGBORO	OGe—Medium to coarse grained hornblende; OGp—Coarse porphyritic and biotite hornblende granite; and M—Migmatite
2	IGBOBURO	OGp—Coarse porphyritic and biotite hornblende granite; M—Migmatite; and OPg—Prophyroblastic gneiss
3	AWON	Su—undifferentiated Schists; Qs—Quartzite Schist; and P—Pegmatite
4	ONIKOKO	OGe—Medium to coarse grained hornblende; M—Migmatite; and OGp—Coarse porphyritic and biotite hornblende granite
5	ODO-OBA	OGb—Coarse prophyritic biotite and biotite muscovite granite; OGh—Coarse ptophyritic hornblende granite; and M—Migmatite
6	AYIN	OGu –Undifferentiated Older Granite; Su—Undifferentiated Schists; OGp—Coarse porphyritic and biotite hornblende granite

Source Olusola (2019)

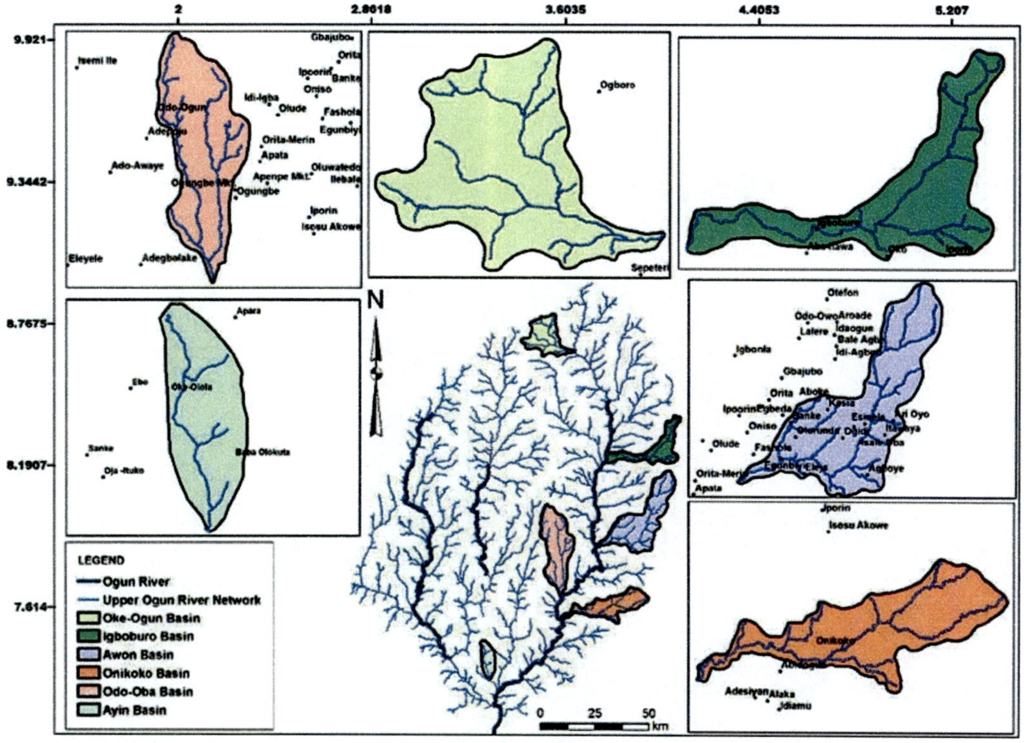

Fig. 21.3 Selected third-order basins. *Source* Adeyemi et al. (2020)

extent, the upper section has witnessed intense weathering of its profiles which is expected because of the humid condition present within the region.

21.4.1.1 Oke-Ogun Basin

Oke-Ogun is located within the northernmost section of the UORB, within latitudes 8° 47' N and 8° 39' N and longitudes 3° 25' E and 3° 40' E (Fig. 21.3). The land area is about 226.68 km^2 (Fig. 21.4). Oke-Ogun Basin cut into the Basement Complex landscape, with outcrops of resistant rocks mainly within the porphyritic and biotite hornblende which occur as inselbergs. Some of the channels within the basin were ephemeral while others are perennial (Olusola 2019). The relief of the basin is low and rises gradually northwards (Fig. 21.5a). The relief across the west–east showed a downwash of the area under the influence of fluvial processes. The entire relief in a west–east transect was dissected by the Ogun River towards the eastern fringe of the basin (Fig. 21.5b). From the north–south, the relief showed pockets of inselbergs punctuating the rolling plains (Fig. 21.5a). The highest point within the basin is 500 m a.s.l, while the lowest point is 323 m a.s.l. Oke-Ogun Basin is underlain by three major lithological units (Fig. 21.4). These are coarse porphyritic and biotite hornblende (OGp), medium- to coarse-grained hornblende (OGe) and Migmatites (M). Evidence of outcrops is visible along the river channel but mostly, within the OGp, isolated hills of great heights can be seen scattered across the landscape. OGp is to the north, while OGe is to the western part and M is towards the south-eastern part of the basin (Fig. 21.4).

21.4.1.2 Igboburo Basin

Igboburo basin is located towards the eastern part of the UORB (Fig. 21.3). Igboburo basin is within latitudes 8° 22′ N and 8° 12′ N and longitudes 3° 47′ E and 4° 50′ E (Fig. 21.6). The land area is about 222.52 km^2. Igboburo basin

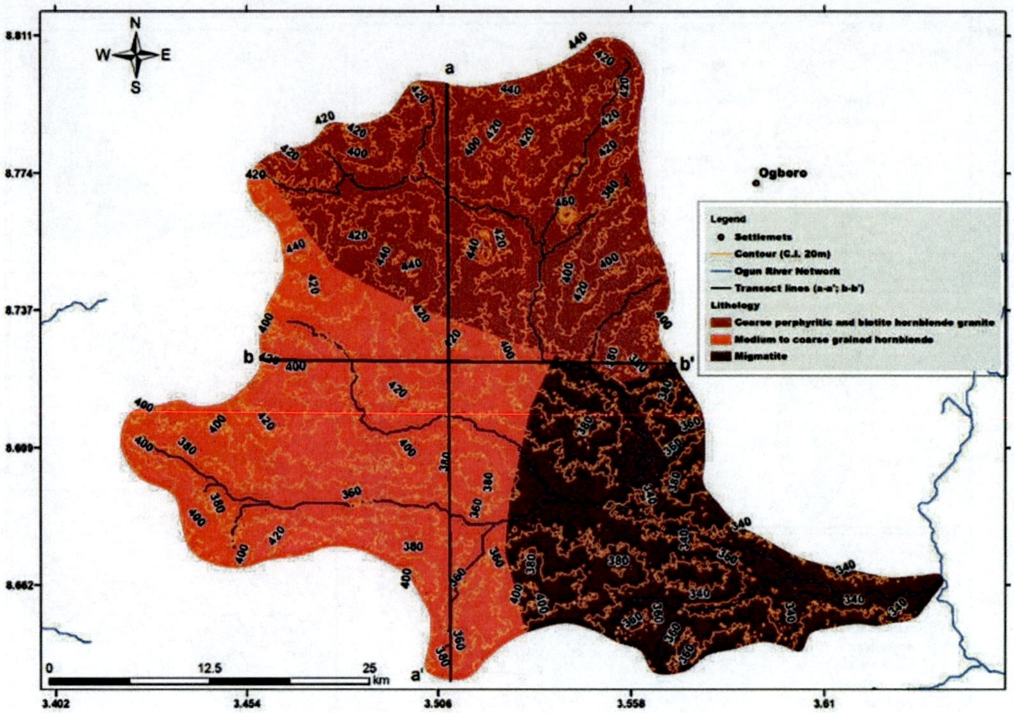

Fig. 21.4 Drainage and topography of Oke-Ogun Basin. *Source* Olusola (2019)

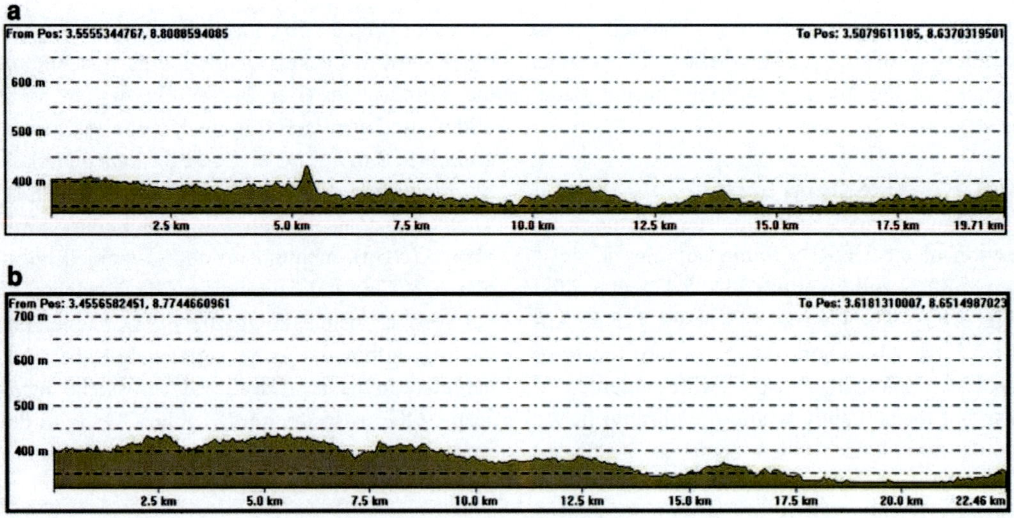

Fig. 21.5 **a** Transect running North–South of Oke-Ogun basin (H.E = 5) (a-a′). **b** Transect running West–East of Oke-Ogun basin (H.E = 5) (b-b′). *Source* Olusola (2019)

cut into the Basement Complex landscape, with outcrops of resistant rocks covering about 50% of the river channel (Olusola 2019). The river channel is mainly ephemeral with pools of water being observed though not flowing during the dry season. The relief of the basin is low and

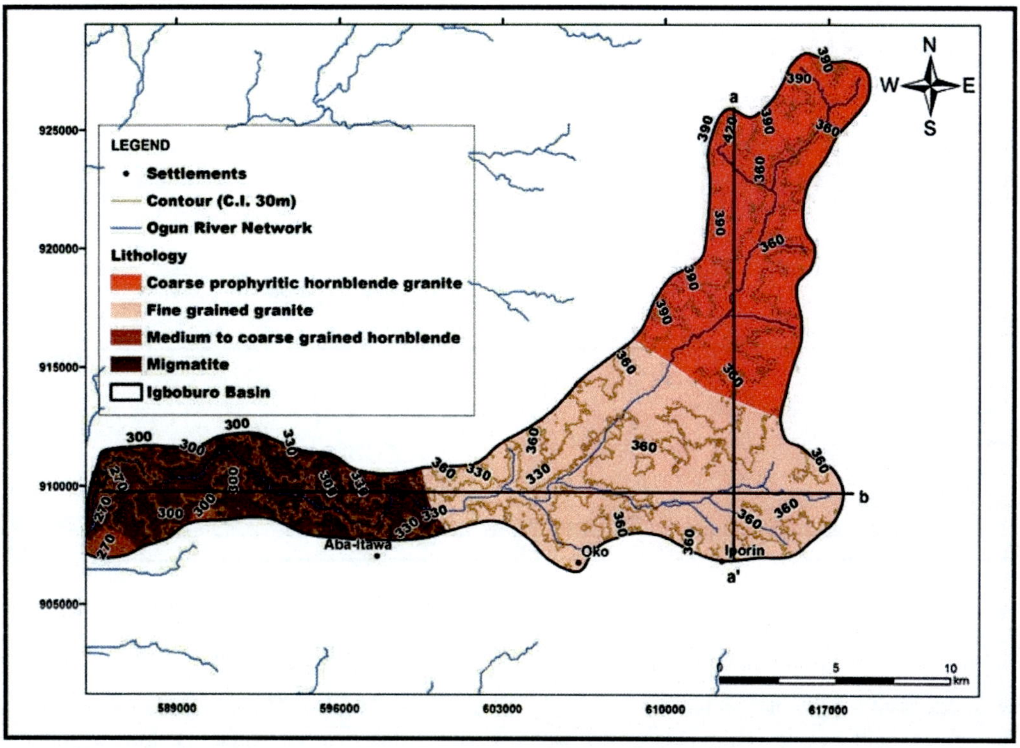

Fig. 21.6 Drainage and topography of Igoburo Basin. *Source* Olusola (2019)

descends gradually southwards (Fig. 21.7a). The relief across the west–east shows a gentle rise towards the east with outcrops of resistant granites (Fig. 21.7b). The highest point within the basin is 455 m a.s.l, while the lowest point is 262 m a.s.l. Igboburo basin is underlain by three lithological units. These are coarse porphyritic hornblende granite (OGh), Migmatites (M) and Quartzite (Q). Mediun- to coarse-grained hornblende (OGe) occurs within a small space towards Itesiwaju LGA. The coverage of OGe is quite small without the river dissecting it. Evidence of outcrops is visible along the M section where it is more pronounced.

21.4.1.3 Awon Basin

Awon catchment is situated south-east of the UORB (Fig. 21.3). It is within latitudes 8° 10′ N and 7° 50′ N and longitudes 3° 37′ E and 4° 50′ E (Fig. 21.8). The land area is about 487.13 km^2. Settlements within and around the basin include communities such as Ari-Oyo, Ogidi, Isale Oba, agboye, Kasia, Olorunda, Egunbiyi, Eleja, etc. Awon basin cut into the Basement Complex landscape of Southwestern Nigeria. The basin is well dissected with very few outcrops of resistant rocks occurring. The channel is mainly an alluvial one (Olusola 2019). The river takes its source from Ori-Ire and flows towards Iseyin where it empties into the Ogun River. The river channel is mainly ephemeral. The relief along the north–south direction of the basin shows gradual descent in height with evidence of fluvial dissections (Fig. 21.9a). The relief across the west–east shows a rolling plain of relatively flat surface dissected only where the river and its tributaries flow across it (Fig. 21.9b). The highest point within the basin is 381 m a.s.l, while the lowest point is 194 m a.s.l. Awon basin is underlain by four lithological units. These are Hornblende biotite granite gneiss (GGh), Migmatites (M), Quartzites (Q) and Undifferentiated Gneiss Complex (Su). The dominant as expected is the M, while the other three are restricted towards

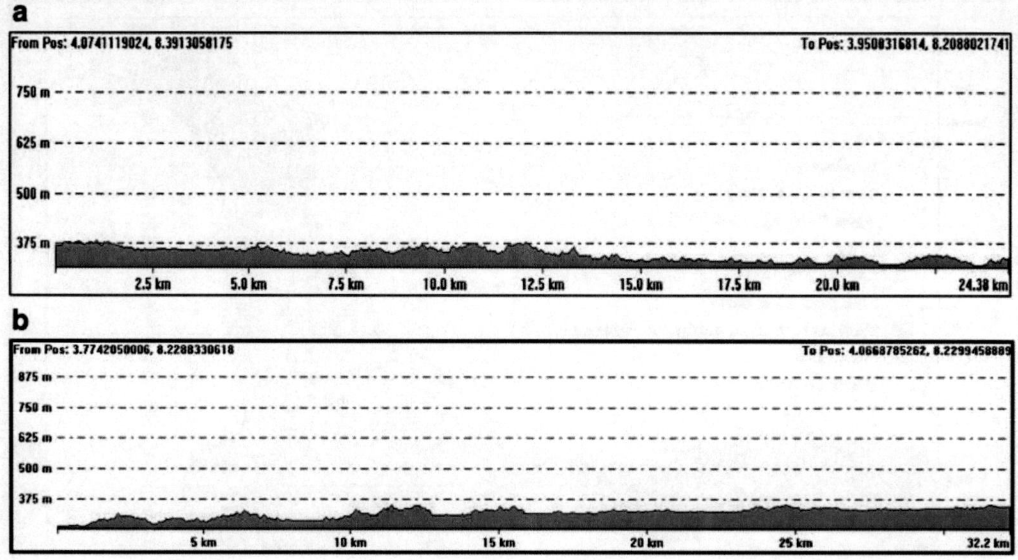

Fig. 21.7 **a** Transect running North–South of Igoburo Basin (H.E = 5) (a-a′). **b** Transect running West–East of Igoburo Basin (H.E = 5) (b-b′). *Source* Olusola (2019)

the south-eastern part of the basin (Fig. 21.8). The greater part of the basin is covered with alluvium of recent deposits washed from hill-slopes into the river channel network.

21.4.1.4 Onikoko Basin

Onikoko river basin is located towards the southern part of the UORB (Fig. 21.3). The land area is about 248 km^2. Onikoko basin cut into the Basement Complex landscape with no observable outcrops of resistant rocks along the river channel. The channel is mainly alluvial and it is perennial. The river takes its source from Afijio and flows towards Aderomi not too far from Lanlate where it empties itself into the Ogun River (Fig. 21.10). The basin traverses peri-urban settlements apart from when it is flowing within forested areas. Anthropogenic imprints are visible within the river basin especially around the Afijio–Onikoko axis. The relief along the north–south direction of the basin shows gradual descent in height as the river approaches the Ogun River (Fig. 21.11a). The relief across the west–east shows a rolling plain gaining height which each distance covered and adequately dissected by tributaries of the rivers within the basin (Fig. 21.12b). The highest point within the basin is 352 m a.s.l, while the lowest point is 122 m a.s.l. Onikoko basin is underlain by four lithological units (Fig. 21.12c). These are Biotite garnet schist (bS), Migmatites (M), Quartzites (Q) and Undifferentiated Gneiss Complex (Su). The dominant is the M, while the other three transects the M at various locations within the river basin.

21.4.1.5 Odo-Oba Basin

The Odo-Oba river basin is located towards the south–central part of the UORB (Fig. 21.3). It is within latitudes 8° 25′ N and 7° 40′ N and longitudes 3° 25′ E and 3° 40′ E (Fig. 21.12). The land area is about 376 km^2. Odo-Oba cut into the Basement Complex landscape with outcrops of resistant rocks observed towards the upper section of the river channel. The channel is largely alluvial and it is perennial (Olusola 2019). The basin traverses peri-urban and urban settlements. The relief along the north–south direction of the basin shows gradual descent in height as the river approaches the Ogun River (Fig. 21.13a). The relief across the west–east shows a rolling plain within the basin (Fig. 21.13b). The highest point

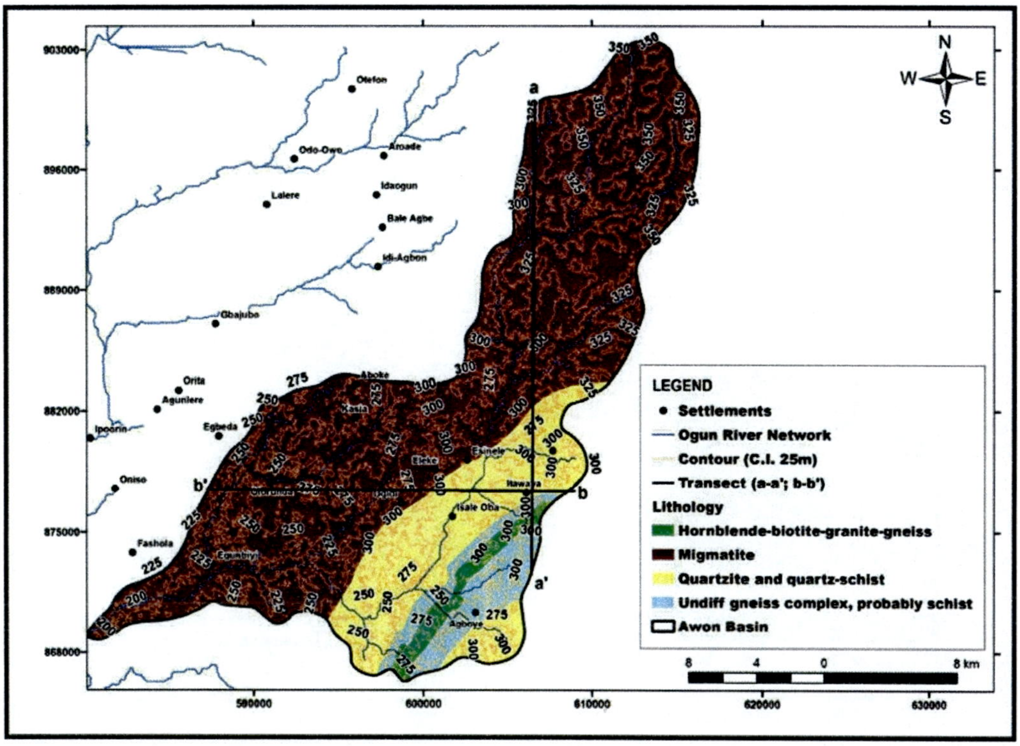

Fig. 21.8 Drainage and topography of Awon Basin. *Source* Olusola (2019)

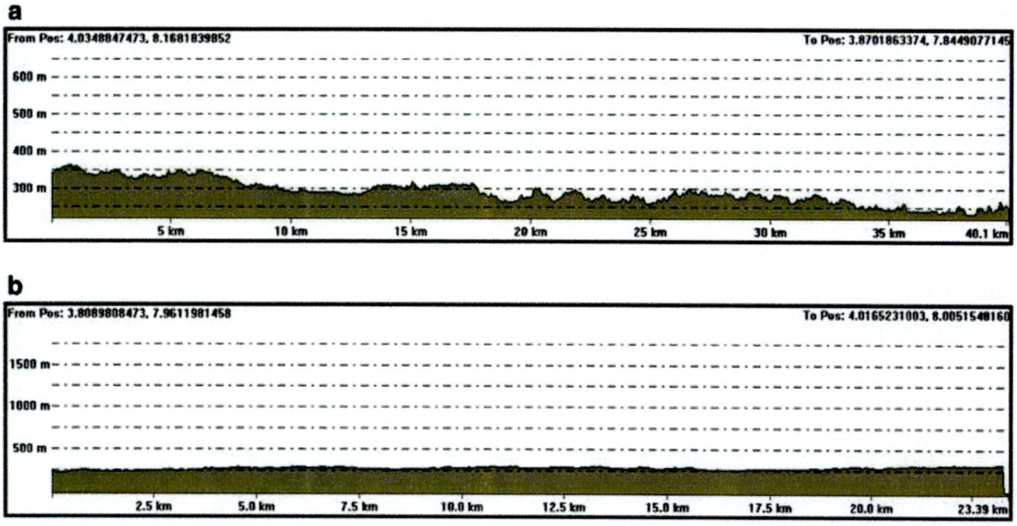

Fig. 21.9 **a** Transect running North–South of Awon Basin (H.E = 5) (a-a′). **b** Transect running West–East of Awon Basin (H.E = 5) (b-b′). *Source* Olusola (2019)

within the basin is 429 m a.s.l, while the lowest point is 146 m a.s.l. Odo-Oba basin is underlain by four lithological formations (Fig. 21.12). These are Biotite garnet schist (bS), Migmatites

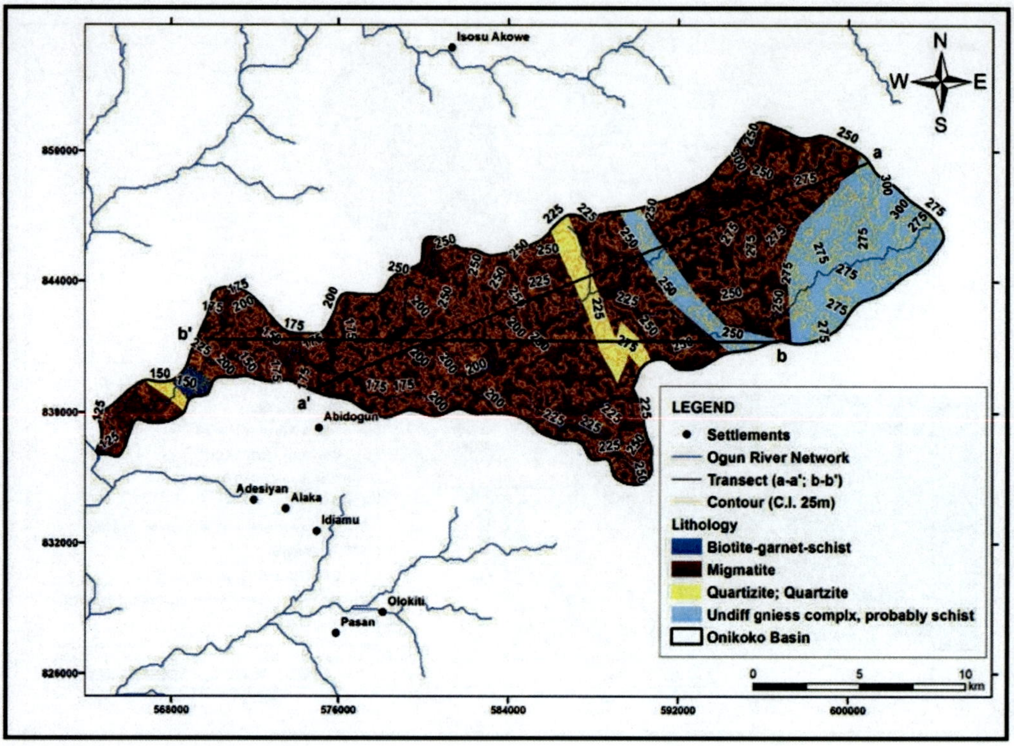

Fig. 21.10 Drainage and topography of Onikoko Basin. 2019 *Source* Olusla ()

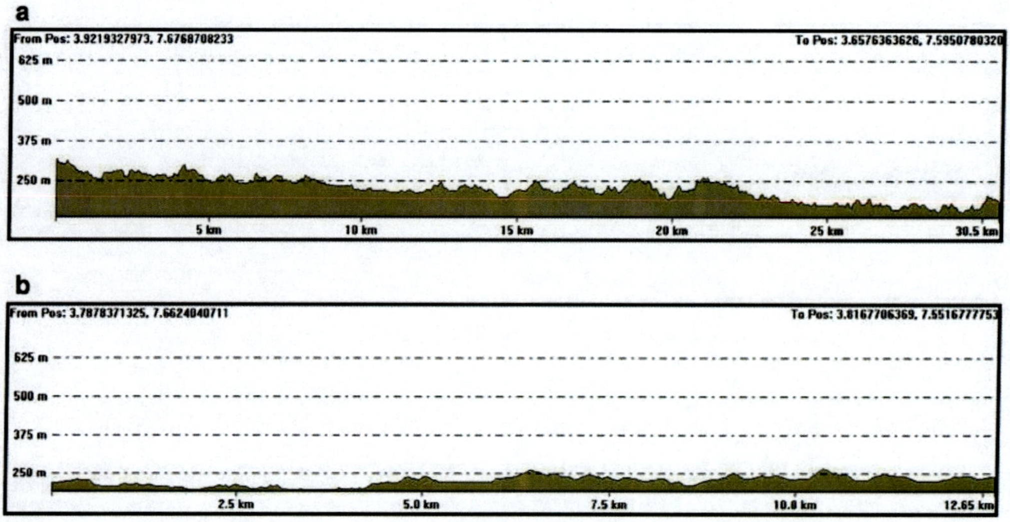

Fig. 21.11 **a** Transect running North–South of Onikoko basin (H.E = 5) (a-a′). **b** Transect running West–East of Onikoko basin (H.E = 5) (b-b′)

(M), Quartzites (Q) and Undifferentiated Gneiss Complex (Su). The dominant is the migmatites, while the other three dissects migmatites at various locations within the basin.

21.4.1.6 Ayin Basin

Ayin river basin is located towards the south-western part of UORB (Fig. 21.3). It is within latitudes 7° 29′ N and 7° 19′ N and longitudes 3° 15′ E and 3° 20′ E (Fig. 21.14). The land area is about 84.45 km^2, the smallest of all the basins considered. The rivers within this basin cut into the Basement Complex landscape, with outcrops of resistant rocks appearing all over towards the upper section of the river channel. The channel contains both an alluvial section and bedrock section and it is largely ephemeral (Olusola 2019). The relief along the north–south direction of the basin shows gradual descent in height with outcrops of hills towering high above the surface as the river approaches the Ogun River (Fig. 21.15a). The relief across the west–east shows an increase in height with increasing

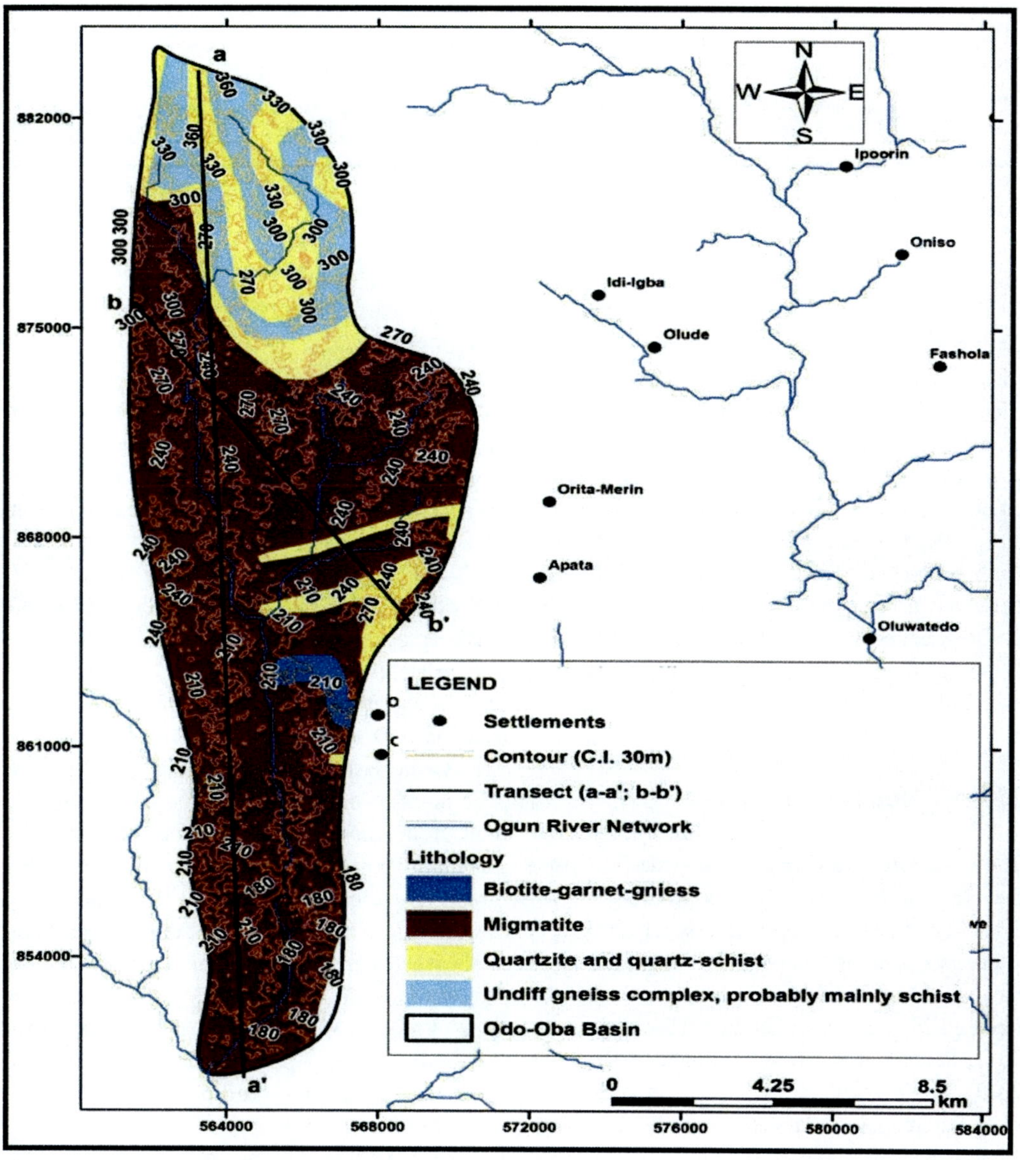

Fig. 21.12 Drainage topography of Odo-Oba Basin. *Source* Olusola (2019)

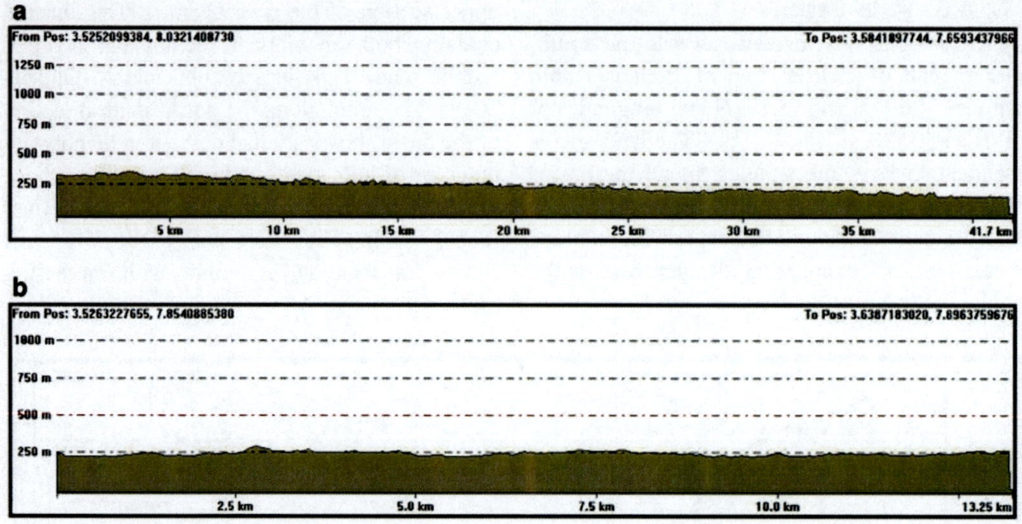

Fig. 21.13 **a** Transect running North–South of Odo-Oba basin (H.E = 5) (a-a′). **b** Transect running West–East of Odo-Oba basin (H.E = 5) (b-b′)

distance from the valleys of River Afo-Ape into the hinterlands and decreases towards the eastern fringes into the valleys of the Opeki River (Fig. 21.15b). The highest point within the basin is 290 m a.s.l, while the lowest point is 88 m a.s.l. Ayin basin is underlain by four lithological units (Fig. 21.13). These are Amphibole Schist (aS), Migmatites (M), Pegmatites (P) and Undifferentiated Gneiss Complex (Su). The dominant is the Amphibole Schist followed by the migmatites. The basin has various outcrops along its river channel especially towards the southern part of the basin.

21.4.2 Slope–Area Curves

The concavity index, an exponent derived from the relationship between slope–area curves (Figs. 21.16, 21.17, 21.18, 21.19, 21.20, 21.21) suggests a rather low concavity though not out of range as these values have been observed and reported in some other studies (Mccleary et al. 2011; Pedrera et al. 2009; Phillips and Desloges 2013). The low values are associated with short, steep drainages influenced largely by mass movements. Even though the area shows several outcrops in some places, the greater part of the basin presents alluvial flow and undulating slope terrain.

Oke-Ogun basin, with steepness (0.08) and concavity index (0.04) presents a point of inflection (Fig. 21.16) that suggests a move from more pronounced debris flow into a fluvial flow as evidenced from the field observation. From Fig. 21.17, the Igboburo basin has a steepness of 0.12 and a concavity index of 0.09. The point of inflection here cannot be fully interpreted as the basin transits from alluvial to bedrock. Hence, the inflection point (Fig. 21.17) is suggesting transition into a transport limited section. The steepness (0.12) and concavity index (0.06) for the Awon basin as shown in Fig. 21.18 is low. The point of inflection suggests a transition pattern from alluvial to bedrock to alluvial. The inflection point here is not too distinct on the graph. Therefore, the pattern of transition cannot be fully determined. One possible reason could be because a greater part of the river flows across migmatite with a pronounced bedrock zone. Onikoko basin with a steepness (0.17) and concavity index (0.1) presents one of the highest concavity among the third-order basins studied (Fig. 21.19).

The transition pattern observed on the field within the basin is from alluvial to bedrock to

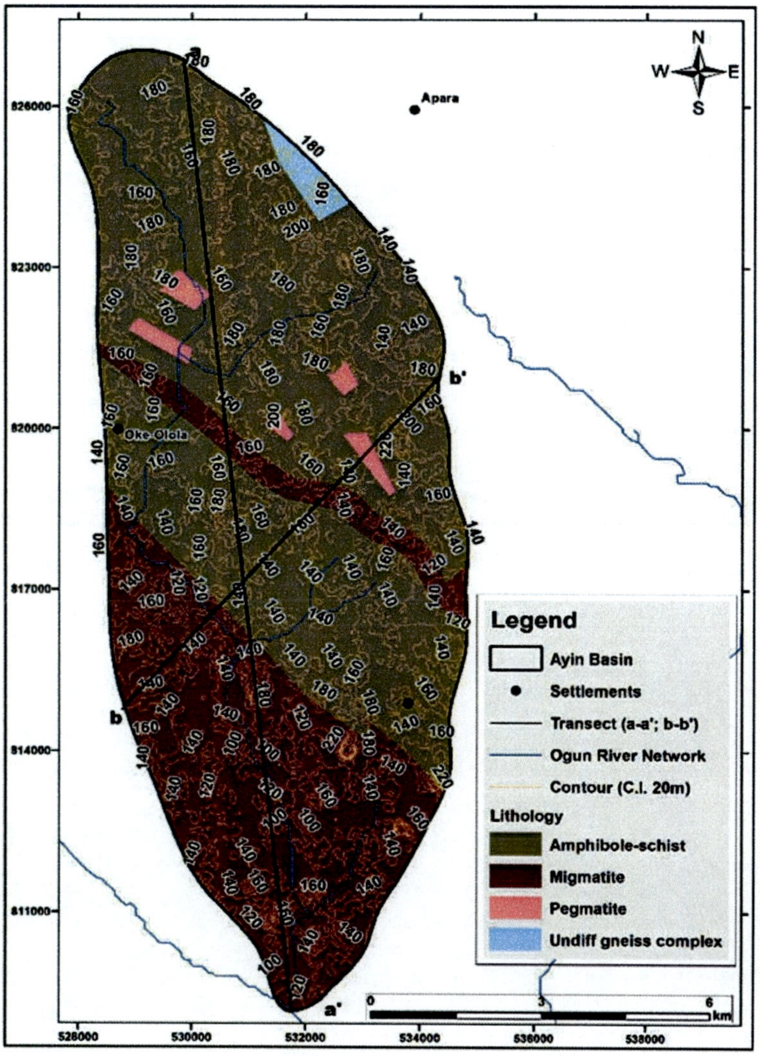

Fig. 21.14 Drainage and topography of Ayin Basin. *Source* Author's Fieldwork (2017)

alluvial. The inflection point here indicates a move into the alluvial section where the river flow is transport limited considering the low shear stress value (Olusola 2019). Odo-Oba presents steepness (0.15) and concavity index (0.08) (Fig. 21.20) that is not quite different from others. The transition pattern as observed within the basin is from bedrock to alluvial. The inflection point here indicates a move from debris flow into a well-defined fluvial flow where transport is non-limited. The steepness (0.25) and concavity index (0.11) for the Ayin basin as revealed from Fig. 21.21 presents the highest for the basin. The transition pattern as observed within the basin is from alluvial to bedrock.

The concavity index (Θ) of the slope as calculated from slope–area (S–A) plots show strong relationships to underlying lithology and drainage area. Most striking are the distinct profiles on rocks traversing different classes of rock types (Olusola and Fashae 2019). The inflection point occurs where the stream sharply steepens in the S–A plot, this is mostly around where the river flows across migmatic bedrock. Since the rivers flowing within the selected third-order basins flows across more than two lithological units,

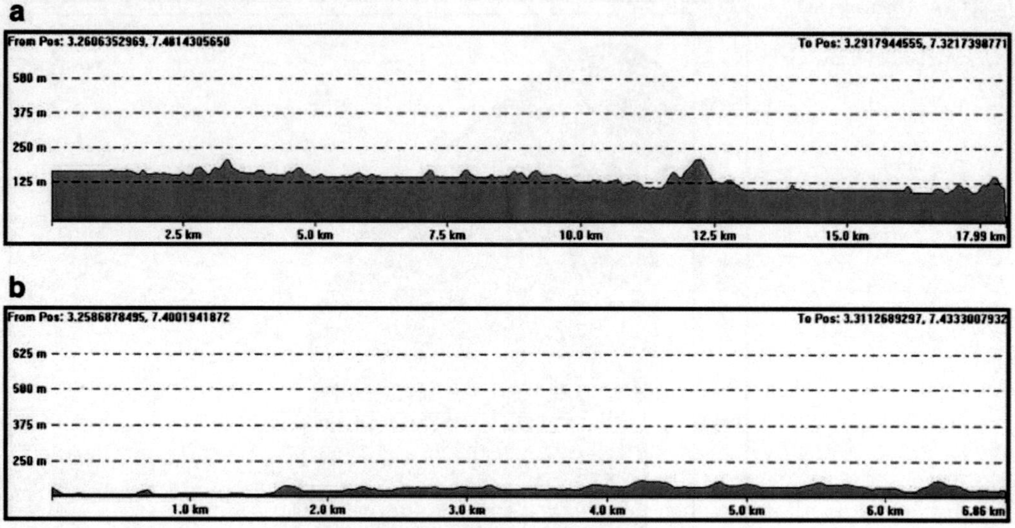

Fig. 21.15 **a** Transect running North–South of Ayin Basin (H.E = 5) (a-a′). **b** Transect running West–East of Ayin Basin (H.E = 5) (b-b′)

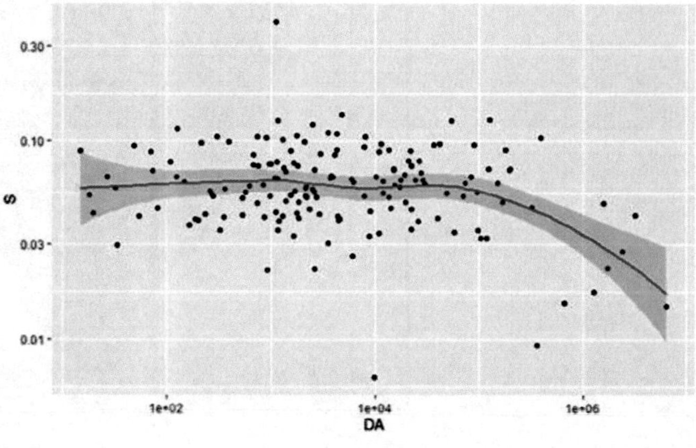

Fig. 21.16 Slope–area plot (Oke-Ogun basin)

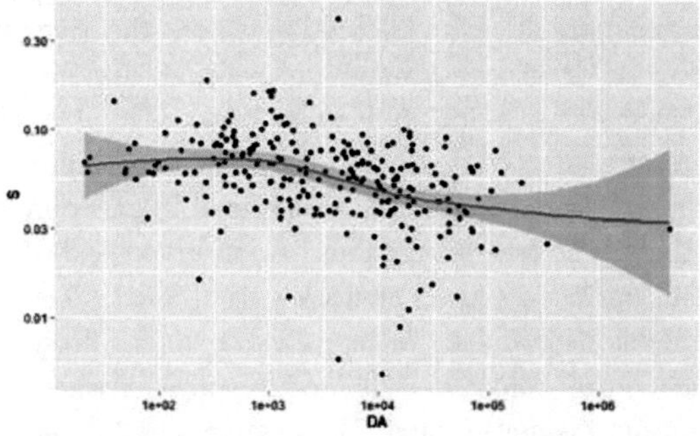

Fig. 21.17 Slope–area plot (Igboburo basin)

there is often a slope break at the constant, especially where the adjoining rocks have varying resistant erosion. Areas showing sharp boundaries between rock types such as the profile

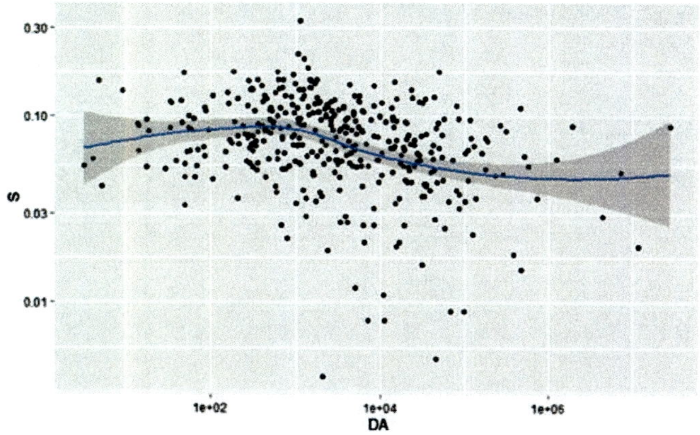

Fig. 21.18 Slope–area plot (Awon basin)

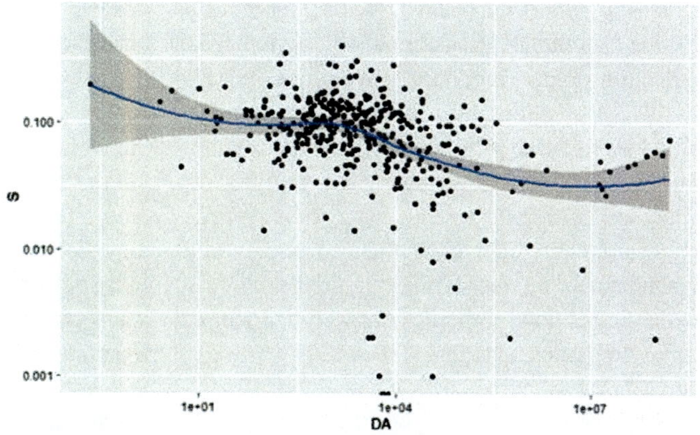

Fig. 21.19 Slope–area plot (Onikoko basin)

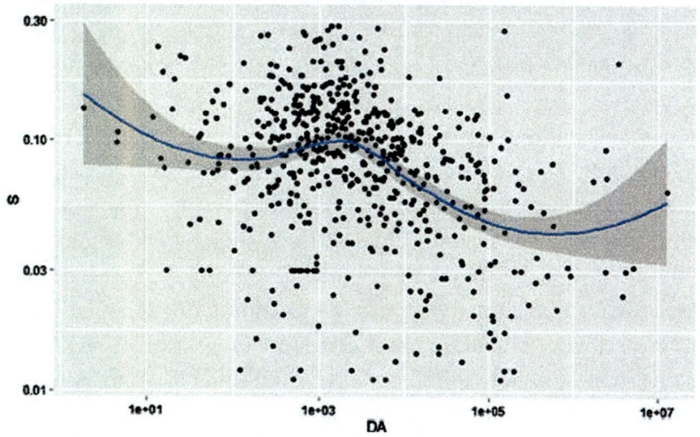

Fig. 21.20 Slope–area plot (Odo-Oba basin)

of Oke-Ogun, Odo-Oba and Ayin, is accompanied by a well-defined convex form. The inflection point here suggests a move into a sediment-limited region (Mccleary et al. 2011; Pedrera et al. 2009; Phillips and Desloges 2013) with a pronounced inflection point. Based on observations from the presented inflection points and the transition pattern as observed from the field, pronounced inflection points are within basins transiting from bedrock to alluvial (Oke-Ogun,

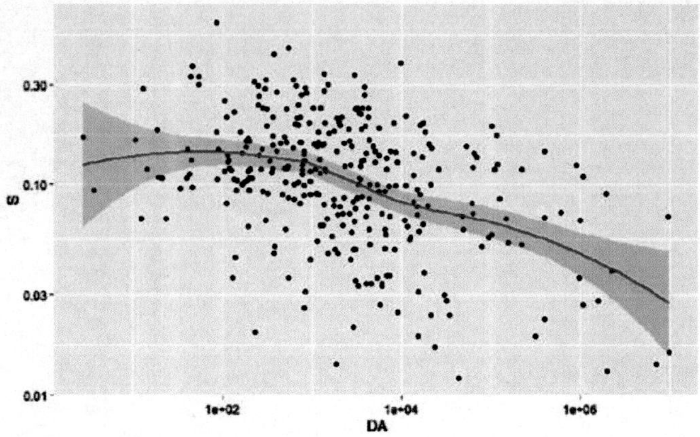

Fig. 21.21 Slope–area plot (Awon basin)

Odo-Oba and Ayin basins), while those basins having multiple transitions presents weak inflection points (Igboburo, Awon and Onikoko basins). Transitions from bedrock channels into alluvial channels could be well defined as alluvial channels are self-formed and over time such channels have been able to mark their environment and carve effectively their channels. Alluvial channels are not sediment-limited and hydrodynamics of flow properties are well defined and follow a particular path.

21.4.3 Channel Network and Lithological Pattern

Environmental factors such as climate, lithology, vegetation, etc., exercise control through their direct and indirect impact on fluvial process and dynamics. Within a homogenous climatic zone, lithology stands to be one of the most important environmental factors to influence river channel morphology. Rose diagrams (Fig. 21.22) present the influence of underlying lithological units on river network orientation and initiation.

Several geomorphologists have highlighted the importance of structure on landform evolution (Faniran et al. 2006). In essence, the impact of underlying structure on channel initiation and evolution cannot be contested. As already pointed out in this work, the settings within which this study was situated is completely a humid tropical environment without any known glacial history but perhaps a cyclical pattern of wet and dry climates since the last glacial maximum (LGM) (Jeje 1980). Therefore, the area is largely stable and alteration to present channels and networks can be closely linked to structural factors. The approach here is to identify lineaments and river networks and compare the pattern. It has been established that known methods of relating lineament patterns to morpho-tectonic subsets are Rose diagrams (see Adeyemi and Faniran 2020).The frequency and length of lineaments (fractures) in the Rose diagram for the study area (Fig. 21.22a-f) shows a bi-modal distribution in most cases along EW–NS and N–S directions, which have corresponding peaks on the length-orientation axis of the Rose diagram of the river channels. Across the entire basins, the lineament pattern and channel network orientation have been shown to follow trends in various directions (see Olusola and Faniran 2020). As observed and discussed in Olusola and Faniran (2020), the Rose diagram for the lineament trends suggests four major trends. These are E–W, N–S, SSE–SSW and NE–SW. Across these four major trends, the N–S and E–W were particularly dominant. Since the general trend of tectonic grains within the Nigerian basement are relatively N–S (Oluyide 1988), the main channels draining the selected basins in Upper Ogun appear to have been controlled along pre-existing weak zones in the country. Therefore, it can be established that these two lineament orientation sets (N–S, E–W)

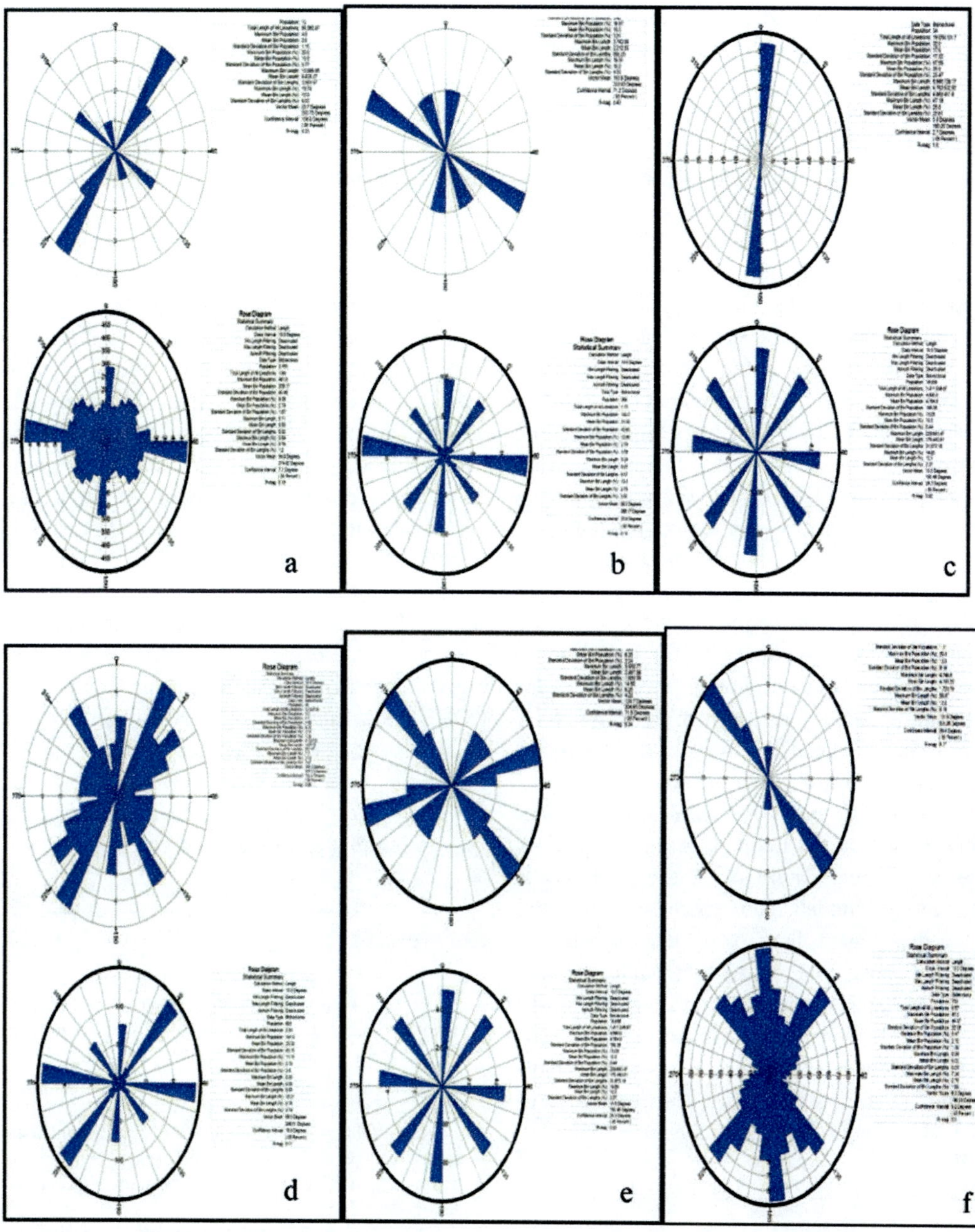

Fig. 21.22 Rose diagram illustrating trends of lineament (upper Rose diagram) and channel network (lower Rose diagram) **a** Oke-Ogun Basin **b** Igboburo Basin **c** Awon Basin **d** Onikoko Basin **e** Odo-Oba Basin **f** Ayin Basin. *Source* Adeyemi and Faniran (2020)

could define the preferred orientation of groundwater occurrence within UORB (UORB) and to a large extent they could also serve as active conduits connecting most springs in the area to the main channel (Adeyemi and Faniran 2020).

In another study, river long profiles and hypsometry of the six basins has been presented

(Olusola and Fashae 2019). It was established that longitudinal profile segmentation across lithological material maintains segmentation typical of rocks of varying resistance, while at the lower reaches, however, the observed falls are closely linked to the influence of base level. As regards the hypsometry of the entire basins, it was concluded that across the entire basin, the transition towards equilibrium will be completed when the area becomes ultimately dissected and remnants of the resistant rock completely eroded (Strahler 1952). The entire six basins can be divided into two groups. Those with hypsometric integrals above 35% and those below 35%. Above 35% are quite high and suggest that greater parts of the basin are still largely uneroded. Those below 35% give the impression of an area tending towards equilibrium or can be termed quasi-equilibrium (Olusola and Fashae 2019). Altogether, these plots together with the slope–area plots show that profiles and segmentations are good geomorphological evidences for revealing process-form dynamics. The S–A graphs have explained roles of underlying lithology, landform evolution and interactions between fluvial and non-fluvial flow, respectively. Based on the findings of the study, S–A plots shows that within the bedrock channels, sediment is limited and processes operating within these channels are sometimes non-fluvial and are most often than not well defined. Therefore, transiting into bedrock channels could produce inflection points that are not quite distinct as slope–area solutions might not have enough to account for the inherent hydrodynamic variations operating therein.

21.5 Conclusion

This study has provided explanations as regards the interactions between lithology and drainage basin attributes across sub-basins within the UORB. The interpretations as presented in the study are needed to develop new approaches into geomorphic inquiry. First and foremost, this work is an effort to contribute to the methodology of geomorphological research by meaningfully relating theory to practice. Meaningful characterization and description are basic to effective interpretation by providing baseline information with which to foresight likely future states. Emphasis has been placed on 'what is there' and 'why it is there'.

In comparison with other locations especially higher latitudes, the study has been able to show areas of difference and areas of intersection. The occurrence and the response of channel form to underlying lithology have been effectively discussed and appropriated. Also, this study has clearly shown that third-order streams within heterogeneous lithologies are still undergoing geomorphic processes. Furthermore, in hydrogeomorphological terms, the lineament trends as presented in the study show ties to ancient tectonic events that have affected the Upper Ogun Basin. The trend draws inference on groundwater flow within the basement complex by establishing the fact that in the Basement Complex of UORB,the occurrence of groundwater follows the path of porosity in fractures and by extension that these major trending lineaments (N–S and E–W) are likely to be the pathway linking most springs in the area to the main channel.

References

Adeyemi OO, Faniran A (2020) Lithology and channel network initiation and orientation: a case study of Upper Ogun River Basin, Southwestern Nigeria. In: Proceedings of geomorphometry conference 2020, June 22–26.

Adeyemi O, Olutoyin F, Olumide O (2020) Downstream hydraulic geometry across headwater channels in Upper Ogun River Basin, Southwestern Nigeria. Afr Geogr Rev 1–16

Burke KC, Dewey JF (1972) Orogeny in Africa. In: Dessauragie AJ, Whiteman TFJ (eds) African Geology. University of Ibadan Press, Nigeria, pp 583–608

De Swardt AMJ, Ogbukagu IK, Hubbard FH (1965) 1:250,000 Geological Map of Sheet No. 60 Iwo. Geol. Survey Nigeria

Douglas I (1977) Humid landforms. Australian National Univ. Press, Canberra, p 288

Faniran A (1982) African landforms. Heinemann, Ibadan

Faniran A, Jeje LK (2000) Humid tropical geomorphology. Heinemann Educational Books (Nigeria) Plc

Faniran A, Jeje L, Ebisemiju F (2006) Essentials of geomorphology. Penthouse Publications, Ibadan, Oyo State, Nigeria
Gregory KJ, Walling DE (1973) Drainage basin form and process: a geomorphological approach. Edward Arnold (Publishers) Ltd.
Hack JT (1973) Stream-profile analysis and stream-gradient index . J Res US Geol Surv 1:421–429
Hayakawa YS, Oguchi T (2006) DEM-based identification of fluvial knickzones and its application to Japanese mountain rivers. Geomorphology 78(1–2):90–106
Huggett RJ (2007) Fundamentals of geomorphology. Routledge, Londan and New York
Jeje LK (1970) Some aspects of the geomorphology of southwestern Nigeria. Unpublished PhD thesis, University of Edinbugh
Jones HA, Hockey RD (1964) The geology of South-western Nigeria. Bull, Geol Surv, 31–100
Jenson SK, Domingue JO (1988) Extracting topographic structure from digital elevation data for geographicinformation system analysis. Photogram Eng Remote Sens 54(11):1593–1600
Mccleary RJ, Hassan MA, Miller D, Moore RD (2011) Spatial organization of process domains in headwater drainage basins of a glaciated foothills region with complex longitudinal profiles. Water Resour Res 47:1–17. https://doi.org/10.1029/2010WR009800
Montgomery DR, Dietrich WE (1988) Source areas, drainage density and channel initiation. Water Resour Res 25:1907–1918
Oguntoyinbo A (1982) The climate of Ibadan. In: Filani MO, Akintola FO, Ikporukpo CO (eds) Ibadan Region. Department of Geography, University of Ibadan. Special Silver Jubilee (NGA) publication, pp 16–40
Olusola AO (2019) Process-form dynamics of upper Ogun river basin, Southwestern Nigeria. Unpublished Ph.D. thesis submitted to the Department of Geography, University of Ibadan
Olusola A, Fashae O, Onafeso O, Adelabu S (2020) Morphologic and hydraulic variability of small bedrock and alluvial channels in relation to lithological controls, Upper Ogun River Basin, Southwestern Nigeria. Phys Geogr 1–21
Olusola AO, Fashae O (2019) River long profiles of selected third-order basins in Basement Complexes. In: Wade S (ed) Earth observations and geospatial science in service of sustainable development goals. Springer Nature Switzerland AG, pp 15–24
Oyawoye MO (1972) The basement complex of Nigeria. In: Dessauvagie TFJ, Whiteman A.J (ed) Africa Geology. Ibadn University Press, pp 66–102
Pedrera A, Pérez-Peña JV, Galindo-Zaldívar J, Azañón JM, Azor A (2009) Testing the sensitivity of geomorphic indices in areas of low-rate active folding (eastern Betic Cordillera, Spain). Geomorphology 105(3–4):218–231.https://doi.org/10.1016/j.geomorph.2008.09.026
Phillips RTJ, Desloges JR (2013) Geomorphology glacially conditioned speci fi c stream powers in low-relief river catchments of the southern Laurentian Great Lakes. Geomorphology. https://doi.org/10.1016/j.geomorph.2013.09.030
Reinfelds I, Cohen T, Batten P, Brierley G (2004) Assessment of downstream trends in channel gradient, total and specific stream power: a GIS approach. Geomorphology 60:403–416
Strahler AN (1952) Hypsometric (area-altitude) analysis of erosional topography. Bull Geol Soc Am 63 (11):1117–1142
Thomas MF (1974) Granite landforms: a review of some recurrent problems of interpretation. Inst Br Geogr, Spec Publ 7:13–37
Thornbury WD (1954) Principles of geomorphology. Wiley, New York
Thorp MB (1970) Climate and geomorphology. Geogr Viewpoint 2:77–95
Twidale CR (1976) On the survival of paleoforms. Am J Sci 276:77–95
Wigwe GA (1966) Drainage composition and valley forms in parts of northern and western Nigeria. Unpublished Ph.D. thesis, University of Ibadan

Check for updates

22 Multi-criteria-based Morphometric Prioritization for Soil Erosion Susceptibility and Denudation Rate Assessment of Purulia District, India

Gour Dolui, Kousik Das, Nilanjana Das Chatterjee, and Rajkumar Bhattacharya

Abstract

The quantitative drainage morphometric analysis is an essential method for understanding the hydrological, geomorphological, geological, and environmental interaction and resultant processes of a particular area or watershed. The present study aims to identify the erosion susceptibility zones and the recent denudational rates by quantitative measurement of morphometric characteristics over the granitic terrain of the Purulia district, West Bengal, India. Erosion susceptibility and denudation rate are measured based on multi-criteria morphometric prioritization and hierarchical anomaly index using remote sensing and GIS techniques. In this study, the ASTER digital elevation data, and Survey of India topographical maps are used to delineate relief, slope, and structure of drainage system such as stream pattern, stream network, stream order, stream length, and others significant linear, relief, and aerial aspects of the sub-watersheds. For the analysis, 27 numbers of sub-watersheds are delineated using digital terrain analysis techniques that cover the entire studied district. Based on the relationship with erodibility, different prioritization ranks are recognized following the estimation of compound values. All the morphometric parameters reveal the interrelations between hydrological processes and their associated geomorphic landforms. Notably, similar drainage pattern and their dominant first-order streams are associated with steep terrain of some sub-watersheds and significantly responsible for more surface erosion. Therefore, high elevated western and southern hilly landscape areas are identified as high erosion susceptibility zone with high-intensity denudation and which also influenced by underlying lithology, associated landforms, and land-use patterns. The rest of the central and northeastern parts of the district have a lesser susceptibility to soil erosion and lower denudation rate which associated with flatted topography with a very gentle slope and hard rock terrain.

Keywords

Drainage morphometry · Remote sensing and GIS · Soil erosion susceptibility · Hierarchical anomaly index · Denudation

G. Dolui (✉)
Department of Geography, Panskura Banamali College (Autonomous), Panskura, Purba Medinipur, West Bengal, India

K. Das · N. Das Chatterjee · R. Bhattacharya
Department of Geography and Environment Management, Vidyasagar University, Midnapore, West Bengal 721102, India
e-mail: nilanjana_vu@mail.vidyasagar.ac.in

P. K. Shit et al. (eds.), *Drainage Basin Dynamics*, Geography of the Physical Environment,
https://doi.org/10.1007/978-3-030-79634-1_22

22.1 Introduction

Basin morphometric analysis, using several arithmetic equations, is basically the measurement of the composition of shapes and extension of the basin hydrological system developed over the surface of the earth (Clarke 1966). Strahler (1964) stated that a quantitative explanation of the basin geometry can be understood completing the morphometric analysis and also revealed the controlling factors of the primary slope, heterogeneity in rock hardness, litho-structural forms, recent diastrophism, and the geological and geomorphological chronosequence over the evolution of drainage basin. Recently, Eze and Efiong (2010) have also reported similar ideas that the dynamic surface runoff which is an essential and responsive factor to the nature of morphometric parameters also controlled by river basin shape and area over the existing geomorphological structure. Nautiyal (1994) analyzed the morphometry of the river watershed based on quantifying the linear character, topographical slope, and drainage network for assessing the characteristics of the basin hydrological system Aouragh and Essahlaoui (2014). In a similar way, the measurement of three fundamental aspects of drainage morphometry (i.e., linear, aerial, and relief), several researchers completed a morphometric analysis based on some basic morphometric parameters such as stream order, drainage density, stream length, basin relief, shape, elongation, etc. (Nag and Chakraborty 2003; Javed et al. 2009; Magesh et al. 2011; Kaliraj et al. 2015; Welde 2016; Rather et al. 2017; Shivhare et al. 2018).

Soil erosion and surface lowering is a common natural process in high elevated topography (Whipple et al. 1999). Erosion of the topsoil of the landscape leads to constant land degradation and erosion and the consequent decline of soil quantity and quality in the upper surface of the topography (Aouragh and Essahlaoui 2018). Thus, erosion susceptibility in the Purulia district became essential for understanding active weathering and erosional processes. This soil erosional intensity can be measured by the prioritization of morphometric parameters and the position and distribution of slope (Chandniha and Kansal 2017; Ameri et al. 2018). Morphometric analysis is a fundamental tool for prioritization of sub-basin by computing several linear and aerial parameters of the drainage catchment (Rekha et al. 2011; Patel et al. 2013; Aher et al. 2014; Gajbhiye et al. 2014; Farhan et al. 2016; Shivhare et al. 2018; Aouragh and Essahlaoui 2018) and this analysis also provides an insight into drainage basin evolution and its function in the development of drainage morphometry (Angillieri 2008). Although the drainage morphometry of a particular basin is the reflective image of the underlying lithology and geological structure, surface geomorphology, landforms, and prevailing climate over there (Horton 1945; Agarwal 1998; Rudraiah et al. 2008), yet drainage morphometric analysis is a significant tool for prioritization of micro-watersheds even without considering the land use, natural vegetation, soil type, and climatic characteristics of the watersheds (Biswas et al. 1999; Jain and Goel 2002; Das 2014). Therefore, to understand the influence of lithology and topography on the erosion and denudation rate in Purulia district both morphometric prioritization and slope prioritization have been considered. Remote sensing and GIS techniques are the best suitable methods for analyzing drainage morphometry and sub-watersheds-based prioritization of the large area because satellite images provide a synoptic view of this area as a useful element for the analysis (Javed et al. 2011; Aouragh and Essahlaoui 2018). This kind of investigation has extensively studied by several researchers (Chopra et al. 2005; Thakkar and Dhiman 2007a, b; Sreedevi et al. 2009; Thomas et al. 2011; Jasmin and Mallikarjuna 2013; Kumar et al. 2015; Singh and Singh 2018) not only for measuring soil erosion but also for watershed management, resource conservation, and sustainable development.

Being a part of Chhotanagpur plateau, Purulia district is a plateau region that consists of mainly granitic rocks in most of the areas and make resistant to weathering. In contrast, the high elevated western parts may have a higher intensity of surface erosion due to its steep slopes. In this perspective, this study aims to identify the erosion susceptibility zones and the present denudational status of Purulia district based on morphometric analysis and sub-watershed-based prioritization using remote sensing and GIS techniques.

22.2 Materials and Methods

22.2.1 Study Area and Database

The entire Purulia district is taken into consideration for morphometric analysis which is a part of Chhotanagpur plateau of eastern India (Fig. 22.1). The area is characterized by undulating topography with rocky and hilly terrain largely in the western part and most southern part. About 150–200 m is a common range of elevation in most of the central portion of the district and its rise roughly with conical and residual hills toward the western margin (Nag and Chakraborty 2003). Most of the areas of the district consist of open hard crust of granitic rocks (Fig. 22.2) and barren land including scattered grassland and croplands. While dense forest is covered in the high elevated hilly terrain of the western and southern parts of the district.

In the present study, several datasets are used such as geological maps and information from Survey of India (SOI) including Advanced Spaceborne Thermal Emission and Reflection Radiometer (ASTER) Digital Elevation Model (DEM) with 30-m spatial resolution. As the quality of ASTER DEM is fundamental to estimate a variety of hydrological and topographical parameters, the DEM has been used to generate a drainage network. A total of 27 sub-watersheds are demarcated by digital terrain analysis (DTA) technique which enclosed the entire area of the district. The quality of the DEM may determine the accuracy for building and measuring the terrain characters such as slope and elevation (Fig. 22.3). As the slope and aspect are very essential input factors in the model of morphometric analysis, the accuracy of these elements can influence the overall performance of the analysis. Therefore, 30-m spatial resolution of this DEM can provide better amounts of a detailed database of topographical and hydrological parameters.

22.2.2 Morphometric Analysis

The efficient algorithms of the ArcHydro tool are used to delineate the drainage network and associate morphometric parameters from ASTER DEM. The basic procedure of drainage extraction has already been provided in detail by Youssef et al. (2011). Flow direction and flow accumulation are properly identified after DEM sinks. Strahler's scheme, which was primarily introduced by Horton and soon after modified by Singh Singh (1980), is applied for stream ordering. In this study, on the basis of fundamental three aspects (linear, relief, and aerial) several important morphometric parameters such as stream order (U), bifurcation ratio (Rb), mean bifurcation ratio (mRb), stream length (Lu), mean stream length (Lsm), stream length ratio (Rl), drainage density (Dd), texture ratio (T), stream frequency (Fs), form factor (Rf), elongation ratio (Re), circularity ratio (Rc), length of overland flow (Lof), compactness coefficient (Cc), shape basin (Bs), basin relief (Bh), relief ratio (Rh), and ruggedness number (Rn) are measured using their standard numerical formula given in Table 22.1.

22.2.3 Prioritization Using Morphometric Parameters

Prioritization of watersheds in diverse scales (i.e., sub-, mini-, and micro-watersheds) had been successfully completed by several researchers using morphometric analysis (Biswas et al. 1999; Thakkar and Dhiman 2013; Javed et al. 2009; 2011; Patel et al. 2013; Gajbhiye et al. 2014; Farhan and Anaba 2016). For sub-watersheds prioritization, several parameters related to erosion potentiality from linear and shape attributes of morphometric analysis is implemented (Patel

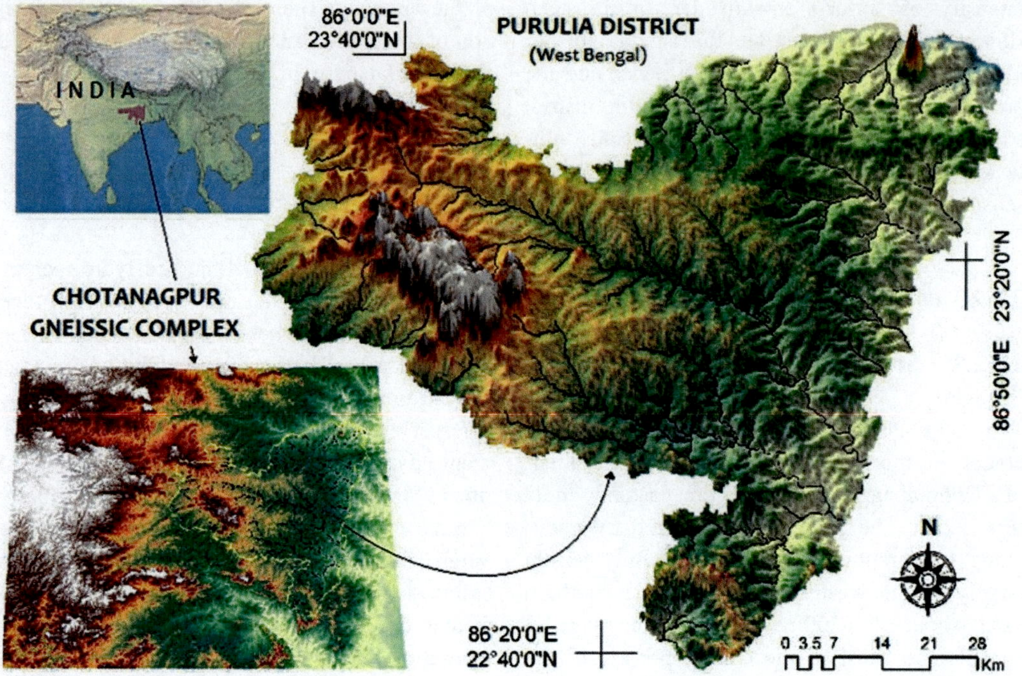

Fig. 22.1 Location of the study area

et al. 2013; Farhan and Anaba 2016). The selected linear attributes are Dd, Fu, Rmb, T, and Lof. The shape parameters are Re, Rc, Rf, Cc, and Bs, including relative relief (m) as a relief attribute. It has been established that in watershed-scale erodibility has a direct relation with linear morphometric parameters and also the same for relief parameter. Therefore, the highest measured values of both linear and relief parameters are assigned with rank 1, the second highest value are assigned with rank 2 and by this process rest of the sub-basins has been assigned rank accordingly (Biswas et al. 1999; Ratnam et al. 2005; Farhan and Anaba 2016). On the other hand, the shape attributes have an opposite correlation with linear attributes; consequently, the lower value signifies the greater erodibility. Therefore, shape factors are assigned with rank 1 with respect to the lowest measured value and second lowest as rank 2 and similarly the rest of sub-watersheds. After that, all assign ranks of linear, shape, and relief parameters have been added to compute the compound value and assign prioritized rank in ascending order of the calculated compound values. Compound value is designated as Cp which is measured by the following formula (Singh and Singh 2018):

$$\mathrm{Cp} = 1/\mathrm{n}\sum\nolimits_{\mathrm{i}=0}^{\mathrm{n}} \mathrm{R_i} \tag{22.1}$$

where Cp = Compound value of a particular watershed,

R_i = Rank of a particular watershed for a parameter,

n = Number of parameters.

Finally, all sub-watersheds grouped into three erosion susceptibility groups based on priority rank of compound values (Ratnam et al. 2005; Farhan and Anaba 2016): (i) High erosion susceptibility (rank 1–9), (ii) Medium erosion susceptibility (rank 10–18), and (iii) Low erosion susceptibility (rank 19–24). In this similar procedure, three erosion susceptibility zones have been demarcated using slope prioritization and ranking of the sub-watersheds (Fig. 22.4).

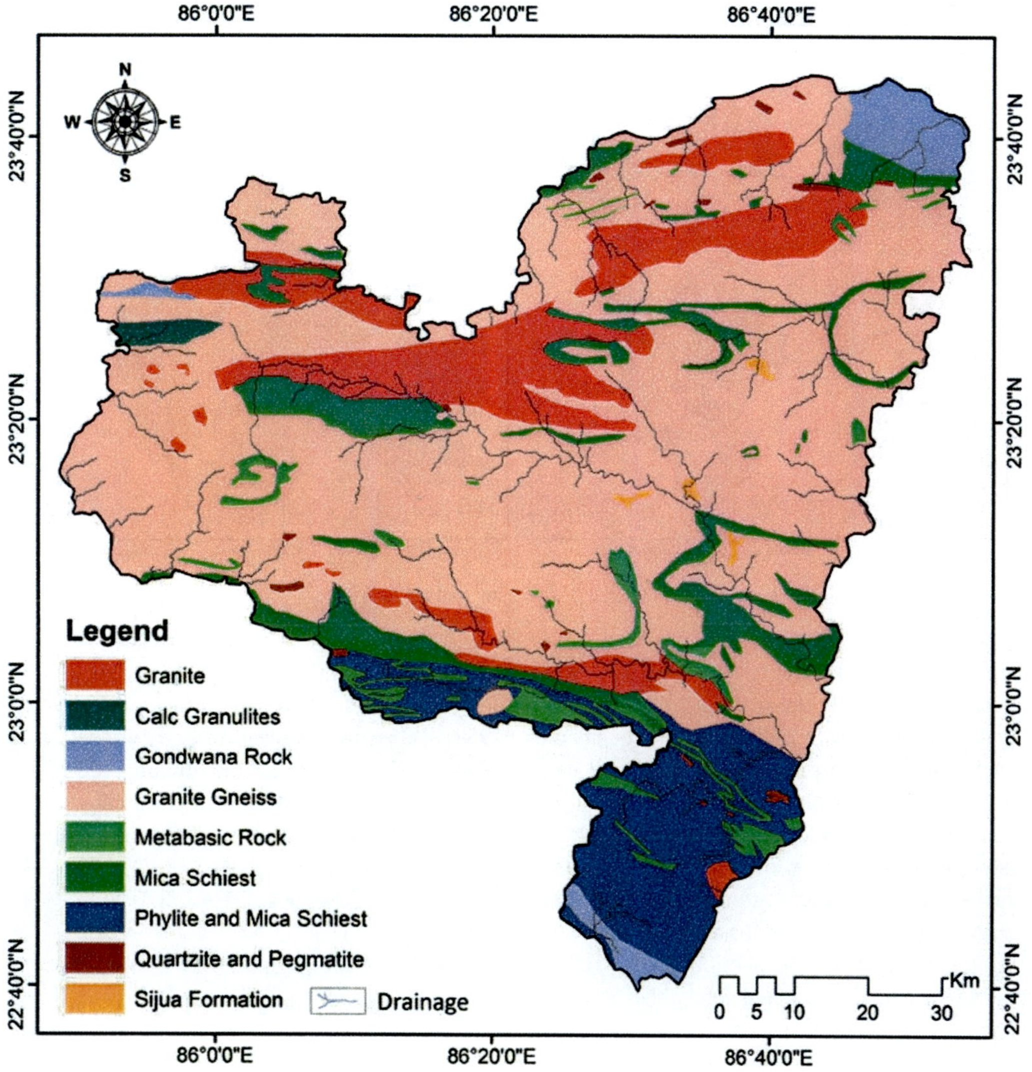

Fig. 22.2 Geological formation of Purulia district (after GSI)

22.2.4 Denudation Rate Estimation

The denudation process and its quantitative analysis explain the geomorphic erosional activities in a drainage basin in terms of terrestrial and climatic processes (Ciccacci et al. 1980). In this study, the denudation rates of the sub-watersheds were measured by morphometric analysis, and some required parameters like basin area, length, stream order, drainage density, and the number of stream segments are obtained by incorporated GIS-based topographic computation techniques. Previously, many researchers in their studies regarding estimation of denudation rate have used those parameters to compute the hierarchical anomaly density (G_a) and hierarchical anomaly index (D_a) (Tokunaga 2000; Del Monte et al. 2002; Della Seta et al. 2007; Gioia et al. 2011; Bahrami 2013; Kaliraj et al. 2015). Naturally, denudation intensity strongly affects by density parameters which illustrate the characteristics of drainage networks such as different stream segments and their areal extension. The ratios of stream length and basin area are used to

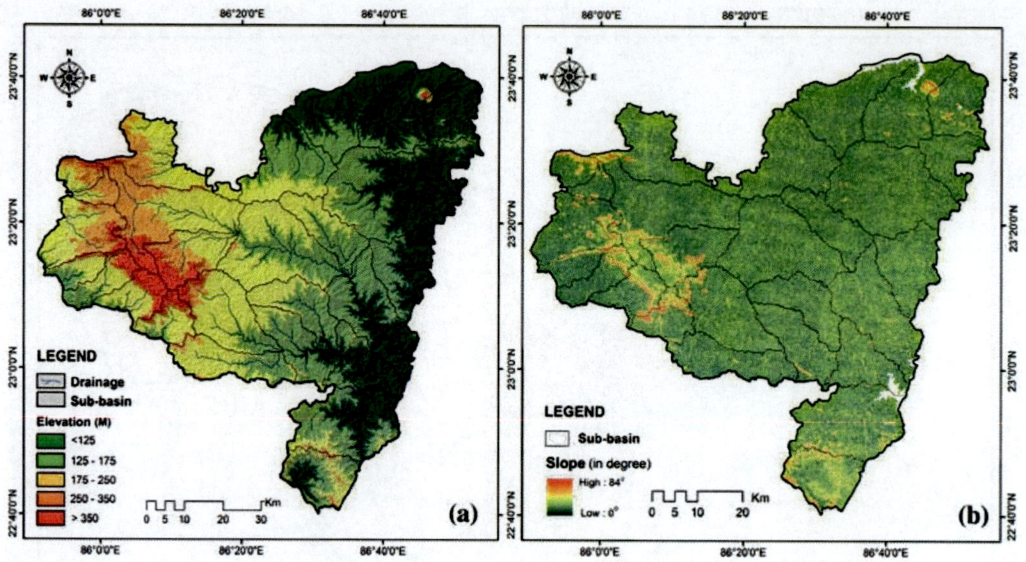

Fig. 22.3 Terrain characteristics of the study area: **a** Elevation and **b** Slope

derive such density parameters (D). To estimate hierarchical anomaly density (G_a), the minimum number of streams in each order is considered which builds an absolutely well-organized dendritic drainage network (Melton 1958). Subsequently, the hierarchical anomaly index (D_a) is measured from the ratio of hierarchical anomaly density and the actual number of first-order streams in the drainage basin (Avena et al. 1967). Ultimately, the denudation rate (T_u) of the drainage basin is estimated using the equation obtain and modified from the previous studies of Ciccacci et al. (1980, 1986, 2003). The hierarchical geometry of a drainage system follows a progression in which a first-order stream meets a second-order stream ($1 \rightarrow 2$), the second order with a third order ($2 \rightarrow 3$), and third- to a fourth-stream order ($3 \rightarrow 4$), and such kind of drainage systems are supposed to have a hierarchical progressive orientation (Bahrami 2013).

To determine the organization's stages of the drainage network, the numerical anomaly of stream segments in a particular drainage system can be fairly understood from a series of indices. Hierarchical anomaly number (Ha) is one of them and it is the smallest number of first-order stream segments which makes the drainage network perfectly hierarchical pattern (well-organized dendritic drainage network) (Melton 1958). When the numbers of stream junction of sub-watersheds increase the stream orders also increase (Bahrami 2013) for example stream order 3, 4, and 5 have different numbers of stream junctions, i.e., ($1 \rightarrow 2$, $1 \rightarrow 3$, $2 \rightarrow 3$), ($1 \rightarrow 2$, $1 \rightarrow 3$, $1 \rightarrow 4$, $2 \rightarrow 3$, $2 \rightarrow 4$, $3 \rightarrow 4$), and ($1 \rightarrow 2$, $1 \rightarrow 3$, $1 \rightarrow 4$, $1 \rightarrow 5$, $2 \rightarrow 3$, $2 \rightarrow 4$, $2 \rightarrow 5$, $3 \rightarrow 4$, $3 \rightarrow 5$, $4 \rightarrow 5$).

$$\mathrm{Ha}_{(i \rightarrow j)} = 2^{(i-2)} - 2^{(i-1)} \quad (22.2)$$

$$\mathrm{Ha}^{(ct)} = \sum \left\{ \mathrm{Ha}_{(i \rightarrow j)} \times Ns_{(i \rightarrow j)} \right\}$$

$$\mathrm{Ha}^{\mathrm{d}} = \frac{\mathrm{Ha}_{(i \rightarrow j)}}{\mathrm{A}} \quad (22.3)$$

$$\Delta_{\mathrm{a}} = \frac{\mathrm{Ha}^{(ct)}}{\mathrm{N1}} \quad (22.4)$$

where (s) is the stream order, ($N^*_{i,r}$) is the number of hierarchical segments of order (i) which influence the segments of order (r), (G_a) is the hierarchical anomaly number, (g_a) hierarchical anomaly density, (Δ_a) is the hierarchical anomaly index, and N1 means total number of stream order (i).

Table 22.1 Methodological formulae for the measuring morphometric parameters

Sl. no	Parameter	Calculation formula	Description	References
Linear aspect				
1	Stream order (U)		Hierarchical rank	Horton (1945)
2	Bifurcation ratio (Rb)	Rb = Nu/Nu + 1	Nu + 1 = No. of segments of next higher order	Schumm (1956)
3	Mean bifurcation ratio (mRb)		mRb = Average Rb of all orders	Strahler 1964
4	Stream length (Lu)		Length of the Stream (km)	Horton (1945)
5	Mean stream length (Lsm)	Lsm = Lu/Nu	Lu = Total stream length of order u; Nu = Total no of stream segment of order u	Horton (1945)
6	Stream length ratio (Rl)	Rl = Lu/Lu-1	Lu-1 = Total stream length of its next lower order	Horton (1945)
Arial aspect				
7	Drainage density (Dd)	Dd = Lu/A	L = Total length of Stream, A = Area of the Watershed	Horton (1945)
8	Texture ratio (T)	T = N1/P	N1 = Total number of first order stream, P = Perimeter of watershed	Horton (1945)
9	Stream frequency (Fs)	Fs = N/A	N = Total number of Stream	Horton (1945)
10	Form factor (Rf)	$Rf = A/Lb^2$	Lb = Maximum Basin length	Horton (1932)
11	Elongation ratio (Re)	Re = 2√(A/ð)/Lb	ð = 3.14	Schumm (1956)
12	Circularity ratio (Rc)	$Rc = 4ðA/P^2$		Miller (1953
13	Length of over land flow (Lo)	Lof = 1/2Dd	Dd = Drainage density	Horton (1945)
14	Compactness coefficient (Cc)	$Cc = 0.2821P/A^{0.5}$		Horton (1945)
15	Shape basin (Bs)	$Bs = Lb^2/A$		Nooka Ratnam et al. (2005)
Relief aspect				
16	Basin relief (Bh)	Vertical distance between the lowest and highest point of watershed		Horton (1945)
17	Relief ratio (Rh)	Rh = Bh/Lb	Bh = Basin Relief, Lb = Basin length	Schumm (1956)
18	Ruggedness number (Rn)	Rn = Bh × Dd	Dd = Drainage density	Moore et al. (1991)
Measurement of denudation				
19	Hierarchical anomaly density (G_a)	G_a	Ga = Number of 1st order streams make the drainage network perfectly hierarchized	Avena et al. (1967)
20	Hierarchical anomaly index (Δ_a)	$\Delta_a = G_a/N_1$	N1 = is total number of 1st order streams	Avena et al. (1967)
21	Denudation rate index (T_u)	$\text{Log } T_u = 1.44780 + 0.32619D + 0.10247\Delta_a$ (if $D \leq 6$)		Ciccaci et al. (1980)

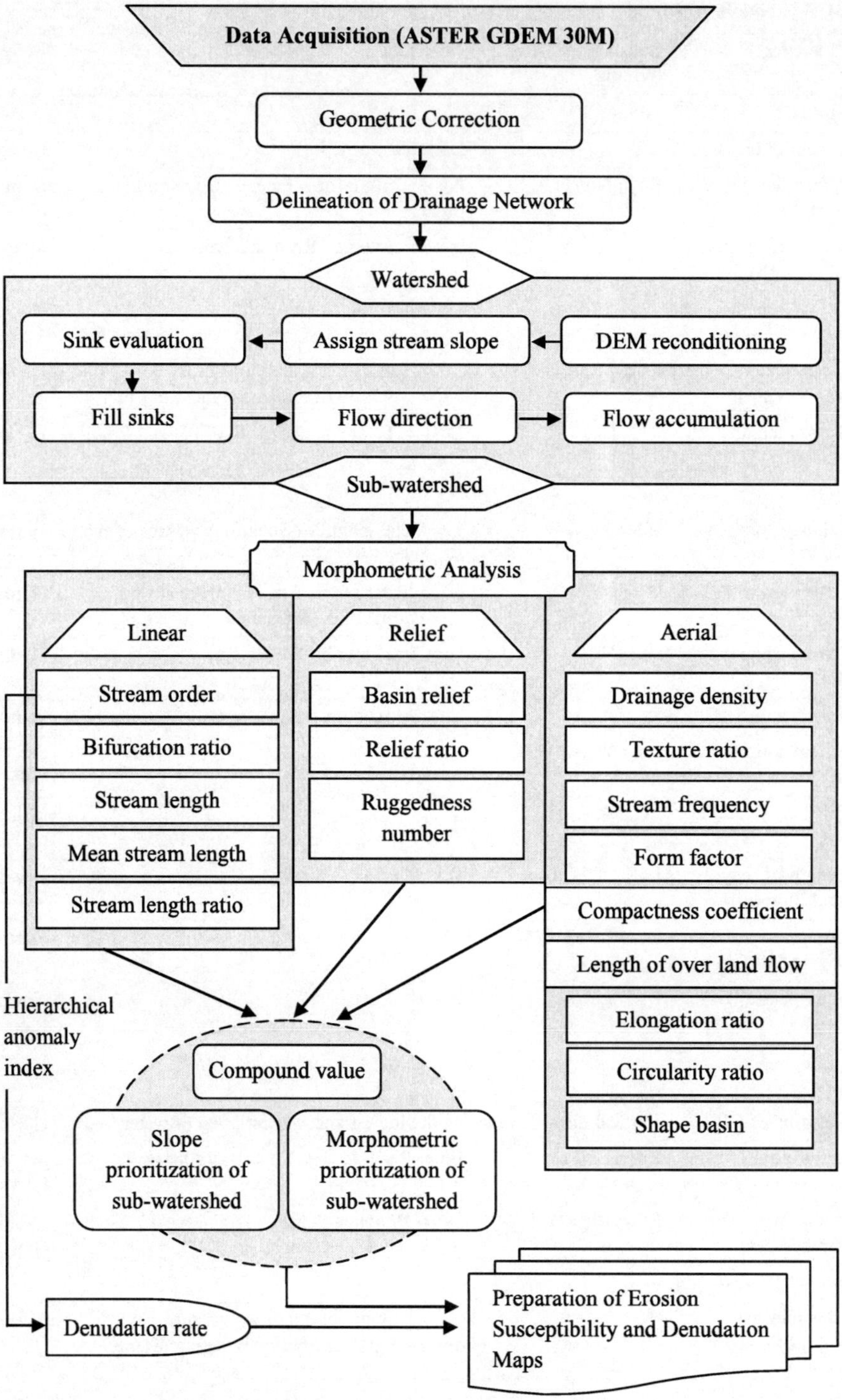

Fig. 22.4 Methodological flow chart for erosion susceptibility mapping

This initial assessment of the drainage network extension and association degree led to the assessment of denudation rate within drainage basins. Finally, some relationships were used which allow to compute the denudation rate index which is expressed as suspended sediment yield (T_u, t/km^2/year), and was calculated as a function of the formerly determined morphometric parameters.

$$\log T_u = 1.05954 + 2.79687 \log D + 0.13985 \Delta_a \text{with} D \geq 6 \quad (22.5)$$

$$\log T_u = 1.44780 + 0.32619 D + 0.10247 \Delta_a \text{with} D < 6 \quad (22.6)$$

where (T_u) is the denudation rate, (Δ_a) is the hierarchical anomaly index, and (D) is the drainage density.

22.3 Results and Discussions

22.3.1 Morphometric Analysis

The morphometric analysis of watersheds provides a detailed explanation of interconnection between processes act over the earth's surface and several earths' system components such as geology, geomorphology, and hydrology (Paul and Bayode 2012). For understanding basin characteristics and erosion susceptibility, analysis of different parameters of drainage morphometry of sub-watersheds is very significant (Javed et al. 2011). The fundamental 18 parameters are analyzed through linear, areal, and relief aspects for each of the sub-watersheds of Purulia district and shown in Tables 22.2 and 22.3. The details of the results of each morphometric parameter have been provided below.

22.3.2 Basic Parameters

22.3.2.1 Sub-Watershed Area (a) and Perimeter (P)

The area of the drainage system is always considered as the foremost hydrological features of a watershed. This area may reflect the total volume of surface water which basically originates from precipitation. In Purulia district out of 27 sub-watersheds, No. 23 sub-watershed shows the maximum areal coverage (512.48 km^2), whereas sub-watershed No. 2 shows minimum areal coverage (54.14 km^2). The basin perimeter explains the length of the line which discriminates the surface divide of the sub-watersheds. The highest and lowest values of the basin perimeter are 51.47 km and 220.25 km for sub-watershed No. 1 and 10, respectively.

22.3.3 Linear Parameters

22.3.3.1 Stream Order (U)

Horton (1945) and Strahler (1964) elaborately described the stream order in the quantitative morphometric analysis. Basically, first-order streams are the initial streams without any tributary which determine by the surface overland flow. After that, second-order streams are generated by the conjunction of two first-order streams with the higher surface flow, and in this way, next order streams have formed with more overland flow (Paul and Bayode 2012). In this study, the application of the stream ordering method through GIS illustrates that the drainage network of the district enclosed as a sixth-order basin (Table 22.2). All the 27 sub-watersheds have first-order streams with significant variation from each other. It ranges from 35 first-order streams (in sub-watershed 2) to 658 first-order stream (in sub-watershed 10). The higher concentration of first-order streams is mainly found in the western and southern parts of the district than the eastern part (Fig. 22.5). The total number of streams (Nu) has occurred significantly higher range (852 to 44 km). Therefore, as expected, sub-watersheds of the eastern part are getting a higher concentration of surface flow from western's sub-watersheds. This variation of stream ordering and basin size is highly determined by the structural and physiographic conditions of the district.

Table 22.2 Computation of basic morphometric parameters for Purulia district

Sub-basin	Basin perimeter	Basin area	Number of streams							Basin length	Stream length
			1st	2nd	3rd	4th	5th	6th	Total		
1	51.475	74.797	49	12	3	1			65	15.220	80.14
2	58.175	54.138	35	8	1				44	12.667	59.86
3	115.800	239.712	393	98	19	3	1		514	29.493	418.56
4	67.400	79.948	48	13	2	1			64	15.807	96.34
5	93.200	118.952	188	44	7	1			240	19.809	207.78
6	91.650	130.500	207	49	13	3	1		273	20.880	226.55
7	84.825	118.904	199	44	12	2			257	19.805	206.23
8	128.800	222.217	144	31	6	1			182	28.251	229.61
9	171.300	269.702	166	36	6	1			209	31.536	266.25
10	220.250	413.520	658	151	35	7	1		852	40.200	734.25
11	109.250	198.370	328	71	13	4	1		417	26.487	355.39
12	160.900	419.309	259	52	11	3	1	1	327	40.519	446.95
13	120.625	176.292	121	24	5	1			151	24.770	181.48
14	148.200	331.625	223	54	12	3		1	293	35.464	353.16
15	135.075	294.477	483	110	31	6	1		631	33.150	514.33
16	125.000	192.329	121	31	6	1			159	26.025	207.17
17	123.225	195.924	127	32	9	3	1		172	26.300	204.55
18	113.350	182.216	113	31	6	1	1		152	25.239	193.66
19	134.400	371.281	243	44	9	2	1		299	37.814	399.45
20	92.475	203.325	329	79	19	5	1		433	26.860	345.4
21	140.749	313.430	208	41	7	1			257	34.345	319.28
22	143.000	310.595	196	40	6	2			244	34.169	301.44
23	166.050	512.486	310	72	12	4	1		399	45.411	548.45
24	130.825	289.669	189	44	13	3	1		250	32.841	325.09
25	92.275	186.364	118	25	7	2	1		153	25.564	208.19
26	104.900	146.662	93	18	5	1			117	22.311	154.38
27	108.050	233.45	151	29	5	1			186	29.053	243.64

22.3.3.2 Total Length of Streams (Lu)

To understand the characteristics of surface runoff, stream length as a hydrological parameter is very significant to measure. Naturally, smaller stream lengths are determined by larger slopes and high elevation where longer stream lengths are characterized by flatted surfaces with very gentle slopes. Therefore, the total length of streams is high basically in first order which is gradually reduced as stream order increases. The total number of streams is counted from each sub-watershed and the total length of these streams is also measured through the GIS tool (Table 22.2). The total stream length is highest (734.25 km) in sub-watershed No. 10. Moreover, the total stream length of the sub-watershed from the western and southern parts is more than the northern and eastern side sub-watersheds. So, it indicates more active

Table 22.3 Result of morphometric analysis for 27 sub-watersheds of Purulia district

Sub basin	Mean Bifurcation ratio	Drainage density	Stream frequency	Circulatory ratio	Form factor	Elongation ratio	Texture ratio	Compactness coefficient	Relief ratio	Length of overland flow
1	2.22	1.07	0.86	0.35	0.32	0.64	0.93	1.67	0.029	0.77
2	2.48	1.10	0.81	0.20	0.33	0.65	0.89	2.23	0.023	0.23
3	3.70	1.74	2.14	0.22	0.27	0.59	3.74	2.10	0.025	0.68
4	2.44	1.20	0.80	0.22	0.31	0.63	0.96	2.12	0.019	0.48
5	3.51	1.74	2.01	0.17	0.30	0.62	3.52	2.41	0.020	0.89
6	3.07	1.73	2.09	0.19	0.29	0.61	3.63	2.26	0.027	0.61
7	2.84	1.73	2.16	0.20	0.30	0.62	3.74	2.19	0.030	0.64
8	3.16	1.03	0.81	0.16	0.27	0.59	0.84	2.43	0.005	0.17
9	3.32	0.98	0.77	0.11	0.27	0.58	0.76	2.94	0.029	0.29
10	4.13	1.77	2.06	0.10	0.25	0.57	3.65	3.05	0.008	0.20
11	3.47	1.79	2.10	0.20	0.28	0.59	3.76	2.18	0.012	0.34
12	3.47	1.06	0.77	0.20	0.25	0.57	0.83	2.21	0.001	0.84
13	2.97	1.02	0.85	0.15	0.28	0.60	0.88	2.56	0.003	0.21
14	2.53	1.06	0.88	0.18	0.26	0.57	0.94	2.29	0.003	0.35
15	3.82	1.74	2.14	0.20	0.26	0.58	3.74	2.22	0.009	0.50
16	3.01	1.07	0.82	0.15	0.28	0.60	0.89	2.54	0.003	0.22
17	2.70	1.04	0.87	0.16	0.28	0.60	0.91	2.48	0.005	0.44
18	3.16	1.06	0.83	0.17	0.28	0.60	0.88	2.36	0.005	0.31
19	3.38	1.07	0.80	0.25	0.25	0.57	0.86	1.96	0.007	0.44
20	3.42	1.69	2.12	0.29	0.28	0.59	3.61	1.82	0.020	0.18
21	3.59	1.01	0.81	0.19	0.26	0.58	0.83	2.24	0.008	0.35
22	2.91	0.97	0.78	0.19	0.26	0.58	0.76	2.28	0.005	0.21
23	3.46	1.07	0.77	0.23	0.24	0.56	0.83	2.06	0.005	0.13
24	3.00	1.12	0.86	0.21	0.26	0.58	0.96	2.16	0.019	0.24

(continued)

Table 22.3 (continued)

Sub basin	Mean Bifurcation ratio	Drainage density	Stream frequency	Circulatory ratio	Form factor	Elongation ratio	Texture ratio	Compactness coefficient	Relief ratio	Length of overland flow
25	2.76	1.11	0.82	0.27	0.28	0.60	0.91	1.90	0.025	0.65
26	2.75	1.05	0.79	0.16	0.29	0.61	0.83	2.44	0.004	0.19
27	3.20	1.04	0.79	0.25	0.27	0.59	0.83	1.99	0.006	0.23

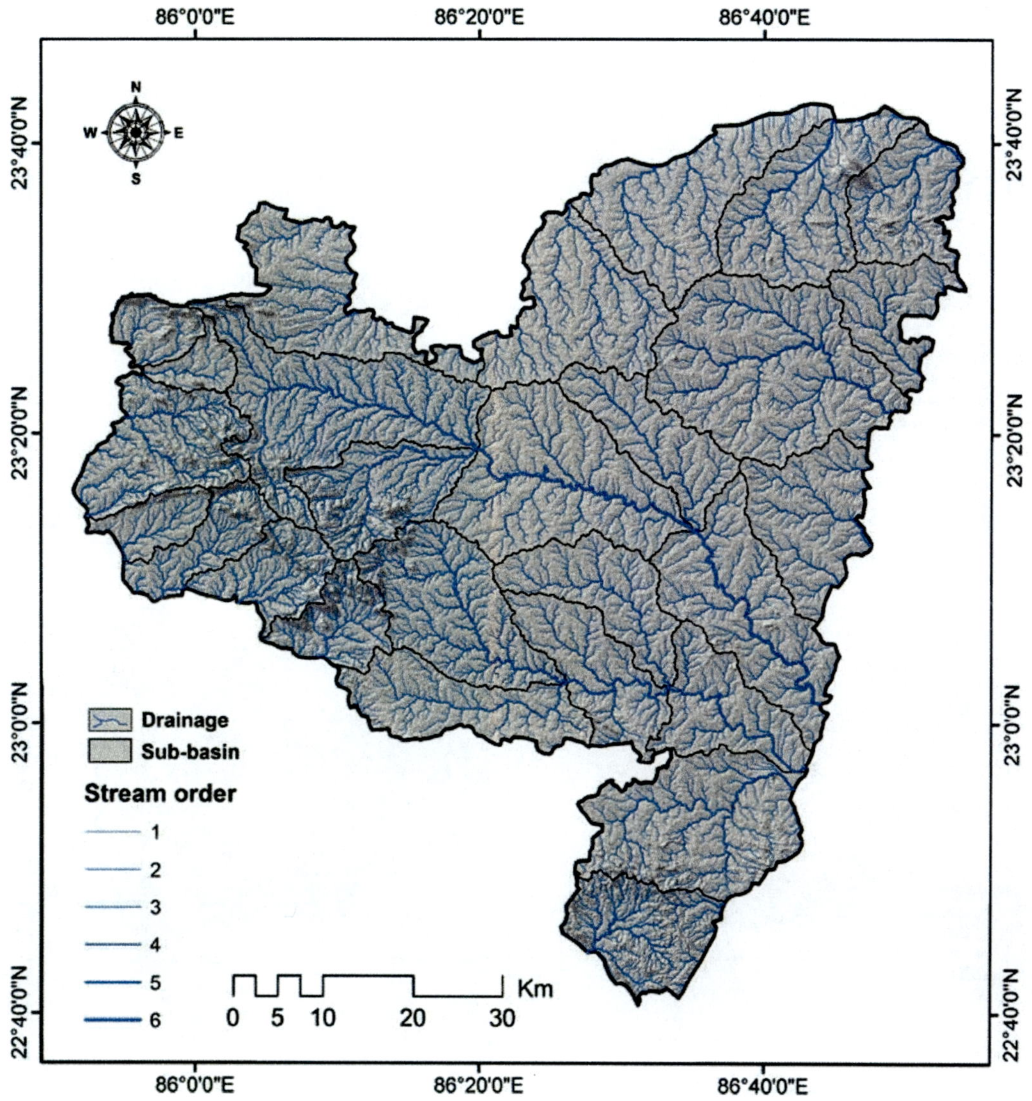

Fig. 22.5 Stream orders of the study area

hydrological action in the western and southern parts compare to other segments of the district.

22.3.3.3 Bifurcation Ratios (Rb) and Mean Bifurcation Ratios (mRb)

The concept "bifurcation ratio" (Rb) was founded by Horton (1932) for expressing the ratio between the number of streams of a particular order and the number in the subsequent lower order. After that in 1964, Strahler defined Rb as the ratio between the number of streams of the certain order and the number of sections of the higher order (Nu + 1). In this work, the mRb considerably varies from 2.22 to 4.13 (Table 22.3), and the mRb of the entire district is 3.13 (Fig. 22.6c). Higher mRbs illustrate complex watershed structure with poor water permeability and therefore, those sub-watersheds indicate potential topsoil degradation with high-intensity surface runoff.

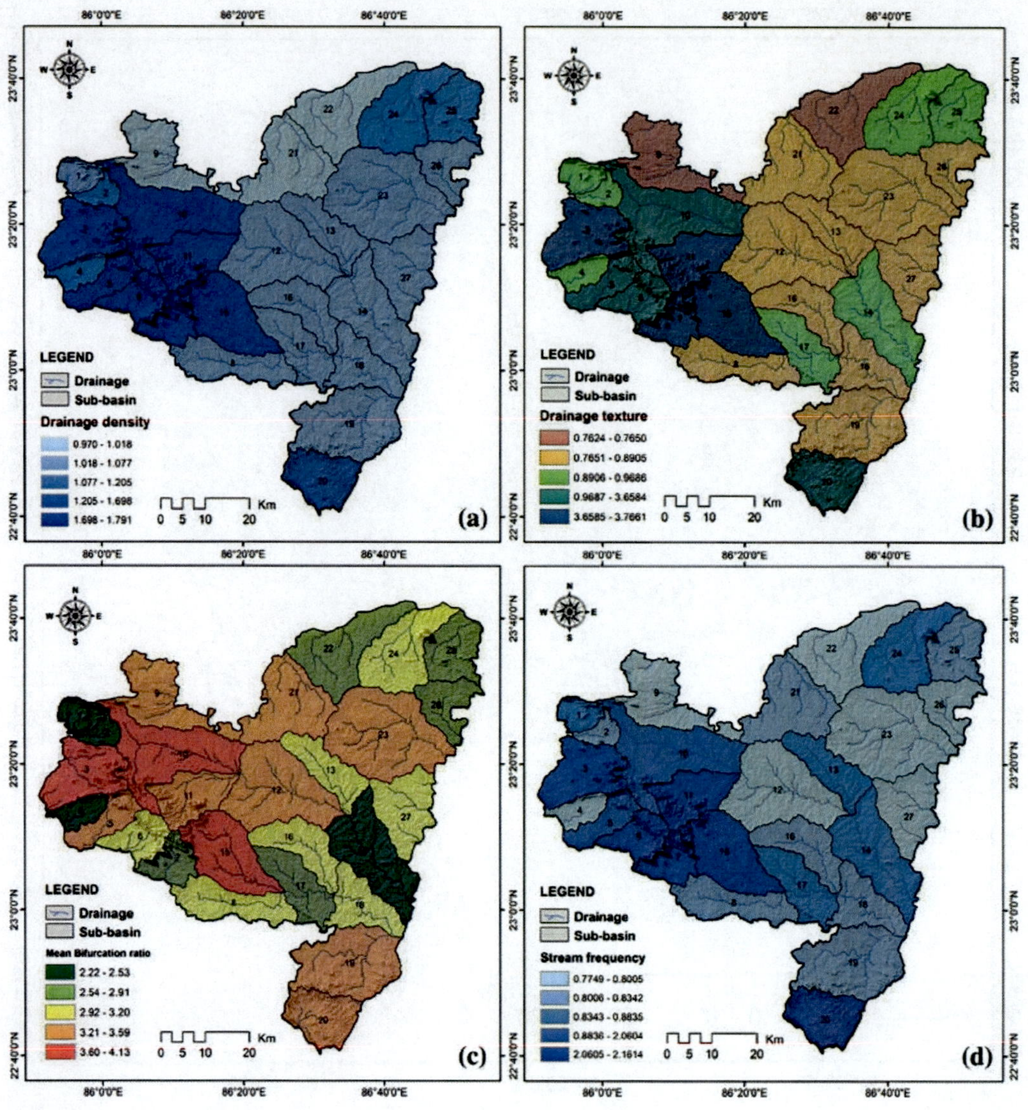

Fig. 22.6 Drainage characteristics: **a** Drainage density, **b** Drainage texture, **c** Mean bifurcation ratio and **d** Drainage frequency

22.3.3.4 Stream Length Ratio (Rl)

Rl delineates the relation between individual stream lengths of a particular stream order with stream lengths of the subsequent order (Horton 1932). Basically, the stream length ratio provides us a common idea regarding the virtual permeability of the lithological structure of the basin (Al-Saady et al. 2016) which also shows the relationship between the discharge of overland flow and erosional stage of the watersheds. The Rls of the study area significantly vary (0.22–592.66) and reveal an inconsistent distinction among successive orders. As RLs illustrate the hydrological character of existing lithology and sediments of a basin, fewer numbers of lengthy streams are originated upon a permeable surface which mostly observed in the eastern portion of the study area. On the contrary, higher numbers of streams with shorter lengths mostly developed on semi-permeable rock and coarse grain sediment surfaces in the central to the western part of the district.

22.3.4 Shape or Arial Parameters

22.3.4.1 Drainage Density (Dd)

According to Horton (1932) Dd is the overall length of the streams (all stream orders) per unit drainage area. Basically, Dd is the function of entire lithological and physiographical characteristics of a particular watershed. Therefore, Dd illustrates the overview of geological formation, tectonic movement, soil types and texture, topography, the shape of the basin, and the existing climate over there. In Purulia, Dd is 1.27 km/km^2 and it significantly varies from 0.97 to 1.79 km/km^2 in the sub-watershed scale (Table 22.3). Spatial distributions of Dd clearly indicate considerable differences in relation to topography. Most of the sub-watersheds with higher Dd located in the western and southern parts and gradually reduce toward the east and north of the district (Fig. 22.6a). The sub-watersheds with higher Dd mainly observed around Ajodhya hills, the highest elevated region of the district. As drainage networks highly influence the Dd, sediment erosion rates are naturally increased on these sub-watersheds.

22.3.4.2 Texture Ratio (T)

T is the ratio between the total number of stream sections (Nu) of all orders and the basin perimeter (P) (Horton 1945). According to him, infiltration capacity is the only significant aspect which influences the T of a particular basin. Geomorphology and drainage morphometry of the study area is influenced by this important factor as it also depends on the underlying bedrock and surface relief of the terrain. For the sub-watersheds in the study area, T ranges from 0.76 to 3.76. The very coarse texture is found in 3, 11, 15, 10, 20 number sub-basin located mostly in the western part and fine texture mostly observed in the northeastern part of the district (Fig. 22.6b).

22.3.4.3 Stream Frequency (Fs)

The ratio between the total number of streams (Nu) in a watershed and the area of this watershed (A) is called stream frequency (Fs). Horton (1945) recognized it as the number of drainages per unit area. From several studies, it is acknowledged that Fs values vary from 3.91 to 9.99 and determine mainly by geological formation lying of the watershed. Therefore, Fs has an influence and correlation with T. Similarly, Fs values have a direct correlation with Dd values of any watershed that means the increasing frequency of streams can increase Dd. Basically, Fs varies with the size of the watershed and thus values of Fs and Dd for large and small size basins are not comparable (Farhan and Anaba 2016). High Fs and Dd meaning higher surface runoff and erosion probability (Sreedevi et al. 2013). The values of drainage frequency are range from 0.77 (SW 9) to 2.16 (SW 7) (Table 22.3). Higher drainage frequency mainly observed in the western hilly region of the district (Fig. 22.6d).

22.3.4.4 Circularity Ratio (Rc)

The ratio between the basin areas to the area of a circle with the same perimeter of a particular sub-watershed is defined as circularity ratio (Miller 1953). In relation to streamflow, circularity ratio is the most helpful measure of shape parameters (Al-Saady et al. 2016). Basically, decreasing circularity ratio is related to the increasing stream order and the results from this analysis show the same thing which is also compatible with Morisawa (1962) work. Therefore, a perfect picture of the basin shape might not be measured by circularity ratio. The Rc values of the sub-watersheds range from 0.10 to 0.35 (Table 22.3). So, it's reflecting that the total area and the sub-watersheds are elongated or less circular with minimum discharge. Miller (1953) reveals that the circularity ratio is controlled by the drainage frequency, stream length, underlying geology, land use types, slope, and relief of the basin. The higher circularity is observed mainly in the southern and eastern parts of the district (Fig. 22.7a).

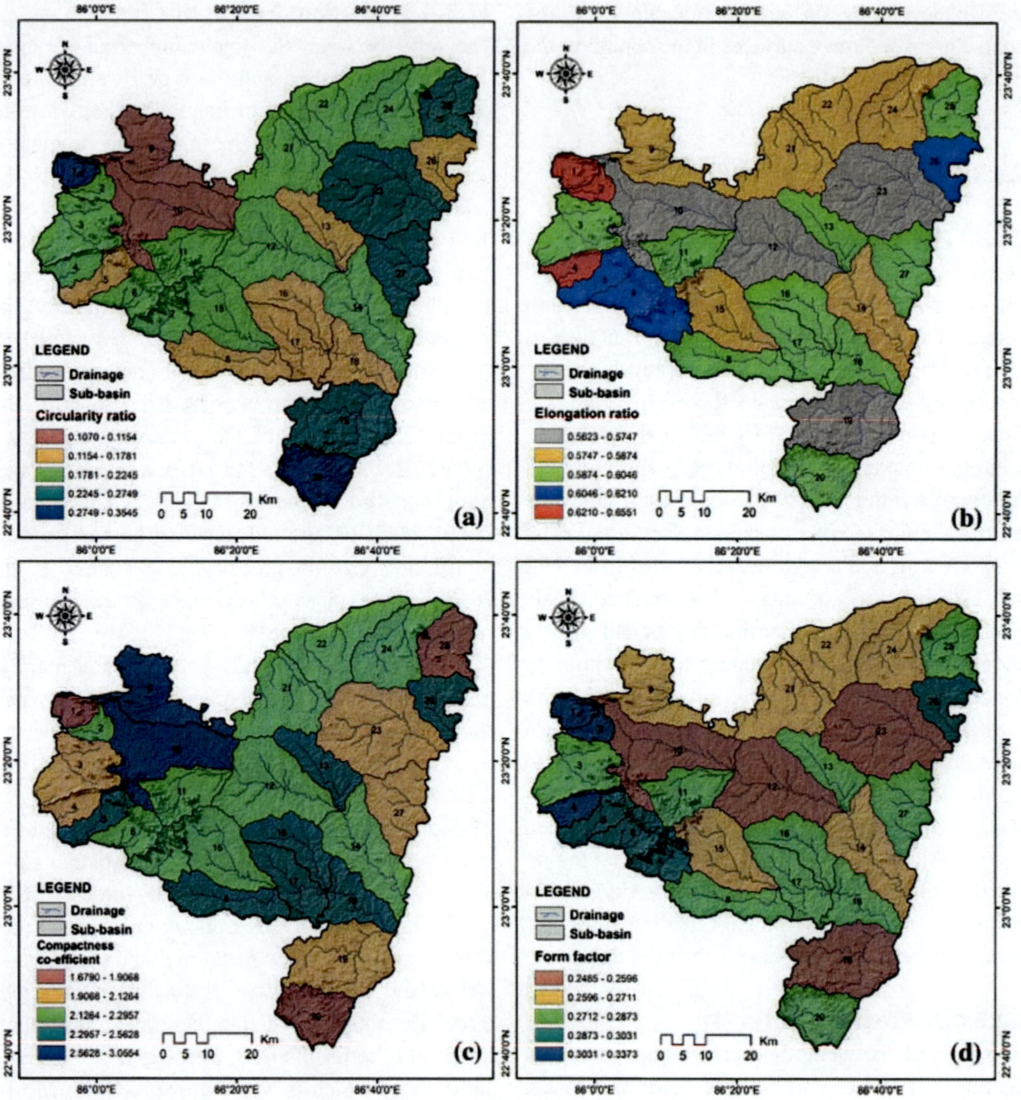

Fig. 22.7 Areal parameters: **a** Circularity ratio, **b** Elongation ratio, **c** Compactness coefficient and **d** Form factor

22.3.4.5 Elongation Ratio (Re)

The ratio between the diameter of the circle of a drainage basin area and the maximum length of the basin is called the Re (Schumm 1956). According to Strahler (1964), if this ratio is between 0.6 and 1.0 then the drainage basin may come under varied climate with heterogeneous geological forms in the beneath. The changing slopes of a watershed have been classified on the basis of the measured index for example: elongated (0.5–0.7), less elongated (0.7–0.8), more elongated (<0.5), circular (0.9–0.10), and oval (0.8–0.9) (Pareta and Pareta 2012). The young stage of basin evolution due to the intense neotectonic movement is identified by elongated shape, whereas the intermediate shape with an inclination toward circularity indicates an early mature stage (Lykoudi and Angelaki 2004). The results show that the elongation ratios for the sub-watersheds are varied from 0.56 to 0.66 (Table 22.3). There is no such apparent symmetry of spatial distribution between Re values and relief of the sub-watersheds (Fig. 22.7b). As most of the basin of central to northern portion is elongated types, runoff and erosion efficiency is little less on this sub-basin (Singh and Singh 1997), and hard lithology influence the infiltration and discharge capacity.

22.3.4.6 Compactness Coefficient (Cc)

The ratio between the actual basin perimeter and the perimeter of a circle of the equivalent area is defined as the Cc (Luchisheva 1950). Generally, a perfect circle shape of the basin is determined by 1 Cc value where, with the increasing value up to 1.128, the basin becomes more squared shape and could be very elongated for 3 Cc values (Zavoianu 1985). The compactness coefficient values of the study area are range from 1.68 to 3.06 for sub-watershed 1 and 10, respectively (Table 22.3). Most of the sub-watersheds are semi-circle to square-shaped as those have more than 2.00 Cc values (Fig. 22.7 c). Therefore, the flow accumulation of those basins has been taken place with a long duration before peak flow occurs.

22.3.4.7 Form Factor (Rf)

The ratio of the basin area with a square of the basin length is defined as form factor (Rf) (Strahler 1957). Basically, Rf varies from 0 to 1 as the highly elongated shape to perfect circular shape (Manu and Anirudhan 2008) and it has been used to predict the basin intensity of a particular area. For a perfectly circular basin, the Rf values always are less than 0.79 where smaller the value (<0.45), the basin will be more elongated. Generally, the watersheds are characterized with greater peak flow with shorter duration belongs to strong Rf, while fewer peak flow and longer duration belong to low Rf. The Rf values of 27 sub-watersheds range from 0.24 to 0.34 (Fig. 22.7d). Therefore, the total region dominance with flatter peak flow for a long time as the Rf has average lower values.

22.3.4.8 Length of Overland Flow (Lof)

Length of Overland Flow (Lof) stands for the duration of water over the surface before it gets concentrated into specific stream channels, and is equivalent to half of the Dd (Horton 1945). The measure of overland flow is very significant because it affects not only the water regime of a drainage system but also the long-term evolution of drainage basins (Zavoianu 1985). It is also a significant and independent variable that affects both hydrographic and hydrologic development of the watershed (Horton 1945). The Lof values varies from 0.14 (SB 23) to 0.89 (SB 5). This variation is due to the spatial variation of underlying lithology, slope, forest cover, and rainfall intensity.

22.3.4.9 Shape Basin (Bs)

The shape of the basin represents the ratio between the square of the basin length and the area of this particular basin. Basically, the shape of the basin is in the inverse ratio to the Rf (Ratnam et al. 2005). This shape factor provides a perception concerning the circular character of the watershed. Surface erosion is responded highly with the greater circular character of the watershed. The shape of the basin ranges from 2.96 to 4.02 for sub-basins 2 and 23, respectively, and indicates that most of the sub-basin shapes are elongated in nature.

22.3.5 Relief Parameters

22.3.5.1 Basin Relief (Bh)

Bh illustrates the difference in elevation between the highest and lowest points from the source region to the downstream region of a particular watershed. The maximum elevation is found in the far western part of the district mainly adjoin part of Ajodhya hill with more than 500 m average height. The lowermost sections mainly observed in the eastern and north-eastern parts of the district associated with less than 150 m height. As the Bh has a direct relation with the drainage network system and erosional processes (Horton 1945), higher erosional potentiality can be identified in the western elevated section of the district. The complex pattern of the dissected plateau and undulating terrain in the western and some parts of most southern sections have been highly susceptible to surface erosion. Bh of the

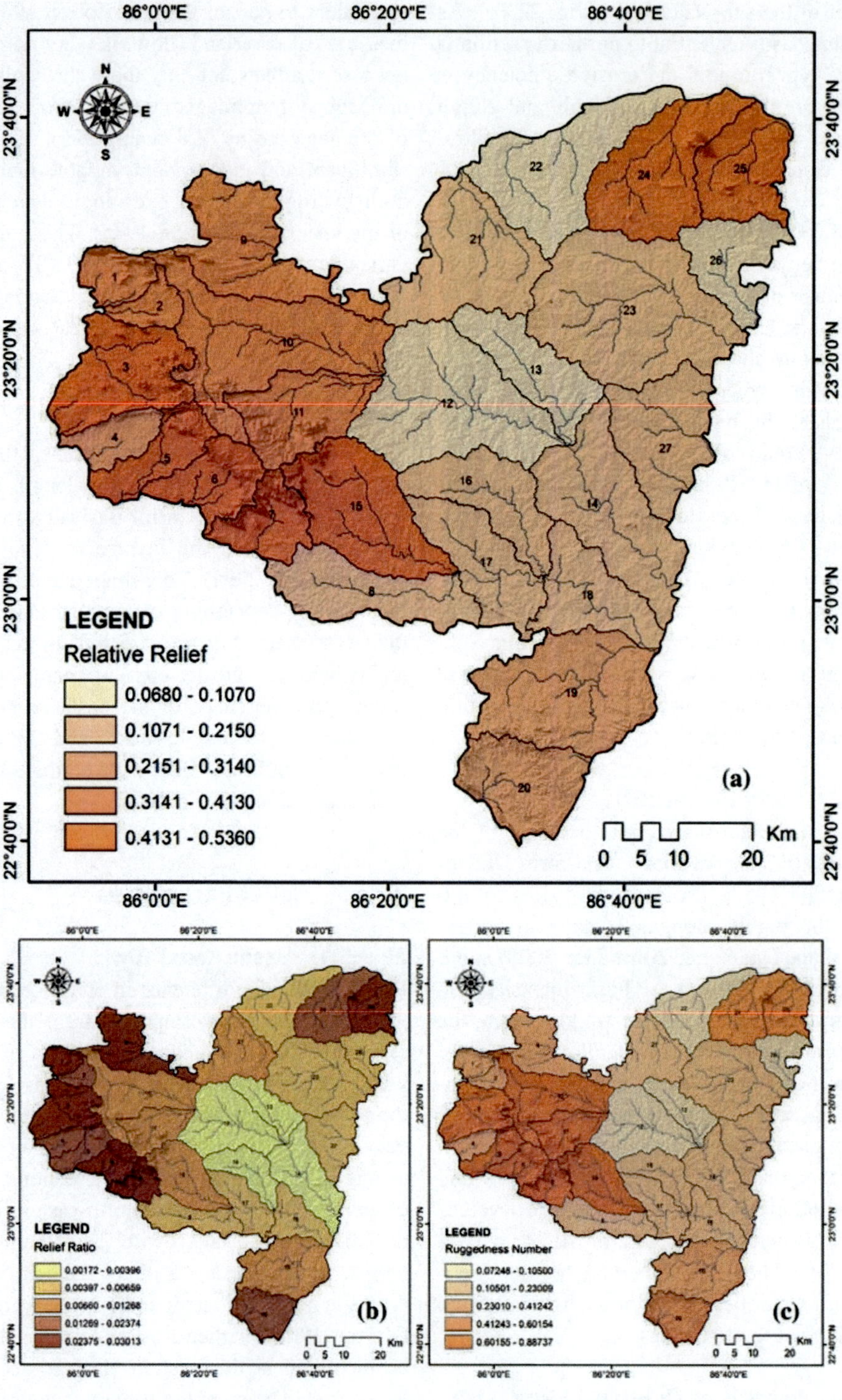

Fig. 22.8 Basic relief parameters of the study area: **a** Basin relief, **b** Relief ratio, **c** Ruggedness number

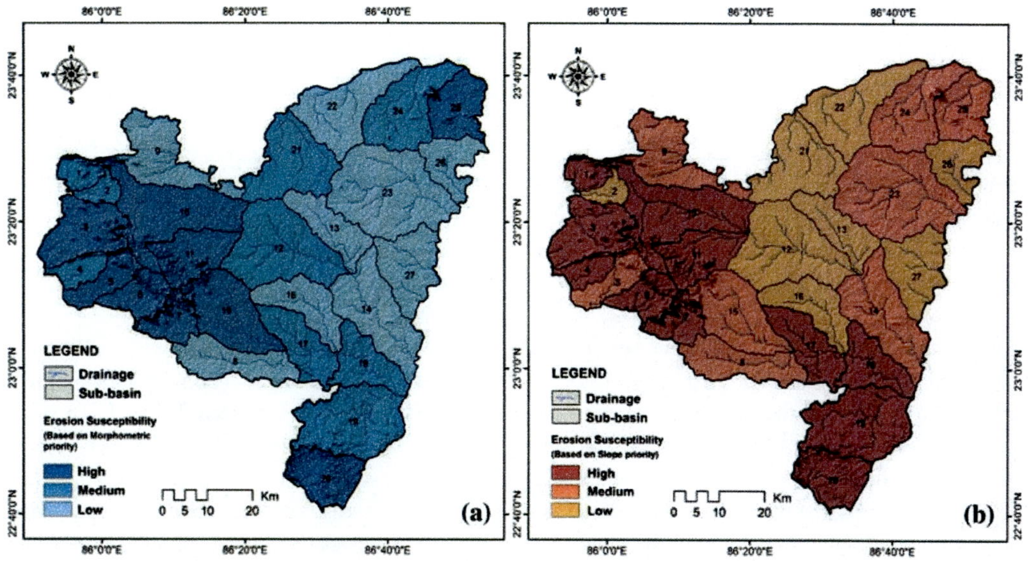

Fig. 22.9 Erosion susceptibility mapping: **a** Based on morphometric prioritization and **b** Based on slope prioritization

total area gradually decreases from western to eastern part parallel to the mainstream direction (Fig. 22.8a).

22.3.5.2 Relief Ratio (Rh)

Rh is defined as the ratio between Bh and the longest width of the basin parallel to the main streamline (Schumm 1956). The value of relief ratio ranges from 1.72 to 30.13 for sub-basins 12 and 7, respectively. Therefore, this large variation in relief ratio demarcates the variation in lithology, geomorphic features, structural units, etc. The intensity of channel gradients is explained by relief ratio and thus, Rh can be recognized as an important factor for measuring erosion processes and to predict the surface runoff intensity (Fig. 22.8b).

22.3.5.3 Ruggedness Number (Rn)

Rn is the outcome of the Dd over the Bh. The structural complexity of the topography can be recognized by the Rn. Ozdemir and Bird (2009) explain the direct relationship between peak discharge and related erosion with Bh and Dd as both increases simultaneously. The results show that the Rn of the study area varies from 0.072 to 0.887 for sub-basins 12 and 5, respectively. Most of the sub-watersheds with higher value located near the western high elevated region and indicates higher erosion propensity (Fig. 22.8c).

22.3.6 Soil Erosion Susceptibility Analysis

Morphometric analysis is employed for assessing and mapping soil erosion susceptibility based on morphometric and slope prioritization. The influences of soil and underlying lithology including the structure of the underlying stratum, depth of impermeable layer below the surface, and land-use pattern are also explained.

22.3.6.1 Assessing the Influence of Morphometric Parameters on Soil Erosion Susceptibility

Sub-basin wise prioritization is performed of selected 18 numbers of morphometric parameters like drainage density, Bh, elongation ratio, circularity ratio, form factor, shape factor, etc. In the last decades, several researchers conducted their study regarding the prioritization of watersheds (Rekha et al. 2011; Kanth and Hassan

2012; Gajbhiye et al. 2014; Rahaman et al. 2015). Some morphometric parameters (Rb, Dd, Dt, and Fs) directly correlated to the soil erosion which is arranged and assigned a rank in respect of their highest and lowest values. Some parameters (Rf, Re, Cc, and Rc) which have negative correlations with erosion, are assigned a rank in the opposite way as the lowest to the highest value.

Therefore, all assigned ranks of 27 sub-watersheds are used to calculate the compound values. After that, the final priority rank has been assigned in respect of the compound values for getting erosion susceptibility zones. Dd, Fs, T, and Rb illustrate the zone of higher concentration mostly in the western hilly region and adjoin areas. As a result, susceptibility to erosion is mostly dominant in this region. On the other hand, other parameters which have negative influence such as Rf, Re, Rc, Cc demonstrate that most of the central to western parts have lower values and indicates the higher intensity of soil erosion efficiency. Therefore, considering all these ranks of those morphometric parameters, an erosion susceptibility map has been prepared (Fig. 22.9a).

22.3.6.2 Assessing the Influence of Slope on Soil Erosion Susceptibility

The erosion susceptibility map is also derived based on slope priority ranks (Fig. 22.9b). As slope has a direct relation with surface erosion potentiality, most of the highly susceptible erosion zones located in far western and some parts of the southern division of the district. Those regions consist of a highly elevated and dissected plateau tract. The major hills of Jhalida and Ajodhya regions are located in the higher section of the district and therefore these regions have a higher intensity of surface erosion.

22.3.6.3 Identify the Influence of Land Use Land Cover, Geology, Soil and Landforms on Soil Erosion Susceptibility

Finally, the erosion susceptibility map of Purulia district has obtained by combining morphometric and slope priority ranks (Table 22.4) of the 27 sub-watersheds (Fig. 22.10a) and this map shows the basic two characteristics as given below:

High Erosion Zone: Western and most Southern part of the district
Basically, high elevated hilly region
Low Erosion Zone: Erosion potentiality gradually reduced from central to northeastern part of the district

Underlying lithology, geomorphic features, land use, and soil characteristics have also influenced the intensity and spatial variation of the erosion susceptibility in Purulia district. As seen in the study area that high erosion-prone zone in the western and most southern part of the district consist of ridges and steep slope landform features, highly fragile phyllite and mica schist including granitic gneiss in bedrocks, dense vegetation covers in the hilly regions and fine to coarse loamy soils in the overlying surface. On the other hand, the intentionally less erosion-prone areas of the central to northeastern parts of the district consist of a gentle slope, mostly hard granite, quartzite, and granitic gneiss rocks beneath the surface, predominantly barren and wasteland and gravel to coarse loamy soil in the overlying surface. Therefore, strong influences of lithology including landforms, topography, and land use are identified in this region.

22.3.7 Denudation Characteristics

The denudation status of all the sub-watersheds of the Purulia district is related to the hierarchical drainage networks and associated geomorphic activities (Gioia et al. 2011). Therefore, the denudation process firmly controlled by slope terrain, external geomorphic processes, and local environment and denotes landscape lowering and evolution. In this study, in the sub-basin scale estimated Dd ranges from 0.97 to 1.79 km/km^2 with an average of 1.27 km/km^2. As most of the area of Purulia district is covered with hard granitic rocks, the denudation rate mostly controlled by the steepness of slope and elevation of

Table 22.4 Slope and morphometric prioritization for mapping erosion susceptibility

Sub basin	Percentage of area in different slope								Slope prioritization			Morphometric prioritization		
	<3	Rank	3–6	Rank	6–9	Rank	> 9	Rank	Total rank	Priority rank		Compound Value	Priority Rank	
1	51.90	24	24.08	8	5.44	8	18.57	4	44	1	High	103	11	Medium
2	73.38	11	18.27	21	1.78	20	6.56	10	62	3	Low	123	14	Medium
3	59.08	21	22.41	12	5.64	6	12.87	7	46	1	High	60	1	High
4	75.01	7	16.28	23	2.86	12	5.85	11	53	1	High	111	12	Medium
5	64.87	20	15.14	25	5.49	7	14.50	6	58	2	Medium	66	2	High
6	51.18	25	21.06	15	9.28	3	18.48	5	48	1	High	80	6	High
7	47.99	26	23.23	10	7.02	4	21.76	2	42	1	High	73	4	High
8	71.67	16	24.08	8	2.73	13	1.51	18	55	2	Medium	141	22	Low
9	69.93	17	25.38	6	2.56	17	2.13	15	55	2	Medium	136	21	Low
10	73.68	10	15.56	24	3.54	9	7.22	8	51	1	High	92	7	High
11	53.90	23	17.95	22	9.56	2	18.59	3	50	1	High	74	5	High
12	75.91	6	22.42	11	1.50	23	0.17	25	65	3	Low	128	18	Medium
13	76.93	5	21.73	13	1.24	24	0.10	26	68	3	Low	141	22	Low
14	72.02	14	24.31	7	2.71	14	0.96	19	54	2	Medium	130	19	Low
15	73.21	12	18.96	18	2.57	16	5.25	12	58	2	Medium	66	3	High
16	72.01	15	25.76	5	2.00	19	0.22	24	63	3	Low	130	19	Low
17	66.20	19	28.57	2	3.51	10	1.72	16	47	1	High	126	17	Medium
18	68.58	18	26.45	4	3.32	11	1.64	17	50	1	High	124	15	Medium
19	57.05	22	29.24	1	6.81	5	6.91	9	37	1	High	114	13	Medium
20	29.40	27	28.30	3	16.08	1	26.21	1	32	1	High	95	9	High
21	79.07	3	19.63	17	1.04	25	0.26	22	67	3	Low	125	16	Medium
22	77.39	4	20.85	16	1.52	22	0.24	23	65	3	Low	160	24	Low
23	74.28	8	23.66	9	1.67	21	0.39	20	58	2	Medium	134	20	Low
24	72.40	13	21.46	14	2.69	15	3.45	14	56	2	Medium	98	10	Medium
25	74.19	9	18.77	19	2.50	18	4.54	13	59	2	Medium	94	8	High
26	85.25	1	14.46	26	0.25	27	0.04	27	81	3	Low	150	23	Low
27	80.35	2	18.55	20	0.82	26	0.27	21	69	3	Low	136	21	Low

the terrain. This study area also notices that the areas with weathered resistant hard rock and semi-impermeable weathered rocky surfaces have third-, fourth- and fifth-order streams with lower drainage density (0.97 to 1.20 km/km^2), whereas the areas with the steep slope and high elevated terrain have a dense number of first-order streams with high Dd (1.79 km/km^2). As a result, it is observed that the streams of all hierarchical orders which involved creating a perfectly ordered tree-shaped structure drainage system are strongly controlled by the slope of the terrain. Additionally, the hierarchical anomaly index (Da) was estimated with respect to the total number of first-order streams (N1) and hierarchical anomaly density (Ga) of each sub-watershed of the entire district. The obtained value of Da expresses the drainage networks which are considerably affected by lithological compositions, particularly along the high elevated region (Ciccacci et al. 2003; Kaliraj et al. 2014). In this study area, the sub-basin wise estimated denudation rates indicate the significant spatial variations of denudation intensity and

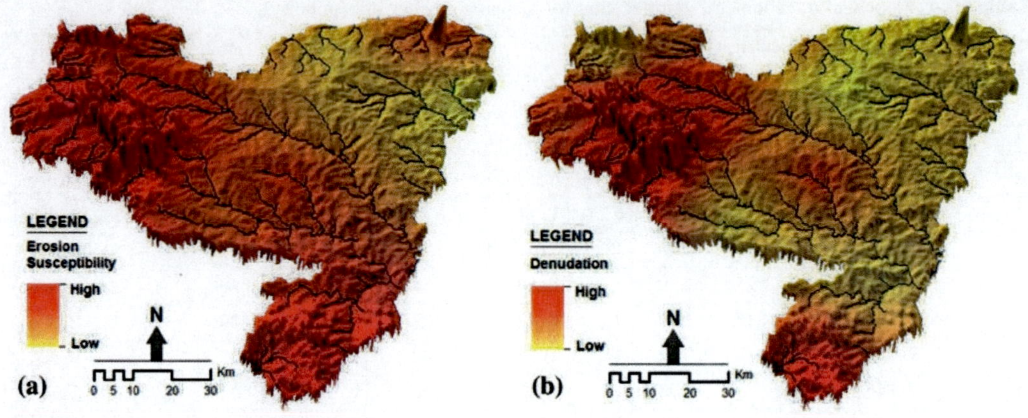

Fig. 22.10 **a** Erosion susceptibility map based on both morphometric and slope priority rank and **b** Denudation intensity map based on hierarchical anomaly density

considerably related to the erosional susceptibility zones of the Purulia district. So, the denudation level is identified more along the surface which is characterized by the steep slope, high elevation, and higher drainage density and is firmly affected by high-intensity weathering and materials alterations (Burbank et al. 2003; Bishop 2007; Kaliraj et al. 2014). Therefore, high-intensity denudation is predominantly observed in the western part of the district surrounding the Ajodhya hills and the most southern hilly tract of the district (Fig. 22.10b).

22.4 Conclusions

This study shows that ASTER DEM and topographical sheets attached with GIS techniques are acceptable tools in the delineation of sub-watersheds and associated drainage networks for geo-morphometric analysis. Several linear, aerial, and relief parameters and soil erosion susceptibility mapping by morphometric analysis are found to be eligible tools for prioritization of the sub-watersheds in Purulia district. All these parameters reveal the interaction between hydrological systems and their associated geomorphological features and topography. Due to the long time crustal evolution, the interaction between surface topography and drainage network system including underlying lithology determines the catchment area, streamflow directions, stream orders, stream length, morphological processes and landforms. Very prominent dendritic drainage patterns are found in most sub-watersheds and the large numbers of first-order streams are observed than the subsequent hierarchical streams. This evidently reveals that large numbers of first-order streams with short flow length are associated with steep slope terrain of some sub-watersheds and significantly responsible for more surface erosion. In this condition, drainage basin prioritization is considered as a realistic method for understanding and mapping the soil erosion susceptibility. The results of prioritization indicate that about 40% of areas of Purulia district are ranked as high priority with more soil erosion potentiality. Sub-watersheds No. 1, 2, 3, 4, 5, 6, 11, 10, 19, and 20 are classified as very high erosion-prone zones, whereas sub-watersheds No. 17, 15, 16, 19, 9, and 12 are categorized as medium erosion-prone areas. The rest of the sub-watershed have the minimum susceptibility to erosion. These areas of high erosion susceptibility are highly influenced by underlying lithology, elevation, and associated landforms and land use land cover. The region with the highly elevated steep slope in the western and southern hilly landscape consisted of the week and fragile bedrocks are directly influenced the erosion propensity of this area. The rest of the central and north-eastern portions of the district have flatted and the topography with a very gentle slope; as a result, these areas have a lesser susceptibility to soil erosion.

The denudation process represents the interaction among drainage networks, surface topography, and the geomorphological landforms of the sub-watersheds. The hierarchical anomaly index value determines the influences of topographical and lithological variation on the development of the drainage network system, mainly along with the high elevated areas of the district. The degree of denudation is more in the area consists of high Dd which leads to more transportation of eroded materials through channels. Therefore, the result of denudational rate shows significant similarities with the erosion susceptibility zones of the district. As the wide variety of denudation processes have determined along with landscape variation, the sub-watersheds are showing continuous changes in the denudation rate. Consequently, this investigation shows that the high-intensity denudation is mainly observed surrounding the Ajodhya hills in the western part and the most southern hilly tract of the district due to the effective controls of slope and elevation of the landscape, associated drainage conditions, and adjoining lithological variation.

References

Agarwal CS (1998) Study of drainage pattern through aerial data in Naugarh area of Varanasi district, UP. J Indian Soc Remote Sens 26(4):169–175. https://doi.org/10.1007/BF02990795

Aher PD, Adinarayana J, Gorantiwar SD (2014) Quantification of morphometric characterization and prioritization for management planning in semi-arid tropics of India: a remote sensing and GIS approach. J Hydrol 511:850–860. https://doi.org/10.1016/j.jhydrol.2014.02.028

Al-Saady YI, Al-Suhail QA, Al-Tawash, BS, Othman AA (2016) Drainage network extraction and morphometric analysis using remote sensing and GIS mapping techniques (Lesser Zab River Basin, Iraq and Iran). Environ Earth Sci 75(18):1243.https://doi.org/10.1007/s12665-016-6038-y

Ameri AA, Pourghasemi HR, Cerda A (2018) Erodibility prioritization of sub-watersheds using morphometric parameters analysis and its mapping: a comparison among TOPSIS, VIKOR, SAW, and CF multi-criteria decision making models. Sci Total Environ 613:1385–1400. https://doi.org/10.1016/j.scitotenv.2017.09.210

Angillieri MY (2008) Morphometric analysis of Colangüil river basin and flash flood hazard, San Juan, Argentina. Environ Geol 55(1):107–111. https://doi.org/10.1007/s00254-007-0969-2

Aouragh MH, Essahlaoui A (2014) Morphometric analysis of a Guigou Sub-watershed, Sebou basin, Middle Atlas, Morocco using GIS based ASTER (DEM) image. Int J Innov Res Sci Eng Technol 3(4):11503–11512

Aouragh MH, Essahlaoui A (2018) A TOPSIS approach-based morphometric analysis for sub-watersheds prioritization of high Oum Er-Rbia basin, Morocco. Spat Inf Res 1–16.https://doi.org/10.1007/s41324-018-0169-z

Avena GC, Giuliano G, Palmieri EL (1967) Quantitative evaluation of the hierarchization and evolution of drainage networks. Bull Soc Geol Italy 86:781–796

Bahrami S (2013) Analyzing the drainage system anomaly of Zagros basins: implications for active tectonics. Tectonophysics. https://doi.org/10.1016/j.tecto.2013.07.026

Bishop P (2007) Long-term landscape evolution: linking tectonics and surface processes. Earth Surf Proc Land 32:329–365. https://doi.org/10.1002/esp.1493

Biswas S, Sudhakar S, Desai VR (1999) Prioritisation of subwatersheds based on morphometric analysis of drainage basin: a remote sensing and GIS approach. J Indian Soc Remote Sens 27(3):155. https://doi.org/10.1007/BF02991569

Burbank DW, Blythe AE, Putkonen J, Pratt-Sitaula B, Gabet E, Oskin M, Barros A, Ojha TP (2003) Decoupling of erosion and precipitation in the Himalayas. Nature 426:652–655. https://doi.org/10.1038/nature02187

Chandniha SK, Kansal ML (2017) Prioritization of sub-watersheds based on morphometric analysis using geospatial technique in Piperiya watershed, India. Appl Water Sci 7(1):329–338. https://doi.org/10.1007/s13201-014-0248-9

Chopra R, Dhiman RD, Sharma PK (2005) Morphometric analysis of sub-watersheds in Gurdaspur district, Punjab using remote sensing and GIS techniques. J Indian Soc Remote Sens 33(4):531. https://doi.org/10.1007/BF02990738

Ciccacci S, Del Monte M, Marini R (2003) Denudational processes and recent morphological change in a sample area of the Orcia River upper basin (Southern Tuscany). Geogr Fis Dinam Quat 26:97–109

Ciccacci S, Fredi P, Lupia Palmieri E, Pugliese F (1980) Contributo dell'analisi geomorfica quantitativa alla valutazione dell'entita`dell'erosione nei bacini fluviali. Boll Soc Geol It 99:455–516

Ciccacci S, Fredi P, Lupia Palmieri E, Pugliese F (1986) Indirect evaluation of erosion entity in drainage basins through geomorphic, climatic and hydrological parameters. Int Geomorphol 2:33–48

Clarke JI (1966) Morphometry from maps. Essays in geomorphology. Elsevier, New York, pp 235–274

Das D (2014) Identification of erosion prone areas by morphometric analysis using GIS. J Inst Eng (India): Series A. 95(1):61–74. https://doi.org/10.1007/s40030-014-0069-8

Del Monte M, Fredi P, Palmieri EL, Marini R (2002) Contribution of quantitative geomorphic analysis to the evaluation of geomorphological hazards. In: Allison R, Chichester J (eds) Applied geomorphology: theory and practice. Wiley, pp 335–358

Della Seta M, Del Monte M, Fredi P, Palmieri EL (2007) Direct and indirect evaluation of denudation rates in Central Italy. Catena 71:21–30.https://doi.org/10.1016/j.catena.2006.06.008

Eze EB, Efiong J (2010) Morphometric parameters of the Calabar river basin: implication for hydrologic processes. J Geogr Geol 2(1):18

Farhan Y, Anaba O (2016) A remote sensing and GIS approach for prioritization of wadi Shueib mini watersheds (Central Jordan) based on morphometric and soil erosion susceptibility analysis. J Geogr Inf Syst 8(01):1. https://doi.org/10.4236/jgis.2016.81001

Farhan Y, Anbar A, Al-Shaikh N, Mousa R (2016) Prioritization of semi-arid agricultural watershed using morphometric and principal component analysis, remote sensing, and GIS techniques, the Zerqa River watershed, Northern Jordan. Agric Sci 8(01):113. https://doi.org/10.4236/as.2017.81009

Gajbhiye S, Mishra SK, Pandey A (2014) Prioritizing erosion-prone area through morphometric analysis: an RS and GIS perspective. Appl Water Sci 4(1):51–61. https://doi.org/10.1007/s13201-013-0129-7

Gioia D, Martino C, Schiattarella M (2011) Long to short-term denudation rates in the southern Apennines: geomorphological markers and chronological constraints. Geol Carpath 62:27–41. https://doi.org/10.2478/v10096-011-0003-1

Horton RE (1932) Drainage basin characteristics. Trans Am Geophys Union 13:350–361

Horton RE (1945) Erosional development of streams and their drainage basins; hydrophysical approach to quantitative morphology. Geol Soc Am Bull 56 (3):275–370. https://doi.org/10.1130/0016-7606(1945)56[275:EDOSAT]2.0.CO;2

Jain SK, Goel MK (2002) Assessing the vulnerability to soil erosion of the Ukai Dam catchments using remote sensing and GIS. Hydrol Sci J 47(1):31–40. https://doi.org/10.1080/02626660209492905

Jasmin I, Mallikarjuna P (2013) Morphometric analysis of Araniar river basin using remote sensing and geographical information system in the assessment of groundwater potential. Arab J Geosci 6(10):3683–3692. https://doi.org/10.1007/s12517-012-0627-1

Javed A, Khanday MY, Ahmed R (2009) Prioritization of sub-watersheds based on morphometric and land use analysis using remote sensing and GIS techniques. J Indian Soc Remote Sens 37(2):261. https://doi.org/10.1007/s12524-009-0016-8

Javed A, Khanday MY, Rais S (2011) Watershed prioritization using morphometric and land use/land cover parameters: a remote sensing and GIS based approach. J Geol Soc India 78(1):63. https://doi.org/10.1007/s12594-011-0068-6

Kaliraj S, Chandrasekar N, Magesh NS (2015) Morphometric analysis of the River Thamirabarani sub-basin in Kanyakumari District, South west coast of Tamil Nadu, India, using remote sensing and GIS. Environ Earth Sci 73(11):7375–7401. https://doi.org/10.1007/s12665-014-3914-1

Kanth T, Hassan Z (2012) Morphometric analysis and prioritization of watersheds for soil and water resource management in Wular catchment using geo-spatial tools. Int J Geol Earth Environ Sci 2:30–41

Kumar A, Samuel SK, Vyas V (2015) Morphometric analysis of six sub-watersheds in the central zone of Narmada River. Arab J Geosci 8(8):5685–5712. https://doi.org/10.1007/s12517-014-1655-9

Luchisheva A (1950) Practical hydrology. Gidrometeoizdat, Leningrad

Lykoudi E, Angelaki M (2004) The contribution of the morphometric parameters of an hydrographic network to the investigation of the neotectonic activity: an application to the upper Acheloos river. In: Proceedings of the 10th international congress, Thessaloniki, April, Bul. of the Geol. Soc. of Greece (vol 36)

Magesh NS, Chandrasekar N, Soundranayagam JP (2011) Morphometric evaluation of Papanasam and Manimuthar watersheds, parts of Western Ghats, Tirunelveli district, Tamil Nadu, India: a GIS approach. Environ Earth Sci 64(2):373–381. https://doi.org/10.1007/s12665-010-0860-4

Manu MS, Anirudhan S (2008) Drainage characteristics of Achankovil river basin, Kerala. J Geol Soc India 71 (6):841

Melton MA (1958) Geometric properties of mature drainage systems and their representation in E4 Phase Space. J Geol 66:35–54. http://www.jstor.org/stable/30056939

Miller VC (1953) A quantitative geomorphologic study of drainage basin characteristics in the Clinch Mountain area, Virginia and Tennessee, Project NR 389042, Technical Report 3. Columbia University Department of Geology, ONR Geography Branch New York

Moore ID, Grayson RB, Ladson AR (1991) Digital terrain modelling: a review of hydrological, geomorphological and biological applications. Hydrol Process 5 (1):3–30

Morisawa ME (1962) Quantitative geomorphology of some watersheds in the Appalachian Plateau. Geol Soc

Am Bull 73(9):1025–1046. https://doi.org/10.1130/0016-7606(1962)73[1025:QGOSWI]2.0.CO;2

Nag SK, Chakraborty S (2003) Influence of rock types and structures in the development of drainage network in hard rock area. J Indian Soc Remote Sens 31(1):25–35. https://doi.org/10.1007/BF03030749

Nautiyal MD (1994) Morphometric analysis of a drainage basin using aerial photographs: a case study of Khairkuli Basin, District Dehradun, UP. J Indian Soc Remote Sens 22(4):251–261. https://doi.org/10.1007/BF03026526

Ozdemir H, Bird D (2009) Evaluation of morphometric parameters of drainage networks derived from topographic maps and DEM in point of floods. Environ Geol 56(7):1405–1415. https://doi.org/10.1007/s00254-008-1235-y

Pareta K, Pareta U (2012) Quantitative geomorphological analysis of a watershed of Ravi River Basin, HP India. Int J Remote Sens GIS 1(1):41–56. https://doi.org/10.1007/s12665-012-2086-0

Patel DP, Gajjar CA, Srivastava PK (2013) Prioritization of Malesari mini-watersheds through morphometric analysis: a remote sensing and GIS perspective. Environ Earth Sci 69(8):2643–2656. https://doi.org/10.1007/s12665-012-2086-0

Paul II, Bayode EN (2012) Watershed characteristics and their implication for hydrologic response in the Upper Sokoto Basin, Nigeria. J Geogr Geol 4(2):147. https://doi.org/10.5539/jgg.v4n2p147

Rahaman S, Ajeez S, Aruchamy S, Jegankumar R (2015) Prioritization of sub watershed based on morphometric characteristics 1243 Page 22 of 23 Environ Earth Sci (2016) 75:1243 123using fuzzy analytical hierarchy process and geographical information system: a study of Kallar Watershed, Tamil Nadu. Aquat Procedia 4:1322–1330

Rather MA, Kumar JS, Farooq M, Rashid H (2017) Assessing the influence of watershed characteristics on soil erosion susceptibility of Jhelum basin in Kashmir Himalayas. Arab J Geosci 10(3):59. https://doi.org/10.1007/s12517-017-2847-x

Ratnam KN, Srivastava YK, Rao VV, Amminedu E, Murthy KSR (2005) Check dam positioning by prioritization of micro-watersheds using SYI model and morphometric analysis—remote sensing and GIS perspective. J Indian Soc Remote Sens 33(1):25. https://doi.org/10.1007/BF02989988

Rekha BV, George AV, Rita M (2011) Morphometric analysis and micro-watershed prioritization of Peruvanthanam sub-watershed, the Manimala River Basin, Kerala, South India. Environ Res, Eng Manage 57(3):6–14. http://www.erem.ktu.lt/index.php/erem/article/view/472

Rudraiah M, Govindaiah S, Vittala SS (2008) Morphometry using remote sensing and GIS techniques in the sub-basins of Kagna river basin, Gulburga district, Karnataka, India. J Indian Soc Remote Sens 36(4):351–360. https://doi.org/10.1007/s12524-008-0035-x

Schumm SA (1956) Evolution of drainage systems and slopes in badlands at Perth, Amboy, New Jersey. Geol Soc Am Bull 67:597–646

Shivhare N, Rahul AK, Omar PJ, Chauhan MS, Gaur S, Dikshit PK, Dwivedi SB (2018) Identification of critical soil erosion prone areas and prioritization of micro-watersheds using geoinformatics techniques. Ecol Eng 1(121):26–34. https://doi.org/10.1016/j.ecoleng.2017.09.004

Singh O, Singh J (2018) Soil erosion susceptibility assessment of the Lower Himachal Himalayan Watershed. J Geol Soc India 92(2):157–165. https://doi.org/10.1007/s12594-018-0975-x

Singh OP (1980) Geomorphology of drainage basins in Palamau upland. Recent Trends Concepts Geogr 1:229–247

Singh S, Singh MC (1997) Morphometric analysis of Kanhar river basin. Natl Geogr J India 43(1):31–43

Sreedevi PD, Owais S, Khan HH, Ahmed S (2009) Morphometric analysis of a watershed of South India using SRTM data and GIS. J Geol Soc India 73(4):543–552. https://doi.org/10.1007/s12594-009-0038-4

Sreedevi PD, Sreekanth PD, Khan HH, Ahmed S (2013) Drainage morphometry and its influence on hydrology in a semi arid region: using SRTM data and GIS. Environ Earth Sci 70(2):839–848. https://doi.org/10.1007/s12665-012-2172-3

Strahler A (1957) Quantitative analysis of watershed geomorphology. Am Geophys Union Trans 38:912–920

Strahler AN (1964) Quantative geomorphology of drainage basins and channel networks. In: Chow VT (ed) Handbook of applied hydrology. McGraw Hill Book Company, New York

Thakkar AK, Dhiman SD (2007a) Morphometric analysis and prioritization of mini-watersheds in Mohr watershed, Gujarat using remote sensing and GIS techniques. J Indian Soc Remote Sens 35(4):313–321. https://doi.org/10.1007/BF02990787

Thakkar AK, Dhiman SD (2007b) Morphometric analysis and prioritization of miniwatersheds in Mohr watershed, Gujarat using remote sensing and GIS techniques. J Indian Soc Remote Sens 35(4):313–321. https://doi.org/10.1007/BF02990787

Thomas J, Joseph S, Thrivikramji KP, Abe G (2011) Morphometric analysis of the drainage system and its hydrological implications in the rain shadow regions, Kerala, India. J Geogr Sci 21(6):1077. https://doi.org/10.1007/s11442-011-0901-2

Tokunaga E (2000) Dimension of a channel network and space-filling properties of its basin. Trans Jpn Geom Union 21(4):431–499

Welde K (2016) Identification and prioritization of subwatersheds for land and water management in Tekeze dam watershed, Northern Ethiopia. Int Soil Water Conserv Res 4(1):30–38. https://doi.org/10.1016/j.iswcr.2016.02.006

Whipple KX, Kirby E, Brocklehurst SH (1999) Geomorphic limits to climate-induced increases in topographic relief. Nature 401(6748):39–43. https://doi.org/10.1038/43375

Youssef AM, Pradhan B, Hassan AM (2011) Flash flood risk estimation along the St. Katherine road, southern Sinai, Egypt using GIS based morphometry and satellite imagery. Environ Earth Sci 62(3):611–623. https://doi.org/10.1007/s12665-010-0551-1

Zavoianu I (1985) Morphometry of drainage basins (developments in water science). Elsevier, Amsterdam

23 Geomorphic Appraisal of Active Tectonics and Fluvial Anomalies in Peninsular Rivers of the Bengal Basin (West Bengal, India)

Sandipan Ghosh

Abstract

The Bengal Basin is one of the largest peripheral collisional foreland basins at the juncture of three converging lithospheric plates (Indian Plate, Eurasian Plate and Burma Plate), carrying signatures of regional scale sesimogenic faults, which is conducive for the frequent and recurrent earthquakes. The recent earthquakes occurred with a magnitude range (M_w) of 3.6–5.7 in the western shelf zone of the Bengal Basin and the seismologists have predicted an event of potential earthquake of M_w 8.2–9.0 due to locked mega thrust. Such activeness of seismic events reflects the vulnerability of densely populated towns and cities located in the Ganga–Brahmaputra–Megna Delta. The main point of research interest is the anomalous fluvial responses to active tectonics in the shelf zone of the Bengal Basin to explore the relative perkiness of regional tectonic uplift or subsidence. Since Palaeogene time, the peninsular river system (viz., Brahmani, Dwarka, Mayurakshi, Ajay, Damodar, Dwarkeswar, Silai and Kasai river basins) was directly influenced by the underlying structure and *en echelon* faults, and several landforms, channel morphology and morphostratigraphical units were distorted and deformed due to seismic shocks. The present study tries to document and understand the significant tectonic elements, geomorphometric anomalies and soft-sediment deformation structures on the alluvial river valleys and Quaternary floodplains using seismic information, proxy data, geomorphic indices of active tectonics, thematic mapping and stratigraphic analysis of depositional facies.

Keywords

Syntectonics · Earthquake · Bouguer anomaly · Seismites · Geomorphic indices of active tectonics · Bengal Basin

23.1 Introduction

Active tectonics is defined as those tectonic processes that produce deformation of the earth's crust and surface on a time scale of significance to human society (Schumm et al. 2002; Keller and Pinter 2002). More recently, much attention has been paid to the role of tectonics in Anthropocene, and active tectonics has become a major concern with much emphasis on earthquake studies (Ouchi 1985; Schumm 1986; Hoolbrok and Schumm 1999; Azor et al. 2002; Jain and Sinha 2005; Kale and Shejwalkar 2008; Sahu et al. 2010; Mahmood and Gloaguen 2014; Anand and Pradhan 2019). The major concern is

S. Ghosh (✉)
Department of Geography, Chandrapur College, Purba Barddhaman, West Bengal, India

P. K. Shit et al. (eds.), *Drainage Basin Dynamics*, Geography of the Physical Environment,
https://doi.org/10.1007/978-3-030-79634-1_23

centred on the seismo-tectonic features, earthquake prediction and fluvial responses to tectonic activity. The western shelf zone of the Bengal Basin is considered as one of the important and structural elements of Indian sub-continent (Valdiya 2016) because the basin geosyncline is one of the largest peripheral collisional foreland basins, due to resistance to subduction below the Eurasian Plate and plate convergence. An obvious question in this regard is whether there is any geomorphic evidence to support the view that the western margin of the Bengal Basin has experienced significant and protracted uplift or subsidence from Tertiary to recent times? Alluvial rivers are the active and sensitive element of the fluvial landscape, because any shifts in the tectonic controls can instigate various rapid geomorphic and sedimentary responses from the fluvial system through different complex responses and resilience (Schumm et al. 2002). According to Jain and Sinha (2005), the primary response of rivers to tectonics is manifested as change in channel slope, while secondary changes are reflected in aggradation/degradation and variations in channel morphology.

The imperceptibility slow and secular as well as episodic crustal movements that have been taking place since the beginning of the Quaternary period some two million years ago are described as neo-tectonic activities. Areas involved in or subject to the impact of continental drift and collision of Indian Plate, Eurasian Plate and Burma Plate are particularly prone to persistent tectonic stability and attendant deformation and displacement (Valdiya 1984). Dennis (1972) and Valdiya (1984) mentioned major indicators of neo-tectonic movements: (a) *historical instances* of sunken or buried archaeological sites, (b) *geomorphic features* such as changing the course of river and streams, abrupt drainage swings, raised floodplain, alluvial terraces and anomalous channel gradient, (c) *structural dislocation and deformation* and recent sedimentary deposits, including the dissection of pediments, fans, warping or tilting of recent layered deposits, movement of ground along newly developed or reactivated older faults and thrusts, (d) *altimetric variation* of elevation determined by geodetic measurements and geomorphic indices of active tectonics and (e) *recurrent seismicity*, implying surface or underground episodic or spasmodic movements or rocks or soft-sediment deformation structures along faults.

For geomorphologists, the important factor is that the deformation is impacting a river, and it is the syntectonic response of the river that is of concern. Syntectonics refers to contemporaneous or coeval deformation and river response, which permits discussion of both active tectonic and neo-tectonic impacts on rivers and floodplains (Schumm et al. 2002). It is a largely ignored aspect of tectonic geomorphology, the study of landforms that result from tectonic processes, which has been involved primarily with earthquake effects and prediction. The notable studies of Ouchi (1985), Leeder and Alexander (1987), Yeats et al. (1997) and Holbrook and Schumm (1999) revealed minutely that tectonic deformation causes a notable change in channel slope, which, in turn, is responsible for variations in channel forms, processes and hydrological behaviour of alluvial rivers (Table 23.1). It is estimated that any deformation of the order of few mm (2 – 3 mm $year^{-1}$) can produce anomalous features in a river basin (Schumm 1986; Schumm et al. 2002). In India, earlier studies on the drainage systems of Ganges River Basin have severally highlighted the manifestation of region E–W thrust as kinck points in the longitudinal profiles of the rivers in north and eastern India (Seeber and Gornitz 1983; Valdiya 1999; Jain and Sinha 2005; Malik and Mohanty 2007). In the western part of the Bengal Basin five important large to medium scale studies of tectonic geomorphology have been still performed by Singh et al. (1998), Sinha and Ghosh (2012), Roy and Sahu (2015), Barman et al. (2019), Roy (2019), and Roy and Bera (2019) showing role of marginal faults on the evolution of laterite uplands, old fluvial/deltaic plains and young fluvial plains and correlation among subsurface lithology, gravity anomaly and anomalous drainage behaviour. Finding the research novelties and gaps of previous works, it is now utmost necessity to correlate the geo-tectonic

Table 23.1 Important works on the tectonic geomorphology of fluvial processes and landforms

Sl. no	Authors	Essential remarks on research articles and books
1	Seeber and Gornitz (1983)	Analyzing the river profiles along the Himalayan arc to understand the relative activeness of tectonics
2	Ouchi (1985)	Experimental understanding of variable fluvial responses to sow active tectonics in the alluvial rivers
3	Holbrook and Schumm (1999)	Field-based understanding and thematic modelling of geomorphic and sedimentary responses of alluvial rivers to tectonic warping or tilting in modern and ancient fluvial sedimentary settings
4	Keller and Pinter (2002)	One of milestone works to inter-relate the earthquakes, uplift or subsidence and landscapes; providing separate chapters on geomorphic indices of active tectonics and fluvial anomalous response to uplift or subsidence
5	Schumm et al. (2002)	One of the important books to cover the various aspects of tectonic geomorphology to understand the typical response of fluvial forms to active tectonics in alluvial rivers
6	Jain and Sinha (2005)	Active tectonics in a basin plays an important role in controlling a fluvial system through the change in channel slope
7	Kale and Shejwalker (2008)	Explaining the merit of geomorphometric analysis in response to ongoing post-rift flexural uplift or neo-tectonic activity in the western Deccan basalt Province
8	Goswami (2012)	Exploring the geomorphic evidences of active faulting syntectonic responses in the north-western Ganga Plain
9	Mahmood and Gloaguen (2014)	Assessment of relative active tectonics on drainage systems of Hindu Kush mountain range using GIS and geomorphic indices
10	Nath et al. (2015)	Assessing the earthquake scenario in West Bengal with emphasis on seismic hazard micro-zonation of Kolkata city
11	Das et al. (2016)	Active tectonic response of dryland fluvial system to tectonic—climatic perturbations of Late Quaternary in Kutch Basin, western India
12	Anand and Pradhan (2019)	Geomorphic understanding of active tectonics using morphometric parameters in the Ganga Basin
13	Dubey and Shankar (2019)	Exploring the impact of relative seismic events on drainage pattern in the Sone Valley
14	Das (2020)	Investigating geomorphic expression of tectonic control in Koyna-Warna shallow seismic region of Deccan Traps

elements of the Bengal Basin with recurrent seismicity (earthquakes) and fluvial signatures of floodplains (to realize current seismicity and active tectonics in the shelf zone of the Bengal Basin) considering the maximum numbers of peninsular rivers (flowing over the western margin of basin). The present study of tectonic geomorphology has made a reconnaissance attempt to understand the surface deformation pattern and fluvial dynamics across and along the subsurface faults and lineaments with the help of gravity anomaly, seismo-tectonic elements, geomorphic indices and soft-sediment deformation structures of fluvial facies.

23.2 Geomorphic Setting of Study Area

The geographical extension of the Bengal Basin ranges between 25° and 20°30′ N latitudes and 87°30′ to 90°30′ E longitudes (Fig. 23.1). The sedimentary basin occupies an area of 89,000 km^2 in total about which 57,000 km^2 on land ad 32,000 km^2 offshore up to 200 m bathymetry (Hossain et al. 2019). In Indian sub-continent, one of the important tectonic and structural elements is the Bengal Basin which is one of the largest peripheral colliosional foreland basins,

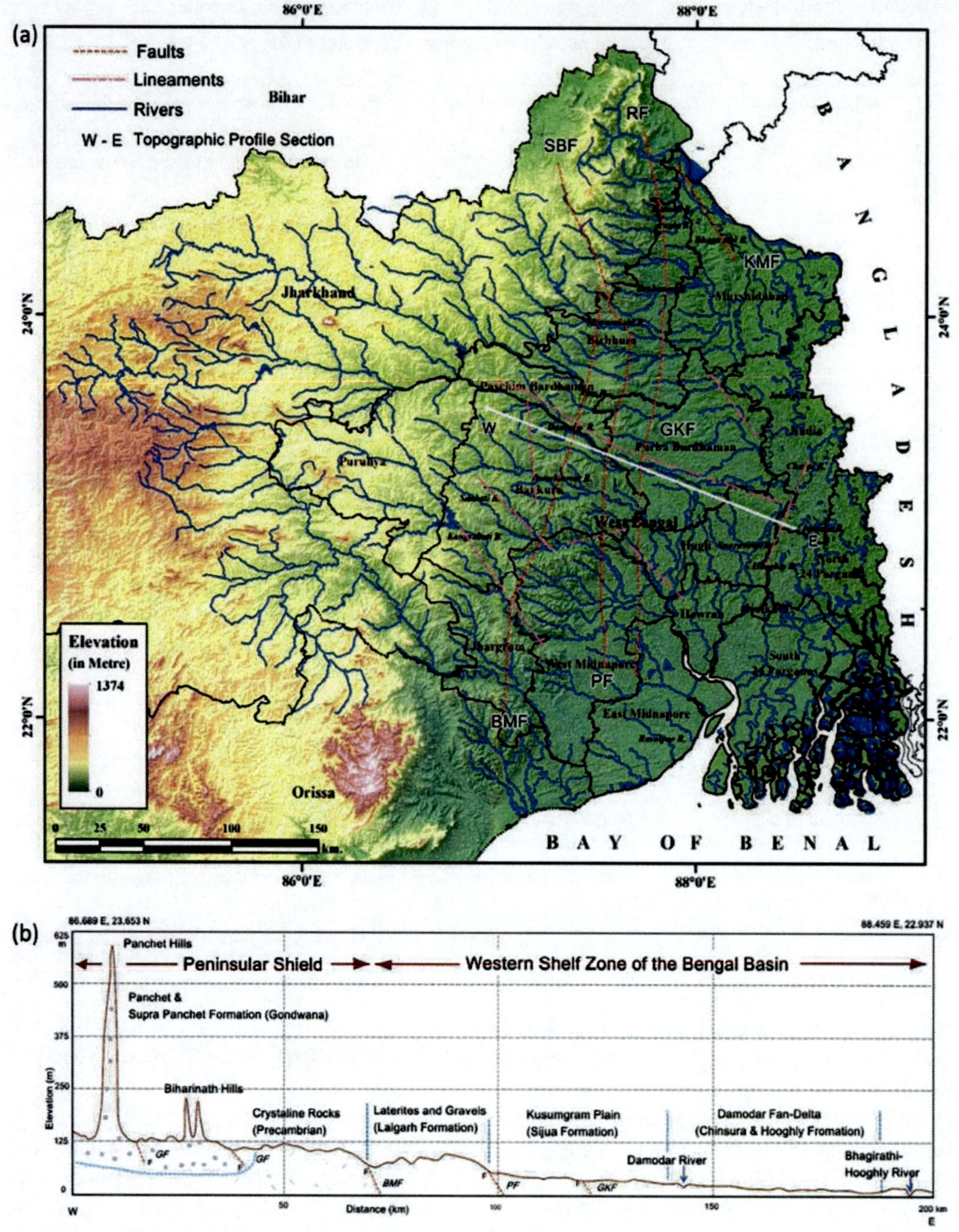

Fig. 23.1 **a** Study area depicting spatial distribution of peninsular rivers in connection with elevation zones and sub-surface faults in the western part of the Bengal Basin, and **b** regional elevation profile W–E showing the occurrences of geomorphic units in association with underlying geology and faults of the Peninsular Shield and Bengal Basin

consisting of Permo-Carboniferous to Mesozoic and Tertiary deposits covered by the Quaternary to Recent alluvium (having ~21 km thick sedimentary succession in West Bengal). From the geomorphic point of view, this basin, occupied by the Ganges-Brahmapurta-Meghna (GBM) Delta, is getting Himalayan and Peninsular sediments at an average rate of 1.1 gigatonne per year (Goodbred and Kuhel 2000). It is found that in these densely populated regions of India and Bangladesh, scientists have found many evidences of active tectonics and the increasing trend of earthquakes due to the resistance to subduction below the Eurasian Plate. The plate convergence, tectonic deformation and seismic activity occur in the intra-plate region, including stable shelf in the western part of the Bengal Basin and results in conjunction with high-angle basement faults of variable trends (NE–SW, E–W, NW–SE) (Roy and Chatterjee 2015).

The Bhagirathi-Hooghly River is the western branch of the GBM Delta and flows more than 500 km from Farraka to Sagar Island. The Bhagirathi-Hooghly River has an undulating catchment area of 66,000 km^2 along the right bank drained by the major tributaries (Fig. 23.1), which are Brahmani, Dwarka, Mayurakshi, Ajay, Kunur, Khari, Damodar, Dwarkeswar, Silai and Kasai (Rudra 2018). These tributaries of peninsular shield area together contribute about 48,410 million m^3 of water annually into the Bhagirathi-Hooghly River. The Ajay (basin area of 6074 km^2), Damodar (basin area of 20,874 km^2) and Mayurakshi Rivers (basin area of 9345 km^2) originate in the Precambrian crystalline metamorphic units of the Chhotanagpur Plateau, and these rivers are flowing west to east direction guiding by the underlying lithology and structure. Other peninsular rivers are originating in the hard rock terrain of plateau fringe and lateritic *Rarh* uplands. A wider extension of the GBM Delta is the 'para-deltas' of western uplands formed by these peninsular tributaries to the Bhagirathi-Hooghly River within its geographic area, and the Damodar Fan Delta is a key morphostratigraphic unit developed through marine transgression and tectonic activity since Pleistocene (Rudra 2018; Mahata and Maiti 2019). Agarwal and Mitra (1991) named that unit as 'palaeo-delta', which occupies the western and northern-most part of the main GBM Delta occurring dominantly west of the Bhagirathi-Hooghly River (Fig. 23.2). To the west, it is bounded by the exposed Pleistocene laterites, Tertiary Formations, Gondwana sediments and Precambrian rocks. The deposition of pebbles and gravels with ferruginous matrix is observed as alluvial fan to fan-delta morphotstratigraphic units of Oligocene to Miocene age, which are assumed to be developed as a result of Basin Margin Fault (BMF) in the interfluves of Ajay–Dmaodar–Dwarkeswar (Mahapatra and Dana 2009; Ghosh 2014). Pleistocene uplands of laterites, elevation range of 40–70 above mean sea level, are carrying main geomorphic signatures of lateritization, neo-tectonics, badlands, complex fluvial incision and ruggedness (Niyogi et al. 1970). Singh et al. (1998) revealed that tectonic evolution of laterite uplands, old fluvial/deltaic plains and young fluvial plains in connection with the subsequent uplift of faulted blocks since ~7 ka in the western margin of GBM Delta.

23.3 Methodology and Database

Tectonic Geomorphology of alluvial rivers may be studied in two ways: (1) the study of landforms and sedimentary features produced by tectonic processes and (2) the application of geomorphic principles and techniques to the solution of tectonic problems (Keller and Pinter 2002). Though geomorphic characterization of the tectonic properties of a landscape is an extremely complex task, but it is now well recognized that the commonly used geomorphic indices of active tectonics (GAT) are powerful methodological tools to evaluate the relationship between tectonics and fluvial hydrogeomorphology at basin scale or reach scale, and to recognize the Quaternary to Recent deformation of fluvial elements and sedimentary structures (Bull and McFadden 1977; Schumm 1986; Holbrook and Schumm 1999; Keller and Pinter 2002; Jain

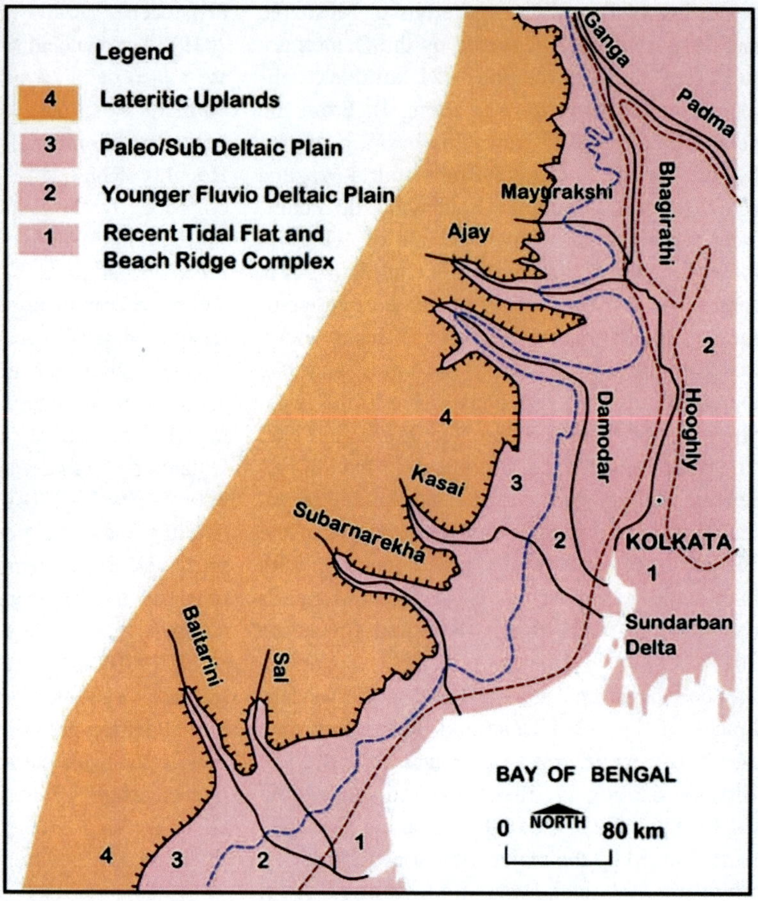

Fig. 23.2 Quaternary palaeogeomorphic units developed in the western shelf zone of the Bengal Basin (modified from Agarwal and Mitra (1991), Singh et al. (1998))

and Sinha 2005; Montenat et al. 2007; Kale and Shejwalkar 2008; Mahmood and Gloaguen 2014; Shanmugam 2016; Anand and Pradhan 2019; Ghosh and Shivakumar 2019; Ayaz and Dhali 2019; Das 2020). These relief, areal, shape and gradient parameters particularly provide useful information about the dynamics of tectonics in the regions underlined by the same rock-type (Keller and Pinter 2002; Kale and Shejwalkar 2008). Quantitative measurements allow geomorphologists objectively to compare different landforms and to calculate less straightforward parameters (GAT) that may be useful for identifying the fluvial response or anomalies (a particular characteristic of an area) to different levels of past or ongoing tectonic activity (Keller and Pinter 2002).

Calculations of a number of geomorphic indices for a large region such as the western shelf zone of the Bengal Basin (from the Rajmahal Basalt Traps to the Kasai River Valley) were made feasible by the analysis of Digital Elevation Model (DEM) of 30-m resolution SRTM (Shuttle Radar Topographic Mission) global data. The spatial scale or unit of geomorphic study is the basin and channel reach of the fluvial system of Peninsular Rivers. Here 12 rivers (viz., Dwarka, Brahmani, Mayurakshi, Ajay, Kunur, Khari, Damodar, Dwarkeswar, Silai and Kasai) and their watersheds (in between the Chhotanagpur Plateau and the Bengal Basin) are selected for GAT analysis. The digital elevation data are used to extract information about the drainage basins, network and profiles using the standard procedures and 3D tools of Global Mapper 21.0 software. A word of caution is added here regarding the longitudinal and cross profiles extracted from the SRTM-DEM data. Because of stepping, anthropogenic interventions and alluvial floodplain of low altitude, the detection of significant breaks and knick zones in the profiles is a not a very simple and straightforward task, especially for low-gradient alluvial rivers. Buffering of path profile (mean elevation error correction using standard deviation) and smoothing of the long profiles using running mean (aka moving average) of 5–11 consecutive elevation values partially reduce the problems but do not eliminate them completely. Next, the extraction of fluvial features is cartographically done from the Survey of India (SOI) topographical sheets of 1: 50,000 scale (viz., 72 P/12, 73 N/1, N/2, N/5 and N/6, 73 M/1, M/6, M/7, M/10, M/12, M/15 and M/16 etc.) using ArcGis 9.3 software. To get micro details and temporal variations of fluvial features, the public domain of Google Earth Pro is used severally. In addition, the sample field studies of selected locations, along the Damodar River Basin, are performed to get soft-sediment deformation features and twisted lithofacies in the alluvial stratigraphy of Holocene to Recent age. To get additional geologic and seismo-tectonic information, the unpublished reports of the Geological Survey of India (GSI) (Ghosh 1992; De et al. 1994; Bhattacharya and Dhar 2003; Shivgotra et al. 2011; Deb and Singh 2011), Bhukosh web portal (http://bhukosh.gis.gov.in/Bhukosh/Public), United States Geological Survey web portal (https://earthquake.usgs.gov/), National Center of Seismology web portal (https://seismo.gov.in/MIS/riseq/earthquake), and important research works (Chandra 1977; Khan and Chouhan 1996; Kayal 2008; Raj et al. 2008; Govindaraju and Bhattacharya 2012; Steckler et al. 2016; Nath et al. 2010; 2014, 2015, 2018; Dey et al. 2019; Singh et al. 2020) are analyzed here.

GAT may detect anomalies in the drainage system or along or across active fault regions. These anomalies are possible due to local changes of topography from tectonic activity resulting from uplift or subsidence. Some of the geomorphic indices most useful in studies of Himalayan active tectonics include (Table 23.2):

(1) The Sinuosity Index (S_I) (Schumm 1956; Mueller 1968) is measured to the deviation of channel from a straight path. Simply it is the ratio of channel length to valley length. Mueller (1968) introduced the topographic and hydrologic sinuosity index to understand the controls of tectonic activity or stream discharge on sinuosity. It is assumed that any tectonic deformation that changes the slope of a river valley may result in a corresponding change in sinuosity to maintain equilibrium channel slope.
(2) The Hypsometric Integral (H_I) (Strahler 1952) is a useful attribute of the hypsometric curve (the area under the curve) to identify the stages of landform development (regarding erosional and depositional sequences), based on a function of total area and total elevation at basins scale. Higher index values of H_I might result from the recent incision of initial landforms formed by deposition in the alluvial rivers.
(3) The Stream Length-Gradient Index (S_L) (Hack 1973) correlates to stream power or energy gradient in response to topographic upliftment or subsidence. It is calculated on the basis of channel slope or gradient of the reach and the total length of channel, measured as horizontal length from the watershed divide to the midpoint of the reach. The S_L index values of the basins are used to discuss the influences of environmental variables on longitudinal river profiles and to test whether the rivers have reached equilibrium or not (Mahmood and Gloaguen 2014).
(4) Drainage Basin Asymmetry Factor (A_F) (Hare and Gardner 1985) is used to recognize the distinct pattern and geometric anomalies of basin in the presence of active tectonics. A_F is a function of the basin area to the right (facing downstream) of the truck stream and the total basin area. The factor is sensitive to tilting perpendicular to the trend of the truck stream. A_F significantly greater or smaller than 50 ($A_F > 50$ implies tilt down to the left of basin, looking downstream) shows influences of active

Table 23.2 Geomorphic indices of active tectonics (GAT) and their calculations

Sl. no	Index	Formula	Variables	Reference
1	Sinuosity Index (S_I)	$S_I = A_L/E_L$	A_L = actual or observed length of stream; E_L = expected length of stream	Schumm (1956); Litchfield et al. (2013)
2	Standard Sinuosity Index (SSI)	SSI = CI/VI, TSI = 100 (VI − 1)/(CI − 1), HSI = 100(CI − VI)/(CI − 1), CI = CL/AL, VI = VL/AL	CI = channel index, VI = valley index, CL = channel length, AL = air length, TSI = topographic sinuosity index, HSI = hydraulic sinuosity index	Mueller (1968); Ghosh and Mistri (2012)
3	Hypsometric Integral (H_I)	$H_I = (E_m - E_{min})/E_{max} - E_{min})$	E_m = mean elevation, E_{max} = maximum elevation, E_{min} = minimum elevation	Strahler (1952); Anand and Pradhan (2019)
4	Stream Length-Gradient Index (S_L)	$S_L = (H_1 - H_2)/(\ln L_2 - \ln L_1)$	H_1 and H_2 are the elevations of each end of a given reach, L_1 and L_2 are the distances from each end of the reach to the source	Hack (1973); Kale and Shejwalkar (2008)
5	Drainage Basin Asymmetry Factor (A_F)	$A_F = 100(A_r/A_t)$	A_r = are of the basin to the right of the trunk stream, A_t = total area of the drainage basin	Hare and Gardner (1985); Mahmood and Gloaguen (2014)
6	Ratio of Valley Floor Width to Valley Height (V_F)	$V_F = V_{fw}/[\{(E_{ld} - E_{sc}) + (E_{rd} - E_{sc})\}/2]$	V_{fw} = width of valley floor, E_{ld} = elevation of the left valley divide, E_{rd} = elevation of the valley floor, E_{sc} = elevation of the valley floor	Bull and MaFadden (1977); Mahmood and Gloaguen (2014)
7	Basin Elongation Ratio (E_R)	$ER = 2 (A/\pi)^{0.5}/L_b$	A = basin area, L_b = length of the basin	Schumm (1956); Kale and Shejwalkar (2008)

tectonics/lithological control or differential erosion, as for example the stream slipping down bedding plains over times (Keller and Pinter 2002; Mahmood and Gloaguen 2014).

(5) The Ratio of Valley Floor Width to Valley Height (V_F) (Bull and McFadden 1977; Bull 1978) is a useful geomorphic index conceived to discriminate between V-shaped ($V_F < 1$) and U-shaped ($V_F > 1$) flat-floored valleys. It is a mathematic function of valley floor width, elevations of left and right valley divides, and elevation of the valley floor. High index values of V_F reflect low uplift rates or stable region and broad valley floors, and low V_F values are associated with local uplift, deep valleys and active incision (Keller and Pinter 2002).

(6) Basin Elongation Ratio (E_R) (Schumm 1956) is basin shape index, which can give information about tectonic deformation of shape with time. It is estimated based on the total area of the basin and maximum length of the basin. Relatively young drainage basins in active tectonics areas tend to be elongated in shape parallel to the topographic slope of a region (Mahmood and Gloaguen 2014).

23.4 Results

23.4.1 Elements and Features of Active Tectonics

23.4.1.1 Geotectonic Settings

Peninsular India lying south of the Indo-Gangetic Alluvial Plain (IGAP) commonly referred to as the Indian Shield reportedly occupied a much wider area than the present triangular shaped region, made up of a diverse mosaic of igneous and metamorphic terrains that have undergone deformation and metamorphism (Hossain et al. 2019). Bengal Basin is situated in the north-western part of Indian Shield, covering West Bengal of India and entire Bangladesh. The Bengal Basin is a classic example of a peripheral foreland basin formed via continent–continent collision (Mukherjee et al. 2009). The basin was initiated at the breakup of Gondwanaland in the late Mesozoic and evolved through the formation of the proto-GBM delta to the present delta starting around 10.5 Ma (Mukherjee et al. 2009). Goodbred and Kuhel (2000) sedimentation in the Bengal Basin was marked by basin-ward subsidence in the middle to Late Eocene causing extensive marine transgression almost over the entire Bengal Basin (Roy and Chatterjee 2015).

The onset of Miocene witnessed a further increase in sediment supply when a huge thickness of fluvial sediments showing alternating and repetitive deposition of sandstone and shale along with swallowing up of the basin from pro-delta to brackish marine with limited marine influence (Roy and Chatterjee 2015). During the Early Pleistocene, shallow marine conditions prevailed only in the deeper part of the eastern Deep Basin. The sea finally receded from the Bengal Basin area possibly in Late Pleistocene. After a brief period of depositional hiatus, the older sediments in the entire basin area were covered completely by a thick mantle of fluvial Holocene alluvium (Roy and Chatterjee 2015). The geological succession of the Bengal Basin (adjacent areas of Peninsular region) is depicted in Table 23.3 and Fig. 23.3.

The main tectonic and structural zones of the basin are identified as (Fig. 23.4): (1) Basin Margin Fault Zone (BMFZ), (2) Stable Shelf Zone (SSZ), (3) Eocene Hinge Zone (EHZ) and (4) Deep Basin (Hossain et al. 2019). The main point of interest is the BMFZ and SSZ, because these geological units are associated with many active faults and west–east flowing peninsula rivers, which have directly influenced the land-form evolution, laterite genesis and badlands. The NNE—trending zone of BMFZ demarcates the western crystalline or metamorphic complex of Precambrian age from the shelf sediments. The fault zone is, apparently, the result of distension and down warping of the shelf region during Early—Late Cretaceous, probably, concomitant with the eruption of Rajmahal basaltic lavas. Western portions of this feature are the exposures of Gondwana sedimentary rocks resting on the Precambrian granitic-Gneissic basement (Kaila et al. 1992; Hossain et al. 2019).

To the northwest, the Bengal Basin is separated from the Rajmahal Hills by the approximately N-S running Rajmahal Fault and the Rajmahal Hills of West Bengal and Jharkhand is a fault-bounded small tectonic element (Rajmahal Fault at east and Saithia-Brahmani fault at west) situated in the western edge of the SSZ of Bengal Basin (Hossain et al. 2019). The Tertiary sedimentary prism thickens towards east and merges with the deep shelf beyond EHZ, which is narrow elongated zone that separates the thick post-Eocene sediments in the east from the shelf zone of the west. At western part of Paschim Barddhaman the subsequent break in the basement slope is very conspicuous, and the L basement slope gently tends towwards east, marked by numerous step faults with small displacements (Ghosh and Guchhait, 2018). The formation of Damodar Fan Delta (DFD) and the associated drainage development were largely influenced by these basement faults since Tertiary times (Mahata and Maiti 2019). The deep seismic profile of Beliator—Khandaghosh zone has encountered two *en-echelon* type faults (Garhmayna—Khandaghosh Fault and Pingla

Table 23.3 Generalized stratigraphic succession of the geological units in the western part of Bengal Basin and eastern part of Peninsular Shield

Era	Period	Formation	Lithology
Cenozoic	Quaternary to Recent	Hooghly Formation/Arambagh Formation/Ganga-Kosi Formation. Katwa Formation	Recent to sub-recent oil/alluvium/sandy clay/loose sand/ferruginous sediment with calcretes
		Panskura Formation/Chinsura Formation/Malda Formation/Bethuadahari Formation	Ferrous shale, mudstone/Calc mud/impure limestone, fine to coarse sandstone with floral remains
		Sijua Formation	Ferruginous sandstone clay/thin pebbles bed/impure limestone/green clay/Carbonaceous clay with rich floral assemblage and plant roots
		Barind Formation/Baikunthapur Formation/Lalgarh Formation/Ausgram Formation	Alternating clay and sand beds
	Tertiary	Siwalik Group	Ferruginous sandstone, pebbly grit, red shale, clay, gravel and fossil wood
Mesozoic	Gondwana supergroup	Durgapur Formation	Compact, thinly laminated quartzite, carbonaceous shale with fossils of algae and foraminifera
		Unconformity	
		Rajmahal Formation/Dubrajpur Formation/Supra-Panchet Formation	Traps with inter-trappeans, ferruginous sandstone, red shale/clay stone
		Unconformity	
Paleozoic		Raniganj Formation	Sandstone and shale with thick coal seams
		Barren Measure Formation	Grey shale nodule of iron ores
		Barakar formation	Sandstone with coal seams
		Talchir Formation	Conglomerate, siltstone, shale and sandstone
Proterozoic	Lower to Middle Proterozoic	Manbhum Granite/Kuilpal Granite	Granite containing phenocrysts of feldspar
		Dalma Volcanics	Ultramafics, mica schist, phyllites, quartzites, tuffs, cherts and calc silicates
	Archean		silicified metamorphics, anopthosites, schist, amrble/cal. Granulites/amphibolites/hornblende schist/composite gneiss, biotite gneiss
	Proterozoic	Chhotanagpur Gneissic Complex	

Source Sengupta (1966), Das Gupta and Mukherjee (2006)

Fault), which has affected the entire sedimentary column starting from the surface to the basement (Kaila et al. 1992). The depth basement of the SSZ varies from 4 to 5 km near Khandaghosh to about 7.5 km near Dhatrigram.

Lithospheric flexure and subsidence of the basin (subduction of the Indian plate below the Eurasian and Burmese plates) are occurring at a rate of 2–4 mm/year in Mukherjee et al. (2009). As a result of the resistance to subduction flow below the Eurasian plate and plate convergence, intense deformation and seismic activity occur in the India intraplate region, including on the stable shelf in the western part of the Bengal

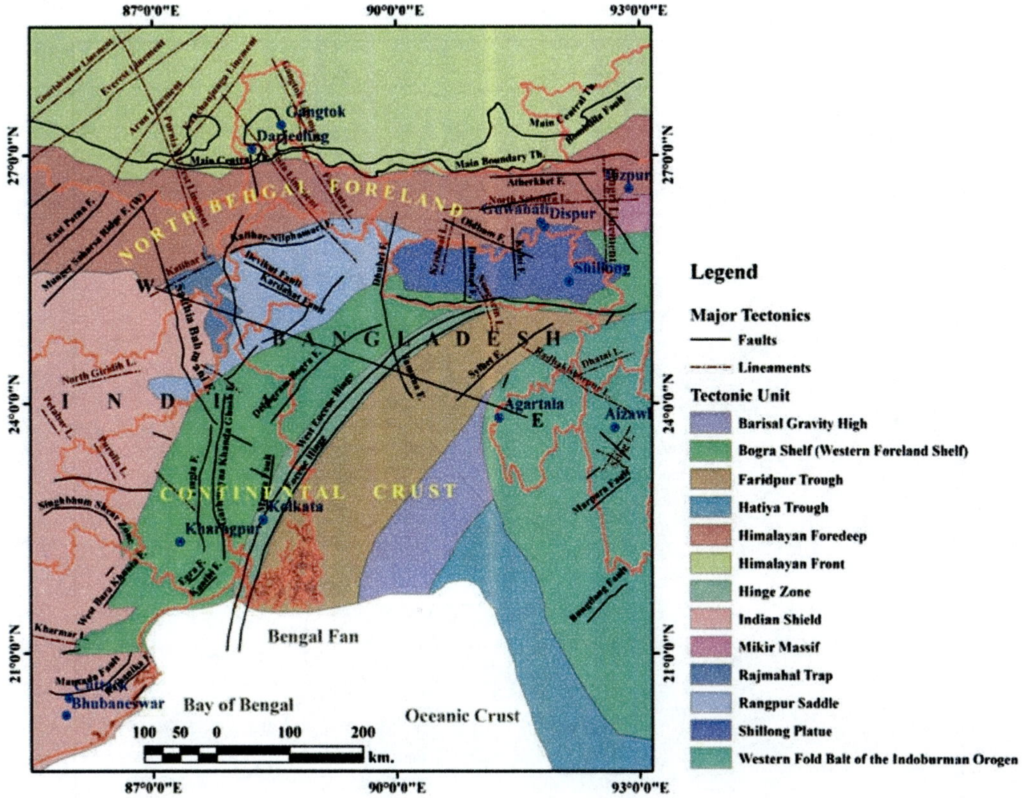

Fig. 23.3 Geological configuration of the Bengal Basin and adjacent areas of India and Bangladesh (modified considering Sengupta 1966; Das Gupta and Mukherjee 2006; Nath et al. 2010, 2015)

Basin (Fig. 23.4), resulting in folding of strata and brittle deformation via high-angle basement faults (Mukherjee et al. 2009; Roy and Chatterjee 2015). The SSZ has further been subdivided into the Baharampur terrace, the Baidyapur depression, the Contai terrace from north to south, the Dinajpur slope, the Rangpur Saddle and the Bogra slope from west to east (Nath et al. 2014). The Bogra slope represents a monocline plunging fold gently sloping towards the southeast of the EHZ and the width of it varies from 60 to 125 km (Nath et al. 2014, 2018).

Thc wcstcrn sub-basin of thc Bcngal Basin (BMFZ and SSZ) shows the occurrence of laterite and lateritic soils, the protoliths of which include ferruginous sandstone, red shale, grit and gravel beds (mainly Lalgarh Formation) containing dicotyledonous fossil woods (Roy and Chatterjee 2015). Moving eastward, red soils change over to grey alluvium along the BMFZ, which is a zone dislocation running along the N15°E–S15°W trend and marked by the crowding of gravity contours (Roy and Chatterjee 2015). Since Palaeogene, it may be expected that the river systems and landforms of West Bengal were directly influenced by the numerous faults and lineaments of Bengal Basin, viz., Garhmayna—Khandaghosh Fault (GKF), Chottanagpur Foor-hill Fault or Basin Margin Fault (BMF), Jangipur-Gaibardha Fault (JGF), Pingla Fault (PF), Debagram—Bogra Fault (DBF), Saithia—Brahmani Fault (SBF), Rajhamal Fault (RF), Malda—Kisanganj Fault (MKF), Purulia Shear Zone (PSZ) etc. The most prominent tectonic feature is the NE–SW trending Eocene Hinge Zone (EHZ), which is 25 km wide and extends to a depth of about 4.5 km below Kolkata city (Nath et al. 2014, 2018). The western margin of

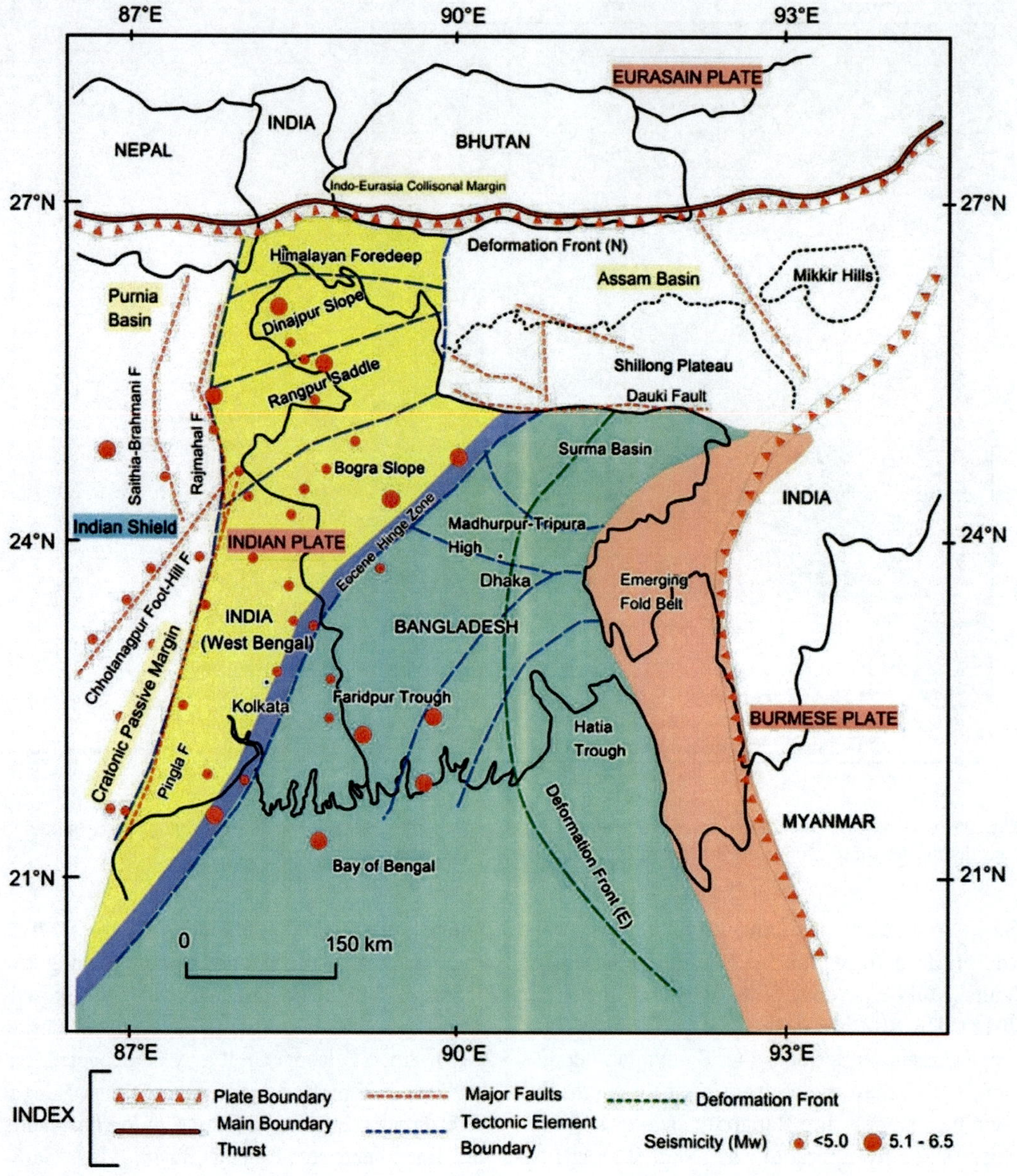

Fig. 23.4 Tectonic elements and divisions of the Bengal Basin with plate boundary, major faults, thrust and seismic events ($M_w > 4.3$) (modified considering Mukherjee et al. 2009; Hossain et al. 2020; Singh et al. 2020)

the Bengal Basin is marked by two interesting features: (1) a series of basement ridges buried under the alluvium bordering the eastern margin of the Precambrian shield area and (2) a series of normal faults and fault-line scarps along the eastern margin of this zone of shallow basement ridges (Sengupta 1966).

23.4.1.2 Bouguer Anomaly

Bouguer anomalies are most sensitive to the mass distributions and hence are influenced by the nature of underlying geological formations. In areas underlain by masses with relatively higher density, Bouguer anomalies are reflected as gravity highs and vice versa (Verma 1985). Bouguer anomalies are negative over elevated areas, showing inverse relationship to topography, i.e. higher the elevation, more negative is the Bouger anomaly. The negative values under elevated areas, therefore, give indirect evidence in support of the root formation. To the east of Singhbhum and Hazaribagh region significant low Bouger anomaly is attributable to large thickness of Tertiary to Quaternary sediments of the Bengal Basin (Verma 1985). The Bouger anomaly map of the Raniganj coalfield shows that the central part of the Gondwana basin is oval shaped. The deepest part of the basin is located few km southwest of Asansol and has a residual anomaly of −32 mgal (Verma 1985; Kayal 2008). Southern part of this coalfield is bounded by a steeply dipping fault (Southern Boundary Fault) and it appears that faulting in the basement is mostly responsible for deposition of sediments in the Bengal Basin. Central Indian Tectonic Zone (CITZ) may have played a role in the postulated collapse of the crust south of the Shillong Plateau, following the outpouring of the Rajmahal lava flows, and in the development of the E-W down-basin faulting in this area (Das Gupta and Mukherjee 2006). Its impact on the West Bengal Shelf is marked by the presence of a depression zone with a monoclonal terrace (viz., Radha Monocline, Baidyapur Depression and Contai Monocline) each to its north and south. From the presence of a number of important riverine alignments, some of these faults may be mid active even to this day (Das Gupta and Mukherjee 2006). The shelf zone is occupied by a wide NNE-SSW gravity low, reflecting structural high or depressions underlying the basin (Reddy et al. 1993; De et al. 1994; Roy and Chatterjee 2015). The most easterly of the alignments links the N-S stretch of the Jalangi River with similar stretches of the Hooghly River Shantipur to Kolkata.

Seismic refraction studies indicated that the alluvium in the Birbhum and Purba-Paschim Barddhaman areas directly overlies the Gondwana sediments and/or Archaeans. In the north of Durgapur the maximum thickness of Gondwana sediments in the extended Raniganj is of the order of 2.8 km. In this area surveyed by De et al. (1994), Bouguer anomaly values range from + 40 mgal in the north of Illambazar to −35 mgal in the east of Bolgona (Fig. 23.5). The ESE-WNW trending gravity low takes a turn towards south near panagarh and continues up to southwest of Bishnupur forming three isolated lows (of the order of −10 mgal) near Sonamukhi, Layakbandh and Bishnupur (in between Basin Margin Fault and Pingla Fault). These gravity lows and crowding are found over the Tertiary sediments in the marginal zone of the Bengal Basin geosynclines, where the thickness of Tertiary columns is expected to about 200 m and is affected by the underlying basement faults. Then, the gravity high structure (from Siuri to Bishnpur) occurs over the shelf zone of the Bengal Basin, reflecting a series of buried basement ridges, aligned NNE – SSW under thick alluvium (De et al. 1994). There is a continuous fall of gravity values over the Bengal Basin shelf province, identified as two prominent lineaments—the marginal zone and fault scarp zone of the Bengal Basin. This fault scarp zone is characterized by a series of normal, down-to-basin, en echelon faults (Basin Margin Fault, Pingla Fault and Garhmayna-Khandaghosh Fault) limiting the western boundary of the Bengal Basin. Faulting probably has occurred at different times during the accumulation of huge thickness of Tertiary sediments of West Bengal (Sengupta 1966).

23.4.1.3 Pattern of Earthquakes

Seismological data of the events that took place in the Bengal Basin during 1918–1989 have revealed an increased frequency of earthquakes in the last 30 years (Khan and Chouhan 1996). The increase in seismic activity is an indication of fresh tectonic activity or propagation of fractures from the adjacent seismic zones (Khan and Chouhan 1996). The Bengal Basin is located at a junction point of the three lithospheric plates viz. the Indian Plate, The Eurasian Plate and the Burma Plate posing high seismic susceptibility in the region. The major earthquakes, that have been affected the Bengal Basin since nineteenth century, are as follows: (1) Cachar earthquake M_W 7.5 (10 January 1869), (2) Bengal

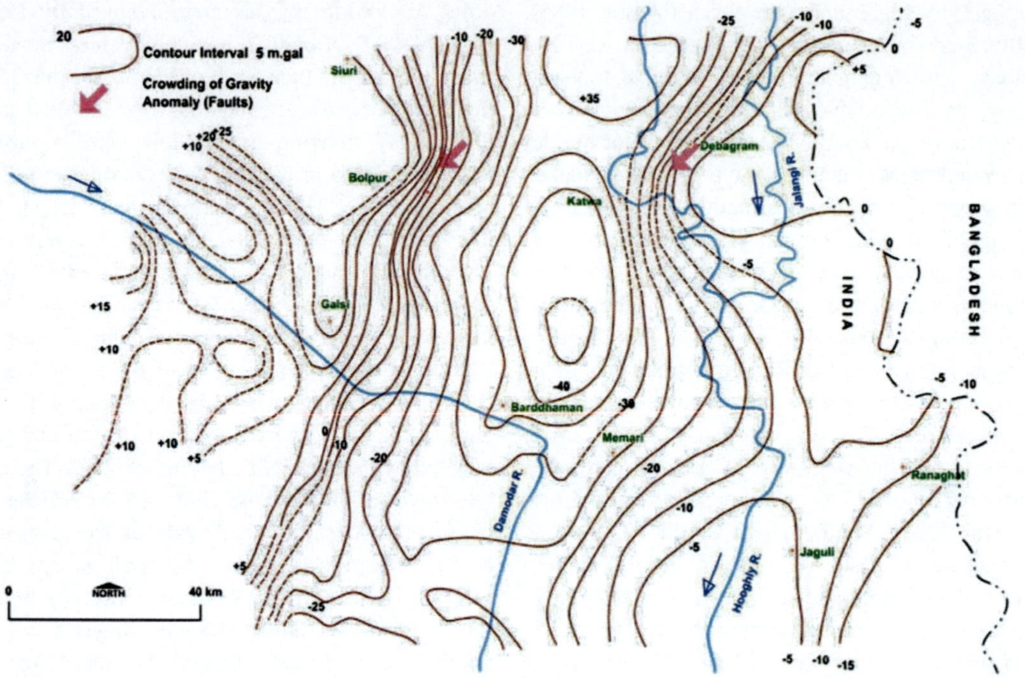

Fig. 23.5 The trend of Bouguer anomaly (interval 5 mgal) in the western part of the Bengal Basin, showing crowding of contours coincided with faults (modified from Das Gupta and Mukherjee 2006; Roy and Chatterjee 2015)

earthquake M_W 7.0 (14 July 1885), (3) Great Assam earthquake M_W 8.7 (12 July 1897), (4) Srimangal earthquake M_W 7.6 (8 July 1918), (5) Dhubri earthquake M_W 7.1 (3 July 1930), (6) Bihar-Nepal earthquake M_W 8.3 (15 January 1934) and (7) Assam earthquake M_W 8.4 (15 August 1950) (Khan and Chouhan 1996). The state of West Bengal has recorded a history of earthquake activity dating back to the past three centuries. Most of the earthquakes occur in Himalayan ranges in the northern part of the state or deep earthquakes within the Bengal Basin. The seismic hazard zonation map published by the Bureau of Indian Standard has classified the whole Indian Territory into four zones (zone II to V), and megacity Kolkata falls in the boundary of zones III and IV, indicating high seismic risk (Govindaraju and Bhattacharya 2012).

Many scholars assume that the shelf zone of the Bengal Basin is stable, but the sediments below the surface are the contributing factor in the ground failure and are capable of amplifying the ground motion, thus enhancing the hazard potential during a strong earthquake (Nath et al. 2015). The 1737 Kolkata earthquake, the 1885 Bengal earthquake of Mw 6.8, the 1897 Shillong earthquake of M_w 8.1, the 1918 Srimangal earthquake of M_w 7.6, the 1934 Bihar-Nepal earthquake of M_w 8.1, the 1935 Pabna earthquake of M_w 6.2 and the 1964 Sagar Islan earthquake of M_w 5.4 caused widespread damage in the Kolkata City and its surroundings (Nath et al. 2015). Magnitude measures the energy released at the source of the earthquake and intensity measures the strength of shaking produced by the earthquake at a certain location. A way to measure the size of an earthquake is to compute how much energy is released. The amount of energy radiated by an earthquake is a measure of the potential for damage to man-made structures using the empirical equation, $E = 10^{1.5Mw + 4.8}$ (where E is energy released in joule) (Fig. 23.6). The magnitude scale portrays energy logarithmically to approximately base 32. For example, a magnitude 6.0 earthquake releases about 32 times as much energy as magnitude 5.0

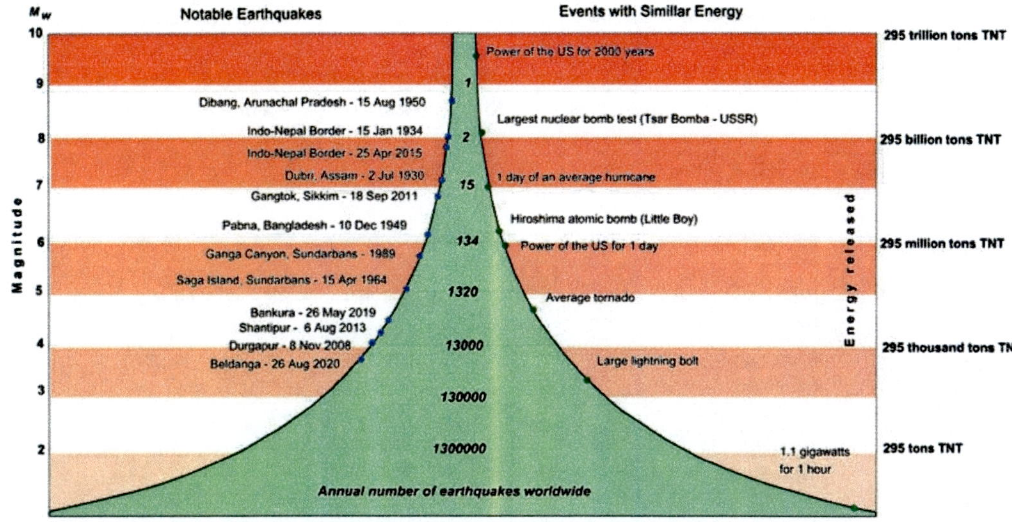

Fig. 23.6 Earthquake magnitudes of the Bengal Basin and energy release (comparing with TNT explosion) and comparison with other natural and man-made events (diagram source USGS earthquake public domain)

earthquake. The table of notable earthquakes is presented here with magnitude, energy equivalent and energy release (Table 23.4). The recent recurrent earthquakes of $M_w > 4.0$ depict that these events are evidence of active tectonics under the thick alluvium of the Bengal Basin and either the GBM delta itself or the substructure of hard strata underlying the Bengal Basin still has the capability of generating occasional earthquakes.

Recently, a mild tremor shook the Bengal Basin region when a shallow-depth earthquake of M_w 4.9 (hypocentral distance 130 km) with epicentre at 23.47° N, 87.12° E (near Bankura town) struck at 11:40 a.m. local time on 6 February, 2008 (Raj et al. 2008). Approximately 2 years ago, an earthquake of M_w 4.0 (hypocentral distance 15 km) was also recorded on 13 December, 2005, with its epicentre located at 22.31° N, 87.64° E (nearer to PF) (Raj et al. 2008). While there is a perceptible vulnerability at least, a bigger earthquake occurs, the recent earthquakes indicate a neo-tectonic activity along the underlying seismogenic source. On 28 February, 2018, another earthquake of Mw 4.8 (hypocentral depth of 7 km) was triggered in the shelf zone of Bengal Basin, having epicentre at 22.6° N, 87.7° E (over PF) (Dey et al. 2019). Strong to moderate ground shaking was felt in the epicentral zone and in surrounding districts of East Midnapore, West Midnapore, Jhargram and Bankura. Again the moderate size of the earthquake was triggered between the PF (to west) and the GKF (to its east). Changes in the stress patterns within the continental interior flexural loading of the thick sediments along the basin margin can reactivate these faults, resulting in significant seismic hazard to nearby regions and cities. Occasional light-to-moderate intra-plate earthquakes have occurred in the past, beneath this region of the western Bengal Basin (Dey et al. 2019). Given the ~7 km focal depth of the earthquake, it originated on the east dipping Pingla Fault (PF). In general observation the nodal planes could be the fault plane, but the similarity in orientation of the NNW-SSE striking nodal plane with PF, suggest this to b the fault plane of this earthquake (Dey et al. 2019). That earthquake ruptured a segment of the Precambrian gneissic basement fault (~4 km^2 fault area) in contact with the eastward thickening sedimentary layer. It is found that that earthquake occurred in response to intra-plate stresses due to the N20°E motion of the Indian plate, and the E-W flexure of the basement due to sediment loading in the Bengal Basin (Dey et al. 2019).

Table 23.4 A summary of major earthquakes that occurred in the tectonic blocks of the Bengal Basin and estimation of earthquake energy magnitude equivalent during trigger

Sl. no	Date	Location	Earthquake Magnitude (M_w)	Energy equivalent to Kilo-watt-hour (kWh)	Compared release of energy during explosion of TNT (kg)
1	2 July 1930	Dhubri, Assam	7.1	808,532,532	1,338,260,744
2	15 January 1934	Indo-Nepal Border	8.0	15,984,442,704	26,457,008,613
3	21 March 1935	Pabna, Bagladesh	6.2	40,897,569	67,692,528.85
4	10 December 1949	Kishoreganj, Bangladesh	6.0	21,071,599	34,877,129
5	15 April 1964	Sagar Island, West Bengal	5.2	1,484,900	2,457,767
6	12 August 1969	Bankura, West Bengal	5.7	7,792,871	12,898,545
7	23 June 1980	Sundarbans, West Bengal	5.0	765,063	1,266,312
8	26 March 1981	Chingrakhali-Bhairabnagar	4.9	549,158	908,951
9	12 June 1989	Sundarbans, West Bengal	5.7	7,792,871	12,898,545
10	20 June, 2002	Jayachari-Rajshahi	5.1	1,065,853	1,764,171
11	20 October, 2003	Purulia	4.3	75,109	124,319
12	6 February, 2008	Beliator, Bankura	4.3	75,109	124,319
13	8 November, 2008	Durgapur	4.2	53,913	89,236
14	5 January, 2009	Bangaon	4.2	53,913	89,236
15	6 August, 2013	Shantipur	4.5	145,779	241,290
16	15 December, 2015	Chas, Jharkhand	4.1	38,698	64,053
17	28 August, 2018	Hooghly	4.7	282,942	468,317
18	28 July, 2019	Kenda, Jharkhand	4.1	38,698	64,053
19	8 April, 2020	Bankura, West Bengal	4.0	27,777	45,977
20	26 August, 2020	Beldanga, West Bengal	3.8	14,311	23,688

Note 1 kWh is equivalent to the energy of 1000 J used for 3600 s; TNT—Trinitrotoluene, $C_6H_2(NO_2)3CH_3$, equivalent is a convention for expressing energy, typically used to describe the energy released in an explosion or earthquakes

Steckler et al. (2016) estimated a potential earthquake of M_w 8.2–9.0 in the eastern part of Bengal Basin. The peak ground acceleration (PGA) and peak spectral acceleration (PSA) are estimated during those earthquakes by the USGS (United State Geological Survey), and the maps are presented in Fig. 23.7 to the observed spatial extension of earthquake shaking in the Bengal Basin.

23.4.1.4 Seismites

To explore the alluvial signatures of active tectonics, the geologists and geomorphologists are getting clues from the seismites which are principally fluvial sedimentary beds and structures deformed by seismic shaking. Seilacher (1969) first proposed the genetic term 'seismites' to interpret earthquakedeformed bed of soft-sediment deformation structures (SSDS). The genetic term 'seismites' have the following characteristics: these units differ from ordinary marine slides by the soupy top layer and by lack of a basal slip surface. It seems more plausible to connect them with seismic shocks acting on gently dipping muds in which compaction gradually increased down from the water–sediment interface (Shanmugam 2016). In this case, the sliding process may not have had time to develop fully so that deformational structure became 'frozen' in an embryonic stage, without resulting in a major lateral transport (Shanmugam 2016). Depending on mud consistency and paleoslope, as well as strength, duration and type of the shock, quite different structures may result. In perfectly horizontal mud layers, or under weaker shocks, for instance, nothing but the liquefied zone would form. The standard vertical alluvial sequences for seismites composed of the following four division: (a) soupy zone (top), (b) rubble zone, (c) segmented zone and (d) undisturbed sediment (bottom) (Seilacher 1969; Suter et al. 2010; Shanmugam 2016). SSDS as water-escape structures, aqueous environments being alluded to because it within these soft-sediment deformations chiefly raised by liquefaction, which involves a change of state from solid like to liquid like in cohesionless grain mass. The principal features of SSDS include load cast, imbricate structure, deformed cross-bedding, pseudonodule, convolute laminates and clastic dyke (Montenat et al. 2007; Suter et al. 2010).

The fluvial facies of floodplains have numerous layers of sandy silt or silty materials, which are the most sensitive sediments with contrasted granulometry, such as alternately homogenous fine sands or silts and argillaceous beds (Montenat et al. 2007). Various fluidization and fluid expulsion phenomena may occur independent of seismicity (e.g. over-pressure of fluid due to loading by sudden deposition of a thick depocenter; channel deposits; mud-or debris-flows etc.); water escapes often associated with gravity deposits, generating convoluted beds (turbiditic sediments). To investigate SSDS, the Quaternary alluvial sections of the Damodar River are taken into consideration. From Rhondia (23°22′ 19″ N, 87°28′ 41″ E) to Palla (23° 09′ 47′ ′ N, 87° 59′ 57″ E), majority of the field evidences were collected and, in these stretch of river valley, the imprints of tectonic activity are indicated by sand dykes, independent mounds, load casts and deformed beds of clay or sandy-silt (Fig. 23.8). A well-preserved sand dyke and sill horizon was observed within a section of Quaternary lithofacies, exposed near Tirat-Harabhanga village (23° 36′ 45″ N, 87° 03′ 33″ E). The feature was observed at the base of >5 m thick section consisting of semi-compact sandy-clay and clay horizon (F_{cl}). The horizon shows the intrusion of the overlying host sandy clay section by sand dykes and sills (1–5 cm width) in varied directions. The sand dykes are water escape structure formed by vertical shear stress during earthquakes. The litho-section (>4 m depth) of Haripur village (23° 10′ 31″ N, 87° 48′ 24″ E) unveiled a very well-preserved seismites section in the study area. The Haripur seismites section with a total thickness of approximately 1 m consists of undeformed clay silt layer at the base. The undeformed clay-silt (F_m) layer is overlain by an enormously deformed clay-silt layer (5 – 30 cm). The sand silt layers at the periphery of this layer are partially oxidized and these too are variably deformed. The deformation within the layer has been observed in the form of crumbled, churned, sometimes detached, contorted, overturned and intricate folded bands of

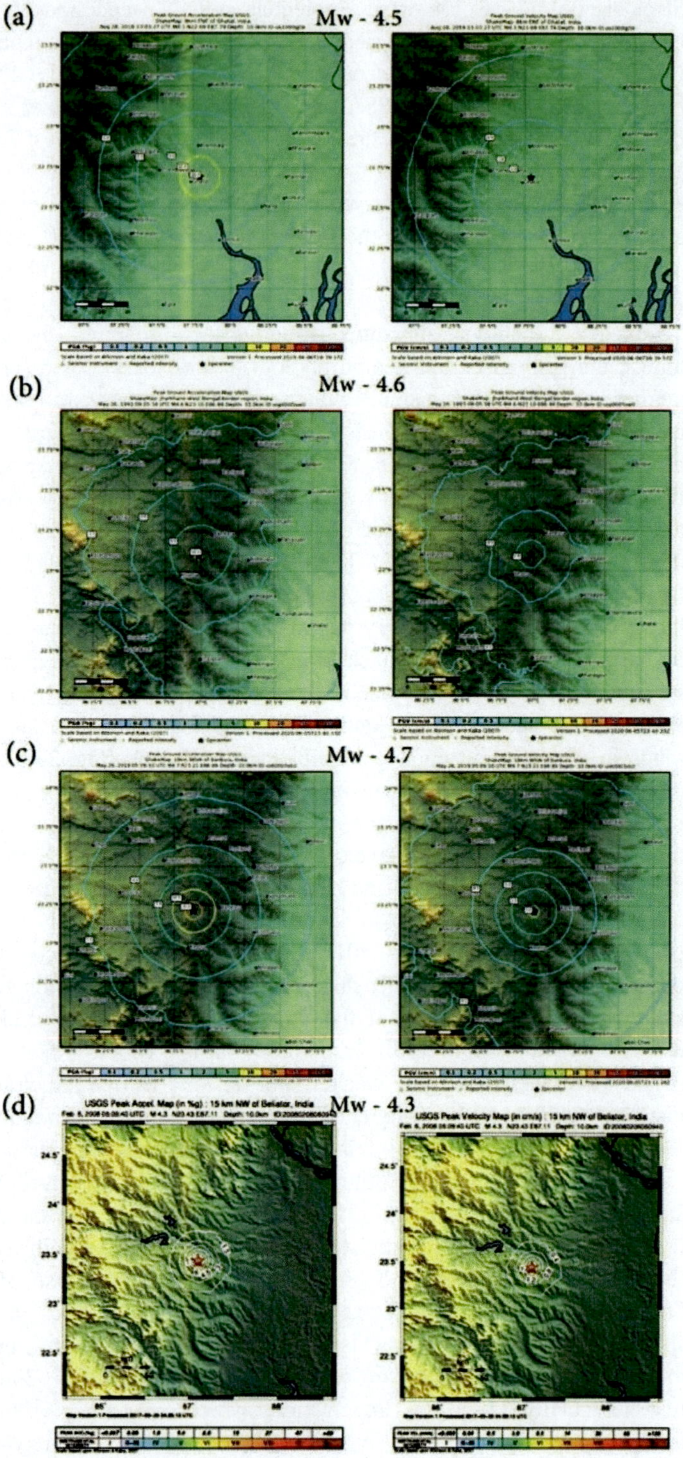

Fig. 23.7 A series of thematic maps showing peak ground acceleration and peak velocity observed during shallow-focus earthquakes in western part of the Bengal Basin: **a** M_w 4.5 (near Ghatal on 28 August, 2018), **b** M_w 4.6 (near Bankura on 16 May 1993), **c** M_w 4.7 (near Bankura 26 May, 2019) and **d** M_w 4.3 (near Beliator 6 February, 2008) (maps derived from USGS earthquake public domain)

clay-silt and fine sand. This deformed layer is again overlain by a thin deformed layer of clay-silt and fine sand approximately 30–40 cm above the previous deformed layer.

The Bhaluksundha (23° 35′ 08″ N, 87° 11′ 05″ E) Quaternary section (>4 m depth) shows yet another type of seismites exposure in the left bank of Damodar. A twisted medium sand dyke intrudes into silty clay horizon along with prominent number of small horizon. Seismites section exposed near Satyanandapur shows a deformed and disrupted clay-silt layer overlain by undeformed sand layer. Lobate discontinuous structures are observed within the deformed layer. Section exposed near Kamalpur area (23° 07′ 55″ N, 88° 00′ 34″ E) displays intricately overturned and folded unconsolidated sand layers. The glimpses of small-scale load structures are formed due to gravitationally unstable density gradient during earthquakes and overloading. Additionally, small-scale slump structures and multiple deformed laminations are formed also by gravitational body force during earthquake. A similar SSDS was observed in Idilpur (23° 13′ 28″ N, 87° 49′ 42″ E), at left bank of Damodar, displaying deformed lamination of lower clay-silt layer and underlain by deformed sand layers. The deformed and folded unconsolidated sandy-silt layers and gravels of various litho-sections suggest profound control of active tectonics on the sedimentary structures of Damodar River Valley (Fig. 23.9).

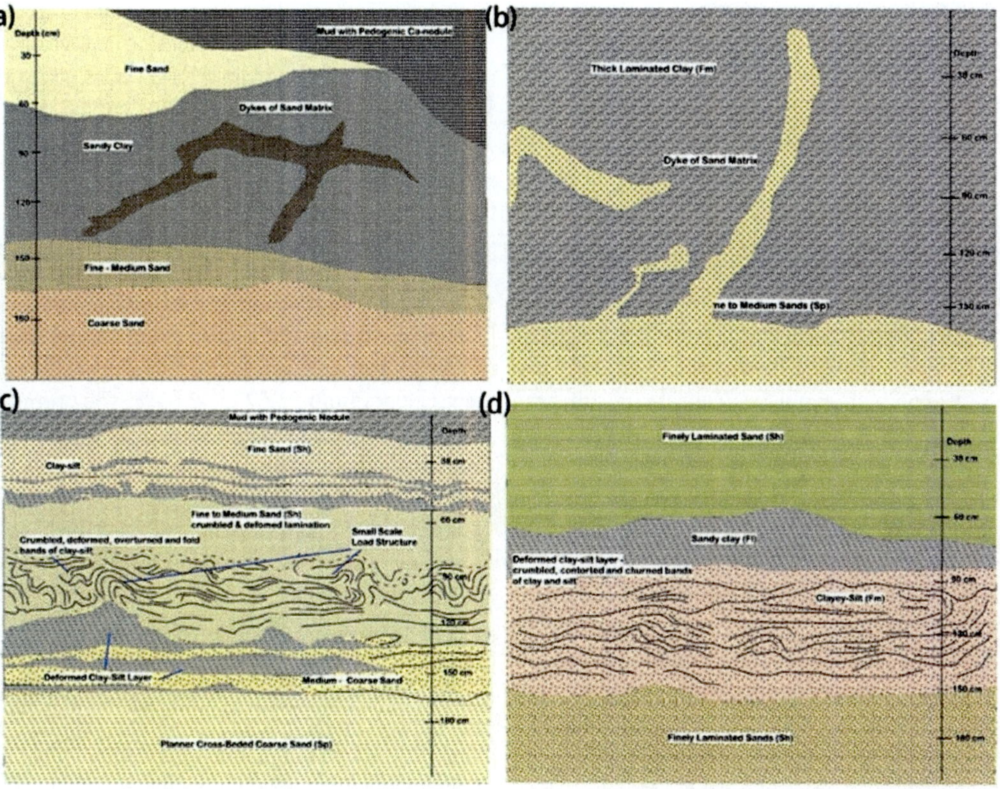

Fig. 23.8 Soft-sediment deformation structures and seismites in Quaternary sections: **a** intrusion of sand dykes in sandy clay (F_{cl}) deposits in varied directions at Tirat-Harabhanga, **b** elongated sand dykes intruded in thick laminated clay-silt deposits (F_m) at Haripur, **c** glimpses of deformed clay layers and small scale load structures formed due to gravitationally unstable density gradient at Kamalpur and **d** glimpse of deformed, overtuned and folded unconsolidated clayey silt and sand layers at Idilpur

Fig. 23.9 Tertiary sediment deformation structures at Hetodoba, Durgapur: **a** deformation and warping of gravel lithofacies and a trace of fault along deformation, **b** occurrence of out-sized clasts in upward finding gravel facies due to seismic shock, and **c** deformation of ferruginous sandy matrix in litho-section of gravels (length of scale 30 cm)

23.4.2 Standard Sinuosity Index

Understanding the syntectonic modification of river systems is one of key assessments in tectonic geomorphology. The fluvial system is very sensitive element of earth surface and a large magnitude of earthquake either ruptures or deforms the surface, or when ground shaking modifies the fluvial forms and processes. Achievement of 'grade' means that a tectonically induced change in an alluvial river can induce changes in other characteristics of the system. Like the longitudinal profiles, local anomalies in channel character can result from the variations in lithology, geological structure, hydrology and so on, but such anomalies can also result from tectonic activity. To understand the tectonic adjustment of river between faults (mainly Basin Margin Fault, Pingla Fault and Garhmayna-Khadaghosh Fault), the geomorphic index of Standard Sinuosity Index (Mueller 1968) is used in the selected stretch of the peninsular rivers. Meandering nature of the river is occurred to maintain a channel slope (minimum expenditure of energy) in equilibrium with discharge and sediment load. As channel slope is controlled by the tectonic deformation, so the variation of meander growth and sinuosity is very obvious in the active floodplains. A river meanders when the straight-line slope of the valley is too steep for equilibrium—the sinuous path of the meander reduces the slope of channel. Any tectonic deformation that changes the slope of a river valley may result in a corresponding change in sinuosity to maintain the equilibrium channel slope (Keller and Pinter 2002). In many cases, it is found that where meandering rivers cross tectonic upwraps, they tend to be less sinuous on the upstream flank and more sinuous on the downstream flank.

The Brahmani River shows very straight course of channel (less sinuous) due to the effect of Rajmahal Fault (RF). Upstream of RF, the SSI is estimated about 1.109 and it reduces to 1.075 at downstream of RF (Fig. 23.10). The hydraulic sinuosity index (HSI) varies from 47.33 to 57.97%, whereas topographic sinuosity (TSI) index reduces from 52.67 to 42.03%. The upstream course of Dwarka River is guided by the Saithia-Brahmani Fault (SBF) along the Gondwana Basin and Rajmahal Basalt Traps (North West–South East orientation of lineament). SSI shows a low value of 1.029, signifying a straight course of tectonic control (TSI—90.24%). While downstream of SBF, SSI increases to 1.123, reflecting hydraulic dominance on sinuosity (HSI—52.67%). The channel reaches of Mayurakshi River between SBF and GKF (Tilpara Barrage to Gunur) shows straighter course (SSI—1.026 to 1.085) due to tectonic control on river. The topographic control on river sinuosity is well proved by high TSI (>73%).

Similarly, the geotectonic system of BSM, PF and GKF has profound impact on channel

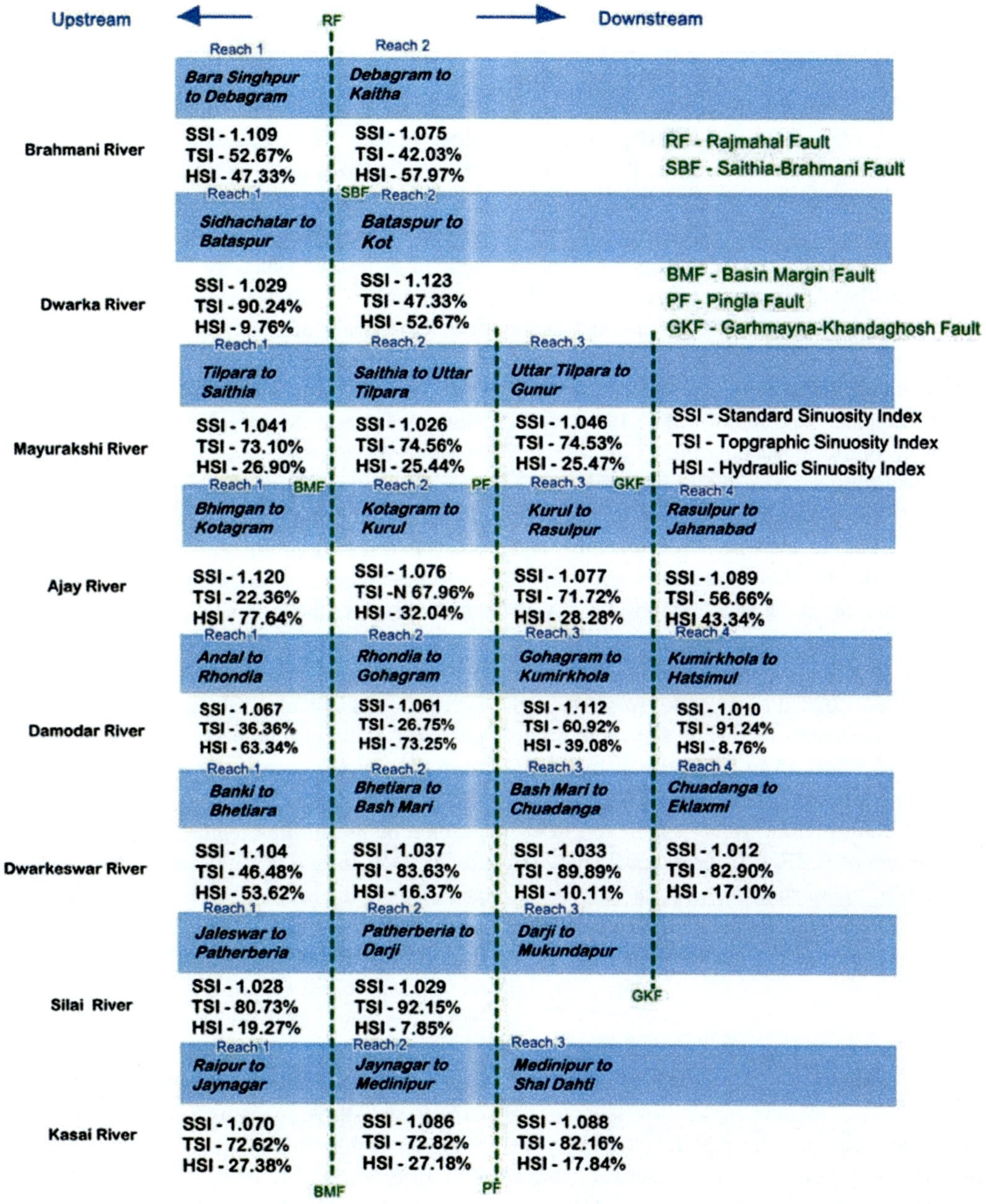

Fig. 23.10 Comparing standard sinuosity index (SSI), topographic sinuosity index (TSI) and hydraulic sinuosity index (HSI) in the alluvial rivers in response to Rajmahal Fault (RF), Saithia-Brahmani Fault (SBF), Basin Margin Fault (BMF), Pingla Fault (PF) and Garhmayna-Khandaghosh Fault (GKF)

sinuosity of Ajay River, showing channel deflection and straightness (Fig. 23.11a). SSI varies from 1.120 to 1.089 at downstream end. Upstream of CF the sinuosity is controlled by hydraulic variables, which reflect in high HSI (77.64%). Downstream of CF, the sinuosity is adjusted by topographic control, as reflected in TSI (56.66–71.72%). The reach between PF and GKF shows more straightness than other reaches (SSI < 1.076). The Damodar River shows prominent deflection of channel course across the faults, maintaining straight alluvial channel (SSI —1.067 to 1.112). The reaches between BMF and PF show more hydraulic control than topographic control, as reflected in HSI greater than 63.64%. While the downstream of PF and GKF (Gohagram to Barddhaman), the sinuosity decreases from 1.112 to 1.010 due to tectonic control (TSI—60.92 to 91.24) (Fig. 23.11b). In this reach, the development of several fluvial terraces, incision, channel narrowness and faults is observed. The channel reach of Dwarkeswar River, between Banki to Eklaxmi, does not show any considerable variation of SSI (1.104–1.012). It may be the resultant effect of BMF, PF and GKF, maintaining a straight course (Fig. 23.11 c). The TSI is estimated as 46.48% at upstream of BMF, but it increases considerably and reaches up to 89.89% at downstream. Similar tectonic control on channel sinuosity is well traced in Silai and Kasai River as well. SSI of Silai River (Jaleswar—Patharberia—Darji) is estimated as 1.028 due to effect of BMF and numerous lineaments, showing very high value of TSI (80.73–92.15%). Very less hydraulic control on sinuosity is observed in Kasai River and SSI varies from only 1.070 to 1.088. Downstream of BMF, TSI increases from 42.03 to 82.16%, marinating a straight course of channel across the terraces of laterites and gravels.

23.4.3 Stream Length—Gradient Index

The Stream Length—Gradient Index (S_L) correlates to stream power. Total stream power available at a particular reach of channel is an important hydrologic variable because it is related to the ability of a stream to erode its bed and transport variable because it is related to the ability of a stream to erode its bed and transport sediment. Total or available stream power is proportional to the product of the slope of the water surface and discharge. The slope of the water surface generally is approximated by the slope of the channel bed, and there is a good correlation between total channel length upstream and bankfull discharge which is thought to be important in forming and maintain rivers (Keller and Pinter 2002). Variations in S_L index reflect the downstream variations in discharge and stream power, but more commonly the lithologic or tectonic controls on channel slope of a given reach. S_L index has been widely applied as proxy to identify areas of anomalous uplift within a landscape. In this study, the reaches of peninsular alluvial rivers (mainly west to east flowing Kasai, Siliar, Damodar, Khari, Ajay, Dwarka and Brahmani Rivers) across or along the Basin Margin Fault (BMF), Pingla Fault (PF), Garhmayna—Khandaghosh Fault (GKF), Saithia—Brahmani Fault (SBF) and Rajmahal Fault (RF) are considered for S_L index analysis and long-profiles.

In the region of Bankura and Paschim Medinipur districts, the fluvial system of Silai and Kasai Rivers has developed well-integrated drainage network and badlands over Quaternary laterites and floodplain alluviums, showing glimpses of fluvial anomalies in channel planform (e.g. more towards straight than meander) and drainage pattern (e.g. annular growth due to tectonic bulge). In this region, the tectonic control of BMF and PF is well traced in the channels and topographic irregularities. The mean channel slope or gradient (S_c) of Silai River varies from 0.00532 to 0.00007 m m^{-1} within the stretch of 64.2 km. In this stretch, the elevation of channel bed varies from 341.96 to 31.18 m and the index values of S_L range between 210.91 and 27.71 m. Across the BMF, the S_L shows high value, i.e. 115.19 to 341.96, which is the reflection of topographic control on high channel gradient. The trace of micro convexity is observed in the long profile curve of river, showing the effect of neo-tectonic uplift. BMF and lineaments have profound impact on the enrichment of kinetic flow energy of incision and low entropy of fluvial system (as regain of power due to uplift). In that stretch of Silai River, maximum concentration of gullies and badlands (at Bagnada and Ganagani) is recognized, showing progressive erosion of relief. Between BMF and PF, the River maintains low energy profile (S_L—27.71 to 42.20 m) with

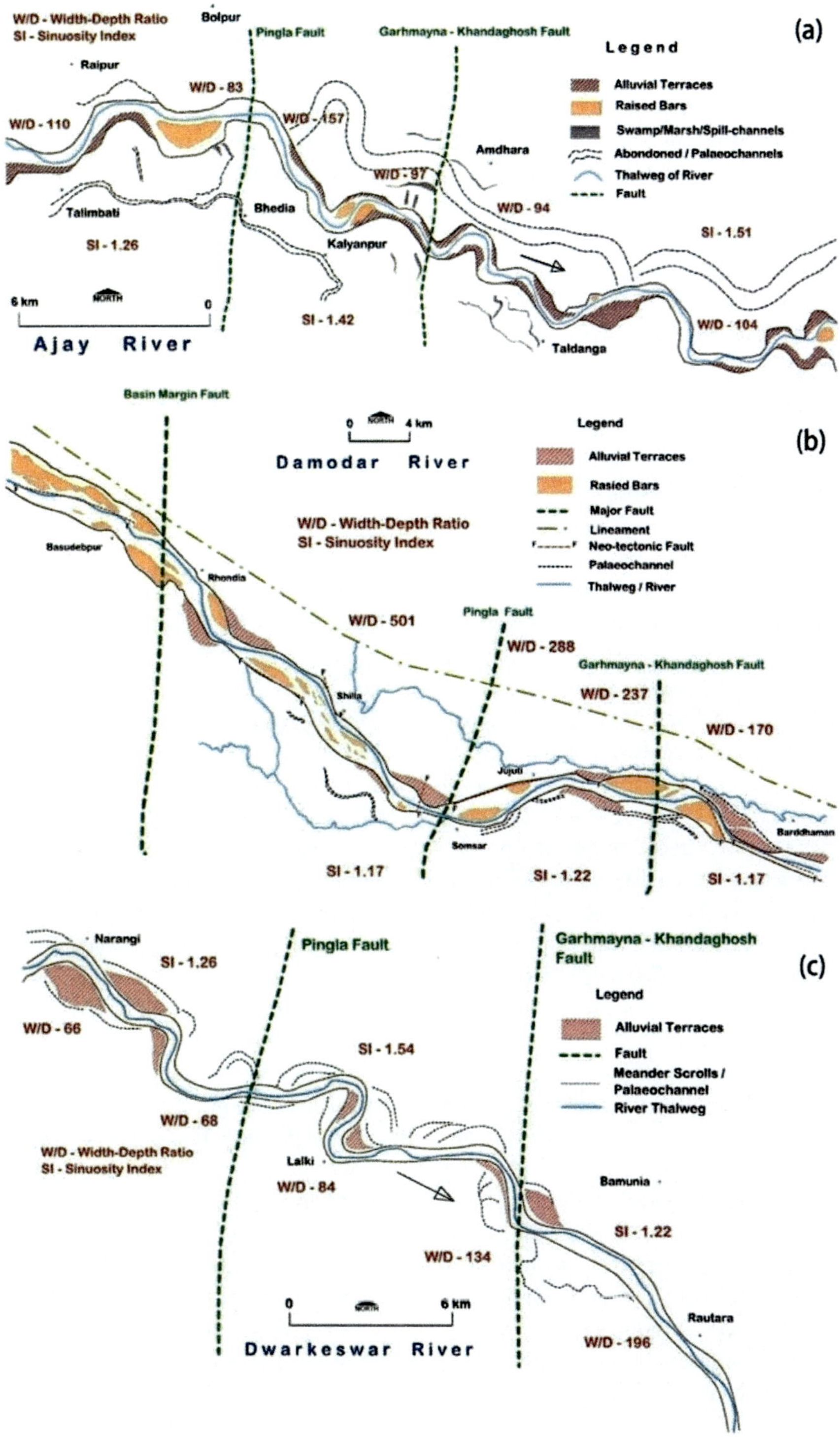

Fig. 23.11 Fluvial responses (viz., width-depth ratio, sinuosity index, alluvial terrace, channel shifting, palaeochannels etc.) to active tectonics across Basin Margin Fault, Pingla Fault and Garhmayna-Khandaghosh Fault in **a** Ajay River, **b** Damdoar River and **c** Dwarkeswar River

ample aggradations (downstream of fault) and valley widening.

Accordingly, Kasai River shows the same kind of signatures, having mean S_c variation of 0.00277 to 0.00058 m m^{-1} within a stretch of 99.2 km. The elevation of channel bed decreases from 136.47 m to 67.86 m and S_L varies from 275.16 to 21.06 m. In the upstream of BMF, the Kasai River is flowing over the Precambrian crystalline complex and Tertiary formations, adjusting the channel profile with underlying geology (Fig. 23.12). Within first 14 km stretch of study reach, the major lineament controls the steep channel gradient, as suggested from high S_L of 147–275 m. Between BMF and PF, the curve of S_L shows four distinct pulse ($S_L > 100$) and convexity of long profile is traced at the locations as a result of occasional channel cut-and-fill sequences (a sign of quasi-equilibrium). The variable change of slope and bed elevation may be the resultant effect of neo-tectonic bulge between BMF and PF. Similar effect is observed in the sub-basins of Silai and Kasai Rivers. Purundar River (a tributary source of Silai) is controlled by a major lineament over laterites, as reflected from high S_L ($S_L > 150$ m). The convexity of profile is well traced in the long profile, showing frequent pulse of S_L towards downstream (S_c variation of 0.00925–0.00007 m m^{-1}). Accordingly, the Tamal River (a tributary source of Kasai) shows bed elevation variation of 78.449–19.84 m within a stretch of 44 km and the index values of S_L decreases from 205.47 to 17.45 m in that stretch. The upstream shows frequent S_L pulse ($S_L > 70$ m) profile convexity and low entropy channel gradient. This river is local base level controlled, showing frequent sign of incision and deposition.

Leaving the sedimentary formations of Gondwana the Damodar River shows considerable variation of channel slope and planform in between Durgapur and Barddhaman (a stretch of 153 km). The mean bed elevation drops from 104.79 to 27.115 m, and S_c varies from 0.00109 to 0.00036 m m^{-1}. Throughout the long profile, S_L index shows frequent high amplitude pulses ($S_L > 100$ m) and profile convexity due to effect of faulting (Fig. 23.13). The reach between BMF and PF shows two separate convexities in long profile, as reflected in S_L of 52.55–110.49 m. There is a considerable increase in channel gradient (0.00033–0.00072 mm^{-1}) between Panagarh and Gohagram. In upstream of GKF, there are considerable topographic expressions (e.g. palaeochannels, raised bars, swamps and abandoned channels) of floodplain subsidence. The subsidence is reflected in low value of S_L, viz., below 50 m. Below GKF, the channel gradient is suddenly increased and convexity of bed elevation profile is clearly visible. It reflects low entropy, high fluvial incision and sudden trigger of high energy due to upliftment along GKF. S_L increases up to 136 m and reduces to 55 m at the vicinity of Barddhaman town. This stretch of Damodar (between Somsar and Belkash) shows increasing straightness of channel ($S_I < 1.1$), unparallel alluvial terraces, narrowness of thalweg and traces of Quaternary faults along the river banks.

Similarly, the tectonic effect of BMF, PF and GKF is observed in the Ajay River, and there is a noticeable variation of S_L (174.91–27.97 m). Five distinct S_L pulses are recognized between BMF and PF, and the convexity of long profile is clearly visible in three parts (Fig. 23.13). These anomalies of channel gradient and stream power are the reflection of tectonic deformation due to these faults. The stretch between PF and GKF shows considerable variation of cut-and-fill sequences, incision of bed (exposure of underlying laterites at Illambazar—Panagarh Bridge) and high energy gradient, forming alluvial terraces of laterite gravels and badlands in the Illambazar Formation. Below the GKF, the river floodplain shows high entropy (S_c variation of 0.00088 to 0.00034 m m^{-1}), subsidence and fluvial aggradations ($S_L < 60$ m), forming abandoned channels, spill channels, swamps, overbank flooding and southern lateral shifting, high sinuosity ($S_T > 1.2$) and raised channel bars. Something differently the Khari River (smaller than Ajay and Damodar) shows the feature of tortuosity, which is a channel property of a curve being tortuous (twisted having many turns) between PF and GKF (Fig. 23.14). It is found that S_L, across PF, shows high S_L ($S_L > 80$ m)

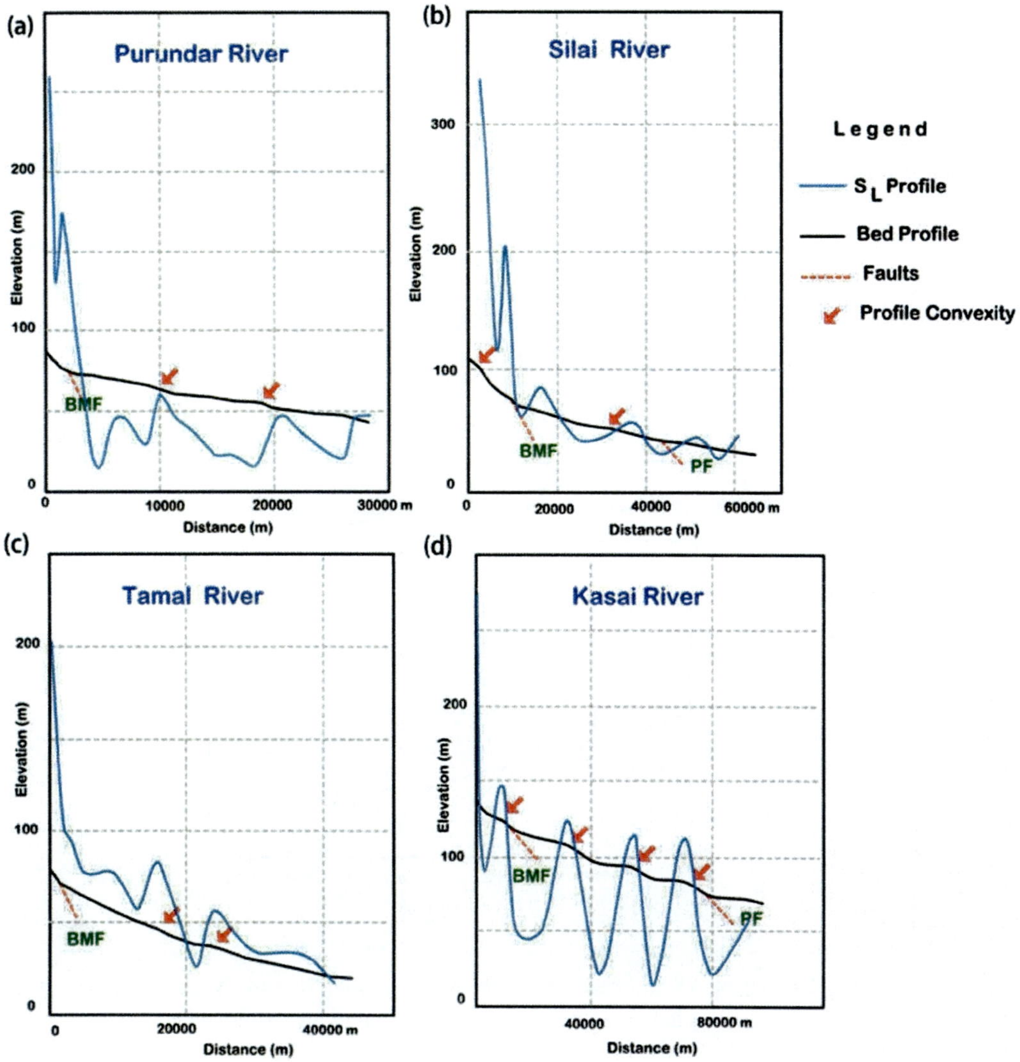

Fig. 23.12 Variations of longitudinal profiles (trace of convexity) and recurrent pulses of S_L values with the effect of Basin Margin Fault (BMF), Pingla Fault (PF) and Garhmayna—Khandaghosh Fault (GKF) in **a** Purundar River, **b** Silai River, **c** Tamal River and **d** Kasai River

and convexity in bed profile, showing the effect of topographic control on channel steep gradient. The development of meander neck-cutoff and numerous fluvial terraces of different topographic levels may be controlled by high fluvial energy (during peak monsoon flow) driven by tectonic uplift. Below GKF, there is a steady decrease of S_L (85–13 m) and the high entropy features of floodplain aggradations are quite prominent, as the development of abandoned channels, high sinuosity ($S_I > 2.0$) and overbank flooding (due to regional subsidence). A similar situation is also observed in Kunur River (Fig. 23.15).

The Dwarka River is flowing along the SBF, crossing the Rajmahal Basalt Traps and Gondwana Formation. The role of SBF is quite prominent, restricting the oscillation of river along a lineament (Fig. 23.16). The channel slope varies from 0.00335 to 00,027 m m^{-1} within the riverbed elevation range of 70.35–

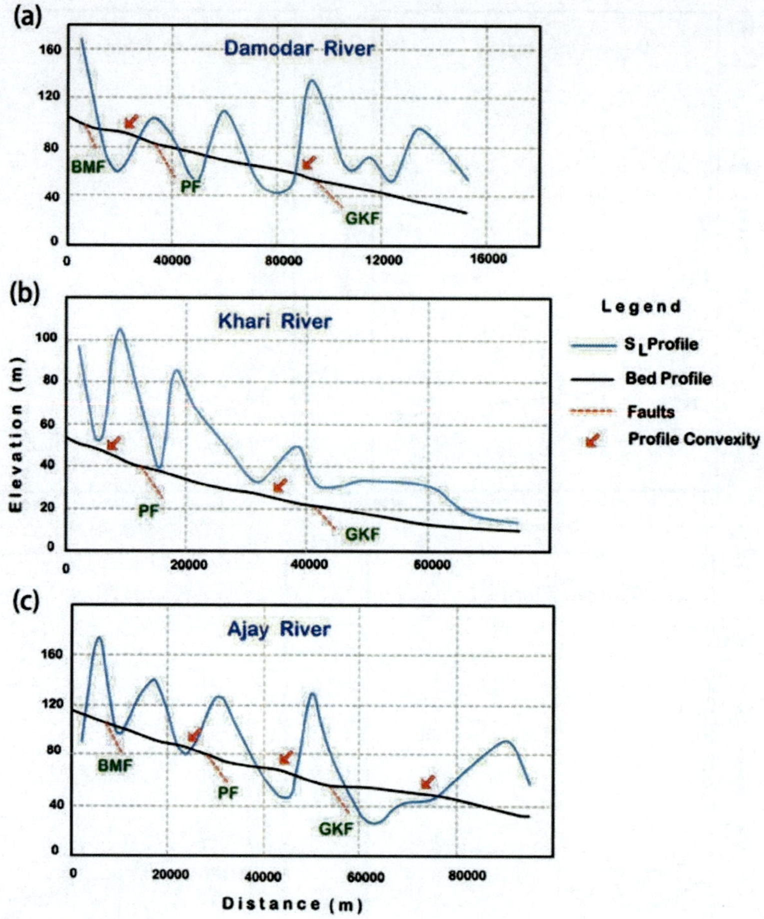

Fig. 23.13 Variations of longitudinal profiles and bed elevations (traces of convexity in relation to zone of deformation by faults) generating recurrent pulses of S_L values with the effect of faults in **a** Damodar River, **b** Khari River and **c** Ajay River

8.37 m. In the initial reach of 15 km yields S_L value of greater than 40 m and it increases up to 71 m. A distinct convexity of elevation profile is traced, reflecting the control of tectonic uplift on high channel gradient and generating several pulses of S_L and cut-and-fill sequences in the channel. The upstream degradation shows narrow and deep valley and alluvial terraces, and downstream degradation shows migrating meandering thalweg, valley widening and raised bars. The influence of basalt trap structure and RF is clearly observed in the Valley of Brahmani River, having S_L range of 354–47 m. The channel slope of Brahmani (bed elevation range of 126.11–46.74 m) reduces from 0.00931 to 0.00036 m m^{-1} within the stretch of 23 km. A prominent convexity, due to uplift along RF, is observed and is traced along the channel elevation profile. Due to this knick point effect, the index value of S_L escalates up to 354 m, indicating low entropy of fluvial system and high fluvial incision with development unparallel terraces. Below RF, the S_L reduces to less than 50 m, developing aggradational landforms, flooding, valley widening and raised channel bars.

23.4.4 Hypsometric Analysis

The distribution of the elevations within a basin provides information on the balance between external processes (which tend to lower the landscape) and internal processes (which tend to create relief). One of the most useful parameters that describes and analyzes the distribution of

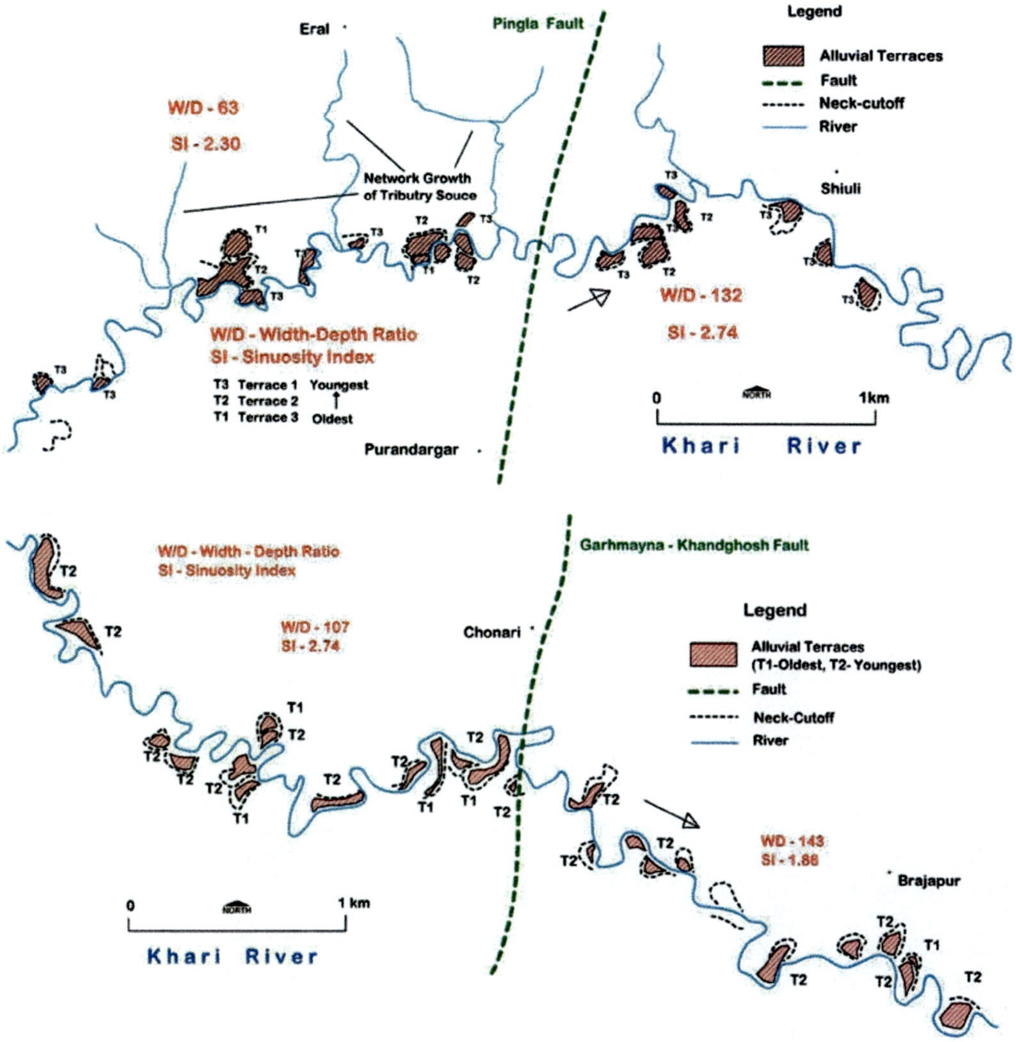

Fig. 23.14 Geomorphic map of Khari River showing upstream and downstream fluvial anomalies (mainly development of terraces and neck-cutoffs) in relation to Pingla and Garhmayna—Khandaghosh Faults

elevations in an area is hypsometry. Hypsometric integral (H_I), a dimensionless parameter, is proposed by Strahler (1952). The advantage of H_I is those users calculate and compare different sins of different areas irrespective of scale. Hypsometric analysis (or area-altitude analysis) is the study of the distribution of horizontal cross-sectional area of a landmass with respect to elevation. The index values of H_I have been used to differentiate between erosional landforms at different stages during their evolution. H_I is an indicator of the present volume as compared with the original volume of the basin. High H_I indicates that most of the topography is high relative to the mean representing a youthful topography stage (Keller and Pinter 2002). Intermediate to low H_I represents more evenly dissected drainage basin, indicating a mature stage of development.

Considering six sub-basins (elevation range of 177–32 m from mean sea level) of the Silai—Kasai drainage system, it is found that the index values of H_I increase from 0.362 to 0.558, with mean of 0.441 (Table 23.5). In addition, the hypsometric analysis of Bagnada and Gangani

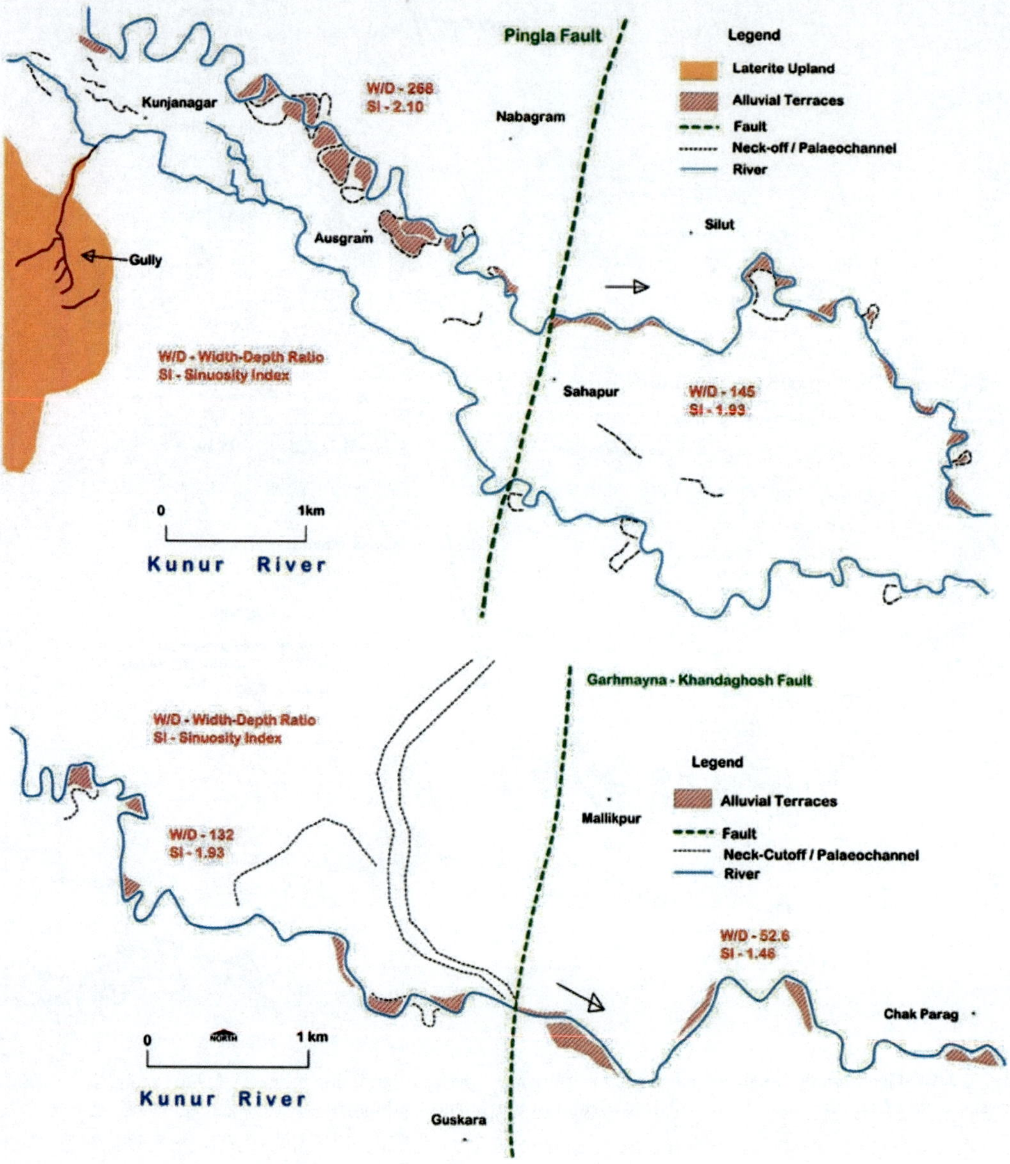

Fig. 23.15 Geomorphic map of Kunur River showing fluvial anomalies (upstream development of alluvial terraces and downstream decrease of sinuosity and width-depth ratio) and in connection with Pingla and Garhmayna—Khandaghosh Faults

badlands suggests relatively high values of H_I, ranging from 0.481 to 0.593. The three sub-basins (elevation range of 120–46 m from mean sea level) of the Dwarkeswar River show H_I range of 0.410–0.451, reflecting mature equilibrium stage of landform development. In the six sub-basins (elevation range of 165–44 m) of the Damodar—Ajay interfluves, H_I varies from 0.396 to 0.619, reflecting mature to youthful stage. The four sub-basins of the Mayurakshi River show H_I range of 0.434–0.480. Similarly in the four sub-basins of the Dwarka—Brahmani interfluves, the index values vary from 0.448 to 0.501, suggesting a mature stage.

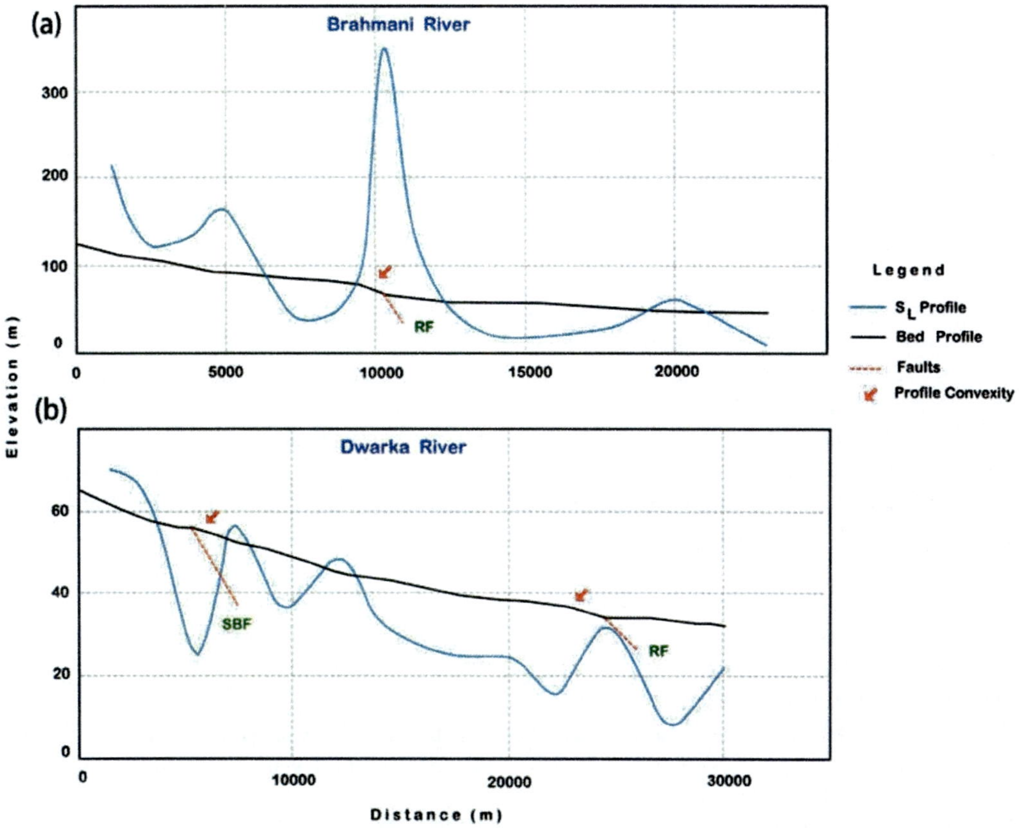

Fig. 23.16 Variations of longitudinal profiles and bed elevations (traces of convexity in relation to zone of deformation by faults) generating recurrent pulses of S_L values in connection with Rajmahal Fault (RF) and Saithia-Brahmani Fault (SBF)

Table 23.5 Summary result showing geomorphic indices of active tectonics in the peninsular rivers of the Bengal Basin

Peninsular river system	SSI range	S_L range (m)	H_I range	A_F range (%)	E_R range
Brahmani—Dwarka River System	1.029–1.123	354–8	0.421–0.5	35.73–53.25	0.530–0.633
Mayurakshi River System	1.026–1.046	165–22	0.433–0.480	63.35–79.24	0.434–0.550
Ajay—Damodar—Dwarkeswar River System	1.012–1.112	175–27	0.396–0.619	43.33–64.58	0.442–0.751
Silai – Kasai River System	1.028–1.088	341–17	0.362 0.578	39.06–70.97	0.367–0.779

Note SSI—standard sinuosity index, S_L—stream length—gradient index, H_I—hypsometric integral, A_F—basin asymmetry factor and E_R—basin elongation ratio

This index is similar to the S_L index in that rock strength or sediment resistance as well as other tectonic factors affect the value. High H_I (>0.5) generally means that not as much of the uplands have been eroded and may propose a younger landscape and dissected relief, possibly produced by active tectonics. The occurrence of gullies and badlands in the lateritic interfluves of these peninsular rivers has proved the active phase of fluvial incision. In this case, the high H_I could also result from recent incision into a young geomorphic surface formed by the deposition. Comparing the H_I values, two groups can be identified with respect to the convexity or concavity of the hypsometric curve: (1) Class A basins with convex hypsometric curves (0.51–0.62) and (2) Class B basins with concave—convex hypsometric curves (0.36–0.50). The range of higher to lower values of hypsometric integral suggests that vast amount of mass is subjected to denudational process while loss of materials has been eroded (Anand and Pradhan 2019). The lower HI values in the basins resulted due to the high kinetic energy of runoff, which dissected the lateritic landscape and floodplain alluvium due to high erosional activities. The values of H_I, less than 0.3, mean a fully stabilized watershed in respect of base level of erosion and the value, greater than 0.6 indicate watershed highly susceptible to fluvial erosion due to regional uplift. In this analysis, the range of H_I ($0.3 \leq H_I \leq 0.6$) reflects that the sub-basins of peninsular rivers are moderately susceptible to incision (control of moderate active tectonics or regional up-warping or tilting), maintaining mature equilibrium to slightly dis-equilibrium stage of landform development in the western shelf zone of the Bengal Basin.

23.4.5 Drainage Basin Asymmetry Factor

To understand the geometric or basin planform change due to active tectonics or lithological control, the factors of drainage basin or valley asymmetry can give some clues about the migration of river or amount of asymmetry in the basin. Using the geomorphic index (viz., asymmetry factor), the users calculate tectonic tilting of the basin and their direction and how much tilting is taking place in comparison to other basin (Fig. 23.17). For most of the stream networks that formed and continued in stable setting, the asymmetry factor (A_F) should equal about 50 (neutral value). The minor rivers of floodplain or pediments or plateau fringe are very sensitive to minimal tectonic up-warping or down-warping. So, A_F is sensitive to tilting perpendicular to the trend of the trunk of high-order river in the basin. Any drainage basin with a flowing trunk stream subjected to a tectonic rotation will most likely have an effect on the tributaries lengths. According to Keller and Pinter (2002), assuming the tectonic activity caused a left to the drainage basin, the tributaries to the left of the main stream will be shorter compared with the ones to the right side of the stream with an asymmetry factor greater than 50, and vice versa.

The four sub-basins (areal coverage of 2.69–9.12 km^2) of the Dwarka—Brahmani interfluves show an anomalous range of A_F, i.e. 33.53–53.25%, depicting the role of RF and SBF in the drainage basin orientation (Table 23.4). The four sub-basins (areal coverage of 8.55–9.85 km^2) of the Mayurakshi River depict A_F range of 38.78–43.37% (values much lower than 50), reflecting tilting of left side of the basins. In the six sub-basins (areal coverage of 17.90–810 km^2) of the Damodar—Ajay interfluves, A_F varies from 43.33% (left tilting) to 64.58% (right tilting). The influence of PF, GKF and BSF makes the trunk streams to tilt more left side of the basins. In the sub-basins (areal coverage of 10.47–159.11 km^2) of Dwarkesar River, it shows anomalous behaviour (complex response) of A_F which varies from 43.18 to 64.25% within the tectonic blocks of PF and GKF. The sub-basins (areal coverage of 52.82–257.54 km^2) of the Silai–Kasai interfluves show again anomalous nature of A_F, which varies from 39.06% (tilting left) to 70.97% (tilting right). In this study, A_F significantly greater or smaller than 50 shows prominent influence of active tectonics or lithological control or differential erosion, as for example, the stream slipping

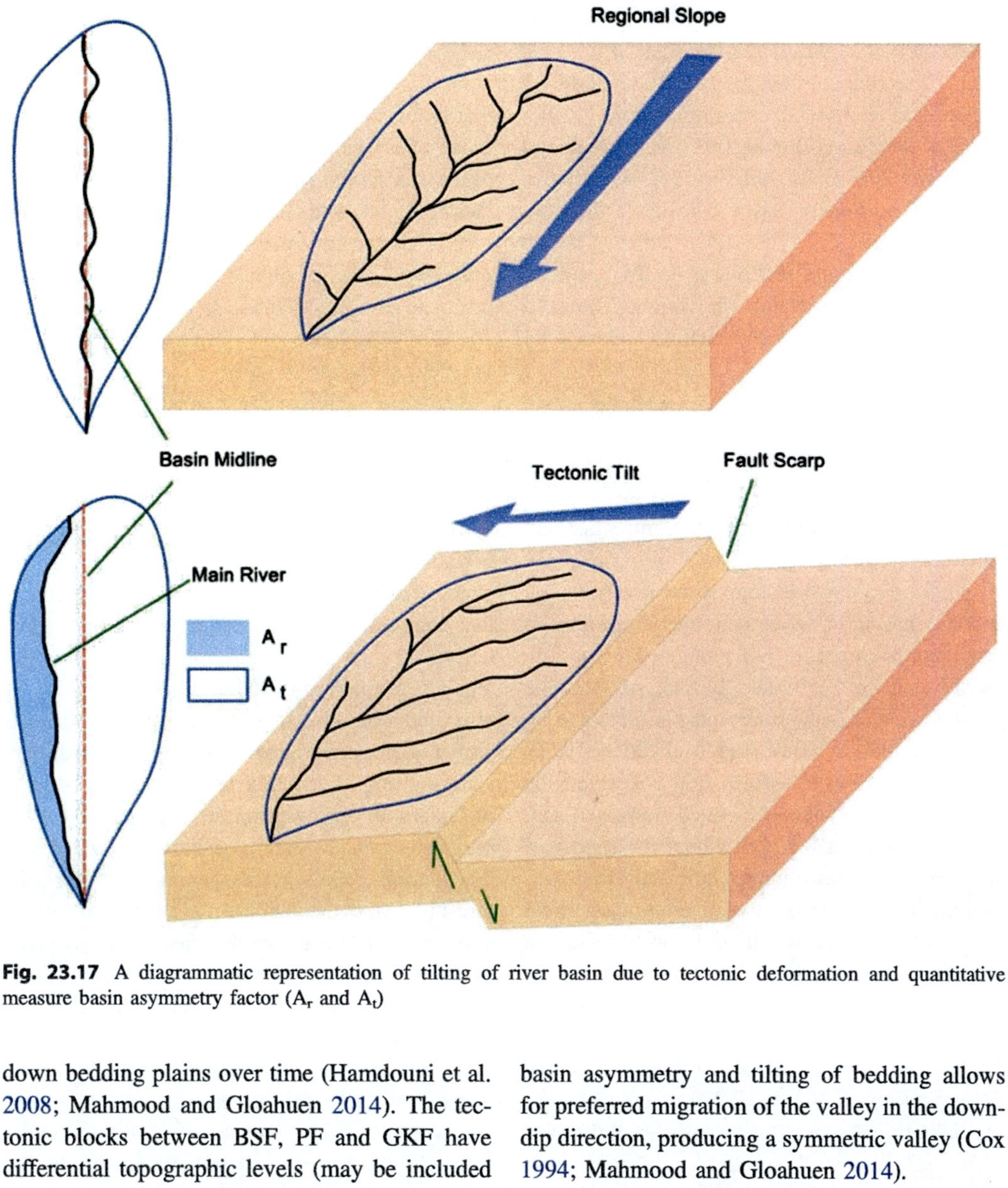

Fig. 23.17 A diagrammatic representation of tilting of river basin due to tectonic deformation and quantitative measure basin asymmetry factor (A_r and A_t)

down bedding plains over time (Hamdouni et al. 2008; Mahmood and Gloahuen 2014). The tectonic blocks between BSF, PF and GKF have differential topographic levels (may be included by fault scarps), occasionally characterized by steep sides and flat floors. The steep sides are created by displacement on faults such that the valley floor moves downs relative to the surrounding margins, or conversely the margins move up relative to the floor. This movement results in basin tilting and causes the river to migrate laterally and deviate from the basin midline. Also, structural control of the orientation of bedding may play a role in the growth of basin asymmetry and tilting of bedding allows for preferred migration of the valley in the down-dip direction, producing a symmetric valley (Cox 1994; Mahmood and Gloahuen 2014).

Valley floor width-to-height ratio (V_F) (Bull and McFadden 1977) is a geomorphic index conceived to discriminate asymmetric valley shapes due to regional tectonic activity. Because uplift is associated with incision, the index is thought to be a proxy for active tectonics where low values of VF are associated with higher rates of uplift and incision. Deep V-shaped valleys (VF < 1) are connected with linear active down-cutting streams distinctive of areas subjected to active tectonics, while flat-floored (U-shaped) valleys (VF > 1) show an attainment of the base level of erosion (may be local), mainly in response to relative tectonics quiescence (Keller 1986; Keller and Pinter 2002; Mahmood and Gloahuen 2014). In this study, a distance between 2 and 4 km is set to measure the geomorphic index along the river valleys, taking more than one cross-sections of DEM upstream of faults for mean value derivation. The sub-basins of the rivers are mainly taken into consideration.

23.4.6 Basin Elongation Ratio

The index of elongation ratio (E_R), introduced by Schumm (1956), is a measure of basin shape included by regional tectonics. Relatively young drainage basins in active tectonics areas tend to be elongated in shape. The elongated shapes are transformed into circular basins, as tectonic activity reduces with time and continued topographic evolution (Bull and McFadden 1977). The reason of this transformation is because the drainage basin widths are much narrower near the active deformation area where the energy of the stream has been directed primarily to downcutting; by contrast, a lack of continuing rapid uplift permits widening of the basins upstream from the active fault zone. The E_R value nearer to 1 suggests highly elongated basin with regional impact of active tectonics. In the sub-basins of the Silai–Kasai interfluves, the index values of E_R vary from 0.367 to 0.779, suggesting an anomalous nature of basin elongation (Table 23.4). The sub-basins of Dwarkeswar River show ER range of 0.409–0.750, reflecting a trend of basin elongation between BMF and PF. In the six sub-basins of Damodar–Ajay interfluves, E_R varies from 0.442 to 0.841, showing control of BMF, PF and GKF on the growth of elongated basin shape. The sub-basins of Mayurakshi River show a trend of semi-circular to elongated basin (upstream of SBF), as reflected from E_R (0.434–0.550). E_R ranges between 0.421 and 0.633 in the Dwarka–Brahmani interfluves, reflecting also semi-circular to more elongated basin shape due to adjoining impact of RF and SBF. It is observed that the basins, situated upstream of PF and SBF (zone of active deformation), show considerable basin elongation, suggesting tectonic uplift of laterite block and the downstream of PF, the inconsistent nature of basin shapes is observed, reflecting quasi-equilibrium fluvial system in response to active tectonics (with regional variations of subsidence and uplift in western shelf of the Bengal Basin).

23.5 Discussion

Where tectonism has been persistent for long periods of time active deformation will produce a channel response that will be superimposed on the long-term tectonic effects. Major valley deformation or total disruption of the river system can be the result of long-term tectonism. It is proposed that stream gradients and valley floor altitudes do not change progressively through geologic time, but rather relatively brief periods of instability and incision are separated by long periods of relative stability (grade). In the terrain, the tectonic forces act as geomorphic extrinsic threshold, and the landform elements of fluvial systems depicts complex response to that event at variable time scale. From the study, it is realized that due to variability of spatial scale of river basins each basin can react differently against same seismic event, producing anomalous landforms in the valleys. For example, when a small drainage basin, like Kunur or Khari, was rejuvenated, the system responded not simply by incision but by hunting for a new equilibrium by incision, aggradation and renewed incision. The rivers of shelf zone do not maintain a steady-state equilibrium, which involves fluctuations about an average (can be traced in stream length–gradient index), but a metastable equilibrium occurs when an external influence, like earthquakes or slow tectonic warping/tilting, carries the fluvial system over some threshold into a next equilibrium regime, forming several landforms (e.g. neck cutoff, alluvial terraces, migration of thalweg, channel shifting, valley narrowness, low sinuosity etc.).

The most commonly cited evidence of fluvial response to deformation is the formation of alluvial terraces in the alluvial valley. The parallel to unparallel terraces are recognized in the rivers of Ajay, Khari, Kunur, Damodar and Kasai at different elevation levels reflecting channel shifting and incision (Figs. 23.14, 23.15 and 23.18). It is found that convex shape of terrace elevation profile between BMF and PF depicts the deformation by uplift, and concave shape between PF and GKF, also to downstream of GKF, signifies regional subsidence. The recurrent occurrences of floods in Ajay, Kunur, Khari, Kasai and Silai Rivers are observed in the region of floodplain subsidence. The uplift occurs between BMF and PF and base level falls, influencing the grade and the cut-and-fill sequences of the sub-basins. The seismic events disrupt the equilibrium of the system, a river may incise through its floodplain in order to reach a new graded profile and begin cutting a new floodplain. The old floodplain becomes a river terrace—an inactive bench stranded above the new level of the river. Repeated episodes of downcutting may preserve several terraces above a river. Faults are as likely to cut the surface of a fluvial terrace as any other landform. The river valley of Damdoar shows several glimpses of Quaternary faults in the terraces between PF and GKF (Fig. 23.11b). The fault cut a given surface (generating steep bank height), and this faulting is known to postdate the age of the terrace surface.

The achievement of grade means that that a tectonically induced change in an alluvial river can induce changes in other characteristics of the system. Longitudinal channel profiles and local anomalies in channel character can result from variations in geology, hydrology and mostly tectonic warping. Where streams flow across tectonic upwraps, erosion across the crest of the warp can generate sufficient sediment to change the local channel pattern from meandering to braided downstream, as reflected in the Damodar River (Fig. 23.19). In addition, because of the dependence on slope, longitudinal titling of just a few tenths of a per cent also can change the pattern of a river (a 0.1% change in slope is equivalent to differential uplift of 1 m over a distance of 1 km). The meandering rivers, like Khari and Kunur, cross-tectonic downwarps between PF and GKF (Ajay—Damodar interfluves), and they tend to be more sinuous on the upstream flank and less sinuous on the downstream flank. It has been suggested that overbank flooding can be anomalously frequent downstream of subsidence. The control of sinuosity is obvious in the peninsular rivers, showing standard sinuosity index nearer to unity, means straight course in between faults. In general, the river meanders (high oscillation in active floodplain) due to minimum expenditure of energy in response to valley slope, but the tectonic deformation of shelf zone increases the valley slope to the extent that alluvial river is confined within the active valley with minimum oscillation, reflecting relatively high energy gradient of Brahmani, Dwarka, Mayurakshi, Ajay, Damodar, Dwarkeswar, Silai and Kasai compared with the floodplains of Bhagirathi-Hooghly River. The domal uplift across Kasai and Silai Rivers shows annular growth of drainage pattern and anomalous nature of sinuosity (Fig. 23.20). There is an increase in sinuosity on the downstream side of uplift as the valley floor is steepened. On the upstream side of the uplift channel, more straightening is observed. In the upstream side of uplift (BMF and PF), the suspended sediment-load rivers, like Kunur and Khari, more development of neck-cutoff and palaeochannels is observed (it aids more degradation), and these will straighten the river downstream.

The Damodar River Valley has witnessed many past seismic events as crossing over the faulted graben of Gondwana sedimentary basin. The deformation of Quaternary sediments (seismites and SSDS) exposed in the banks and nearby areas has sufficient indications and evidences of active seismic activity. The general and almost linear NW-SE and WNW-ESE trend (sinuosity nearer to unity) of the main channel is possibly controlled by tectonic units between BMF, PF and GKF. It is observed that below

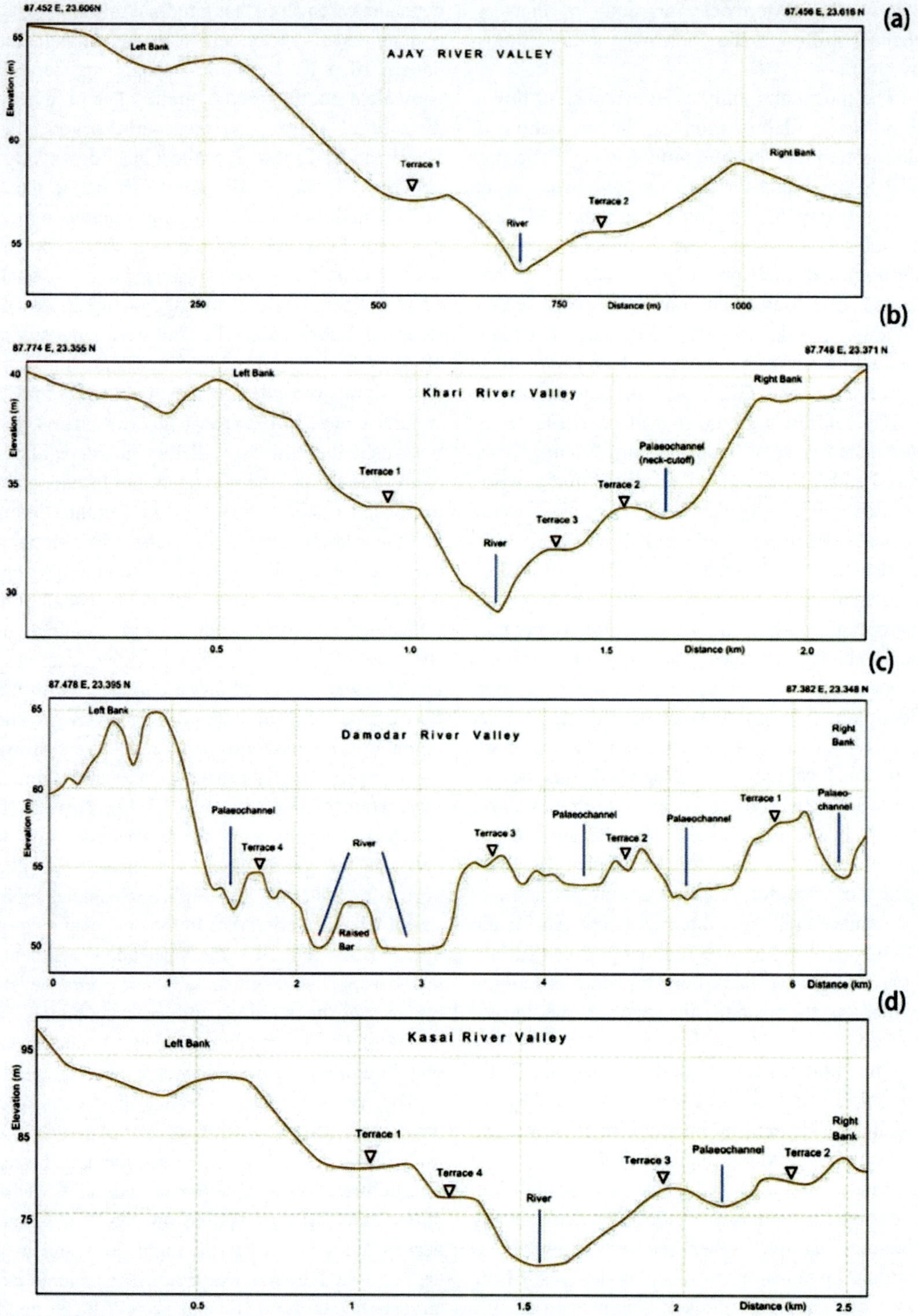

Fig. 23.18 Representation of channel cross-section to depict the alluvial terraces of different topographic levels in **a** Ajay River (near Deul), **b** Khari River (near Asinda), **c** Damodar River (near Shilla) and **d** Kasai River(near Dubli)

	Upstream ←	Axis of Deformation	→ Downstream
Uplift	Aggradation, frequent channel shifts, indistinct channel, sediment deposition raised bars	Degradation, terrace fromation, low sinuosity	Sinuosity increase, bank erosion, point bar growth
Subsidence	Sinuosity increase, bank erosion, point bar growth	Aggradation, braided pattern growth	Flooded, indistinct channel, sediment depostion, palaeochannels
		Zone of Deformation	

Fig. 23.19 Possible fluvial response to active deformation (uplift or subsidence) observed in the study areas

Fig. 23.20 Elevation map showing tectonic control (faults and lineaments) on drainage development in the Silai–Kasali interfluves

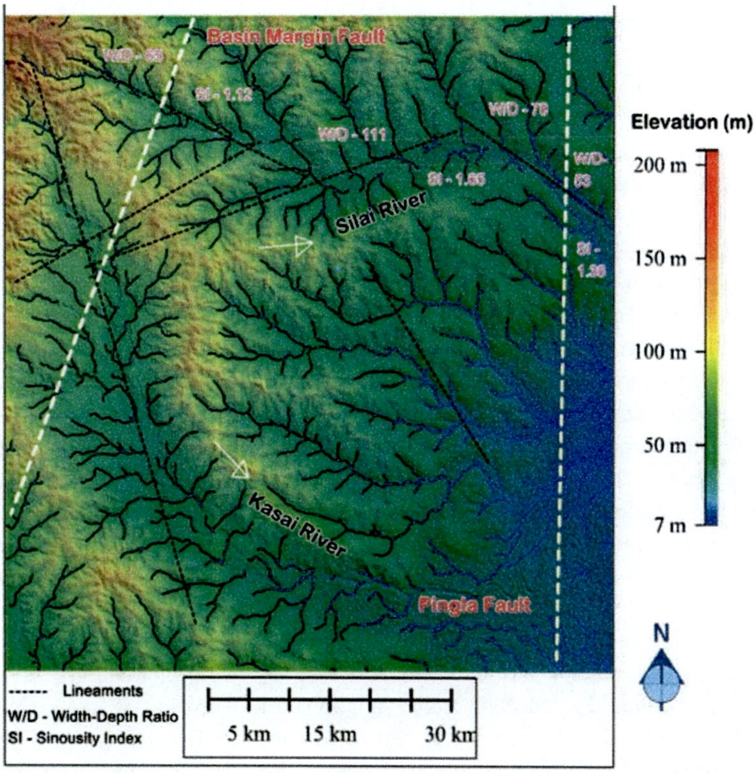

BSF, the width-depth ratio (>100) of the main channel increases from Durgapur to Jujuti, showing bank erosion and raised bars possibly due to recurrent seismic activity. The floodplain signatures of palaeochannels, abandoned channels, low lying depressions, swamps and channel shifting reflect the evidence of subsidence between Rhondia and Jujuti. These features might have been caused due to reactivation of PF and GKF, as reflected from the recurrent earthquakes of M_W 3.6–4.7. These seismic shocks can develop seismites (sand dykes) and SSDS (deformation of clay and sandy-silt beds) in the fluvial facies. The course of Damodar while traversing through Barddhaman becomes WNW-ESE trending and the width–depth ratio also steadily decreases with maintaining linearity up to Palla. From Palla locality, the river takes a sharp and southerly dip and continues to flow further in SSW direction up to Jamalpur. The linearity of course and sudden perpendicular bend might be because it is flowing on a definite tectonic plane, which had a definite role in the evolution of Damodar Fan Delta. Local faulting or dislocations within the Quaternary lithofacies as observed near Rhondia, Gohagram, Haripur and Idilpur are direct evidence of prevalence of active seismicity in the shelf zone of the Bengal Basin. The anomalous development of valley narrowing and linearity, alluvial terraces, sudden rise in bank height, swamps, and raised point and mid-channel bars can be recognized as syntectonic response of Damdoar fluvial system.

23.6 Conclusion

Alluvial rivers have multiple responses (rather more complex responses) to active tectonics of the Bengal Basin, and some are closely interrelated. The concentration here is on individual alluvial rivers' responses, which include those rivers that flow through their own sediments and are not bedrock confined. Apart from seismicity and tectonic elements of the Bengal basin shelf zone, considerable information and documentation of tectonic geomorphology are gained by examining drainage patterns, channel morphology and soft-sediment deformation. Epeirogenic tilting on the ancient deposits or underlying Tertiary Formations has a profound effect on channel slope as reflected from variations of stream length—gradient index between subsurface faults. Deflections of river across the faults are evident in Mayurakshi, Ajay, Damodar and Dwarkeswar Rivers, showing frequent changes of active channel and linear orientation over the floodplains. Low-gradient rivers (mainly suspended-sediment load) are especially sensitive to tectonic movements, such that even very subtle deformation can alter the courses (abandoned channels) and forms (meander cut-offs and terraces) of rivers, like Kunur and Khari. The value of standard sinuosity index is estimated nearer to unity (<1.15) in all peninsular rivers, showing the dominance of topographic control (TSI > 70%) on river sinuosity between the tectonic blocks. The index values of S_L (> 100) and H_I (0.4 to 0.6) show recurrent pulses of stream gradient due to uplift and mature to youthful stage of landform development in the sub-basins of peninsular rivers. Other parameters of basin symmetry factor and basin elongation ratio show also the seismic effect on fluvial landforms, signifying relatively moderate tectonic activity over the basin margin. The recurrent occurrences of earthquakes in and around the Bengal Basin prove the relative activeness of fault planes or scarps, which triggered low to high magnitude earthquakes in BMF, PF, GKF and EHZ in Anthropocene.

The present study has tried to explore syntectonic responses of alluvial rivers to active tectonics in the western shelf zone of the Bengal Basin. The direct implication of the above is that the Quaternary terrain that forms >75% of the total study area is vulnerable (possibility of ample liquefaction) to the effects of a medium to high magnitude earthquakes. The major towns or cities, like Asansol, Bankura, Durgapur, Bardhaman, Bolpur, Medinipur and Kolkata, are situated in the close vicinity of tectonic elements and earthquake epicentres. Apart from that, currently, DGPS (differential global positioning system)-based geodetic surveys and DDS (deep seismic sounding) seismic refraction surveys are strongly recommended in the active deformed zones of alluvial river basins. Seismic microzonation of important towns and cities can provide better information of seismic vulnerability in contagious areas and faulted blocks. Suitable measures should be taken to regularly monitor recurrent seismic activity by installing seismograph on or near the dams and barrage, which are situated adjacent to basin margin faults. There are needs for more interdisciplinary fundamental research to comprehend the effect of fault scarps on alluvial rivers or deformation of alluvial terraces and sediments using sophisticated techniques and instruments. These studies can provide indication and information about the relative activeness of seismicity in the Bengal Basin.

References

Agarwal RP, Mitra DS (1991) Paleogeographic reconstruction of Bengal Delta during Quaternary period. In: Vaidyanadhan R (ed) Quaternary deltas of India, Geological Society of India, Bangalore

Anand AK, Pradhan SP (2019) Assessment of active tectonics from geomorphic indices and morphometric parameters in part of Ganga Basin. J Mountain Sci 16 (8):1943–1961

Ayaz S, Dhali MK (2019) Longitudinal profile and geomorphic indices analysis on tectonic evidence of fluvial form, process and landform deformation of Eastern Himalayan Rivers, India. Geol, Ecol Landscapes 4(1):11–22

Azor A, Keller EA, Yeats RS (2002) Geomorphic indicators of active fold growth: South Mountain—Oak Ridge anticline, Ventura basin, Southern California. Bull Geol Soc Am 114(6):745–753

Barman SD, Islam A, Das BC, Mandal S, Pal SC (2019) Historical evidences in the identification of Palaeochannels of Damodar River in Western Ganga-Brahmaputra Delta. In: Das BC, Ghosh S, Islam A (eds) Quaternary geomorphology in India. Springer, Switzerland, pp 105–126

Bhattacharaya AK, Dhar N (2003) A report on geoenvironmental appraisal in Durgapur—Panagrah urban area and its environments for developmental activities within Asansol-Durgapur Development authority area, District Burdwan, west Bengal. Geological Survey of India, Calcutta

Bull WB (1978) Geomorphic tectonic activity classes of the south front of the San Gabriel Mountain. Geosciences Department, University of Arizona, Arizona, California

Bull WB, McFadden LD (1977) Tectonic geomorphology north and south of the Garlock fault, California. In: Doehring DO (ed) Geomorphology in Arid Regions. State University of New York, Binghampton, pp 115–138

Chandra U (1977) Earthquakes of peninsular India—a sesismotectonic study. Bull Seismol Soc Am 67 (5):1387–1413

Cox PT (1994) Analysis of drainage-basin symmetry as a rapid technique to identify areas of possible Quaternarytilt-block tectonics: An example from the Mississippi Embayment. GSA Bulletin 106(5):571–581

Das A, Bhattacharya F, Rastogi BK, Chauhan G, Ngangom M, Thakkar MG (2016) Response of a dryland fluvial system to climatic – tectonic perturbations during the Late Quaternary: evidence from Rukmawati River Basin, Kutch, western India. J Earth Syst Sci 125(6):119–1138

Das S (2020) Koyna-Warna shallow seismic region, India: is there any geomorphic expression of active tectonics? J Geol Soc India 96:217–231

Das Gupta AB, Mukherjee B (2006) Geology of N.W. Bengal Basin. Geol Soc India, Bangalore

De S, Ghosh D, Ghatak TK, Chakraborty K (1994) Compilation of Bouguer anomaly map in degree sheet 73M. Geological Survey of India, Calcutta

Deb K, Singh M (2011) Report on the PGRS input to the thematic mapping in parts of Bankura-Bardhaman districts, West Bengal for ground characterization of lineaments / faults in the context of neotectonic activity in the area keeping in view the recent seismic activity in the area. Geological Survey of India, Kolkata

Dennis JG (1972) Structural geology. McGraw-Hill Kogakusha, Tokyo

Dey S, Powali D, Chaudhury J, Ghosh M, Mondal R, Kanaujia J, Mitra S (2019) 28 August 2018 (Mw 4.5) Bengal Basin earthquake highlights active basement fault beneath the sediments. Curr Sci 116(10):1633–1636

Dubey RK, Shankar R (2019) Drainage pattern and its bearing on relative active tectonics of a region: a studyin the son valley, Central India. J Geol Soci India 93:693–703

Ghosh BK (1992) Quaternary geological and geomorphological mapping in the interfluves area of Damodar and Rupnarayan rivers in parts of Bankura and Midnapur districts, West Bengal. Geological Survey of India, Calcutta

Ghosh S (2014) Palaeogeographic significance of ferruginous gravel lithofacies in the Ajay—Damodar interfluve, West Bengal, India. Int J Geol, Earth Environ Sci 4(3):81–100

Ghosh S, Mistri B (2012) Hydrogeomorphic significance of sinuosity index in relation to river instability: a case study of Damodar River, West Bengal, India. Int J Adv Earth Sci 1(2):49–57

Ghosh S, Guchhait SK (2018) Modes of formation, palaeogene to early quaternary palaeogenesis and-geochronology of laterites in rajmahal basalt traps and Rarh Bengal of Lower Ganga Basin. In: Das BC, GhoshS, Islam A (eds), Quaternary Geomorphology in India. Springer, Switzerland, pp 25–60

Ghosh S, Shivakumar R (2019) An assessment of geomorphometric anomalies and their significance on the seismotectonic activity through geoinformatics. J Earth Syst Sci 128:178. https://doi.org/10.1007/s12040-019-1175-9

Goodbred SL, Kuhel SA (2000) The significance of large sediment supply, active tectonism, and eustasy on margin sequence development: late Quaternary stratigraphy and evolution of the Ganges-Brahmaputra delta. Sed Geol 133(3–4):227–248

Goswami PK (2012) Geomorphic evidences of active faulting in the northwestern Ganga Plain, India:

implications for the impact of basement structures. Geosci J 16(3):289–299
Govindaraju L, Bhattacharya S (2012) Site-specific earthquake response study for hazard assessment in Kolkata city, India. Nat Hazards 61:943–965
Hamdouni RE, Irigaray C, Fernández T, Chacón J, Keller EA (2008) Assessment of relative active tectonics,southwest border of the Sierra Nevada (southern Spain). Geomorphology 96:150–173.
Hack JT (1973) Stream profile analysis and stream gradient index. J Res United States Geol Surv 1 (4):421–429
Hare PH, Gardner TW (1985) Geomoprhic indicators of vertical neotectonisim along converging plate margins, Nicoya Peninsula, Costa Rica. In: Morisawa M, Hack JT (eds) Tectonic Geomorphology. Allen and Unwin, Boston, pp 75–104
Holbrook J, Schumm SA (1999) Geomorphic and sedimentary response of rivers to tectonic deformation: a brief review and critique of a toll for recognizing subtle epeirogenic deformation in modern and ancient settings. Tectonophysics 305:287–306
Hossain MS, Khan MSH, Chowdhury KR, Abdullah R (2019) Synthesis of the Tectonic and Structural Elements of the Bengal Basin and its surroundings. In: Mukherjee S (ed) Tectonics and structural geology: Indian context. Springer Geology, Switzerland, pp 135–218
Hossain MS, Xiao W, Khan MSH, Chowdhury KR, Ao S (2020) Geodynamic model and tectono-structural framework of the Bengal Basin and its surroundings. J Maps 2:445–458
Jain V, Sinha R (2005) Response of active tectonics on the alluvial Baghmati River, Himalayan foreland basin, eastern India. Geomorphology 70:339–356
Kaila KL, Reddy PR, Mall DM, Venkateswarlu N, Krishna VG, Prasad ASSSRS (1992) Crustal structure of the west Bengal Basin, India from deep seismic sounding investigations. Geophys J Int 111(1):45–66
Kale VS, Shejwalkar N (2008) Uplift along the western margin of the Deccan Basalt Province: is there any geomorphometric evidence? J Earth Syst Sci 117 (6):959–971
Kayal JR (2008) Microearthquake seismology and seismotectonics of South Asia. Springer, Dordrecht
Keller EA, Pinter N (2002) Active tectonics; earthquakes, uplift and landscape. Prentice Hall, New Jersey
Keller EA (1986) Investigation of active tectonics: use of surficial earth processes. Act Tectonics. Impact on Society, National Academy Press, Washington DC, 136–147
Khan AA, Chouhan RKS (1996) The crustal dynamics and the tectonic trends in the Bengal Basin. J Geodyn 22(3–4):267–286
Leeder MR, Alexander J (1987) The origin and tectonic significance of asymmetrical meander-belts. Sedimentology 34(2):217–226
Litchfield NJ, Van Dissen R, Sutherland R, Barnes PM, Cox S, Norris R et al. (2013). A model of active faulting inNew Zealand: fault parameter descriptions. GNS Science Report 2012/19, retrieved from:https://www.gns.cri.nz/static/pubs/2012/SR%202012-019.pdf
Mahapatra S, Dana RK (2009) Lateral variation in gravelly sediments and processes in an alluvial fan-fan-delta setting, north of Durgapur. J Geol Soc India 74:480–486
Mahata HK, Maiti R (2019) Evolution of Damodar fan delta in western Bengal Basin, West Bengal. J Geol Soc India 93:645–656
Mahmood SA, Gloaguen R (2014) Appraisal of active tectonics in Hindu Kush: insights from DEM derived geomorphic indices and drainage analysis. Geosci Front 3(4):407–428
Malik JN, Mohanty C (2007) Active tectonic influence on the evolution of drainage and landscape: geomorphicsignatures from frontal and hinterland areas along the Northwestern Himalaya, India. J Asian Earth Sci 29:604–618
Montenat C, Barrier P, Estevou PO, Hibsch C (2007) Seismites: an attempt at critical analysis and classification. Sed Geol 196:5–30
Mueller JE (1968) An introduction to the hydraulic and topographic sinuosity indexes. Ann Assoc Am Geogr 58(2):371–385
Mukherjee A, Fryar AE, Thomas WA (2009) Geologic, geomorphic and hydrologic framework and evolution of the Bengal Basin, India and Bangladesh. J Asian Earth Sci 34:227–244
Nath S, Adhikari MP, Devaraj N, Maiti SK (2015) Seismic vulnerability and risk assessment of Kolkata city, India. Nat Hazards Earth Syst Sci 15:1103–1121
Nath S, Adhikari MP, Maiti SK, Devaraj N, Srivastava N, Mohapatra LD (2014) Earthquake scenario in west Bengal with emphasis on seismic hazard microzonation of the city of Kolkata, India. Nat Hazards Earth Syst Sci 14:2347–2375
Nath SK, Srivastava N, Ghatak C, Adhikari MD, Ghosh A, Sinha Ray SP (2018) Earthquake induced liquefaction hazard, probability and risk assessment in the city of Kolkata, India: its historical perspective and deterministic scenario. J Seismol 22:35–68
Nath SK, Thingbaijam W, Vyas SC, Sengupta P, Dev SMSP (2010) Macroseismic—drive site effects in the southern territory of West Bengal, India. Seismol Res Lett 81(3):480–487
Niyogi D, Sarkar SK, Mallick S (1970) Geomorphic mapping in plains of West Bengal, India. In: Chatterjee SP, Das Gupta SP (eds) Selected papers (vol. 1)—Physical Geography 21st International Geographical Congress. National Committee for Geography, Calcutta, pp 89–94
Ouchi S (1985) Response of alluvial rivers to slow active tectonic movement. Geogr Soc Am Bull 96:504–515
Raj A, Nath SK, Thingbaijam KKS (2008) A note on the recent earthquakes in the Bengal Basin. Curr Sci 95 (9):1127–1129
Reddy PR, Venkateswarlu N, Prasad ASSSRS, Rao K (1993) Crustal density model across West Bengal basin, India: an integrated interpretation of seismic

and gravity data. Proc Indian Acad Sci—Earth Planet Sci 102(3):487–505

Roy AB, Chatterjee A (2015) Tectonic framework and evolutionary history of the Bengal Basin in the Indian subcontinent. Curr Sci 109(2):271–279

Roy S (2019) Influence of faulting on the extra-channel geomorphology of the Ajay—Damodar interfluves in Lower Ganga Basin. In: Das BC, Ghosh S, Islam A (eds) Quaternary geomorphology in India. Springer, Switzerland, pp 79–88

Roy S, Bera S (2019) Geophysical control on the channel pattern adjustment in the Kunur River Basin of western part of Lower Ganga Basin. In: Das BC, Ghosh S, Islam A (eds) Quaternary geomorphology in India. Springer, Switzerland, pp 89–104

Roy S, Sahu AS (2015) Quaternary tectonic control on channel morphology over sedimentary low land: a cae study in the Ajay—Damodar interfluves of eastern India. Geosci Front 6:917–946

Rudra K (2018) Rivers of the Ganga—Brahmaputra—Meghna Delta. Springer, Switzerland

Sahu S, Raju NJ, Saha D (2010) Active tectonics and geomorphology in the Sone-Ganga alluvial tract in mid-Ganga Basin, India. Quat Int 227(2):116–126

Schumm SA (1956) The evolution of drainage system and slopes in badlands at Perth Amboy, New Jersey. Bull Geol Soc Am 67:597–646

Schumm SA (1986) Alluvial river response to active tectonics. In: Wallace RE (ed) Active Tectonics. National Academic Press, Washington D.C., pp 80–94

Schumm SA, Dumont JF, Holbrook JM (2002) Active tectonics and alluvial rivers. Cambridge University Press, Cambridge

Seeber L, Gornitz V (1983) River profiles along the Himalayan arc as indicators of active tectonics. Tectonophysics 92:335–367

Seilacher A (1969) Fault-graded beds interpreted as seismites. Sedimentology 13:155–159

Sengupta S (1966) Geological and geophysical studies in western part of Bengal basin, India. Bull Am Assoc Pet Geol 50(5):1001–1017

Shanmugam G (2016) The Seismite problem. J Palaeogeogr 5(4):3108–3362

Shivgotra V, Deb K, Singh M (2011) Thematic mapping in parts of Bankura—Burdwan districts West Bengal for ground characterization of lineaments/faults in the context of neo-tectonic activity in the area keeping in view the recent seismic activity in the area. Geological Survey of India, Kolkata

Singh LP, Prakash B, Singh AK (1998) Evolution of lower Gangetic Plain landforms and soils in West Bengal, India. CATENA 33(2):75–104

Singh R, Khan PK, Singh AP (2020) Earthquake source dynamics and kinematics of the Eastern Indian Shield and adjoining regions. Acta Geophys 68:337–355

Sinha R, Ghosh S (2012) Understanding dynamics of large rivers aided by satellite remote sensing: a case study from Lower Ganga plains, India. Geocarto Int 27(3):207–219

Steckler MS, Mondal DR, Akhter SH, Seeber L, Feng L, Gale J, Hill EM, Howe M (2016) Locked and loading megathrust linked to active subduction beneath the Indo-Burman Ranges. Nat Geosci 9:615–618

Strahler AN (1952) Hypsometric (area-altitude) analysis of erosional topography. Bull Geol Soc Am 63:1117–1142

Suter F, Martinez SI, Velez MI (2010) Holocene soft-sediment deformation of the Santa Fe-Sopetran basin, northern Colombian Andes; evidence for pre-historic seismic activity. Sed Geol 235(3):188–199

Valdiya KS (1984) Aspects of tectonics. Tata McGraw-Hill Publishing Company Limited, New Delhi, Focus on South-Central Asia

Valdiya KS (1999) Why does river Brahmaputra remain untamed? Curr Sci 76(10):1301–1305

Valdiya KS (2016) The making of India; geodynamic evolution. Springer, Switzerland

Verma RK (1985) Gravity field. D. Reidel Publishing Company, Dordrecht, Seismicity and tectonics of the Indian Peninsula and the Himalayas

Yeats RS, Sieh K, Allen CR (1997) The geology of earthquakes. Oxford University Presses, Oxford

Check for updates

24 Morphological Landscape Mapping of the Bhagirathi Flood Plains in West Bengal, India, Using Geospatial Technology

Ismail Mondal and Jatisankar Bandyopadhyay

Abstract

The Bhagirathi, Mathabhanga-Churni, and Ichamati rivers in West Bengal are characterized by frequent riverbank erosion and planform changes along with their courses. This study examines the morphodynamic changes of river Bhagirathi, Churni, Mathabhanga, and Ichamati considering the relationship between land use and land cover and geomorphology during the last 43 years' time span using geospatial technology. The result shows a remarkable change in bank line positions and their geometry over time. The result also portrays that the bank line is unstable and subject to migration incessantly. The degree of sinuosity increases from its previous river course (1970) to the present course for all rivers, with an abnormal nature. Besides, the study also found that among the river channels under consideration, the Bhagirathi River is drastically changing the aspects and shifted to the river course toward the east. Also, river channels are subjected to excessive meandering and migration owing to various natural and anthropogenic phenomena including LULC changes along with the river courses. The outcome of this study obtained from processing satellite imagery and Survey of India (SoI) Toposheets is helpful to comprehend the channel planform dynamics and to formulate policy recommendations for future development works.

Keywords

Channel shifting · Sinuosity index · Land use and Land cover · Geographic information science · Nadia rivers

24.1 Introduction

The fluvial system is taken into account as the most prominent geomorphic system on the earth's surface, which is continuously reshaping the earth's surface (Morisawa et al. 1985) and evolving the morphology of rivers and flood plains. Meandering is the most common natural process in a flood plain and alluvial river system, which may be a driver to environmental change, and causing significant changes within the riverine marginal areas, which is considered as obvious signs of evolution in river shape (Yousefi et al. 2018; Ahmed and Fawzi 2009), Meandering leads to the lateral migration process to succeed morphological adjustments to accumulate the range of flows and sediment loads

I. Mondal (✉)
School of Oceanographic Studies, Jadavpur University, Kolkata-700032, India

J. Bandyopadhyay
Department of Remote Sensing and GIS, Vidyasagar University, Midnapore 721102, India

P. K. Shit et al. (eds.), *Drainage Basin Dynamics*, Geography of the Physical Environment,
https://doi.org/10.1007/978-3-030-79634-1_24

from upstream, which may cause catastrophic local or regional changes. When rivers reach the mature stage, they become sluggish and form meander bends. These oscillations cause massive riverbank erosion (Debnath et al. 2017). Most of the world's rivers are the issues of meandering due to their natural and also human actions and that they migrate to river course through the basin topography (Ayman et al. 2009; Bag et al. 2019), which may alter surrounding land use and land cover characteristics. Changes in land use due to river dynamics may escort the socio-economic discrepancy by plummeting crop yield, infrastructural damage, and severe loss of livelihood of the poor within the society and socio-economic inequality (Bhunia et al. 2016). Change in land use reflects the history and, therefore, the way forward for human decisions on watershed systems (Yousefi et al. 2018). The similar tune has been echoed in the works of Islam et al. (2021). They have portrayed the change in LULC of an island in Bangladesh using geo-spatial techniques. The study of Miah et al. (2020) also reflects the landscape morphology mapping for preparing the coastal vulnerability map in Bangladesh using GIS techniques.

The Bhagirathi-Hooghly Riverine system is the western-most of major distributaries within the Ganga delta (Bag et al. 2019). On the eastern side, the other distributaries, i.e. Bhairab, Sialmari, Jalangi, and the Mathabhanga-Churni-Ichamati rivers' system, are located. The rivers of Bhagirathi, Jalangi, and Mathabhanga-Churni-Ichamati are commonly referred to as "Nadia Group of Rivers" up to the Bangladesh border (Mondal et al. 2016a; Bag et al. 2019). The course of Bhagirathi and its tributaries river within the study area is characterized by dynamic and irregular meandering and lateral divergence. For example, Islam and Guchhait (2017a) showed that the lower stretch of the Bhagirathi River extending from Katwa to Nabadwip is the most unstable stretch detected through the rapid changes of the planform characteristics especially the meander geometry. Moreover, Islam (2010) found the role of fluctuating river regime due to the controlled hydrology of the Farakka Barrage Project (FBP) in the lean months in channel instability and bank erosion of the Bhagirathi. Similarly, Guchhait et al. (2016) further extended the study and explained the changes in channel and flood characteristics not as the sole response to the construction of the FBP, but the contribution of the Ajay-Mayurakshi system in the form of huge monsoonal discharge and coarse-grained sediment and the differential flood plain sediment (e.g. hard clay on the western bank of the Bhagirathi and erosion permitting sand and silt on the eastern bank). Besides, Islam and Guchhait (2020) established that channel instability of this region is the outcome of the transport limited condition of the Bhagirathi River contributed by the higher sediment load compared with the lesser amount of discharge. Thus, due to this the channel instability, adjacent areas are subjected to incessant erosion, which often causes significant damage to cropland, riverside land, livestock, and also the human livelihood (Islam and Guchhait 2017b; Islam and Guchhait 2018). However, integrating channel instability with the changes in the basin-scale dynamics of the LULC and river basin management is a vital issue that has not been explored adequately in the context of the Bhagirathi flood plains.

Considering the character of the river course and the importance of surrounding land cover changes due to channel changes, this study focuses on the inter-relationship between land use and land cover and geomorphology vis-a-vis shifting river bank line of the study area. Besides, this study also intends to carry out a deep investigation of the morphological aspects such as changes in channel geometry, and channel instability with reference to river width, depth, velocity, and discharge with an eye to propose a rational supervision for reducing erosion hazards and protecting the riverbank. Since rivers play such an important role in morphological activities, the main goal of this research is to look into the morphological aspects of the Bhagirathi River. In addition, the changing courses of rivers and their effects have been analyzed. The main objectives of the study are to investigate the trend of river channel migration and its unpredictability over time and space and to find out the major changes in the Ichamati, Mathabhanga, Churni, and Bhagirathi rivers in terms of bank line positions, impact of channel dynamics on the land use and land cover of the flood plain since historical past to present. Finally, using remote sensing and GIS techniques, this study has been carried, which will serve as a model for the long-term management of bank erosion hazard.

24.2 The Study Area

24.2.1 Locational Attributes

The study area lies in the Ganga-Brahmani interfluves area in the Krishnanagar subdivision of Nadia district, India. The entire study area is extended between 23^0 0′ N to 49° N latitude and 88° 9′ to 88° 48′ E longitude and includes the area ~701 sq. miles (Fig. 24.1). The present study area is a portion of the Bengal Basin (Mukherjee 1938, 2012; Bag et al. 2019), deliberated under fluvial landscape dominated by active and older floodplain (Bagchi et al. 1944, 1972) and characterized by a gentle slope and also the low elevation in <5–35 m, migrating the riverine channels, scrolls meander, natural leaves, floodplain formation and also oxbow lakes, etc. In the West, Bhagirathi is considered as the westernmost major distributary channel of the Ganga delta and interconnected via the Jalangi river links to the Bhagirathi River opposite of Nabadwip and lower part of Santipur. In the eastern riverine system and its major distributaries, i.e. Bhairab, Sialmari, Jalangi, and also the Mathabhanga-Churni-Ichamati rivers' system is (generally known as "Nadia Group of Rivers") up to the Bangladesh border (Mondal et al. 2016a; Bag et al. 2019).

24.2.2 Hydro-Geomorphic Settings

The Bhagirathi river has conversant a figure of fluctuations in the situation of branch-off facts, apart from the parental Ganga river, between Suti and Jangipur, Geria, and Mithipur. The earlier river channel of Bhagirathi, the present area has been represented through the mature meander and scrolls. In another branch-off fact perhaps wherever in the eastern side of the Jangipur Farakka Barrage, while, it remains not so well well-defined as a large- scale of human intrusion has demolished the present landforms (Basu 1965, 1968). Though, the present research studies on palaeogeographic and morphological development of the Ganges delta (Niyogo 1975; Umitsu 1993; Acharyya et al. 1999) specify that the Bhagirathi-Hooghly river system has been shifted in the direction of the eastern side through space and time by the progress of the estuaries of the peninsular riverine systems (i.e. Damodar, Ajay, and Mayurakshi) with coupled of neotectonic actions. Therefore, a tapered strip of the low-lying landscape (Matured Alluvial Plain) has developed toward the western side of the present river boundary mature of Alluvial Plain. The sparse marshy landformsare located in the north-south direction (cyclical water bodies and also wetlands, meander scrolls, cut-off meandering bend), "*bils/jhils*" at the western portion of the current river of Bhagirathi is an alluvial lowland are in the similar position of the current N-S conventional river course of the river of Damodar (additional towards downstream of the Bhagirathi-Hooghly River) it can be measured as a broad lineament feature and with all prospect, it has the ancient river course of the Ganga river at all period earlier 800 (Roy 1952). Meandering (scroll) zone serpentine of the Bhagirathi-Hooghly River among the Samudragarh-Dhatrigram expanse of Burdwan district and its adjacent portion of the district of Nadia might be the image in the variation in riverine discharge and the tidal river flow of the Hooghly estuary (Basu 1965), and surface expression of the normal fault is consecutively over the Burdwan to Jalangi alluvial plain area (Sengupta 1966). The Bhagirathi-Hooghly River of the eastern part is a low-lying part of the alluvial region (The Ichamati river interfluves among the Bhagirathi-Hooghly riverine system (Mondal et al. 2016b) by the flowing physiognomies of northern to southern part is exhausted via the "Nadia Group of Rivers," i.e. Churni, Mathabhanga, Bhairab, Jalangi, and Ichamati, etc. The middle part is exhausted in the Jamuna River, a distributary channel of the Bhagirathi River and its tributary

Fig. 24.1 Study area

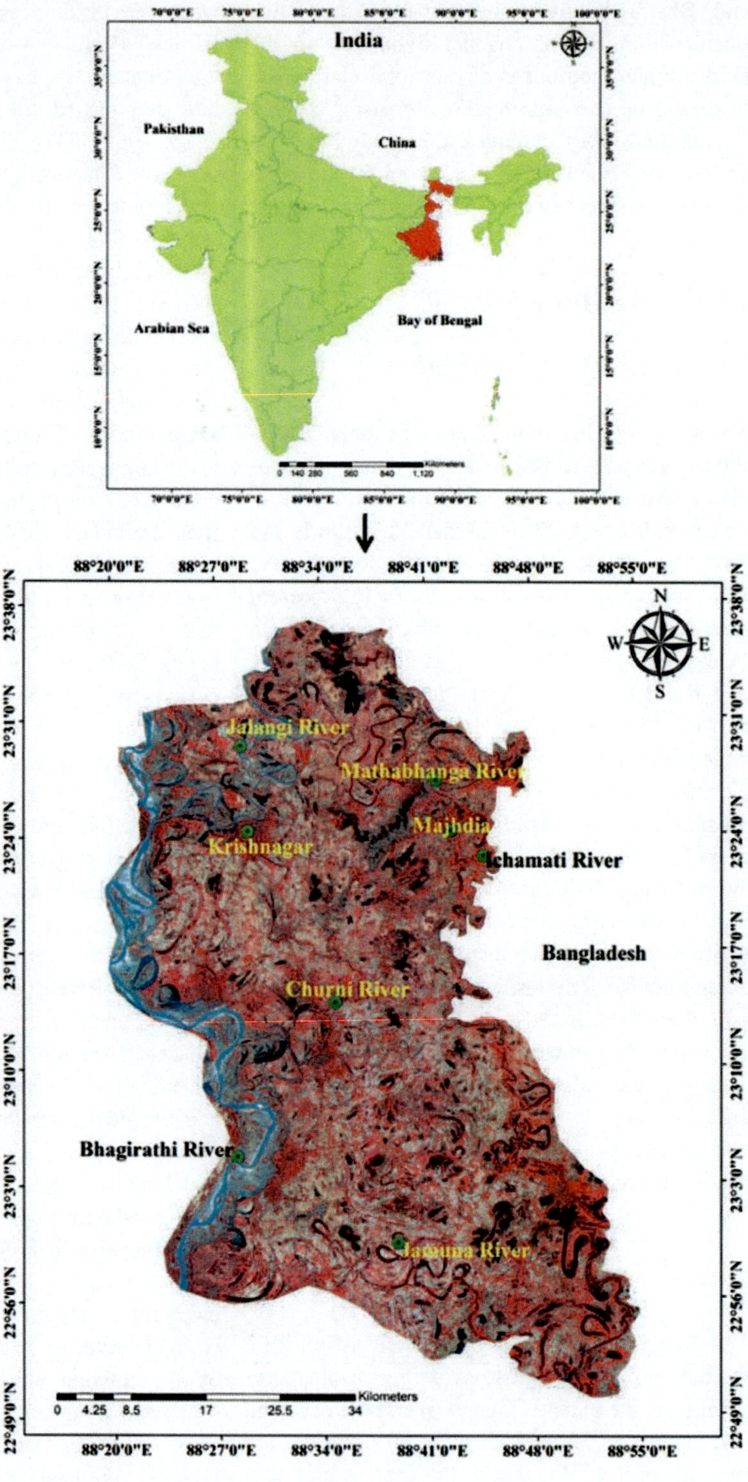

to the Ichamati River (Mondal et al. 2016b). The River Mathabhanga and Churni system have jointly produced the Ichamati River section near Majdiah (Nadia District) and flow through the delta plain of Ganga to meet to the southward direction and fallen of Bay of Bengal (Fig. 24.2).

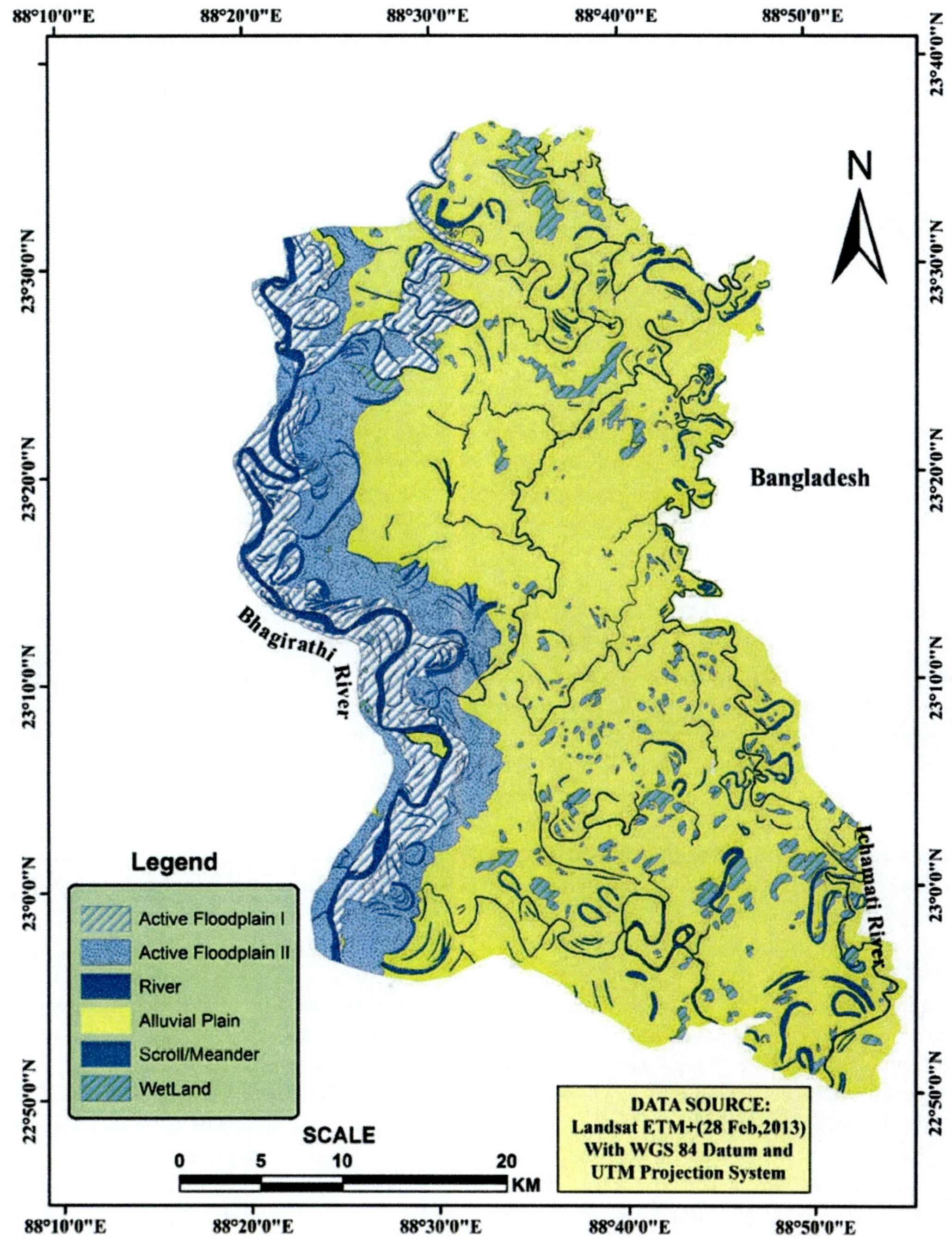

Fig. 24.2 Geomorphology map

The whole district is a network of moribund rivers and streams, but the Bhagirathi, Jalangi, and Matabhanga Rivers have been for more than an era, and it is still being characteristically acknowledged as the "Nadia groups of Rivers." Till the introduction of the Indian railways, these waterways afforded the consistent resources of communication among the upper basin of the Ganges and the seacoast. Forever subsequently the British (East India Company) work of the era, considerable effort has been knowledgeable in observance of them exposed for the riverine navigation during the decade. In 1781, Chief Rennel documented that they were not typically navigable in the dry (summer) period. Captain Colebrook in a description highlighted the ancient course of the Ganges (1797), inscribes so "the Bhagirathi, Mathabanga, Jalangi, and Ichamati are not navigable during the dry time." There have been reports of all rivers remaining exposed in turn throughout the dry season. The Jalangi, Churni, Mathabanga, and Ichamati were used earlier to be navigable throughout the entire or greater portion of the year (Garrett 1910).

The river Bhagirathi was navigable in the dry period of 1796 (Rennel 1781). The Mathabhanga River when surveyed in1975 by the Survey of India was navigable during the dry period for vessels of a moderate problem. In the initial nineteenth century, Mathabhanga seems to have been more effortlessly navigable than, moreover, the Bhagirathi to have sustained open each year from 1809 to 1818 (Garrett 1910). In 1813, measures were taken towards progressing its meander channel, and a toll was established to discharge the expenditure of the effort. It seems to have caused by the exertion than complete, in 1818 "the obstacles had developed so various and hazardous, as to the reason the collision of numerous vessels, and to need heavy victims on the interpretation of demurrage paid for the dis-chargeof boats to come predictable loads (Garrett 1910; Rennell 1764–66;1910).

The suppliers of Calcutta in that year immediately appealed government that stages should be taken for alleviating a wicked from that which of the commercial concern hurt too rigorously." The present study belongs to the tropical monsoon climate zone (June to August). The major soil of this region is the mature alluvial source, transported down by the Ganga, and its distributaries channels are the Bhagirathi, Jalangi, and the western rivers of Bhagirathi, Ajay, etc. The river embankment is characterized by sandy soil where the mixed clay layer exceeds the non-synthetic sand layer, which may increase the vulnerability of the risk of riverbank erosion. The morphology of the river system has been studied from 1970 to 2013 with a total span of time of 43 years to identify the major changes along with the course of the stream.

Now, the soils of the Bhagirathi floodplains especially of Nadia and North 24 Parganas of West Bengal are found in a different grouping of soils including alluvial soils and the coastal soils. The study area is an alluvial region primarily dependent on the agricultural pursuit. The present study area portrays the nine soil types according to NBSS and LUP guidelines. They include W050 to W061 (Fine, Fine loamy, Fine silty, Fluventic Ustifluvents, Aeric Haplaquepts, and Aquic Ustifluvents). Basically, the soils of this region are characterized by deep, poorly drained, fine soils occurring on the level to nearly level meander plain with loamy surface and moderate flooding associated with very deep, poorly drained, and fine and coarse loamy soils (Fig. 24.3).

24.3 Materials and Methodology

24.3.1 Database

The remotely sensed data have been used, which provided a synoptic view of the larger regions over different periods of time, and with better accuracy. It has been possible to study river channel changes and their effects on land use in a less time consuming and cost-effective manner. The Landsat MSS, TM, and ETM + time series data are used in the present study (Table 24.1).

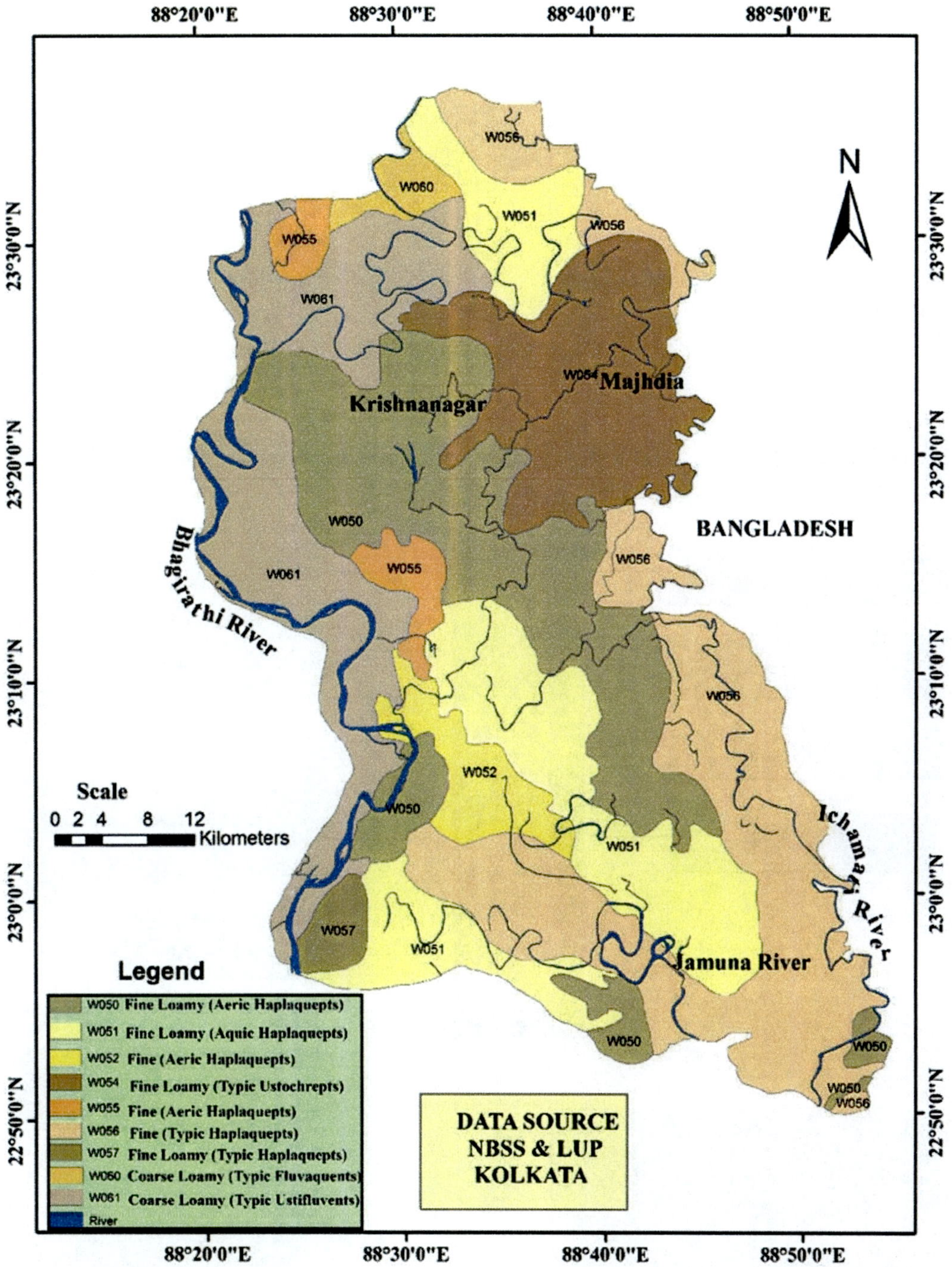

Fig. 24.3 Soil map

The time period for his study is 43 years, and all quantitative measurements have been done during this time period. Survey of India (SOI) Toposheet and Google earth image also used to be considered for the baseline map (Fig. 24.4).

Table 24.1 Satellite data information

Data	Date of acquisition	Satellite/Sensor	Path/Row/Scale
Collateral Data	1970	SoI Toposheet	1:50,000
Remote Sensing Data	28 February 1975	Landsat-2 MSS	138/43 and 44
	27 February 1990	Landsat-5 TM	138/43 and 44
	28 February 2013	Landsat-7 ETM+	138/43 and 44
	28 February 2013	Google earth image	2.5 m

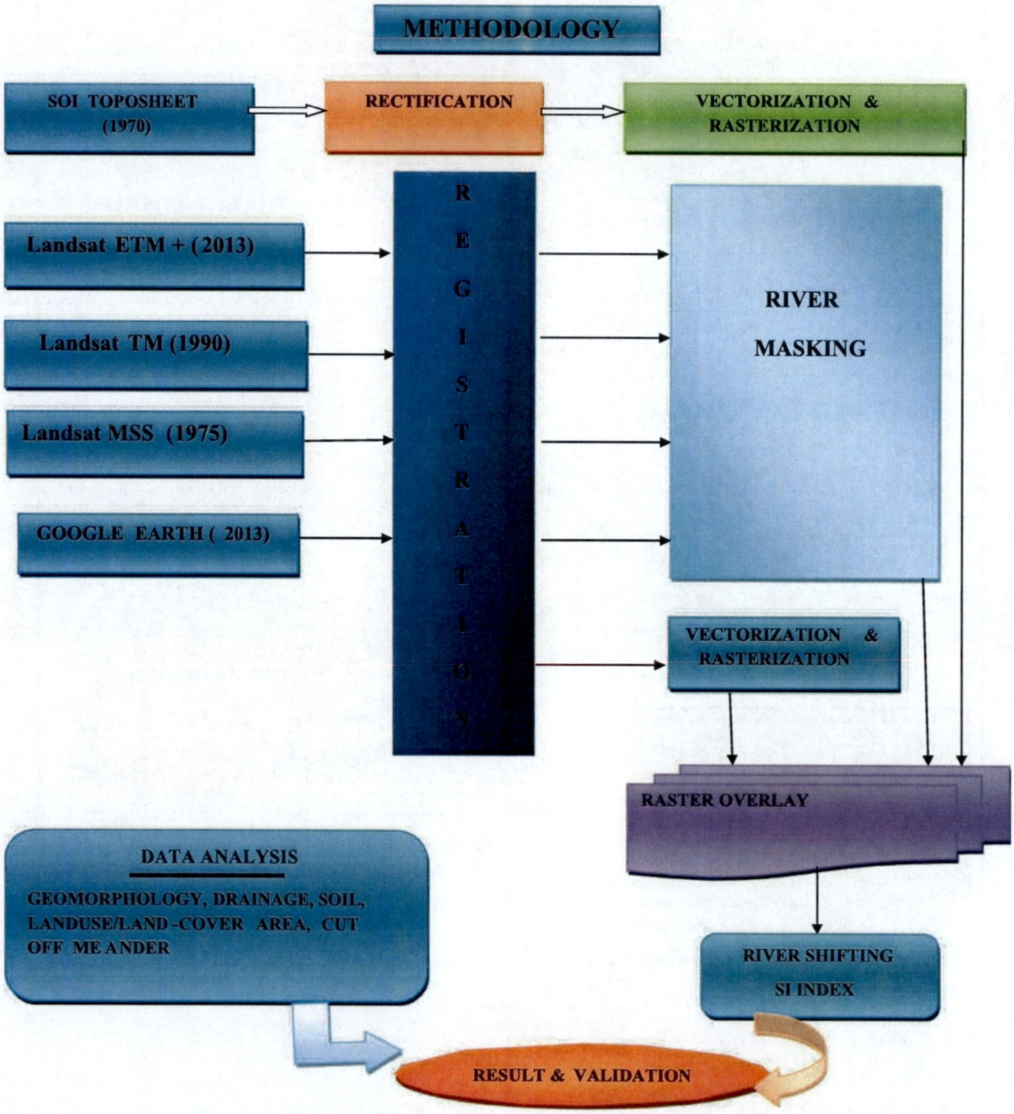

Fig. 24.4 Flowchart of the methodology

24.3.2 Methodology

The present study has been carried out using a systematic methodology from collecting the data to validating the results (Fig. 24.4).

24.3.2.1 Software and Data Processing

To avoid the river regime effects, all the satellite images were collected for the same month for better estimation. The Toposheet, Google earth image, and block map have been rectified using ArcGIS 10.5 software and then all the satellite images are registered with respect to the SOI Toposheet (1: 50,000) and ground-truthing information. All the satellite images were geo-referenced based on Universal Transverse Mercator (UTM) projection system with the World Geodetic System (WGS) datum and North 45 zone (Table 24.1). Then the resampling and geometric correction have been done in order to the same scale using ARC GIS and ERDAS Imagine software was used for the multi-layering map and modeling. The second-order polynomial transformation with the nearest neighbor resampling technique has been used with RMSE of the transformation, 0.26 was attained (<1-pixel error).

24.3.2.2 Channel Change Detection and Analysis

The river course has been extracted from the orthorectified satellite imageries and Google earth image based on the water boundary (as denoted the edge of the catchment) and save as a vector layer (Geodatabase files) using ArcGIS 10.5 software. Then the superimposed bank lines provide the complete channel migration configuration of the rivers, based on selected cross-section points, which was produced across various cross-sections of the river courses. Consequently, the measurement of bank erosion was estimated as the distance between the new bank line and the old bank line and then standardized by channel width. Computation of the Sinuosity index (SI) has been done using the method of Mullar's (1968). SI deals with the meandering nature of the river, it has defined whether a river channel is a conventional or meandering migration. It is a ratio between the actual length and the straight length of the river.

$$\text{SI} = \text{Channel length}/\text{Air length}$$

where the channel length is the actual path of a stream, and air length is the expected straight path of a stream. The river channel is basically divided into four types—1. Linear (SI = 1), 2. Regular (SI = 1.00–1.25), 3. Irregular (SI = 1.25–1.50), and 4. Meander (SI = > 1.50).

24.3.2.3 Land Use and Land Cover Mapping and Analysis

For the analysis of LULC change, the entire study area's landscape has been classified into cropland, a settlement with homestead orchards, sand splay, vegetation, aquaculture land, wetland pond, river, scroll/meander, and sand. The LULC characteristics of the study area were delineated based on a supervised classification method with maximum likelihood algorithm in ERDAS Imagine 2014 software. The SOI Toposheet, Google earth image, field verification, and native knowledge about the area were employed for conveying LULC classes. To execute accuracy for quantifying and checking the validity of the classification, we have taken 410 random sample points or reference points using stratified random sampling in ERDAS Imaging software. For a better accuracy assessment, a minimum of 30 reference points for each class and 250 for a whole are required (Debnath et al. 2017). A combined table was created with a combination of these reference points and the classified map. Based on this combined table, confusion matrix table was developed, which is the relationship between the classified map and also the reference Google earth data précised in an error matrix. Then omission, commission, producer accuracy, and user's accuracy were calculated based on this matrix. The overall accuracy, kappa coefficient, was measured for the 2013 LULC map. Kappa coefficient continuously ranges between 0 and 1. Then to identify the LULC pattern within the historical migration area, the temporal river course was overlaid on the LULC maps of 2013 (Fig. 24.5).

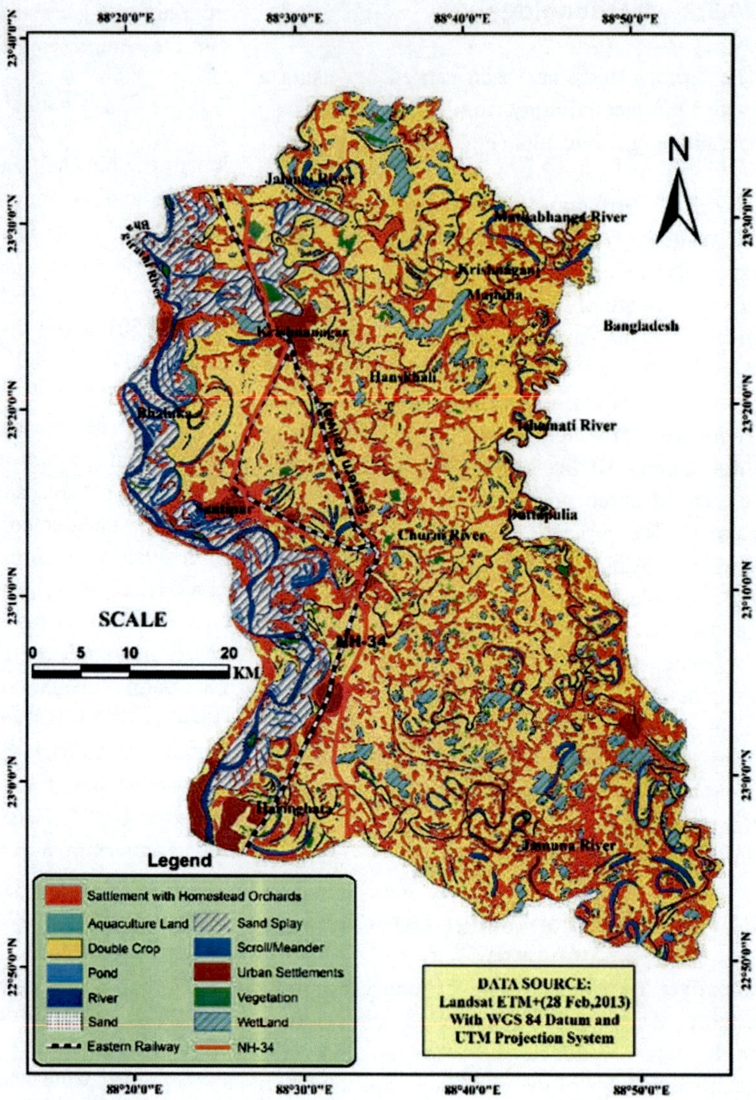

Fig. 24.5 Land use and Land cover of the study area

24.4 Results and Discussion

24.4.1 Morphodynamic Change Analysis of Bhagirathi, Churni, Mathabhanga, and Ichamati Rivers

24.4.1.1 Bhagirathi-Hooghly River

The satellite images and topographical maps show that the length of the Bhagirathi-Hooghly River is 110 km in 2013 within the study area, which was 82 km in 1970. The course of Bhagirathi is changing over time. The sinuosity index was 2.49 in 1970 and 3.46 in 2013, which indicates that the river is steadily increasing its meandering character and transforming into a meander river course. Besides the channel area also significantly changed during the study period that indicates rapid sedimentation due to increasing sinuous character. Also reveals that the lateral erosion has become increase because of the high rate of sedimentations over a 43-year time span. In the study reach, the river has widened as well as the river got in the mature stage. However, reach wise sinuosity index varies significantly. From Bankara to Piarinagra reach, river course registered maximum sinuosity is 3.47 and the minimum recorded is 1.35 in Char Sundalpur to Gosir reach. However, the remaining study section/reach has been recorded as 2.33 (Simpla to Kuturia), 2.46 (Char Sultanpur to Rashulpur Char) 1.5 (Rudrapara to Nabadwip), 1.48 (Gramkhalna to Piarinagar) during 2013 (Figs. 24.6a, 24.7a, 24.8).

24.4.1.2 River Churni

From the local history, most probably Churni was an artificial channel, not an exact river. During the seventeenth century, Maharajah Krishna Chandra the king of Nadia district was excavated this inland waterway. The length of the channel is almost 56 km and flows through Shibnivas, Hanskhali, Beranger, Haringhata, Ranaghat, and finally joins River Bhagirathi near Chakdaha. Before, there was an alternative imperative river is Anjana origin from the Jalangi River and its confluence with the River of Bhagirathi. On the present form distributary, the river channel flowed out from the Anjana adjacent in Yatrapur and also its confluence of the Ichamati River. The lower part of Mathabhanga at that time was known as Ichamati. The fluctuation of Anjana and its distributary channel rose with the inland waterway. Later on, this distributary channel was occupied artificially, and the alluvial

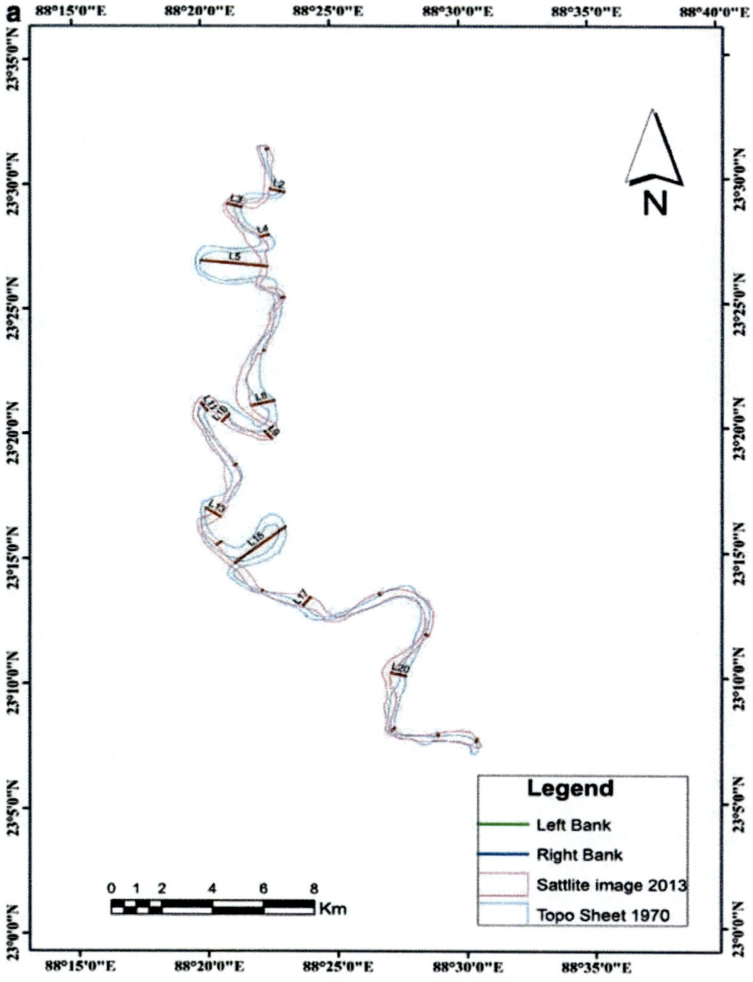

Fig. 24.6 a Location-wise cross-sections across the river Bhagirathi. **b** and **c** Location-wise cross-sections across the river Churni **(6b)** and Mathabhanga **(6c)** **d** Location-wise cross-sections across the river Ichamati

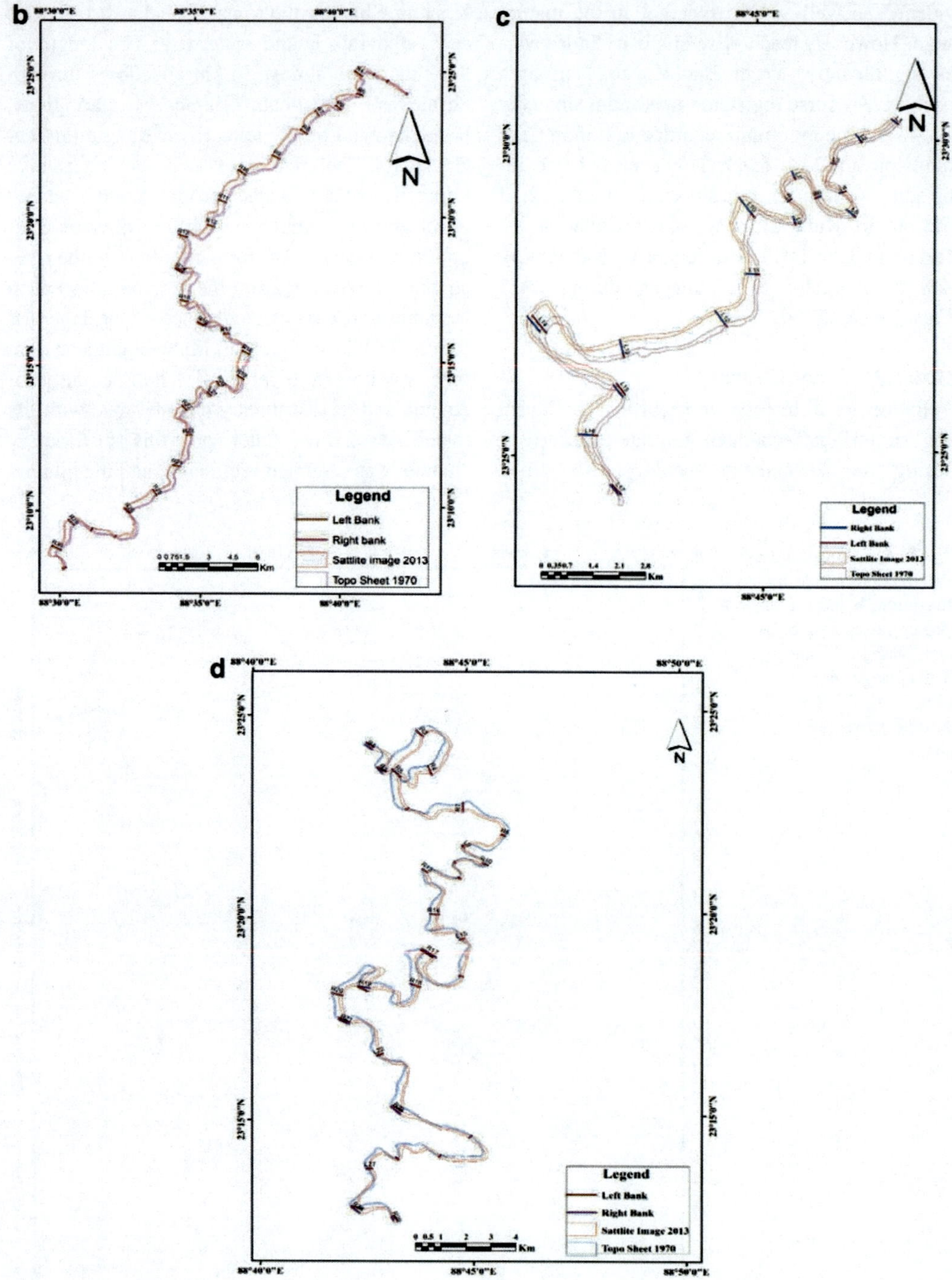

Fig. 24.6 (continued)

Fig. 24.7 a Meander river course of Bhagirathi. b Meander river course of Churni c Meander river course of Mathabhanga d Meander river course of Ichamati

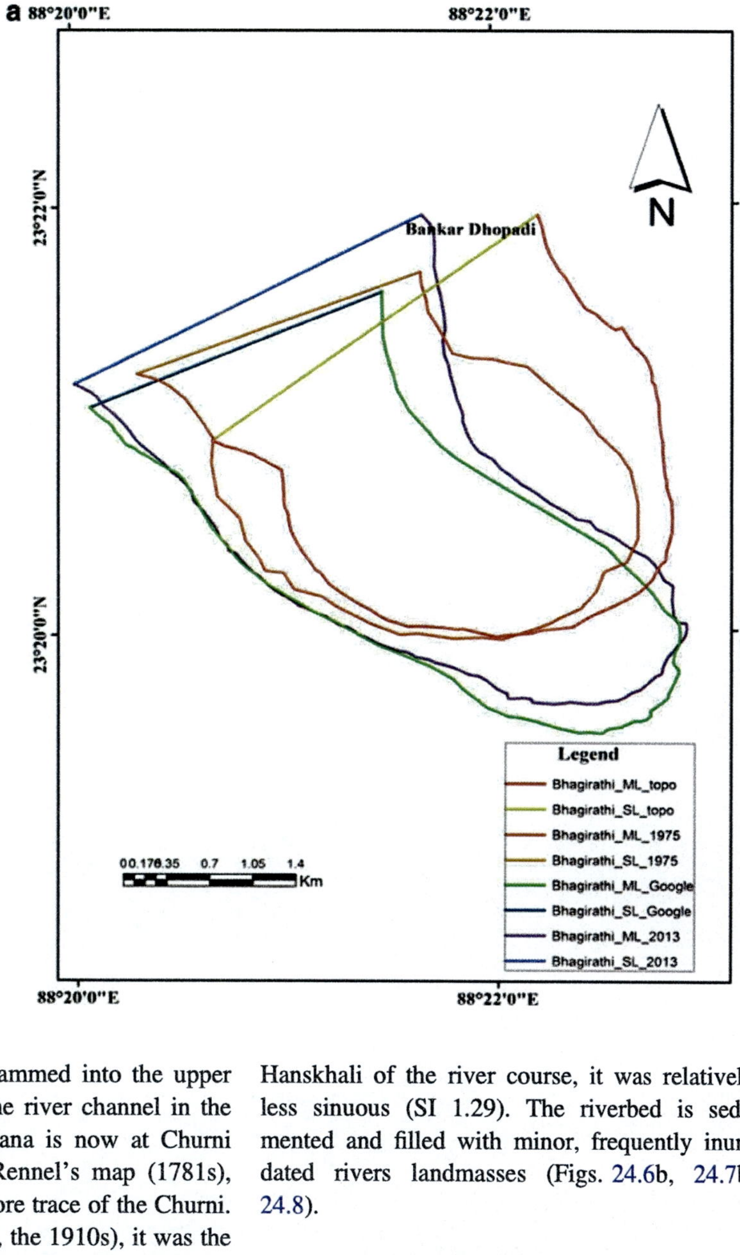

deposited sedimentation jammed into the upper portion of the Anjana. The river channel in the lower portion of the Anjana is now at Churni River. However, in the Rennel's map (1781s), there is situated and no more trace of the Churni. 31 years before (Runnel's, the 1910s), it was the ancient period major trade way privileged undivided Bengal. Present-day, the Churni has lost its navigability due to high sedimentation. The sinuosity index is 2.64 in 2013, which was 2.19 1970, retaining a meandering river course throughout the study period. Figures 24.6b, 24.7b, 24.8 shows reach-wise sinuosity index of river Churni. The river course near Byaspur registered a maximum SI value (3.25), and near Hanskhali of the river course, it was relatively less sinuous (SI 1.29). The riverbed is sedimented and filled with minor, frequently inundated rivers landmasses (Figs. 24.6b, 24.7b, 24.8).

24.4.1.3 Mathabhanga

The Mathabhanga River or Hauling leaves of the Padma almost 10 miles under the adjacent point where the Jalangi River separates from it. The river flows first in a southeastern course as it concerns Hat Boalia, where it is diverging, and one of a branch, that is subsequently known as Kumar or Pangasi, proceeds (Mathabhanga River 2017), by the way in the same direction in

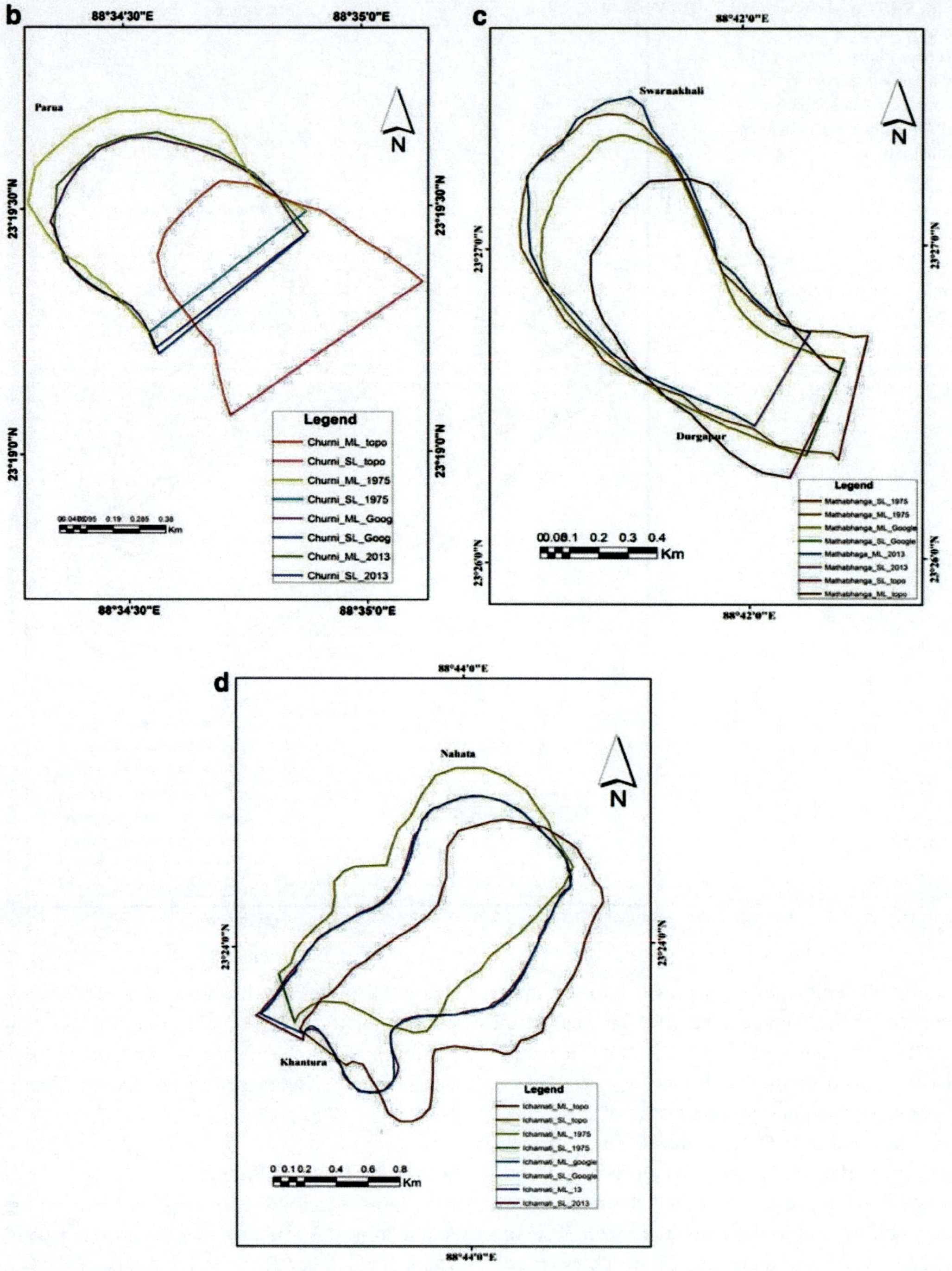

Fig. 24.7 (continued)

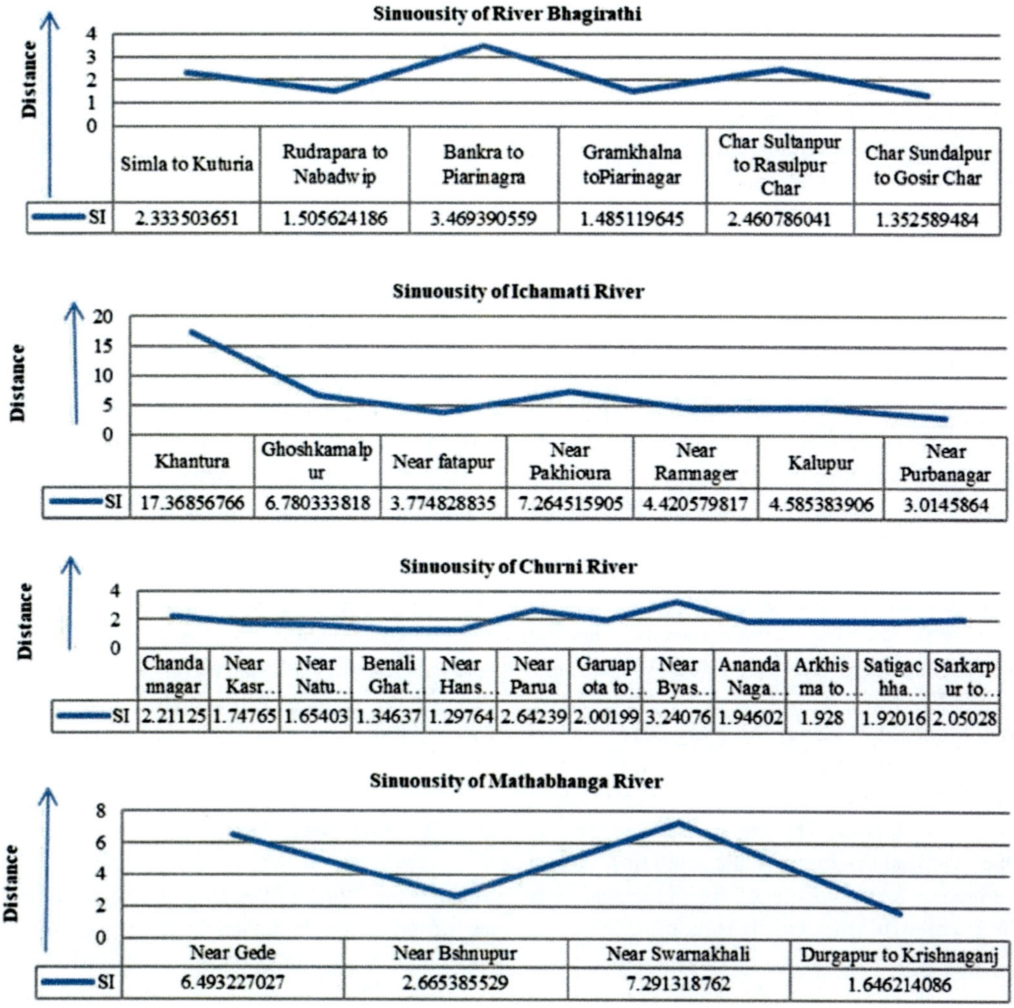

Fig. 24.8 Sinuosity index of river Bhagirathi, Ichamati, Churni, and Mathabhanga

historical Alamdanga up to the river boundary of the district that it forms for a few miles part it is passed into the Jessore, whereas the other subdivision pursuers in a very evasive course, in general movement into the south section, till, later on, transitory of huadanga, it influences of Kissengunge, east of the Krishnagar subdivision, where another divergence in to take place, and there is two consequential tributary existence known as the Churni and Ichamati interfluve, now the name of the parental river existence is lost. The SI index range from 5.9 to 7.29 during the period of 1970 to 2013 indicates that the sinuous nature of the river course increases gradually. However, the reach-wise sinuosity values vary with time. Near the Sarnakhali, the river course is very highly sinuous and is estimated as SI = 7.29, which was 5.89 in 1970, whereas from Durapur to Krishangang, river course registered minimum sinuosity (SI = 1.64) (Figs. 24.6c, 24.7c, 24.8).

24.4.1.4 Ichamati River

The Ichamati River origins from Majdia, Nadia district. The Ichamati River is a transinternational boundary that flows over India and Bangladesh and it forms into the two international boundaries among these two countries (Mondal et al. 2014,

2016a, b, c, Mondal and Bandyopadhyay 2016). Presently, the river has been faced due to the problem of a huge amount of siltation and leading to a tinny flow of water bodies in the summer season and also the inundation level is very in the rainy season. According to Mondal et al. 2014, 2016a, b, c, Mondal and Bandyopadhyay 2016), the survey report is highlighted in the present situation, and also corrective problems were studied. The Ichamati river is divided into three sections: the extended portion flows into the Mathabhanga river, a distributary channel of the Padma river, and after flowing for 208 km, it links the Kalindi river in near Hasnabad, 24 Parganas (N), another it linksto Debhata in Satkhira District, Bangladesh, and the third is the most prominent river on the western side of Dhaka (Ahmed 2012; Hirst et al. 2002). After the Ichamati River of Dinajpur in Rennel's map (1764–66), it shows the previous two ancient rivers simultaneously. According to Rennel's map (1764–66) and Mondal et al. 2014, 2016a, b, c, Mondal and Bandyopadhyay 2016), survey report is highlighted the three Ichamati rivers in the past scenario where it is a single inland waterway channel. The other river was noticeable overhead originates in the south of Jafarganj conflicting to the mouth of the Hoorsagar near the Nathpur factory and its runs headed for the Joginighat near Munshiganj. On the other hand, Joginighat is located at the confluence point of Jamuna and Ichamati. The lower portion of the Ichamati River is a divided channel from the Mathabhanga River near the Majhdia, Nadia. After its crossing in a length of the river 19.5 km on the Indian continental side, here the Ichamati moves into Bangladesh adjacent to Mubarakpur. Its movements form 35.5 km on the Bangladesh side and again it moves in India near the Duttaphulia of Nadia district. The Ichamati River forms in the two international borders of India and Bangladesh from 21 km of Angrail to Kalanchi, and it again from the Goalpara along with the Kalindi and Raimangal estuary outflow into the Bay of Bengal. Here, the Ichamati river bed share is 17 feet in and around whereas it is higher than that of Mathabhanga, Churni is also lower than the Mathabhanga by seven inches. The water level of Mathabhanga during the lean period is higher than that of the river Padma. Therefore, the Ichamati river presently faces the problem of no enters the water during the dry and summer season. Here, one of the major causes of siltation, so the river needs the construction of a guard wall for any railway and road flyover bridge. The Ichamati River's average bed shear stress is 15.6 (N/m^2) (Mondal et al. 2019), so, therefore, its requirement to be immediately excavated in direction to that there is water flow during the lean season. Later on, this is necessary to be set up a complete solution for both two India and Bangladesh, and there is a requirement for consensus on this opinion. The course of Ichamati is highly meandering in nature. The SI index is 17.36 in 2013, which was 15.42 in 1970. However, the reach-wise SI values vary significantly. Near the Kantura, about 76.90 km river course sows highly sinuous (SI = 17.20), whereas near the Purbanagar river course retain a steady-state channel (SI = 3.20) (Figs. 24.6d, 24.7d, 24.8).

24.4.2 Channel Migration

Channel migration of the river course shifting is one of the major significant characteristics of the alluvial river system (Debnath et al. 2017; Das 2013). The bank line of a river changes over time due to the sedimentation problem of the channel, which also causes the widening of a river and bank erosion. The study showing the actual river shifting of right and left banks of Bhagirathi, Churni, Mathabhanga, and Ichamati rivers during the 43 years. According to Fig. 24.9 maximum shifting of the Bhagirathi, river course was recorded by the cross-section R8/L8 (1600 m), R13/13 (927 M), R20/L20 (1000 m), and R3/L3 (849.69 m), and right bank shifting recorded at cross-section R20/L20 (1253 m), R8/L8 (967.72 m), R13/L13 (927.26 m). However, cross-sections R1/L1 (168.5 m), R7/L7 (196 m), R22/L22 (40 m) recorded minimum left bank shifting during the assessment period.

Figure 24.9 shows the left and right migration trend of the Churni River during the assessment

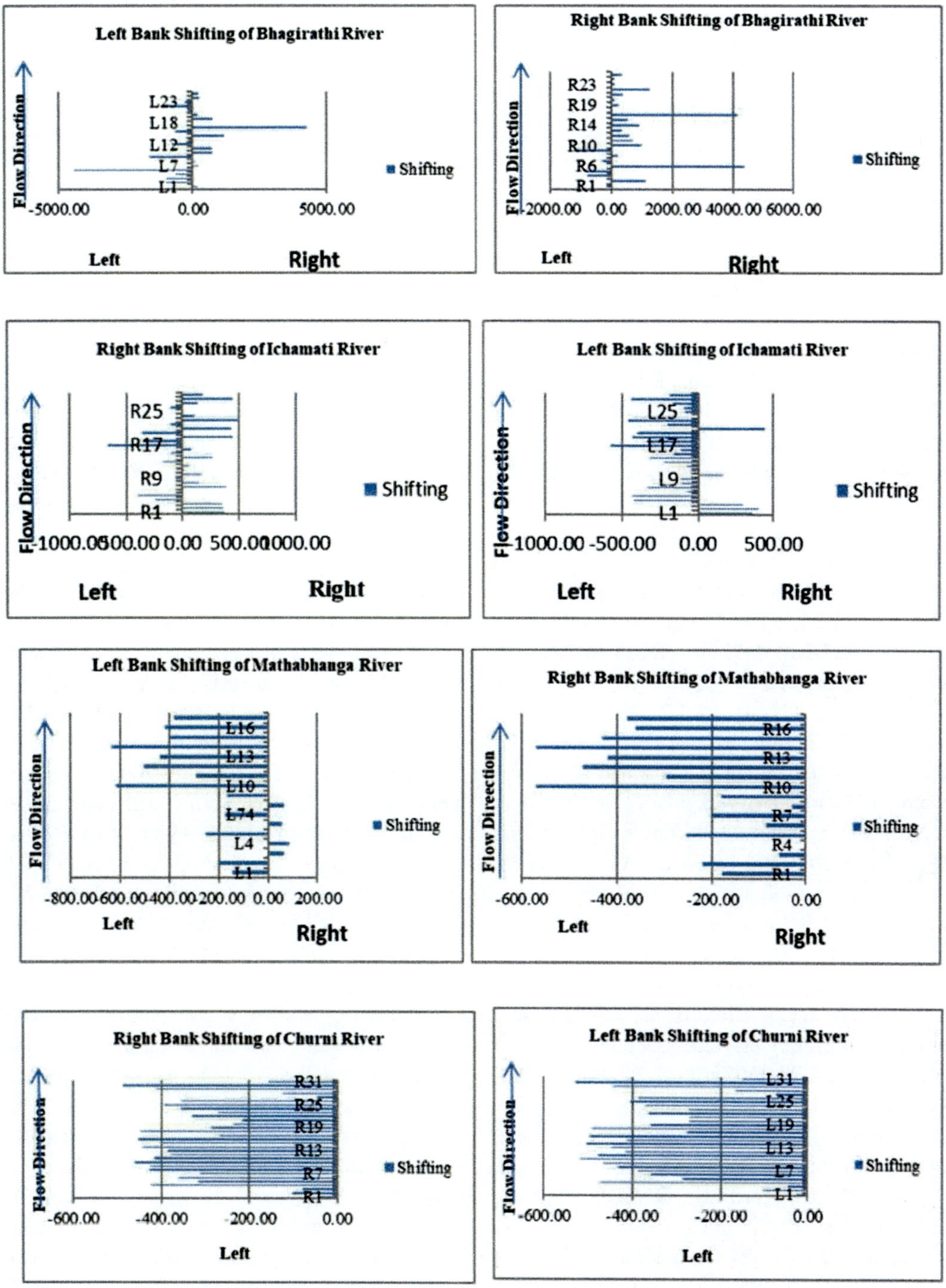

Fig. 24.9 Shifting of left and right bank of river Bhagirathi, Ichamati, Mathabhanga, and Churni (1970–2013)

period. The shifting rate of the river course was high from 1970 to 2013 and observed 529.4 m at cross-section R31/L31, 516.26 m at cross-section R9/L9 cross-section, 490 m at cross-section R17/L17, 490.94 m at R15; whereas minimum recorded at cross-sections R27/L27 (72.28 m), R1/L1 (47.04 m), R3/L3 (45.14 m). However, for the right bank, the maximum shifting recorded at section R 18/L18 (448 m), R31/L30 (486 m), R16/L16 (451 m), R14/L14 (412 m), R10/L10 (439 m), and minimum recorded at cross-section R1/L1 (47 m), R3/L3 (78 m), R31/L31 (156 m), R28/L28 (46 m), R29/L29 (146 m). Whereas the maximum shifting of the left bank in Mathabhanga river course was at cross-section R14/L14 (632 m), R10/L10 (614 m), R12/L12 (502 m), R16/L16 (415 m), and the minimum was R8/L8 (65 m), R6/L6 (56 m), R3/L3 (66 m), R1/L1 (145 m) (Fig. 24.9). Whereas, for the right bank, maximum recorded R10/L10 (570 m), R14/L14 (569 m), R12/L12 (472 m), R16/L16 (361 m) and minimum recorded R8/L8 (28 m), R6/L6 (85 m), R3/L3 (57 m), R7/L7 (200 m), R1/L1 (178 m) and R9/L9 (167 m). However, maximum left bank shifting of Ichamati river was recorded 575 m at cross-section R17/L17, 461 m at cross-section R23/L23, 446 m at cross-section R28/L28, 434 m at cross-section R21/L21, 441 m at cross-section R19/L19, 394 m at cross-section R7/L7,392 m at cross-section R5/L5, and minimum at cross-section R24/L24 (3.4 m), R11/L11 (46 m), R6/L6 (71 m), R8/L8 (107 m), R22/L22 (200 m); Whereas maximum right bank shifting recorded at cross-section R17/L17 (575 m), R23/L23 (507 m), R28/L28 (446 m), R21/L21 (434 m), R19/L19 (441 m), R7/L7 (394 m), R5/L5 (392 m), R3/L3 (337 m), and minimum recorded at cross-sections R25/L25 (35 m), R16/L16 (78 m), R11/L11 (12 m), R9/L9 (54 m), R6/L6 (34 m) (Fig. 24.9). This quantitative analysis reveals that the shifting rate is high at the Bhagirathi River followed by Ichamati, Mathabhanga, and Churni rivers. The course of the Bhagirathi River is getting huge water from Farakka braze throughout the year, which enhances erosion and deposition activity in the river channel causes maximum shifting above the river courses and bank line erosions. Similarly, for Ichamati, Churni, and Mathbhanga River courses, it is minimal due to the problem of *siltation* leading to a thin flow of water in the dry season. Therefore, urgent management is required.

24.4.3 Channel Dynamics Impacts on Flood Plain LULC

Land use/land cover plays an important role in soil erosion, siltation of river bed, and riverbank erosion. Generally, thick forest coverer and grassland along the river bank prevent riverbank erosion (Ahmad et al. 2017; Mondal et al. 2016b). Land use/land cover map of the study area prepared using Landsat (ETM +) data of 2013 by the supervised image classification method. LULC map (Fig. 24.5) shows that the study area is characterized by cropland, the settlement with the homestead orchards, vegetation, aquaculture land, sand splay, sand, river, wetland pond, and scroll/meander. Among the LULC classes, 55% areas (926,924.64 sq.km) covered by double cropland, 13% (228,709.033 sq.km) areas covered by settlement with homestead orchards, 12% area (205,761.698 sq.km) covered by sand splay categories, and only 5% (88,561.802 km^2) land categories as vegetation. 0.9% (1621. 09 sq.km) area covered by aquaculture land (4.0%) of 65,341.315 sq.km covered by wetland pond, 4% area covered by river, 4% scroll/meander, and only 1% sand (Fig. 24.10). The present study has some limitations in the amendment for the cropping pattern due to the effect on heavy saltwater with soil texture, solid drainage system, and also more significantly lacks suitable irrigation amenities (Mondal et al. 2014, 2016b). However, this study has considered the over-generalized, monotonous cropping categorization conquered by the paddy field. The study is mainly alluvial plain, so, this soil category region crop production is very high, the Aman rice crop has to rest on impulsive in monsoonal precipitation. Throughout the Rabi season, there is very little enthusiasm in Boro agronomy. Between oilseeds and other cropland,

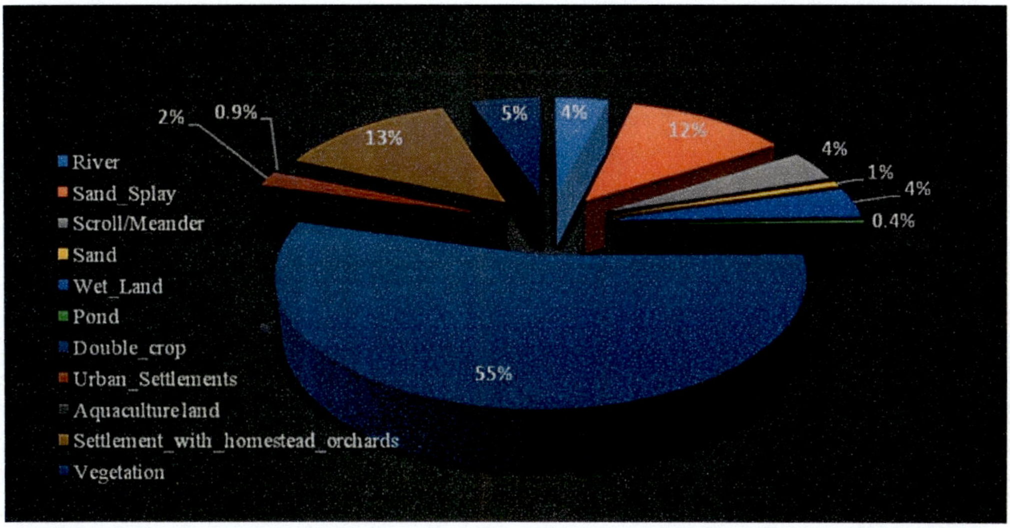

Fig. 24.10 Percentage of land use and land cover

mustard dominates in harvesting, and an existing year with a tiny space has also been assumed for sunflowers, etc.

24.5 Conclusion

24.5.1 Major Findings

In the present study, we focused on the lateral migration and morphodynamics changes of the Bhagirathi, Churni, Mathabhanga, and Ichamati river course considering the relationship between land use and land coverand geomorphology visa-a-viz shifting river bank lineduring the 43 years usingthe geospatial technology. According to the results, the banks of the Bhagirathi and Mathabhanga rivers are severely eroded. The study shows that all the river courses have shifted their bank in both sides, however, the pattern of shifting is not the same, i.e. displays an irregular migration pattern. However, the Bhagirathi Riveris highly dynamic than Churni, Mathabhanga, and Ichamati. The overall sinuosity index is increasing for all rivers; however, the degree of sinuosity is not the same. In 1970, the Bhagirathi was wide and straight, and during the 43 years, river course has significantly changed. The Ichamati river course was recorded in highest sinuosity values (SI = 17.3), followed by Mathabhanga (SI = 7.29), Bhagirathi (SI = 3.46), and Churni (SI = 2.64). Generally, river morphologyand the LULC In the study area, morphology is interrelated with both banks of the river course erosion that threatens adjacent settlements and infrastructure in the study area. However, it's a continuous shifting of the river Bhagirathi that has been observed towards the left or east during the entire assessment period. This may be due to excessive sedimentation, high discharge, and neo-tectonic activity (as the Bengal basin tilted towards the eastward).The sinuosity index also demonstrated that the river course/channel has subjected toexcessive sedimentation, which designates the meandering commotion of the river. Throughout the study showed that river stream changes owing to various natural and manmade phenomena. We can identify the river bank moving along with his risk zone area using features and land use/land cover, altering landforms in connection to geomorphology, and accuracy. we can identify the river bank shifting along with his risk zone area.

Now the Bhagirathi, Churni, Mathabhanga, and Ichamati rivers' catchment contains various inland waterway, channel, Khal, Bill, boar, wetland, oxbow, etc. lost their capacity of water volume due to effect on our unawareness after the long time of period. The overall result has been highlighted if abundant in both banks of that's waterbody overflowing during the rainy season and its effect due to flooding of adjacent areas. The rate of siltation is very high in all river channels,*Khal*, and*bil*in the same condition. According to our previous study, Mondal et al. (2016a) have estimated the rateof sedimentation of the Ichamati river 16,328.82 m^3/month, and the silt deposition rate is high in the monsoonal period, and the other season, the rate is less (11,109 m^3). Here, the study is due to the effect on the high rate of siltation in the depth of the waterbody, so it parallels assessment near to the agriculture field. In the winter and summer seasons, the water body is observing in quite dry, and other hands, monsoonal periods quite different huge inundation after getting water from the raining and flooded the adjacent villages and also the agriculture field areas. Here, the siltation process of the Bhagirathi, Churni, Mathabhanga, and Ichamati Rivers has played a significant role in overflow in surrounding areas, and also on both sides of the river banks. In the study, we observed that a huge amount of illegal work by the brickfields industry, mostly precisely carelessly accumulating silt-soil from the river areas, and also illegal building of riverbank sites near the brickfields zones, etc., consecutively free of obligation, siltation is an alternative imperative reason for the death of the riverine stage. Finally, we suggested four river basinsif the present current of water has been flowed out, silt transported in the Bay of Bengal by the current. Then discontinuation with the source of all those rivers currently being transformed into closed water bodies.

24.5.2 Managements Issues

- The land acquisition suggestions regarding the potentials for decrease or prevention in the assessment of the probable effect on agriculture land, forest cover, wetland, etc.
- The alluvial nature of the bed and bank materials is easily erodible by the destructive episodes of the tidal cycle.
- Fresh muds consolidate in the slack water period where marsh encroachment is significant and as the current increase in the next stage of the tide and erosion is not sufficient to remove all of the material deposited.
- Faster moving currents of tidewater along the deep channels created mid-channel deposition ridge in the estuary.
- Diversity of the surface level by silt-laden inundation and silt-free inundation.
- Natural levees and open tidal floodplains (unprotected areas) are building up regularly by alluvial sedimentation to match the increased height of seasonal river floods,
- At the same time, the depth of seasonal flooding by rainwater (silt-free) is increasing over the saucer-shaped lowland in which the high tidal inflow of river flooding is prevented by coastal embankments.

References

Acharya SK, Chakraborty P, Lahiri S, Raymahashay BC, Guha S, Bhowmik A (1999) Arsenic poisoning in the Ganges Delta. Nature 401:545

Ahmed T (2012) Ichamati River. In: Islam S, Jamal AA (eds) Banglapedia: national encyclopedia of Bangladesh (2nd edn). Asiatic Society of Bangladesh

Ahmad F, Goparaju L (2017) Land evaluation in terms of agroforestry suitability, an approach to improve livelihood and reduce poverty: a FAO based methodology by geospatial solution: a case study of Palamu district, p 25. Jharkhand, India. https://doi.org/10.12775/EQ.2017.006

Ahmed AA, Fawzi A (2009) Meandering and bank erosion of the River Nile and its environmental impact on the area between Sohag and El-Minia. Egypt Arab J Geosci 2011(4):1–11. https://doi.org/10.1007/s12517-009-0048-y

Bag R, Mondal I, Bandyopadhyay J (2019) Assessing the oscillation of channel geometry and meander migration cardinality of Bhagirathi River, West Bengal, India. J Geogr Sci 29(4):613–634

Bagchi KG (1944) The Ganges Delta. Calcutta University press, Calcutta, p 157

Bagchi KG, Munsi SK, Bhattacharyya R (1972) The Bhagirathi-Hooghly Basin. In: Proceedings of the interdisciplinary symposium. Sri Sibendranath Kanjilal, Calcutta

Basu SR (1965) A study of the member belt of the Bhagirathi river. Calcutta University, M.A Thesis (unpublished)

Basu SR (1968) Landscape evaluation-a methodology. Geo. Rev. of India., vol XXX, no 4

Bhunia GS, Shit PK, Pal DK (2016) Channel dynamics associated with land use/cover change in Ganges river, India, 1989–2010. Spatial Inf Res 24(4):437–449

Das B (2013) Risk reduction management of flood by Bhagirathi River a case study of Agradweep of Bardwan district in Gangetic Delta. Int J Eng Res Appl 3:567–576

Debnath J, Pan ND, Ahmed I, Bhowmik M (2017) Channel migration and its impact on land use/land cover using RS and GIS: a study on Khowai River of Tripura, North-East India. The Egypt J Remote Sens Space Sci 20(2):197–210

Garrett JHE (1910) Bengal District Gazetteers: Nadia. Indian Civil Service, Published: West Bengal District Gazetteers Department of Higher Education Government of West Bengal, Kolkata, 2001

Guchhait SK, Islam A, Ghosh S, Das BC, Maji NK (2016) Role of hydrological regime and floodplain sediments in channel instability of the Bhagirathi River, Ganga-Brahmaputra Delta, India. Phys Geogr 37(6):476–510

Hirst MFC (2002) Director of Surveys, Bengal and Assam, Report on the Nadia Rivers 1915, first published in 1916 by the Bengal Secretariat Book Depot, reproduced in Rivers of Bengal, vol III, p 27, West Bengal District Gazetteers, Higher Education Department, Government of West Bengal, 2002

Islam A (2010) Variability of stream discharge and bank erosion—a case study on the river Bhagirathi. J River Res Inst-River Behav Control 31:55–66

Islam A, Guchhait SK (2017a) Analysing the influence of Farakka Barrage Project on channel dynamics and meander geometry of Bhagirathi river of West Bengal, India. Arab J Geosci 10(11):1–18

Islam A, Guchhait SK (2017b) Search for social justice for the victims of erosion hazard along the banks of river Bhagirathi by hydraulic control: a case study of West Bengal, India. Environ Dev Sustain 19(2):433–459

Islam A, Guchhait SK (2018) Analysis of social and psychological terrain of bank erosion victims: a study along the Bhagirathi river, West Bengal, India. Chinese Geogr Sci 28(6):1009–1026

Islam A, Guchhait SK (2020) Characterizing cross-sectional morphology and channel inefficiency of lower Bhagirathi River, India, in post-Farakka barrage condition. Nat Hazards 103(3):3803–3836

Islam MS (2021) Assessing the dynamics of land cover and shoreline changes of Nijhum Dwip (Island) of Bangladesh using remote sensing and GIS techniques. Regional Stud Marine Sci 41:101578

Mathabhanga River (2017) Trans-boundary Rivers—Banglapedia. en.banglapedia.org. http://www.wikizero.com/en/Mathabhanga_River. Accessed 13 July 2017

Miah J, Hossain KT, Hossain MA, Najia SI (2020) Assessing coastal vulnerability of Chittagong District, Bangladesh using geospatial techniques. J Coastal Conserv 24(6):1–18

Mondal I, Bandyopadhyay J (2014) Environmental change of trans international boundary Indo-Bangladesh border of Sundarban Ichamati River catchment area using geoinformatics techniques, West Bengal, India. Universal J Environ Res Technol 4 (3):143–154

Mondal I, Bandyopadhyay J (2016) Physicochemical analysis of Ichamati River and estimation of soil parameters using geospatial technology, West Bengal, India. J Inst Eng (India): Series E Springer 97(2):151–158. https://doi.org/10.1007/s40034-016-0086-4,

Mondal I, Bandyopadhyay J, Paul AK (2016a) Estimation of hydrodynamic pattern change of Ichamati River using HEC RAS model. West Bengal, India. Model Earth Syst Environ 2: 25. https://doi.org/10.1007/s40808-016-0138-2

Mondal I, Bandyopadhyay J, Paul AK (2016b) Recent trend for water resource management of sustainable development through river-interlinking using remote sensing and GIS technology: a case study on Sundarban River system, West Bengal, India. Central Ground Water Authority & Central Ground Water Board Eastern Region, Kolkata, 21–22nd May 2016, pp 153–164

Mondal I, Bandyopadhyay JP, Paul AK. (2016c) Water quality measurement for seasonal fluctuation of Ichamati River, West Bengl, India. Model Earth Syst Environ, Springer. https://doi.org/10.1007/s40808-016-0153-3,2:1-13

Mondal I, Thakur S, Bandyopadhyay J (2019) Delineating lateral channel migration and risk zones of Ichamati River, West Bengal, India. J Cleaner Prod, Elsevier. https://doi.org/10.1016/j.jclepro.2019.11874244: 118740

Morisawa M (1985) Rivers: form and process. In: Clayton KM (ed) Longman, Long- man Group Limited, London and New York, p 222

Mueller JE (1968) An introduction to th e hydraulic and topographic sinuosity indexes. Ann Assoc Am Geogr 58(2):371–385

Mukherjee RK (1938) The changing face of Bengal. Calcutta University Press, Calcutta, p 288p

Mukherjee I (2012) Impact of hydrological characteristics and fluvial morphology on land utilization in the Bhagirathi and Jalangi Floodplains of East central West Bengal [D]. The University of Burdwan

Niyogi D (1975) Quaternary geology of the coastal plain in West Bengal. Indian J Earth Sci 2:51–61

Rennell J (1764–66) https://en.wikipedia.org/wiki/James_Rennell

Rennell J (1781) The Ganges and Brahamaputra Rivers: Royal socity. London Phylosophical Transactions, vol. LXXI

Rennell J (1910) The journals of Major James Rennell. Baptist Mission Press and Asiatic Society, Calcutta
Roy NR (1952) Bangalir Itihas (In Bengali). Book Emporium, Calcutta, p 923
Sengupta S (1966) Geological and geophysical studies in western part of Bengal Basin, India. Bull Am Assoc Petrol Geol 50:1001–1017
Umitsu M (1993) Late quaternary sedimentary environments and landforms in the Ganga Delta. Sediment Geol 83:177–186
Yousefi S, Moradi HR, Pourghasemi HR, Khatami R (2018) Assessment of floodplain landuse and channel morphology within meandering reach of the Talar River in Iran using GIS and aerial photographs. Geocarto Int 33(12):1367–1380

Index

P. K. Shit et al. (eds.), *Drainage Basin Dynamics*, Geography of the Physical Environment,
https://doi.org/10.1007/978-3-030-79634-1